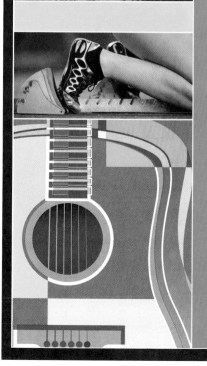

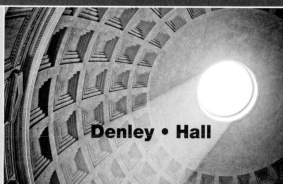

Viewing Life Mathematically

A Pathway to Quantitative Literacy

Denley • Hall

Executive Editor: Susan Fuller

Co-Editor: Barbara Miller

Assistant Editors: Margaret Gibbs, Robin Hendrix

Executive Project Manager: Kimberly Cumbie

Vice President, Research and Development: Marcel Prevuznak

Editorial Assistants: Danielle C. Bess, Claudia Vance, Nina Waldron

Review Coordinators: Lisa Young, Jessica O'Leary

Senior Graphic Designers: Jennifer Moran, Tee Jay Zajac

Digital Production Editors: Robert Alexander, Doug Chappell

QSI (Pvt.) Ltd: E. Jeevan Kumar, D. Kanthi, U. Nagesh, B. Syam Prasad

Art: Jennifer Moran

Cover Design: Jennifer Moran

A Division of Quant Systems, Inc.
546 Long Point Road, Mount Pleasant, SC 29464

Library of Congress Control Number: 2014948152

Printed in the United States of America
10 9 8 7 6 5 4

ISBN: 978-1-935782-05-6
Textbook and Software Bundle: 978-1-935782-06-3

Table of Contents

The Mathematics of Growth

Geometry

Probability

Statistics

Personal Finance

Preface

Purpose and Style

We created *Viewing Life Mathematically: A Pathway to Quantitative Literacy* as an alternative college-level mathematics pathway for students whose majors are neither calculus-based nor statistics-based. It is assumed that students taking the course will have sufficient knowledge of basic algebraic manipulation, ability to graph functions, and ability to solve algebraic equations. The material includes a variety of fields within mathematics that allow the course to be tailored to an institution's specific objectives. Each topic engages students in developing mathematical skills and techniques that can be applied in their everyday lives and their own programs of study. The unifying theme is the introduction of concepts and ideas that encourage students to view aspects of life by thinking mathematically.

We strive to make these applications of mathematics interesting to students and accessible through a clear, conversational writing style that leads students through the mathematics. The content is neatly organized, utilizing text boxes to emphasize the location of important definitions, theorems, and formulas. Numerous visual displays and diagrams are provided to support the quantitative reasoning process. We often present alternate ways to approach a problem and clearly explain the thinking process behind the steps, so that students can choose the process that works best for them. To further develop analytical skills, we encourage students to use technological resources, including calculators, spreadsheet programs, and online resources, to perform calculations for them, so that they are able to focus on interpreting the output and understanding the contextual implications of the mathematics.

Our goal for this text is to develop the quantitative reasoning skills that students majoring in liberal arts disciplines need to succeed. With these students in mind, *Viewing Life Mathematically: A Pathway to Quantitative Literacy* focuses on building practical knowledge and problem-solving skills in everyday real-world contexts. The ability to analyze and interpret data is essential for all students to become informed citizens of a complex, technological world, regardless of their chosen discipline.

Special Features

Chapter Opener

Each chapter begins with a list of sections and a list of chapter objectives to prepare students for the topics that will be covered. The objectives list also helps students identify the most important concepts in the chapter, enabling students to focus their time and effort appropriately.

Chapter Introduction

An application of the chapter content is introduced on the first page of each chapter to inspire student interest and provide an understanding of how the topics to be studied are useful.

Examples

Examples are presented in a step-by-step manner that is easy for students to follow. Titles are given to alert students to the concept they will be learning throughout the example. Examples make use of tables, diagrams and graphs, and technology where applicable, giving clarification to the mathematical skill being presented.

TECH TRAINING

Tech Training boxes appear throughout the text, providing step-by-step instructions for using a calculator, spreadsheet program, or online resource to solve problems. Calculator instructions are given for a TI-83/84 Plus graphing calculator and a TI-30 scientific calculator with the Equation Operation System (EOS). All spreadsheet directions are written for Microsoft Excel 2010. The online resource used is Wolfram|Alpha (Wolfram Alpha LLC. 2009. Wolfram|Alpha. http://www.wolframalpha.com). Wolfram|Alpha is a free resource that can be utilized by any student with internet access. Students may use other types of technology, but may need to consult a user manual to troubleshoot any differences.

Definitions and Formulas

Definitions and formulas are presented in highly visible boxes for easy reference.

Skill Check

Skill Check questions allow students to test their knowledge as they progress through a chapter to ensure that they understand the material. Solutions to these problems are located at the end of the section in which they appear before the section exercises.

Think Back

Think Backs are just-in-time information located in the margins of the text to remind students of important concepts that they may have forgotten from previous math courses or previous chapters in this course.

☞ Helpful Hint

Helpful Hints appear in the margins of the text and provide students with extra insightful pieces of information to help them avoid common mistakes and understand concepts more deeply.

Fun Fact

Fun Facts appear in the margin with additional information to pique student interest and make mathematics more relatable.

MATH MILESTONE

Math Milestones appear in the margins to give more information about mathematicians and the history of mathematics to help students connect further with the material.

Section Exercises

Each section includes a variety of exercises to give students practice applying and reinforcing the skills they learned in the section. The exercises exhibit a wide range of difficulty levels and applications, often using real-world data.

Chapter Summary

Each chapter features a Chapter Summary that highlights important definitions, properties, processes, and formulas given in the chapter. Material is listed by section in the order that it appears, allowing students to quickly find the information they need when studying for a test or doing homework assignments.

Chapter Exercises

At the end of each chapter, a set of Chapter Exercises provides extra problems for students to practice and to identify strengths and weaknesses before taking an exam.

Answer Key

The Answer Key in the back of the book contains the answers for odd-numbered Section Exercises and all answers to Chapter Exercises. This allows students to check their work to ensure they are accurately applying the methods and skills that they have learned.

Content

Chapter 1: Critical Thinking and Problem Solving

For most students in college, thinking mathematically means thinking algebraically. This chapter aims to introduce thinking mathematically as creative problem solving through reasoning, processes, and techniques.

Chapter 2: Set Theory

From Twitter followers to pizza toppings; from understanding surveys to organizing a closet; from factors to prime numbers, looking at the way items are grouped provides a solid mathematical foundation. This chapter uses a systematic approach to organizing and analyzing information and groups of items in order to solve problems involving sets. Formal set theory is introduced, as well as its definitions and notation, as a means to express, analyze, and manipulate categories and groupings.

Chapter 3: Logic

Logic is the basic structure of thinking mathematically. This chapter introduces formal mathematical logic, along with its definitions and notation. The ultimate goal of the chapter is for the reader to create and critique valid and invalid logical arguments.

Chapter 4: Rates, Ratios, Proportions, and Percentages

Rates, ratios, proportions, and percentages are universal, from credit cards and mortgages to retail sales and tipping. This chapter develops the skills for understanding and calculating these important values.

Chapter 5: The Mathematics of Growth

Mathematical models use information about the past to create images of the future. In this chapter, methodology is developed to create, manipulate, and analyze how quantities change over time. The focus on real-world applications involves linear, quadratic, exponential, and logarithmic growth.

Chapter 6: Geometry

From GPS navigation to rocket trajectories, it is important to be able to model situations geometrically. In this chapter, the mathematics behind the spatial relationships that exist between objects is developed in a way that engages students in elementary problem solving in Euclidean and non-Euclidean geometry.

Chapter 7: Probability

Little in life is certain. Behavioral economists suggest that many mistakes in everyday life stem from massively overestimating or underestimating chance. Being able to express and manipulate the uncertainties of random events is a crucial skill in life. This chapter introduces formal probability, as well as its definitions and notation, as a means to express, analyze, and calculate chance.

Chapter 8: Statistics

Being able to understand statistical concepts is an important part of mathematical literacy in every field of study at the college level. This chapter introduces the fundamental concepts behind collecting and analyzing statistical data.

Chapter 9: Personal Finance

Wise financial planning both for the short term and the long term is an important life skill. This chapter introduces the basic skills of mathematical finance for effective budgeting and long-term decision making. Topics include budgeting with Excel as well as computing sales prices, interest, compound interest, effective interest, and monthly payments.

Chapter 10: Voting and Apportionment

Making the right choice can be challenging. This chapter introduces the mathematics behind the way individual preferences affect the decisions groups make and how to ensure fair outcomes. Topics include apportionment, voting methods, and voting paradoxes.

Chapter 11: The Arts

Art, architecture, and music are not necessarily known for their mathematical content by the novice observer. This chapter uses geometry and proportions to explore the mathematics of such beauty.

Chapter 12: Sports

Just as mathematics provides a language to describe and analyze other facets of life, the sporting world is no exception. Each sport in this chapter is covered in a way that will allow every student to learn how mathematics affects the decisions made in those sports, and how mathematics is used in the physical participation of each sport. This chapter explores how the world of sports incorporates problem solving skills, ratios, proportions, probability, and statistics into decision making.

Chapter 13: Graph Theory

This chapter includes a thorough coverage of introductory topics to graph theory and how they relate to real-world applications. The structure of networks including Facebook and the interstate highway system provide powerful mediums to connect people. This chapter looks at the mathematics behind these networks.

Chapter 14: Number Theory

From ISBN numbers to internet security, the properties of numbers plays an often unnoticed role throughout modern life. This chapter provides a gentle introduction into the power of number theory, covering topics from prime and composite numbers to modular arithmetic, as well as applications of public-key encryption.

Hawkes Learning Systems: Courseware

Hawkes Learning Systems specializes in interactive courseware with a unique mastery-based approach to student learning. The courseware is designed to help you develop a solid foundation of skills and has been proven to increase your overall success. Within each homework lesson you will find three learning modes: Learn, Practice, and Certify.

Learn: Learn is a multimedia presentation that includes the information you need to successfully answer each question in your assignment. Each lesson includes definitions, rules, properties, and examples, along with instructional videos.

Practice: Practice gives you unlimited opportunities to practice the types of problems you will receive in Certify. In Practice, you have access to learning aids through the Interactive Tutor. Step-By-Step breaks a problem down into smaller steps; Solution offers guided solutions to every problem; and Explain Error gives targeted feedback specific to your mistake.

Certify: This is the credit component of your homework! You will answer your problem set by using your knowledge and the foundation you built in Learn and Practice. You will have the opportunity to try again with no penalty if you do not demonstrate Mastery in your initial attempt(s). Pay close attention to any due dates or benchmarks assigned by your instructor.

Video: View instructional videos anytime, anywhere at HawkesTV.com.

Feel free to contact support for questions or technical help.

Support Center: support.hawkeslearning.com

Chat: chat.hawkeslearning.com

E-mail: support@hawkeslearning.com

Phone: 843.571.2825

Acknowledgements

As we breathe life into this book, we would be amiss to fail to recognize that truly good teaching and writing does not happen in a vacuum. It takes a team of talented individuals with different points of view to critique and hone the thoughts that appear on the page. We are grateful for those who partnered with us in this endeavor. To Jim Hawkes, Emily Cook, Marcel Prevuznak, and Kim Cumbie, who believed in us and this project enough to branch out into a new area for Hawkes, we are indebted. To our editors who never seemed to tire: Margaret Gibbs, Susan Fuller, Barbara Miller, Robin Hendrix, and all those working at Hawkes Learning, thank you seems too small, but it is heartfelt. To our reviewers, who helped not only shape the content, but also fine-tune the language, we are appreciative.

Iana Anguelova, College of Charleston

Dr. Edna Bazik, National Louis University – Chicago

Dr. Shari Beck, Navarro College – Corsicana Campus

Jennifer Briney, MacMurray College

Joan Brown, Eastern New Mexico University

David Busch, Iowa Central Community College

John Callaghan, Gateway Community College

Dr. Judith Covington, Louisiana State University – Shreveport

Dr. John Dawson, Kirkwood Community College – Iowa City

Dr. Sarah Duffin, Southern Utah University

Dr. Johnny Duke, Georgia Highlands College

Kevin Dyke, Georgia Highlands College

Kay Geving, Belmont University

Dr. Kim Harris, University of North Carolina – Charlotte

Carla Hill, Marist College

Dr. Peggy Hohensee, Fletcher Technical Community College/Kaplan University

Dr. Liz Jurisich, College of Charleston

Dr. Dorothy Kerzel, Mississippi University for Women

Dr. Chris Mattingly, Martin Methodist College

Tonya Meisner, University of Wisconsin – Marinette

Janette Miller, University of Wisconsin – Sheboygan

Kristi Peters, Enterprise State Community College

Robin Rufatto, Ball State University

Dr. Tsvetanka Sendova, Bennett College for Women

Jim Sheff, Spoon River College

Melody Shipley, North Central Missouri College

Lymeda Singleton, Texas A&M University – Commerce

Joseph Szurek, University of Pittsburgh – Greensburg

John Thoo, Yuba College

Dr. Alfredo Vaquiax–Alvarado, Our Lady of the Lake University

Dr. Catherine Whatley, York Technical College

Dr. Nancy Wyshinski, Trinity College

Mike would like to thank his wife, Glinda, for her undying support. Without her wit, intellect, and influx of ideas, this book would never have been completed. To Grayson and Jack, thank you for your support, your patience, and your encouragement in its completion.

Lastly, Kim would like to thank Tristan. This book would never even exist if it wasn't for your vision, drive, and loving support! And to Emma and Chloë, thank you for not only being patient but also helpful and encouraging throughout. I love you all!

About the Authors

Kim Denley has been involved in mathematics reform for more than a decade. As part of a nationally acclaimed course redesign for undergraduate mathematics at the University of Mississippi, she began her commitment to helping students realize the utility of mathematics in their everyday lives. This book builds on the earlier trajectory of her statistics text, *Beginning Statistics*, to provide an intentionally readable book for non-STEM majors. Kim loves living in Music City—Nashville, Tennessee—with her husband and daughters, where she volunteers in the community, offering mathematical help to women seeking their GED.

Mike Hall has served as a mathematics educator for nearly two decades, having taught at the University of Mississippi and Arkansas State University. His research interests have focused on mathematics self-efficacy, learner autonomy, and instructional technology. Having spent many years researching the content that students learn in undergraduate mathematics courses, Mike discovered a true need for a textbook that allowed students to explore mathematics in the context of daily activities, and that was approachable for all students, regardless of their mathematical background. Simply put, this book is the labor of his passion for learning mathematics. Mike lives in Jonesboro, Arkansas, with his wife and sons, where he volunteers in local youth baseball programs.

Chapter 1
Critical Thinking and Problem Solving

Objectives

- Understand mathematical reasoning
- Distinguish between inductive and deductive reasoning
- Identify arithmetic and geometric sequences
- Understand Pólya's problem-solving process
- Apply problem-solving strategies
- Find estimates

Critical Thinking and Problem Solving

According to the United States Bureau of Labor Statistics, the average cost of a new house in 1980 was about $68,000. The average cost of a gallon of milk was about $1.70, the average cost of a gallon of gas was about $1.25, and the average cost of a year of tuition at a public four-year institution was about $2400. These numbers seem almost outrageously low by today's standards. The average price of a new house in 2011 was about $214,000, the average cost of a gallon of milk was about $2.65, the average cost of a gallon of gas was about $3.53, and the average yearly cost of tuition at a public four-year institution was about $12,500.

Table 1			
Price Differences between 1980 and 2011			
	1980 Cost (Dollars)	2011 Cost (Dollars)	Difference (Dollars)
House	68,000	214,000	+146,000
Milk	1.70	2.65	+0.95
Gas	1.25	3.53	+2.28
College Tuition	2400	12,500	+10,100
Source: US Bureau of Labor Statistics, http://www.bls.gov			

It might appear as if the cost of goods has grown quite large. However, if we look more closely and take into account the rate of inflation over the 30-year period from 1980 to 2010, the costs of the products selected here are similar to those in 1980. How can this be? The ability to understand the cost of goods and be an informed consumer is an invaluable skill when making decisions regarding purchases and saving for the future. This chapter will focus on the use of mathematical concepts that can be used to analyze the world around us and, in turn, help us make informed decisions.

Thinking mathematically requires that we consider mathematical concepts when making decisions. Whether we are making a decision in a logical argument or deciding which freeway to take on the way to school or work, mathematics plays a key role in the choices we make and the problems we solve. Though most of us do not think in mathematical terms when solving daily problems, mathematical models such as graphs, tables, and other visual representations give us a visual approach to finding solutions and decision making.

Figure 1

1.1   Thinking Mathematically

Mathematical thinking is important for decisions we all make every day. Moreover, possessing the ability to think mathematically makes one a better problem solver for all occasions. Consider the idea that the product of two odd numbers will always be odd. Is this always true? Take a moment to consider this. It is, in fact, always true. Although you may not spend time thinking of mathematical facts like this every day, everyone needs to have an understanding of mathematics in order to make responsible decisions each day. For example, you need to have an understanding of addition and subtraction with numbers and their attributes in order to maintain a checking account.

Reasoning

Reasoning is defined by the Merriam-Webster dictionary as the drawing of inferences or conclusions through the use of statements offered as explanation or justification.

We use reasoning in making decisions every day. For example, when driving home from school on a given day, you probably notice patterns in traffic that alert you to take a given route based on the time of day, weather conditions, month of the year, etc. Logical reasoning is also found in political debates, formal writing, medical decisions, and predicting the weather, just to name a few. Throughout the text, we will consider many types of reasoning in an effort to become better problem solvers and mathematical thinkers.

We begin our discussion of reasoning with inductive and deductive reasoning.

Inductive Reasoning

"Doing" mathematics is a process that is far more than just calculations; it involves observing patterns, testing conjectures, and estimating in order to arrive at an answer that best fits the information given. To that end, the ideas of mathematical argument require us to exercise the use of the ideas of **inductive** reasoning.

Inductive Reasoning

Inductive reasoning is a line of reasoning that arrives at a general conclusion based on the observation of specific examples. Inductive reasoning can be considered a *generalization*.

Example 1: Using Inductive Reasoning

Consider the following argument.

In New York City, it snowed 30 inches during January 2010 and 35 inches during January 2011. Therefore, New York City will receive at least 30 inches of snow every January.

Does this argument use inductive reasoning?

Solution

Notice that in this argument, the example is specific to January 2010 and January 2011, and then a very general conclusion is made. Obviously, the likelihood of it snowing exactly the same amount each January is not very practical, but nevertheless, the argument here is inductive.

Let's try another example that is more mathematical.

Example 2: Using Inductive Reasoning

Consider the following sequence of numbers.

$$1, 4, 9, 16, 25, \ldots$$

If the number pattern continues, can you conclude what the next number will be? What about the 15th number?

Solution

If we look closely, we can see that the numbers are simply the squares of the natural numbers: $1^2 = 1$, $2^2 = 4$, $3^2 = 9$, etc. Therefore, the next number in the pattern would be $6^2 = 36$, then $7^2 = 49$, and so on. That means the 15th number in the pattern would be 15^2, or 225. Notice again that we begin with a specific example and continue to a generalized answer.

Think Back

Recall from earlier courses that natural numbers are often called the "counting numbers" and are defined as the set of numbers $\{1, 2, 3, \ldots\}$.

Although we use inductive reasoning often, there are a few drawbacks to this type of reasoning. First, since we are using specific examples to draw generalized conclusions, we can only conjecture that our conclusion is true based on the information and examples we have. For instance, after eating out in your area, you may conclude that Friday night is a good night to be a waiter because the number of people eating out is larger than any other night of the week. Unless you can consider every restaurant that exists, you cannot truly be sure that this statement is true. Second, if we can find even one situation that does not satisfy our conclusion, then the argument is invalid. This contradictory example is called a **counterexample**. It is important to note that one counterexample is enough to prove that a line of reasoning is false, but one positive example is never enough to prove that it is true. Third, since there is no way to know *every* case, we must always assume patterns will continue in the same manner. Recall from Example 2 that our assumption was that the established pattern would continue. Therefore, using the assumption that each successive number would be the next perfect square, we could reason that the 15th number would be 15^2.

Example 3: Counterexamples

Consider the statement:

You must have a degree in computer science to become wealthy in the computer industry.

Is this a valid argument?

Solution

For this example, we need to look no further than two of the most famous people in the computer industry: Steve Jobs (cofounder of Apple Computers) and Bill Gates (cofounder of Microsoft). Neither Jobs nor Gates have a degree in computer science, yet both became quite wealthy in the computer industry. Hence, we have found a *counterexample* to our conclusion, and the argument is invalid.

Example 4: Reasoning with Sequences

Identify a pattern in each of the following sequences of numbers, then use the established pattern to find the next term in the sequence.

a. 4, 9, 14, 19, ____

b. 2, 6, 18, 54, ____

c. 5, 6, 8, 11, ____

Solution

When trying to identify a pattern in a sequence of numbers, there is not a set method to follow. We will introduce you to several techniques in this example. The more practice you have with these techniques, the easier it will become to identify patterns.

a. When considering the sequence 4, 9, 14, 19, ____, we should try to determine whether the difference between the consecutive terms is constant (that is, if the difference is the same for each consecutive pair of terms) or if the difference between the consecutive pairs of terms varies.

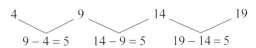

We can see here that there is a common difference between each number in the sequence. The common difference is 5. Thus, the next term in the sequence will be $19 + 5 = 24$.

b. The sequence 2, 6, 18, 54, ___ does not have a common difference between terms as in part **a.** Since there is no common difference, we need another approach. Instead of a common difference, perhaps there is a common ratio between the consecutive pairs of terms.

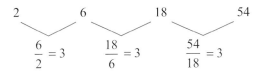

We can see that each successive number is the product of the previous number and 3. So, $2 \cdot 3 = 6, 6 \cdot 3 = 18, 18 \cdot 3 = 54$, etc. This means that the next term would be $54 \cdot 3 = 162$.

c. For the sequence, 5, 6, 8, 11, ____, there is no common difference or common ratio between the numbers. We need to ask ourselves, *What do we have to do to the first term in order to obtain the second?* In this case, we have to add 1 to 5 in order to get 6. For the next term, we see that the sum is $2 + 6 = 8$. We can now see a pattern developing that we need to investigate further. We added 1 to the first term and 2 to the second term. If we continue in this manner by adding 3 to the third term, 4 to the fourth term, etc., does the pattern continue? For the next term, $8 + 3 = 11$, so indeed it does. Thus, to find the next term in the sequence, we add 4 to 11 to get 15.

This example illustrates two common types of sequences. Sequences where the common difference between any two consecutive terms in a number sequence is the same—called an **arithmetic sequence**—and sequences where the ratio between any two consecutive terms in a number sequence is the same—called a **geometric sequence**.

Arithmetic Sequences

When the common difference between any two consecutive terms in a number sequence is the same, we call this an **arithmetic sequence**.

Geometric Sequences

When the common ratio between any two consecutive terms in a number sequence is the same, we call this a **geometric sequence**.

Example 5: Reasoning with Patterns

A certain type of human cell reproduces in the following manner: 1 cell, 4 cells, 9 cells, 16 cells. Determine the number of cells present on the next production of cells.

Solution

The second example we did concerning inductive reasoning asked us to consider the sequence of numbers 1, 4, 9, 16, ... Notice that the first iteration contains 1 cell and the second iteration contains $2^2 = 4$ cells. The third and fourth iterations contain $3^2 = 9$ cells and $4^2 = 16$ cells, respectively. So, the next iteration in the sequence will contain $5^2 = 25$ cells.

Skill Check # I

Consider the given figures. To build the figures, the first figure requires 16 line segments, the second figure requires 28 line segments, and the third figure requires 40 line segments. How many line segments would be required

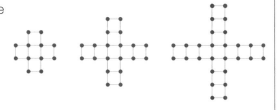

a. in the 8th figure?   **b.** in the nth figure?

Deductive Reasoning

Now that we have discussed inductive reasoning, we will turn our attention to **deductive reasoning**.

Deductive Reasoning

Deductive reasoning is a process that begins with commonly accepted facts and logically arrives at a specific conclusion.

Recall that inductive reasoning involves making conclusions that are generalizations based on specific examples. So, inductive reasoning moves from specific to general. On the other hand, deductive reasoning uses the idea that the commonly accepted facts guarantee the truth of a conclusion. In a deductive argument, the premises are general statements that support a specific conclusion.

Example 6: Using Deductive Reasoning

Consider the following statement.

If you are a mammal, then you have lungs.

How can this statement be evaluated as a deductive argument?

Solution

This statement has two parts: determining if you are a mammal and whether or not you have lungs. Breaking down the individual components of the statement, we can observe the following.

   a. I am a mammal (premise is true).
   I have lungs (conclusion is true).

   b. I am a mammal (premise is true).
   I do not have lungs (conclusion is false).

Notice that, in instance **b.**, when the conclusion is false, the argument does not seem to follow "logical" thought. There is a much more detailed discussion of deductive reasoning in Chapter 3.

Example 7: Inductive versus Deductive Reasoning

To illustrate the difference between inductive and deductive reasoning, consider the following process.

Choosing a positive integer, multiplying it by 2, and adding 1 to the product will result in an odd number.

   a. Evaluate this conclusion as an inductive argument.

   b. Evaluate this conclusion as a deductive argument.

Solution

   a. We will begin by looking at some specific examples.

Table 1		
Process Examples		
Positive Integer	**Arithmetic**	**Result**
3	$3 \cdot 2 + 1$	7
6	$6 \cdot 2 + 1$	13
8	$8 \cdot 2 + 1$	17

If we make the conclusion that choosing a positive integer, multiplying it by 2, and adding 1 to the product will result in an odd number simply by looking at these three examples, we are using inductive reasoning. We are making a *generalization* based

1

on three specific examples. Although we know this to be valid, whenever we make a generalization, we are taking a risk that the generalized form will not always hold and therefore will be an invalid conclusion.

b. Now we will let the variable x represent the chosen integer and use deductive reasoning to show that we have indeed made a valid conclusion.

> Number: x
>
> Multiply by 2: $2x$
>
> Add 1: $2x + 1$

This means that, in algebraic form, the problem may be expressed as $2x + 1$ (multiplying the chosen integer by 2 and adding 1). Using deductive reasoning, we know that when we multiply any number by 2, we will always get an even number. Thus, whenever we have an even number, as in $2x$, and add 1, we will always get an odd number. Therefore, we can assert that, *given any positive integer x, when x is multiplied by 2 and we add 1, the result will be an odd number*. We can then use this general statement (which we have just proven to be true) and apply it to specific numbers.

Skill Check #2

Consider the following process. *Select a number and multiply the number by 10. Now subtract 25 from that product. Then divide by 5. Finally, subtract the original number from the result.* Can you determine the general solution using deductive reasoning? Test the general solution with specific numbers.

To conclude this section, it is important to note that thinking mathematically is not just a skill for mathematics courses, but a skill for life. Being able to reason through information allows each of us to make informed decisions on a daily basis. Whether it is determining which route to take to school, deciphering a debate between political candidates, or solving a mathematics problem, thinking mathematically (that is, using the fundamentals of reasoning) allows each of us to reach conclusions that are well informed and based on the principles of thought outlined in this section. Chapter 3 includes a more involved discussion of logical thought that will further help you to develop your reasoning skills.

Skill Check Answers

1. **a.** 100   **b.** $4 + 12n$

2. To determine the general solution: Select a number: x; Multiply it by 10: $10x$; Subtract 25 from the product: $10x - 25$; Divide by 5: $2x - 5$; Subtract the original value: $2x - 5 - x$; This gives that the general result is $x - 5$.
   Now we test the general solution with specific numbers.
   Select three numbers: 2, 10, 15; Multiply them by 10: 20, 100, 150; Subtract 25 from the product: −5, 75, 125; Divide by 5: −1, 15, 25; Subtract the original value: −3, 5, 10; Result is always 5 less than the original number.

1.1   Exercises

Find a counterexample to each statement.

1.   Every Tuesday is an even day of the month.

2.   The product of two numbers is always even.

3.   If the difference between two numbers is even, then the numbers are both even.

4.   If a geometric figure has 4 sides, then the figure is a square.

5.   If $a > b$ and $a > c$, then $b > c$.

Find the missing terms of each sequence and determine if the sequence is arithmetic, geometric, or neither. If the sequence is arithmetic, state the common difference; if the sequence is geometric, state the common ratio.

6. $1, 3, 5, 7,$ _____ , _____ , _____

7. $1, 5, 9, 13,$ _____ , _____ , _____

8. $15, 10, 5, 0,$ _____ , _____ , _____

9. $5, 10, 20, 40,$ _____ , _____ , _____

10. $1, \dfrac{1}{2}, \dfrac{1}{4}, \dfrac{1}{8},$ _____ , _____ , _____

11. $3, 5, 8, 12, 17,$ _____ , _____ , _____

12. $10, 100, 1000, 10,000,$ _____ , _____ , _____

Find the next term of each sequence and determine if the sequence is arithmetic, geometric, or neither. If the sequence is arithmetic, state the common difference; if the sequence is geometric, state the common ratio.

13.

14.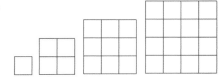

1

Identify whether each statement is an example of inductive or deductive reasoning.

15. The sides of all squares are proportional. The three quadrilaterals shown are all squares. Therefore all of their sides are proportional.

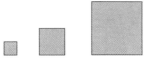

16. If Jessica plays basketball and makes 7 out of every 12 free throws, then she should make 35 out of every 60 free throws.

17. I have an 8:00 a.m. math class on Tuesdays and Thursdays. Each class day, I leave for class in my car at 7:30 a.m. Every day that the drive to campus takes 15 minutes, I arrive to class on time. Therefore, if I leave for class at 7:30 a.m. today and the drive to campus takes 15 minutes, I will be on time.

18. My grade on the first test in Quantitative Reasoning was 85, so I will make a B in the course.

19. If you live in New York, you are a resident of the United States.

20. All squares are rectangles, and all rectangles have four sides. Therefore all squares have four sides.

21. All known planets travel about the sun in elliptical orbits; therefore all planets travel about the sun in elliptical orbits.

Solve each problem.

22. Figurate numbers are numbers that can be represented by any regular geometric figure. (A regular geometric figure is a figure such as a triangle, pentagon, or hexagon where all sides have the same length.)

   a. Triangular Numbers: Triangular numbers are numbers that can be represented by an equilateral triangle. The first three triangular numbers, 1, 3, and 6, are shown in the figure. Find the next two triangular numbers and represent them with dots in the shape of a triangle.

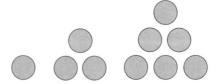

   b. Square Numbers: Square numbers are numbers that can be represented by a square. We discussed the first five square numbers, 1, 4, 9, 16, and 25, in Examples 2 and 5. Determine the 6th and 7th square numbers and represent them using squares in the shape of a square.

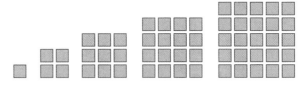

23. Consider the following pattern.

$$1 \cdot 12 = 12$$
$$11 \cdot 12 = 132$$
$$111 \cdot 12 = 1332$$
$$1111 \cdot 12 = 13{,}332$$

a. Based on this pattern, predict the product $111{,}111 \cdot 12$.

b. Check your prediction by finding the product.

c. What type of reasoning did you employ?

24. Consider the following information.

Multiplication Patterns		
Multiplication	**Repeated addition**	**Sum**
$4 \cdot -2$	$-2 + (-2) + (-2) + (-2)$	-8
$3 \cdot -7$	$-7 + (-7) + (-7)$	-21
$5 \cdot -6$	$-6 + (-6) + (-6) + (-6) + (-6)$	-30

What can you conclude about the sign of the final sum based on multiplication?

25. Jessica has decided to start doing push-ups as part of her exercise routine. She is keeping track of the number of push-ups done each day using the given bar chart.

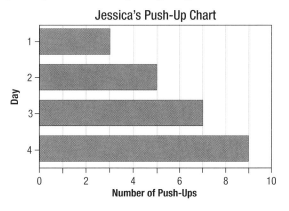

a. How many push-ups will Jessica do on day 15?

b. How many push-ups will Jessica do on day n?

c. Jack says since he does 20 push-ups a day as part of his exercise routine, he does more push-ups than Jessica. Is Jack correct? Explain

1.2 Problem Solving: Processes and Techniques

To begin this section, we need to establish the difference between a *problem* and *problem solving*. In mathematics, a **problem** is a situation where the answer, or even the methodology to arrive at the answer, is not immediately known. **Problem solving** is the actual act of performing mathematical operations, such as addition, multiplication, division, and subtraction, as well as mathematical reasoning, to arrive at a solution to a problem. The processes of problem solving is learned over an extended period of time through one's own persistence in finding a solution to a given problem. In short, once taught the processes of problem solving, individuals must exert the effort required to actually solve problems.

It should also be noted here that there is also a distinction between the *answer* and the *solution* to a problem. These two ideas are often mistakenly considered to be the same. They are not. The **answer** to a problem such as *What is the product of 4 and 6?* is simply 24; while the **solution** to this problem could be much more complex, since it includes all the reasoning and computation done to arrive at the answer. For example, a solution to this problem could be that "product" implies multiplication, so we have $4 \cdot 6 = 24$. Understanding both of these ideas will make you a better problem solver and mathematical thinker. The primary difference between the two is being able to distinguish between computational processes, or algorithms, and the process of understanding *how* to arrive at the solution to a problem. In essence, the problem-solving process cannot be reduced to a simple algorithm (which, by the way, is why so many people struggle with problems that require mathematical thought). In this section, we are going to spend a considerable amount of time discussing the problem-solving process and how to use this process to solve problems.

Pólya and Problem Solving

George Pólya established a four-step problem-solving process that can be employed in any situation, but is especially helpful when solving problems involving mathematical concepts. The process consists of the following steps.

1. Understand the Problem

Before you are able to solve a problem, you must first understand what is being asked. The steps to understanding the problem include reading and identifying pertinent information, discerning what is being asked, and assigning names or *variables* to unknown quantities. It is important to note that reading while problem solving means being an active reader; in other words reading slowly, writing down given information, drawing diagrams, and rereading if necessary.

2. Develop a Plan

Developing a plan is the part where most people struggle to solve problems. The secret to developing a plan that is effective in solving a problem is to determine which problem-solving strategy best fits the situation. The selection of a strategy is not exact since some problems may require multiple strategies and multiple attempts at a solution. Choosing a strategy will become easier with practice and experience.

Problem-Solving Strategies
Recognizing the steps and strategies used in problem solving will assist a great deal in becoming a better problem solver. The following strategies are great problem-solving methods.

- Find a pattern
- Adopt a new point of view
- Solve a simpler problem
- Work backwards

- Draw a picture
- Guess and test
- Account for all possibilities
- Use logical reasoning
- Use a variable

These are just a few of the possible strategies that may be employed when solving problems.

It should also be noted that this is the point in problem solving where many of us feel that topics seem easy in class, yet we struggle to set up problems in such a way that the problem-solving method is obvious when attempting problems at home. The reason for this is usually tied to the amount of practice each of us puts into attempting problem solutions and the amount of personal perseverance we have for working through problems to achieve a final solution. Each of these can be remedied by practice and focus.

3. Carry Out the Plan

Carrying out the plan involves the use of mathematical operations to determine the solution. In other words, this is the step in the problem-solving process where we use the algorithms we know to actually perform the operations as set out in our plan from Step 2.

4. Look Back

An essential part of problem solving is having the ability to be reflective in thought. We should ask ourselves the following questions after we arrive at an answer.

- Is my answer feasible and correct?
- What worked?
- What didn't work?
- Will the solution to this problem assist me the next time I see a similar problem?

Answering these questions after arriving at an answer will assist in all future attempts to solve a problem. Remember that problem solving is a skill that can take a significant amount of time and effort to master. However, with practice, everyone can become a better problem solver. The following figure is a visual representation and overview of the problem-solving process.

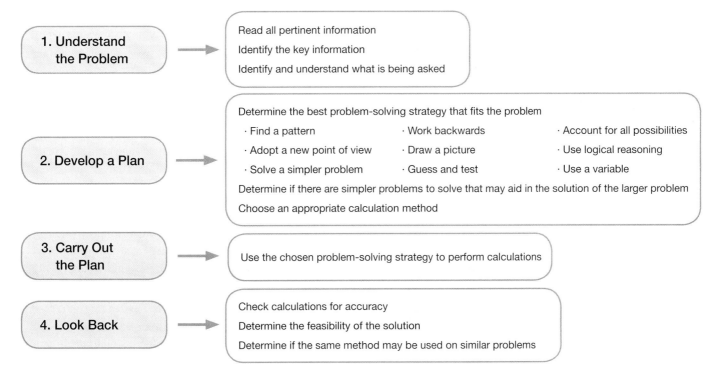

Figure 1: The Problem-Solving Process

1

Becoming a good problem solver does not happen overnight; it takes time to fully achieve. The more you practice, the easier problem solving will become. A problem that might be difficult for one person may seem trivial for you. For example, if you were asked to find the product $12 \cdot 11$, you would almost immediately arrive at a strategy that would allow you to find the solution. However, a student who is just learning the concept of multiplication might struggle quite a bit with a solution strategy. But with practice and by employing the concepts in this section, you can become an effective problem solver. The most common mistake made concerning problem solving centers around the inability to formulate a good plan for arriving at a solution. Since there are no algorithms per se for problem solving, the process can be more challenging. The more experience you have with different types of problems, the easier it will be to develop an effective plan. This is the time to try as many problems as you can and experiment with different types of strategies to understand how they work and when they are most useful.

Example 1: Working Backwards

While three watchmen were guarding an orchard, a thief slipped in and stole some apples. On his way out, the thief met the three watchmen, one after another, and to each in turn he gave one-half of the apples he had, plus two more in addition to that. In this way, he managed to escape with one apple. How many apples had he stolen originally?

Solution

1. Understand the Problem

After reading the problem, we can identify the pertinent information needed to solve the problem. The thief had one apple when he escaped the orchard. We are asked to determine how many apples he started with if he gave each of the three watchmen $\frac{1}{2}$ of the apples in his possession, plus two more.

2. Develop a Plan

Choosing a strategy that best reflects the situation often makes problem solving a little easier. In this problem, it seems it might be best to work backwards. Notice that we have the ending value of one apple already. Using this fact, we can find the solution by working backwards using the information given. When using the method of working backwards, note that all of the operations of arithmetic are done in reverse.

3. Carry Out the Plan

We have decided to work backwards. Recall from the initial problem that the thief is *dividing* his apples in half and then *subtracting* two more from the number of apples in his possession each time he meets a watchman. To work backwards, we must reverse the operations each time. So instead of subtracting 2 apples, we add 2. Instead of dividing his apples in half (that is, dividing by 2), we multiply by 2. Knowing that the thief escaped with only 1 apple, we use this idea to know that he must have had $1 + 2 = 3$ apples, and then twice that amount, or 6 apples, when he met the last watchman. Using the same logic, we can continue to work backwards to calculate the number of apples he had as he met each watchman in turn.

☞ Helpful Hint

One reason we may find mathematics challenging is that the language we use casually every day may have precise meanings in mathematics—perhaps different from its everyday meanings. Here, for example, the following all have the same meaning in mathematics:

dividing in half;
dividing in 2;
dividing by 2.

Be careful, dividing by ½ is NOT the same thing.

Table 1	
Number of Apples	
Escaped With:	**1 Apple**
3rd Watchman:	(1 apple + 2 apples) · 2 = 6 apples
2nd Watchman:	(6 apples + 2 apples) · 2 = 16 apples
1st Watchman:	(16 apples + 2 apples) · 2 = 36 apples

Thus, the thief originally stole 36 apples.

4. Look Back

The last step in the process requires that we consider the answer to determine if it is feasible. First check the answer to see if it is correct. We can use our solution of 36 apples to determine how many apples the thief has after meeting each watchman.

He meets the 1ˢᵗ watchman and gives him $\frac{1}{2}$ of his apples and 2 more.

$$36 \text{ apples} - \frac{1}{2}(36 \text{ apples}) - 2 \text{ apples} = 16 \text{ apples}$$

He meets the 2ⁿᵈ watchman and gives him $\frac{1}{2}$ of his remaining apples and 2 more.

$$16 \text{ apples} - \frac{1}{2}(16 \text{ apples}) - 2 \text{ apples} = 6 \text{ apples}$$

He meets the 3ʳᵈ watchman and gives him $\frac{1}{2}$ of his remaining apples and 2 more.

$$6 \text{ apples} - \frac{1}{2}(6 \text{ apples}) - 2 \text{ apples} = 1 \text{ apple}$$

When looking at our process, we should ask ourselves if what we have done makes sense and is feasible. The answer to both of these is yes. So, our answer is that the thief originally stole 36 apples.

Example 2: Guess and Test

Arrange the numbers 1, 2, 3, 4, 5, and 6 in the circles of the given triangle such that the sum along each side is 12.

Solution

1. Understand the Problem

Since we are given a figure to consider, understanding the problem must include understanding the figure. In this case, we are being asked to place the numbers 1 through 6 in the circles such that each side of the triangle has a sum of 12.

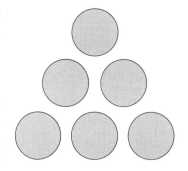

2. Develop a Plan

To devise a plan, we need to consider the possibilities of how to carry out the plan. Unless the answer is obvious to you, we most likely need to use the strategy of guess and test. Guess and test is an effective strategy for solving problems where you understand what is being asked and you are given enough information to "test" your results immediately. Guess and test is sometimes the best strategy for helping us to determine where to begin in solving a problem, and it allows us to eliminate erroneous solutions quite rapidly, thus leading to a proper solution more quickly. As you use this strategy more often, you will become better at developing an idea of where to start in finding the solution to a problem.

3. Carry Out the Plan

Our first guess will be to place the numbers 1 through 6 in any order on the circles and add the numbers along each side to see what happens.

```
      1
    6   2
   5  4  3
```

From our initial guess, we can see that the sum along each side ($1 + 2 + 3 = 6$, $3 + 4 + 5 = 12$, $5 + 6 + 1 = 12$) is not 12. We can use this information to formulate a better second guess.

Consider the possibilities of acquiring a sum of 12 when using the numbers 1 through 6. We can see through our first guess that $3 + 4 + 5 = 12$ and that $6 + 5 + 1 = 12$. The only mistake we had was with the third side. In fact, the only possible sums of 12 using the numbers 1 through 6 are the following.

$$1 + 5 + 6 = 12$$
$$2 + 4 + 6 = 12$$
$$3 + 4 + 5 = 12$$

Looking at our three sums, we can see that the numbers 4, 5, and 6 are each used in two of the three sums, while the numbers 1, 2, and 3 are used only once. This indicates that the numbers 4, 5, and 6 should be placed in the corner circles of our triangle.

Notice that having a well-designed plan makes the process that leads to a solution clear and simple.

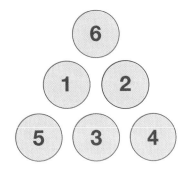

4. Look Back

To look back, we need to check to see that the sums of all of the sides are indeed 12. Now we can check our answer.

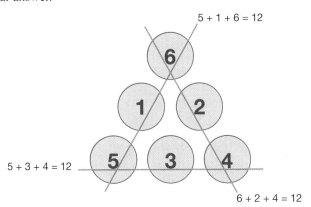

So, the sum of each side of the figure is 12. Therefore, we have a solution.

Skill Check # I

A magic square is a square arrangement of numbers in which the sums of all the rows, columns, and diagonals are equal. Using the numbers 2, 7, 12, 17, 22, 27, 32, 37, and 42 only once, determine the placement of the missing numbers in the following magic square.

17		7
12	22	
		27

Example 3: Solve a Simpler Problem

Find the sum of the numbers 1 through 1000.

Solution

1. Understand the Problem

In this problem, we are asked to add the numbers 1 through 1000 to find the total. In other words, we are asked to compute $1 + 2 + 3 + \ldots + 998 + 999 + 1000$. The sum will be quite large and using only paper and pencil to solve this problem may be tedious.

2. Develop a Plan

One plan we could use is to get out some scratch paper and start calculating. Another strategy could be to use a calculator with a memory key (M+) and press the key 1000 times. We could also just find the sum for each 100 numbers and then find the total sum.

This problem is an example of a classic problem and solution developed by Carl Gauss over 200 years ago. Did he solve the problem using a calculator? Of course not! There were none at the time.

While there are many ways to approach this problem, we are going to use the strategy of solving a simpler problem. Consider just the first few terms, 1 through 10. If we find the sum of these we get

$$1 + 2 + 3 + 4 + 5 + 6 + 7 + 8 + 9 + 10 = 55, \text{ or } 5 \cdot 11.$$

How is this helpful? Well, if you notice from our simpler problem, if we start with the outside numbers and work our way toward the middle, there are 5 "pairs" of 11.

$$1 + 10 = 11$$
$$2 + 9 = 11$$
$$3 + 8 = 11$$
$$4 + 7 = 11$$
$$5 + 6 = 11$$

Using this knowledge, we should be able to carry out a similar plan and find a solution.

3. Carry Out the Plan

Using the ideas we developed in devising the plan, we can now carry out the plan to find the sum of the numbers 1 through 1000. Applying the same logic to our problem, we can identify the following "pairs" of 1001.

$$1 + 1000 = 1001$$
$$2 + 999 = 1001$$
$$3 + 998 = 1001$$
$$\vdots$$
$$500 + 501 = 1001$$

This means we will have 500 pairs of 1001. Using our calculators or pencil and paper, we can easily obtain the final answer of $500 \cdot 1001 = 500,500$.

4. Look Back

We need to determine if the answer makes sense and is feasible. Have we committed any type of mathematical errors? Well, by solving the simpler problem, we have used deductive reasoning and know the given pattern will always follow as long as the numbers in the sequence remain consistent. So, we have indeed found the solution to be 500,500. (Note that we could have also used the strategy of finding a pattern to solve this problem.)

Fun Fact

Example 3 is based on a famous story about the mathematician Carl Freidrich Gauss. The story goes that as a child in primary school, after the young Gauss misbehaved, his teacher J.G. Büttner gave him a task to add a list of integers in arithmetic progression (1 to 100). To the teacher's astonishment, young Gauss produced the result in a matter of seconds.

Gauss found the solution by a method of realizing that pairwise addition of terms from opposite ends of the list yielded identical intermediate sums: $1 + 100 = 101$, $2 + 99 = 101$, $3 + 98 = 101$, and so on until $50 + 51 = 101$, for a total sum of $50 \cdot 101 = 5050$. However, the details of the story are at best uncertain. Some authors question whether it ever happened.

Source: http://en.wikipedia.org/wiki/Carl_Friedrich_Gauss

Example 4: Find a Pattern

Your uncle gives you $1 on the day you were born, $2 on your 1st birthday, $4 on your 2nd birthday, and so on, doubling the amount on each birthday. How much money does he give you on your 18th birthday?

Solution

1. Understand the Problem

The statement of the problem is quite clear. You are to find the amount of money your uncle gives you on your 18th birthday if the established pattern continues. The pattern indicates that the amount of money will double each year.

2. Develop a Plan

Since there is an established pattern from the information we are given, the selection of the strategy of finding a pattern seems most appropriate. While it is true that we could solve this problem in many different ways, we will try to find a pattern. We can see that the established pattern shows that the amount of the gift will double each year. The first six gifts (ending on your 5th birthday) would be the following amounts: $1, $2, $4, $8, $16, and $32. Do you notice anything about the amounts? These amounts are powers of 2, or "doubles" of the previous term. If you consider that you were 0 years old on the day you were born, then $2^0 = 1$. On your first birthday, you receive $2^1 = 2$ dollars, and so on. So it appears that on your nth birthday you will receive 2^n dollars.

3. Carry Out the Plan

Now that we have established what the pattern entails, we need to carry out the plan to find the amount received on your 18th birthday. From our knowledge formed in developing the plan, you should receive 2^{18} dollars, or $262,144. What a generous uncle!

4. Look Back

Looking back, we need to determine if the answer makes sense and is feasible. Have we committed any type of mathematical errors? At first glance, it seems almost improbable that the answer is feasible. After all, that is a lot of money. However, what we see here is the power of doubling. Think of it like this. If you have $100 and double it, you have $200. Double it again, you have $400. Continue to double, and you have $800, $1600, $3200, $6400, $12,800, and so forth. You can see how quickly these numbers grow. Considering that we understand what is happening in the problem, we can be reasonably certain that the solution is accurate and correct.

Think Back

Recall that exponential notation can be represented by a^x, where a is called the base and x is called the exponent. The exponent indicates the number of times the base should be multiplied by itself. That is, 3^4 indicates that the base 3 should be multiplied by itself 4 times, so $3^4 = 3 \cdot 3 \cdot 3 \cdot 3 = 81$.

Skill Check #2

Suppose you could spend $20 every minute of every day for a year. How much money would you spend in 365 days?

Example 5: Draw a Diagram

A baseball league is forming in which each of the teams will play four games against each of the other teams. There are five teams in the league: Raiders, Jackals, Blazers, Warriors, and Eagles. Determine how many total games will be played.

Solution

1. Understand the Problem

The problem asks us to consider five teams playing each other in a baseball league where they will all play each other four times. This means that each team will play 16 games. By this logic, it would appear that the total number of games played will be 80. We need to actually solve the problem, however, before settling on an answer.

2. Develop a Plan

As with other problems, there are many ways to approach a solution. For this example, we will use a diagram to solve the problem. Our plan is to let each team be represented by their team name, and use arrows to connect each team to the other teams.

3. Carry Out the Plan

Carrying out the plan here involves creating a visual representation of the problem. We begin with a figure that represents the games the Raiders will play against the Blazers (and consequently, the games the Blazers will play against the Raiders). Because each team plays each other four times, the line represents four games.

Next, we consider all games that will be played by the Raiders against all the teams (and the games they all play against the Raiders).

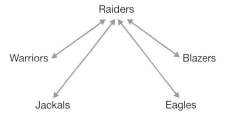

Continuing in this manner, we arrive at all of the possible game match-ups.

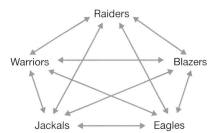

From the figure, we can see that there are 10 lines. Remembering that each line represents four games, we conclude that a total of 40 games will be played.

4. Look Back

Recall from our understanding of the problem that we thought there might be 80 total games played. Why is there a difference in our actual solution? The answer lies in the fact that, although the Raiders play the Blazers four times and the Blazers play the Raiders 4 times, there is a total of only four games between the two teams. With our initial thinking, when each team plays another team, we counted that game as a game for both teams. So, in our preliminary discussion of the problem, we counted each game twice.

Example 6: Make a List

How many ways can you make change for $0.50 out of quarters, dimes, or nickels?

Solution

1. Understand the Problem

We have been asked to determine the number of ways to make change for $0.50 using only quarters, dimes, or nickels. That means we need to look at all of the possibilities for using these coins to arrive at $0.50. Some obvious choices are 2 quarters, 5 dimes, or 10 nickels. But there are many more.

2. Develop a Plan

To solve this problem, we are going to make a list of all possible ways to make $0.50 out of quarters, dimes, or nickels. We will use a table here, but you can organize your list in any way you would like.

3. Carry Out the Plan

By making a list of the possible combinations of coins to obtain $0.50, we can see that there are a total of 10 ways to make change for $0.50 using only quarters, dimes, or nickels. How do we know we've exhausted all the possibilities? We know all of the possibilities have been exhausted because any other combination of quarters, dimes, or nickels would be a repeat of the ones already listed.

Table 2			
Possible Ways to Make Change for $0.50			
Quarters	**Dimes**	**Nickels**	**Amount**
2	0	0	$0.50
1	2	1	$0.50
1	1	3	$0.50
1	0	5	$0.50
0	5	0	$0.50
0	4	2	$0.50
0	3	4	$0.50
0	2	6	$0.50
0	1	8	$0.50
0	0	10	$0.50

4. Look Back

Looking back, we need to determine if the answer makes sense and is feasible. Have we committed any mathematical errors? In fact, we did not make any mathematical errors and our answer is quite feasible for the problem given. Therefore, we can feel confident that our answer of 10 combinations is correct.

Example 7: Use a Variable

A number is multiplied by 8 and the product is added to 6. If the sum is 30, what is the number?

Solution

1. Understand the Problem

We are being asked to multiply an *unknown* number by 8. Once we find this product, we are adding the product to 6. The sum is 30. We need to find the missing number.

2. Develop a Plan

Our plan is going to revolve around finding the missing number. Since the number multiplied by 8 is an unknown, we are going to use the variable x to represent the unknown number, then use algebra to solve for the unknown number.

3. Carry Out the Plan

We let the unknown value be x. Using the given information, we multiply our variable by 8 to get $8x$. Then we add 6 to get $8x + 6$. We know this sum is 30. So we have $8x + 6 = 30$. Solving for x, we need to subtract 6 from both sides, then divide by the coefficient of x.

$$8x + 6 = 30$$
$$8x + \cancel{6} - \cancel{6} = 30 - 6$$
$$8x = 24$$
$$\frac{\cancel{8}x}{\cancel{8}} = \frac{24}{8}$$
$$x = 3$$

We can see that the solution is $x = 3$.

4. Look Back

The primary objective of looking back is to check the answer to make sure it is correct. A number, 3, is multiplied by 8 and the product is added to 6 to get a sum of 30.

$$8(3) + 6 = 30$$

Skill Check #3

The sum of 3 consecutive odd natural numbers is 39. Find the three numbers.

Skill Check Answers

1.

17	42	7
12	22	32
37	2	27

2. $10,512,000

3. 11, 13, and 15

1.2   Exercises

Solve each problem using Pólya's problem-solving process.

1. Complete the following magic square using the numbers 1 through 16, where the sum of every row, column, and diagonal is the same.

16	2		
5	11		8
	7	6	
4	14		1

2. A farmer looks over a field and sees 37 heads and 98 feet. Some are goats, some are chickens. How many of each are there?

3. Jessica needs to mail a USB drive to her friend. She uses a combination of 41-cent stamps and 8-cent stamps to pay $1.71 in postage. How many of each stamp did Jessica use?

4. The Alpha Zeta Math Club is having a pizza party. The president of the club decides to have fun cutting the pizza and challenges the members to cut a pizza into 11 pieces with only 4 straight cuts. Show a way that this could be done.

5. A president and vice president for a sorority are chosen from five people. How many different president/vice-president combinations are possible?

6. Jeff is one year older than Erica, and Erica is one year older than Alden. If the sum of their ages is 75, how old is each person?

7. Holly needs to tile the floor in her new apartment. How many 8-inch square floor tiles are required to cover a rectangular floor that measures 15 feet by 18 feet?

8. The product of two whole numbers is 196 and their sum is 35. What are the two numbers?

9. The product of two whole numbers is 902 and their sum is 63. What are the two numbers?

10. Twice the difference of a number and 1 is 4 more than that number. Find the number.

11. There are eight teams in the intramural basketball league. Each of the teams will play each other five times. What is the total number of games played in the league?

12. There are six girls to every five boys in a psychology course. If there are 385 students enrolled in the course, how many are girls?

13. There are 8 boys to every 13 girls majoring in business. If there are 377 girls in the business degree program, what is the total number of students?

14. Tim bought a pair of Zeus running shoes on sale that were marked down 45% to $46.75. What was the original price of the shoes?

15. If a pound of coffee costs $8, how many ounces can be bought for $2.25?

16. Ali wants to buy three large pizzas. Marco's Pizza Emporium is having a special deal: Buy two large pizzas and get a third for half price. A large pizza costs $11.50 and Ali has $30. Does she have enough money to buy all three pizzas?

17. Arrange the numbers 1, 2, 3, 4, 5, and 6 in the given circles so that the sum of the numbers of each side is equal to 9.

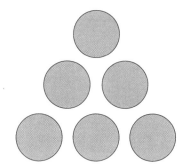

18. A number is multiplied by 8, and that product is added to 3. The sum is equal to the product of 5 and 7. Find the number.

19. Place the digits 1, 2, 3, 4, and 5 in the circles so that the sums across (horizontally) and down (vertically) are equal.

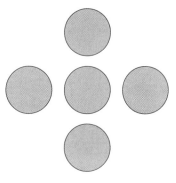

20. The houses on Elm Street are numbered 1 through 200. How many of the house numbers contain at least one digit that is a 9?

21. Three apples and two pears cost $0.78. Two apples and three pears cost $0.82. What is the total cost of one apple and one pear?

22. a. Place the numbers 1, 2, 3, 4, 5, 6, 7, 8, and 9 in the magic square so that the sum of the numbers in each row, column, and diagonal is equal to 15.

b. Use the digits 0, 1, 2, 3, 4, 5, 6, 7, and 8 to create another magic square so that the sum of the numbers in each row, column and diagonal is equal to 12.

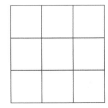

23. A rectangle has an area of 72 square inches. Its length and width are whole numbers.

a. What are the possible dimensions of the rectangle?

b. Which of those dimensions yield a rectangle with the smallest perimeter?

24. Find the sum of the whole numbers from 1 to 900.

25. Find the sum of the even numbers from 1 to 500.

26. Find the sum of the odd numbers from 1 to 700.

Solve each problem.

27. One of the most interesting and useful patterns in all of mathematics is called Pascal's triangle, as shown. It is named after the French mathematician Blaise Pascal (1623–1662) who showed that these numbers play an important role in the theory of probability.

$$
\begin{array}{c}
1 \\
1 \quad 1 \\
1 \quad 2 \quad 1 \\
1 \quad 3 \quad 3 \quad 1 \\
1 \quad 4 \quad 6 \quad 4 \quad 1 \\
1 \quad 5 \quad 10 \quad 10 \quad 5 \quad 1 \\
1 \quad 6 \quad 15 \quad 20 \quad 15 \quad 6 \quad 1
\end{array}
$$

a. Copy the array and fill in the next three rows of numbers using the pattern(s) that you notice.

b. Compute the sum of the elements in each row of Pascal's triangle for the first 6 rows.

c. Look for a pattern in the sums of the first 6 rows. See if you can predict the sum of the elements for row 7, then check by adding. Do the same for rows 8 and 9. State the general rule for the sum of the elements in words.

28. Looking at each row of figures in the given diagram, determine the total number of rectangles found in each row. Use the pattern for the first four rows to determine the number of rectangles in the last row.

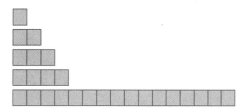

29. Write down the next three rows to continue this sequence of equations.

$$1 = 1 = 1^3$$
$$3 + 5 = 8 = 2^3$$
$$7 + 9 + 11 = 27 = 3^3$$
$$13 + 15 + 17 + 19 = 64 = 4^3$$

30. What is the ones digit in the number 3^{2034}? (**Hint:** Start with smaller exponents to find a pattern.)

1.3 Estimating and Evaluating

Along with the reasoning and problem-solving skills covered in Sections 1.1 and 1.2, critical thinking also involves the ability to estimate and evaluate. These skills are part of a bigger picture of being *numerate*; that is, having the ability to think about and communicate with numbers, just as being *literate* is having the ability to function in the world of words. The beauty is that numeracy and literacy overlap more often than we think. Consider questions like the following.

- *Did you get charged the correct amount when you purchased your shirt at the mall?*

- *Does your final average in chemistry reflect the work you did throughout the semester?*

- *How much time will you require to finish a project?*

- *Is it reasonable to say that 95% of Americans live a middle class lifestyle?*

Reading these questions requires literacy while answering them requires numeracy. We'll continue our critical thinking skills by applying reasoning and the steps of problem solving to estimation and evaluation.

Estimating

To **estimate** means to make an approximate calculation, or to roughly judge the value of a quantity. It is important to note that when making these rough calculations, the estimate needs to be an *accurate* estimation, not just any "shot in the dark" guess. It's actually quite easy to make any approximation, but one that has accuracy requires more thought. Consider, for instance, the moment when your waiter brings your bill for the evening's meal. How do you know the waiter correctly calculated your total amount due? Usually, a computer handles the actual addition, but a human is responsible for the input of the items into the computer. This is where errors tend happen. Was everything you ordered on the ticket? Are there charges for items you did not order? If you have an estimated total in mind for your bill before the ticket is brought, you'll be poised to notice when there might be an error.

> ### Estimate
>
> To **estimate** is to make an approximate calculation or to roughly judge the value of a quantity.

Consider for a moment the following job advertisement posted online at Monster.com.

> Top 100 Homebuilder located in Columbus, Georgia, is looking for an estimator. Computer proficiency and general construction knowledge is required. Degreed individuals are encouraged to apply. Timberline Estimating and Builder MT experience is helpful but not essential. Candidates will be judged on the following criteria.
>
> - Analytical skills leading to competent decision making
>
> - Task-oriented proven work performance with emphasis on self-starting
>
> - Team-oriented attitude
>
> - Strong written and oral communication skills
>
> Salary will be commensurate with experience. Competitive benefit package. [1]

Are you qualified for this job? Chances are that you might answer "Of course not. I don't have any construction experience." But look again; yes, construction knowledge is required, but so are the skills of estimation and critical thinking.

There are, of course, sophisticated software programs that analyze data and calculate estimates, such as the one mentioned in the job advertisement. However, since estimation and rounding are skills needed for ordinary, daily activities as well as for opening up more job opportunities, it is useful to know effective methods for estimating and rounding.

We've all been asked to round numbers to specific decimal places before. Instead of the "how to," we'll focus more on the decision process of determining an appropriate estimate given a certain situation. In most situations, there most likely will not be someone looking over your shoulder telling you to which place you should round your estimation.

Remember, estimates are an *educated guess*, not an exact amount. So, first, you'll need to decide how precise you want your estimate to be. If you are estimating a food bill, rounding to the nearest dollar is probably sufficiently detailed. However, if you're estimating the cost of your grocery bill, rounding to the nearest ten dollars might be adequate.

Another consideration for rounding is *when* should it occur. In other words, should you round the numbers you are dealing with before any calculations are done, or wait and round at the end of all calculations? Because we are working at making a rough calculation, choose the method where the numbers are easiest to work with, keeping in mind the particular situation. For instance, when making purchases, it might actually be better to round up at the beginning and overestimate the total price rather than underestimate it and be short on money at the register. However, overestimating the amount of income you'll earn from working six hours of overtime next week could be rather disappointing on payday.

Using Pólya's problem-solving steps along with reasoning skills can help us tackle estimation in an informed manner. Let's take a look at an example.

Think Back

Estimations are approximations. The symbol ≈ means "is approximately equal to."

Example 1: Using Pólya's Problem-Solving Steps in Estimation

Suppose you are planning a wedding reception for 100 people. This is your first time planning such an event. Estimate the cost of the reception.

Solution

Let's use Pólya's steps to help us with the estimation.

1. Understand the Problem

If we are estimating the cost of the reception, then we need to account for all of the possible expenses we might encounter and add the approximate expenses to find an estimate for the total.

2. Develop a Plan

There are several factors that go into the cost of the reception. Breaking the overall cost into smaller segments by listing as many of these as possible is a great place to start. This is a way to account for all possibilities while solving simpler problems.

3. Carry Out the Plan

How many cost factors can you name? If you're not in the reception business, you might seek the help of an online planner or other resource so that you don't leave anything out. Suppose an online search for "planning a reception checklist" produced the following.

Venue cost

Food

Décor

Entertainment

First of all, you need a venue big enough to comfortably hold 100 people. Since this cost varies based on location, it is best to get a few rough estimates—again, this can be done online. The Wedding Report for 2011 lists the average cost of reception venues as $3228. Since all we need is an estimate, we'll use this average. However, keep in mind that some venues include the cost of food in the total cost of the venue, while other venues do not include this and may still cost tens of thousands of dollars. [2]

Food for a reception is usually based on a "per person" amount. As a result, changing the number of guests could drastically change the total cost of the food. Several wedding sites give total cost averages in the range of $4750. Again, keeping in mind that we're looking for an estimate and not an exact figure, this will help keep us moving forward on finding the estimate and keep us from being bogged down in finding the least expensive price per head.

Decorations can be anything from very simple to spectacular! However, the average reported cost is in the $900 range for reception decorations and flower arrangements. So, we'll use this for our estimate.

Last, assuming entertainment is a desired feature of the reception, we'll need to make an initial decision based on some quick research. Having a DJ or single musician ranges from $300 to $700, while having a live band ranges from $900 to $1600. It might seem like averaging the two of these would give a good general idea, but stop and think for a minute. An average of the lowest and highest amounts would give us an amount of $950. This amount is either far too much for the first category, or not really enough for the higher category. So instead of being so far off in either direction, we will simply chose one of the categories and mention it in our estimation. For now, choosing the lower end of having a DJ or single musician, means that entertainment cost is about $\frac{\$300 + \$700}{2} = \$500$.

Now, we're prepared to combine all of the individual estimates in order to arrive at a total estimate for a wedding reception.

Venue cost	$3228
Food	$4750
Decor	$900
Entertainment	$500
	$9378

Think Back

Only consider the digit directly to the right of the place you are rounding to. Round down for digits less than 5. Round up for all other digits.

Since the estimation price is $9378, now might be a good time to consider if rounding should play a part, and if so, how precise it should be. Because this is an estimate, giving prices that end in $78 might appear to be more accurate than we were anticipating. Since we are new to the game of planning a reception, friends, relatives or whomever the estimation is for might assume that enough research went in to be confident enough to estimate to the nearest dollar. Although we did do the research, this is unlikely. Rounding to the nearest hundred dollars or the nearest thousand dollars might paint a better picture of the estimate. Rounding to the nearest hundred dollars gives $9400 while rounding to the nearest thousand dollars gives $9000.

Consider the scale of the project. A $400 difference in price is meaningful in a nine-thousand-dollar project. Therefore rounding to $9400 for the estimation is the better choice.

Let's stop for a moment and consider when a difference of $400 might not make such a large impact. Suppose the estimate was more like $32,378. When considering a budget near thirty-two thousand dollars, $400 is no longer likely to be a game-changer.

4. Look Back

The last step in the process requires that we consider the answer to determine if it is feasible. If you were just beginning the process of planning a wedding reception, this amount could either cause sticker shock or be quite a nice surprise, depending on your past experience with weddings. Since we based our intermediate figures on national averages, it is safe to assume that this is a middle-of-the-road figure. What we have learned in the process is that there are several areas where we might be able to save money, but also places where we might want to splurge, for example, hiring a live band to dance the night away!

Skill Check #1

Using the estimation prices given in Example 1, what is the estimated cost for the same reception if the number of guests was only 50? Or 200 (assuming that all venues in our previous calculations could accommodate 200 people)? Think carefully about how you might arrive at the estimates.

Example 2: Using Deductive Reasoning in Estimation

A public university is expanding its campus by building a new Math and Computer Science Center. Estimate the cost of the new center by using the cost specifications of the most recently constructed building on campus. The previous classroom building had 20,000 square feet (sq ft) of assigned space, with a construction cost of $6,000,000. The new center needs to contain 40,000 sq ft of assigned space.

Solution

The problem clearly asks for an estimate of the building cost for the new center based on previous figures. This is both Step 1 and Step 2 in our problem-solving process. We understand the problem and are given a plan of action. We can then use deductive reasoning to carry out that plan. State-funded universities build classroom buildings on roughly the same level of quality. Therefore, since we are given the construction cost of the first building, we can reasonably conclude that the new center will cost approximately the same amount per square foot. Because we are doubling the square footage of assigned space in the new center, we will need to double the construction cost. Therefore, using the information we are given, a good estimate for the cost of the new Math and Computer Science Center is $12,000,000.

Let's take a moment to appreciate the value of accuracy when estimating. There is not really any such thing as a "wrong" estimate. Any estimate made using the original values is a viable estimate. However, some estimates are more useful than others. What then is considered a good estimate? Is being within 10% good enough? 15%? The correct answer might be "it depends." Think about the two examples that we've discussed: a wedding reception costing $9400 and a building costing $12,000,000. A 10% difference on each of these estimates is as follows.

Wedding reception: 10% of $9400 = $940

University building: 10% of $12,000,000 = $1,200,000.

It's now a little more obvious why the accuracy of an estimate depends on the scenario. $940 is a reasonable amount of accuracy for the cost of a wedding reception, but missing the mark by $1.2 million in a construction estimate is certainly less useful.

Example 3: Estimating Using Graphs

A 2011 study of student graduation data by Complete College America showed that on average, students end up taking significantly more credits than required to complete their college degree.[3] According to the graph, students graduating with a certificate are required to take 30 credit hours.

Students Are Wasting Time On Excess Credits

| | | | = 5 Needed Credits |
| | | | = 5 Excess Credits |

Certificate
Should take **30** credits
Students take **?** credits

Associate
Should take **60** credits
Students take **79.0** credits

Bachelor's
Should take **120** credits
Students take **136.5** credits

a. Notice that the number of credits that students receiving a certificate take is missing. Estimate the number of credits students receiving a certificate actually take.

b. If the average cost of a credit hour is $287 when seeking a certificate, estimate the total cost for courses that students *should* pay for a certificate.

c. If the average cost of a credit hour is $287 when seeking a certificate, estimate the cost for courses that students end up paying if they take the average number of credits for a certificate. Use your answer from part **a.**

Solution

a. We can see that under the Certificate heading, we are told that students should take 30 credits to complete the degree. The legend explains that the green books represent the 30 credits needed for the certificate and the blue books represent the excess credits that the students take while pursuing a certificate. We can compare the two stacks and see that they represent about the same number of credits. In other words, on average students take twice as many credits as they need to get a certificate. So a good estimate for the number of credits students actually take when receiving a certificate is 30 + 30 = 60. In reality, the actual figure for the average number of credits students take when receiving a certificate is 63.5.

b. Students should take 30 hours to receive a certificate based on the graph. For our estimate, we can round the price of a credit hour to make our estimation easier. Since the cost is $287 on average, we can round to $300. Multiplying these two together gives us the following.

$$30 \cdot \$300 = \$9000$$

So, on average, students receiving a certificate should pay approximately $9000.

   c. Based on part **a.**, we know that students take twice as many credit hours as needed for a certificate. Therefore, we can double the cost that they should pay in order to find the cost they actually pay.

$$\$9000 \cdot 2 = \$18,000$$

Skill Check #2

Use the graph in Example 3 and estimate the number of excess credits students take when earning a bachelor's degree.

Whether you consciously realize it or not, you are constantly being asked to intake information, problem solve, and make a judgment. Whether it be commercials asking you to buy their product, insurance agents asking you to trust their judgment, politicians expecting you to believe their view of the world, or newspapers presenting their side of the story for you. One aspect of being a good citizen and member of society is thoughtful evaluation of what others might present to us and problem solving in a thoughtful manner. Remember that a critical thinker is active, not passive.

Skill Check Answers

1. The only amount likely to change is the cost of food since it is the only one affected by the number of guests. Therefore, a reasonable estimate would be halving the cost of food for 50 people, or doubling it for 200. Therefore, $7003 or $7000 rounded to the nearest hundred dollars is an appropriate estimate for 50 guests, and $14,128 or $14,000 for 200.

2. On average, students take approximately 17 extra credit hours.

1.3   Exercises

Round each number as indicated.

1. 817 to the nearest ten

2. 588 to the nearest hundred

3. 12,399 to the nearest thousand

4. 10,048 to the nearest thousand

5. 109,999 to the nearest hundred thousand

Estimate each product. Show your process of estimation.

6. $391 \cdot 22$

7. $45 \cdot 33$

8. $14,581 \cdot 22,912$

9. $5247.11 \cdot 418$

10. $2.88 \cdot 55$

1

In each scenario, estimate the required amounts. Show your process of estimation.

11. Suppose your project needs one project manager, two user-experience professionals, and three app developers. The rates for each of these team members are as follows.

Project manager: $118.50 per hour

User-experience professionals: $88 per hour

App developers: $125 per hour

The project requires the project manager for 100 hours, each user experience professional for 50 hours, and each app developer for 100 hours. Estimate the total cost of the project.

12. In his first week at college, Chase spent the following.

Entertainment: $28

Apparel: $24

Travel (gas): $18

If there are 12 weeks in the semester, estimate the amount of money Chase will spend on these things throughout the semester.

Solve each problem.

13. The given line graph shows the federal hourly minimum wage over time, adjusted for inflation. [4]

a. Estimate the overall increase in wages from 1935–2010.

b. Estimate when the highest adjusted minimum wage occurred.

c. Over what time periods did the adjusted minimum wage decrease?

14. As the administrative assistant in the provost's office, Allie has to keep supplies stocked. She placed a stationary order for the office. She purchased 3 boxes of office paper for $89.97, pens for $43.96, and clips/rubber bands for $24.97. The total bill was $169.98. Is this a reasonable amount if tax were added? Explain your answer.

15. Trying to prepare a budget for his first few months in his new place, Darren has done research and found the following average costs per month: health insurance $183, phone/internet/cable $140, electric $179. If Darren will be earning $1700 per month, estimate the amount of money he will have after paying these bills each month.

16. According to TrueCar.com, the average transaction price of light vehicles in the United States for April 2012 was $30,303. Recent statistics show that one-third of car buyers sign up for a six-year loan at an average interest rate of 2.49%. With no down payment, the monthly payment is $454. Estimate the actual cost of the average new car after six years. Estimate the amount of excess money you paid for the new car by financing it over six years.

17. Jane had a dream to complete an Ironman triathlon. The race consists of a 2.4-mile swim, a 112-mile bike ride, and a marathon run of 26.2 miles. In training, Jane can swim at an average pace of 1.9 mph. She rides her bike at an average of 20.8 mph, and runs at 4.1 mph. Estimate how long it will take her to complete the Ironman triathlon.

18. Amelia is planning a road trip from Nashville, Tennessee, to Washington, D.C. The distance from Nashville to Washington is 651 miles and will take about 36 gallons of gas to complete.

    a. If gas costs an average of $3.745, estimate the fuel cost for the one-way trip.

    b. Use your answer for part a. to estimate the total cost of fuel for the round trip.

19. Because Elias is self-employed, he is advised to pay quarterly estimated federal tax payments since his income tax withholding will not fully cover next year's tax liability. Online advice instructs him to determine last year's tax, minus any withholding, and divide by four. Estimate the quarterly taxes Elias should pay if his total tax from last year was $23,741 and his total withholding was $7500.

20. Debbie needs to purchase some printer paper and USB jump drives for the department where she works. Each ream of paper costs $4.55 and each pack of 4 jump drives costs $27.85. She needs nine reams of paper and six packages of jump drives. Approximate to the nearest dollar the amount of money Debbie will spend to make the purchase.

21. Marcel has decided to paint his new apartment. He estimates the cost of paint will be $385. Which estimate should Marcel use for the job so that his budget includes a contingency factor of 8% in addition to the cost of paint?

    a. $38

    b. $432

    c. $395

    d. $800

22. Three students estimate the product $57 \cdot 159$. Explain which of the following estimations will be the most precise and why: $50 \cdot 100, 60 \cdot 200, 60 \cdot 160$.

23. The following line graph represents the amount of financial aid in the form of loans, grants, and work study funds at an institution for the years 2000–2010. Answer each of the following questions.

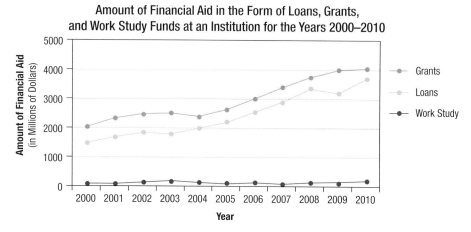

Amount of Financial Aid in the Form of Loans, Grants, and Work Study Funds at an Institution for the Years 2000–2010

a. Estimate the share of aid from grants for the year 2003.

b. Estimate the total financial aid from these three funding sources in the year 2000.

c. Estimate the total financial aid from these three funding sources in the year 2007.

d. Estimate which year represents the greatest difference between grants and loans.

24. As part of planning for a pool party, Jamaal has $155 to spend. He wants to invite at least 23 people. Estimate the maximum amount of money he can spend on average for each person in attendance.

Chapter 1 Summary

Section 1.1 Thinking Mathematically

Definitions

Reasoning

Reasoning is the drawing of inferences or conclusions through the use of statements offered as explanation or justification.

Inductive Reasoning

Inductive reasoning is a line of reasoning that arrives at a general conclusion based on the observation of specific examples.

Counterexample

A counterexample is one situation that does not satisfy the conclusion of an argument, and thus proves that the argument is invalid.

Arithmetic Sequence

An arithmetic sequence is when the common difference between any two consecutive terms in a number sequence is the same.

Geometric Sequence

A geometric sequence is when the common ratio between any two consecutive terms in a number sequence is the same.

Deductive Reasoning

Deductive reasoning is a process that begins with commonly accepted facts and logically arrives at a specific conclusion.

Section 1.2 Problem Solving: Processes and Techniques

Definitions

Problem

In mathematics, a problem is a situation where the answer or even the methodology to arrive at the answer is not immediately known.

Problem Solving

Problem solving is the actual act of performing mathematical operations, such as addition, multiplication, division, and subtraction, as well as mathematical reasoning to arrive at a solution.

Solution

The solution to a problem includes all the reasoning and computation done to arrive at the answer.

Pólya's Problem-Solving Process

1. Understand the problem

2. Develop a plan

3. Carry out the plan

4. Look back

Section 1.3 Estimating and Evaluating

Definition

Estimate

To estimate is to make an approximate calculation or to roughly judge the value of a quantity.

Chapter 1 Exercises

Find a counterexample to each statement.

1. The Super Bowl is played on the first Sunday in February.

2. The product of an even number and an odd number is odd.

3. If the sum of two numbers is odd, then both numbers are odd.

4. A number multiplied by itself is always even.

Find the missing terms of each sequence and determine if the sequence is arithmetic, geometric, or neither. If it is an arithmetic sequence, state the common difference; if it is a geometric sequence, state the common ratio.

5. 4, 7, 10, _____ , _____ , _____

6. −14, −7, 0, _____ , _____ , _____

7. 3, 6, 18, 36, _____ , _____ , _____

8. 4, 12, 36, 108, _____ , _____ , _____

9. $\dfrac{1}{3}, \dfrac{1}{9}, \dfrac{1}{27},$ _____ , _____ , _____

10. 7, 5, 9, 7, 13, _____ , _____ , _____

Use inductive reasoning to predict the next line of each pattern. Complete the computations to verify.

11. Consider the following pattern.

$$1 \cdot 8 + 1 = 9$$
$$12 \cdot 8 + 2 = 98$$
$$123 \cdot 8 + 3 = 987$$
$$1234 \cdot 8 + 4 = 9876$$
$$12{,}345 \cdot 8 + 5 = 98{,}765$$
$$123{,}456 \cdot 8 + 6 = 987{,}654$$
$$1{,}234{,}567 \cdot 8 + 7 = 9{,}876{,}543$$

12. Consider the following pattern.

$$999,999 \cdot 1 = 999,999$$
$$999,999 \cdot 2 = 1,999,998$$
$$999,999 \cdot 3 = 2,999,997$$
$$999,999 \cdot 4 = 3,999,996$$

Solve each problem.

13. A pet store employee counts 23 heads and 68 feet on the animals in the store. The store currently only has kittens and parakeets left in the store. How many of each are there?

14. A box office sells tickets for a concert at two different prices: $35 and $55. The box office sold a total of 95 tickets at a value of $4325 today. How many of each of the different tickets did the box office sell?

15. A president and vice president for a club must be chosen from four people. How many different president/vice-president combinations are possible?

16. The product of two whole numbers is 234 and their sum is 31. What are the two numbers?

17. Jessica bought a coat on sale that was marked down 65% to $68.25. What was the original price of the coat?

18. Place the numbers 1 through 9 in the given circles so that the sum of the numbers on each side of the triangle is 20.

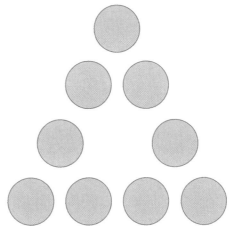

19. A rectangle has an area of 120 square inches. Its length and width are natural numbers.

   a. What are the possible dimensions of the rectangle?

   b. Which of those dimensions yield a rectangle with the smallest perimeter?

20. Find the sum of the whole numbers from 1 to 200.

21.   How many line segments are there in each of these figures?

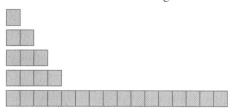

22.   What is the ones digit in the number 3^{341}?

23.   If there are 4 million childbirths per year in the United States, how many children are born every minute on average? (Assume there are 365 days in a year.)

24.   A car travels 27 miles in 20 minutes, maintaining a constant speed. How fast is the car traveling in miles per hour?

Bibliography

1.3

1. Monster, s.v. "Estimator Construction Job." Accessed September 11, 2012. http://jobview.monster.com/Estimator-Construction-Job-columbus-GA-US-114041998.aspx

2. The Wedding Report. "2011 Cost Update." Accessed September 2012. http://www.theweddingreport.com/bz/index.php/2011-wedding-cost-update-3-4-decrease-from-2010-spending/

3. Complete College America. "Time is the Enemy." September 2011. http://www.completecollege.org/docs/Time_Is_the_Enemy_Summary.pdf

4. US Department of Labor, Wage and Hour Division (WHD). "History of Federal Minimum Wage Rates Under the Fair Labor Standards Act, 1938-2009." Accessed January 24, 2012. http://www.dol.gov/whd/minwage/chart.htm

   US Department of Labor, Bureau of Labor Statistics (BLS). "CPI Inflation Calculator." Accessed June 18, 2012. http://data.bls.gov/cgi-bin/cpicalc.pl

Chapter 2
Set Theory

Sections

Objectives

- Develop an understanding of set operations
- Use Venn diagrams to represent sets
- Use Venn diagrams to solve problems with sets
- Solve problems that involve survey analysis

Set Theory

When you are hanging out with your friends and you are all trying to determine a place to go to dinner, the decision is usually based on some common attributes of the restaurant you each prefer. One reason you might choose a restaurant is because several members of the group enjoy the milkshakes at that particular place or the restaurant may be close to where everyone works. Finding the ways in which the group opinions and choices overlap is part of the solution to finding a restaurant that makes everyone happy.

The same concept applies when formal surveys are used. The people analyzing the survey need techniques to help them understand where responses are the same, or overlap. For instance, suppose a school nutritionist is conducting a survey of student eating habits. He asks all of the students to fill out the following survey.

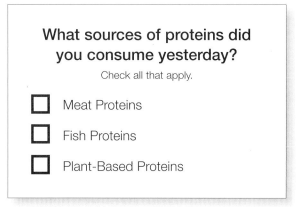

Figure 1

Using the methods in this chapter, the results of the student surveys can be analyzed and displayed visually to help the nutritionist understand student preferences on proteins for planning future meals in the cafeteria.

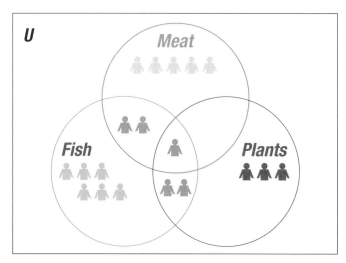

Figure 2

2.1 Set Notation

Have you ever wondered how a pizza restaurant organizes all of their toppings in a manner that allows them to efficiently create pizzas to order? How is it possible that a supermarket keeps all of the products organized in a manner that allows shoppers to search for and find products? The answer is quite simple: the use of mathematical sets. From the ways in which you select clothes in your closet to how you organize the apps on your smartphone, the use of sets allows us to have a systematic way to analyze groups of items and information.

A flock of sheep, a litter of kittens, and an ensemble of musicians are all sets of things with living members. The alphabet is a set of letters; a line is a set of points. No matter what the set consists of, the mathematical concept of sets will be the building block for many areas of mathematics we will look at throughout this book.

Set

A **set** is a collection of objects made up of specified elements, or members.

To begin, we need to explain some notation for describing a set. Mathematical sets are commonly represented using capital letters and the elements of the set are represented by lowercase letters. We can express the member x as being an element of a set G symbolically by $x \in G$ (read "x is an element of G"). If y is not an element of G, we write $y \notin G$ (read "y is not an element of G").

One of the most common ways to describe a set is to simply make a list all of the elements in the set. This notation is called the **roster method**. The common way to write a set using the roster method is for the elements of the set to be surrounded by braces and separated with commas. For example, the following sets A, B, C, and D are described using the roster method.

$$A = \{1, 2, x, y, z\}$$

$$B = \{x, 1, 2, y, z\}$$

$$C = \left\{\frac{1}{2}, \frac{2}{3}, \frac{3}{4}, \frac{4}{5}\right\}$$

$$D = \{\text{Gwen, King Charles spaniel, Zoë, taxi}\}$$

Roster Method

Roster notation is a way to describe a set by listing all of the elements in the set.

Notice that set D does not contain any numerical elements at all. Sets are not limited to numbers. In fact, they can contain anything—real or imaginary. An element of a set may be a number, but it could also be a griffin, a lychee, or a caterpillar. The only thing the elements of a set might have in common is that they are all members of that set. In fact, did you wonder what the elements in D all have in common when you first looked at that set? The only obvious thing that they have in common is that they are all members of the set D.

In set theory, sets themselves can even be elements of other sets. Consider the set G, which is a set consisting of sets of paired numbers.

$$G = \{\{1,2\}, \{3,4\}, \{5,6\}, \{7,8\}, \{9,10\}\}$$

Fun Fact

The griffin is a mythical creature considered to be the king of all creatures, known for guarding valuable and precious articles.

Heraldic guardian griffin at Kasteel de Haar, Netherlands

Native to the Guangdong and Fujian provinces of China, the cultivation of the tropical fruit lychee can be traced back to at least 2000 BC.

Lychees for sale at a Malaysian fruit stall.

The individual numbers 1 through 10 are not members of the set G; however the sets of paired numbers such as $\{1, 2\}$ are.

It's important to point out that there is no concept of order in a set. Thus, the two sets A and B that we introduced previously are really the same set because they contain the same elements. We've listed them again so that you can check for yourself.

$$A = \{1, 2, x, y, z\}$$

$$B = \{x, 1, 2, y, z\}$$

When two sets contain exactly the same elements, the sets are said to be equal. So, we can write $A = B$.

Equal Sets

Two sets are said to be **equal** if they contain exactly the same elements. If sets A and B are equal, we write $A = B$.

Sets come in all shapes and sizes: some have infinitely many elements and some have a finite number of elements. Recall that a set having infinitely many elements means that a list of the elements would go on forever; whereas a finite set has a specific number of elements. The roster notation for an infinite set often includes a series of three dots, called an ellipsis, that indicates that the set continues on without ceasing. For instance, the set of positive integers can be written as $\mathbb{Z} = \{1, 2, 3, 4, \ldots\}$. An ellipsis can also be used to indicate when a certain pattern is continued within a list of set members. For instance, when describing the even numbers between 2 and 100, we could write $L = \{4, 6, 8, \ldots, 98\}$.

☞ **Helpful Hint**

Some textbooks may use $n(\ )$ to indicate the cardinal number of a set.

In this text, we will focus most of our discussion on finite sets. Since finite sets have a specific number of elements, we can discuss their cardinal numbers. The cardinal number of a finite set is the number of elements contained in the set, or in other words, the size of the set. The cardinal number is denoted by $|\ \ |$. For example, if $A = \{a, b, c, d, e\}$, then $|A| = 5$.

Cardinal Number

The number of elements contained in a finite set is called the **cardinal number**, or **cardinality**. The cardinal number is denoted by $|\ \ |$.

Example 1: Using the Roster Method to Represent a Set

Use the roster method to represent S, the set of states in the United States that begin with the letter M. Then, find $|S|$.

Solution

There are eight states in the United States that begin with the letter M. Therefore,

$S = \{\text{Maine, Maryland, Massachusetts, Michigan, Minnesota, Mississippi, Missouri, Montana}\}$

Because S contains eight elements, $|S| = 8$.

Note that although we listed the elements in S in alphabetical order for the ease of the reader, this is not necessary when using the roster method. Remember that the order of the elements does not matter in a set.

When two finite sets have the same cardinal number, that is, the same number of elements (no matter if the elements are of the same type or not), the sets are said to be **equivalent**. For instance, sets C and D both contain 4 elements, so C is equivalent to D. We've listed the sets again so that you can check for yourself.

$$C = \left\{ \frac{1}{2}, \frac{2}{3}, \frac{3}{4}, \frac{4}{5} \right\}$$

$$D = \{\text{Gwen, King Charles spaniel, Zoë, taxi}\}$$

Symbolically, we write $C \sim D$, which is read "C is equivalent to D."

Equivalent Sets

Sets are **equivalent** if they have the same cardinal number; that is, the same number of elements. If sets C and D are equivalent, we write $C \sim D$.

Example 2: Determining Equal and Equivalent Sets

Determine if the given pairs of sets are equal, equivalent, or neither.

a. $A = \{$Public Health, International Studies, Mechanical Engineering, Music Education, Political Science$\}$

$B = \{$Tim, Gloria, Alan, Warren, Karen$\}$

b. $X = \{2, 4, 6, 8, 10, 12, 14, 16, 18, 20\}$

$Y = \{20, 18, 16, 14, 12, 10, 8, 6, 4, 2\}$

Solution

a. $|A| = 5$ and $|B| = 5$, therefore, the sets are equivalent to one another, and we can write $A \sim B$. However, because they do not contain the same elements, they are not equal to one another.

b. $|X| = 10$ and $|Y| = 10$. Even though the order of the elements is different, both contain the even numbers from 2 to 20. Therefore, the sets are equal, and we can write $X = Y$. When two sets are equal, they are by definition also equivalent.

Think Back

$\mathbb{R}$ represents the set of real numbers, $\mathbb{N}$ represents the set of natural numbers, $\mathbb{Z}$ represents the set of integers, and $\mathbb{W}$ represents the set of whole numbers. All of these sets are infinite sets.

Describing a set using the roster method is very convenient for small sets, however it quickly loses its usefulness for big sets. An alternative when the members of a set all share a certain property is to describe the elements using **set-builder notation**. For instance, the following describes the integers, $\mathbb{Z}$, using set-builder notation.

$$\mathbb{Z} = \{n \mid n \text{ is an interger}\}$$

It is read "$\mathbb{Z}$ is the set of all n such that n is an integer."

Set-Builder Notation

Set-builder notation is used to describe a set when the members all share certain properties.

Set-builder notation is extremely helpful when representing an infinite set of numbers like the natural numbers. It is impossible to list out every member of an infinite set, but by describing it in this manner, set-builder notation encompasses the entire set.

☞ **Helpful Hint**

Set-builder notation may also use a colon instead of a vertical line to represent "such that." For example,

$Y = \{x : x \text{ is an even integer}\}$.

Example 3: Using Set-Builder Notation to Represent a Set

Use set-builder notation to represent Y, the set of all even integers.

Solution

$Y = \{x \mid x \text{ is an even integer}\}$

Note that there are many correct ways to write the solution. Alternative solutions include $Y = \{x \mid x \in \mathbb{Z} \text{ and } x \text{ is even}\}$ and $Y = \{2x \mid x \text{ is an integer}\}$.

Skill Check #1

Represent the set J of all natural numbers less than 10, using both the roster method and set-builder notation.

There are a couple of very special types of sets that we need to define. The first of these is the **empty set**, also called the **null set**. Think for a moment about the set A, which is the set of states that border Hawaii. Since Hawaii is a made up of islands that do not border any other state, A has no elements and is an empty set. If a set is empty, we denote this symbolically by writing $A = \varnothing$, or sometimes $A = \{\ \}$. The cardinality of the empty set is 0. The empty set should not be confused with the set $\{\varnothing\}$ or the set $\{0\}$, both of which contain a single element. The first is the set whose only element is an empty set and the second is the set containing the number 0, both of which have a cardinality of 1.

Empty Set

The **empty set**, or **null set**, is the set that contains no elements. If set A is empty, we write $A = \varnothing$.

Example 4: Determining Empty Sets

Determine if the following sets are empty sets.

a. The set A of negative numbers less than 100.

b. The set B of any state that contains the letter q in its name.

Solution

a. Since it is the case that all negative numbers are less than 100, the set A is not empty. In fact, it is a set with infinitely many elements.

b. Since there are not any states whose names contain the letter q, this set is empty. That is, $B = \varnothing$.

Skill Check #2

Give an example of an empty set.

The second special set to mention is the **universal set**, *U*. The universal set is the set of all elements that are being considered in any particular situation. For instance, suppose you are buying a car. The universal set could be the set of all new cars available, it could be the set of all used cars available, or it might be both—the set of all new and used cars available. Choosing a universal set "sets the scene" for those elements that will be considered.

Universal Set

The set of all elements being considered for any particular situation is called the **universal set** and is denoted by *U*.

The universal set is especially useful in that it allows us to express everything that is not in a particular set. For instance, let's go back to our car example. Suppose the universal set is the set of all new and used cars that are available. Assume you can afford a car under $25,000. Then the set of cars you can afford is $A = \{c \mid c \in U \text{ and cost} < \$25{,}000\}$. Since not everyone has the luxury of choosing from every possible make and model of car in the world, not all cars are included in *A*. For instance, in most situations, a $4,000,000 Lamborghini would not be up for consideration, and therefore would not be in the set *A*. It would be in the **complement** of *A*, denoted *A′*. In other words, it is an element of the universal set that is not in the set *A*. In our car example, $A' = \{c \mid c \in U \text{ and cost} \geq \$25{,}000\}$.

☞ Helpful Hint

The complement of *A* can be denoted by *A′*, A^c, or $\overline{A}$.

Complement

The **complement** of *A* consists of all the elements in the given universal set that are not contained in *A*. The complement of *A* is denoted *A′*.

The set-builder notation for the complement of set *A* is $A' = \{x \mid x \in U, x \notin A\}$.

Consider for a moment two special sets: the universal set and the empty set. One contains all of the elements being considered and the other contains no elements. Hence, the universal set and the empty set will always be complements of one another.

$$U' = \varnothing \quad \text{and} \quad \varnothing' = U$$

Example 5: Determining the Complement of a Set

Twitter is a type of online social media where registered users can post "tweets" that are up to 140 characters long. In the world of Twitter, you can keep up with what other users post by "following" them and in turn, other users can "follow" you and your tweets.

Let

$U = \{x \mid x \text{ is a registered user on Twitter}\}$

$A = \{x \mid x \text{ is a registered user on Twitter and } x \text{ follows you on Twitter}\}$

$B = \{x \mid x \text{ is a registered user on Twitter and you do not follow } x \text{ on Twitter}\}$

Determine the complements of *A* and *B*.

Solution

Since A is the set of all Twitter users who follow you on Twitter, the complement of A is the set of all registered Twitter users who do not follow you on Twitter. We write the complement of A in the following manner.

$$A' = \{x \mid x \text{ is a registered user on Twitter and } x \text{ does not follow you on Twitter}\}$$

Since Twitter currently has more than half a billion users, it is likely that A' is much larger than A.

B is the set of all registered users who you do not follow on Twitter; so, the complement of B is the set of registered users who you do follow on Twitter. We can write the complement of B as

$$B' = \{x \mid x \text{ is a registered user on Twitter and you follow } x \text{ on Twitter}\}.$$

Skill Check #3

Let $U = \{x \mid x \text{ is a book published in the US in 2014}\}$ and $A = \{x \mid x \in U \text{ and you read } x \text{ as an e-book}\}$. Find A'. Is it possible for you to have read a book in A'?

Skill Check Answers

1. Roster notation: $J = \{1, 2, 3, 4, 5, 6, 7, 8, 9\}$;
   Set-builder notation: $J = \{x \mid x \in \mathbb{N}, x < 10\}$

2. Answers will vary. Examples may include $A = \{x \mid x \in \mathbb{N}, x < 0\}$;
   $B = \{x \mid x \text{ is a US President before Barack Obama and } x \text{ is a woman}\}$

3. $A' = \{x \mid x \in U, \text{ and you did not read } x \text{ as an e-book}\}$; Yes, but not as an e-book.

2.1   Exercises

Determine whether each statement is true or false. If the statement is false, explain why.

1. It is always possible to list every element of a set using the roster method.

2. $\varnothing \sim \{0\}$

3. Let $U = \{\text{set of all students enrolled at Xavier University}\}$ and $A = \{x \mid x \in U \text{ and } x \text{ is a student with less than 30 earned credit hours}\}$. Then $A' = \{x \mid x \in U \text{ and } x \text{ is a student with at least 30 earned credit hours}\}$.

4. $0 \in \varnothing$ 5. $x \in \{x, y, z\}$

6. $\{x\} \in \{x, y, z\}$ 7. $|\varnothing| = 0$

8. Let $Y = \{$Tim, Emilia, Aleesa, Whit$\}$, then $|Y| = 4$.

9. Let $A = \{$Sounds, Angels, Ravens, Titans$\}$ and $B = \{201, 38, 46, 23\}$, then $A \sim B$.

Write each set using the roster method.

10. A is the set of months of the year that have exactly 30 days.

11. B is the set of states whose names begin with the letter N.

12. C is the set of positive numbers smaller than 100 that have 2 digits that are the same.

13. D is the set of last names of people who teach this course at your university.

14. E is the set of planets in our solar system.

15. F is the set of weekdays.

Solve each problem.

16. Give an example of a set that cannot be represented using the roster method.

17. Describe 2 sets of which you are a member.

Write each set using set-builder notation.

18. Let G be the set of whole numbers.

19. Let H be the set of natural numbers less than or equal to 50.

20. Let U be the set of all the states in the United States of America and J be the set of all states that border an ocean.

21. Let U be the set of all students enrolled at a public school of higher education and K be the set of all collegiate athletes.

Write a description for each set. It is possible for more than one description to be correct.

22. $J = \{-11, -9, -7, -5, -3, -1\}$ 23. $K = \{\$1, \$2\ \$5, \$10, \$20, \$50, \$100\}$

24. $M = \{$Saturday, Sunday$\}$ 25. $N = \{$January, June, July$\}$

Write each set using the roster method.

26. $A = \{x \mid x \text{ is an odd number and } 20 < x < 30\}$

27. $B = \{x \mid x \leq 15 \text{ and } x \text{ is a positive multiple of } 3\}$

28. $C = \{x \mid 2x = 4\}$

29. $D = \{x \mid x \text{ is a state that shares a common border with Colorado}\}$

The table shows the total number of official Olympic medals for all recorded time (1896 through the Winter games of 2014) for the top 10 countries, some of which are no longer countries. Let the universal set consist of the 10 countries listed. Solve each problem.

Top 10 All-Time Olympic Medal Winning Countries				
Team	**Gold**	**Silver**	**Bronze**	**Combined Total**
United States (USA)	1072	860	749	2681
Soviet Union (URS)	473	376	355	1204
Great Britain (GBR)	246	276	284	806
Germany (GER)	252	260	270	782
France (FRA)	233	254	293	780
Italy (ITA)	235	200	228	663
Sweden (SWE)	193	204	230	627
China (CHN)	213	166	147	526
Russia (RUS)	182	162	177	521
East Germany (GDR)	192	165	162	519

Source: Wikipedia, s.v. "All-time Olympic Games medal table," accessed July 2014, http://en.wikipedia.org/wiki/All-time_Olympic_Games_medal_table

30. Let X equal the set of countries who have won more than 1000 medals overall. Write the set X using the roster method.

31. Let Y equal the set of countries who have won between 500 and 1000 medals overall. Write the set Y using the roster method.

32. Let Z equal the set of countries who have won less than 500 medals overall. Write the set Z using the roster method.

33. Let G equal the set of countries who have won more than 200 Gold medals. Write the set G using the roster method.

34. Is $X = G$? Explain.

35. Is $X \sim G$? Explain.

Use the given sets to answer each question.

P = {lasagna, rotini, orzo, tortellini, penne}

Q = {x | x is a pasta shape}

R = {penne, tortellini, orzo, rotini, lasagna}

S = {marinara, pesto, alfredo, Bolognese, carbinara}

36. Is $P = Q$? Why or why not?

37. Is $P = R$? Why or why not?

38. Is $P = S$? Why or why not?

39. Are any of P, Q, R, and S equivalent? Explain.

Use the given sets to answer each question.

A = {Business, Physics, Psychology, Kinesiology, Graphic Design, History}

B = {History, Graphic Design, Kinesiology, Physics, Business}

C = {Art History, Education, Nursing, Biology, Statistics}

D = {x | x is a university major}

40. Is $A = B$? Why or why not?

41. Is $B = C$? Why or why not?

42. Is $A = D$? Why or why not?

43. Are any of A, B, C, and D equivalent? Explain.

Determine the cardinal number of each set.

44. $W = \{3, 4, 5, 6, 7, 8, 9, 10, 0\}$

45. $X = \{x \mid x \in \mathbb{Z}, x \text{ is even, and } |x| < 10\}$

46. The empty set

47. $Y = \{x \mid x \text{ is a United States president, past or present}\}$

Use the set $A = \{b, a, s, k, e, t\}$ to solve each problem.

48. Find $|A|$.

49. If $U = \{a, b, c, d, \ldots, x, y, z\}$, find A'.

50. If $U = \{a, b, c, d, \ldots, x, y, z\}$, find $|A'|$.

51. If $U = \{a, b, c, d, \ldots, x, y, z, A, B, C, D, \ldots, X, Y, Z\}$, find A'.

52. If $U = \{a, b, c, d, \ldots, x, y, z, A, B, C, D, \ldots, X, Y, Z\}$, find $|A'|$.

Use the set $B = \{1, 2, 3, 4\}$ to solve each problem.

53. Find $|B|$.

54. If $U = \{0, 1, 2, 3, 4, 5, 6, 7, 8, 9\}$, find B'.

55. If $U = \{0, 1, 2, 3, 4, 5, 6, 7, 8, 9\}$, find $|B'|$.

56. If $U = \{-9, -8, -7, -6, \ldots, 6, 7, 8, 9\}$, find B'.

57. If $U = \{-9, -8, -7, -6, \ldots, 6, 7, 8, 9\}$, find $|B'|$.

Solve each problem.

58. Let A be a set in which one of the elements is President Barack Obama. List at least three different universal sets for A.

59. Let X be a set in which one of the elements is π. List at least 3 different universal sets for X.

2.2 Subsets and Venn Diagrams

MATH MILESTONE

John Venn (1834–1923) was British born and educated. He came from a long line of Anglican priests, and indeed became one himself. His ultimate passions led him to become a lecturer at Cambridge University, studying and teaching logic and probability theory.

An important part of interacting with mathematical concepts is finding ways to express them visually. One way to visualize the relationships between sets is in the form of a **Venn diagram**. Venn diagrams were first introduced by British logician John Venn. These diagrams are used to help conceptualize relationships in many fields, including set theory, logic, probability, statistics, and computer science. In a Venn diagram, the sets are represented by circles (or ovals) contained within a rectangular region representing the universal set

Venn Diagram

A Venn diagram is a way to visualize the relationships between sets. In a Venn diagram, the sets are represented by circles (or ovals) contained within a rectangular region representing the universal set.

The ovals are often labeled with the names of the sets they represent, and the elements of a set can be listed within their set oval.

Figure 1 shows a Venn diagram that represents the set $A = \{1, 2, 3\}$ within the universal set $U = \{1, 2, 3, 4, 5, 6, 7, 8, 9, 10\}$. Notice how the elements of A and U are positioned in their respective regions in the diagram. It's also worth noting that neither the size of the oval nor the size of the rectangle have any significance in Venn diagrams.

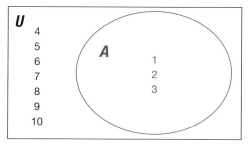

Figure 1

Example 1: Interpreting Venn Diagrams

The following Venn Diagram represents the sets S, T, and V within the universal set $U = \{x \mid x \in \text{English alphabet}\}$. Use the diagram to answer the following questions.

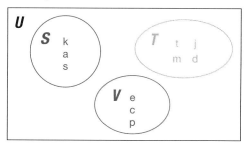

a. List the elements of the sets S, T, and V in roster form.

b. Find $|S|$, $|T|$, and $|V|$.

c. Find T'.

d. Is $S = V$? Is $S \sim V$? Explain your answers.

2

Solution

a. $S = \{k, a, s\}$

$T = \{t, m, j, d\}$

$V = \{e, c, p\}$

Remember, that the order in which the elements are listed is not important. Therefore, it is also correct to list the elements of each set in a different order.

b. In order to find the cardinal number of each set, S, T, and V, we simply need to count the number of elements in each set. Therefore $|S| = 3$, $|T| = 4$, and $|V| = 3$.

c. Recall that T' contains all of the elements in the universal set that are not in T. Be careful to list all elements not in T and not just the elements in the other sets. Therefore $T' = \{a, b, c, e, f, g, h, i, k, l, n, o, p, q, r, s, u, v, w, x, y, z\}$.

d. For $S = V$, they would have to have exactly the same elements. Since this is not the case, $S \neq V$. However, since S and V both have a cardinality of 3, they are equivalent to one another. That is, $S \sim V$.

Example 2: Interpreting Venn Diagrams

There is an increasing number of electric vehicles on the roads in the United States. The number of charging stations available for these cars varies from state to state. The states with the most public and private electric charging stations are California, Florida, Oregon, Texas, and Washington.

Let

$U = \{x \mid x \in$ all public and private electric charging stations in the United States$\}$

$C = \{x \mid x \in U, x \in$ all public and private electric charging stations in California$\}$

$F = \{x \mid x \in U, x \in$ all public and private electric charging stations in Florida$\}$

$O = \{x \mid x \in U, x \in$ all public and private electric charging stations in Oregon$\}$

$T = \{x \mid x \in U, x \in$ all public and private electric charging stations in Texas$\}$

$W = \{x \mid x \in U, x \in$ all public and private electric charging stations in Washington$\}$

The following Venn diagram depicts the top five states with the most electric charging stations. The cardinal number for each set is shown inside the appropriate oval.[1]

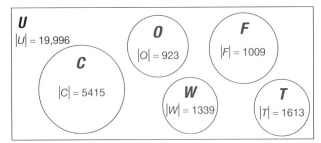

1 US Department of Energy: Alternative Fuels Data Center, http://www.afdc.energy.gov

Use the Venn diagram to answer the following questions. Assume all questions refer to both public and private stations.

a. Which state has the most electric charging stations?

b. Which state has the second highest number of charging stations?

c. How many electric charging stations do the top five states have all together?

d. How many charging stations are in the United States, but not in one of the top five states in the diagram?

Solution

a. Using the Venn diagram, we can see that California has the most charging stations with 5415 in the state.

b. Don't be fooled by the size of the circles in the diagram. Remember that the size of a circle is of no consequence in a Venn diagram. Looking at the size of the sets by their numbers, we see that Texas, with 1613 stations, has the second highest number of charging stations.

c. In order to determine how many combined charging stations there are in the top five states, we need to add together the size of all five sets.

$$\text{Total Number} = |C| + |T| + |W| + |F| + |O| = 5415 + 1613 + 1339 + 1009 + 923 = 10{,}299$$

So the top five states have a combined total of 10,299 electric charging stations.

d. In order to find the total number of electric charging stations in the United States that are not in one of the top five states, we need to subtract the answer we found in part **c.** from the total number of charging stations in the universal set. We know the cardinal number of U is 19,996. Therefore, we have $19{,}996 - 10{,}299 = 9697$ electric charging stations that are in the United States, but not in one of the top five states.

Notice in the previous example that none of the circles in the Venn diagram overlap. Since it is impossible for a single electric charging station to be in two states at once, the circles will be completely separate and not overlap in the diagram.

Let's consider a situation where the circles in a Venn diagram do overlap in a particular way. At State University, a graduating senior with a GPA of at least 3.8 is always chosen at random to introduce the commencement speaker at graduation. Consider what the Venn diagram would look like to illustrate this situation. Let the universal set be all students at State University. Then, let set A consist of the graduating seniors. In order to be chosen to introduce the commencement speaker, you must be a graduating senior who also has a GPA of at least 3.8. The set of all possible students that can be chosen to introduce the speaker is a smaller set B contained within A, referred to as a **subset**.

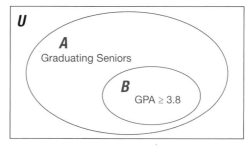

Figure 2

Subset

If A and B are sets, B is a **subset** of A if every element of B is also an element of A. We write $B \subseteq A$.

Example 3: Drawing a Venn Diagram with Subsets

Let

$U = \{x \mid x \text{ is a student at State University}\}$

$W = \{x \mid x \text{ is a student at State University majoring in Computer Science with a minor in Business}\}$

$Y = \{x \mid x \text{ is a student at State University majoring in Computer Science}\}$

Draw a Venn diagram to represent the sets U, W, and Y at State University.

Solution

Begin by drawing a rectangle representing the universal set of all students at State University. This rectangle can be any size you like.

Next, we need to decide how the sets W and Y are to be drawn. Notice the set Y contains all students majoring in computer science and the set W is a more specific group of students who not only are majoring in computer science, but also are minoring in business. Therefore $W \subseteq Y$. So the Venn diagram will look like the following.

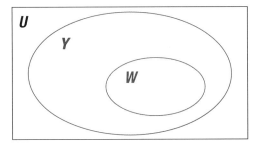

Although, your diagram may look a little different than the one shown here, it should resemble the structure of this one. In other words, the oval representing W should be completely contained within the oval representing Y.

Skill Check #1

Draw a Venn diagram of the following.

$U = \{x \mid x \text{ is a computer}\}$

$A = \{x \mid x \text{ is an iPad}\}$

$B = \{x \mid x \text{ is a tablet computer}\}$

A more restricted type of subset is the proper subset.

Proper Subset

When $B \subseteq A$, and A contains at least one element that is not contained in B, B is said to be a **proper subset** of A, and is written $B \subset A$.

Another way to think of a proper subset is that B is a subset of A, but B is not equal to A. It is actually "properly contained" within A. Consider our example about State University and computer science majors. If all computer science majors at State University were also minoring in business, then Y and W would actually be the same set and W would not be a proper subset of Y. In Example 3, W was a proper subset of Y.

Helpful Hint

$\not\subseteq$ = "not a subset of"

$\not\subset$ = "not a proper subset of"

As another example, consider the set $X = \{a, b, c, d, e, f\}$. The sets $Y = \{a, b, f\}$ and $Z = \{c\}$ are both proper subsets of X. However, the sets $V = \{a, b, c, d, e, f\}$ and $W = \{a, b, h\}$ are not proper subsets of X; V is not a proper subset because it contains every element of X, and W is not a subset at all. It's important to note that the empty set will always be a proper subset of any set, except in the case of the empty set itself. Think about that for a moment. Since a proper subset is *any* combination of elements within the set that does not include the entire set, then this must also include the empty set, which contains no elements at all. However, the empty set cannot be a proper subset of itself because it is equal to itself. In that same vein, every set is a subset, but not a proper subset, of itself.

Example 4: Identifying Proper Subsets

Let $X = \{1, 2, 3\}$. List all the proper subsets of X.

Solution

All proper subsets of X must exclude at least one member of X. In our example, this means that each proper subset can have at most two elements in it. In fact, the proper subsets may contain two elements, one element, or no elements. A table listing out the possible proper subsets in order will help us.

Table 1		
Proper Subsets with Precisely 2 Elements	**Proper Subsets with Precisely 1 Element**	**Proper Subsets with Precisely 0 Elements**
$\{1, 2\}$	$\{1\}$	$\varnothing$
$\{1, 3\}$	$\{2\}$	
$\{2, 3\}$	$\{3\}$	

So, there are 7 proper subsets of the set $\{1, 2, 3\}$.

Skill Check #2

List all of the proper subsets of the set $\{a, b, c, d\}$.

Notice that in Example 4, the number of subsets that X contains would be the same no matter what the elements are. It would not matter if the elements were apple, banana, and pineapple instead of 1, 2, and 3. Any set with three elements will always have 7 proper subsets. Then how many subsets are in a set containing 10 elements? Or 100? Determining the number of proper subsets by listing the subsets seems easy enough when there are only 3 elements in the set; however, this method quickly becomes tedious with a set containing 100 elements. Luckily, there is no need to resort to such measures. Knowing the cardinal number of a set is enough to determine how many subsets and proper subsets are contained within a set.

Number of Subsets and Proper Subsets of a Set

If the cardinal number of a set is n, then there are 2^n subsets and $2^n - 1$ proper subsets contained in the set.

We'll leave the verification of why there are 2^n subsets as an exercise. But, we can see that since the definition of a proper subset requires that we exclude at least one member of the set, the only subset that is not proper is the set itself. Therefore the number of subsets and proper subsets differ by only one.

Let's consider Example 4 again. The set $x = \{1, 2, 3\}$ has a cardinal number of 3. We can use the formula to confirm that we noted all of the proper subsets. Using the formula, the number of proper subsets is as follows.

$$2^3 - 1 = 8 - 1$$
$$= 7$$

Thinking back, the only subset that we did not include in the proper subsets was the set itself, that is, the set containing all three elements. If we add this one to the list of subsets, we get $2^3 = 8$, which is the total number of subsets.

Example 5: Determining the Number of Subsets

Dr. Williams is eating at China Buffet one afternoon and notices a sign that says:

"So many possibilities—you could spend a lifetime eating at China Buffet and never have the same meal twice!"

He wonders how many different plates he could make from the 16 items on the buffet. He can have all, none, or some of the items. Help Dr. Williams determine the number of different plates he could make at the buffet.

Solution

When Dr. Williams makes a plate, he chooses a subset of the items from the buffet. The number of different plates that Dr. Williams could put together is the number of subsets of the 16 items on the buffet. Since there is no requirement on the number of food items he needs to have on a plate, we can use the formula for the number of subsets to determine how many different plates he could make. Using the formula for the number of subsets, we have

$$\text{number of plates} = 2^{16} = 65,536.$$

This means that if Dr. Williams ate at China Buffet once a day every day, he could eat for almost 180 years without duplicating a meal. Certainly, it would seem the sign is not misleading.

Skill Check Answers

1.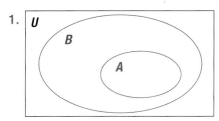

2. $\varnothing$, {a}, {b}, {c}, {d}, {a, b}, {a, c}, {a, d}, {b, c}, {b, d}, {c. d}, {a, b, c}, {a, b, d}, {a, c, d}, {b, c, d}

2.2 Exercises

Use the Venn diagram to solve each problem.

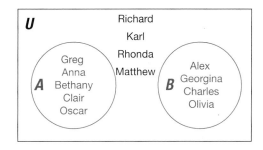

1. List A and B using the roster method.

2. Find A'.

3. List U using the roster method.

Draw each Venn diagram.

4. Let U consist of all artists. Draw a Venn diagram to represent the two sets violinist and musicians.

5. Let U consist of all four legged animals. Draw a Venn diagram to represent the two sets male dogs and female dogs.

6. Draw a separate Venn diagram to illustrate each of the following.

   a. $W \subseteq Y$

   b. $Y \subseteq W$

7. Let $U = \{x \mid x \in \mathbb{R}\}$, $W = \{x \mid x \text{ is a counting number less than } 20\}$, and $Y = \{2, 4, 6, 8, 10\}$. Draw a Venn diagram to represent U, W, and Y with the elements in the proper regions.

8. Let $U = \{$Red, Orange, Yellow, Green, Blue, Indigo, Violet$\}$, $A = \{$Green, Orange, Yellow$\}$, and $B = \{$Indigo, Violet$\}$. Draw a Venn diagram to represent U, A, and B with the elements in the proper regions.

Solve each problem.

9. Let $A = \{0, 1, 2, 3, 4, 5\}$ and $B = \{5, 4, 3, 2, 1\}$. Is $B \subseteq A$?

10. Let $A = \{\{a, b\}, c, d, e\}$ and $B = \{a, b, c\}$. Is $B \subseteq A$?

11. Explain why Red $\not\subset \{$Red, Blue, Green$\}$.

12. The set B contains the names of three of the most expensive paintings ever sold. If $B = \{$*The Card Players* by Cézanne, *No. 5 1948* by Pollock, *Woman III* by de Kooning$\}$, list all the subsets of B.

13. The set C contains the names of the top three grossing films of all time as of Summer 2014. If $C = \{$*Avatar*, *Titanic*, *Marvel's The Avengers*$\}$, list all of the proper subsets of C.

Determine whether each statement is true or false. If the statement is false, explain why.

14. The set $\{$s, t, u, v$\}$ has exactly 16 subsets and 17 proper subsets.

15. A set can have an even number of proper subsets.

Solve each problem.

16. Given $A = \{x \mid x \in \text{ positive whole numbers less than } 100\}$ and $B = \{x \mid x \in \text{ even non–negative integers less than } 100\}$, is either set a subset of the other? Explain.

17. W contains The New York Times' top five fiction books for 2013. If $W = \{$*The Goldfinch*, *Americanah*, *The Flamethrowers*, *Life After Life*, *Tenth of December*$\}$, how many subsets does W contain? How many proper subsets?

18. Given that $B = \{$♪, ✦, ♡, ☺, ◎, ☎, ♨, ◌$\}$. How many subsets does B have? How many proper subsets does B have?

19. Determine the number of subsets contained in Y if $Y = \{x \mid x$ is an odd positive integer and $x < 25\}$.

20. A set has 32 subsets. How many elements are in the set?

21. A set has 127 proper subsets. How many elements are in the set?

22. A new keyless car security system allows the owner to create a "handprint" to start the car by selecting up to five places on the six-space keypad on which to place their fingers. How many possible handprints can the user choose from if at least one space must be selected?

23. Pizza Jet promotes a special lifetime offer—once you order all possible pizza combinations from Pizza Jet, you can get free pizza for life! Pizza Jet offers a selection of 12 toppings (including cheese) for their pizzas. How many different pizzas must you order from Pizza Jet to qualify for their lifetime offer? Is this possible? Explain.

24. The national fast food chain Wendy's advertises that there are 256 ways to personalize a Wendy's hamburger. How many condiments must Wendy's carry for their customer to have this many choices?

25. Fill out the table to verify that the number of subsets of a set is 2^n and the number of proper subsets of a set is $2^n - 1$. Assume the elements of the sets are the numbers $0, 1, 2, \ldots, n$.

Number of Elements in the Set	List the Proper Subsets	Number of Proper Subsets	Number of Subsets
0			
1			
2			
3			
4			
n			

2.3 Operations with Sets

In the previous sections, the introduction to sets and set notation provided us with the foundation for discussing operations with sets. In this section, we will use set operations to compare the manner in which two separate sets are related.

Consider the members of a mathematics class as the universal set. A teacher wants to conduct a survey of her students who are also taking English and biology courses. Suppose the teacher defines the set E of students taking English and the set B of students taking biology. We can use a Venn diagram, as shown in Figure 1, to represent how the two sets of students are related. The teacher is interested in surveying students that are common to both sets B and E. We call this commonality

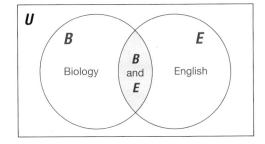

Figure 1

between the two sets the ***intersection*** of the sets B and E. In Section 2.2, we saw that it's possible for sets of data to contain the same elements. When some, but not all, of the elements of one set are contained in the other, the sets are represented as overlapping circles in a Venn diagram. In Figure 1 the overlapping region represents the intersection of the two sets B and E.

Intersection

The **intersection** of two sets A and B is the set of all elements common to both A and B. We denote the intersection of A and B as $A \cap B = \{x \mid x \in A \text{ and } x \in B\}$.

Example 1: Determining the Intersection of Sets

Find the intersection of the sets $A = \{n, u, m, b, e, r, s\}$ and $B = \{r, u, l, e\}$.

Solution

Since the intersection of two sets consists of all of the elements that appear in *both* sets, we can see that the intersection of A and B consists of the elements r, u, and e.

$$A \cap B = \{n, u, m, b, e, r, s\} \cap \{r, u, l, e\}$$
$$= \{r, u, e\}$$

We can also use a Venn diagram to find the intersection.

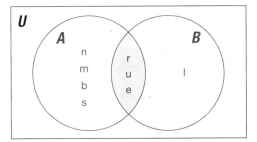

Notice that any elements in the intersection of the Venn diagram are only listed once.

Some diagrams display the number of elements in each region instead of listing the elements themselves. The following example illustrates this.

Example 2: Using a Venn Diagram to Find the Intersection

The given Venn diagram represents the number of students that participated in certain activities while on a spring break trip. Determine the number of students that went both hiking and skiing over spring break.

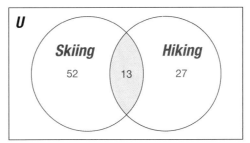

Solution

Using the Venn diagram, we can see that 52 students went skiing exclusively on their break, 27 students went hiking exclusively, and 13 students went both skiing *and* hiking.

Big's Diner is famous for their hot dogs and chili. They have hot chili, mild chili, chili with beans, chili without beans, vegan chili, and their specialty, Texas chili. Their hot dog selection consists of jumbo dogs, turkey dogs, bratwurst, Vienna, and Cumberland. You can even order a chili dog, if you like. If we draw a Venn diagram of all of the offerings at Big's Diner, we get the Venn diagram in Figure 2.

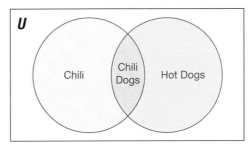

Figure 2

We can form the complete menu offered at Big's Diner from these two sets of dishes. We do so by combining the elements of the two sets into one set. When we combine the elements of two or more sets together, we call this the union.

Union

The **union** of two sets A and B is the set of all elements in A or in B. We denote the union of A and B as $A \cup B = \{x \mid x \in A \text{ or } x \in B\}$.

Example 3: Determining the Union of Sets

Let $A = \{1, 2, 3, 4, 5\}$, $B = \{2, 4, 6, 8\}$, and $C = \{1, 3, 5, 7, 9\}$. Find the following unions.

a. $A \cup B$

b. $A \cup C$

c. $B \cup C$

Solution

☞ **Helpful Hint**

Note that we do not include any element more than once in the union.

a. To find the union of sets A and B, we need to find the elements that appear in either set A or set B. The elements of set A are 1, 2, 3, 4, 5 and the elements of set B are 2, 4, 6, 8. We simply combine the items together to create a new set to represent the union. Therefore,

$$A \cup B = \{1, 2, 3, 4, 5\} \cup \{2, 4, 6, 8\}$$
$$= \{1, 2, 3, 4, 5, 6, 8\}.$$

The Venn diagram shown gives us a visual illustration of the union of the elements in set A and in set B.

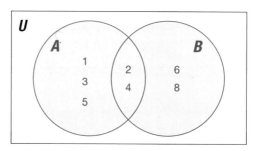

Note that $A \cap B = \{2, 4\}$.

b. Finding the union of sets A and C is done in a similar manner.

$$A \cup C = \{1, 2, 3, 4, 5\} \cup \{1, 3, 5, 7, 9\}$$
$$= \{1, 2, 3, 4, 5, 7, 9\}$$

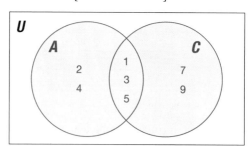

Note that $A \cap C = \{1, 3, 5\}$.

c. Finding the union of sets B and C is also done in a similar manner.

$$B \cup C = \{2, 4, 6, 8\} \cup \{1, 3, 5, 7, 9\}$$
$$= \{1, 2, 3, 4, 5, 6, 7, 8, 9\}$$

Note that the intersection of these two sets is empty. Therefore, $B \cap C = \varnothing$.

In addition to finding the intersection and the union of two sets, we can also combine the operations of intersection and union. Let's try an example of combining the operations.

Example 4: Combining Intersection and Union

Let $U = \{a, b, c, d, e, \ldots, z\}$, $M = \{m, a, t, h\}$, $N = \{m, o, n, e, y\}$, and $K = \{i, n, v, e, s, t, o, r\}$. Find

a. $M \cup (N \cap K)$

b. $M \cap (N \cup K)$

Solution

a. In order to find the solution when combining the operations of intersections and unions of sets, it might be best to describe the set first. $M \cup (N \cap K)$ is the set of all elements that are in set M or are in N and in K. Let's do each part and see what we get.

$$M \cup (N \cap K) = \{m, a, t, h\} \cup \left(\{m, o, n, e, y\} \cap \{i, n, v, e, s, t, o, r\} \right)$$

Just as in order of operations with numbers, we need to perform the operation in parentheses first. So,

$$N \cap K = \left(\{m, o, n, e, y\} \cap \{i, n, v, e, s, t, o, r\} \right) = \{o, n, e\} \text{ and}$$

$$\begin{aligned} M \cup (N \cap K) &= \{m, a, t, h\} \cup \{o, n, e\} \\ &= \{m, a, t, h, o, n, e\} \end{aligned}$$

b. Similarly, we can find $M \cap (N \cup K)$

$$\begin{aligned} M \cap (N \cup K) &= \{m, a, t, h\} \cap \left(\{m, o, n, e, y\} \cup \{i, n, v, e, s, t, o, r\} \right) \\ &= \{m, a, t, h\} \cap \{m, o, n, e, y, i, v, s, t, r\} \\ &= \{m, t\} \end{aligned}$$

Skill Check #1

Let, $M = \{m, a, t, h\}$, $N = \{m, o, n, e, y\}$, and $K = \{i, n, v, e, s, t, o, r\}$.
Find $K \cap (M \cup N)$.

Each of the examples involving the intersection and union of sets had elements that were common to both sets except Example 3c., where the intersection was the empty set. Just as in Example 3c., it is possible that two sets could have no common elements. For instance, if you were to draw a single card out of a standard deck of playing cards, the card will be either a red card or a black card. It is impossible for the card to be both red and black at the same time. In this case, we say the set of red cards and the set of black cards are **disjoint**.

> ### Disjoint
>
> Two sets A and B are **disjoint** if there are no elements in set A that are also contained in set B. In the case where two sets are disjoint, the intersection of those two sets is the **null set**, or empty set, denoted by $\varnothing$. Therefore, $A \cap B = \varnothing$ when sets A and B are disjoint.

It might be easier to visualize two sets that are disjoint by creating a Venn diagram. Using our example of playing cards (all playing cards are either red or black, but not both), we can see that two sets with no elements in common are disjoint. See Figure 3.

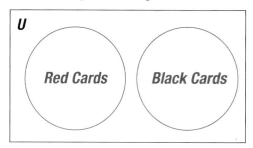

Figure 3

Example 5: Identifying Disjoint Sets

Let $U = \{\text{all students}\}$, $A = \{\text{students with GPA} < 2.5\}$, and $B = \{\text{students with GPA} > 3.0\}$. Determine if sets A and B are disjoint and draw a Venn diagram to illustrate the relationship between sets A and B.

Solution

Since it would be impossible for a student to have a GPA that is less than 2.5 and a GPA greater than 3.0 at the same time, sets A and B are disjoint. We can illustrate the relationship between sets A and B using a Venn diagram. The universal set is represented as all students' grade point averages of 0.0 to 4.0. Therefore, when sets A and B are described as students with GPAs less than 2.5 and greater than 3.0, respectively, there are no common GPAs. In addition, any student with a GPA between 2.5 and 3.0 is not represented in either set A or set B.

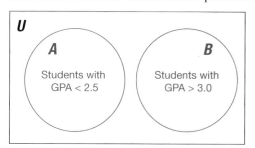

Up to this point, we have defined the intersection and union of sets as well as what it means for two sets to be disjoint. It is also helpful to discuss the items that are *not* part of a given set.

In Section 2.1, we defined the complement of a set A as the set that consists of all the elements in the given universal set that are not contained in A. Recall that the complement of A is denoted A'. We can use Venn diagrams to give a visual illustration of the complement.

Figure 4 illustrates the intersection of sets A and B as we have seen previously.

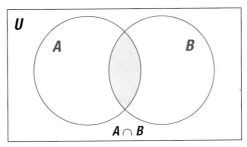

Figure 4: Intersection of Sets A and B

Figures 5 and 6 represent the complements of sets A and B, respectively. Notice that everything in the universal set U, **except** what is in set A, is in the complement of A and everything in the universal set U, except what is in set B, is in the complement of B.

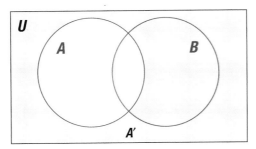

Figure 5: Complement of Set A Figure 6: Complement of Set B

Fun Fact

De Morgan's Laws are named after the mathematician Augustus De Morgan (1806–1871). The laws are based on the notion that the logic set forth by the Greek philosopher Aristotle was somewhat restrictive in its approach to the logical argument.

Now that we have visual representations of the complements of sets, we can begin to analyze the components of the union and intersection of the sets. To make understanding each of these a little easier, we can use De Morgan's Laws. De Morgan's Laws allow us to relate the three operations of sets—intersection, union, and complement—in an effort to analyze sets.

De Morgan's Laws

Let A and B be sets. Then,

$$(A \cup B)' = A' \cap B'$$

and

$$(A \cap B)' = A' \cup B'.$$

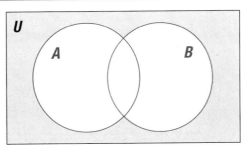

Figure 7: De Morgan's Law $(A \cup B)' = A' \cap B'$

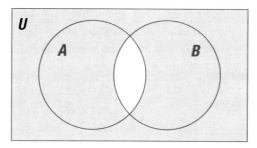

Figure 8: De Morgan's Law $\left(A \cap B \right)' = A' \cup B'$

Example 6: Using De Morgan's Laws

Given sets $U = \{a, b, c, d, \ldots, z\}$, $A = \{h, o, u, n, d\}$, and $B = \{r, o, c, k\}$, verify that $\left(A \cup B \right)' = A' \cap B'$.

Solution

The universal set U consists of all of the letters of the alphabet.

Since $A \cup B = \{h, o, u, n, d\} \cup \{r, o, c, k\} = \{r, o, c, k, h, u, n, d\}$, we know that $\left(A \cup B \right)' = \{a, b, e, f, g, i, j, l, m, p, q, s, t, v, w, x, y, z\}$.

Similarly,

$A' = \{a, b, c, e, f, g, i, j, k, l, m, p, q, r, s, t, v, w, x, y, z\}$ and

$B' = \{a, b, d, e, f, g, h, i, j, l, m, n, p, q, s, t, u, v, w, x, y, z\}$,

so $A' \cap B' = \{a, b, e, f, g, i, j, l, m, p, q, s, t, v, w, x, y, z\}$.

Notice that $\left(A \cup B \right)' = A' \cap B'$. Hence we have verified De Morgan's law for A and B.

Skill Check #2

Given sets $U = \{a, b, c, \ldots, z\}$, $A = \{h, o, u, n, d\}$, and $B = \{r, o, c, k\}$, verify $\left(A \cap B \right)' = A' \cup B'$.

Section 2.1 introduced the cardinal number, which is the notion of the number of items in a given set. Recall that a set $A = \{a, b, c\}$ has 3 elements, and we denote this by $|A| = 3$.

Example 7: Determining the Cardinal Number of a Union

Let $A = \{1, 2, 3, 4, 5\}$ and $B = \{2, 4, 6, 8\}$. Find $|A \cup B|$.

Solution

In Example 3**a.**, we found that $A \cup B = \{1, 2, 3, 4, 5, 6, 8\}$. Therefore, the number of elements in the set $A \cup B$ is $|A \cup B| = 7$.

You might notice that, in order for us to find the cardinality of the union of two sets, we had to be careful not to count an element more than once. In the case of Example 7, the elements 2 and 4 appear in both sets.

> ## Inclusion-Exclusion Principle
>
> The **inclusion-exclusion principle** states that the number of elements in the union of two sets A and B is calculated by adding the number of elements in set A to the number of elements in set B, less the number of elements that appear in both sets. We denote this by $|A \cup B| = |A| + |B| - |A \cap B|$.

The definition of the inclusion-exclusion principle means that when we take $|A| + |B|$, we are counting the total elements in both sets, which means that certain elements are counted twice. To correct this, we must subtract off the number of elements that appear in the intersection of the sets.

Example 8: Applying the Inclusion-Exclusion Principle

A standard deck of playing cards has 52 cards (26 of which are red and 26 of which are black) divided into 4 suits (clubs, spades, diamonds, and hearts), where there are 13 of each suit (Ace, 2, 3, 4, 5, 6, 7, 8, 9, 10, Jack, Queen, King). Of these cards, 12 are considered face cards (4 Kings, 4 Queens, and 4 Jacks). Find the number of cards in a standard deck that are either clubs or face cards.

Solution

We start the solution by writing what we are looking for using set notation.

$$|\text{clubs} \cup \text{face cards}| = |\text{clubs}| + |\text{face cards}| - |\text{clubs} \cap \text{face cards}|.$$

If the set A consists of clubs and the set B consists of face cards, then this is equivalent to

$$|A \cup B| = |A| + |B| - |A \cap B|$$

There are 13 clubs in the deck and there are 12 face cards. However, there are 3 face cards that are also clubs (King of clubs, Queen of clubs, and Jack of clubs). Therefore,

$$|A \cup B| = 13 + 12 - 3 = 22.$$

So, the number of cards that are either clubs or face cards is 22.

> ## Skill Check #3
>
> Find the number of playing cards that are either even $(2, 4, 6, 8, 10)$ or are diamonds.

Skill Check Answers

1. $\{n, e, t, o\}$

2. $A \cap B = \{o\}$, so $(A \cap B)' = \{a, b, c, d, e, f, g, h, i, j, k, l, m, n, p, q, r, s, t, u, v, w, x, y, z\}$.
   $A' = \{a, b, c, e, f, g, i, j, k, l, m, p, q, r, s, t, v, w, x, y, z\}$ and
   $B' = \{a, b, d, e, f, g, h, i, j, l, m, n, p, q, s, t, u, v, w, x, y, z\}$, so
   $A' \cup B' = \{a, b, c, d, e, f, g, h, i, j, k, l, m, n, p, q, r, s, t, u, v, w, x, y, z\}$. This gives that
   $(A \cap B)' = A' \cup B'$

3. 28

2.3   Exercises

Use the given sets to solve each problem.

$U = \{1, 2, 3, \ldots, 20\}$

$A = \{1, 2, 3, 4, 5, 6, 7, 8\}$

$B = \{2, 4, 6, 8, 10, 12\}$

$C = \{5, 7, 9, 11, 13, 15\}$

1. Find $A \cup B$.

2. Find $A \cup C$.

3. Find $B \cap C$.

4. Find $A \cap C$.

5. Verify $(A \cup B)' = A' \cap B'$.

6. Verify $(A \cap B)' = A' \cup B'$.

Use the given sets to solve each problem.

$U = \{a, b, c, d, \ldots, z\}$

$A = \{n, u, m, b, e, r, s\}$

$B = \{r, u, l, e\}$

7. Find $A \cup B$.

8. Find $A \cap B$.

9. Find $|A \cap B|$.

10. Verify $(A \cup B)' = A' \cap B'$.

11. Verify $(A \cap B)' = A' \cup B'$.

Use the given sets to solve each problem.

$U = \{A, B, C, D, \ldots, Z\}$

$A = \{I, C, E\}$

$B = \{C, U, B, E\}$

12. Find $A \cup B$.

13. Find $A \cap B$.

14. $|A \cap B|$

15. Verify $(A \cup B)' = A' \cap B'$.

16. Verify $(A \cap B)' = A' \cup B'$.

Use the given sets to solve each problem.

$U = \{A, B, C, D, \ldots, Z\}$

$A = \{F, A, C, T, O, R\}$

$B = \{P, R, O, D, U, C, T\}$

17. $A \cup B$

18. $A \cap B$

19. $|A \cap B|$

20. Verify $(A \cup B)' = A' \cap B'$.

21. Verify $(A \cap B)' = A' \cup B'$.

22. How many subsets does set A have?

23. How many subsets does U have?

Use the given sets to solve each problem.

$U = \{a, b, c, d, \ldots, z\}$

$M = \{b, r, i, d, g, e\}$

$N = \{g, a, t, o, r\}$

$K = \{b, a, l, i, s, t, e, r\}$

24. Find $N \cap (M \cup K)$.

25. Find $N \cup (M \cap K)$.

26. Find $K \cup (N \cap M)$.

27. Verify $(M \cup K)' = M' \cap K'$.

28. Verify $(M \cap N)' = M' \cup N'$.

Solve the problem.

29. A grocery store found that 275 of its customers use push carts to shop, 185 used a carry basket to shop, and that 145 used both a push cart and a carry basket. How many customers use only a push cart or a carry basket? Draw the Venn diagram.

Use the Venn diagram to solve each problem.

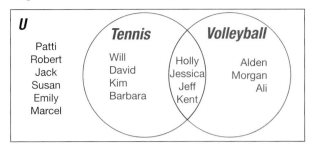

30. Which students played only tennis?

31. Determine which students played tennis or volleyball.

32. Determine which students played tennis and volleyball.

33. Find the number of students that play tennis or volleyball.

Solve each problem.

34. Determine the number of playing cards in a standard deck that are red cards or face cards.

35. Determine the number of playing cards in a standard deck that are odd numbered cards or black cards.

Show that each pair of sets is equal by drawing a Venn diagram of each set.

36. $A \cap B$ and $B \cap A$

37. $A \cup B$ and $B \cup A$

38. $(A \cap B) \cap C$ and $A \cap (B \cap C)$

39. $(A \cup B) \cup C$ and $A \cup (B \cup C)$

40. $A \cup \varnothing$ and A

Use set notation to represent each shaded region.

41.

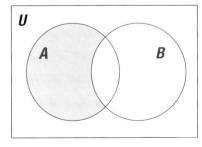

42.

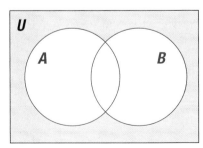

2.4 Applications and Survey Analysis

An interesting application of Venn diagrams is their usefulness in representing the relationships between two or more sets of information. For example, if a teacher surveyed a class to determine which students have taken algebra, statistics, or trigonometry, it might be helpful to represent the results in a format where each set of students can be represented in a visual manner. It is certainly the case that some of the students might have taken more than one of the courses. That is where Venn diagrams allow us to illustrate visually how we can organize the responses.

Assume a tour guide would like to know what language(s) a group of tourists speak while trying to plan an excursion. The possible languages are English, Spanish, and Italian. An example of the number of tourists that speak each language is represented in Figure 1.

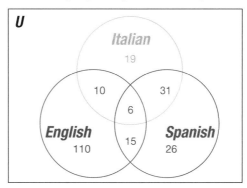

Figure 1

Think Back

Recall from Section 2.3 that "and" implies the intersection of sets and "or" implies the union of two sets.

The Venn diagram shows that 110 students speak only English, 26 speak only Spanish, 19 speak only Italian, 10 speak English and Italian but not Spanish, 31 speak Italian and Spanish but not English, 15 speak English and Spanish but not Italian, and 6 speak all three languages.

Example 1: Interpreting a Venn Diagram of Three Sets

The given Venn diagram contains the number of elements that belong to the three sets A, B, and C.

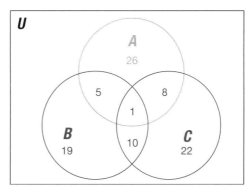

Use the information in the diagram to determine

a. $|A \cap B \cap C|$.

b. $|A \cap B|$.

Solution

a. To find $|A \cap B \cap C|$, we need to consider that the intersection of two sets, such as A and B, represents the set of elements that are in both A and B. Therefore, to find the intersection of the three sets A, B, and C, we are looking for the number of elements that are common to all three sets (the elements in the triangular middle section of the Venn diagram). We observe in the Venn diagram that there is only one element represented in the intersection of sets A, B, and C. Thus, $|A \cap B \cap C| = 1$.

b. Working with three sets and their intersections, we need to be careful and make sure we are considering the proper intersection. For $|A \cap B|$, we need to determine the number of elements in the area where sets A and B overlap. Referring to the Venn diagram, we can see that set C intersects sets A and B as well. There are 5 elements in $A \cap B$ that are not in C and 1 element in $A \cap B$ that is also in C. Therefore, $A \cap B$ contains $5 + 1 = 6$ elements, giving $|A \cap B| = 6$.

Skill Check #1

Use the Venn diagram in Example 1 to find the following.

a. $|A \cap C|$

b. $|B \cap C|$

Example 2: Constructing a Venn Diagram of Three Sets

Consider the universal set $U = \{a, b, c, \dots, z\}$. Given subsets $A = \{a, e, i, o, u\}$, $B = \{a, b, c, d, e, f, g, h, i, j, k, l\}$, and $C = \{a, l, u, m, n, i\}$, draw a Venn diagram to represent the relationships between the sets.

Solution

When there are three sets under consideration, getting started can be confusing because there are eight possible areas to place elements in the Venn diagram. Each of the elements in the universal set may only be placed in one of these eight areas. The eight areas are: (**1**) set A only, (**2**) set B only, (**3**) set C only, (**4**) sets A and B but not C, (**5**) sets A and C but not B, (**6**) sets B and C but not A, (**7**) sets A and B and C, and (**8**) the universal set but not sets A or B or C.

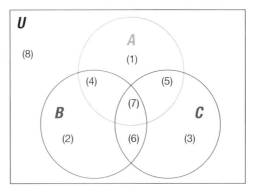

There are many approaches to determining the represented sets. One way is to begin with the intersection of all three sets, area (**7**). So, comparing the elements of sets A, B, and C, we can determine the elements common to all three sets are a and i. This gives $A \cap B \cap C = \{a, i\}$.

We must repeat this process for each of the other seven areas while making sure to not count any element more than once.

The next step is to determine the elements that belong in the different parts of set $A = \{a, e, i, o, u\}$. We compare the sets and determine that set A consists of all of the vowels of the alphabet. When compared to sets B and C, we find that the only vowel not in either B or C is o. So, the only element in area (**1**), set A only, is o. Similarly, the element in area (**4**), $A \cap B$ only, is e and the element in area (**5**), $A \cap C$ only, is u.

For set B only, we consider set $B = \{a, b, c, d, e, f, g, h, i, j, k, l\}$. Some of the elements, a, i, of set B have already been accounted for in the intersection of all three sets. By comparing set B to set A and set C, set A also contains element e and set C contains element l. This means the only elements in area (**2**), set B only, are b, c, d, f, g, h, j, k. Similarly, the element in area (**4**), $A \cap B$ only, is e and the element in area (**6**), $B \cap C$ only, is l.

For set C only, we consider set $C = \{a, l, u, m, n, i\}$. Some of the elements, a, i, of set C have already been accounted for in the intersection of all three sets. By comparing set C to set A and set B, set A also contains element u and set B contains element l. This means the elements in area (**3**), set C only, are m, n.

The elements of the alphabet not accounted for in any of sets A, B, or C are the elements p, q, r, s, t, v, w, x, y, z, which go in area (**8**). We are now ready to draw our Venn diagram.

We find the solution by placing each of the letters from each set in the appropriate areas within a Venn diagram. $A \cap B \cap C = \{a, i\}$, set A only $= \{o\}$, set B only $= \{b, c, d, f, g, h, j, k\}$, set C only $= \{m, n\}$, set $A \cap B$ only $= \{e\}$, set $A \cap C$ only $= \{u\}$, and set $B \cap C$ only $= \{l\}$. Then the solution is shown in the following diagram.

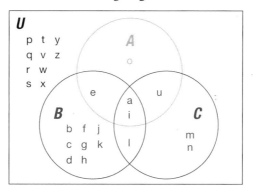

Fun Fact

One of the most popular polling services in the United States is Gallup, Inc. Started in 1935 by statistician George Gallup, the Gallup Organization conducts hundreds of surveys each year on topics that range from political opinions to shoppers' tastes in clothing. Known for its accuracy (the Gallup Poll has accurately predicted every presidential winner since 1936, except in 1948 when Harry Truman defeated Thomas Dewey), the Gallup Organization helps report public opinion and keep the populace informed.

When conducting political or fact-finding polls, the polling organization usually asks questions in a manner that is inclusionary, meaning that any participant in the survey may fall into one or more categories. For instance, if you asked college students to name their favorite food, they could respond that they liked pizza, burgers, and/or chicken. They could like one, two, three, or none of those options. The analysis of responses to a list of questions is called **survey analysis**.

Survey Analysis

Survey analysis is the analysis of responses to a list of questions.

2

Example 3: Drawing a Venn Diagram for Survey Analysis

A survey of 500 students showed that 350 listen to jazz, 300 listen to classical, and 200 listen to both. Draw a Venn diagram to illustrate this survey.

Solution

We are given that the number of students that listen to jazz and classical music is 200 students. We must notice that each of the students that like both types of music are also counted in the categories for their respectful types of music. For instance, although 350 students listen to jazz, 200 of those listeners also listen to classical. Recall from the previous section the inclusion-exclusion principle, which states that we must subtract the students that have been counted twice.

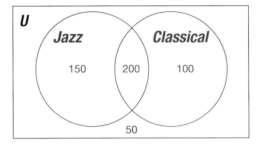

The resulting Venn diagram shows that 150 students like only jazz, 100 students like only classical, 200 students like both types of music, and 50 students don't like either type of music.

Skill Check #2

A survey of 400 customers at an ice cream shop showed that 225 customers like chocolate ice cream, 300 customers liked vanilla, and 200 customers liked both. Draw a Venn diagram to illustrate this survey.

Example 4: Drawing a Venn Diagram for Survey Analysis

Students majoring in international relations are polled on whether they take courses in any of three languages: French, German, and Russian. No student who took French also took Russian, but 39 who took French also took German. Eighty-four who took German also took Russian. All together, 55 reported taking French, 141 reported taking German, and 92 reported taking Russian. Draw a Venn diagram to illustrate this poll.

Solution

Recall that there are eight possible areas to place elements in the Venn diagram when there are three sets of consideration. Each of the student responses may only be placed in one of the eight areas. The eight areas are students taking: (**1**) French only, (**2**) German only, (**3**) Russian only, (**4**) French and German but not Russian, (**5**) French and Russian but not German, (**6**) German and Russian but not French, (**7**) French and German and Russian, and (**8**) None.

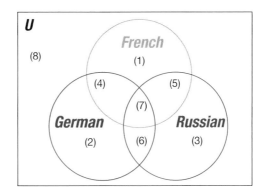

We will start with the intersection of all three sets to begin. We are told that no student who took French also took Russian. Therefore, we know that no students could have taken all three languages.

The next step is to determine the elements that belong in the each of the groups.

We are given that 39 students took French and German. Since we are given that 55 students took French, of which 39 also took German, we can remove the students that have been counted twice to determine that $55 - 39 = 16$ students took only French.

There were 141 students that took German. Of those students, 84 students also took Russian and 39 also took French. Therefore, $141 - 84 - 39 = 18$ students took only German.

Lastly, 92 students reported taking Russian. Of those students, 84 also took German and 0 students took French and Russian. Thus, $92 - 84 = 8$ students took only Russian.

We can place our information into a Venn diagram for a visual representation of the solution.

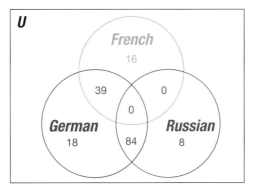

Skill Check #3

A survey of shoppers at a grocery store found that 225 shoppers like bananas, 198 like apples, and 180 like grapes. Twenty-five shoppers like all three fruits. There are 110 shoppers that like bananas and apples, 58 that like apples and grapes, and 55 that like bananas and grapes. Draw a Venn diagram to illustrate this poll.

Skill Check Answers

1. a. 9   b. 11

2.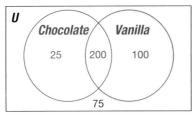

3.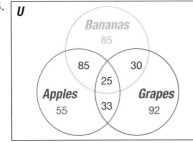

2.4   Exercises

Use the given sets to solve each problem.

$A = \{1, 2, 3, 4, 5, 6, 7, 8\}$

$B = \{2, 4, 6, 8, 10, 12\}$

$C = \{5, 7, 9, 11, 13, 15\}$

1. Find $(A \cup B) \cup C$.

2. Find $(A \cap B) \cap C$.

3. Find $(A \cup B) \cap C$.

4. Find $(A \cap B) \cup C$.

Use the given Venn diagram to shade each solution.

5. $(A \cup B) \cup C$

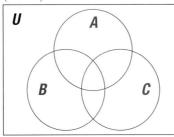

6. $A \cap (B \cup C)$

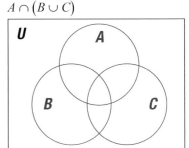

Draw each Venn diagram.

7. Construct a Venn diagram illustrating the following sets: $A = \{$apple, orange, grape, peach$\}$, $B = \{$grape, banana, apple, kiwi$\}$, and $C = \{$kiwi, apple, peach, banana$\}$ if $U = \{$apple, orange, peach, grape, banana, kiwi$\}$.

8.   Construct a Venn diagram illustrating the sets: $A = \{1, 2, 3, 4\}$, $B = \{2, 4, 6, 8, 10\}$, and $C = \{3, 4, 6\}$ if $U = \{1, 2, 3, 4, 5, 6, 7, 8, 9, 10\}$.

9.   A survey of 400 college freshmen showed that 200 drink soda, 300 drink juice, and 150 drink both. Draw a Venn diagram to illustrate this survey.

10.   A survey of 350 students showed that 225 listen to rap and 200 listen to rock and 135 listen to both. Draw a Venn diagram to illustrate this survey.

Create a Venn diagram with the given information.

11. $|A| = 26$ $|B| = 32$ $|C| = 23$ $|A \cap B| = 6$
$|C \cap B| = 12$ $|A \cap C| = 9$ $|A \cap B \cap C| = 5$ $|U| = 65$

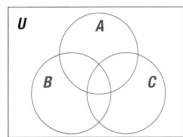

Use the given information and Venn diagram to answer each question.

There are 43 students in the University Travel Club. They discovered that 17 members have visited Mexico, 10 have been to England, 28 have visited Canada, 8 have been to Mexico and Canada, 3 have only been to England, and 4 have only been to Mexico. No student has been to only England and Canada. Two students have been to all three countries. Some of the club members have not been to any of the three.

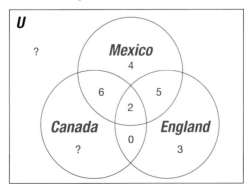

12.   How many students have been to all three countries?

13.   How many students have been only to Canada?

14.   How many have been to Mexico or Canada but not England?

15.   How many have been to none of the countries?

Solve each problem.

16. A researcher collecting data on 100 households finds that 47 have a DVD player; 52 have only streaming video, and 27 have both. Determine the answer to the following questions.

    a. How many do not have video streaming?

    b. How many have neither video streaming nor a DVD player?

    c. How many have a DVD player but not video streaming?

17. A survey of 125 freshman business students at a large university produced the following results:

    35 read *Money*;
    25 read *The Wall Street Journal*;
    32 read *Fortune*;
    21 read *Money* but not *The Wall Street Journal*;
    11 read *The Wall Street Journal* and *Fortune*;
    13 read *Money* and *Fortune*;
    9 read all three.

    Use this information to answer the following questions:

    a. How many students read none of the publications?

    b. How many read only *Fortune*?

    c. How many students read *Money* and *The Wall Street Journal*, but not *Fortune*?

18. A survey of 600 workers yielded the following information: 417 belonged to the Auto Workers Union, 275 were Democrats, and 215 of the Auto Workers Union were Democrats.

    a. How many workers belonged to the Auto Workers Union or were Democrats?

    b. How many workers belonged to the Auto Workers Union but were not Democrats?

    c. How many workers were Democrats but did not belong to the Auto Workers Union?

    d. How many workers neither belonged to the Auto Workers Union nor were Democrats?

19. Imagine Dragons, One Direction, and Bruno Mars toured the United States. A large group of teenagers were surveyed and the following information was obtained: 825 saw One Direction, 1033 saw Imagine Dragons, 1247 saw Bruno Mars, 211 saw all three, 514 saw none, 240 saw only Bruno Mars, 677 saw Bruno Mars and Imagine Dragons, and 201 saw Imagine Dragons and One Direction but not Bruno Mars.

    a. What percent of the teenagers saw at least one band?

    b. What percent of the teenagers saw exactly one band?

20. There are three types of blood antigens that determine blood type: A, B, and Rh+. An individual's blood type is determined by the specific combination of these antigens. In order to receive a blood transfusion, you can't receive blood from a donor who has an antigen that you don't have yourself. That means that people with AB+ blood can receive a transfusion from ANY donor, since they have all of the possible antigens. They can only donate to other people with AB+. People with O– blood (none of the antigens) can only receive type O– blood, since all other blood types have at least one of the antigens. However, they can donate their blood to anyone, since their blood does not have any of the antigens. A laboratory looked at blood samples for 200 patients and found the following information provided in the table. How many patients were classified as O–? Explain your reasoning.

Blood Antigen Survey Results	
Number of Samples	**Antigen in Blood**
80	A
36	B
82	Rh
10	A and B
62	A and Rh
22	B and Rh
4	A, B, and Rh

Chapter 2 Summary

Definitions

Set

A set is a collection of objects made up of specified elements, or members.

Roster Notation

Roster notation is a way to describe a set by listing all of the elements in the set.

Equal Sets

Two sets are said to be equal if they contain exactly the same elements. If sets A and B are equal, we write $A = B$.

Cardinal Number

The number of elements contained in a finite set is called the cardinal number. The cardinal number is denoted by $|\ |$.

Equivalent Sets

Sets are equivalent if they have the same cardinal number; that is, the same number of elements. If sets C and D are equivalent, we write $C \sim D$.

Set-Builder Notation

Set-builder notation is used to describe a set when the members all share certain properties.

Empty Set

The empty set, or null set, is the set that contains no elements. If set A is empty, we write $A = \varnothing$.

Universal Set

The set of all elements being considered for any particular situation is called the universal set and is denoted by U.

Complement

The complement of A consists of all the elements in the given universal set that are not contained in A. The complement of A is denoted A'.

Definitions

Venn Diagram

A Venn diagram is a way to visualize the relationships between sets. In a Venn diagram, the sets are represented by circles (or ovals) contained within a rectangular region representing the universal set.

Subset

If A and B are sets, B is a subset of A if every element of B is also an element of A. We write $B \subseteq A$.

Proper Subset

When $B \subseteq A$, and A contains at least one element that is not contained in B, B is said to be a proper subset of A and is written $B \subset A$.

Number of Subsets and Proper Subsets of a Set

If the cardinal number of a set is n, then there are 2^n subsets and $2^n - 1$ proper subsets contained in the set.

Section 2.3   Operations with Sets

Definitions

Intersection

The intersection of two sets A and B is the set of all elements common to both A and B. We denote the intersection of A and B as $A \cap B = \{x \mid x \in A \text{ and } x \in B\}$.

Union

The union of two sets A and B is the set of all elements in A or in B. We denote the union of A and B as $A \cup B = \{x \mid x \in A \text{ or } x \in B\}$.

Disjoint

Two sets A and B are disjoint if there are no elements in the set A that are also contained in the set B. In the case where two sets are disjoint, the intersection of those two sets is the null, or empty, set $\varnothing$. Therefore, $A \cap B = \varnothing$ when A and B are disjoint.

Formulas

De Morgan's Laws

Let A and B be sets. Then $(A \cup B)' = A' \cap B'$ and $(A \cap B)' = A' \cup B'$.

Inclusion-Exclusion Principle

The inclusion-exclusion principle states that the number of elements in the union of two sets A and B is calculated by adding the number of elements in the set A to the number of elements in the set B, less the number of elements that appear in both sets. We denote this by $|A \cup B| = |A| + |B| - |A \cap B|$.

Section 2.4   Applications and Survey Analysis

Definition

Survey Analysis

Survey analysis is the analysis of responses to a list of questions.

Chapter 2 Exercises

Determine whether each statement is true or false. If the statement is false, explain why.

1. $\{3\} \in \{1, 2, 3, 4, 5, 6\}$

2. $1 \in \{x \mid x \text{ is an integer}\}$

3. $\{3\} \subseteq \{1, 2, 3, 4, 5, 6\}$

4. $\{1\} \not\subset \{x \mid x \text{ is an integer}\}$

5. Let $A = \{\text{red, yellow, blue}\}$. Then $|A| = 3$.

6. Let $B = \{-2\}$. Then $|B| = 2$.

7. $|\varnothing| = 1$

8. $\left|\{\varnothing\}\right| = 1$

Write each set as indicated.

9. Let the set A consist of the even counting numbers less than 14. Write A using the roster method.

10. Use the roster method to write the set B that consists of the seasons of the year.

11. Use set-builder notation to write the set C that consists of the set of real numbers between 100 and 1000.

12. Use set-builder notation to write the set D that consists of the months of the year that have 30 days.

Use the given sets to solve each problem.

$A = \{\text{Felix, Amber}\}$

$B = \{\text{moral, social, civil}\}$

13. Find $|A|$ and $|B|$.

14. List all the subsets of A and subsets of B.

15. List all the proper subsets of A.

16. Is $A = B$? Why or why not?

17. Is $A \sim B$? Why or why not?

Use the given sets to solve each problem.

$G = \{\text{I, II, III}\}$

$F = \{\text{love, joy, peace}\}$

18. Find $|G|$ and $|F|$.

19. List all the subsets of G and subsets of F.

20. List all the proper subsets of F.

21. Is $G = F$? Why or why not?

22. Is $G \sim F$? Why or why not?

Determine the number of proper subsets of each set.

23. $\{\alpha, \beta, \chi, \delta, \varepsilon, \phi, \mu, \pi\}$

24. $\varnothing$

Draw a Venn diagram to illustrate each group of sets. A universal set is not given, so choose one that fits and define it.

25. Parents and their children

26. Sculptors and Artists

27. Men and Women

28. $A = \{x \mid x \in \mathbb{R}\}$ and $B = \{x \mid x \text{ is an integer}\}$

Use the given sets to write each set in roster notation.

$U = \{1, 2, 3, 4, 5, 6, 7, 8, 9, 10\}$

$A = \{1, 2, 3, 4, 5\}$

$B = \{1, 3, 5, 7\}$

29. $A \cap B$

30. $A \cup B$

31. $A' \cap B$

32. $A' \cup B'$

33. $|A \cap B|$

34. $\left| (A \cup B)' \right|$

Use the given Venn diagram to write each set in roster notation.

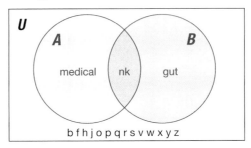

35. A

36. B

37. $A \cap B$

38. $A \cup B$

39. $(A \cup B)'$

40. $(A \cap B)'$

41. U

42. $|A \cup B|$

43. $|A \cap B|$

44. $|(A \cup B)'|$

Draw a Venn diagram to illustrate each group of sets.

45. $U = \{$Mike, Kim, Susan, Marcel, Jay, Darren, Barbara$\}$

$A = \{$Mike, Kim, Susan$\}$

$B = \{$Susan, Barbara, Marcel$\}$

$C = \{$Susan, Mike, Darren, Barbara$\}$

46. $U = \{x \mid x$ is a whole number$\}$

$A = \{2, 3, 6, 8, 10\}$

$B = \{3. 4, 5, 6, 7, 8\}$

$C = \{1, 3, 5, 7\}$

Solve each problem.

47. A school gym teacher is trying to determine what sports students enjoy the most. She collected information on 250 students and found that 150 like volleyball, 110 like soccer, and 65 students like both.

  a. Draw a Venn diagram to represent the findings of the teacher.

  b. How many students like only volleyball?

  c. How many students like only soccer?

  d. How many students like neither soccer nor volleyball?

48. A camp counselor is planning activities for the summer and wants to know what campers would enjoy. He asks 650 campers and finds that 457 enjoy swimming, 250 enjoy tennis, 223 enjoy jogging. He finds that 176 enjoy swimming and tennis, 75 enjoy swimming and jogging, 105 enjoy tennis and jogging, and 45 enjoy all three.

    a. Draw a Venn diagram to represent the survey results.

    b. How many campers enjoy only swimming?

    c. How many campers enjoy only tennis?

    d. How many campers enjoy only jogging?

    e. How many campers enjoy only swimming and tennis?

    f. How many campers enjoy only tennis and jogging?

    g. How many campers enjoy only swimming and jogging?

    h. How many campers enjoy none of the three?

49. A study found that 25% of a certain population has blue eyes, 20% of the population has blonde hair, and 12% of the population has blonde hair and blue eyes. Estimate the percentage of the population that has blue eyes or blonde hair?

Bibliography

2.2

1. US Department of Energy: Alternative Fuels Data Center. "Alternative Fueling Station Counts by State." Accessed January 21, 2014. http://www.afdc.energy.gov/fuels/stations_counts.html

Chapter 3
Logic

Objectives

- Construct statements using logic symbols
- Construct truth tables
- Determine the validity of formal arguments
- Identify common fallacies in arguments

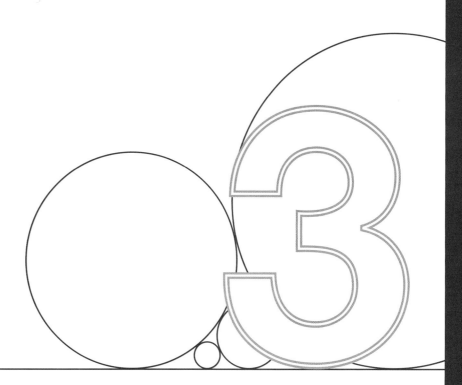

3 Logic

In their 1941 film *In the Navy*, Abbott and Costello show by multiplication, division, and repeated addition that $13 \cdot 7 = 28$. What?!? Wait a second. . . How can that be? Here's their method of creative math.

Division

A. "7 in to 2. . . 7 won't go into 2, so I gotta take the 2 from here and put it down there. 7 into 8? 1 time"

B. "Now I'm gonna carry the 7 from here and put it under the 8. 7 from 8, 1"

C. "Now I have 21, 7 into 21? 3 times SO $28 \div 7 = 13$!"

A.
$$7 \overline{)\cancel{2}\,8} \searrow 2$$

B.
$$7 \overline{)\cancel{2}\,8} \searrow 21$$

C.
$$7 \overline{)\cancel{2}\,8}\,^{13} \underset{21 /\!/ 7}{}$$

Multiplication

A. "7 times 3, 21"

B. "7 times 1, 7"

C. "7 and 1 is 8 and then bring this 2 down IT'S 28!
$7 \cdot 13 = 28$!"

A.
$$\begin{array}{r} 13 \\ \times \; 7 \\ \hline 21 \end{array}$$

B.
$$\begin{array}{r} 13 \\ \times \; 7 \\ \hline 21 \\ 7 \end{array}$$

C.
$$\begin{array}{r} 13 \\ \times \; 7 \\ \hline 21 \\ + \; 7 \\ \hline 28 \end{array}$$

Repeated Addition

A. "Put down 13 seven times"

B. "Add all the 3's 3, 6, 9, 12, 15, 18, 21"

C. "Add all the 1's 22, 23, 24, 25, 26, 27, 28 YOU GET 28!
$13 + 13 + 13 + 13 + 13 + 13 + 13 = 28$!"

A.
$$\begin{array}{r} 13 \\ 13 \\ 13 \\ 13 \\ 13 \\ 13 \\ + 13 \end{array}$$

B.
$$\begin{array}{rl} 13 & 21 \\ 13 & 18 \\ 13 & 15 \\ 13 & 12 \\ 13 & 9 \\ 13 & 6 \\ + 13 & 3 \end{array}$$

C.
$$\begin{array}{rrl} 22 & 13 & 21 \\ 23 & 13 & 18 \\ 24 & 13 & 15 \\ 25 & 13 & 12 \\ 26 & 13 & 9 \\ 27 & 13 & 6 \\ 28 & + 13 & 3 \\ \hline & 28 & \end{array}$$

JUST TO BE CLEAR... Division: $28 \div 7 = 4$

Multiplication: $13 \cdot 7 = 91$

Repeated Addition: $13 + 13 + 13 + 13 + 13 + 13 + 13 = 91$

Abbott and Costello's famous skit managed to show by division, multiplication, and repeated addition that $13 \cdot 7 = 28$. So it must be true, right? This makes us laugh because we can all see the errors that he makes right before our eyes. What is not so funny is that we often jump to the same types of misguided conclusions in our everyday lives by using similarly faulty logic.

So, what was wrong in Costello's conclusion that $13 \cdot 7 = 28$? The problem is that his conclusion cannot be logically supported by the rules of addition, multiplication, and division. Although this famous error in his conclusion is obvious to us here, what if you are faced with a conclusion and you don't know if you can trust it? In other words, you are not sure if the conclusion is right or wrong, or if the argument is valid or faulty.

Consider the following observation a student made in class one day.

Teacher: Before the end of the week, you will have a surprise quiz.

Student: That can never happen. Because, if you waited until Friday to give it, we would all realize it was coming and it would no longer be a surprise. So, the quiz can't be given on Friday. Similarly, since it's not on Friday, it can't be on Thursday. If you waited until after Wednesday, we already know it can't be on Friday. So, we would know it must be on Thursday and hence it would not be a surprise. We could make the same argument for Wednesday and Tuesday. So, that means the quiz is today, Monday. We now know it is happening today, so it's not a surprise quiz and therefore you can't give it.

Although all his classmates were initially excited at the thought that they were not going to have a surprise quiz that week, they were surprised when the next morning they heard the words, "Clear your desks and take out a blank sheet of paper."

As we move through the chapter, not only will we show you ways to identify common errors in logic, like those in the Costello scene and from the student hoping for no quiz, we will also help you better understand how to make strong arguments for yourself.

3.1 Logic Statements and Their Negations

Lots of claims similar to those made by Abbot and Costello or the hopeful student mentioned in the chapter introduction are made everyday. "If you run for office, you will surely win." "9 out of 10 dentists choose this toothpaste." "It always rains the first day of school." Our ordinary English language encompasses not only these types of statements but is also littered with opinions, sarcasm, riddles, commandments, and the list goes on. Because of this, it is often difficult to determine the validity of many of the things we hear day to day. However, there are times when we want to determine with certainty if statements are not only factually true, but also logically true. Mathematical logic provides us with a way to do just that. It provides a consistent framework in which to evaluate claims for logical truth.

All claims that we can logically evaluate through mathematics are made up of what are called statements. A mathematical **statement** is a complete sentence that asserts a claim that is either true or false, but not both at the same time. Any statement has exactly one of two possible **truth values** at any particular moment: true (T) or false (F). If a statement is true, then we say the statement has a true truth value. If a statement is false, then its truth value is false. If it is not possible to assign a truth value to a sentence, then it is not a mathematical statement.

> **Statement**
>
> A **statement** is a complete sentence that asserts a claim that is either true or false, but not both at the same time.

☞ Helpful Hint

Logic statements are most commonly represented by lower case letters.

The following sentences are all examples of statements. They are represented by lowercase letters, as is the practice in mathematical logic.

a: The car is blue and the cat is black.

b: The first even number is 2.

c: Bill Haslam is the governor of Tennessee.

d: Charlie Brown was the first president of the United States.

These four sentences are indeed statements because they have a clear true/false value at any given time, even though we may not know which it is. For instance, sentence a might be true or false depending on what car and which cat are being described. However, sentence b is always true and sentence d is always false. Regardless of their truth value, all four of the sentences are examples of mathematical statements. Decide for yourself if the truth value of statement c is either always true, always false, or dependent on something.

It has to be noted that not all complete sentences in the English language are statements. That is, they don't assert a claim that is either true or false. Consider the following sentences.

s: This is the best movie ever made.

t: Stop!

u: Are we going to the movies tonight?

v: Chair sees box cat telephone.

w: This sentence is false.

At first sight, some of these sentences might seem to vie for statement status. Note that just because they are labeled with a lowercase letter, we cannot assume that they are statements with truth values. The insistence that a statement is either true or false, and cannot be both at the same time, is crucial here. For instance, take the first sentence *s*. It is a matter of opinion which is the best movie ever made, and therefore the sentence cannot really be said to be true or false. Matters of opinion are not mathematically legitimate statements.

Neither the exclamation "Stop!" nor the question "Are we going to the movies tonight?" make any claim to be tested, and are therefore also not statements. As for sentence *v*, this is just a random and meaningless string of words and quite obviously not a statement.

That leaves us with example *w*, which is a little more perplexing. It looks like a statement, but is it true or false? From our definition, we know that to be a mathematical statement it must be one or the other, but not both at the same time. When reading the definition of a statement, you might have wondered how a sentence could be both true and false at the same time. To see how that might happen, let's suppose that sentence *w* is actually true. Then just as it asserts, the sentence must also be false. So now we have a sentence that is both true and false at the same time. On the other hand, what if we assume the sentence is false? Then it is false that "this sentence is false," so it is true. Somehow it is both true and false yet again! This type of sentence is an example of a **paradox**. Paradoxes are not allowed to be statements because they have no truth value. At first, it might seem that this example is a little contrived, but once you start looking, you will see that paradoxes are more common than you think.

Paradox

A **paradox** is a sentence that contradicts itself and therefore has no single truth value. A paradox cannot be a mathematical statement.

Example 1: Identifying Statements

Determine if the following sentences are statements.

a: It is raining outside.

b: Beaches are the most beautiful place to vacation.

c: Today is Monday.

d: Today is Monday and tomorrow is Friday.

e: I lie all the time.

Solution

Sentence *a* is a statement because it can be assigned a truth value depending on the weather outside. However, sentence *b* is an opinion, and therefore not a statement. Sentence *c* is a statement since it can be either true or false, depending on the current day the statement is read. Sentence *d* is a statement even though it is always false. And finally, sentence *e* is a paradox and not a statement since it contradicts itself and therefore has no truth value.

Skill Check # I

Write down two statements of your own: one that is always true and one that is always false.

Negation

Sometimes, it is the case that we want the opposite truth value of a statement, or its negation. The **negation** of a statement is the logical opposite of that statement, or its denial. Negations always have the opposite truth value of the original statement. In other words, if a statement is true, its negation is always false. Likewise, if a statement is false, its negation is always true. Although there are a variety of different ways to express the idea of negation in the English language, mathematical negations of statements can be written with the symbol $\sim$, read as "not." For instance, the negation of statement a is $\sim a$.

> ## Negation
>
> The **negation** of a statement is the logical opposite of that statement, or its denial. Negations always have the opposite truth value of the original statement.

Consider the following statement.

e: 5 is a prime number.

Since 5 is actually a prime number, statement e is true. Therefore, $\sim e$ is false. The most common way to write a negation of a statement is to insert the word "not" into the statement. For instance, we could write $\sim e$ as "5 is not a prime number." However, there are many ways to negate statements with words. Here is another way to negate e.

$\sim e$: It is not true that 5 is a prime number.

Sometimes it takes a bit more thought to negate a statement in English. This is true when the statement contains words that are **quantifiers**, such as *all*, *some*, *none*, or *no*. In the English language, we can negate the word "all" with "not all." However, English does not allow us to put the word "not" in front of every quantifier. For instance, it is incorrect to say "not some" or "not none." Instead, we use other ways to negate the words some and none. Table 1 gives us ways to negate these quantifiers.

Table 1	
Negating Quantifiers	
Quantifier	**Negations**
All are	Not all are; Some are not; At least one is not
Some are	None are
Some are not	All are
None are	There is at least one that is; Some are

Notice that the negation of *all* is not the word *none* (or *no*) since this does not give an opposite truth value to a statement. For instance, the statements "all leaves are green" and "no leaves are green" are both false, and therefore, cannot be negations of one another.

Example 2: Negating a Statement

Negate the following statements.

a: Melony is wearing a red raincoat.

b: The door is not closed.

c: None of the tourists brought raincoats.

d: I run less than Cara.

Solution

~*a*: Melony is not wearing a red raincoat.

~*b*: The door is not not closed. However, as we noted in the helpful hint, when we negate a negation, we are back to no negation at all. So we more commonly say, "The door is closed."

~*c*: Some of the tourists brought raincoats.

~*d*: I do not run less than Cara. Notice that we could also write, "I run the same as or more than Cara." We need both parts since the opposite of "less than" is "more than or equal to."

Skill Check #2

Negate the following statement.

Some of the students completed their assignments.

Compound Statements

Recall that mathematical logic seeks to evaluate claims systematically. There is rarely a need to mathematically evaluate a single statement like the ones we have looked at so far since they are inherently either true or false. More often, we need to evaluate the truth value of two or more statements combined together using connecting words such as *and*, *or*, or *implies*. We call these types of statements **compound statements**.

Compound Statement

A **compound statement** is composed of two or more statements joined together using connective words such as *and*, *or*, or *implies*.

Let's look at how to write compound statements symbolically. For instance, consider the following two statements *p* and *q*.

p: It is raining.

q: It is sunny outside.

We can combine these two *simple statements* using the connecting word *and* to form the following compound statement.

r: It is raining and it is sunny outside.

When combining two or more statements together to form a compound statement using the word *and*, the symbol $\wedge$ is used between the lower case letters. To write the compound statement *r* symbolically, we write $r = p \wedge q$.

Example 3: Using Logic Symbols for Compound Statements Involving *and*

Use the following simple statements *a* and *b* to symbolically write the given compound statement *c*.

a: Snow is falling.

b: The sun is shining.

c: Snow is falling and the sun is shining.

Solution

$$c = \text{Snow is falling and the sun is shining.}$$

$$= (\text{Snow is falling}) \text{ AND } (\text{The sun is shining})$$

$$= (\text{Snow is falling}) \wedge (\text{The sun is shining})$$

$$= a \wedge b$$

Let's turn to the connecting word *or*. There are two ways to think about the word *or*. For instance, take the statement "Your painting is so inspirational that you are very talented or have had lots of training." You could interpret this statement to mean that you are either very talented or had lots of training, but not both. This is referred to as an **exclusive** *or*, meaning *one or the other, but not both*.

However, it could mean that both are true; in other words, you are talented and have had lots of training. This type of or, meaning either or both of the options can be true, is known as the **inclusive** or and is what is used in mathematical logic. To represent the inclusive or, we use the symbol $\vee$.

Example 4: Using Logic Symbols for Compound Statements Involving *or*

Use the following statements *p* and *q* to symbolically write the given compound statement *r*.

p: He will go to the movies tonight.

q: He will stay home to give the dog a bath tonight.

r: He will go to the movies tonight or he will stay home to give the dog a bath tonight.

Solution

$$r = \text{He will go to the movies tonight or he will stay home to give the dog a bath tonight.}$$

$$= (\text{He will go to the movies tonight})$$
$$\text{OR}$$
$$(\text{He will stay home to give the dog a bath tonight})$$

$$= (\text{He will go to the movies tonight})$$
$$\vee$$
$$(\text{He will stay home to give the dog a bath tonight})$$

$$= p \vee q$$

The last way we will consider combining simple statements into compound statements is with implications. Two statements can be joined together using the sentence structure "if *a*, then *b*." We call this type of combination an **implication** because statement *a* implies statement *b*. Both "*a* implies *b*" and "if *a*, then *b*" have identical meanings in the English language, but sometimes one sounds more natural than the other. Mathematically, we use $a \Rightarrow b$ to symbolically represent "if *a*, then *b*."

Here are a few of the many ways to convey $p \Rightarrow q$.

"p implies q"

"if p, then q"

"p is sufficient for q"

"q is necessary for p"

"p will lead to q"

"q if p"

"q whenever p"

"p only if q"

Example 5: Using Logic Symbols for Compound Statements Involving Implications

Use the following statements s and t to symbolically write the given compound statement q.

s: The water temperature on Saturday is below 76.2°.

t: You are allowed to wear a wetsuit in the triathlon.

q: If the water temperature on Saturday is below 76.2°, then you are allowed to wear a wetsuit in the triathlon.

Solution

q = If the water temperature on Saturday is below 76.2°, then you are allowed to wear a wetsuit in the triathlon.

= If (the water temperature on Saturday is below 76.2°), then (you are allowed to wear a wetsuit in the triathlon).

= If (s), then (t).

= $s \Rightarrow t$

Table 2 summarizes the logic symbols discussed in this section.

Table 2 Logic Symbols	
Symbol	**Read**
$\wedge$	And
$\vee$	Or
$\sim$	Not
$\Rightarrow$	Implies

Skill Check #3

Write the following compound statements mathematically given the simple statements a, b, and c.

a: I am hungry.

b: I am tired.

c: I am in college.

1. I am hungry and tired.

2. I am hungry or I am in college.

3. I am tired and not in college.

Skill Check Answers

1. Answers will vary.

2. None of the students completed their assignments.

3. **1.** $a \wedge b$
   2. $a \vee c$
   3. $b \wedge \sim c$

3.1   Exercises

Decide if each sentence is a mathematical statement.

1.   My computer is fast.

2.   The car in front of me is turning around.

3.   Who are you voting for in the election?

4.   I always lie.

5.   Running is fun.

6.   Are you cold today?

7.   I received 35 e-mails today, half of which ended up in my spam folder.

8.   Get out of my room!

9.   This computer's processor is at least 1.8 GHz.

10.   Either we are going to the beach or the mountains for vacation this year.

11.   No one goes to that mall anymore; it's too crowded.

Use the given simple statements to write each compound statement in words.

a: Driving makes me smile.

b: I grill more often than I bake.

c: It is sunny.

12. $b \wedge a$ **13.** $a \vee c$ **14.** $c \wedge \sim b$

15. $c \Rightarrow b$ **16.** $\sim c \Rightarrow \sim b$

Use the given simple statements to write each compound statement in words.

> *p*: My video reached 1000 views on YouTube.
>
> *q*: I have 1000 Facebook friends.
>
> *r*: The home page of my website has a bounce rate of less than 20%.

17. $p \wedge q$ **18.** $\sim q \Rightarrow \sim r$ **19.** $q \vee p$

Reword each conditional statement using the specified alternate phrasing.

20.   Use the form "*q* whenever *p*" in the statement "If I flip the switch, the lights turn on."

21.   Use the form "*p* is sufficient for *q*" in the statement "If I peddle faster, then the wheels turn more slowly."

22.   Use the form "*p* implies *q*" in the statement "If I complete more homework exercises, then I get a higher grade."

23.   Use the form "*q* is necessary for *p*" in the statement "If the class is 50% male, then 50% of the class must be female."

24.   Use the form "*p* will lead to *q*" in the statement "If you complete college, then better jobs await you."

Negate each statement.

25.   Kelsey's website had more than 50,000 visits yesterday.

26.   Three hundred twenty-nine people applied for the same job I did.

27.   I did not get the job.

28.   Austin slept until 8:00 a.m. this morning.

29.   None of the Christmas tree lights are not working.

30.   Every student is volunteering at the food pantry this year.

Using the letters given to represent simple statements and the proper logic connectives, express each compound statement in symbolic form.

31. It is not true that Miranda likes both art and history.

m: Miranda likes art.

n: Miranda likes history.

32. Miranda does not like art or Miranda does not like history.

m: Miranda likes art.

n: Miranda likes history.

33. If I don't eat meat, then I don't get sick.

a: I don't eat meat.

b: I get sick.

34. I don't get sick or I don't eat meat.

a: I don't eat meat.

b: I get sick.

35. If Mrs. Walker is a teacher, then she is not a rocket scientist.

t: Mrs. Walker is a teacher.

r: Mrs. Walker is a rocket scientist.

36. If Mrs. Walker is not a rocket scientist, then she is a teacher.

t: Mrs. Walker is a teacher.

r: Mrs. Walker is a rocket scientist.

37. Right angles are formed by the lines if the lines are perpendicular.

w: The lines are perpendicular.

z: Right angles are formed by the lines.

38. The lines are not perpendicular or right angles are formed by the lines.

w: The lines are perpendicular.

z: Right angles are formed by the lines.

Use the given statements to write each compound statement in words.

p: The moon is full.

q: I don't know if it's cloudy or bright outside.

r: I've lost my glasses.

39. $p \wedge r$

40. $(\sim p) \vee q$

41. $\sim(q \wedge r)$

42. $r \Rightarrow q$

43. $\sim r \Rightarrow \sim q$

3.2   Truth Tables

As we stated in Section 3.1, mathematical logic deals in part with deciding the truth value of different types of statements. While some are easy to determine, others require more thought. For instance, consider the compound statement $a \wedge d$ given below.

$a \wedge d$: Today is Monday **and** it is sunny outside.

Is this statement true or false? To decide the truth of this compound statement for today, it would be enough to consult the calendar to determine the day of the week while you look out the window to check the weather. For instance, suppose the portion "Today is Monday" is true and the portion "It is sunny outside" is also true. Then the compound statement in which the two pieces are put together as one is obviously true as well. However, what truth value would the statement have if today is not Monday, but it is sunny outside? It can often get quite complicated to keep up with the truth values for compound statements, especially when they have more and more pieces to them. To help us determine the truth value of more complex compound statements, we can use a **truth table**. A truth table is a table used to orderly and systematically determine the truth value for compound statements. It has a row for each possible combination of truth values of the individual statements that make up the compound statement.

> ### Truth Table
>
> A **truth table** is a table that has a row for each possible combination of truth values of the individual statements that make up the compound statement.

Let's look at constructing the truth table for the compound statement we were just discussing: Today is Monday and it is sunny outside, $a \wedge d$. To construct a truth table, begin by creating a column for each part of the compound statement. In this case, we have the two simple statements a and d. Then list all the possible combinations that could occur for the truth values. For example, both a and d could be true meaning that it actually could be Monday and sunny outside. The first row of Table 1 shows this. The remaining 3 rows in the truth table list the other possible combinations for a and d. Take a moment to consider why there needs to be 4 rows in the table.

Table 1
Truth Table

a	d
T	T
T	F
F	T
F	F

Next, add a column for each piece of the compound statement that is needed. In this example, we need another column for $a \wedge d$, as shown in Table 2. As the statements get more complicated, you will begin to see the natural progression of difficulty in the columns leading up to the last column, which always contains the original compound statement.

Table 2
Truth Table

a	d	a ∧ d
T	T	
T	F	
F	T	
F	F	

How do we complete the last column, which combines the two statements? When two simple statements are joined with the word *and*, the compound statement is said to be a **conjunction**. A conjunction is true only when both individual parts are true; otherwise, it is false.

> ## Conjunction
>
> If *a* and *b* are statements, then "*a* and *b*" is a compound statement called a **conjunction**. A conjunction is true only when both statements are true; otherwise it is false.

Now, we can fill in the last column of our truth table. Looking across each row, if both columns *a* and *d* are true, then the last column $a \wedge d$ will be true; otherwise it is false.

Table 3
Truth Table

a	*d*	*a* ∧ *d*
T	T	T
T	F	F
F	T	F
F	F	F

With our truth table filled in, we are fully armed to answer the question, "Is the statement $a \wedge d$, *Today is Monday and it is sunny outside*, true or false?" The actual answer is, "It depends." We can look at the table and see that there are four possible truth value combinations in our situation. A better question might be, "Under what circumstances is the compound statement $a \wedge d$, *Today is Monday and it is sunny outside*, true?" We can see that it is true only when it is both Monday **and** it is sunny. Sounds simple enough, so let's consider another example.

Suppose we want to make a truth table for *q*: It is raining or it is sunny outside. Before we can make the truth table we need to know the truth values for an *or* compound statement. A compound statement containing the word *or,* is said to be a **disjunction**. In contrast to a conjunction where there is only one true value, a disjunction will always be true unless both simple statements are false.

> ## Disjunction
>
> If *a* and *b* are statements, then "*a* or *b*" is a compound statement called a **disjunction**. A disjunction will always be true unless both statements are false.

The truth table for *q*: It is raining or it is sunny outside, written mathematically as $c \vee d$, will be as shown in Table 4.

Table 4
Truth Table

c	*d*	*c* ∨ *d*
T	T	T
T	F	T
F	T	T
F	F	F

Remember that in mathematical logic, we will assume that disjunctions are *inclusive*. That is to say that the statement is true if *c* or *d* or both are true.

Finally, we need to look at the truth values for a compound statement in which one simple statement implies another. The statement structure of "If *a*, then *b*" can best be described by thinking of a promise. If you do *a*, then I promise to do *b*. The only time the promise is FALSE, or broken, is when you do *a* and I don't do *b*. Formally, we say the **conditional statement** "if *p*, then *q*," symbolized $p \Rightarrow q$, is true in all cases except the case in which *p* is true and *q* is false. Think of everything before the *implies* arrow as the *if* part of the statement, and everything after the arrow as the *then* part.

Conditional

If *a* and *b* are statements, then "if *a*, then *b*" is a compound statement called a **conditional.** A conditional will always be true unless *a* is true and *b* is false.

☞ **Helpful Hint**

Note that the order of the statements in a conditional *does* matter. $p \Rightarrow q$ is NOT the same as $q \Rightarrow p$.

Example 1: Constructing a Truth Table for a Conditional Statement

Consider the conditional statement

> If it rains, then we will stay home tonight.

Let the following statements represent *w* and *z*.

   w: It rains.

   z: We will stay home tonight.

Construct the truth table for $w \Rightarrow z$.

Solution

Begin by making a column for each simple statement *w* and *z* and filling in all possible truth combinations.

Table 5
Truth Table

w	*z*	
T	T	
T	F	
F	T	
F	F	

The last column will contain the conditional $w \Rightarrow z$. Recall that the "promise" is only broken when the first part is true and the second part is false. Therefore, the truth values for the conditional are as follows.

Table 6
Truth Table

w	*z*	$w \Rightarrow z$
T	T	T
T	F	F
F	T	T
F	F	T

Having looked at the different types of compound statements, we are now armed to construct a more involved truth table piece by piece.

Example 2: Constructing a Truth Table for a Disjunction

Construct the truth table for the following compound statement: $a \lor \sim b$.

Solution

First, we need to decide on the beginning pieces for the table. We certainly need columns for the simple statements a and b. Before adding the column for the disjunction, we need a column for the negation of b as well. Begin by completing the first two columns of the table so that we have all four possible combinations of truth values for a and b. Notice that the last column is the compound statement we were given.

Table 7
Truth Table

a	b	$\sim b$	$a \lor \sim b$
T	T		
T	F		
F	T		
F	F		

Filling in the negation of b column, we have the following. Remember that the negation has the opposite truth value of the original statement.

Table 8
Truth Table

a	b	$\sim b$	$a \lor \sim b$
T	T	F	
T	F	T	
F	T	F	
F	F	T	

Finally, we need to fill in the column for the disjunction. Remember that the disjunction is true unless both pieces are false. Comparing the 1st and 3rd columns, we can fill in the remainder of the truth table.

Table 9
Truth Table

a	b	$\sim b$	$a \lor \sim b$
T	T	F	T
T	F	T	T
F	T	F	F
F	F	T	T

Table 10
Summary of Logic Statements

	Notation	Read	Truth Value Rule
Negation	$\sim p$	not p	opposite of truth value of p
Conjunction	$p \land q$	p and q	true only when both p and q are true
Disjunction	$p \lor q$	p or q	false only when both p and q are false
Conditional	$p \Rightarrow q$	if p, then q	false only when p is true and q is false

Now, we're ready to tackle a more challenging example. Consider the following statements from our original examples at the beginning of the section.

a: Today is Monday.

d: It is sunny outside.

e: 5 is a prime number.

Suppose we would like to examine the following compound statement.

$(a \vee d) \Rightarrow \sim e$: **If** today is Monday **or** it is sunny outside, **then** 5 is **not** a prime number.

Notice that, in this example, rather than using the word *implies* we have replaced it with an "if *a*, then *b*" construction. To form the truth table for this compound statement, we need several columns. We need each original statement *a*, *d*, and *e*, but we'll also need the negation of *e* and a few other pieces. Let's begin our truth table with columns for the simple statements first. The first two columns are fairly easy to fill in.

Table 11			
Truth Table			
a	*d*	*e*	*~e*
T	T		
T	F		
F	T		
F	F		

☞ Helpful Hint

For each individual statement within a compound statement, the number of rows required for the truth table increases by a factor of 2 for each simple statement. For example, a compound statement with 4 simple statements needs $2 \cdot 2 \cdot 2 \cdot 2 = 2^4 = 16$ rows.

Now, before just absentmindedly filling in the column for *e*, stop and think about it for a minute. We don't actually have enough rows for all of the possible outcomes with the three statements. Just consider the first row. If *a* and *d* are both true, then *e* can either be true or false. We need an identical row in our table for each possible truth value of *e*. Expanding our table to accommodate this, we can then complete the last two columns. Recall that the negation of *e* will have the opposite truth value of *e*.

Table 12			
Truth Table			
a	*d*	*e*	*~e*
T	T	T	F
T	T	F	T
T	F	T	F
T	F	F	T
F	T	T	F
F	T	F	T
F	F	T	F
F	F	F	T

☞ Helpful Hint

When constructing truth tables, the order of the columns is not crucial. Some tables show all of the simple statements of the compound statement first, while others follow a progression of the compound statement reading left to right.

Now that we have the columns for the simple statements filled in, we need the column for the compound piece *a* ∨ *d*.

Table 13				
Truth Table				
a	*d*	*e*	*~e*	*a* ∨ *d*
T	T	T	F	T
T	T	F	T	T
T	F	T	F	T
T	F	F	T	T
F	T	T	F	T
F	T	F	T	T
F	F	T	F	F
F	F	F	T	F

The final column of the table will contain the conditional statement, that is, the "if a, then b" or "a implies b" part. So as we fill in the last column, we will be focused on the two columns that contain our "if a, then b" parts. Remember, the only time we will place a false truth value is when the *if* part is true and the *then* part is false.

			"If" ↓	"Then" ↓	
					Table 14
					Truth Table
a	d	e	$a \vee d$	$\sim e$	$(a \vee d) \Rightarrow (\sim e)$
T	T	T	T	F	F
T	T	F	T	T	T
T	F	T	T	F	F
T	F	F	T	T	T
F	T	T	T	F	F
F	T	F	T	T	T
F	F	T	F	F	T
F	F	F	F	T	T

Now, the truth table is complete for $(a \vee d) \Rightarrow \sim e$: **If** today is Monday or it is sunny outside, **then** 5 is **not** a prime number.

Let's try another example.

Example 3: Constructing a Truth Table from Words

Construct the truth table for the following compound statement.

If you're not making mistakes, then you're not doing anything.—John Wooden, member of the Basketball Hall of Fame, both as a player and a coach.

Solution

This statement might seem rather easy to write down at first glance, but we will take a moment to list the simple statements without using the negations. Let statements a and b be the following.

> ***a*:** You are making mistakes.

> ***b*:** You are doing something.

So, our conditional statement is

> **If** you are **not** making mistakes, **then** you are **not** doing anything: $\sim a \Rightarrow \sim b$.

A conditional statement is false only when the *if* part is true and the *then* part is false. The truth table will look like the following.

		"If" ↓	"Then" ↓	
				Table 15
				Truth Table
a	b	$\sim a$	$\sim b$	$\sim a \Rightarrow \sim b$
T	T	F	F	T
T	F	F	T	T
F	T	T	F	F
F	F	T	T	T

From the truth table, we can see that the compound statement is true in all but one of the cases; the statement is false when *you are not making mistakes* and *you are doing something*.

Skill Check #1

Construct a truth table for the conditional statement in Example 3. Let statements *a* and *b* be the following.

a: You're not making mistakes

b: You're not doing anything

Does your truth table have the same truth values as the table in Example 3? Should it?

For some compound statements, the values in the last column of the truth table are all true. When this happens, the statement is called a **tautology**. A tautology will have a *true* truth value in all possible circumstances. Table 15 from Example 3 does not represent a tautology because there is one instance where the statement is false. Let's take a look at a statement that *is* a tautology.

Tautology

A **tautology** is a statement that is true in all possible circumstances.

Example 4: Constructing a Truth Table for a Tautology

Construct the truth table for the following compound statement.

Next year, Imre can take physics or he cannot take physics.

Solution

Let *c* represent "Imre can take physics next year."

Then statement *c* is the first part of the compound statement. To express the entire statement, we need to symbolize the part after the *or* in the statement—"Imre cannot take physics next year." Notice that this is simply the negation of *c*. So, our entire compound statement can then be expressed by $c \vee {\sim} c$. The truth table is then represented by the following.

Table 16		
Truth Table		
c	${\sim}c$	$c \vee {\sim}c$
T	F	T
F	T	T

Hence, "Next year, Imre can take physics or he cannot take physics" is a tautology, since all truth values for the disjunction in the last column are true.

Helpful Hint

Because the English language is prone to ambiguities, sometimes you will need to use your judgment to insert any necessary connectives. In mathematics, it is wise to always be as precise as possible.

Consider the following response opinion on immigrant fishermen being jailed in Davidson County printed in *The Tennessean*.

> *If they come here illegally then commit more offenses while they are here,
> no matter how small . . . they will eventually commit greater offenses.*

Construct the truth table for the statement given in the newspaper.

Solution

As we said before, English is a rich and complicated language. We need to be careful with simply seeing the words "if . . . , then" without looking at the intent behind the statement. In the original quotation from the newspaper, the words "if . . . , then" appear, but are not used in the same mathematical way we have been talking about. Also, the word *and* was implied, but not written. Here's a rewording of the actual quotation using the logical "if *a*, then *b*" compound statement.

> *If they come here illegally **and** they commit more offenses while they are here,
> no matter how small, **then** they will eventually commit greater offenses.*

Let's break down this compound statement and write out its truth table.

Let *p*, *q*, and *r* represent the following simple statements.

p: They come here illegally.

q: They commit more offenses while they are here, no matter how small.

r: They will eventually commit greater offenses.

That gives us the following mathematical statement.

$(p \wedge q) \Rightarrow r$: **If** they come here illegally **and** commit more offenses while they are here, no matter how small, **then** they will eventually commit greater offenses.

When we build the truth table, we need to include all the parts that will eventually build up our final conditional statement. Remember to include enough rows for three simple statements, that is, $2 \cdot 2 \cdot 2 = 8$.

The first part of the table should look like Table 17.

Table 17
Truth Table

p	*q*	*r*	*p* ∧ *q*	(*p* ∧ *q*) ⇒ *r*
T	T	T		
T	T	F		
T	F	T		
T	F	F		
F	T	T		
F	T	F		
F	F	T		
F	F	F		

We find it easier to list all of the simple statements of the compound statement in order, and then copy a column over again for clarity if needed. In Table 18, we've duplicated the column for *r* after the $p \wedge q$ column for easy reference when completing the last column.

Next, complete the column for the conjunction (that is, the *and* statement). Remember that a conjunction is true only if both of the individual statements are true.

Table 18					
Truth Table					
p	*q*	*r*	*p ∧ q*	*r*	*(p ∧ q) ⇒ r*
T	T	T	T	T	
T	T	F	T	F	
T	F	T	F	T	
T	F	F	F	F	
F	T	T	F	T	
F	T	F	F	F	
F	F	T	F	T	
F	F	F	F	F	

Finally, fill in the conditional column using the columns that contain the *if* and *then* parts.

			"If" ↓	**"Then"** ↓	
Table 19					
Truth Table					
p	*q*	*r*	*p ∧ q*	*r*	*(p ∧ q) ⇒ r*
T	T	T	T	T	T
T	T	F	T	F	F
T	F	T	F	T	T
T	F	F	F	F	T
F	T	T	F	T	T
F	T	F	F	F	T
F	F	T	F	T	T
F	F	F	F	F	T

The truth table crystallizes what would need to be proven for this implication to be a true model for social policy. Since there is only one instance where the statement is false, the original speaker would have to prove that particular instance never happens in reality. In other words, one would have to show that *illegal immigrants **never** commit more crimes without increasing their seriousness*, or equivalently, *illegal immigrants who continue to commit crimes **always** become more serious criminals*. The quantifiers *never* and *always* in these sentences should make you hesitant to draw conclusions when they are used as part of a logical argument.

Skill Check Answer

1. No, the truth tables do not look exactly the same. The truth tables have the same meaning, but do not look identical since *a* and *b* are defined differently.

Truth Table		
a	*b*	*a ⇒ b*
T	T	T
T	F	F
F	T	T
F	F	T

3.2   Exercises

Complete each truth table.

1. $a \wedge {\sim} b$

Truth Table			
a	**b**	**~b**	**a ∧ ~b**
T	T		
T	F		
F	T		
F	F		

2. ${\sim} w \vee {\sim} z$

Truth Table				
w	**z**	**~w**	**~z**	**~w ∨ ~z**
T	T			
T	F			
F	T			
F	F			

3. $c \Rightarrow {\sim} d$

Truth Table			
c	**d**	**~d**	**c ⇒ ~d**
T	T		
T	F		
F	T		
F	F		

4. $\left(a \vee b \right) \vee c$

Truth Table				
a	**b**	**c**	**a ∨ b**	**(a ∨ b) ∨ c**
T	T	T		
T	T	F		
T	F	T		
T	F	F		
F	T	T		
F	T	F		
F	F	T		
F	F	F		

5. $(p \vee r) \Rightarrow q$

Truth Table				
p	**r**	**q**	**$p \vee r$**	**$(p \vee r) \Rightarrow q$**
T	T	T		
T	T	F		
T	F	T		
T	F	F		
F	T	T		
F	T	F		
F	F	T		
F	F	F		

6. $(\sim a \wedge \sim b) \wedge \sim c$

Truth Table							
a	**b**	**c**	**$\sim a$**	**$\sim b$**	**$\sim c$**	**$\sim a \wedge \sim b$**	**$(\sim a \wedge \sim b) \wedge \sim c$**
T	T	T					
T	T	F					
T	F	T					
T	F	F					
F	T	T					
F	T	F					
F	F	T					
F	F	F					

Use a truth table to determine if each statement is a tautology.

7. $(\sim p \vee \sim q) \vee (p \wedge q)$

8. $w \Rightarrow v$

9. $(m \wedge n) \vee (\sim n)$

10. $(a \wedge \sim b) \vee (a \Rightarrow b)$

Convert each compound statement into variables and then construct the truth table for the compound statement.

11. We will buy a new smartphone or we will buy a new computer.

12. I plan to go to the movies and go out to eat this weekend.

13. Ella is not on the dance team and Coleman is on the soccer team.

14. If it rains and you don't put the convertible hood up, the inside of your convertible will get wet.

15. If Meg doesn't pass, then she will lose her scholarship and drop out of school.

3.3 Logical Equivalence and De Morgan's Laws

In Section 3.2, we talked about compound statements using the following.

Conjunctions: *The zoo is open, but it's not a nice day.* (Think *and*)

Disjunctions: *Congress made a bill that the president must sign or veto.* (Think *or*)

Conditional statements: *Laugh and the world laughs with you.* –Anthony Burgess (Think "if *a*, then *b*")

As we noted before, switching the order of the simple statements in conjunctions and disjunctions makes no difference to the truth value. That is to say, $p \wedge q$ has the same meaning as $q \wedge p$ and $p \vee q$ has the same meaning as $q \vee p$. However, this is not the case when we switch the order in conditional statements. When the order of the statements is rearranged, new conditions are set forth. Given a conditional statement, we can subsequently consider the converse, inverse, and contrapositive of the original conditional statement. Note that these new variations are conditional statements themselves. Each of these, along with the biconditional statement is listed symbolically in Table 1 along with the way we read them so that you can see the relationship that each has with conditional statements.

☞ Helpful Hint

Note that the converse, inverse, and contrapositive are also conditional statements.

Table 1		
Variations on Conditional Statements		
Name	**Symbols**	**Read**
Conditional	$p \Rightarrow q$	If p, then q
Converse of Conditional	$q \Rightarrow p$	If q, then p
Inverse of Conditional	$\sim p \Rightarrow \sim q$	If not p, then not q
Contrapositive of Conditional	$\sim q \Rightarrow \sim p$	If not q, then not p
Biconditional	$p \Leftrightarrow q$	p if and only if q

Consider the following quote by Marilyn Monroe.

> *Give a girl the right shoes, and she can conquer the world.*—Marilyn Monroe

Although Ms. Monroe used the word *and,* she was trying to convey the conditional statement

> **If** *you give a girl the right shoes,* **then** *she can conquer the world.*

In this conditional statement, assuming *a* identifies the *if* statement and *b* identifies the *then* statement, we would have the following.

> *a*: Give a girl the right shoes.
>
> *b*: She can conquer the world.
>
> $a \Rightarrow b$: **If** you give a girl the right shoes, **then** she can conquer the world.

Another way to express what Ms. Monroe was saying is that for a girl to conquer the world, it is sufficient for her to have the right shoes. Obviously there are other ways to conquer the world, so the right shoes are not necessary; but if you are a girl with the right shoes, that's enough.

Example 1: Writing Variations of a Conditional Statement

Write the converse, inverse, contrapositive, and biconditional variations of this famous conditional statement by Marilyn Monroe.

$a \Rightarrow b$: *If you give a girl the right shoes, then she can conquer the world.*

Solution

Converse: $b \Rightarrow a$: If she can conquer the world, then the girl was given the right shoes.

Inverse: $\sim a \Rightarrow \sim b$: If you do not give a girl the right shoes, she cannot conquer the world.

Contrapositive: $\sim b \Rightarrow \sim a$: If she cannot conquer the world, then the girl was not given the right shoes.

Biconditional: $a \Leftrightarrow b$: Give a girl the right shoes if and only if she can conquer the world.

Example 2: Writing the Contrapositive of a Conditional Statement

Given the following conditional statement, write the contrapositive.

If I cannot find my phone, then it is in the car.

Solution

Let the following represent a and b.

a: I can find my phone.

b: My phone is in the car.

The original conditional statement is then $\sim a \Rightarrow b$. The contrapositive of the conditional statement $\sim a \Rightarrow b$ is $\sim b \Rightarrow a$. Now we need to translate this compound statement into words. Therefore, the contrapositive to the conditional statement given is:

If my phone is not in the car, then I can find it.

Example 3: Writing a Biconditional Statement

Let the following statements represent c and d.

c: I am breathing.

d: I am alive.

Write the biconditional statement $c \Leftrightarrow d$ using words.

Solution

Biconditional statements use the phrase "if and only if." Using statements c and d, the biconditional statement is:

I am breathing if and only if I am alive.

Skill Check #1

Write the inverse to the statement "If I cannot find my phone, then it is in the car."

We can find the truth values for each of these four variations, as well as the original statement, by using a truth table. Remember that a conditional statement is false only when the *if* statement is true and the *then* statement is false. The truth values for the conditional statements is shown in Table 2.

				Table 2				
				Truth Table				
a	*b*	~*a*	~*b*	Conditional $a \Rightarrow b$	Converse $b \Rightarrow a$	Inverse $\sim a \Rightarrow \sim b$	Contrapositive $\sim b \Rightarrow \sim a$	Biconditional $a \Leftrightarrow b$
T	T	F	F	T	T	T	T	
T	F	F	T	F	T	T	F	
F	T	T	F	T	F	F	T	
F	F	T	T	T	T	T	T	

Before we fill in the last column, we need to define the truth values of **biconditional statements**.

Biconditional Statement

Biconditional statements, read "if and only if," are true only when each component of the statement has the same truth value; that is, either both are true or both are false.

Let's fill in the last column now.

				Table 3				
				Truth Table				
a	*b*	~*a*	~*b*	Conditional $a \Rightarrow b$	Converse $b \Rightarrow a$	Inverse $\sim a \Rightarrow \sim b$	Contrapositive $\sim b \Rightarrow \sim a$	Biconditional $a \Leftrightarrow b$
T	T	F	F	T	T	T	T	T
T	F	F	T	F	T	T	F	F
F	T	T	F	T	F	F	T	F
F	F	T	T	T	T	T	T	T

Notice that the conditional statement and its contrapositive have the same truth values: T, F, T, T. Because of this, the statements are said to be **logically equivalent**. We write this mathematically using the symbol ≡. So, we have $(a \Rightarrow b) \equiv (\sim b \Rightarrow \sim a)$.

Logically Equivalent Statements

Logically equivalent statements are statements that have exactly the same truth values in all situations. We write this mathematically using the symbol ≡.

Skill Check #2

Find the other statements that are logically equivalent in Table 3.

Looking back at our original statement by Marilyn Monroe, the following two statements are logically the same statement. In other words, they have the same meaning.

If you give a girl the right shoes, then she can conquer the world,

and

If she cannot conquer the world, then the girl did not get the right shoes.

Now, we will look at some other places where we find logical equivalence. Consider the following truth table.

	Table 4	
	Truth Table	
p	$\sim p$	$\sim(\sim p)$
T	F	T
F	T	F

From this, we can see that $p \equiv \sim(\sim p)$. For example, the following statements have the same meaning.

I am going to the party tonight,

and

There is no way I'm not going to the party tonight.

Does this sound familiar? Think back when you might have heard someone throw a double negative into a conversation. Sometimes it seems that conversations could be made a lot clearer by avoiding negatives altogether.

Augustus De Morgan, an English mathematician and logician, formally defined two famous equivalent negations that show how to negate *and* statements and *or* statements. De Morgan's Laws show how a negation sign is "distributed" across compound statements.

☞ Helpful Hint

Notice that in De Morgan's Laws, one side of the equivalence has a negation sign on the outside of a set of the parentheses while the other side has no parentheses at all.

De Morgan's Laws

1. $\sim(p \wedge q) \equiv \sim p \vee \sim q$

2. $\sim(p \vee q) \equiv \sim p \wedge \sim q$

You can verify De Morgan's Laws for yourself by constructing their truth tables from scratch, which we will do as an exercise at the end of this section. We use De Morgan's Laws in our next example.

Example 4: Applying De Morgan's Laws

Consider the following compound statement.

It is not true that Jack and Jill went up the hill.

From the given statements, choose the statement that is logically equivalent.

a: Jack did not go up the hill and Jill did not go up the hill.

b: It is not true that Jack and Jill did not go up the hill.

c: Jack went up the hill or Jill went up the hill.

d: Jack did not go up the hill or Jill did not go up the hill.

Solution

To determine which statement is logically equivalent, let us first write the original compound statement in symbolic form.

Let the following statements represent p and q.

> p: Jack went up the hill.
>
> q: Jill went up the hill.

Then our statement "It is not true that Jack and Jill went up the hill" can be written logically as

$$\sim(p \wedge q).$$

By De Morgan's Laws, we know that a compound statement that reads $\sim p \vee \sim q$ is logically equivalent.

Write each of our choices symbolically as well.

> a: Jack did not go up the hill and Jill did not go up the hill: $\sim p \wedge \sim q$
>
> b: It is not true that Jack and Jill did not go up the hill: $\sim(\sim p \wedge \sim q)$
>
> c: Jack went up the hill or Jill went up the hill: $p \vee q$
>
> d: Jack did not go up the hill or Jill did not go up the hill: $\sim p \vee \sim q$

Therefore, the logically equivalent statement is d: Jack did not go up the hill or Jill did not go up the hill.

Skill Check #3

Determine which of the statements from Example 4 is equivalent to "Neither Jack nor Jill went up the hill."

Negating Conditional Statements

We have now seen how to negate statements involving both *and* and *or*. To close this section, we will look at how to negate conditional statements. Remember that a negation of a mathematical statement has the opposite truth value as the original statement. Recall that you can think of conditional statements as "the promise was broken." It might appear that one of the variations that we have looked at for implications would be the negation of "if a, then b" statements. However, take another look at the truth tables for each of these.

Table 5								
Truth Table								
a	b	$\sim a$	$\sim b$	**Conditional** $a \Rightarrow b$	**Converse** $b \Rightarrow a$	**Inverse** $\sim a \Rightarrow \sim b$	**Contrapositive** $\sim b \Rightarrow \sim a$	**Biconditional** $a \Leftrightarrow b$
T	T	F	F	T	T	T	T	T
T	F	F	T	F	T	T	F	F
F	T	T	F	T	F	F	T	F
F	F	T	T	T	T	T	T	T

You can see from the table that none of these variations (converse, inverse, contrapositive, or biconditional) has the opposite truth value pattern of the conditional statement. As it happens, the negation of a conditional statement is not itself a conditional statement. It is a compound statement. The following rule shows that the negation of the conditional statement $p \Rightarrow q$ is $p \wedge \sim q$. Think of the conditional promise "If you make an A in the class, I'll give you a reward" and the broken compound promise "You made an A in the class and I did not give you a reward."

> ### Negation of Conditional Statements
> $$\sim(p \Rightarrow q) \equiv p \wedge \sim q$$

Now we can use this rule to negate a conditional statement.

Example 5: Writing the Negation of a Conditional Statement

Write the negation of the following conditional statement.

If I go to Moss' Diner, then I get the triple stack pancakes.

Solution

First, write each piece of the conditional statement symbolically.

Let the following statements represent *a* and *b*.

> **a:** I go to Moss' Diner.
>
> **b:** I get the triple stack pancakes.

Our original conditional statement is then written symbolically as $a \Rightarrow b$.

We know from the rule for the negation of conditional statements that the negation of a conditional statement $a \Rightarrow b$ is the compound statement $a \wedge \sim b$. So, we can write the negation of the conditional statement as the compound statement

I go to Moss' Diner and I do not get the triple stack pancakes.

Example 6: Writing the Negation of a Conditional Statement

Negate the following conditional statement by using the rule of negation of conditional statements along with De Morgan's Laws. Remember that the solution will be a compound statement.

$$a \Rightarrow (c \wedge d)$$

Solution

From the negation rule we know that the negation of the conditional statement is

$$a \wedge \sim(c \wedge d).$$

We can then use De Morgan's Laws to negate the latter half of the statement $\sim(c \wedge d)$ to be $\sim c \vee \sim d$. So, we have that the negation of $a \Rightarrow (c \wedge d)$ is $a \wedge (\sim c \vee \sim d)$.

You can see that using the negation rule along with De Morgan's Laws is far less time consuming than building truth tables for large conditional statements.

Skill Check Answers

1. If I can find my phone, then it is not in the car.

2. The converse is logically equivalent to the inverse.

3. Jack did not go up the hill and Jill did not go up the hill: $\sim p \wedge \sim q$

3.3   Exercises

Write the converse, inverse, contrapositive, and biconditional for each conditional statement.

1.   If I go to the movies, then I will get popcorn.

2.   Getting an 89 on the final is sufficient for me to get an A in biology.

3.   It is dark whenever I get out of my last class.

4.   Seeing a puppy will lead to me smiling.

Find each conditional variation.

5.   Find the inverse of $p \Rightarrow q$.

6.   Find the converse of $\sim q \Rightarrow p$.

7.   Find the contrapositive of $\sim p \Rightarrow q$.

8.   Find the inverse of $\sim q \Rightarrow \sim p$.

Complete each truth table for the given tautology and then decide which two statements within the compound statement are logically equivalent, if applicable.

9. $(p \Rightarrow \sim q) \Leftrightarrow \sim(p \wedge q)$

Truth Table						
p	q	$\sim q$	$p \Rightarrow \sim q$	$p \wedge q$	$\sim(p \wedge q)$	$(p \Rightarrow \sim q) \Leftrightarrow \sim(p \wedge q)$
T	T					
T	F					
F	T					
F	F					

10. $\sim(p \Rightarrow q) \Rightarrow (p \vee q)$

Truth Table					
p	**q**	**p⇒q**	**~(p⇒q)**	**p∨q**	**~(p⇒q)⇒p∨q**
T	T				
T	F				
F	T				
F	F				

11. $(\sim p \wedge \sim q) \Leftrightarrow \sim(p \vee q)$

Truth Table							
p	**q**	**~p**	**~q**	**~p∧~q**	**p∨q**	**~(p∨q)**	**(~p∧~q)⇔~(p∨q)**
T	T						
T	F						
F	T						
F	F						

12. $\left[p \wedge (\sim p \vee q)\right] \Leftrightarrow (p \wedge q)$

Truth Table					
p	**q**	**~p**	**~p∨q**	**p∧(~p∨q)**	**p∧q**	**[p∧(~p∨q)]⇔(p∧q)**
T	T					
T	F					
F	T					
F	F					

Show that each pair of statements is logically equivalent or explain why they are not.

13. **a.** $a \Rightarrow b$ **b.** $b \Rightarrow a$

14. **a.** $n \Rightarrow m$ **b.** $m \vee \sim n$

15. **a.** $p \wedge \sim q$ **b.** $\sim(\sim p \vee q)$

16. **a.** $r \wedge \sim s$ **b.** $\sim(r \Rightarrow s)$

17. **a.** $w \Rightarrow z$ **b.** $\sim w \Rightarrow \sim z$

18. **a.** $p \vee q$ **b.** $p \vee (q \wedge \sim p)$

De Morgan's Laws state that $\sim(p \wedge q) \equiv \sim p \vee \sim q$ and $\sim(p \vee q) \equiv \sim p \wedge \sim q$. Construct a truth table to show that each pair of compound statements is logically equivalent.

19. $\sim(p \wedge q)$ and $\sim p \vee \sim q$ **20.** $\sim(p \vee q)$ and $\sim p \wedge \sim q$

Use De Morgan's Laws to write an equivalent statement without using parentheses.

21. $\sim(\sim p \wedge q)$

22. $\sim(\sim p \vee \sim q)$

Use De Morgan's Laws to write an equivalent statement using parentheses.

23. $\sim p \wedge q$

24. $p \vee q$

Use De Morgan's Laws to write an equivalent statement.

25. There is a space available in the 8:00 a.m. Biology class or I am not able to make the perfect schedule for next semester.

26. In Biology the nucleotide bases adenine and thymine pair together in DNA and the bases do not pair together in RNA.

Negate each conditional statement.

27. If Brooke comes to my room before 2:00 p.m. then I do not finish my homework before 10:00 p.m.

28. My computer goes to sleep whenever I leave it alone for 15 minutes.

29. Having a ticket from the Sunday paper is sufficient to get a free ice cream at Dairy Dip.

30. Being in Charleston, South Carolina, implies that I am on Eastern Standard Time.

31. $\sim p \Rightarrow a$

32. $a \Rightarrow (c \vee d)$

33. $(c \wedge d) \Rightarrow b$

34. $(w \vee z) \Rightarrow (w \wedge z)$

3.4 Valid Arguments and Fallacies

Valid Arguments

How does one build up a good logical argument? We have looked at tools to help us determine the truth value of statements, but what about constructing an argument of your own?

When we refer to an argument in logic, we are not talking about a quarrel or shouting match; instead, we are referring to the process of following a line of reasoning that is meant to support a particular belief or idea. Formally, an **argument** consists of a series of logical statements. Statements at the beginning of the argument are called **premises** (or *hypotheses* or *assumptions*), which are used to support a particular ending statement called a **conclusion**.

Argument

An **argument** consists of a series of logical statements.

Premise

A **premise** is a statement at the beginning of an argument used to support the conclusion.

Conclusion

A **conclusion** is the ending statement of an argument.

The terms *premise* and *conclusion* are relative terms. They infer a position in an argument. No statement taken out of the context of the argument can be either a premise or a conclusion on its own; they work together. To illustrate this, consider a quote by former presidential candidate Mitt Romney, given in an article on CNN.com during the day after the first presidential debate in October 2012.

> "There's no question in my mind if the president is reelected, you'll continue to see a middle-class squeeze," Romney said. [1]

From Section 3.3, we know that Mr. Romney's compound statement contains two simple statements.

> *a*: The president is reelected.

> *b*: You will continue to see a middle-class squeeze.

Because statement *a* is at the beginning of the argument, that is, the *if* part of the conditional statement, it is considered the premise of the argument. Statement *b* is the conclusion, the *then* component.

Premise: The president is reelected.

Conclusion: You will continue to see a middle-class squeeze.

Notice that if you just considered one of these statements by Mr. Romney on its own, it could be either a premise or the conclusion for any argument, depending on the order of the statements. Having the statements together in context tells you which one they are.

---

1 CNN, http://www.cnn.com

Here's another example from that same debate, stated by President Obama.

> ". . . And so the question is does anybody out there think that the big problem we had is that there was too much oversight and regulation of Wall Street? Because if you do, then Governor Romney is your candidate." [2]

Here are the premise and conclusion:

Premise: (You think) there was too much oversight and regulation of Wall Street.

Conclusion: Governor Romney is your candidate (for the next President).

Again notice that either statement could be a premise or a conclusion if considered by itself. In speech, we often leave out some of the implied words in an argument as in these examples from the election debate. However, there are particular words that often indicate conclusions or premises are to follow. The following table gives an overview of some of the more common words or phrases you might come across.

Table 1	
Premise and Conclusion Indicators	
Premise Indicators	**Conclusion Indicators**
as	accordingly
as indicated by	consequently
because	entails that
due to the fact that	hence
for	implies that
for the reason that	it must be that
given that	it follows that
if	so
in as much as	then
in that	therefore
may be concluded from	thus
may be inferred from	we may conclude that
seeing that	we may infer that
since	whence
the reason that	wherefore

Here's a rather lengthy, but fun, set of premises leading to a conclusion. Consider the following reply from Sherlock Holmes when he was questioned on how he knew at a glance that a man was a retired Marine sergeant in *A Study in Scarlet* by Sir Arthur Conan Doyle.[3]

> *. . . I could see a great blue anchor tattooed on the back of the fellow's hand. That smacked of the sea. He had a military carriage, however, and regulation side whiskers. There we have the marine. He was a man with some amount of self-importance and a certain air of command. You must have observed the way in which he held his head and swung his cane. A steady, respectable, middle-aged man, too, on the face of him—all facts which led me to believe that he had been a sergeant.*

Mr. Holmes gave a large set of premises to support his conclusion.

Skill Check #1

Identify the premise and conclusion in each of the following.

a. Our technology will deliver a better product, so you will have fewer complaints.

b. Your employees will feel better and be more productive if you build in communal break times to the work day.

---

2 *The New York Times*, http://www.nytimes.com

3 Doyle, *A Study in Scarlet*

When making good formal arguments, we basically use two types of reasoning—inductive and deductive—that we discussed in Chapter 1. Recall that **inductive reasoning** is a line of argument that takes specific examples as its premises and then draws a general conclusion from these. In other words, the statements given are all examples that are observed to be in support of a specific conclusion. For example, if you observe something to be true many times (or even once) and conclude that it must be true in all instances, you have used inductive reasoning. Sherlock Holmes used inductive reasoning for his argument to identify the Marine sergeant.

Inductive Reasoning

Inductive reasoning is a line of argument that takes specific examples as its premise and then draws a general conclusion from these.

Here's an example of inductive reasoning.

Every fall, the leaves have fallen from the trees. I'll be raking leaves
again this November.

While arguing in this way can seem persuasive because we become convinced by the weight of the evidence, it is not logically foolproof. Consider the following inductive arguments.

- *Since $0^2 = 0$ and $1^2 = 1$, all numbers squared must equal themselves.* We all know that this is not a true mathematical statement simply because $2^2 = 4 \neq 2$.

- *The state senator voted "No" on every bill to fund higher education. Therefore, the senator does not value higher education.* Without asking the senator if he actually values higher education, we do not know this conclusion to be true.

- *I failed the test. My three friends in the class failed the test. So, I know everyone failed the test.* It does not necessarily follow that the conclusion is true. The only way we could know is by checking everyone's grade.

- *It didn't snow in Texas this year or last year. It never snows in Texas.*

You can quickly get the idea that it would take only one case that does not follow the observed pattern to make the reasoning faulty. Think about the fourth example. Since we don't have a crystal ball to look into the future and know if it will ever snow in Texas, we cannot be sure about the truth of this "never" statement. It is usually improbable that we can or will observe all possible cases in a scenario and, therefore, we can never use inductive reasoning to prove anything. Hence, inductive reasoning builds a weak argument logically. Inductive reasoning relies on a collection of supporting evidence, rather than logic, to be persuasive. Consequently, we can see that inductive reasoning must be used with extreme caution as a means of building logical arguments.

Since inductive reasoning doesn't help us build strong logical arguments, we'll focus on the second type of reasoning—**deductive reasoning**. In contrast to inductive reasoning, **deductive reasoning** takes statements and logically combines them. It uses statements that are commonly accepted as facts to make a case for certain conclusions. Suppose that you believe that all US presidents were effective leaders. Since Zachary Taylor was the 12th US president, by deductive reasoning, you must also believe he was an effective leader whether or not you know anything about his presidency.

Deductive Reasoning

Deductive reasoning uses statements that are commonly accepted as facts to make a case for a certain conclusion.

Fun Fact

How much do you know about our 12th president, Zachary Taylor? Taylor served as the president of the United States from 1849–1850. He was known as "Old Rough and Ready" because of his leadership skills during his 40-year military career. He was the last president to hold slaves, although he took a moderate stance on the issue. He was the president who served the third-shortest term in office, as he died 16 months into his term, and was succeeded by his vice president, Millard Fillmore.

While everyone might not agree with your original premise, your argument is said to be valid if the conclusion follows logically from the premise. That is, a **valid argument** is a deductive argument where the conclusion is guaranteed from the premise. It is important to realize that having an argument that is valid implies no judgment on the truth of the premise. The validity of an argument is concerned only with the logical connection *between* the premise and the conclusion.

> ### Valid Argument
>
> A **valid argument** is a deductive argument where the conclusion is guaranteed from the premises.

Here are some examples of valid arguments.

1. *All men are mortal.*

   Socrates is a man.

   Therefore, Socrates is mortal.

2. *Taylor Swift makes great music.*

   Her new CD was just released. It must be great!

3. *Cutting out sweets from your diet makes you lose weight.*

   He hasn't eaten sweets in months, so he must be losing weight.

You will have an opportunity to check for yourself that the conclusions follow from the statements made, but first we will look at how to do that.

There are many valid ways to combine logical statements that are always logically consistent. All of the examples use the same basic form of argument. They rely on having a true conditional statement such as $p \Rightarrow q$, along with a true premise p, and then use logic to conclude that q must also be true. Using a truth table, we can see how this works.

Table 2
Truth Table

p	q	$p \Rightarrow q$
T	T	T
T	F	F
F	T	T
F	F	T

Notice that row 1 contains the only combination where p is true and the conditional compound statement $p \Rightarrow q$ is also true. Notice that when this is the case, row 1 shows us that q must also be true.

Let's explore valid argument **1.** about Socrates in more detail to see what is happening.

Fun Fact

Historians believe Socrates followed in his father's footsteps working as a mason for many years before devoting his life to philosophy.

Example 1: Determining the Validity of an Argument

Show that the following argument is valid.

> *All men are mortal.*
> *Socrates is a man.*
> *Therefore, Socrates is mortal.*

Solution

The first line of the argument, "All men are mortal," can be read as the conditional statement "if one is a man, then he is mortal," which can be expressed with notation as $p \Rightarrow q$. This means that we have the following simple statements.

p: One is a man.

q: One is mortal.

If we assume $p \Rightarrow q$ is true, we can consider the case for Socrates. Since we know that Socrates is a man, we know that Socrates satisfies p. Consider the truth table for $p \Rightarrow q$.

Table 3		
Truth Table		
p	**q**	**p ⇒ q**
T	T	T
T	F	F
F	T	T
F	F	T

We are assuming that both p and $p \Rightarrow q$ are true. According to the truth table, if these two statements are true, then q must also be true. Consequently, we conclude that q is true of Socrates, and so Socrates must indeed be mortal.

Skill Check #2

Show that arguments **2.** and **3.** on the previous page are valid by using the same method given in Example 1.

If the conclusion is not always guaranteed from the premises, the argument is said to be **invalid**.

Invalid Argument

An **invalid argument** is an argument where the conclusion is not always guaranteed from the premises.

Example 2: Determining the Validity of an Argument

Let's change argument **3.** from the valid examples given previously.

The original argument was

> *Cutting out sweets makes you lose weight. He hasn't eaten sweets in months, so he must be losing weight.*

Switch the last two portions so that the argument becomes

> *Cutting out sweets makes you lose weight. He's losing weight, so he must be cutting out sweets.*

Show that the modified argument is invalid.

Solution

Despite the fact that we often hear people say things like this, the argument is invalid because the conclusion (although it might be true) does not necessarily follow from the premises stated. Let's see why the conclusion does not follow directly from the premises.

Identifying the premises and conclusion is always a good place to start. In this case, it's easy to spot the conclusion because it follows "so" in the argument. The conclusion is "he must be cutting out sweets." Thus, "he is losing weight" must be the premise.

Let a and b be the following.

   a: He is cutting out sweets.

   b: He is losing weight.

You can think of the first part of the new argument as "**if** you cut out sweets, **then** you lose weight." This can be expressed symbolically as $a \Rightarrow b$.

Assume $a \Rightarrow b$ is true. The argument also asserts b is true; he is losing weight. Next, we consult the truth table for $a \Rightarrow b$ to find the rows where both are true.

	Table 4	
	Truth Table	
a	b	$a \Rightarrow b$
T	T	T
T	F	F
F	T	T
F	F	T

Unlike our previous example, notice that there are two combinations where both $a \Rightarrow b$ and b are true: the first and third rows. The first row has a true value for statement a and the third row has a false value for statement a. Since there is more than one possible truth value for statement a when both $a \Rightarrow b$ and b are true, we cannot conclude that a is in fact true always. It does not follow logically from the truth table. The weight loss might be a result of some dietary restriction, exercise, or any number of things. So this argument is invalid, even if the conclusion is true.

Skill Check #3

Consider whether the following is a valid argument.

All organisms with wings can fly. Since penguins have wings, they must be able to fly.

When faced with an argument that is valid, even though the conclusion is not true (as is the case in the previous Skill Check example), the problem lies in the premise, not the argument. In these instances, there must be an error in the beginning statement. Making the best argument requires that our argument is **sound**. That is, we want a logically valid argument and true premises from which to prove logically that the conclusion follows from the premises. To have a sound argument is the most confident we can be.

Sound Argument

A **sound argument** is a valid argument using true premises.

Example 3: Determining Sound Arguments

Use a truth table to show that the following is a sound argument.

If the President of the United States is unable to carry out his or her constitutional role, then the Vice President becomes the President of the United States. In November 1963, President John F. Kennedy was fatally shot. Therefore, without an election, Lyndon Johnson became President of the United States.

Solution

Begin by identifying the simple statements within the compound statement.

p: The President of the United States is unable to carry out his or her constitutional role

q: The Vice President becomes the President of the United States.

The truth table for a conditional compound statement is given in Table 5.

Table 5 Truth Table		
p	q	$p \Rightarrow q$
T	T	T
T	F	F
F	T	T
F	F	T

You can see that the implication is true on the first, third, and fourth lines. For a sound argument, we must start with a true statement. Therefore, we'll only consider the first two rows of the table. We are told that in 1963, President John F. Kennedy was fatally shot. In other words, statement p is true. We also are told that statement q is true in this argument because the Vice President at the time, Lyndon Johnson, became President of the United States. From the truth table, we know that given statements p and q, which are true, the implication $p \Rightarrow q$ is also true. So we know that this argument is both valid and sound.

We consider one more example that is slightly more involved.

Example 4: Determining Sound Arguments

Determine if the following argument is sound.

We'll go skiing this weekend if it snows at least 2 inches, and if we go skiing I'll get some new boots. It snowed 3 inches, so I must have new boots!

Solution

Let's begin by identifying the simple statements within the compound statement. We can write our statements symbolically as follows.

p: It snows at least 2 inches.

q: We go skiing.

r: I get new boots.

This argument is actually a double implication. Ultimately, we want to show that $p \Rightarrow r$, by showing that $p \Rightarrow q$ and $q \Rightarrow r$. The truth table for this compound implication is given in Table 6.

			Table 6		
			Truth Table		
p	*q*	*r*	*p* ⇒ *q*	*q* ⇒ *r*	*p* ⇒ *r*
T	T	T	T	T	T
T	T	F	T	F	F
T	F	T	F	T	T
T	F	F	F	T	F
F	T	T	T	T	T
F	T	F	T	F	T
F	F	T	T	T	T
F	F	F	T	T	T

We begin with the premise that it snowed 3 inches, which ensures the truth of *p*. We know from the truth table for implications that if we have a true *p* and a true *p* ⇒ *q*, we can conclude *q*, that we do indeed go skiing. Then going one more step, it also allows us to combine this new conclusion, *q*, with *q* ⇒ *r*, to conclude *r* and ensure that I get new boots. Therefore the argument is sound.

Skill Check #4

Use a truth table to show that the following is a sound argument.

Speeding is a crime. I occasionally drive over the speed limit. I am a criminal.

Creating chains of inferences in this way is a basic technique in logical reasoning and often continues for multiple layers. Logical arguments depend both on the truth of their basic premises and on the logical validity of the ways in which those premises are put together. As we have seen, manipulating logical arguments in ways that seem plausible, but are actually invalid, happens more often than you may have initially thought.

Fallacies

Not all arguments make solid and compelling cases for their conclusions. Arguments often contain **fallacies**. Loosely speaking, a fallacy is a lapse in logic or an error in reasoning, often referred to as a faulty argument. It is reasoning that leads to an invalid argument.

Fallacy

A **fallacy** is an error in reasoning that leads to an invalid argument.

If you search for "fallacy" on the internet, you will find literally dozens of categorizations of fallacies. Since our aim here is to have you think critically about logical arguments, and not memorize the names of every fallacy known, we will introduce a few of the most prevalent fallacies you might run across. Evident by their Latin titles, people have been loose with their logic for many, many centuries.

As we consider each of the following eight fallacies, we want to point out that when we speak, arguments involve implications that are often not necessarily verbalized in the argument itself. Although, mathematically speaking, some of the scenarios we will look at are simply statements and not arguments at all because they omit the implied conclusion, we will consider such implications and note the unspoken conclusion since this is often the case in the world in which we live.

1. *Post Hoc, Ergo Propter Hoc* (After This, Therefore Because of This)

My wallet had money in it before you went to get gum from my purse.

The brakes worked just fine before you drove the car.

A black cat crossed my father's path. The next day, he had a heart attack.

This title comes from the Latin meaning, "after this, therefore because (on account) of this," that is, since one thing happened first, it must have *caused* the other to happen. The statement "We never had a problem with the plumbing before you moved in" from your new landlord might send you into a tizzy. By implying that, as a new tenant, you are the reason for the plumbing issues, the landlord might wish to have an excuse for the bad plumbing already in existence, or have you conveniently pick up the cost for any damage that might need repaired. Even if the premise of this statement were true (that there were no problems with the plumbing before you arrived and now there are problems with the plumbing), it does not follow that you as the new tenant *caused* the plumbing to be problematic. Try not to let your landlord use this fallacy on you.

Formally, the argument is completed with a conclusion that is merely inferred, as in these examples. Here is another example, but with a good spin. "Once I started eating chocolate, my weight started to drop. Therefore, my slimming diet should include chocolate at least once every day." *Post hoc, ergo propter hoc*—I think not. Oh, if only this were a sound argument!

2. *Dicto Simpliciter* (Hasty Generalization)

People from Chicago are just rude. We had terrible service at our restaurant and the hotel staff were so unhelpful.—A statement made on the first day of a vacation.

I don't like coffee. It's too sweet.—A statement made after the first taste of a white chocolate mocha with an extra shot of syrup.

I've decided to buy a lottery ticket every day this year. This is easy money.—A statement made after winning $50 on his first lottery ticket.

We often make broad sweeping conclusions based on only a few occurrences. These hasty generalizations are often because the size of the sample is too small to make a reasoned conclusion. Recall that this type of reasoning is inductive and can be thrown off its course by one counterexample. For example, take Liam who said the following after being introduced to two Germans who were vegetarians: "I'm never going to visit Germany. They don't eat meat there." Of course we all know this not to be the case!

Consider the scenario of someone having seen a cyclist nearly knocked off their bike at the corner of 1st and Main Street saying, "That intersection is deadly. It needs wider lanes and longer traffic lights." Although there might actually be an issue that needs to be addressed at the intersection, it will take more than one close call, along with sufficient evidence, to establish a cause. If we generalize a rule based on very few occurrences, we are guilty of committing the fallacy of hasty generalization.

3. *Ad Hominem* (Personal Attack)

How can you expect anything good from such a bleeding-heart politician?—The response from Candidate B after Candidate A outlined the five points of change he would make if elected.

Do we really want to believe what a divorced man has to say about our School Board policies?—A school board member responding to a parent voicing concern over lack of dress code policies.

From the Latin meaning "to the man," the ad hominem fallacy refers to attacking someone's person, character, or motives, instead of addressing the premise of the argument. As another example, a lawyer might say to a jury, "You cannot believe anything this witness says because he is a convicted criminal." By pointing out the witness's character flaws, the lawyer hopes you will not consider his eye-witness account of the crime. An attack on the person and not their argument simply changes the focus. Using the ad hominem attack does not disprove their argument.

4. *Petitio Principii* (Circular Reasoning)

You should eat healthy meals. They're good for you.

She's really happy right now because she's in a good mood.

I think, therefore I am.—René Descartes

Meaning to "beg the question," this form of fallacy is circular reasoning, that is, the premise and conclusion state the same thing. The speaker uses the conclusion that he is trying to prove as the evidence to prove it. He just uses different wording to restate the same thing. As the audience, we often have a hard time dissecting this one because there seems to be no obvious beginning or end to the argument. "Tim Tebow is the most successful college quarterback this decade because he's the best we've ever seen." Do not be fooled by the insertion of the word "because" here. There is no supporting evidence following it. Both parts of the sentence are premises (or opinions, if you like) until they have some evidence to back them up. Of course, if you believe that *thinking* and *being* are not synonymous, then Descartes may not have been fallacious in his statement after all.

5. *Non Sequitur* (It Does Not Follow)

If you wear this outfit, you're sure to get a date.

By giving to this overseas charity, you are neglecting the needs of your local community.

She lives on Main Street. She must be very wealthy.

Non sequitur comes from Latin, meaning literally "it does not follow." When a conclusion has nothing to do with the premise, we say a diversion or "red herring" was introduced into the argument. Often political endorsements are non sequiturs. Consider the 2008 presidential election. Chuck Norris, film's on-screen "tough guy," had this endorsement for former presidential candidate Mike Huckabee.

> "... I believe the only one who has all of the characteristics to lead America forward into the future is ex-Arkansas Gov. Mike Huckabee." [4]

Whether or not you supported Huckabee for president in 2008, we would hardly say that he must make a good president because he is endorsed by Chuck Norris.

6. *Straw Man* (Exaggerated or Distorted View of the Opponent)

We can't have national health care. No one wants a government death panel deciding their loved-one's fate.

I'm against extending tax cuts for the wealthy. Why should we be worried about Bill Gates losing a few pennies?

Consider the candidate who makes a speech about his opponents at a rally where the opponents are not there to defend their positions. You could easily imagine hearing statements such as these.

Like the straw man who is easily blown over, this fallacy involves the buildup of a distortion of someone's ideas or beliefs so that they can easily be knocked down. The premise is mostly arguing on a poor representation of the truth. It's easy to make ourselves look good when we can effortlessly tear down someone else. This fallacy includes any attempt to "prove" an argument by overstating, exaggerating, or oversimplifying the arguments of the opposing side.

MATH MILESTONE

René Descartes was a famed philosopher who thought that science and mathematics could be used to explain everything in nature, and wrote many works explaining his reasoning. At the age of 41, he published *La Geométrie*, which gave birth to Cartesian geometry that is widely studied today. Sadly, he died of pneumonia at the age of 54.

Fun Fact

Eldridge Cleaver, ran as a candidate in the 1968 presidential election under the Peace and Freedom Party. He received 36,571 votes, even though some states held that he did not meet the constitutional requirement for age and therefore could be excluded from the ballot. The Constitution requires that the US President be at least 35, but does not specify when he must have reached that age. Cleaver would not have been 35 until more than a year after the inauguration day in 1969.

Fun Fact

Argumentum ad Populum

Consider this famous sketch from the 1975 film *Monty Python and the Holy Grail*.

As Bedevere tries to calm the angry village crowd who wants to burn one of the locals as a witch, he says, "There are ways of telling whether she is a witch. Tell me, what do you do with witches?"

The crowd replies, "Burn, burn them up!"

Bedevere asks, "And what do you burn apart from witches?"

"Wood!" they cry.

He leads them on with, "So, why do witches burn?"

"B— 'cause they're made of . . . wood?"

"Good!," he commends.

He then uses similar lines of questioning to come to the conclusion that because she weighs as much as a duck, she must be made of wood, and must be a witch; therefore she can be burned at the stake! Now that's some powerful deductive reasoning!

7. *False Dilemma* (Illusion of Limited Choice)

You're either part of the solution or part of the problem.—Eldridge Cleaver during his 1968 presidential campaign.

Either you can afford this new car or you can decide to walk around for the next year.

If you aren't for us, you're against us.

"Life isn't always black or white" would be a great thing to keep in mind when faced with this type of fallacy. As we well know, rarely are there ever simple solutions with only two choices. Yet, the false dilemma fallacy relies on your thinking just that. It occurs when we build an argument around the assumption that only two possible choices are available. For instance, "If we don't commit to sweeping changes in the law, the American way of life will be lost forever."

Users of this type of fallacy hope for you to ignore the idea that other options might exist in the dilemma.

8. *Argumentum ad Populum* (Appeal to the People)

9 out 10 doctors prefer . . .

You should watch Reality Takes a Leap . . . the number 1 show on television.

I have hundreds of letters supporting his good character. He couldn't have committed that crime. No one would believe it.

When we want to use the fact that the majority, or even just a large number, of people are doing something as the basis for supporting a conclusion, we are appealing to the "everybody does it" idea. The Latin meaning is literally "argument to the people."

This type of fallacy is used extensively in propaganda. My young daughter trying to convince me to buy a puppy: "Zoe has a new puppy, Mae Mae has a new puppy, and Isabella has a new puppy. Everyone is getting a new puppy." This is also a hasty generalization, or dicto simpliciter. Not *everyone* is getting a new puppy! My mother used to counter this argument of mine with her own version of logic—"If everyone else jumped off a bridge, would you?"

Or how about this argument: "You can't give me a ticket for speeding. I was just going as fast as everyone else on the interstate." The fact that everyone was speeding on the interstate has nothing to do with an individual getting a ticket for breaking the law. The offender hopes to divert the attention away from his or her lawlessness and focus it on the fact that "everyone was doing it."

Example 5: Identifying Fallacies

Identify the type of fallacy used in each of the following statements.

a: If public spending is not reduced, our economy will collapse.

b: Heather just had a new fuel pump put in her car. Forty-eight hours later, her car won't start. Heather's father thinks the mechanic must not have installed the fuel pump correctly.

Solution

a: False dilemma; it is possible that both can be true at the same time, that is, the economy can be strong and public spending remain the same.

b: Post hoc, ergo propter hoc; there are many other things that could cause Heather's car to not start, including a lack of fuel!

Skill Check #5

Name the fallacy in the following statement: Chloë drives an expensive car. She must have money to spare!

Table 7

Summary of Fallacies

Name (Latin)	Name (translated)	Description	Example
Post Hoc, Ergo Propter Hoc	After This, Therefore Because of This	Because one thing happened first, it caused the other to happen.	My favorite baseball team won because I wore my lucky cap to the game.
Dicto Simpliciter	Hasty Generalization	Making broad, sweeping generalizations based on a few specific occurrences.	My state senator was caught in a scandal. This just proves that all politicians are untrustworthy.
Ad Hominem	Personal Attack	Attacking someone's person, character, or motives instead of addressing the premise of the argument.	Robert said the car accident wasn't his fault, but I'm sure he is lying; he stole money from me once.
Petitio Principii	Circular Reasoning	The premise and conclusion state the same thing.	Government-run health insurance is dangerous because it is socialistic.
Non Sequitur	It Does Not Follow	The conclusion has nothing to do with the premise.	Tens of thousands of Americans have seen lights in the night sky that they could not identify. The existence of life on other planets is fast becoming certainty.
Straw Man	Exaggerated or Distorted View of the Opponent	The buildup of a distortion of someone's ideas or beliefs so that they can easily be knocked down.	People who accept evolution believe that humans and monkeys are the same. This is clearly not true.
False Dilemma	Illusion of Limited Choice	Argument that rests on the assumption that there are only two choices as a solution.	We can do this the easy way or the hard way.
Argumentum ad Populum	Appeal to the People	Stating that the majority, or even just a large number, of people are doing something as the basis for supporting a conclusion.	Two out of the three people in my car think that turning right at the stop sign is the way to get to the concert. So a right turn is the correct way to go.

This is in no way an exhaustive list of the fallacies that we encounter in our daily lives. However, it does give us a basis for building our critical thinking skills when we are next confronted with a statement or argument that does not sit quite right with us. If nothing else, it has hopefully raised your awareness of the ways in which language and ideas can be manipulated.

Skill Check Answers

1. a. **Premise:** our technology will deliver a better product; **Conclusion:** you will have fewer complaints

   b. **Premise:** you build in communal break times to the work day; **Conclusion:** your employees will feel better and be more productive

2. Argument 2: The first line can be expressed as $p \Rightarrow q$ (If it was made by Taylor Swift, then it is good music). Since we know that Taylor Swift made her new CD, p is satisfied. If we take $p \Rightarrow q$ to be true, the truth table of $p \Rightarrow q$ ensures that when p is true, q is true. Consequently, we can conclude that q is true of Taylor Swift's new album, and that it is great.

   Argument 3: The first line can be expressed as $p \Rightarrow q$ (If you cut sweets out of your diet, then you will lose weight). Since we know that he has not eaten any sweets (that is, he has cut out sweets), p is satisfied. If we take $p \Rightarrow q$ to be true, the truth table of $p \Rightarrow q$ ensures that when p is true, q is true. Consequently, we can conclude that q is also true, and that he is losing weight.

3. If we represent "the organism has wings" as p and "the organism can fly" as q, then the first sentence can be expressed as $p \Rightarrow q$. Since we know that penguins do have wings, p is true of penguins. If we take $p \Rightarrow q$ to be true, the truth table of $p \Rightarrow q$ ensures that when p is true, q is true. Consequently, we can conclude that q is true of penguins, and that they can fly. Of course, this conclusion is not true. However, because we assume the first conditional statement to be true, and because our premise p is true of penguins, this argument is logically valid.

4. Speeding is a crime can be expressed as $p \Rightarrow q$, where
   p: One drives above the speed limit.
   q: One commits a crime.
   Speeding is a crime can be expressed as $p \Rightarrow q$, and the truth table is as follows.

Truth Table		
p	q	$p \Rightarrow q$
T	T	T
T	F	F
F	T	T
F	F	T

For a sound argument, we must start with a true statement. Therefore, we'll only consider the first two rows of the table. We are told that "I occasionally drive over the speed limit." In other words, statement p is true. We also are told that statement q is true in this argument because "I am a criminal." From the truth table, we know that given true statements p and q, the implication $p \Rightarrow q$ is also true. So we know that this argument is both valid and sound.

5. Non sequitur

3.4   Exercises

Fill in each blank with the correct term.

1. An ending statement is called a/an _____.

2. A/An _____ is a series of statements used to support an idea or theory.

3. Statements at the beginning of an argument are called _____.

4. A faulty argument is another name for a/an _____.

5. The _____ fallacy gives the illusion of limited choice in a situation.

6. _____ is a fallacy in which an event is concluded because of a first event happening.

7. The _____ fallacy makes an appeal to the public.

8. The _____ fallacy tends towards an exaggerated or distorted view of the opponent's view.

9. An argument that seems to have no starting point, but continues with circular reasoning, is referred to as _____.

10. The fallacy that tries to focus the argument on an opponent's person rather than the issue at hand is referred to as _____.

11. Making an argument based on a quick sweeping assumption about an issue is called the _____ fallacy.

12. Often referred to as a "red herring," the _____ fallacy has a conclusion that does not follow from the premise.

Determine whether each argument uses inductive or deductive reasoning, and whether the reasoning is valid.

13. The number pi (π) begins with 3.14159. Four is the only even digit when you write out the number π.

14. We have put up our Christmas tree the day after Thanksgiving every year for the last 5 years. A photograph from 2 years ago shows our Christmas tree in the background, so it couldn't have been taken before Thanksgiving.

15. We have put up our Christmas tree the day after Thanksgiving every year for the last 5 years. It's Black Friday today, so we must be putting up the Christmas tree.

16. All of my friends have smartphones. They must be the most popular type of cell phone.

17. In a survey of 200 adults, 5 out of 7 people said they carry smartphones as their cell phone. With 95% confidence, I can say that most people carry a smartphone as their cell phone.

Determine whether each argument is valid or invalid.

18. All pigs are pink. I am pink, therefore I must be a pig.

19. All professors at the university received a 3% raise this year. My dad is a psychology professor at the university, so he must have gotten a raise.

20. To renew a US passport for a child under 16, you need to demonstrate that he or she is a US citizen with either a US birth certificate or a valid US passport. I am renewing my daughter's passport, so I need her birth certificate.

21. Most people who live in California are Democrats. Stephanie lives in California, so she must be a Democrat.

22. All dogs bark. I have a dog. Therefore, my dog barks.

23. If it rains, I will need an umbrella. I do not need my umbrella so it must not be raining.

24. Most people who live in Kansas are Republicans. Jim lives in Kansas. Therefore, Jim is a Republican.

Determine the missing piece of information needed to make each argument valid.

25. I can only carry a bottle of liquid that is at most 3.4 fl oz in my carry-on luggage while flying. A larger bottle will be confiscated by airport security. My hand sanitizer got taken away when I went through airport security.

26. I was born on US soil. Therefore, I am a US citizen.

27. Whenever I play folk music, I tune my guitar to an open D tuning. This evening I am playing some folk music.

28. I am tired, so it must be late.

29. If you are not over 16, then you cannot apply for a driver's license in the state of Tennessee. Emma applied for a driver's license.

30. To get into the local state university, you must have a cumulative ACT score of at least 21. Ben's cumulative ACT score is less than 21.

31. If you buy a new car, then you cannot pay your student loan bills. You are able to pay your student loan bills.

Identify the premise and conclusion in each argument.

32. "If the owners truly cared about the game and the fans, they would lift the lockout and allow the season to begin on time while negotiations continue," NHLPA Executive Director Donald Fehr said. [5]

33. "Even if we stopped burning fossil fuels today, there is enough carbon dioxide in the atmosphere—and it is such a persistent, lasting gas—that temperatures will continue to rise for a few hundred years." [6]

34. The sea level is rising because melting glaciers and ice sheets are adding more water to the oceans.

5 BBC, http://www.bbc.co.uk
6 BBC, http://www.bbc.com

35. Fast food obesity has strikingly increased in many countries because of the easy availability of fast food in the grocery shops, gas stations and dispensers everywhere. So it is difficult to escape from the lure of these delicious advertisements and showcases. [7]

36. The current US population is approximately 314 million, about half of which are males, so if 2 percent of the 157 million American men suffer from one of these severe disorders (schizophrenia, major depression, or psychopathy), this results in a figure of 3,140,000. [8]

37. "If a man is struck down by a heart attack in the street, Americans will care for him whether or not he has insurance." [9]

38. An excerpt from the TV show *The West Wing*:[10]

    Press secretary, C. J., says, "...*USA Today* asks you (the president) why you don't spend more time campaigning in Texas and you say because you don't look good in funny hats."

    "It was *big hats*," Sam corrects her.

    "The point is we got whomped in Texas," C. J. states.

    "We got whomped in Texas twice," Josh adds.

    "We got whomped in the primary and we got whomped in November," says C. J.

    "I think I was there," replies the president.

    "And it was avoidable. Sir."

    "C. J. on your tombstone, it's going to read, 'Post hoc, ergo propter hoc.'"

39. A cartoon by Randy Glasbergen.[11]

40. "Dear Friend, a man who has studied law to its highest degree is a brilliant lawyer, for a brilliant lawyer has studied law to its highest degree." Oscar Wilde, *De Profundis*.

41. All potatoes have skin. I have skin. Therefore, I must be a potato.

7 ygoy, http://www.ygoy.com

8 *The Chronicle of Higher Education*, http://www.chronicle.com

9 Fox News, http://www.foxnews.com

10 Sorkin, *The West Wing*

11 Courtesy of Randy Glasbergen

42. During the fall of 2011, the movement "Occupy Wall Street" spread almost overnight into a nationwide/worldwide phenomenon. Many famous personalities voiced their opinions both for and against the movement. One film star claimed, " . . . they're a bunch of posers, propping up the same corporate giant they're trying to take down. I find it a little ironic that most of these kids not only own all the latest technology, but use it constantly while pretending to 'take a stand.' They're a major player in the consumerism that goes on in this country."

Solve the problem.

43. Explain why the following argument given by the spokesman is faulty.

On the presidential campaign trail, when asked, candidates are always in favor of raising taxes on everybody in the nation's top earning brackets. Unless, of course, they're the ones being asked to pay more, in which case, forget about it.

We know this because, time and again, when the records of their tenure in office are scrutinized, they are shown to consistently vote to keep taxes on the wealthy at a lower rate than the majority of Americans.

"Why do you ask about this?" one spokesman for a candidate said this week. "He has made the same decision as 99 percent of his fellow legislators."

44. Another type of valid argument is called modus tollens. It has the form if p implies q and q is false, then p is false. Explain why modus tollens is a valid argument.

Identify the type of fallacy being used in each statement.

45. "Some people believe the answer to this problem is to wall off our economy from the world," he said this month in India, talking about migration of US jobs overseas. "I strongly disagree."

46. You must be an atheist. You never go to church on a Sunday morning.

47. On a discussion of the Treaty of Versailles in a history class, a student responds, "You said this happened five years before Hitler came to power. Why are you so fascinated with Hitler? Are you anti-Semetic?"

48. After hitting the side of the television, the picture on Emma's old television goes back into focus. Emma tells Karan that hitting the television fixed it.

Chapter 3   Summary

Section 3.1   Logic Statements and Their Negations

Definitions

Statement

A statement is a complete sentence that asserts a claim that is either true or false, but not both at the same time.

Paradox

A paradox is a sentence that contradicts itself and therefore has no single truth value. A paradox cannot be a mathematical statement.

Negation

The negation of a statement is the logical opposite of that statement, or its denial. Negations always have the opposite truth value of the original statement.

Negating Quantifiers	
Quantifier	**Negations**
All are	Not all are; Some are not; At least one is not
Some are	None are
Some are not	All are
None are	There is at least one that is; Some are

Compound Statement

A compound statement is composed of two or more statements joined together using connective words such as *and*, *or*, or *implies*.

Logic Symbols	
Symbol	**Read**
∧	And
∨	Or
~	Not
⇒	Implies

Section 3.2   Truth Tables

Definitions

Truth Table

A truth table is a table that has a row for each possible combination of truth values of the individual statements that make up the compound statement.

Conjunction

If *a* and *b* are statements, then "*a* and *b*" is a compound statement called a conjunction. A conjunction is true only when both statements are true; otherwise it is false.

Disjunction

If *a* and *b* are statements, then "*a* or *b*" is a compound statement called a disjunction. A disjunction will always be true unless both statements are false.

Conditional

If a and b are statements, then "if a, then b" is a compound statement called a conditional. A conditional will always be true unless a is true and b is false.

Summary of Logic Statements			
	Notation	Read	Truth Value Rule
Negation	$\sim p$	not p	opposite of truth value of p
Conjunction	$p \wedge q$	p and q	true only when both p and q are true
Disjunction	$p \vee q$	p or q	false only when both p and q are false
Conditional	$p \Rightarrow q$	if p, then q	false only when p is true and q is false

Tautology

A tautology is a statement that is true in all possible circumstances.

Section 3.3   Logical Equivalence and De Morgan's Laws

Definitions

Biconditional Statement

Biconditional statements, read "if and only if," are true only when each component of the statement has the same truth value; that is, either both are true or both are false.

Variations on Conditional Statements		
Name	Symbols	Read
Conditional	$p \Rightarrow q$	If p, then q
Converse of Conditional	$q \Rightarrow p$	If q, then p
Inverse of Conditional	$\sim p \Rightarrow \sim q$	If not p, then not q
Contrapositive of Conditional	$\sim q \Rightarrow \sim p$	If not q, then not p
Biconditional	$p \Leftrightarrow q$	p if and only if q

Logically Equivalent Statements

Logically equivalent statements are statements that have exactly the same truth values in all situations. We write this mathematically using the symbol $\equiv$.

De Morgan's Laws

1. $\sim(p \wedge q) \equiv \sim p \vee \sim q$

2. $\sim(p \vee q) \equiv \sim p \wedge \sim q$

Negation of Conditional Statements

$\sim(p \Rightarrow q) \equiv p \wedge \sim q$

Section 3.4   Valid Arguments and Fallacies

Definitions

Argument

An argument consists of a series of logical statements.

Premise

A premise is a statement at the beginning of an argument used to support the conclusion.

Conclusion

A conclusion is the ending statement of an argument.

Inductive Reasoning

Inductive reasoning is a line of argument that takes specific examples as its premise and then draws a general conclusion from these.

Deductive Reasoning

Deductive reasoning uses statements that are commonly accepted as facts to make a case for a certain conclusion.

Valid Argument

A valid argument is a deductive argument where the conclusion is guaranteed from the premises.

Invalid Argument

An invalid argument is an argument where the conclusion is not always guaranteed from the premises.

Sound Argument

A sound argument is a valid argument using true premises.

Fallacy

A fallacy is an error in reasoning that leads to an invalid argument.

Summary of Fallacies		
Name (Latin)	**Name (translated)**	**Description**
Post Hoc, Ergo Propter Hoc	After This, Therefore Because of This	Because one thing happened first, it caused the other to happen.
Dicto Simpliciter	Hasty Generalization	Making broad, sweeping generalizations based on a few specific occurrences.
Ad Hominem	Personal Attack	Attacking someone's person, character, or motives instead of addressing the premise of the argument.
Petitio principii	Circular Reasoning	The premise and conclusion state the same thing.
Non Sequitur	It Does Not Follow	The conclusion has nothing to do with the premise.
Straw Man	Exaggerated or Distorted View of the Opponent	The buildup of a distortion of someone's ideas or beliefs so that they can easily be knocked down.
False Dilemma	Illusion of Limited Choice	Argument that rests on the assumption that there are only two choices as a solution.
Argumentum ad Populum	Appeal to the People	Stating that the majority, or even just a large number, of people are doing something as the basis for supporting a conclusion.

Chapter 3 Exercises

Negate each statement.

1. The puppy could not keep her eyes open after 10:00.

2. Jour means "the day" in French.

3. All houses have fireplaces.

4. Every senior that graduated had a job offer.

Construct the truth table for each compound statement.

5. $(a \lor \sim b) \Rightarrow c$

6. $w \land (x \land \sim y)$

7. If you and Kathy both go to the movies, then I will go as well.

8. I don't eat dessert and do not have a cup of coffee.

Write the converse, inverse, contrapositive, and biconditional for each conditional statement.

9. If my grade in this course is an A, then I can enroll in the next course.

10. If my car runs out of gas, then my car will not start.

Solve each problem.

11. Complete the truth table for $(p \land (p \Rightarrow q)) \Rightarrow q$.

Truth Table				
p	q	$p \Rightarrow q$	$p \land (p \Rightarrow q)$	$(p \land (p \Rightarrow q)) \Rightarrow q$
T	T			
T	F			
F	T			
F	F			

12. If the bakery has fresh muffins, then I will buy some for my roommates. If I buy fresh muffins then my roommates will be happy. What can you deduce from these statements?

13. Which of the following is an example of inductive logic?

   a. If you liked all of Richie Rock's movies, then you will also like his latest movie.

   b. If there is an accident on the freeway, then we will be late for the play.

   c. If I get a raise next week, then I can buy a new car.

   d. I have to leave work early if the bank closes at 5:00 today.

Determine whether each arguments is valid or invalid.

14. Eagles have feathers.
    Sparrows have feathers.
    Ducks have feathers.

15. All birds have wings.
    Ducks are birds.
    Ducks have wings.

16. Some birds have feathers.
    Sparrows are birds.
    Sparrows have feathers.

17. All birds can fly.
    Eagles are birds.
    Eagles can fly.

Identify the premise and conclusion in each statement.

18. "If the Egyptian government had been doing what the IDF is doing during the Arab Spring, it would have been a very different picture," he said, referring to the activists who used social media to organize protests that toppled Egypt's Hosni Mubarak last year. [1]

19. "The most prominent advocates of global warming aren't scientists," said Heartland's president, Joseph Bast. "They are Charles Manson, a mass murderer; Fidel Castro, a tyrant; and Ted Kaczynski, the Unabomber. Global warming alarmists include Osama bin Laden and James J. Lee (who took hostages inside the headquarters of the Discovery Channel in 2010)." [2]

---

1 CNN, http://www.cnn.com

2 The Heartland Institute, http://www.heartland.org

Identify each type of fallacy being used.

20. If you believe that a man who can cheat on his wife of 38 years will be honest with you on all levels, than you deserve what you get for voting for him.

21. "The only question is if Washington has the courage to seize this opportunity. If they do, this moment won't be remembered as a drawn-out partisan showdown over spending, but as the launching pad from which we began to sail into a healthier, more prosperous future." [3]

3 CNN, http://www.cnn.com

Bibliography

3.4

1. Cohen, Tom. "Obama accuses Romney of dishonesty in debate." CNN. http://www.cnn.com/2012/10/04/politics/debate-main/index.html?hpt=hp_t1

2. *The New York Times*. "Transcript of the First Presidential Debate." http://www.nytimes.com/2012/10/03/us/politics/transcript-of-the-first-presidential-debate-in-denver.html

3. Doyle, Sir Arthur Conan. *The Project Gutenberg EBook of A Study in Scarlet*. http://www.gutenberg.org/files/244/244-h/244-h.htm

4. Norris, Chuck. "My Choice for President." WND. http://www.wnd.com/news/article.asp?ARTICLE_ID=58255

5. BBC. "NHL cancels first two weeks over labour dispute." http://www.bbc.co.uk/news/world-us-canada-19840142

6. Vince, Gaia. "Sucking CO2 from the skies with artificial trees." BBC. http://www.bbc.com/future/story/20121004-fake-trees-to-clean-the-skies

7. ygoy. "Fast Food Obesity." http://obesity.ygoy.com/fast-food-obesity/

8. Barash, David. "The Mathematical Argument for Gun Control." *The Chronicle of Higher Education*. http://chronicle.com/blogs/brainstorm/the-mathematical-argument-for-gun-control/50717

9. Kohn, Sally. "5 reasons ObamaCare is already good for you." Fox News. http://www.foxnews.com/opinion/2012/03/28/5-reasons-obamacare-is-already-good-for/#ixzz28R2qZdRn

10. Sorkin, Aaron. "Post Hoc, Ergo Propter Hoc." *The West Wing*, season 1, episode 2, directed by Thomas Schlamme, aired September 29, 1999.

11. Randy Glasbergen. "goldie39." http://www.glasbergen.com

Chapter Exercises

1. Sutter, John D. "Will Twitter war become the new norm?" CNN. http://www.cnn.com/2012/11/15/tech/social-media/twitter-war-gaza-israel/index.html

2. Lakely, Jim. "'Do You Still Believe in Global Warming?' Billboards Hit Chicago." The Heartland Institute. http://heartland.org/press-releases/2012/05/03/do-you-still-believe-global-warming-billboards-hit-chicago

3. Jones, Van and Phaedra Ellis-Lamkins. "To end the fiscal showdown, tax carbon." CNN. http://www.cnn.com/2012/11/16/opinion/jones-carbon-tax/index.html?hpt=op_t1

Chapter 4
Rates, Ratios, Proportions, and Percentages

Sections

Objectives

- Write rates as fractions
- Solve proportional equations
- Calculate unit rates
- Write and interpret ratios
- Calculate proportions and percentages
- Identify and calculate percentage increase and percentage decrease
- Identify common mistakes with percentages

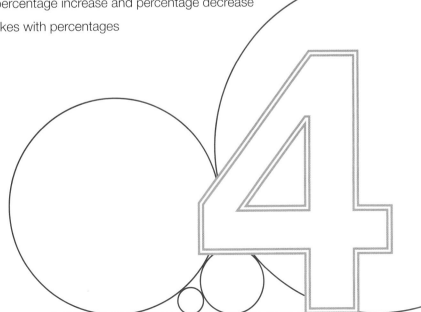

4 Rates, Ratios, Proportions, and Percentages

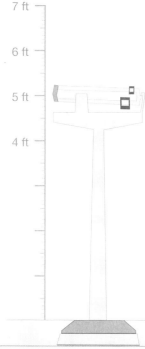

Your BMI, or body mass index, is a relationship between your height and your weight. Developed in the 1800s, this notion of looking at a comparison of height and weight has impacted the world of fitness and health. You can easily find charts and BMI calculators on the Internet to help you calculate your BMI. Simply put, BMI is a fraction that compares your height and weight so that you, along with your physician, can determine a diet an fitness plan that works for your current BMI and your health goals.

Consider the question "Is a person who weighs 150 pounds overweight?" The correct answer is, "That depends." If a 6-foot-6-inch-tall adult weighs 150 pounds, not only are they *not* overweight, they're close to being moderately underweight. However, if a 4-foot-10-inch-tall adult weighs 150 pounds, they probably need to look at monitoring their exercise and diet, since their BMI labels them as being obese. A comparison like the BMI helps to take into account more information than just weight before drawing a conclusion.

Throughout this chapter, we'll look at ways fractions can help with understanding comparisons. Rates, ratios, proportions, and percentages are all forms of fractions that are used in slightly different ways.

Table 1

Body Mass Index

	Normal						Overweight							Obese								Extreme Obesity					
BMI	19	20	21	22	23	24	25	26	27	28	29	30	31	32	33	34	35	36	37	38	39	40	41	42	43	44	45
Height (inches)													**Body Weight** (pounds)														
58	91	96	100	105	110	115	119	124	129	134	138	143	148	153	158	162	167	172	177	181	186	191	196	201	205	210	215
59	94	99	104	109	114	119	124	128	133	138	143	148	153	158	163	168	173	178	183	188	193	198	203	208	212	217	222
60	97	102	107	112	118	123	128	133	138	143	148	153	158	163	168	174	179	184	189	194	199	204	209	215	220	225	230
61	100	106	111	116	122	127	132	137	143	148	153	158	164	169	174	180	185	190	195	201	206	211	217	222	227	232	238
62	104	109	115	120	126	131	136	142	147	153	158	164	169	175	180	186	191	196	202	207	213	218	224	229	235	240	246
63	107	113	118	124	130	135	141	146	152	158	163	169	175	180	186	191	197	203	208	214	220	225	231	237	242	248	254
64	110	116	122	128	134	140	145	151	157	163	169	174	180	186	192	197	204	209	215	221	227	232	238	244	250	256	262
65	114	120	126	132	138	144	150	156	162	168	174	180	186	192	198	204	210	216	222	228	234	240	246	252	258	264	270
66	118	124	130	136	142	148	155	161	167	173	179	186	192	198	204	210	216	223	229	235	241	247	253	260	266	272	278
67	121	127	134	140	146	153	159	166	172	178	185	191	196	204	211	217	223	230	236	242	249	255	261	268	274	280	287
68	125	131	138	144	151	158	164	171	177	184	190	197	203	210	216	223	230	236	243	249	256	262	269	276	282	289	295
69	128	135	142	149	155	162	169	176	182	189	196	203	209	216	223	230	236	243	250	257	263	270	277	284	291	297	304
70	132	139	146	153	160	167	174	181	188	195	202	209	216	222	229	236	243	250	257	264	271	278	285	292	299	306	313
71	136	143	150	157	165	172	179	186	193	200	208	215	222	229	236	243	250	257	265	272	279	286	293	301	308	315	322
72	140	147	154	162	169	177	184	191	199	206	213	221	228	235	242	250	258	265	272	279	287	294	302	309	316	324	331
73	144	151	159	166	174	182	189	197	204	212	219	227	235	242	250	257	265	272	280	288	295	302	310	318	325	333	340
74	148	155	163	171	179	186	194	202	210	218	225	233	241	249	256	264	272	280	287	295	303	311	319	326	334	342	350
75	152	160	168	176	184	192	200	208	216	224	232	240	248	256	264	272	279	287	295	303	311	319	327	335	343	351	359

Source: Adapted from *Clinical Guidelines on the Identification, Evaluation, and Treatment of Overweight and Obesity in Adults: The Evidence Report.* Bethesda (MD): National Heart, Lung, and Blood Institute; 1998 Sep. Report No.: 98-4083

4.1   Rates and Unit Rates

The fraction that we will look at first is called a rate. A **rate** is a fractional comparison between two quantities that are not necessarily in the same units. For instance, miles per gallon when driving a car $\left(\frac{23\text{ miles}}{1\text{ gallon}}\right)$, price of rice $\left(\frac{\$7.99}{5\text{ pounds}}\right)$, heart rate $\left(\frac{73\text{ beats}}{1\text{ minute}}\right)$, and minimum wage rate $\left(\frac{\$7.25}{1\text{ hour}}\right)$ are all examples of rates. Notice that in each of the rates the units are written as part of the rate. The word "per" is used to denote the division bar when reading rates. For instance, the rates given previously are read as "23 miles per gallon," "\$7.99 per 5 pounds," "73 beats per minute," and "\$7.25 per hour."

Rate

A **rate** is a fraction used to compare two quantities that are not necessarily in the same units.

Example 1: Writing Rates

Suppose you and your friends are deciding where to spend Friday night. Laser tag offers the following deals: \$13.00 per person for 2 games or \$18.00 per person for 3 games. However, the county fair is offering \$20.00 wristbands allowing all-you-can-ride access for the night. Since both options sound appealing, you let your friends decide where to spend the evening. To make the decision easier, you would like to present them with rates for the evening's choices based on time. Write down a rate for each choice based on time if it takes 30 minutes to play a game of laser tag or if you could stay at the fair from 8 p.m. until 10 p.m.

Solution

Since we are asked to write rates based on time, we will write each option as a fraction with price in the numerator of the fraction and time in the denominator.

$$\frac{\text{price}}{\text{time}}$$

Consider the laser tag option first. If it takes 30 minutes to play one game of laser tag, the denominator is calculated by multiplying the number of games played by 30.

$$\text{Laser Tag Deal 1: } \frac{\$13}{2\cdot 30\text{ min}} = \frac{\$13}{60\text{ min}}$$

$$\text{Laser Tag Deal 2: } \frac{\$18}{3\cdot 30\text{ min}} = \frac{\$18}{90\text{ min}}$$

For the fair, the rate will be based on staying at the fair from 8 p.m. until 10 p.m., which is 2 hours of unlimited rides.

$$\text{Fair: } \frac{\$20}{2\text{ hours}}$$

Notice that we were careful to write all of the rates in the same fraction form; that is, the numerator was the cost and the denominator was the time. For another purpose it may be more suitable to consider these quantities in the opposite order (this is also a rate). It is important to know which rate makes the most sense for the problem in order to make a true consistent comparison.

Think Back

Recall that fractions are of the form

$$\frac{\text{numerator}}{\text{denominator}}.$$

4

By placing all the rate options in fraction form, your friends can begin to weigh their options for the evening. If we really wanted to make things as clear as possible for them, we would make sure that all the rates were in the same units of time, either all in minutes or all in hours. We'll look at how to do that later in the chapter. But for now, think about which option you would choose based on the rates we calculated.

Example 2: Identifying Rates

Foods in Bulk offers 30 pounds of candy for $64.49 on its website. Which of the following fractions represents the rate given by their website?

a. $\dfrac{1}{\$64.49}$ **b.** $\dfrac{30 \text{ lb}}{\$64.49}$ **c.** $\dfrac{30 \text{ lb}}{1}$ **d.** $\dfrac{\$64.49}{30 \text{ lb}}$

Solution

There are actually two correct ways to write the rate from the website: answers **b.** and **d.** You can either write the fraction in pounds of candy per price in dollars like answer **b.**, or price in dollars per pounds of candy like answer **d.** Both **a.** and **c.** are incorrect because neither compares the amount of candy to the price of the candy.

Helpful Hint

When written out, the word "per" indicates that the denominator amount of a rate is about to follow.

Let's consider another familiar rate example, miles per gallon (mpg). The "per gallon" phrase implies that the comparison is for one gallon of gasoline. Every new car trying to capture the attention of economical customers boasts of its fuel efficiency by posting its mpg rate on the window sticker. Miles per gallon is a comparison between the distance a vehicle travels and the amount of gasoline it consumes (that is, 25 mpg means that a car can drive 25 miles for each gallon of fuel). If you see a new car that boasts 36 mpg on the highway but only 26 mpg in the city, what does that mean? Like most cars, the rate at which the car uses fuel differs depending on the type of driving that is being done. For instance, 36 mpg for highway driving means that for every gallon of fuel, the car can be driven 36 miles on the highway. However, the decrease to only 26 mpg for city driving is in large part due to more frequent stopping and starting that occurs when driving in a city, which consumes fuel more quickly. Traditionally, society has moved away from writing this as a fraction, such as $\frac{36 \text{ miles}}{1 \text{ gallon}}$, and has instead moved to using the mpg notation, such as 36 mpg.

Example 3: Working with Miles per Gallon

As a college graduation present, Katie's parents helped her buy her first new vehicle. She chose one that claims a fuel efficiency of 29 miles per gallon (mpg) on the highway. Check the actual mpg on Katie's new car if she drove 309 miles on 11 gallons of gas.

Solution

Begin by setting up a ratio of the actual miles Katie drove on 11 gallons of gas. We'll put the gallons in the denominator so that we can have a comparison of the actual mpg to the manufacturer's mpg.

$$\frac{309 \text{ miles}}{11 \text{ gallons}}$$

The question asks how many miles did she drive on 1 gallon? We can simply divide here to find that, as follows.

$$\frac{309 \text{ miles}}{11 \text{ gallons}} \approx 28.09 \text{ mpg}$$

Whether or not Katie drove all 309 miles on the highway, she still managed to get close to the manufacturer's mpg rate of 29 mpg.

Skill Check #1

If rice costs $7.99 for 5 pounds, determine the cost per pound of the rice.

Think Back

Recall that a variable is a symbol to represent an unknown quantity.

Often it's useful to be able to compare two rates to one another. For instance, suppose that a certain brand of coffee costs $1.58 per pound and you would like to know how much $4\frac{1}{2}$ pounds of coffee would cost. Using algebra, you can create an equation and solve for the unknown amount to help you find the answer. When you set two rates equal to one another, the resulting equation is called a **proportional equation**. Solving a proportional equation for the unknown amount is often referred to as "solving a proportion."

Proportional Equation

A **proportional equation** consists of two rates set equal to one another.

Example 4: Using Proportional Equations

How much gasoline should Katie have used after driving 243 miles in her new car that boasts 29 mpg? Set up and use a proportional equation to find the answer.

Solution

Begin by setting up the miles per gallon rate that was claimed by the manufacturer as a fraction. Remember to put miles in the numerator and gallons in the denominator.

$$\frac{29 \text{ miles}}{1 \text{ gallon}}$$

We need to find an equivalent fraction that has 243 miles in the numerator rather than 29 miles. We can use a variable, such as x, to represent the number of gallons of gasoline used for 243 miles. That gives us the rate of

$$\frac{243 \text{ miles}}{x \text{ gallons}}.$$

Now we can set the two rates equal to one another to form a proportional equation. Note that both rates are in the same form, with miles in the numerator and gallons in the denominator.

$$\frac{29 \text{ miles}}{1 \text{ gallon}} = \frac{243 \text{ miles}}{x \text{ gallons}}$$

To solve a proportional equation like this, recall that we use algebra to isolate the unknown quantity x. In other words, we want to get x on one side of the equation by itself. Begin by multiplying both sides of the equation by each of the denominators in order to remove the fractions. In this case, multiply by $1x$, or simply x.

4

$$\frac{29}{1} = \frac{243}{x}$$

$$\frac{29}{1} \cdot x = \frac{243}{\cancel{x}} \cdot \cancel{x}$$

$$29x = 243$$

$$\frac{\cancel{29}x}{\cancel{29}} = \frac{243}{29}$$

$$x \approx 8.38$$

So, Katie should have used approximately 8.38, or about $8\frac{1}{3}$, gallons of gas after driving 243 miles.

Helpful Hint

To clear a proportional equation of fractions, multiply both sides of the equation by the product of the denominators.

When we compare things to a single unit, such as miles per one gallon as we did in Example 4, it is referred to as a **unit rate**. In other words, the denominator of a unit rate is one unit. For instance, when looking around for the "best deal" on a hotel for your vacation, what do you consider first? Some would say amenities such as a pool, computer center, or workout room. Some might place high priority on the location of the hotel. Still others might consider things like free breakfast, complimentary goodies, or free Internet access. We would venture to guess, though, that most people include the price in their decision. So, knowing the room rate, or price for a room for one night, at the different hotels you are considering would be beneficial when making your decision. The same scenario could happen for the price per day for a rental car or the cost per game at a bowling alley. These are all examples of unit rates.

> ### Unit Rate
>
> A **unit rate** is a rate comparing two measured quantities, one of which is a single unit written in the denominator of the fraction.

Finding Unit Rates

Shopping at grocery stores really lends itself to unit rates. Suppose you want to know which is the better deal—buying the regular size box of cereal or the larger family size. One store sells the smaller box at $2.99 for 9.8 oz, while a different store sells a 14 oz box for $4.49. How can we compare the cost of two options? By writing each rate as a unit rate, we can find either the price per ounce or the number of ounces per dollar of each box and make an informed decision about the best deal. For comparison sake, you can find either unit rate. However, it's important that you find the same unit rate for each box of cereal. For our example, we will use the unit rate price per ounce. In other words, the price will be in the numerator and one ounce will be in the denominator.

Take the first box of cereal and write the rate given as a fraction, price per number of ounces.

$$\frac{\$2.99}{9.8 \text{ oz}}$$

In order for the denominator to have a unit of one and be a unit rate, we need to divide it by 9.8. The rules of algebra say that if we divide the denominator by a number, we must also divide the numerator by that same number to keep the fractions equivalent. Remember that dividing both the numerator and denominator of a fraction by the same number is equivalent to dividing by 1, which does not affect the value of the fraction at all. So, dividing both the numerator and denominator in this example by 9.8, we have:

$$\frac{\$2.99 \div 9.8}{9.8 \text{ oz} \div 9.8} \approx \frac{\$0.305}{1 \text{ oz}} \text{ or approximately } 30.5\cent \text{ per ounce.}$$

This is the unit rate of price per ounce for the smaller box of cereal.

Now calculate the unit rate for the larger box of cereal. Make sure that you write the fraction in the same form as we did previously so that both unit rates give the price per ounce. Then, divide both the numerator and denominator by the value of the denominator.

$$\frac{\$4.49 \div 14}{14 \text{ oz} \div 14} \approx \frac{\$0.321}{1 \text{ oz}} \text{ or approximately } 32.1\cent \text{ per ounce}$$

We can now see that the larger box is actually more expensive per ounce than the smaller one. Buyer beware! Bulk size isn't always the best deal.

In reality, you can easily compare the unit prices for items in the grocery store without having a calculator on hand. The shelving stickers will list the unit price underneath the container price in most stores. Take the extra time to check for yourself and make the consumer-savvy buy.

Figure 1

Example 5: Finding Unit Rates

Find the unit rate for each of the following types of wood mulch at the local garden center.

a. Pine bark mulch: $3.77 for 3 cubic feet

b. Aromatic cedar mulch: $3.98 for 2 cubic feet

c. Evergreen red mulch: $3.33 for 2 cubic feet

Solution

To find the unit rates for each mulch, we first set up each rate as a fraction; that is, $\frac{\text{price}}{\text{amount}}$.

a. Pine bark mulch: $3.77 for 3 cubic feet $= \dfrac{\$3.77}{3 \text{ cubic feet}}$

b. Aromatic cedar mulch: $3.98 for 2 cubic feet $= \dfrac{\$3.98}{2 \text{ cubic feet}}$

c. Evergreen red mulch: $3.33 for 2 cubic feet $= \dfrac{\$3.33}{2 \text{ cubic feet}}$

Now that we have a rate for each type of mulch, we need to convert them to unit rates. To do this, we divide both parts of the fraction by the amount in the denominator.

a. Pine bark mulch: $\dfrac{\$3.77 \div 3}{3 \text{ cubic feet} \div 3} \approx \dfrac{\$1.26}{1 \text{ cubic foot}}$ or $1.26 per cubic foot

b. Aromatic cedar mulch: $\dfrac{\$3.98 \div 2}{2 \text{ cubic feet} \div 2} = \dfrac{\$1.99}{1 \text{ cubic foot}}$ or $1.99 per cubic foot

c. Evergreen red mulch: $\dfrac{\$3.33 \div 2}{2 \text{ cubic feet} \div 2} \approx \dfrac{\$1.67}{1 \text{ cubic foot}}$ or $1.67 per cubic foot

Once converted to unit rates, we can see that the pine bark mulch is the least expensive mulch in our list.

Skill Check #2

Helium gas (He) was placed in a container fitted with a porous membrane. The helium effused through the membrane at the rate of 1.5 L/24 hr. Find the unit rate of effusion per hour for the helium.

Example 6: Using Unit Rates

Emma is traveling to Italy for spring break. She is preparing herself for all the shopping she has planned by looking at the exchange rate between dollars and euros. To ensure that she does not overspend while there, she wants to have in mind the equivalent dollar amounts for euro prices. Table 1 shows a section of the exchange rates that Emma found listed on the Internet.

Table 1		
Exchange Rates		
EURO vs.	**1 EUR**	**in EUR**
American Dollar	1.351	0.740192
Argentine Peso	5.62251	0.177856
Brazilian Real	2.2542	0.443617
British Pound	0.8419	1.18779
Canadian Dollar	1.3328	0.7503

a. What is the unit exchange rate for American dollars to euros?

b. What is the unit exchange rate for euros to American dollars?

c. If Emma is considering taking $200 for spending money, approximately how many euros will she have to spend?

d. Once in Italy, Emma falls in love with a leather bag priced at €200. She knows that it is over her budget, but wants to know by how much. Use the exchange rate to find out approximately how much the bag would cost in American dollars.

Solution

a. This format is very similar to many others you will find online. Reading it correctly is half the battle. The column labeled "**1 EUR**" gives the amount of corresponding currency that would be equivalent to one euro. If we want a rate of American dollars to euros, the unit rate for euros should be in the denominator.

$$\frac{\text{dollars}}{\text{euros}} = \frac{?}{1}$$

Reading down the first column, we can see that there are 1.351 dollars for each euro. We have

$$\frac{\text{dollars}}{\text{euros}} = \frac{\$1.351}{€1}.$$

b. This time we need the rate for euros to American dollars, so the dollar amount will be in the denominator. The second column labeled "**in EUR**" provides the number of euros equivalent to one unit of currency listed. Therefore, we have

$$\frac{\text{euros}}{\text{dollars}} = \frac{€0.740192}{\$1}.$$

c. To answer this question, we will set up a proportional equation to solve. We know that Emma has $200 and wants to know the equivalent amount of euros. Writing a rate with the unknown amount of euros as x gives us

$$\frac{\$200}{€x}.$$

We can use the dollars/euros unit rate we found in part **a.** to set up a proportional equation.

$$\frac{\$200}{€x} = \frac{\$1.351}{€1}$$

Solving for x, we have the following.

$$\frac{200}{\cancel{x}} \cdot \cancel{x} = \frac{1.351}{1} \cdot x$$
$$200 = 1.351x$$
$$x \approx 148.04$$

So, Emma will have approximately €148 to spend on her trip to Italy.

d. In this situation, we want to change from euros to dollars. We can set up a proportional equation as we did in part **c.** and solve it. This time, however, we will use the unit rate found in part **b.**, euros per dollar, since we are given euros to begin with. The unit rate found in part **b.** is

$$\frac{€0.740192}{\$1}.$$

Setting up a proportional equation using €200 and x as the equivalent amount in dollars gives us

$$\frac{€0.740192}{\$1} = \frac{€200}{\$x}.$$

Solving for x we have the following.

$$\frac{0.740192}{1} \cdot x = \frac{200}{\cancel{x}} \cdot \cancel{x}$$
$$0.740192x = 200$$
$$x \approx 270.20$$

This means that the leather bag costs $270.20 (in American dollars). Since her original budget was $200, buying this bag would put her approximately $70 over her budget.

More often than not, the rates we encounter are unit rates. However, it is not always the case that we can use unit rates in a meaningful manner. Consider for a minute Example 1 about choosing Friday night's entertainment. Recall that the friends had a choice of the following entertainment options given in rates.

Option 1: Laser tag $\frac{\$13}{60 \text{ min}}$

Option 2: Laser tag $\frac{\$18}{90 \text{ min}}$

Option 3: Fair rides $\frac{\$20}{2 \text{ hours}}$

Notice that the rates we found are not unit rates. They all are in differing amounts of time. Although

4

we could technically find unit rates per hour for each of the three options, it would be misleading to present them to your friends because each option is not available at the unit rate. For instance, consider Option 2 of laser tag for 1.5 hours. As a unit rate, it would cost $12 per hour ($18 divided by 1.5 hours). However, we know that purchasing only one hour of laser tag actually costs $13 as shown in Option 1. It is not possible to play laser tag at the unit price found in Option 2. Similarly, although in theory, $10 is what one hour of rides would cost at a unit rate at the fair in Option 3, it's actually impossible to purchase a single hour's worth of rides—you have to buy the $20 pass. So, instead of looking at unit rates, we use the mixed rates to decide how much time and money we would like to spend to have fun.

Skill Check Answers

1. $1.60 per pound

2. 0.0625 L/1 hr

4.1 Exercises

Write each rate described.

1. Google Voice advertises its competitive price for international calling on mobile phones as 15 cents per minute from Mexico.

2. A freelance online editorial company offers document indexing for $3.50 for 12 indexable printed pages.

3. The Department of Veterans Affairs offers tuition assistance for active military personnel under a new GI Bill that pays up to $1473.00 per month for full-time status at a higher education institution.

4. The local grocery store offers 10 cans of tomato sauce for $10 in the weekly sales ad.

5. Grayson needs to work 65 hours every three weeks to meet the quota for his school internship.

Choose the correct form of the rate given.

6. To advertise their weekly special, Carol needs a sign made. One pen costs $3.18, but she is going to offer 3 pens for the price of one. Which is the correct sign?

   a. $3.18/pen

   b. 3 pens/$3.18

   c. $9.54/3 pens

7. Tyson can run 4 km in 30 minutes.

    a. 4 km/30 min

    b. 7.5 km/min

    c. 2 km/hour

    d. 1 hour/2 km

Find each unit price. Round your answer to the nearest hundredth.

8. It costs $130.11 per week to rent a standard-size car in Nashville, TN. Find the price per day it costs to rent this car.

9. It costs $2.01 to drive 25 miles in a new 2012 Hybrid car. How much does it cost per mile to drive 25 miles?

10. It costs $4.78 to drive 25 miles in a 2012 sedan. How much does it cost per mile to drive 25 miles?

11. If your car can go 433 miles on a full tank of gas, find the miles per gallon your car gets if the tank holds 17.2 gallons.

12. Ron gets paid $51,630 per year. If he works 2087 hours in a year, find his hourly rate of pay.

13. If a 4-pound bag of sugar costs $2.99, what is the unit cost for a pound of sugar?

14. The advertised special is $9.99 for 16 ounces of frozen shrimp. Find the unit cost of the shrimp.

15. Eggplants are 4 for $5.00. What is the cost per eggplant? How much eggplant do you get per dollar?

Solve each problem. Round your answer to the nearest hundredth when necessary.

16. Use unit rates to determine whether spending $3.97 for the 14 oz box of cereal or $4.23 for the 16 oz box of cereal is the better value for the money. Explain your reasoning.

17. Which is the better buy, 3 batteries for $4.80 or 12 batteries for $14.76? Explain your reasoning.

4

18. Carla was curious if the bulk buy store in town really was cheaper than the store where she shops. She made a list of 5 items she buys on a regular basis and found the prices for each item at each store. Use unit prices to decide which is the best place for Carla to shop if she plans to buy the 5 items.

Price Comparison		
Items	**Bulk Buy Store**	**Carla's Regular Store**
Eggs	18 count for $3.56	$2.69 per dozen
Skim Milk	$2.60 per gallon	$2.99 per gallon
Romaine Lettuce	6 heads for $3.98	$0.99 per head
Potatoes	10 pounds for $3.98	10 pounds for $3.99
Cheddar Cheese	2 pounds for $5.98	$2.99 for 16 ounces

19. You want to buy a used truck and are interested in the fuel consumption rate for each of the three trucks you are considering. Using the information given, decide which of the used trucks gets the best gas mileage.

    a. 2008 Chevy Silverado: 500 miles on 29.4 gallons

    b. 2006 Toyota Tundra: 428 miles on 26.4 gallons

    c. 2007 Ford F-150: 250 miles on 15.6 gallons

20. Suppose at bulk rate that a box of crackers costs 30.5 cents per ounce.

    a. If a box costs $4.49, how many ounces are in the box of crackers?

    b. What would the box of crackers cost if it contained 14 oz?

21. The following list of ingredients for a sugar cookie recipe makes 2 dozen cookies. What amounts of butter and half-and-half would you need to make 3 dozen cookies?

    3 cups sifted all-purpose flour

    $1\frac{1}{2}$ teaspoons baking powder

    $\frac{1}{2}$ teaspoon salt

    1 cup white sugar

    1 cup butter (softened at room temperature)

    1 egg, lightly beaten (egg should be at room temperature)

    3 Tablespoons half-and-half

    2 teaspoons vanilla extract

22. Suppose we assume the cost of living in the United States is the same as the cost of living in the United Kingdom. It would then be reasonable to assume that a fair exchange rate of dollars to pounds would simply convert the money, without any added inflation. In other words, the ratio of cost of goods in the United States to the cost of goods in the United Kingdom should simply equal the exchange rate.

United States Vs. United Kingdom Price Comparisons		
Item	US Average Cost	UK Average Cost
Bread, white	$2.91 loaf	£1.37 loaf
Ground beef, 100% beef	$2.88 per pound	£2.38 per pound
Eggs	$1.87 per dozen	£2.90 per dozen
Milk, whole	$3.62 per gallon	£3.20 per gallon
Bananas	$0.61 per pound	£0.93 per pound

a. Given the costs for several of the same grocery items in both countries, find the rate of total costs for these items in the United States to that of total cost in the United Kingdom.

b. If the exchange rate is $\dfrac{\$1}{£0.6470}$, determine if the exchange rate reflects the cost of living in the two countries.

23. An entry level accountant's average annual pay for the United States is $44,036 based on 52 weeks per year. Due to the economy, a firm is having to cut back on the number of weeks that it employs its accountants. If the firm cuts the work year to 48 weeks, but keeps the same rate of pay (which is equivalent to the national average), how much should an accountant expect his annual pay to decrease? Round your answer to the nearest dollar.

4.2 Ratios

When the measured quantities being compared are in the same units or of the same type, we call the fraction a **ratio**. The units in a ratio are of the same kind—people, distance, objects, animals—and are units of identical dimension. For instance, a ratio of men to women is a comparison where both are counts of people. If there were 12 people in a room, where 7 were men and 5 were women, the ratio of men to women would be 7 to 5. Or in cooking, when a recipe calls for a 1 to 2 ratio of fat to flour, it means that there should be twice as much weight of flour as there is of fat. In fact, the relationship that a ratio represents must hold true no matter what units of measure are used. In our ingredients example, there will always be twice as much flour as fat whether you measure in ounces, grams, or even in pounds (assuming you were cooking for an army!).

> ### Ratio
>
> A **ratio** is a fraction used to compare two measured quantities whose units are the same type.

Notice that, although we could have, we did not use fractional notation to write the ratios above. In fact, in mathematics, there are three accepted ways to write ratios.

- written with words, "1 to 2"
- written with a colon, "1 : 2"
- written as a fraction, "$\frac{1}{2}$"

Notice that the units are not written in ratios like they are in rates. Although all three notations for ratios are equally valid, there is no need to write all three. In this text, we use the colon method most often in our notation.

When we are identifying the quantities being compared in a ratio, the first number given refers to the first item in the list. For instance, when we say the phrase, "7 to 5, men to women," the 7 refers to the number of men and the 5 refers to the number of women. Consequently, order is very important in ratios.

Intentionally or not, Jan and Dean made ratios a hit in the early 1960s with the lyrics to their number one song, "Surf City," citing *two girls for every boy*." Certainly, life for the California surfers would have been very different if there were two boys for every girl, instead of two girls for every boy!

Example 1: Writing Ratios

On a university campus, the fall enrollment shows that the new freshman class consists of 321 women and 214 men. Write the ratio of men to women.

Solution

Since the ratio we are writing is *men : women*, the number of men must come first. So, the ratio of men to women on the campus is

$$214 : 321 \quad \text{or} \quad 214 \text{ to } 321 \quad \text{or} \quad \frac{214}{321}.$$

Often, it is the case that ratios are written in a reduced form, which is usually the simplest form of the fraction. Since both numbers are divisible by 107 here, the ratio of men to women is equivalent to $2 : 3$.

Equivalent Ratios

Equivalent ratios are ratios that express the same relationship.

In the previous example, the ratio 214 : 321 was shown to be equivalent to the reduced ratio 2 : 3 by dividing both numbers by 107. We could also say the ratio 214 : 321 is equivalent to 428 : 642 by multiplying both numbers by 2. In fact, every ratio has an unlimited number of equivalent ratios that can be found by using either multiplication or division.

Skill Check # I

A local radio station says that out of every 10 listeners, 6 are women and 4 are men. Write the ratio of women to men listeners for the radio station in lowest terms.

Example 2: Writing Ratios

According to *Fortune*'s 2011 list of the top "40 Under 40," the up-and-coming top 40 adults in America includes 15 women. [1]

a. Write the ratio of the number of women on the list to the total number of adults on the list in all three notations.

b. Write the ratio of women to men on the list.

Solution

a. We were asked to write the ratio of *women : total number of adults*, so we write the number of women first. We were also asked to provide the ratio in all three notations. They are the following.

with words: 15 to 40

with a colon: 15 : 40

with a fraction: $\dfrac{15}{40}$

All three are read the same way, "fifteen to forty."

The ratios tell us that 15 out of every 40 adults on the list are women. We can write an equivalent ratio by dividing each number by 5. The simplified ratio is then 3 to 8, which tells us that 3 out of every 8 adults on the list are women.

b. To calculate the ratio of women to men, we need to know the number of men and the number of women on the list. The study states that there were a total of 40 people on the list, of which 15 were women. This means that 25 of them were men. Remember that, since we were asked to write a ratio of *women : men*, we write the number of women first, followed by the number of men.

with words: 15 to 25

with a colon: 15 : 25

with a fraction: $\dfrac{15}{25}$

1 Fortune, http://www.fortune.com

Notice that the total number of people on the list (40) does not appear in the ratio of women to men at all.

In this example, both 15 and 25 are divisible by 5, so we can once again simplify our ratio to an equivalent one by dividing both numbers by 5. The simplified ratio of women to men is $3:5$.

Take a moment and carefully consider the fraction notation of ratios. In the previous example, we calculated the ratio $3:5$ or $\frac{3}{5}$. We need to be careful about what the fraction $\frac{3}{5}$ represents. It does not mean that three-fifths of the 40 people on the list are women, or men, for that matter! It means that the number of women is three-fifths of the number of men. Putting it another way, for every 3 women on the list, there are 5 men. We could also think of reversing the ratio to be *men* : *women*. In this case we have that the men outnumber women in the ranking by a ratio of 5 to 3. This means that for every 5 men in the group there are 3 women. Apparently, the business world is a far cry from the 1960s surfing days in California!

Example 3: Writing and Interpreting Ratios

Table 1 shows a study by an online health journal claiming that, although fish has long been a food of choice for helping to control heart disease, farm-raised tilapia might actually be dangerous to the heart. Omega fatty acid ratios in farm-raised tilapia actually give undesirable amounts of omega-6 acids. In farm-raised tilapia, the ratio of these potentially detrimental long-chain omega-6 fatty acids to the beneficial long-chain omega-3 fatty acids averaged about $11:1$. Table 1 shows a sample of the number of fatty acids in a tablespoon of different types of fish oil. Write a ratio for each so that they can be compared to the given ratio for farm-raised tilapia.

Table 1

Number of Omega Fatty Acids in a Tablespoon of Fish Oil

Fish Oil	Long-Chain Omega-6 Fatty Acids	Long-Chain Omega-3 Fatty Acids
Herring	39	1509
Salmon	92	4657
Sardine	239	3096
Cod Liver	127	2557
Menhaden	159	3624

Source: EFA Education. "Essential Fats in Food Oils." http://efaeducation.org/essential.html

Solution

In the information given, we were told that the ratio $11:1$ for farm-raised tilapia was the ratio of long-chain omega-6 fatty acids to that of long-chain omega-3 fatty acids. So, for comparison we need to list the omega-6 fatty acids first in each of the new ratios. For each ratio, all we need to do is write the numbers in the order they appear in the table, since the omega-6 values are listed first.

Table 2

Ratio of Omega Fatty Acids in Fish Oil

Fish Oil	Long-Chain Omega-6 Fatty Acids	Long-Chain Omega-3 Fatty Acids	Ratio
Herring	39	1509	39:1509
Salmon	92	4657	92:4657
Sardine	239	3096	239:3096
Cod Liver	127	2557	127:2557
Menhaden	159	3624	159:3624

Now that we've written the ratios, to be able to compare them to the ratio of 11 : 1 for farm-raised tilapia, we need to reduce them all to a common form. Since the farm-raised tilapia has a "1" on the right side of the ratio, we'll divide each ratio by the right-hand number so that it also will have a 1. You can think of this as being in the same vein as converting to a unit rate.

Table 3			
Unit Ratio of Omega Fatty Acids in Fish Oil			
Fish Oil	**Ratio**		**Unit Ratio**
Herring	39 : 1509	(39 / 1509) : (1509 / 1509)	0.0258 : 1
Salmon	92 : 4657	(92 / 4657) : (4657 / 4657)	0.0198 : 1
Sardine	239 : 3096	(239 / 3096) : (3096 / 3096)	0.0772 : 1
Cod Liver	127 : 2557	(127 / 2557) : (2557 / 2557)	0.0497 : 1
Menhaden	159 : 3624	(159 / 3624) : (3624 / 3624)	0.0439 : 1
Tilapia			11 : 1

Now that all the ratios are in a common form, we can see that there is at least a hundred-fold difference between the fatty acid ratios of tilapia and the other fish.

Another common use of ratios is to scale quantities either up or down. For example, the distance on a map between your house and the beach is given in inches. In order to find the distance in terms of miles, you can scale it up using the key shown on the map. We have already seen that we can reduce ratios to equivalent ones by dividing both quantities by the same number. It is also the case that we can multiply both quantities by a number to make an equivalent ratio.

Example 4: Using Ratios for Scaling

Using the given map, answer the following questions.

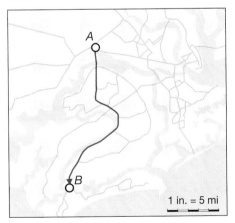

a. If the distance on the map between point A and point B measures 3 inches, what is the actual distance between these points?

b. If you wanted to take an 11-mile walk one day, what distance would that be on the map, in inches?

4

Solution

a. Because the distance on the map is 3 inches, and the scale of the map is 1 inch = 5 miles, we can calculate the actual distance x between points A and B by setting up a proportional equation. In the equation, we'll put inches in the numerator and miles in the denominator of each ratio. Note that it makes no difference which unit we choose to place in the numerator.

$$\frac{1 \text{ inch}}{5 \text{ miles}} = \frac{3 \text{ inches}}{x \text{ miles}}$$

We can now solve the equation for x by multiplying each side of the equation as shown.

$$\frac{1}{5} = \frac{3}{x}$$
$$\frac{1}{\cancel{5}} \cdot \cancel{5}x = \frac{3}{\cancel{x}} \cdot 5\cancel{x}$$
$$x = 15$$

Therefore, the actual distance between the points is 15 miles.

b. Again, we use the scale of 1 inch = 5 miles to find the distance on the map. The proportional equation will have the variable in the numerator this time since we are looking to find the distance in inches the map would show.

$$\frac{1 \text{ inch}}{5 \text{ miles}} = \frac{x \text{ inches}}{11 \text{ miles}}$$
$$\frac{1}{\cancel{5}} \cdot \cancel{5} \cdot 11 = \frac{x}{\cancel{11}} \cdot 5 \cdot \cancel{11}$$
$$11 = 5x$$
$$\frac{11}{5} = \frac{\cancel{5}x}{\cancel{5}}$$
$$2.2 = x$$

Thus, the distance of your walk would measure 2.2 inches on the map.

Example 5: Scaling Ratios

Suppose we know that the distance from Chicago, Illinois, to Albany, New York, is 816 miles. On the map, the distance from Chicago to New York is 1.4 inches. Use this ratio to determine how far it is from Chicago to St. Louis, Missouri, if the distance on the map between these two cities is 0.5 inches.

Solution

We can use the distances given in the scenario to help us set up ratios in an equation that we can solve. First of all we are told that the actual distance between Chicago and Albany is 816 miles and that the map distance is 1.4 inches. So the ratio for Chicago to Albany is

$$\frac{816 \text{ miles}}{1.4 \text{ inches}}.$$

We can then fill in the ratio for Chicago to St. Louis based on what we know.

$$\frac{x \text{ miles}}{0.5 \text{ inches}}$$

Notice we were careful to put the actual distances between cities in the numerator of each ratio while the map distances are in the denominator. We chose this orientation so that the distance we are looking to find is once again in the numerator. The proportional equation is then

$$\frac{816 \text{ miles}}{1.4 \text{ inches}} = \frac{x \text{ miles}}{0.5 \text{ inches}}.$$

We can solve this proportional equation as follows.

$$\frac{816 \text{ miles}}{1.4 \text{ inches}} = \frac{x \text{ miles}}{0.5 \text{ inches}}$$

$$\frac{816}{\cancel{1.4}} \cdot 0.5 \cdot \cancel{1.4} = \frac{x}{\cancel{0.5}} \cdot \cancel{0.5} \cdot 1.4$$

$$408 = 1.4x$$

$$291.43 \approx x$$

Therefore, the distance from Chicago to St. Louis is approximately 291 miles.

So far, we have only seen ratios involving two components. However, there is a shorthand for ratios using any number of parts. For instance, the ratio $3:5:7$ is read "three to five to seven." Three 2-component ratios are inferred by this notation as follows.

$$3:5:7$$

Ratio 1: $3:5$

Ratio 2: $5:7$

Ratio 3: $3:7$

Example 6: Using Multiple Ratios

Raw mixed nuts make a healthy and delicious snack. An online store promises raw cashews, raw almonds, and raw Brazil nuts in a mixed bag of nuts. They go even further to say that the ratio of cashews, almonds, and Brazil nuts is $2:3:5$ in every bag. Suppose that there are 5 ounces of cashews in a bag. How many ounces of almonds and Brazil nuts will the bag have?

Solution

We are told that the ratio $2:3:5$ is that of *cashews : almonds : Brazil nuts*, and that there are 5 ounces of cashews. Using the first part of the ratio, *cashews : almonds*, we can find the amount of almonds that will be in the bag.

The *cashews : almonds* ratio is $2:3$, or $\dfrac{\text{cashews}}{\text{almonds}} = \dfrac{2}{3}$.

Setting up a proportional equation using the variable x for the unknown amount of almonds gives us the following proportional equation to solve.

$$\frac{2}{3} = \frac{5}{x}$$

Using algebra, we get x on one side of the equation by itself as follows.

$$\cancel{3}x \cdot \left(\frac{2}{\cancel{3}}\right) = \left(\frac{5}{\cancel{x}}\right) \cdot 3\cancel{x}$$
$$2x = 15$$
$$x = 7.5$$

So, there are 7.5 ounces of almonds in the bag.

Now, we can use the *cashews* : *Brazil nuts* ratio of $2:5$ to find the amount of Brazil nuts in the bag. Use the same method to find x.

$$\cancel{5}x \cdot \left(\frac{2}{\cancel{5}}\right) = \left(\frac{5}{\cancel{x}}\right) \cdot 5\cancel{x}$$
$$2x = 25$$
$$x = 12.5$$

So, there are 12.5 ounces of Brazil nuts in the bag. In fact, we now know that, for this particular bag, the ratio $2:3:5$ of *cashews* : *almonds* : *Brazil nuts* is the same as the ratio $5:7.5:12.5$.

Example 7: Using Ratios to Find Quantities

In a hardware store, bags of miscellaneous nails are sold at a discount price. If a bag containing 300 nails has a $4:5:6$ ratio of nails in decreasing lengths (long, medium, short), how many of each nail should the bag have?

Solution

In this situation, the ratio $4:5:6$ indicates how to partition every set of $4 + 5 + 6 = 15$ nails. Each set of 15 nails should have 4 long, 5 medium, and 6 short nails, as shown in the following ratios.

$$\text{Long nails}: \quad \frac{4}{15}$$
$$\text{Medium nails}: \quad \frac{5}{15}$$
$$\text{Short nails}: \quad \frac{6}{15}$$

We can use these ratios to set up a proportional equation for each length of nail and solve to find the amounts needed. We then have the following 3 proportional equations.

Long nails: $\qquad \dfrac{4}{15} = \dfrac{x}{300}$

$$\frac{4}{\cancel{15}}^{20} \cdot \cancel{300} = \frac{x}{\cancel{300}} \cdot \cancel{300}$$
$$80 = x$$

Medium nails: $\qquad \dfrac{5}{15} = \dfrac{x}{300}$

$$\frac{5}{\cancel{15}}^{20} \cdot \cancel{300} = \frac{x}{\cancel{300}} \cdot \cancel{300}$$
$$100 = x$$

Short nails:

$$\frac{6}{15} = \frac{x}{300}$$

$$\frac{6}{\cancel{15}} \cdot \cancel{300}^{20} = \frac{x}{\cancel{300}} \cdot \cancel{300}$$

$$120 = x$$

Note that as a check, 80 long nails + 100 medium nails + 120 short nails adds up to 300, the total number of nails in each bag.

Skill Check Answer

1. $3:2$

4.2   Exercises

Fill in each missing value in the given equivalent ratios.

1.   The ratio of 1.2 to 32 is the same as the ratio of 3.6 to____.

2.   **a.** $1:7 = \underline{\quad}:14$

   b. $1:7 = \underline{\quad}:21$

   c. $1:7 = 4:\underline{\quad}$

   d. $1:7 = 5:\underline{\quad}$

3.   **a.** $2:5 = 4:\underline{\quad}$

   b. $2:5 = \underline{\quad}:35$

4.   **a.** $10:3 = \underline{\quad}:9$

   b. $10:3 = 5:\underline{\quad}$

5.   **a.** $5:9 = 2.5:\underline{\quad}$

   b. $5:9 = \underline{\quad}:13.5$

Write a ratio to represent each situation. Be sure to simplify each ratio.

6.   If 9 of the 21 members of a tour group are children, what is the ratio of children to adults on the trip?

7.   Triangles A and B are *similar* triangles. In other words, triangle B has exactly the same shape as Triangle A, just a different scale. Triangle A has sides 45 cm, 54 cm, and 99 cm. If Triangle B has sides of lengths 90 cm, 108 cm, and 198 cm, what is the ratio of the lengths of the sides of Triangle A to the lengths of the sides of Triangle B?

8.   A bag contains the following marbles: 8 black marbles, 17 blue marbles, 7 brown marbles, and 14 green marbles.

   a. What is the ratio of brown marbles to black marbles?

   b. What is the ratio of brown marbles to all the marbles in the bag?

9.   A recent study showed that in American universities there were 18 million college students and 1.3 million faculty. Write the overall student to faculty ratio for universities across America in lowest terms.

10.   Christopher had acquired 12 of the 20 items requested by his supervisor. Write the ratio of acquired items to nonacquired items in lowest terms.

Solve each problem. Be sure to simplify your answer to lowest terms. Round your answer to the nearest tenth when necessary.

11.   If the ratio of tourists to locals is 2 : 3 and there are 40 tourists at the opening of a new museum, how many locals are at the opening?

12.   A powdered drink mix calls for 3 scoops powder to 8 ounces water. How much powder do you need if you have 32 ounces of water?

13.   It was recently estimated that uninsured vehicles outnumber insured ones by about six to five. If there are 2002 vehicles in a county, how many of them are uninsured?

14.   One group (A) contains 125 people. One-fifth of the people in group A will be selected to win $15.00 grocery cards. There is another group (B) in a nearby town that will receive the same number of grocery cards, but there are 585 people in that group.

   a. What will be the ratio of nonwinners in group A to nonwinners in group B after the selections are made?

   b. What is the ratio of winners to people for group B?

15.   A recipe for trail mix calls for a ratio of 4 to 1 of M&M's to sunflower seeds by weight. How many ounces of sunflower seeds will there be in this mixture if there are 2 pounds of M&M's? (**Hint:** There are 16 ounces in 1 pound.)

16.   The local election is over and a new city council member has been elected. The new councilwoman received three votes for every vote received by her opponent. The new councilwoman received 2058 votes. How many votes did her opponent receive?

17.   In a certain desert environment there are a lot of small rodents. There also happen to be many snakes that feed on the rodents. The ratio of rodents to rodent-eating snakes is 13 to 5. If there are 4000 snakes in the area, how many rodents are there?

18. If the ratio of difficult math problems to easy math problems in a book is 4 to 3, how many easy problems are there if there are 360 problems in the book? Round your answer to the nearest whole number.

19. Find the actual width of a building if the model of the building is 5 cm wide by 58.7 cm long and the actual length of the building is 140.9 m.

20. The scale of a map is 1 in. = 11.9 miles. How many inches will be drawn on the map to represent 30.9 miles?

21. The following is the list of ingredients for a sugar cookie recipe that yields 2 dozen cookies. Convert the entire sugar cookie recipe to one that makes 5 dozen cookies.

3 cups sifted all-purpose flour

$1\frac{1}{2}$ teaspoons baking powder

$\frac{1}{2}$ teaspoon salt

1 cup white sugar

1 cup butter (softened at room temperature)

1 egg, lightly beaten (egg should be at room temperature)

3 tablespoons half-and-half

2 teaspoons vanilla extract

22. The ratio of outdoor swimmers to indoor swimmers at the recreation center is $3:2$. If the center has 55 swimmers altogether, how many of them are indoor swimmers?

23. Andrew wins 4 fencing bouts for every 3 that he loses. If he had 35 bouts during the last year, how many bouts did he win?

24. At a recent education conference, one-fourth of the total participants came to learn about online courses. The rest came to learn about other educational practices. If 92 participants showed up, how many were *not* there to learn about online courses?

25. Chloe practices piano an average of 8 hours per week. Patricia practices figure skating an average of 1.2 hours for every hour Chloe practices. How many hours a week does Patricia practice?

26. Minimundus, in Klagenfurt, Austria, is an outdoor museum that is home to 150 miniature models of prominent architecture from all over the globe. The museum has everything from the Eiffel Tower to St. Peter's Basilica to a US southern plantation home. Each replica is a scaled version of the original with a ratio of $1:25$. Whenever possible, the same building materials used for the actual building are used in all of the models.

a. Paris' Eiffel Tower is 1063 ft tall. How tall is the model tower in Minimundus?

b. The model of the Sagrada Familia Basilica at Minimundus stands 22.4 ft tall. Find the height of the actual basilica in Barcelona, Spain.

27. An online website created a model of our solar system. Their page says that "Unlike most models, which are compressed to a single page for viewing convenience, the planets are shown at their true-to-scale average *distances* from the sun. This makes the website page rather large—on an ordinary 72-dpi monitor, it is just over half a mile wide, making it possibly one of the largest pages on the web." [2]

    a. If the sun's diameter is 4.875 inches on the screen and 870,000 miles in reality, what is the scale ratio of the website?

    b. If Saturn is represented as 0.375 inches on the website, what is its diameter in miles?

28. A Cuda Dragster car model has a scale of $1:25$. If the tire height of the model is 1.06 inches, what is the height of the life-size tire?

29. The Clarksville football team outscored its opponents $5:3$ last season. If their opponents scored 78 points, how many points did Clarksville score?

30. In music, when going up an octave you double the frequency. For instance, if one note has a frequency of 400 hertz (Hz), the note an octave above it is at 800 Hz, and the note an octave below it is at 200 Hz.

    a. What is the ratio of frequencies of two notes an octave apart?

    b. If a note has a frequency of 246.94, what is the frequency of the note an octave above it? An octave below it?

31. Instructions for a chemical procedure state to mix salt, baking soda, and water in a $4:5:21$ ratio by mass. How many grams of water would be required to make a mixture that contains 60 grams of salt?

2 Phrenopolis, "Solar System Perspective," http://www.phrenopolis.com/perspective/solarsystem/

4.3 Proportions and Percentages

The word "proportion" has a statistical meaning as well as an algebraic one. As we noted in the first section of this chapter, a proportional equation in algebra relates two rates or ratios. That is, when we set two rates or ratios equal to each other, we call the resulting equation a proportional equation. Solving a proportional equation helped us with ratios. Let's shift gears a little and look at what a proportion means in a statistical sense. Statistically, the term proportion refers to a fraction of a whole.

> ## Proportion
> A **proportion** is a fraction of a whole.

Contemporary media often refer to what fraction of the population votes for one party or another, buys a certain car, runs a marathon, or even develops an illness.

Only 1 in 200 Americans Have Run a Marathon

Flu Outbreak Affects 98% of Americans

Congressman's Re-election Bid In Trouble

Mazda's *Mazda 3* Best Selling Car in America

When we considered ratios in the previous section, we were concerned with comparing quantities to each other. But with proportions, we are comparing a portion of a group to the larger whole. For instance, consider the following information regarding car sales in April 2014.

Table 1	
April 2014 Car Sales	
Car Type	**# of Cars Sold**
Midsize	311,328
Small	267,651
Luxury	97,007
Large	288
Total	**676,274**
Source: The Wall Street Journal, "What's Moving: U.S. Auto Sales," accessed May 2014, http://online.wsj.com/mdc/public/page/2_3022-autosales.html	

Consider the car type labeled "Midsize." We can calculate both ratios and proportions involving midsize car sales. One of many ratios that can be calculated from the table is the ratio of midsize car sales to that of luxury car sales, which is given by the comparison of the two car types, namely, $311{,}328 : 97{,}007$. However, the *proportion* of midsize cars sold in April is the fraction that compares the number of midsize cars to the total number of cars sold in April 2014.

The proportion of midsize car sales is therefore $\dfrac{311{,}328}{676{,}274}$.

You can immediately see that both of the numbers in the proportion are divisible by 2, so the fraction can be rewritten as $\dfrac{155{,}664}{338{,}137}$, its simplest form.

Think Back

Recall that a percentage is a rate per 100.

Proportions like the one above are often written as percentages. Percentages give us a common way to compare fractions. When fractions don't have the same denominator, it is sometimes difficult to

4

compare them. However, with percentages the denominator is always 100, lending itself easily to comparison. To convert a fraction to a percentage, we divide the numerator by the denominator and multiply by 100%. Fractions, percentages, and decimals are all ways of expressing proportions. The fraction $\frac{3}{8}$ is the same as the decimal 0.375, which is the same as the percentage 37.5%.

Example 1: Changing Fractions to Percentages

Change the fraction of midsize car sales $\dfrac{155,664}{338,137}$ to a percentage.

Solution

Divide the numerator by the denominator and multiply by 100% to find the percentage.

$$\frac{155,664}{338,137} \cdot 100\% \approx 46.04\%$$

This tells us that in Table 1, 46.04% of the cars sold were midsize cars. In other words, about 46 out of every 100 cars sold were midsize cars.

Skill Check # I

Change the fraction $\frac{75}{234}$ to a percentage.

Example 2: Finding Proportions and Percentages

The following sample of adults between the ages of 25 and 34 reported whether they had a college degree or not. What proportion of these adults reported having a college degree? Convert the proportion to a percentage.

Table 2	
Adults Between the Ages of 25 and 34	
	Have College Degree?
Yes	373
No	629
Total	**1002**

Solution

The proportion of adults age 25 to 34 with a college degree is $\dfrac{373}{1002}$.

To write this proportion as a percentage, we calculate

$$\frac{373}{1002} \cdot 100\% \approx 37.23\%.$$

Example 3: Using Proportions to Find Quantities

The proportion of people who are left-handed is approximately $\frac{1}{11}$. If there are 45 people in a room, estimate how many of them are left-handed.

Solution

We want to know what $\frac{1}{11}$ *of* 45 is. Remember that *of* in mathematics implies multiplication. So, we can simply multiply the two quantities together.

$$\frac{1}{11} \cdot 45 = 4.\overline{09}$$

In other words, we would expect there to be about 4 left-handed people in a room of 45.

We can easily move between ratios and proportions by thinking carefully about what ratios convey. Remember that a ratio of $4:7$ means that every 11 units $(4 + 7)$ are comprised of 4 of the first kind and 7 of the second kind. In the next example, we will consider finding a proportion from a given ratio.

Example 4: Working with Ratios and Proportions

☞ **Helpful Hint**

In order to change a percentage to a decimal number, drop the percent sign and divide by 100. This can be done by moving the decimal point two places to the left. For example, 78.1% = 0.781.

Recall Example 1 in Section 4.2 about the university campus, which had a new freshman class with a ratio of $2:3$, men to women. University-wide, 53% of the campus population is men. Is the proportion of men in the freshman class different from that of men campus-wide?

Solution

In order to solve this problem we need to compare the percentage of men in the freshman class to that of the entire campus. We know the entire campus has 53% men. To begin, we need to find the proportion of freshmen men. We can do this using the number of men and the total number of students in the freshman class, or we can do this with ratios. Because we know the ratio of men to women in the freshman class is $2:3$, we know that out of every $2 + 3 = 5$ students in the freshman class, 2 of them are men. Therefore, the proportion of men in the freshman class is

$$\frac{2}{5} \text{ or } 40\%.$$

Now we can compare the percentage of men in the incoming freshman class to the percentage of men campus-wide, which is 53%. It is now clear that the proportion of men is smaller in the freshman class.

We can also take percentages or proportions that we know and convert back to the original amounts.

Example 5: Using Proportions and Percentages

The Earth consists of three parts: crust, mantle and core. The Earth's core makes up 31.5% of the Earth's mass. The mantle makes up 68.1% of its mass. Although 20 to 30 miles thick, the crust of the Earth only makes up a small proportion of its mass—the remaining 0.4%. If the total mass of the Earth is 5.9742×10^{24} kg, what is the mass of the Earth's crust?

4

For very large or very small numbers, an alternative method to express decimals, called scientific notation, is often used. Scientific notation requires that we rewrite a number to be between 1 and 10 and then multiply the number by a power of 10 equal to the number of decimal places represented. So, 3,658,000,000,000,000 can be rewritten in scientific notation as 3.658×10^{15}. Small numbers less than 1 will have negative powers of 10. For example, 0.0000223 is written 2.23×10^{-5}.

Numbers in scientific notation can be multiplied by multiplying the respective parts of the notation together and using the product rule for exponents.

For example, the two numbers above are multiplied together as follows.

$(3.658 \times 10^{15})(2.23 \times 10^{-5})$

$= (3.658 \times 2.23)(10^{15} \cdot 10^{-5})$

$= 8.15734 \times 10^{10}$

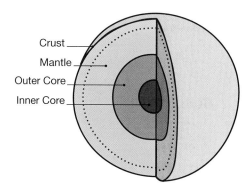

Solution

We are given the fact that the Earth has a total mass of 5.9742×10^{24} kg, and its crust is 0.4% of the overall mass. We simply need to multiply the two quantities together to find the amount of mass that is crust. Because we are given the crust as a percentage, we need to first write it as a decimal before multiplying the two together.

$$\left(5.9742 \times 10^{24} \text{ kg}\right)(0.4\%) = \left(5.9742 \times 10^{24} \text{ kg}\right)(0.004)$$
$$= \left(5.9742 \times 10^{24} \text{ kg}\right)\left(4 \times 10^{-3}\right)$$
$$= 23.8968 \times 10^{21} \text{ kg}$$
$$= 2.38968 \times 10^{22} \text{ kg}$$

So, the Earth's mass is comprised of 23,896,800,000,000,000,000,000 kg of crust, which is a mere 0.4% of its total mass!

Example 6: Finding Proportions and Ratios

Suppose you earned a B on your last History 102 test by getting 80% of the questions correct.

a. If there were 60 equally weighted questions on the test, how many did you answer correctly?

b. What was your ratio of correct to incorrect answers on the test?

Solution

a. Since there were 60 questions on the test, we need to find 80% of 60. To do this, we first change 80% to a decimal, and then multiply.

$$(80\%)(60) = (0.80)(60)$$
$$= 48$$

Therefore, you answered 48 questions correctly on the test.

b. Since we know from part **a.** that you answered 48 questions correctly, we can subtract this from the total number of questions to find the number of questions that you answered incorrectly.

$$60 - 48 = 12 \text{ incorrect questions.}$$

Therefore, the ratio *correct : incorrect* would be written $48 : 12$ or $4 : 1$.

Example 7: Using Percentages

Elaine was very excited to find that the watch she has wanted was on sale for $66.50. The sale price indicated that it was 30% off of the original price. Find the original price of Elaine's watch.

Solution

If the watch is discounted 30%, then Elaine will only pay 70% of the original price of the watch. We can set up a proportional equation to find the original price x of the watch.

$$\frac{\$66.50}{x} = \frac{70}{100}$$

By solving this equation, we can find the original price.

$$\frac{\$66.50}{x} = \frac{70}{100}$$

$$\frac{\$66.50}{x} \cdot 100\,x = \frac{70}{100} \cdot 100\,x$$

$$\$6650 = 70x$$

$$\$95 = x$$

So the watch was originally $95.00.

Skill Check Answer

1. 32.05%

4.3 Exercises

Write each proportion in lowest terms.

1. If there are 48 students in a class and 32 are seniors, what proportion of students are *not* seniors?

2. In a survey, 99 out of 100 people are *not* left-handed. What proportion of people are left-handed according to this survey?

3. Nearly two-in-ten American women end their childbearing years without having borne a child. What proportion of American women end their childbearing years with having borne a child?

4. A survey indicated that 11 out of 100 students studying at a university reported having a disability. What proportion of students did *not* report having a disability?

Change each fraction to a percentage. Round your answer to the nearest hundredth when necessary.

5. $\dfrac{20}{45}$ **6.** $\dfrac{1}{5}$

7. $\dfrac{2}{3}$ **8.** $\dfrac{61}{212}$

Answer each question.

9. 312 is 65% of what number?

10. Suppose 12% of students chose to study French their freshman year, and that meant that there were 21 such students. How many students chose *not* to take French their freshman year?

11. Which of the following is equal to 95%?

 a. $\dfrac{19}{20}$

 b. $\dfrac{18}{20}$

 c. $\dfrac{18}{25}$

12. If two out of every seven individuals in a population carry a gene for a defective enzyme, what ratio of individuals carry the normal gene?

13. There are four hundred participants in the local Thanksgiving 5K race. If 51% of the participants are female, what is the ratio of female participants to male participants?

14. If the drinks portion of a bill costs $18 and the total bill was $30, what percentage of the total was the cost of the drinks?

15. Which is greater, half of 22 or one-fifth of 95?

16. Which is greater, a third of 36 or one-sixth of 138?

17. Thirty-four is 20% of what number?

18. Sixty percent of 340 is what number?

19. There is a herd of cattle out on the range. Fifty-two percent of the cattle are male. If the herd consists of 1175 animals, how many are female?

20. Seventeen percent of the plants in the greenhouse are broad leaf plants and the rest are grasses. What percentage of the plants are grasses?

21. Seventy percent of the animals in a zoo are herbivores. If there are 821 animals in the zoo, how many are *not* herbivores? Round your answer to the nearest whole number.

22. A sample of ore is found to be 0.0022% gold and 0.059% silver. What is the percentage of matter in the ore that is neither gold nor silver?

23. Three-fourths of the students at the local community college live in the county. If there are 2004 students enrolled, how many do *not* live in the county?

24. A website reports that 14% of the world's population is Hindi and 33% are Christian. If there are approximately 6,840,507,000 people in the world, how many are Hindi and how many are Christian according to the website?

4.4   Using Percentages

When purchasing a new stereo or a new pair of shoes, everyone loves to hear a salesperson say, "that item is 20% off today." Have you ever been in a store that was having a 50% off sale and was also advertising a sale with an additional 50% off? Does that mean the item would be free? Of course not! Stores cannot afford to give things away. Having a firm grasp of the concepts of percentages makes being a smart consumer much easier. In addition, being able to navigate through a world of percentages in a confident manner is an important step in becoming a successful member of society. Just like learning to play an instrument or becoming a top athlete requires constant practice and repetition, one of the best ways to learn how to work with percentages or fractions is to practice using them over and over. In this section, the primary focus will be to improve your skills with percentages by working through several examples involving percentages and by pointing out some useful tips along the way.

Service Tipping

Think Back

Remember, to find 10% of an amount, move the decimal point one place to the left.

The "Worldwide Tipping Guide" suggests that not every country in the world considers it a good thing to tip servers at restaurants.[1] In fact, tipping is seen as an insult in some countries. However, according to the guide, the United States has the highest expectations for tipping of any other country at 15-20%. Even though there are new phone applications that allow you to calculate a tip quickly, it is smart to have a general idea of what the amount should be in case you inadvertently type in the wrong amount. Suppose you want to always tip 20%. The easiest way to quickly calculate this amount is to start with 10% of the bill and then double it. Remember from earlier math courses that to find 10% of a number, you move the decimal point one place to the left.

For instance,

$$10\% \text{ of } \$25.00 = \$2\underset{\smile}{.}5.00 = \$2.50$$
$$10\% \text{ of } \$349.00 = \$34\underset{\smile}{.}9.00 = \$34.90$$
$$10\% \text{ of } \$1275.00 = \$127\underset{\smile}{.}5.00 = \$127.50.$$

Once you know 10% of the total amount, you can find twenty percent by doubling the amount. So, we have

$$20\% \text{ of } \$25.00 = 2 \cdot \left(\$2.50\right) = \$5.00$$
$$20\% \text{ of } \$349.00 = 2 \cdot \left(\$34.90\right) = \$69.80$$
$$20\% \text{ of } \$1275.00 = 2 \cdot \left(\$127.50\right) = \$255.00.$$

The last bill is quite large, so 20% seems rather much for a tip. Instead, you decide for this party to tip only 15%. No problem. You still know your base amount of 10%, so to find 15% we need to add half of that amount again.

$$15\% \text{ of } \$1275.00 = \overset{(10\%)}{\$127.50} + \frac{\overset{(5\%)}{\$127.50}}{2}$$
$$= \$127.50 + \$63.75$$
$$= \$191.25$$

A similar idea works for 5% tips as well. Start with 10% of the amount and then halve it to find 5% of an amount quickly. Since tipping is usually in increments of 5%, you'll be able to mentally calculate tips quicker than your friend using their cell phone. You'll soon become the one everyone wants at their table when the bill comes.

---

1 Magellan's, http://www.magellans.com

Skill Check #1

Suppose your lunch bill is $9.84. Approximately how much would a 15% tip be? 20%?

Sales Tax

Sales tax on goods in the United States is not typically included in the advertised sticker price. Instead, it is added on at the checkout register. In 2014, state sales tax ranged from 2.9% in Colorado to 7% in several states. [2]

Unlike tipping amounts, sales tax isn't usually in manageable increments of 5% that you can easily calculate in your head. One way to find the price including tax is to calculate the tax amount and then add it back on to the original price. However, since we're probably using some type of calculator to do the calculation, there is an easier method. Suppose the tax that is to be added on is 4.2%. You can simply calculate $100\% + 4.2\% = 104.2\%$ of the original price to arrive at the final price including tax. Finding the total price of a $25.00 item with 4.2% sales tax using both methods gives the following equivalent results.

Calculating sales tax and then adding it on:

$$(\$25.00)(0.042) = \$1.05$$

$$\$25.00 + \$1.05 = \$26.05$$

Calculating 104.2% of the item:

$$(\$25.00)(1.042) = \$26.05$$

Skill Check #2

Find the total price of a new 32" TV that sells for $219.99 if purchased in Georgia, where the sales tax is 4%, and if purchased in Kentucky, where the sales tax is 6%.

Sale Prices

When finding sale prices you can either think of it in terms of the "tipping" method or the "sales tax" method we mentioned earlier. Since sale discounts usually occur in increments of 5%, we like to think of it in the same manner as the "tipping" method—just subtracting instead of adding at the end!

End of Season
40% OFF SALE

For example, suppose you found the perfect sofa for your new apartment. The store is having its annual *End of Season* sale and offering 40% off on its entire stock. Assuming you are in the store without a calculator, estimate the discounted price of your dream sofa if the original price was $496.00.

We start just as we did with tipping and find our base amount of 10% by moving the decimal point one place to the left.

$$10\% \text{ of } \$496.00 = \$49.60$$

Since this is an in-store estimate for yourself, we can simplify the calculation by rounding this discount to $50. So, if 10% is approximately $50 off, 40% would equate to four $50 discounts, giving you $200 off! Your dream sofa now only costs approximately $296.00—a bargain you can't afford to miss. Another way to estimate the discount is to round the cost of the sofa from $496.00

☞ Helpful Hint

Finding 40% off of a number is equivalent to finding 60% of the number. Sometimes it is easier to find the final cost directly rather than finding the amount of the discount and then subtracting.

2 Sales Tax Institute, http://www.salestaxinstitute.com

to $500.00. Then, 10% of $500.00 would be $50.00. A discount of 40% would then equate to four $50.00 discounts, or $200.00 off. Either way, the discount is approximately $200.

Suppose the sales assistant offers you an additional 10% off if you pay cash in the store that day. What would your final cost be then?

What you need to keep in mind here is that the additional discount is 10% off the sale price, *not* 10% off the original price. Let's calculate 10% of the estimated sale price by moving the decimal point.

$$10\% \text{ of } \$296.00 = \$29.60, \text{ or almost } \$30$$

Alternatively, we can round $296.00 to $300, and then calculate 10% to estimate the discount.

$$10\% \text{ of } \$300 = \$30$$

So, the additional discount means that if you pay cash, the price of your dream sofa would be about

$$\$296.00 - \$30 = \$266.00, \text{ or about}$$

$$\$300.00 - \$30 = \$270.00.$$

Be careful not to be misled into thinking that 40% off the original price, plus an additional 10% off is the same as receiving 50% off the original price. If the sofa was 50% off to begin with, you could just halve the original price to find the sale price.

50% off the original price:

$$\frac{\$496.00}{2} = \$248.00.$$

However, we've just seen that 40% off plus an additional 10% off is approximately $266.00. It is important to remember that in this situation you cannot just add the percentages together.

Skill Check #3

Find the sale price of a new sweater that is 40% off the original price of $88.00.

Just as it is important to know that in certain situations you can't add percentages together, as in the last example, it's worth noting that you can't always subtract percentages either. For instance, suppose a store determines the retail price for items by marking them up 50% over the wholesale price. Later they decide to put the items on sale, and the store manager thinks that if he reduces the retail price by 50% at least he will come out even. It might be tempting to believe that by adding 50% and then subtracting 50%, the store is right back at the wholesale price again. However, this is not the case. In our example, a $100 wholesale item costs $150 retail, since the price is marked up by 50%. If the sale price is 50% off, then the $150 item will be on sale for $75. That's less than the wholesale price the store paid originally! The store manager might be looking for a new job if he doesn't understand percentages. Beware of believing that a percentage increase followed by the same percentage decrease leads back to the original amount. We will look more closely at percentage increases and decreases next.

Percentage Increase and Percentage Decrease

The three topics we just covered (tipping, sales tax and sale prices) can also be thought of as *percentage increases* or *percentage decreases*. In other words, we actually changed the original amount by either adding or subtracting a certain percentage of that amount. In each case, we were told by what percentage we should either increase or decrease the amount. However, sometimes we might be given two amounts, an original price and a sale price, and want to calculate the percentage change. You can use the following formulas to find any percentage increase or decrease.

Think Back

The absolute value of any number is the distance that number lies from zero without regard to sign. For example, the absolute value of 8 is 8, and the absolute value of –8 is also 8. To indicate absolute value symbolically, we use a vertical bar on both sides of the number, that is, $|8| = 8$ and $|-8| = 8$.

Absolute Change

The **absolute change** between two amounts is the absolute value of the difference between the two numbers.

$$\text{absolute change} = |\text{new amount} - \text{original amount}|$$

Percentage Change

$$\text{percentage change} = \frac{\text{absolute change}}{\text{original amount}} \cdot 100\%$$

Example 1: Finding the Percentage Change

An on-campus organization is trying to increase its monthly attendance.

a. If after a big recruiting push, the meeting attendance increased from 25 to 38 people, what was the percentage change in attendance?

b. During the third meeting, only 31 people attended. What was the percentage change from the second meeting to the third?

c. For advertising purposes, the organization decides to promote the overall increase in attendance for the two-month span since it has increased from the original number of attendees. Calculate the percentage change from meeting 1 to meeting 3.

Solution

a. To find the percentage change, we need to know the *absolute change* in attendance between the first meeting and the second meeting. Use subtraction to find this value.

$$|38 - 25| = 13 \text{ more people came to the second meeting}$$

Next, substitute the appropriate values into the equation to find the percentage change.

$$\frac{\text{absolute change}}{\text{original amount}} \cdot 100\% = \frac{13}{25} \cdot 100\% = 52\%$$

Since the attendance had an increase from the original amount, we say there was a 52% increase in attendance.

b. Now, if at the next meeting only 31 people attend, what is the percentage change from meeting 2 to meeting 3?

This time, the absolute change in attendance is as follows.

$$\text{absolute change} = |\text{meeting 3} - \text{meeting 2}| = |31 - 38| = 7$$

Substituting this into our equation, we have

$$\frac{7}{38} \cdot 100\% \approx 18.42\%.$$

Since this was a decline in attendance, we say there was an 18.42% decrease from meeting 2 to meeting 3.

c. Recall that, in meeting 1, there were 25 people in attendance and in meeting 3, there were 31 people. So we need to subtract these numbers to find the absolute change in attendance.

$$\text{absolute change} = |\text{meeting 3} - \text{meeting 1}| = |31 - 25| = 6$$

So, our percentage change equation becomes

$$\frac{6}{25} \cdot 100\% = 24\%.$$

Since this was an increase in attendance from meeting 1 to meeting 3, we say there was a 24% increase overall.

Let us note something that *cannot* be done in the situation from Example 1. It might be tempting to rationalize that since there was a 52% increase the first meeting followed by an 18.42% decrease the second meeting, we could just subtract the two percentages to find the overall percentage change. However, we have just shown that not to be the case. Remember that you cannot always manipulate percentages by adding or subtracting them. It is such a misinterpretation of percentage change that we are not even going to write down the subtraction, lest someone should mistakenly see it written as an accepted form of mathematics!

Percentage Faux Pas

We have seen that percentages are a very useful and informative tool to navigate through today's world. However, percentages are commonly used incorrectly and lead people down the wrong path. As you become more adept at working with percentages, you can be careful to avoid making these mistakes yourself, and you can avoid being fooled by abuses of percentages in the future. The following is a look at some of the more common misuses involving percentages.

Comparing Percentages of Groups of Very Different Sizes

When using percentages to compare things, it is easy to make errors if we are not careful and only look at the percentages and not at the size they represent. Suppose you were told that 75% of people living in Alaska favor free skis for school children while only 40% of Californians do. While it is true that a greater proportion of Alaskans favor free skis, it would be wrong to conclude from these percentages that more Alaskans are in favor of the free skis than Californians. To find out why, let us consider what the percentages are referring to.

Alaska's population in 2010 was around 710,000, while California's population was 37,000,000. Given that information, we can find the actual number of people in favor of free skis in both states.

Alaska: 75% of 710,000 = $(0.75)(710,000) = 532,500$

California: 40% of 37,000,000 = $(0.40)(37,000,000) = 14,800,000$

Helpful Hint

This is another situation where we should avoid adding percentages together. In the ski example, if we add the percentages of people in both states who favor free skis it would appear that 115% of people favor the idea!

So, we can see that 75% of Alaskans is a much, much smaller number than 40% of Californians. Comparing 40% to 75% when the populations are not even in the same ballpark is misleading, to say the least. In fact, 2% of California's population is already greater than the entire population of Alaska.

Comparing percentages of different groups when the sizes of the groups are very different can often lead to misleading conclusions. However, when the groups are roughly the same sizes, these issues disappear. For instance, it would be reasonable to compare percentages from the populations of Kentucky (4,300,000 in 2010) and Colorado (5,000,000 in 2010) since their population sizes are more similar.

Skill Check #4

Even if the total populations are very different sizes, there are times when percentages/proportions are more appropriate for comparison. Can you think of some examples?

Averaging Percentages

Another "no-no" when manipulating percentages is trying to average them. Remember our previous example about comparing percentages of groups with unequal sizes? Well, the same is true for when averaging percentages from groups of unequal sizes. In general, averaging percentages almost always yields incorrect answers. There are exceptions, but to avoid confusion or misuse, stay away from averaging percentages. The next example is an extreme case to help you see this point.

Consider our previous example about approval for free skis to school children in Alaska and California. What percentage of people in both states favor the idea? Since there are two states, what is wrong with simply averaging the two percentages? If we did that, we would have

$$\frac{75\% + 40\%}{2} = \frac{115\%}{2} = 57.5\% \text{ favor free skis.}$$

However, if we add all of the people in favor of the idea together, we still have less than half of the population of California. So, the average percentage of 57.5% cannot possibly be correct.

Here is the correct method.

$$\frac{\text{AK pop. in favor} + \text{CA pop. in favor}}{\text{AK pop.} + \text{CA pop.}}$$

So, the percentage of people in both states in favor of the free ski idea would be

$$\frac{532,500 + 14,800,000}{710,000 + 37,000,000} = \frac{15,332,500}{37,710,000} \approx 0.40659 \approx 0.41 \text{ or } 41\%.$$

Shifting Reference Points

The last potential misuse of percentages that we will consider involves shifting reference points. It is best explained by an example.

Example 2: Shifting Reference Points

Campus police report that car break-ins on campus decreased by 25% this year compared to only a 20% decline the previous year. They promoted the fact that their law enforcement efforts reduced the number of break-ins more this year than in the previous year.

The following is a breakdown of the campus report, including a column to show the percentage change from year to year.

Table 1 Campus Car Break-In Report		
Year	Car Break-Ins	Percentage Change
2009	10	
2010	8	20% decrease
2011	6	25% decrease

Consider the underlying numbers rather than the percentages to confirm the assertion made by the campus police.

Solution

We can see that, just as the police department claims, car break-ins are decreasing by a larger percentage each year. However, notice that the absolute reduction in the number of car break-ins remains exactly 2 fewer each year. Since the number of break-ins reduces from year to year, the absolute difference of 2 becomes a larger percentage of the shrinking overall number of break-ins. In other words, the reference point that we are comparing the difference to changes each year, causing the percentage decrease from year to year to become larger.

It is always wise to have an eye on the underlying numbers before comparing percentages so that you do not fall victim to shifting reference points.

Skill Check Answers

1. $1.48, $1.97

2. Georgia: $228.79, Kentucky: $233.19

3. $52.80

4. Answers may vary. An example is the percentage of voters in a demographic area.

4.4   Exercises

Solve each problem. Round your answer to the nearest hundredth when necessary.

1.   Your dinner bill was $18.00. If you leave a 20% tip, how much will the tip be?

2.   A shirt is on sale for 40% off, and you have an additional 20% off coupon. True or false: The shirt will ultimately be 60% off the original price?

3.   If you wish to leave a 15% tip and your bill is $9.50, how much tip will you leave?

4.   The bill for dinner was $35.90. You left $8.62 for the tip. What percentage of the original bill was the tip?

5. A local store is having a sale. All waterproof shoes are 40% off. If a pair of rain boots normally costs $13.72 plus 7% sales tax, how much will it cost to buy the rain boots during the sale? (**Hint:** Discounts are applied before sales tax is added.)

6. An online company is advertising a Presto Pizza Oven on sale for 45% off the original price of $69.99. What is the sale price for the oven?

7. A new wireless printer costs $99.99 in the store. What would your total cost be if sales tax is 9.5%?

8. Get Lucky is having a clearance sale of their winter items. All sale items are marked an additional 10% off the sale price. If a coat that originally costs $110 is on sale for 25% off, what is the new sale price? Is this a reduction of 35% off the original price?

9. A ticket to see Pink in concert costs $117.80 at regular price. An online ticket company is offering tickets at a 25% discount. Your friend bought a ticket at the discounted price but found out he can't go. He's offering the ticket to you at an additional 10% off what he paid. What price could you get the concert ticket for if you bought it from your friend?

10. In Louisiana, a pack of cigarettes costs about $4.82 plus 4% sales tax. How much would a person spend on cigarettes in a year in Louisiana if he smoked one pack of cigarettes on each of the 365 days of the year?

11. You owe $257.43 on your credit card. Because you have not paid your bill on time, you have to add a past due charge of $25, an over limit fee of $40, and a service charge of 9% of the amount owed on the credit card (without the past due charge and over limit fees). If you wish to send a check for the full amount immediately, how much should you write the check for?

12. A particular species of bird weighs about twenty-eight grams when born. If its weight increases by ten percent a day, how much will it weigh at the end of the third day of its life?

13. If you are selling your house with a local realtor who requires a 6% commission fee, what can you expect to pay the realtor if your house sells for $139,000?

14. Suppose you sell your house for $215,000 but still owe $175,000 on the mortgage. If your real estate agent gets a 6% commission, what is the amount of money you'll make on this sale?

15. You are about to list your house as For Sale By Owner. Suppose you need to allow for a 3% commission for the buyer's realtor, and you need to receive a minimum of $92,500 on the sale. What is the minimum price you can list the house at to accommodate both of these requirements? Round to the nearest hundred dollars.

16. The wholesale cost of a new sofa is $240. The price of the sofa is marked up 40% before putting it on the showroom floor. During a showroom sale, the sofa was marked down 40%. What is the price of the sofa now?

17. The stock price for KTD was $54.56 in January 2014. In February the stock had increased by 6% but in March the price fell by 6%. What was the price of KTD stock in March 2014?

4

Find the absolute change and the percentage change for each situation.

18. 270 is increased to 1134

19. 14 is increased to 70

20. 150 is decreased to 39

21. 475 is decreased to 152

22. 39 is increased to 69.42

23. 62 is decreased to 42.78

Solve each problem. Round your answer to the nearest hundredth when necessary.

24. If a laptop computer regularly costs $879 and is on sale for $599, what percentage discount is being given?

25. A poll in Russia looked at the support of Vladimir Putin's United Russia party. At that time, United Russia held 315 of the 450 seats in the State Duma, but the poll predicted this would fall to 252. If this was the case when the elections happened, calculate the percentage change in the number of seats held by the United Russia party.

26. Suppose that a local newspaper reported that Cityville and Beautytown both had 50 murders last year. However, 5 years ago, Cityville had 42 murders while Beautytown had just 29. Calculate the absolute change and the percentage change for each city over the five years. Could the newspaper make the claim that crime had gone up by the same amounts in each city? Explain.

27. A news outlet reported that fifty million Americans visited online retail sites on Black Friday, representing an increase of 35 percent versus a year ago. How many Americans visited online retail sites on Black Friday last year?

28. If Apple stock fell 0.93% from $363.57 per share, what was the absolute change in stock price?

29. Two companies report a 15% gain in online sales over the last year. Suppose Company Q had 25 online sales and Company Z had 150 online sales last year. Do you believe their growth was the same? Justify your answer.

30. The United States population was estimated to be 312,681,446 in 2011, estimated to be 285,102,075 in 2001, and estimated to be 77,584,000 in 1901. Calculate the absolute change and percentage change between each of the figures given (that is, 1901 to 2001, 1901 to 2011, and 2001 to 2011).

Chapter 4   Summary

Section 4.1   Rates and Unit Rates

Definitions

Rate

A rate is a fraction used to compare two measured quantities that are not necessarily in the same units.

Proportional Equation

A proportional equation consists of two rates set equal to one another.

Unit Rate

A unit rate is a rate comparing two measured quantities, one of which is a single unit written in the denominator of the fraction.

Section 4.2   Ratios

Definitions

Ratio

A ratio is a fraction used to compare two measured quantities whose units are the same type.

Equivalent Ratio

Equivalent ratios are ratios that express the same relationship.

Section 4.3   Proportions and Percentages

Definition

Proportion

A proportion is a fraction of a whole.

Section 4.4   Using Percentages

Formulas

Absolute Change

The absolute change between two amounts is the absolute value of the difference between the two numbers.

$$\text{absolute change} = \left| \text{new amount} - \text{original amount} \right|$$

Percentage Change

$$\text{percentage change} = \frac{\text{absolute change}}{\text{original amount}} \cdot 100\%$$

Chapter 4 Exercises

Write each proportion described. Be sure to simplify your answer.

1.   If there are 26 students in a class and 12 are seniors, what proportion of students are *not* seniors?

2.   Out of 1000 people surveyed, 850 said they like soda. What proportion of the people surveyed said they do *not* like soda?

Answer each question. Round your answer to the nearest hundredth when necessary.

3.   If six out of every ten individuals in a population carry a gene for a defective enzyme, how many individuals carry the normal gene in a population of 900?

4.   The ratio of apples to oranges in a basket is 3 to 4. What fraction of fruit in the basket consists of apples?

5.   A 180-pound student burns 120 calories every half hour of playing ultimate Frisbee. How many calories would he burn in

   a. 10 minutes?

   b. One hour and 45 minutes?

6.   Your dinner bill was $30.00. If you leave a 15% tip, how much will this be?

7.   If you wish to leave a 20% tip and your bill is $58.50, how much tip will you leave?

8.   The original bill after dinner was $49.55. You left $10.90 extra for the tip. What percentage of the original bill was the tip?

9.   You are about to list your house as For Sale By Owner. Suppose you want to allow for a 6% commission for the buyer's realtor, and you need to make a minimum of $110,500 on the sale. What is the minimum price you can list the house for to accommodate both of these requirements? Round your answer to the nearest hundred dollars.

10.   Jessica's annual salary decreased by $3800, which was a 9% decrease. What was Jessica's original salary?

11.   Last year, a new computer sold for $1600. This year, the same computer can be purchased for $1200. What percentage discount is being given from the previous year?

12.   Suzanne's salary grew from $40,000 in 2000 to $56,000 in 2012. During the same time period, Mike's salary grew from $38,000 to $53,000. Whose salary grew more in absolute change? In percentage change?

Write each rate described and then convert it to a unit rate.

13. The university reported that there were 42 freshmen in the entering class for every 3 advisors.

14. Coach Burge wanted the entire class to be able to do 90 sit-ups in 2 minutes by the end of the semester.

Solve each problem. Round your answer to the nearest hundredth when necessary.

15. Which is the better value for the money, $5.99 for a package of 8 cold remedy capsules or $2.85 for 2 capsules? Why?

16. Suppose at bulk rate, instant potato flakes cost 12.04 cents, or $0.1204, per ounce.
    a. If a box costs $77.05, how many pounds of potato flakes are in the box? (**Hint:** There are 16 ounces in a pound.)

    b. What would a 2-pound box of potato flakes cost?

17. Suppose that the ratio of males to females in the sales field of a large company is $4:5$. If there are 81 sales representatives in the company, how many are male and how many are female?

18. If Whitt wins 3 out of every 4 games of Solitaire that he plays on his computer, how many games would he need to play so that he wins 51 games?

19. The following instructions to make mashed potatoes are found on a box of instant potato flakes. Convert the recipe so that it makes 20 servings of potatoes.

    $1\frac{1}{2}$ cups water

    2 tablespoons butter

    $1\frac{1}{2}$ cups milk

    2 cups potato flakes

    $\frac{1}{2}$ teaspoon salt

    Combine water, milk, salt, and butter. Bring to a boil. Remove from heat, add potato flakes, stir, and let stand 30 seconds to 1 minute until moisture is absorbed. Fluff with fork. Makes 8 servings.

20. A sculpture for the lawn in front of the new art gallery is 38.5 feet long. A model is being made for a display inside the gallery. The model is 14.2 inches long. Based on this scale, how long would an actual bench be on the lawn if the bench in the scale model is 2 inches?

Bibliography

4.2

1. Fortune. "Fortune's 40 Under 40." http://archive.fortune.com/magazines/fortune/40-under-40/2011/full_list/

2. Phrenopolis. "Solar System Perspective." http://www.phrenopolis.com/perspective/solarsystem/

4.4

1. Magellan's. "Worldwide Tipping Guide." Accessed December 2011. http://www.magellans.com/store/article/367

2. Sales Tax Institute. "Sales Tax Rates." http://www.salestaxinstitute.com/resources/rates

Chapter 5
The Mathematics of Growth

Objectives

- Demonstrate an understanding of functions, function notation, domain, and range
- Demonstrate an understanding of linear functions and linear growth
- Demonstrate an understanding of quadratic functions and quadratic growth
- Demonstrate an understanding of exponential functions and exponential growth
- Demonstrate an understanding of logarithmic functions and logarithmic growth
- Model data with linear, quadratic, exponential, and logarithmic functions

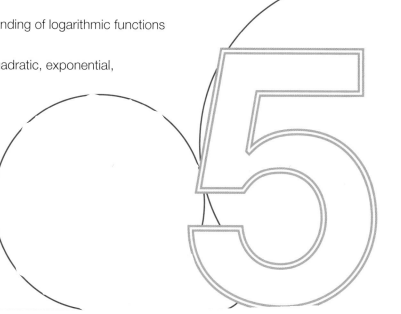

The Mathematics of Growth

According to the Centers for Disease Control and Prevention, the influenza virus is a contagious respiratory illness that infects the nose, throat, and lungs. The spread of the virus can occur when an infected person comes within 6 feet of someone not infected. Assuming that a given person infected with the virus could come in contact with 10 uninfected people each day, the spread of the virus would occur very rapidly if all 10 uninfected people actually become infected with the virus. If this pattern were to continue, a single person could infect 10 others. Those 10 people could, in turn, infect 10 more each. If that pattern were to continue, then after only one week, there would be 1,000,000 people infected. After two weeks, the number infected would be greater than the entire population of the world. Thankfully, not everyone is susceptible to the virus on first contact, while others take a vaccination to avoid actually contracting the virus. Situations such as this are carefully studied by biologists, in conjunction with mathematicians, to model the growth of diseases in an effort to best address the needs of a population.

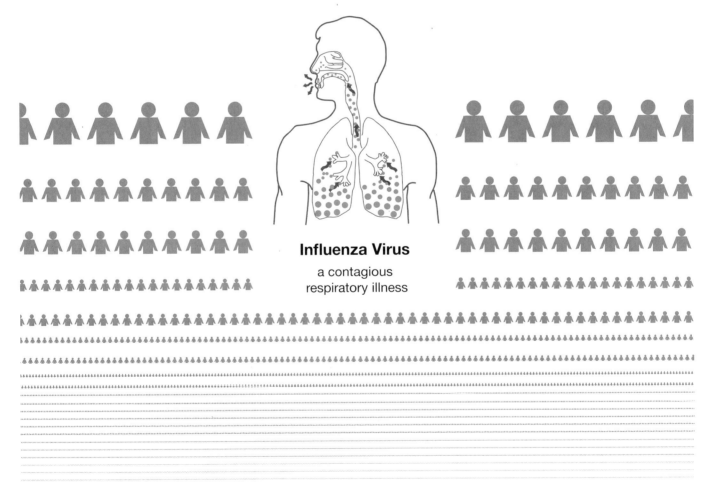

Influenza Virus

a contagious
respiratory illness

Figure 1

5.1   The Language of Functions

The central concept to understanding how mathematics models the world around us is the notion of functions. In general, a **function** is a mathematical process that relates two or more quantities to one another. This relationship between two or more quantities allows us to represent unknown amounts of one quantity with known amounts of another. For instance, consider the example of being paid for an hourly job: the more hours you work at the job, the more money you earn. Thus, there exists a relationship between the amount of money you earn and the number of hours you work. Another example could be the amount of money deposited into a savings account. Can you determine which quantity depends on the other? These are just two examples of mathematical relationships that will we discuss throughout this chapter.

Consider the situation where we have a bank account that does not earn interest. If you made a monthly deposit into the account, as shown in Figure 1, then the total amount in the account grows larger with each deposit. The relationship between the accumulated amount of money in the account and the number of months shows us that a line can represent the graph as long as the monthly deposit is the same amount and there is no appreciated interest.

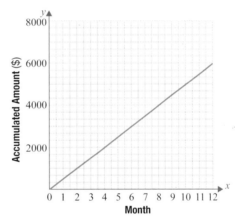

Figure 1: Bank Account Balance

What about other types of relationships that exist between quantities? The amount of pressure in a balloon can be determined by the volume of the balloon, as illustrated in Figure 2. As we can see, the relationship between pressure and volume is much different than that of depositing the same amount of money into an account each month.

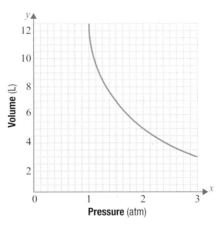

Figure 2: Pressure vs. Volume

5

Skill Check #1

What would the graph of the number of sweaters sold at a department store from January 1 until March 31 look like?

For one last example, let's consider the graph of the path of a basketball as it flies through the air. As we can see from the graph in Figure 3, the height that the basketball reaches in meters is related to the amount of time that has elapsed since the basketball was released.

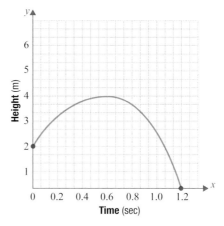

Figure 3: Height of a Basketball

What was the maximum height of the basketball? How long did the basketball travel in the air? What was the height of the basketball after half of a second? These are just a few of the questions that can be answered once we understand more about the relationships that can exist between quantities.

Functions

Mathematical models are a way for us to represent relationships between two quantities. Consider the situation of purchasing gasoline priced at $3.25 per gallon. If we interpret this in terms of an equation of two variables, then the price of filling up your car can be represented by $y = 3.25x$, where x represents the number of gallons of gas you purchase and y represents the total cost of the gas purchase. In this case, we say that the total cost is a **function** of the cost per gallon. In general, we call unknown quantities **variables** since the value of these quantities can change, or vary. In all situations of functions, one variable is dependent on another. For instance, when purchasing gasoline at the pump, the price we pay is **dependent** on how many gallons of gas we buy. In this situation, the amount of gas purchased would be the **independent** variable and the total cost would be the dependent variable. In a function, each value of the dependent variable is uniquely associated with a value of an independent variable.

Function

A **function** is a mathematical equation that describes the relationship between the dependent and independent variables for a given situation.

Dependent Variable

The value of the **dependent variable** changes with respect to the value of the independent variable.

Independent Variable

The value of an **independent variable** does not rely on the values of the other variables in an expression or function, and its value determines the values of the other variables.

The notation we will use to represent y as a function of x is $y = f(x)$ and is read as "y equals f of x." This notation indicates that for a given value of x, we substitute that value into the expression to find a value for the function y. Thus, we create an input (independent variable x) and an output (dependent variable y) of the values that give us coordinate pairs (x, y).

For our needs in this textbook, a basic understanding of how functions work will be sufficient; that is, how functions represent relationships between variables and the evaluation of those relationships. The process of evaluating a function means to substitute a given value into the function for x, the independent variable, to determine the value of the dependent variable y. Let's work a couple of examples to help clarify the process.

Example 1: Evaluating Functions

Evaluate the following functions for the given values.

a. Find $f(5)$ for $f(x) = 2x - 5$.

b. Find $f(0)$ for $f(x) = x^3 - 4$.

c. Find $f(-3)$ for $f(x) = x^2 - 2x - 5$.

Solution

To find the solution, we need to evaluate each function at the given value; that is, substitute the appropriate number in for x in each function.

a. $f(5)$ means $x = 5$, so substitute 5 in for the variable x. If $f(x) = 2x - 5$, then
$f(5) = 2(5) - 5 = 10 - 5 = 5$.

b. $f(0)$ means $x = 0$, so substitute 0 in for the variable x. If $f(x) = x^3 - 4$, then
$f(0) = (0)^3 - 4 = 0 - 4 = -4$.

c. $f(-3)$ means $x = -3$, so substitute -3 in for the variable x. If $f(x) = x^2 - 2x - 5$, then
$f(-3) = (-3)^2 - 2(-3) - 5 = 9 + 6 - 5 = 10$.

Skill Check #2

Evaluate the following functions for the given values.

a. Find $f(3)$ for $f(x) = 3x + 4$.

b. Find $f(-2)$ for $f(x) = 2x^2 + 5$.

c. Find $g(4)$ for $g(x) = 2x^2 - 3x + 1$.

Domain and Range

Recall the example about filling your car with gasoline. The amount of gasoline you put in your car will depend on several things. First, how large is the gas tank? Second, how much gas is already in the tank? Third, how much money do you have to purchase gas? We have already discovered that the relationship between the cost of a gallon of gas, the number of gallons we buy, and the total cost of the gas can be related with a function. We need to think about which values for gallons and total cost make sense in this situation. This will help us understand how functions work in the real world.

Assume that a gallon of gas costs \$3.25. We determined previously that we could represent the relationship between the cost of a gallon of gas, amount of gas purchased, and the total cost of the gas purchase with the function $f(x) = 3.25x$. First, notice that the total amount of gas we are able to purchase can only be 0 or a positive number. Any boundaries or restrictions that are placed on the independent variable, such as these, are factors in defining the **domain**. Once we know the values for the domain, we can determine the output values for the function. Notice that the total cost of gas $f(x)$ can only be 0 or positive. This is because we are only allowed to purchase positive amounts of gas and the cost per gallon is a positive value. We call the values for the output of a function the **range**.

Domain

The **domain** of a function is the set of input values (values of the independent variable) for which a function is defined. You can only evaluate a function at values that are in the domain.

Range

The **range** of a function is the set of output values (values of the dependent variable) that correspond to the domain values.

In general, when we are given a mathematical relationship, we can use the following steps to begin making sense of the information given.

1. Determine the independent and dependent variables and what they represent.

2. Determine the domain of the function based on the context of the function.

3. Determine the range of the function based on the domain.

4. Graph the function to get a "picture" of the mathematical relationship. (**Note:** Graphing can be done with either a graphing utility or by hand.)

Example 2: Exploring Mathematical Relationships

A baseball team wishes to purchase shirts for their team members and families. The Print Perfect T-Shirt Company charges \$200 to set up the design of the shirts and then charges \$12 per shirt purchased.

a. Write a function to represent the relationship between the number of shirts purchased and the total cost.

b. Determine the domain and range of the function for the total cost of the purchased shirts.

c. Graph the function using a graphing utility or by hand.

Solution

a. To create a function that represents the relationship, we need to first determine the independent and dependent variables. Since the total cost is based on the initial setup cost of \$200 and the \$12 for each shirt purchased, the independent variable would be the number of shirts purchased and the dependent variable would be the total cost. With this information, we can now write a function expression that represents the relationship as $f(x) = 12x + 200$.

b. We can now determine the domain and range of the function $f(x) = 12x + 200$. Recall that the domain is determined from input values that are allowed for the function. In this case, the domain consists of all nonnegative integer values since you cannot purchase a negative number of shirts. Notice that it makes little sense to include noninteger values since we are only purchasing whole shirts and not pieces of shirts. The range can now be determined to be all values \$200 and larger.

c. We can graph the function to illustrate the relationship between the number of shirts purchased and the total cost of the shirts. This can be accomplished by hand using several values from the domain, say $x = 0$, $x = 5$, $x = 10$, and evaluating the function at those values.

Total Cost of Shirts Purchased	
x	$f(x) = 12x + 200$
0	$f(0) = 12 \cdot 0 + 200 = 200$
5	$f(5) = 12 \cdot 5 + 200 = 260$
10	$f(10) = 12 \cdot 10 + 200 = 320$

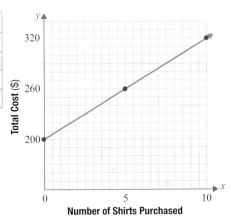

TECH TRAINING

To graph the function $f(x) = 12x + 200$ using a TI-83/84 Plus calculator, follow these steps.

1. Select ⎡ Y= ⎤ from the calculator menu.

2. Move the cursor to Y1:.

3. Enter 12x + 200.

4. Select ⎡GRAPH⎤ from the calculator menu. You may have to adjust the window settings under ⎡WINDOW⎤ in order to properly view the graph.

To graph the function $f(x) = 12x + 200$ using Microsoft Excel, open a new workbook. In Cell A1, type "# of Shirts" and in Cell B1, type "Total Cost". Enter several (3 or 4) values for the number of shirts in Cells A2, A3, A4, etc. Enter the corresponding total cost values into Column B. For example, when $x = 0$, total cost is 200. Select Charts and choose Scatter-Straight Lined Scatter.

Skill Check Answers

1. A graph that decreases from left to right.

2. **a.** 13
   b. 13
   c. 21

5.1   Exercises

Graph each equation by evaluating integer values of *x* from −2 to 2, plotting the resulting points, and then connecting the points with a smooth, unbroken line.

1. $y = 3x$

2. $y = 2x + 1$

3. $y = -2x + 3$

4. $y = \dfrac{1}{4}x - 2$

5. $y = x^2 - 2$

6. $y = x^3 + 1$

7. $y = x^2 - 2x + 1$

8. $y = -\dfrac{3}{2}x$

9. $y = -x^2 - 5x - 7$

10. $y = 2x^2 - 7x$

Evaluate each function for the given values of *x*.

11. $f(x) = x^2 - 3x + 1$

x	$f(x) = x^2 - 3x + 1$
−3	
−2	
−1	
0	
1	
2	
3	

12. $f(x) = -3x + 1$

x	$f(x) = -3x + 1$
−3	
−2	
−1	
0	
1	
2	
3	

13. $g(x) = (2x - 3)^2$

x	$g(x) = (2x - 3)^2$
−3	
−2	
−1	
0	
1	
2	
3	

14. $h(x) = x^3 - 4x + 1$

x	$h(x) = x^3 - 4x + 1$
−3	
−2	
−1	
0	
1	
2	
3	

Write a function to represent each situation and be sure to mention any domain restrictions. Then, answer each question.

15. A computer purchased in 2005 had an initial value of $1500. If the computer depreciates at $150 per year, write a function in x that represents the depreciation in value. What is the value of the computer after 5 years?

16. The cost to rent a car for a day is an initial cost of $42 and $0.45 per mile driven. Write a function in x that represents the total cost of the car rental. What is the total cost if you drove 95 miles?

17. The cost of tuition at Hawkes University per semester is $15,000 for the first 12 hours of coursework. Each hour of credit taken at Hawkes after 12 hours is an additional $575. Write an equation that represents the cost of attending Hawkes, where x represents the number of credit hours of coursework over 12 hours. If a student attending Hawkes took 18 hours of credit, what is his or her cost of attendance?

18. The Wilson Water Company charges a monthly fee of $15 for service and an additional $2.50 for every 250 gallons of water used. Write an equation that represents the total cost of monthly service.

Use Graph A to answer each question.

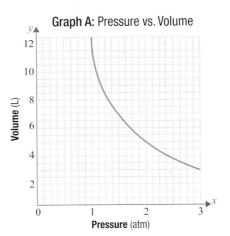

Graph A: Pressure vs. Volume

19. When the pressure is 3 atmospheres, what is the volume?

20. As the pressure decreases from 3 to 2 atmospheres, does the volume increase or decrease?

21. If the volume is 6 liters, what is the pressure in atmospheres?

Use Graph B to answer each question.

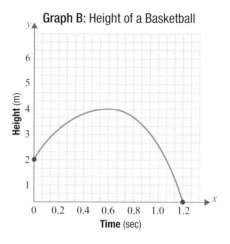

Graph B: Height of a Basketball

22. How long did the basketball travel in the air?

23. What was the height of the basketball after 0.6 seconds?

5.2   Linear Growth

In the previous section, the concept of functions was introduced to show the relationship between dependent and independent variables. Now, we will turn our attention to specific types of functions and the situations that dictate which type of function best represents the relationship between two variables.

We must begin our exploration of mathematical growth with a couple of questions: *What is mathematical growth?* and *Why does understanding growth matter when understanding mathematics and the mathematical world in which we live?* To answer each of these questions, we need to first have a better understanding of "growth" in the context of mathematics. In introductory algebra courses, we learn that relationships exist between mathematical components in equations. In this course, we can expand our understanding of these relationships to discuss applications that represent growth. In general, when we discuss mathematical growth, we are interested in how the dependent variable changes, or "grows," when the independent variable increases or decreases. This understanding allows us to create mathematical models that illustrate growth.

Mathematical growth is defined by the relationship between two variables. We can use a function to represent this relationship and then use graphing methods to visualize and analyze the relationship.

Let's begin with a familiar situation. Assume you must travel 600 miles for a trip in your car. If you travel at a constant rate of 60 miles per hour (mph) it should take you 10 hours to reach your destination. This equation is easily solved using the following formula.

$$\text{distance traveled} = \text{rate of travel} \cdot \text{time of travel}$$

$$\text{or}$$

$$d = r \cdot t$$

You may recall some of the models in this chapter from previous math courses. This is particularly true of mathematical situations that grow linearly. Mathematical models with linear growth consist of situations where one variable *varies directly* with the other. The relationships between quantities that vary at the same rate, or vary directly, are known as **proportions** and can be written as linear functions.

When two expressions are proportional, we say that there is a **direct variation** among the expressions. In other words, this means that as the value of one expression increases, the value of the other expression also increases. In our car example, the distance you travel is directly proportional to the speed you drive and the amount of time you drive at that speed.

Let's begin with an example to help us understand linear functions.

Example 1: Using Direct Variation

Jessica is interested in purchasing a new video game that costs $60. If she has a job that pays $7.50 per hour, find a linear equation for her income and determine the number of hours she must work in order to have enough money to buy the game.

Solution

You might need to ask yourself what we are looking for as the final answer. We are looking for the number of hours Jessica must work in order to earn $60. Let Jessica's total pay be represented by the variable p. We can determine the value of p by taking the number of hours worked w and multiplying it by the pay per hour s, which is $7.50 per hour. Translating this situation into an algebraic equation, we get the following.

$$\text{income} = \text{hourly wage} \cdot \text{hours worked}$$

or

$$p = s \cdot w$$
$$60 = 7.5w$$

Let's start with a chart to organize our thoughts. The chart consists of the number of hours Jessica worked and her income based on the number of hours worked.

Table 1
Amount Earned

Hours Worked (w)	Income in Dollars (p)
1	7.50
2	15.00
3	22.50
4	30.00
5	37.50
6	45.00
7	52.50
8	60.00

Continuing the chart of Jessica's pay tells us that she must work 8 hours to make $60.00. That is the solution to our problem. We have translated the idea that hourly pay multiplied by the hour number of hours worked yields the total pay into an algebraic expression where the total pay is directly proportional to the number of hours worked at a given pay rate. Take a look at the following graph to see a visual representation of the relationship between the two variables.

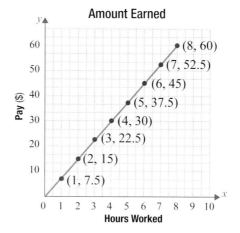

Amount Earned

We can see from both the table and the graph that Jessica must work 8 hours to earn enough money to buy the video game.

Skill Check #1

If Andrew earns $6.75 for every dozen oysters that he catches, how many dozen oysters must he collect to accumulate $215? Find a linear equation to determine the number of dozen oysters Andrew needs to catch.

As you can see from the graphical representation in Example 1, the model is represented by a line. This means that the amount of money earned in comparison to the number of hours worked is **constant**. So, for each hour worked, the increase in pay is the same. In a linear growth model, we call this constant of variation the **slope**. The *slope* of the line represents the *rate of change* of the expression. This means that the rate of change of the model is based on the ratio of change in the *y*-values to the change in the *x*-values. Thus, the primary component to understanding when mathematical growth produces a linear relationship is based on the idea that the rate of growth is constant.

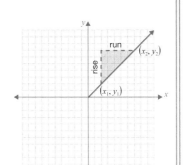

Slope

The **slope** of a line is defined as the ratio of the change in the *y*-values ("rise") divided by the change in the *x*-values ("run") between two points. The slope of a line is defined by the following formula.

$$\text{slope} = m = \frac{\text{rise}}{\text{run}} = \frac{\text{change in } y\text{-values}}{\text{change in } x\text{-values}}$$

When given two points (x_1, y_1) and (x_2, y_2), the slope of the line connecting them is calculated using the following formula.

$$m = \frac{y_2 - y_1}{x_2 - x_1}$$

Let's try to determine the slope of our model from Example 1 by choosing any values for the hours worked *w*. Let's find the slope between 3 hours worked and 8 hours worked. The corresponding points (w, p) on the graph would be (3, 22.5) and (8, 60). Remember that in our example, *w* represents the values on the *x*-axis and *p* represents the values on the *y*-axis. Calculating the slope, we see that

$$m = \frac{60 - 22.50}{8 - 3} = \frac{37.50}{5} = 7.50.$$

A slope of $m = 7.50$ means that for every additional unit increase in the *x*-direction, there is a 7.50-unit increase in the *y*-direction. So for each additional hour Jessica works, her total pay increases by $7.50. This example establishes that the graph of a relationship that creates a line with a constant rate of change is a **linear function**.

Let's try an example to explore this idea further.

Example 2: Applying Linear Functions

Assume you have $20,000 in a bank account that you will use to pay your college tuition. If the cost of your college tuition is $2500 per semester, use a linear function to find the number of semesters it would take for the account to have a balance of $0. Assume that this bank account does not earn interest.

Solution

First, determine what the variables are. You should notice that the number of semesters that you take classes can vary. Unlike the previous problem, you should recognize that when we are starting at time 0, we have a value of $20,000. How does this compare to the previous example? Let's use a table again to find the equation that models this scenario.

Table 2	
College Tuition Cash	
Time (t) (in Number of Semesters)	Account Value (V) (in Dollars)
0	20,000
1	17,500
2	15,000
3	12,500
4	10,000
5	7500
6	5000
7	2500
8	0

As you can see in the table, the bank account will reach a value of $0 after 8 semesters. Also, the rate at which the value of the money in the account is decreasing is constant because we are removing the same amount, $2500, each semester. Therefore, this is a linear function where the value V of the account is based on the time t that has passed.

Since the value of the bank account is decreasing at a constant rate of $2500 per semester, the slope should be negative. In the previous example, we established that when the slope is positive, the values of both variables are increasing. When the slope is negative, as in this example, the value of one variable is increasing while the other is decreasing. We call this association a negative association and measure this association using the slope.

Now let's look at the graph of the value of the account as it relates to the number of semesters.

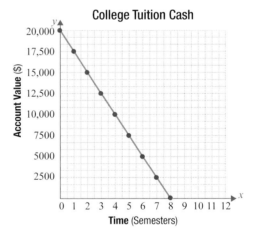

We can see from the graph that the account will be empty in 8 semesters.

In Example 2, the slope, or rate of change, is −2500. The slope is negative because each semester, the amount in the bank account decreases by $2500. Since you are starting with $20,000, this value represents the value of y when $x = 0$. So, 0 semesters of tuition paid implies that $0 is spent. We can obtain the following mathematical model for the amount of money in the account after t semesters.

$$y = -2500t + 20,000 \text{ for } 0 \le t \le 8$$

Equations such as this are called **linear equations in two variables**. We may also use the conventional and useful **function notation**, $f(t) = -2500t + 20,000$. Recall that function notation allows us to better associate which variable is independent and which variable is dependent. This means that the independent variable in Example 2 is the number of semesters attended and the dependent variable is the amount left in the account after t semesters.

To further develop our understanding of functions and their graphs, we need to discuss intercepts. The **x-intercept** is the point where the graph of the function intersects the *x*-axis (horizontal axis) and the **y-intercept** is the point where the graph of the function intersects the *y*-axis (vertical axis). The intercepts may be found by replacing one of the variables with 0 and solving for the remaining variable. For instance, to find the *y*-intercept in the equation $y = 2x - 5$, we substitute $x = 0$ into the equation to obtain $y = 2(0) - 5 = -5$. This gives that the *y*-intercept is $(0, -5)$. Similarly, to find the *x*-intercept, substitute $y = 0$ into the equation and solve for *x*. This gives that the *x*-intercept is $(2.5, 0)$. The intercepts can be useful when graphing a linear function.

If we have the slope and the *y*-intercept of a line, we can use them to write the equation of the line. This equation is called the slope-intercept form.

Slope-Intercept Form of a Linear Equation

The **slope-intercept form** of a linear equation is

$$y = mx + b,$$

where *m* represents the slope of the line and *b* is the value where the graph crosses the *y*-axis. In other words, the point $(0, b)$ is the *y*-intercept.

Let's try another example. A commonplace use of linear growth is the relationship between Fahrenheit (F) and Celsius (C). Let's use what we know about the two measures of temperature to find the equation of the form $y = mx + b$, where *x* is the temperature in degrees Fahrenheit, *y* is the temperature in degrees Celsius, *m* is the slope, and *b* represents the temperature in Fahrenheit when the temperature in Celsius is 0 degrees. Using the common values for the freezing and boiling points of water, in both Fahrenheit and Celsius, we can find the slope of the line that defines the function.

The freezing point of water is $0\,°C$ and $32\,°F$ while the boiling point of water is $100\,°C$ and $212\,°F$.

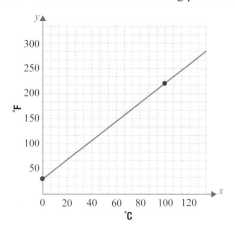

Figure 1: Relationship Between Celsius and Fahrenheit

Looking at the graph, we can see that the line intersects the °F-axis (*y*-axis) when the Celsius temperature is $0\,°$, and the Fahrenheit temperature is $32\,°$. This means that *b* will be 32 in our equation. The next value we have for the relationship between the two measurements is when $x = 100$ and $y = 212$. Let's use these two pairs of values to find the slope.

$$m = \frac{212 - 32}{100 - 0} = \frac{180}{100} = \frac{9}{5}.$$

Now we have a formula, $y = \frac{9}{5}x + 32$, that allows us to find the temperature in either Celsius or Fahrenheit when we know only one measure of temperature.

Example 3: Finding Linear Equations

Assume Jeffrey has an uncle that puts away $10 in a shoebox on the day he is born and adds $5 to the shoebox on his birthday every year. If this trend continues, use a linear equation to find the amount of money in the shoe box on Jeffrey's 18th birthday.

Solution

Let's begin our solution with the notion of what happens each year by looking at a chart of values.

Table 3	
Jeffrey's Shoebox Money	
Age (Years)	Accumulated Amount ($)
0	10
1	15
2	20
3	25
4	30

} $5
} $5
} $5
} $5

Now, let's find the equation of the line. Since the beginning amount is $10 at time $t = 0$, we know this is our y-intercept. The only thing we need now is the slope of the line. We can determine the slope by recognizing that the amount of change in the accumulated amount each year is constant. This tells us the slope is constant. So, $m = 5$. Knowing this, the equation of the line is

$$y = 5x + 10.$$

If we want to know what the amount will be when Jeffrey is 18, we can evaluate the function when $x = 18$.

$$y = 5x + 10$$
$$= 5(18) + 10$$
$$= 100$$

So Jeffrey will have $100 when he turns 18. Looking at the graphical representation, we see that this relationship is linear since the change from year to year is constant.

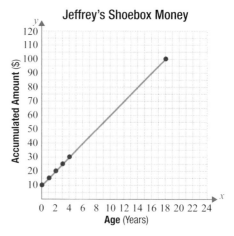

Skill Check #2

Cindy is starting a business selling T-shirts. Each T-shirt in an order costs $9 and there is an artwork fee of $12.75 per order. Find a linear function that could be used to represent the cost of each T-shirt order.

Skill Check Answers

1. 32 dozen; $y = 6.75x$

2. $c = 9t + 12.75$

5.2   Exercises

Find the slope and y-intercept of each linear equation.

1. $y = 2x - 7$

2. $f(x) = 2x - 5$

3. $y = -4x$

4. $y = -\dfrac{3}{2}x + 3$

5. $y = -x - 7$

6. $y = -5x + 8$

7. $2x + 4y = -5$

8. $-x + 2y = 3$

9. $h(x) = x - 4$

10. $g(x) = \dfrac{5}{4}x - 7$

Write a function to represent each situation and answer each question.

11. Opal works a job that pays $12 per hour. Find a function that represents her pay based on the number of hours worked.

12. Assume your car costs $2000 a year to maintain and $0.35 per mile to drive (for fuel). What function represents the total cost of operating the car each year?

13. A company manufactures sets of skis where each set costs $450. Each set of skis sells for $600. What function represents the profit from selling the manufactured skis?

14. The water depth in a pond is initially 48 inches and decreases at a rate of 0.5 inches per day through evaporation. Write a function that represents the depth of the pond at a given time. What will be the depth of the pond after 8 days?

15. Glinda rents a television at Rentals 4 U for an initial charge of $25. The cost of the rental is $17.50 per week. Write a function that represents Glinda's rental cost.

16. A television was purchased for $2750 in 2012 and the value depreciates at a rate of $225 per year. Write a function to model the value of the television based on the number of years of depreciation. What will be the value of the television after 6 years?

17. The price of a computer is $1100 today. The price rises at a constant rate of $25.00 per year. Write a function to model the value of the computer based on the number of years of appreciation. How much will the computer cost in 3.5 years?

18. The cost to rent a moving van for a day is an initial cost of $29 plus $0.55 per mile driven. Write a function to model the cost of the moving van rental based on the initial cost and miles driven. What is the total cost if you drive 625 miles?

19. You can purchase a car for $11,500 or lease it for $275 per month with a down payment of $500. Find a function that models the cost of the lease based on the number of months the car is leased. How long can you lease the car before the amount of the lease is more than the cost of the car?

20. The cost to attend a school is a one-time fee of $3500 and an annual cost of $15,000. Find a function that models the cost of attendance at the school based on the number of years attended. What is the cost for attending the school for 4 years?

21. The yearly cost of tuition at Supernatural University is $4750 for the first 15 hours of coursework. Each hour of credit taken at Supernatural after 15 hours is an additional $850. Write a function that represents the cost of attending Supernatural, where x represents the number of additional hours of coursework over 15 hours. If a student attending Supernatural takes 17 hours of credit, what is his or her cost of attendance?

22. A population of mold decays at a rate of 20 mold spores per day. Write a function to model the decay rate of the mold spores based on the number of days that have passed. If the initial population of mold spores is 1290 spores, how many days will it take for the population to be less than 500 mold spores?

5.3 Discovering Quadratics

In the previous section, we learned that a linear function has the form $f(x) = mx + b$ and that the graph of a linear function is a straight line with a slope of m and a y-intercept of $(0, b)$. We also discussed the fact that linear functions have a rate of change that is constant. This means that linear functions are quite useful at describing phenomena that change at a constant rate, such as distance traveled at a constant speed. However, many phenomena do not change at a constant rate and are, therefore, not linear functions. So, we will need a new type of function for these situations.

For our next type of function, let's consider the paths created by a bouncing ball, a water fountain, and a suspension bridge. The type of function that models these paths is one of the most useful types of functions that change at a nonconstant rate. To begin exploring this new type of function, we construct Table 1 and plot the points to see the difference in growth between the two function representations.

Table 1		
Growth of Functions		
x	**$f(x) = x$**	**$g(x) = x^2$**
0	0	0
1	1	1
2	2	4
3	3	9
4	4	16
5	5	25
6	6	36

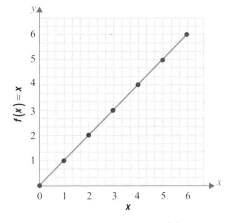

Figure 1: Graph of $f(x)$

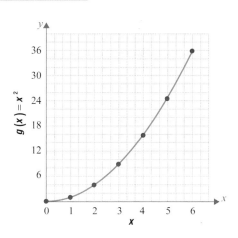

Figure 2: Graph of $g(x)$

The function $g(x)$, which has a power of 2, in Figure 2 is a **quadratic function** and grows more quickly than the linear function $f(x)$, in Figure 1.

Quadratic Function

A **quadratic function** is a function of the form $f(x) = ax^2 + bx + c$, where $a \neq 0$. The common name for the graphical representation of a quadratic function is **parabola**.

Here Are a Few Attributes of Parabolas

- If $a > 0$, the parabola will open up.

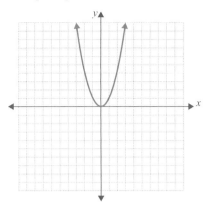

Figure 3: Parabola with $a > 0$

- If $a < 0$, the parabola will open down.

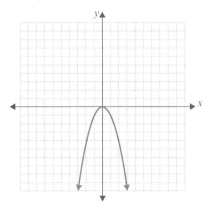

Figure 4: Parabola with $a < 0$

- Every parabola has a maximum or minimum point called the **vertex**. Parabolas that open down have a maximum point (see Figure 5) and parabolas that open up have a minimum point (see Figure 6).

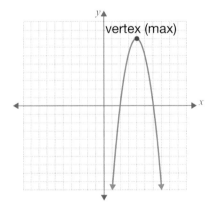
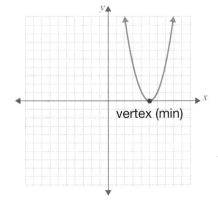

Figure 5: Parabola with Maximum   Figure 6: Parabola with Minimum

- Parabolas are symmetrical about a vertical line passing through the vertex. This vertical line is called the **axis of symmetry** (see Figure 7).

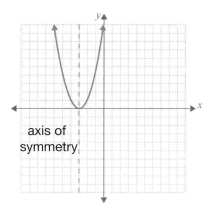

Figure 7: Parabola with Axis of Symmetry

Great examples of common situations that are modeled by quadratic functions include fireworks and forensic science. For instance, the flight of a firework, such as a bottle rocket, can be described by a quadratic function and the height to which it flies can be adjusted by manipulating the coefficients. If we want a firework to reach a maximum height of 300 ft in the air, then the function representing what the flight of the firework might look like could be defined by

$$f(x) = -25x^2 + 75x + 250.$$

If you graph the function using a graphing utility, or by plotting points on a graph and connecting the points, you will see that the y-value of the vertex is at 300.

Forensic scientists use quadratics to determine the total distance it takes for a car to stop after traveling at a certain speed, called the stopping distance. To find this value, the scientists start with a quadratic function that describes the distance a car travels after the brakes are applied. However, the stopping distance must also include the distance traveled during the reaction time, the split second before the brakes are applied. Combining these two quantities gives us the equation for the stopping distance. The distance d required for a car to stop can be modeled by the function

$$d(x) = \frac{x^2}{20} + x,$$

where x represents the speed of the car when the brakes are initially applied. Notice that in this example, the equation has a constant term of 0. This equation can be used by accident investigators to analyze the distance a car traveled before coming to a stop in order to determine the speed the car was traveling.

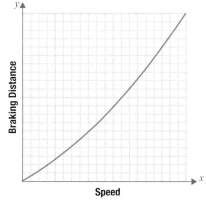

Figure 8: Braking Distance vs. Speed

☞ Helpful Hint

When finding the vertex of the parabola, determine the x-coordinate first by using the formula $x = \frac{-b}{2a}$. Then, substitute the x-coordinate into the function to determine the y-coordinate of the vertex.

In Section 5.1, we covered the concept of functions and how to evaluate functions for a given value of x. The vertex of a parabola with equation $f(x) = ax^2 + bx + c$ occurs when $x = \frac{-b}{2a}$. Using this information, we can find the y-coordinate of the vertex by substituting $x = \frac{-b}{2a}$ into the equation of the parabola and evaluating.

Let's try graphing a couple of examples before modeling with quadratics.

Graphing Quadratic Functions

The graph of $f(x) = ax^2 + bx + c$ can be graphed using the following steps.

1. Establish the shape of the parabola as opening up $a > 0$, or opening down $a < 0$.

2. Find the vertex of the parabola: $\left(\frac{-b}{2a}, f\left(\frac{-b}{2a} \right) \right)$.

3. Determine the x-intercepts (if they exist) by substituting 0 for y or $f(x)$ and solving for x.

4. Determine the y-intercept by substituting 0 for x and solving for y.

5. Plot the vertex and the intercepts.

6. Connect the points to give a graph of the parabola.

Example 1: Graphing Quadratic Functions

Graph the quadratic function $y = x^2 - 6x - 7$.

Solution

We will go through our steps to graph the function.

1. Since $a = 1$ and $a > 0$, the parabola opens up.

2. To find the vertex, we find the x-coordinate using $x = \frac{-b}{2a}$.

   Based on the equation $y = x^2 - 6x - 7$, $a = 1$, $b = -6$ and $c = -7$.

   $$x = \frac{-b}{2a} = \frac{-(-6)}{2(1)} = \frac{6}{2} = 3$$

   So, the x-coordinate of the vertex is 3. The y-coordinate is found by substituting 3 for x in the quadratic function.

   $$\begin{aligned} y &= x^2 - 6x - 7 \\ &= (3)^2 - 6(3) - 7 \\ &= 9 - 18 - 7 \\ &= -16 \end{aligned}$$

   This gives that the vertex is $(3, -16)$. We know that the vertex in this case is a minimum, since the parabola opens up.

3. Find the x-intercepts by substituting 0 for y and solving for x.

   $$\begin{aligned} 0 &= x^2 - 6x - 7 \\ 0 &= (x - 7)(x + 1) \\ x - 7 = 0 \ &\text{or} \ x + 1 = 0 \\ x = 7 \ \text{or} \ \ \ & \ \ \ x = -1 \end{aligned}$$

   So, the x-intercepts are $(7, 0)$ and $(-1, 0)$.

4. The y-intercept is found by substituting 0 for x and solving for y.

   $$\begin{aligned} y &= x^2 - 6x - 7 \\ &= (0)^2 - 6(0) - 7 \\ &= -7 \end{aligned}$$

   So, the y-intercept is $(0, -7)$.

5. We will combine the last two steps. Step 5 is to plot the points we have found and Step 6 is to connect them to form a graphical representation.

Graph of $y = x^2 - 6x - 7$

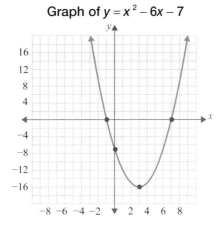

Note: In the previous example, finding the x-intercepts was quite easy. This will not always be the case, as many quadratic functions cannot be solved by factoring. When we are unable to factor easily, the quadratic formula should be used. For your convenience, recall that the quadratic formula is

$$x = \frac{-b \pm \sqrt{b^2 - 4ac}}{2a}.$$

TECH TRAINING

There are many graphing utilities available that may be used to find a representation of a function. For instance, utilities including graphing calculators (such as a TI-83/84 Plus, TI-Nspire, Casio FX9750, etc.), Microsoft Excel, Wolfram|Alpha, or even Google are readily available to most students.

To illustrate the ease of use of Wolfram|Alpha, here is the procedure to graph a function. Go to www.wolframalpha.com and type the function into the input bar of the website, then click the = button. When the results are returned, Wolfram|Alpha will automatically graph the function for you. For example, to graph $2x^2 + 3x - 1$ type "2x^2+3x-1" into the input bar. Then, click the = button. Wolfram|Alpha will return the graph of the function as well as the roots, or x-intercepts. [1]

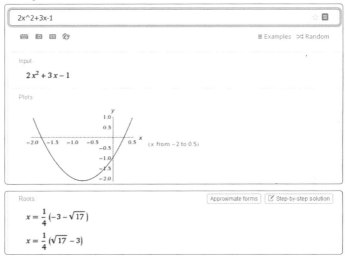

Modeling with Quadratics

Now we transition into modeling with quadratics. As we discovered earlier, quadratic functions are useful for modeling relationships we see all around us.

Example 2: Application of Quadratic Functions

We can borrow from physics that if an object is projected straight upward at time $t = 0$ from a point y_0 feet above ground, with an initial velocity of v_0 feet per second, then its height above the ground after t seconds is given by $h(t) = -16t^2 + v_0 t + y_0$.

Suppose a projectile is fired vertically upward from a height of 400 feet above the ground, with an initial velocity of 750 ft/sec. (That's 375 mph, if you're wondering.)

a. Write a quadratic model for its height $h(t)$ in feet above the ground after t seconds.

b. During what time interval will the projectile be more than 6000 feet above the ground? (That is, when will the height be greater than 6000 feet?)

c. How long will the projectile be in flight?

Solution

a. Given $y_0 = 400$ ft and $v_0 = 750$ ft/sec, $h(t) = -16t^2 + 750t + 400$.

b. To determine when the projectile will be more than 6000 feet above the ground, we need to find for which values of t we get $h(t) > 6000$. So, $-16t^2 + 750t + 400 > 6000$. Using a graphing utility, the following picture can be achieved by graphing both $h(t)$ and $y = 6000$ on a viewing window of $0 \le t \le 50$ and $0 \le y \le 10{,}000$.

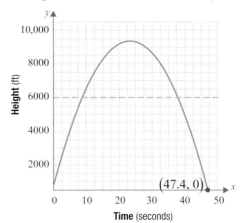

Also, we can determine that the intersection yields values of $t = 9.3$ and $t = 37.6$, indicating that the projectile is above 6000 feet when $9.3 < t < 37.6$.

c. To determine when the projectile will hit the ground, we need to understand when this occurs. The projectile will have a height of 0 when $y = 0$ or when $h(t) = 0$.

Using $h(t) = -16t^2 + 750t + 400$ and setting $h(t)$ equal to zero, we have $-16t^2 + 750t + 400 = 0$. We can use either the graphing utility or the quadratic formula to determine the value of the solution.

Using the quadratic formula, we can find the roots of function $h(t) = -16t^2 + 750t + 400$ as follows.

Quadratic formula: $x = \dfrac{-b \pm \sqrt{b^2 - 4ac}}{2a}$

Based on the function, $a = -16$, $b = 750$, and $c = 400$.

$$t = \frac{-750 \pm \sqrt{750^2 - 4(-16)(400)}}{2(-16)}$$

$$= \frac{-750 \pm \sqrt{562500 + 25600}}{-32}$$

$$= \frac{-750 \pm \sqrt{588100}}{-32}$$

$$\approx \frac{-750 \pm 766.877}{-32}$$

$$t \approx \frac{-750 + 766.877}{-32} \text{ or } t \approx \frac{-750 - 766.877}{-32}$$

This means that $t \approx -0.527$ seconds or $t \approx 47.4$ seconds. Since time can't be negative, the only logical answer would be $t \approx 47.4$ seconds.

TECH TRAINING

The solution to Example 2 part **c.** can be found using Wolfram|Alpha. Go to www.wolframalpha.com and type the function "−16t^2 + 750t + 400" into the input bar. Then, click the = button. The given result graphs the function. The roots in exact and approximate form can be found by scrolling down the page. [2]

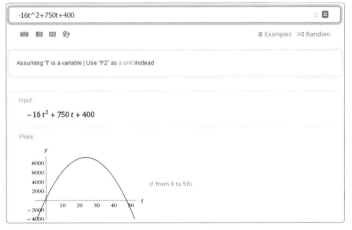

Using Wolfram|Alpha gives an answer of $t \approx 47.4$ seconds. The graph also confirms this value.

Skill Check #1

Suppose a rocket is fired vertically upward from a height of 1200 feet above the ground, with an initial velocity of 620 ft/sec. Write a quadratic model for its height $h(t)$ in feet above the ground after t seconds and determine how long the projectile will be in flight.

Recall the discussion of forensic scientists using quadratics to determine stopping distance. Suppose an experiment was performed to determine the stopping distances of a vehicle under various conditions. The results are shown in Table 2, which displays the number of feet required to stop a vehicle traveling at a given speed (in miles per hour) under various conditions. The distance to stop includes the braking distance and the reaction distance. The driving conditions considered were driving in normal conditions, driving while using a cell phone, and driving on a wet road.

Table 2			
Speed and Feet Required to Stop Under Various Conditions			
Speed (mph)	**Normal (ft)**	**Cell Phone (ft)**	**Wet Road (ft)**
15	10	17	12
20	15	24	18
25	20	32	25
30	26	40	33
35	32	48	42

If we make a plot of this data and draw a curve to fit the data, we can see the shape of the graph for each braking situation. See Figures 9 through 11.

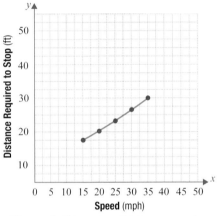

Figure 9: Stopping Distance under Normal Conditions

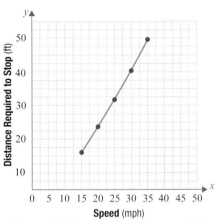

Figure 10: Stopping Distance while Using a Cell Phone

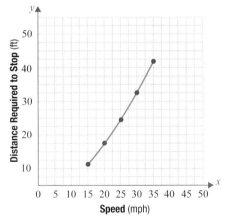

Figure 11: Stopping Distance on a Wet Road

With these relationships in mind, what effect do you think cell phones and wet roads have on stopping distance as opposed to normal driving conditions?

According to the National Highway Traffic Safety Administration, the relationship for stopping distance in each of the conditions is modeled as follows.[3]

Stopping Distance under Normal Conditions: $N(x) = 0.0086x^2 + 0.671x - 1.97$

---

3 National Highway Traffic Safety Administration, http://www.nhtsa.gov/

Stopping Distance while Using a Cell Phone: $C(x) = 0.0086x^2 + 1.11x - 1.37$

Stopping Distance on a Wet Road: $W(x) = 0.02x^2 + 0.5x$

Using these functions, we can see in Figure 12 that the distance required to stop while talking on a cell phone versus normal driving conditions is greater at all speeds, thus making it not safe to talk on a cell phone while driving. Notice that the stopping distance for driving on a wet road is similar to that of talking on a cell phone.

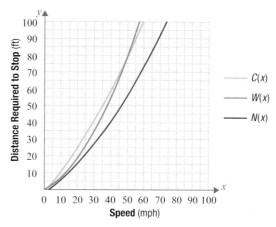

Figure 12: Speed and Feet Required to Stop
Under Various Conditions

Now we have developed two types of functions that model growth: linear functions and quadratic functions. Each of these types of growth is quite common in everyday situations. Next, we will change our focus to functions that grow much differently than what we have seen so far.

Skill Check #2

A car gets its best gas mileage when the speed of the car is kept low and constant. We can model the gas mileage of a car by the function $M(x) = -\frac{1}{28}x^2 + 3x - 31$, where x is the speed of the car in miles per hour and M is measured in miles per gallon. At what speed will the car attain its maximum number of miles per gallon?

Skill Check Answers

1. $h(t) = -16t^2 + 620t + 1200$; 40.6 seconds

2. 42 miles per hour

5.3   Exercises

Graph each parabola using the five-step method for graphing quadratics.

1. $f(x) = x^2 + 2x - 3$ 2. $f(x) = -x^2 + 4x - 3$ 3. $g(x) = x^2 - 9$

4. $h(x) = x^2 - 6x - 7$ 5. $f(t) = -t^2 - t + 6$ 6. $f(x) = 2x^2 - 3x - 1$

7. $f(x) = -x^2 - 2x + 5$

Answer each question.

8. The function $C(x) = 0.0086x^2 + 1.11x - 1.37$ represents the stopping distance in feet while talking on a cell phone and driving at a speed of x mph. What distance will it take you to stop while talking on a cell phone if you are driving 65 mph? 75 mph?

9. The function $W(x) = 0.02x^2 + 0.5x$ represents the stopping distance in feet on a wet road when driving at a speed of x. What distance will it take you to stop while driving on a wet road if you are driving 65 mph? 35 mph?

Use the given information and quadratic equation to answer each question.

Whispertown has a population growth model of $P(t) = at^2 + bt + P_0$ where P_0 is the initial population. Suppose that the future population of Whispertown t years after January 1, 2012, is described by the quadratic model $P(t) = 0.8t^2 + 6t + 24{,}000$.

10. What is the population of Whispertown on January 1, 2020?

11. In what month and year will the population reach 30,000?

12. How long will it take for the population to double to 48,000?

Use the given information and quadratic equation to answer each question.

A projectile is fired vertically upward from a height of 200 feet above the ground, with an initial velocity of 1100 ft/sec. Recall that projectiles are modeled by the function $h(t) = -16t^2 + v_0t + y_0$.

13. Write a quadratic equation to model the projectile's height $h(t)$ in feet above the ground after t seconds.

14. During what time interval will the projectile be more than 8000 feet above the ground? Round your answer to the nearest hundredth.

15. What is the total flight time of the projectile? Round your answer to the nearest hundredth.

Solve each problem. Round your answer to the nearest hundredth when necessary.

16. The population (in thousands) for Tylersville t years after January 1, 2010, is modeled by the quadratic function $P(t) = 0.7t^2 + 12t + 3450$. In what month and year does Tylersville's population reach twice its initial (1/1/2010) population?

17. A ball is thrown straight up, from ground zero, with an initial velocity of 55 feet per second. Find the maximum height attained by the ball and the time it takes for the ball to return to ground zero.

18. From the top of a 250-foot tall building, a ball is thrown straight up with an initial velocity of 25 feet per second. Find the maximum height attained by the ball and the time it takes for the ball to hit the ground.

19. A ball is dropped from the top of a 1250-foot tall building. How long does it take the ball to hit the ground? (Note that $v_0 = 0$.)

20. Grayson drops a rock into a well in which the water surface is 275 feet below ground level. How long does it take the rock to hit the water surface? (Note that $v_0 = 0$.)

5.4   Exponential Growth

So far, we have studied growth that occurs at a constant rate with linear functions and growth that occurs with projectile-type motion, or quadratic growth. Now we turn our attention to relative growth. Using an example from the previous section, let's compare two different types of growth.

Recall, from Section 5.2, the example that was done regarding the uncle that saved money for Jeffrey in a shoebox. The equation that modeled the amount of money in the shoebox for any given year was found to be $y = 5x + 10$. The graph of this function is shown in Figure 1.

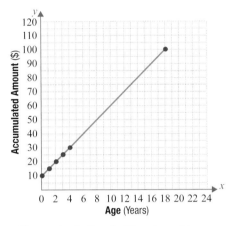

Figure 1: Jeffrey's Shoebox Money

It has already been established that the money in the shoebox grows at a constant rate of $5 per year, and Jeffrey would receive a total of $100 on his 18th birthday.

Now assume that Jeffrey had an aunt that started a similar savings plan, but doubled the amount she gave Jeffrey each year. The next example looks at the way the money grows in this case.

Example 1: Introducing Exponential Growth

On the day Jeffrey was born, his aunt put $1 in a shoebox for him. Each year on his birthday, she decides she wants to put away double the amount from the previous year. So she puts in $2 on his first birthday and $4 on his second birthday. On his third birthday, she continues the trend and doubles what she put away on his second birthday, so she puts $8 in the shoebox. If she continues this trend, how much money will she place in Jeffrey's shoebox on his 18th birthday?

Solution

Here is a chart to represent the data.

Table 1		
Jeffrey's Shoebox Money		
Age (Years)	Amount on Birthday ($)	Accumulated Amount ($)
0	1	1
1	2	3
2	4	7
3	8	15
4	16	31
5	32	63
6	64	127

Now, let's look at a graph of the data.

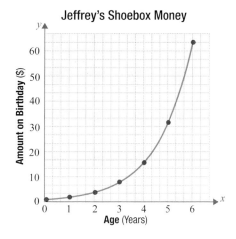

Jeffrey's Shoebox Money

What is different about the change in this situation compared to the example in Section 5.2? What do we need to know in order to develop a formula that models this growth?

In order to establish a formula for the pattern, consider the change taking place. Instead of the change being constant each year, the change each year is dependent on the amount from the previous year. In other words, *each year's growth is proportional to the previous year's growth*.

We begin building the function by starting with the information given. Since the initial amount of money at time $t = 0$ is \$1, this is the *y*-intercept. We recognize that this amount will double each year. So, the amount put away in any given year is represented as a double of the previous year. This can be done mathematically by looking at "doubles" such as:

$$1, 2, 4, 8, 16, 32, \ldots \text{ or } 2^0, 2^1, 2^2, 2^3, 2^4, 2^5, \ldots$$

Notice that the change in value (in years) occurs in the exponent. So, the function for the amount the aunt adds to the shoebox on Jeffrey's x^{th} birthday can be built with a function where the exponent is the variable.

$$f(x) = 2^x$$

To answer our original question, we can insert the value of 18 for the value of *x* to find how much money would be placed into the shoebox on Jeffrey's 18th birthday.

$$f(18) = 2^{18} = \$262,144$$

Notice that the amount deposited on Jeffrey's 18th birthday alone is \$262,144 while the *total* amount deposited by the uncle after 18 years was only \$100.

Skill Check # 1

Assume that on the day you were born your parents deposited \$1 in a shoebox for you. On your 1st birthday, they deposit \$3. On your 2nd birthday, they deposit \$9. On your 3rd birthday, they deposit \$27. Write a function to represent this situation. How much would they deposit on your 12th birthday?

Of importance here is the power of functions where the exponent is a variable. We call these exponential functions.

> **Exponential Function**
>
> An **exponential function** is a function of the form $f(x) = b^x$, where $b > 0$, $b \neq 1$, and x is any real number.

The following example is adapted from a famous problem developed by Professor Bartlett to illustrate the growth power of exponential functions.

Example 2: Application of Exponential Functions

A certain bacteria reproduces in such a way that every minute, the number of bacteria doubles. So at 8:00 a.m., or time $t = 0$, suppose we place a single bacterium in a jar. At 8:01 a.m., there are 2 bacteria in the jar and at 8:02 a.m., there are 4 bacteria in the jar. Assume the bacteria continue to reproduce in this manner, with no loss in the number of bacteria, and that the jar is full of bacteria at 8:30 a.m.

a. At what time will the jar be half full of bacteria?

b. Using what you already know, can you develop a formula to model this data?

Solution

Let's take a look at a table to see the pattern of growth. In the table, time $t = 0$ represents the starting time of 8:00 a.m. At 8:01 a.m., one minute of time will have elapsed. So, at 8:01 a.m., $t = 1$. This continues for the entire 30 minutes of the example.

Table 2
Bacteria Growth

Time (Minutes)	Number of Bacteria
0	1
1	2
2	4
3	8
4	16
5	32
⋮	⋮

a. If you take the time to plot the growth on a graph, you will quickly see the pattern again does *not* grow linearly or as a quadratic, but rather **exponentially**. It is a good question to ask why this example is exponential and not quadratic. The answer is that, since the growth is proportional to the previous amount, the growth must be exponential.

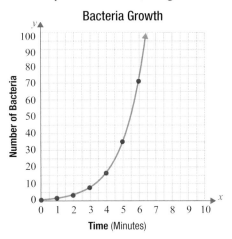

Bacteria Growth

Let's explore a little more about the pattern and what it represents.

At what time will the jar be half full of bacteria?

Careful here. Many will jump to the conclusion that the jar will be half full halfway through the time period, at 8:15 a.m. This is incorrect. Since the bacteria double every minute, the jar is actually half full at 8:29 a.m. Also, the jar is $\frac{1}{4}$ full at 8:28, $\frac{1}{8}$ full at 8:27, $\frac{1}{16}$ full at 8:26, etc.

b. The data can be modeled by the function $f(t) = 2^t$. This is very similar to the previous example. You should notice that the base of each is 2 because the amount doubles every unit of time and that at time $t = 0, f(0) = 2^0 = 1$.

In general, much of what has been discussed regarding exponential functions illustrates how exponential functions can be used to model growth that increases very rapidly, much like that of populations, bank account interest, etc. It is important to note that when something grows in an exponential manner, we should be cautious of whether the growth could continue infinitely. Situations where a population grows exponentially is a great example. If the population reaches a number that is too large, there may not be enough resources to sustain the growth. In this text, we focus on the concept of exponential growth as it relates to understanding how functions grow and how they can be used to represent situations.

General Form of an Exponential Function

The **general form of an exponential function** is

$$y = a \cdot b^x \quad \text{or} \quad f(x) = a \cdot b^x,$$

where b, the base, is a positive number not equal to 1, a is the initial value, and x is any real number.

For example, when b is 2, the function doubles with every unit increase in x, when b is 3, the population triples with every unit increase in x.

Example 3: Application of Exponential Functions

The sales data for a certain company for 8 years can be modeled by the data in Table 3.

Table 3

Sales Data

Year	1	2	3	4	5	6	7	8
Total Sales ($)	55	162	300	580	1050	1872	3454	5165

A plot of the data with an exponential model fitted to the data shows that the function that best represents the sales of the company is $f(x) = 40.2358(1.88346)^x$, where x represents the number of years.

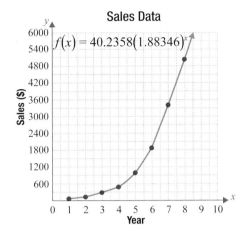

Sales Data

What are the projected sales in year 15?

Solution

The function that represents sales for the company is $f(x) = 40.2358(1.88346)^x$. So, the number of sales in year 15 would be the following.

$$f(15) = 40.2358(1.88346)^{15}$$

So, the projected sales in year 15 would be $535,745.66.

Another common illustration of exponential growth happens as populations grow. Consider a population that initially consists of 100 people and then grows at 5% each year. If the population increased by 5% in the first year, then the new population would be the initial population of 100 plus an additional 5% of 100, or 5 people. So, the new population after one year would be 105. The population for the first 4 years is shown in Table 4.

Table 4	
Population Growth	
Time (Years)	**Population**
0	100
1	105
2	110.25
3	115.7625
4	121.550625

Note that it is impossible to have a part of a person in a population, so if the estimated population were 110.25, we would round to 111 people.

Unlike our previous example where the amount doubled each year, when there is a percentage change each year, the amount of change for each iteration is relative to the previous amount. So, the rate of change alters the exponential as follows:

$$y = a(1+r)^x \text{ or } f(x) = a(1+r)^x,$$

where r is the rate of change, a is the initial amount, and x is the number of time intervals. Note that if the population is decreasing, r is negative.

Example 4: Percent Increase

A certain town had a population of 2000 in 2001. If the population is increasing at a rate of 2% per year, what will the population be in 2021?

Solution

Using our formula for the percentage change in a population, $f(x) = a(1+r)^x$, where $r = 0.02$, $a = 2000$, and $x = 20$, we can find the population.

$$f(x) = a(1+r)^x$$
$$f(x) = 2000(1+0.02)^x = 2000(1.02)^x$$
$$f(20) = 2000(1.02)^{20} \approx 2971.89$$

This means that the population after 20 years would be about 2972 people.

In this section, we discovered a new type of growth, where any given value of the function is proportional to the previous value. This attribute is the primary concept that differentiates exponential functions from linear or quadratic functions. In the next section, we will discuss another type of function that represent the world around us.

It should be noted that much of the exponential growth that we have discussed in this section assumes that a given population can grow indeterminately large. This, however, is not the case when populations grow in reality. For living things to exist, there must be enough resources to sustain their survival. When a given population grows exponentially, the survival resources are used up more and more rapidly, creating a point where the population simply becomes unsustainable.

Skill Check Answer

1. $f(x) = 3^x$; $531,441

5.4   Exercises

Answer the question thoughtfully.

1.   What is the primary difference between exponential and linear functions?

Solve each problem.

2.   The balance owed on your credit card triples from $500 to $1500 in 9 months.

   a. If your balance is growing linearly, how long will it take your balance to reach $6000?

   b. Check that the exponential equation $f(x) = 500(1.13)^x$ also describes the balance on the credit card after x months.

   c. Using the exponential equation in part **b.**, and your answer from part **a.**, find the balance owed on your credit card at that time if it grew exponentially rather than linearly.

3.   Bacteria in a bottle are quadrupling every minute. If the number of bacteria in the bottle at noon is 1, how many bacteria are in the bottle at 12:10 p.m.?

4.   The population of Greene Hills is decreasing at a rate of 2% per year. If the population is 20,000 today, what will the population be in 10 years?

5.   The price of a gallon of milk is increasing at a rate of 1% per year. If a gallon of milk costs $3.50 in 2014, how much will a gallon of milk cost in 2017?

Use the given information and exponential equation to answer each question.

The current population of a town in California is modeled by the function $f(x) = 1066.058(1.0434)^x$.

6.   If January 1, 2012, is when the initial population was measured (that is, $x = 0$), what was the initial population?

7.   What will the population be on January 1, 2015?

8.   What will the population be on January 1, 2030?

Use the given information to answer each question.

A checkerboard has 64 squares that alternate black and white. A grain of rice weighs 0.03 grams. Assume 1 grain of rice is placed on the first square, 2 grains are placed on the second square, 4 grains are placed on the third square, and so on.

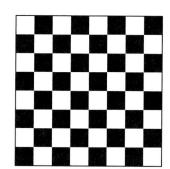

9.   How many grains will be placed on the 14th square?

10.   How many grains will be placed on the 25th square?

11.   How much does the rice on the 20th square weigh?

12.   How much does the rice on the 30th square weigh?

Solve the problem.

13.   In a given 30-day month, would you rather get a penny on day 1, two pennies on day 2, and continue to double that amount each day from the previous day, or start with a quarter on day 1, 2 quarters on day 4, 4 quarters on day 7 and continue to double the amount from the previous amount every 3rd day?

5.5   Logarithmic Growth

In October 2010, an earthquake with a magnitude of 7.7 on the Richter scale struck off the coast of Sumatra, triggering a tsunami that killed about 200 people, injured thousands more, and caused billions of dollars worth of damage. The Richter scale is an example of a type of function that allows us to understand, compare, and study quantities that vary from quite large to quite small.

The function we will study in this section is called a **logarithm**. Logarithms allow us to convert numbers so that we may compare large and small numbers alike. Consider that the earthquake that struck off the coast of Sumatra measured 7.7 on the Richter scale and that an earthquake that struck Conway, Arkansas, in February 2011 measured 4.5 on the Richter scale. How can we relate these earthquakes in a meaningful way? When we notice that the difference between the two is only 3.2 units on the scale, what does that mean? What does each unit represent? What is the difference between an earthquake with a magnitude of 7.7 and an earthquake with a magnitude of 6.7 or 8.7 or 9.7? Since the Richter scale is a logarithm of base 10, if we understand the base of the logarithm, then we can compare the intensities of each earthquake in an effort to compare the destructive power of each. In the case of the Richter scale, for each increase in one unit on the scale, there is an increase of 10 times in the intensity of the earthquake.

Here is an example of a logarithm: $\log_{10} 100$. (Read "log of 100 to base 10" or "log base 10 of 100.")

Simply put, in order to find the value of $\log_{10} 100$, we must answer the following question.

What power of 10 is 100?

If you can keep this idea in mind when working with logarithms, you will always be on the right track.

If we rearrange the expression to match our question, we ask ourselves,

10 raised to what power is 100?

Or $10^x = 100$.

There are many ways to proceed from here. One way is to use the definition of logarithms. The word **logarithm** implies *exponent* or *power*.

> ### Logarithm
>
> If $b^x = a$, then the **logarithm** with base b of a is x. Symbolically, we can express this exponential as an equivalent logarithm
>
> $$\log_b a = x,$$
>
> where $b > 0$, $b \neq 1$, and $a > 0$.

Using this definition, we can revisit our original question: *What power of 10 is 100?* Since $10^2 = 100$, we get $\log_{10} 100 = 2$.

How about if I asked you: *What power of 2 is 8?* Hopefully, you would reply with 3.

The corresponding logarithm is

$$\log_2 8 = 3. \text{ Equivalently, } 2^3 = 8.$$

This means that the power of 2 that is equal to 8 is 3.

Here's another one: *What power of 2 is 32?* The answer is 5.

One way to approach the solution is to examine the powers of 2.

$$2^1 = 2, \ 2^2 = 4, \ 2^3 = 8, \ 2^4 = 16, \ 2^5 = 32$$

Here's the logarithmic statement that says the same thing.

$$\log_2 32 = 5. \text{ Equivalently, } 2^5 = 32.$$

So, the power of 2 that is equal to 32 is 5.

Skill Check # I

Interpret what $\log_5 25 = 2$ means.

Example 1: Equivalent Logarithmic and Exponential Functions

Write each of the following in its equivalent exponential form.

a. $2 = \log_4 x$

b. $4 = \log_b 81$

c. $x = \log_5 125$

Solution

a. $4^2 = x$

b. $b^4 = 81$

c. $5^x = 125$

Example 2: Equivalent Logarithmic and Exponential Functions

Write each of the following in its equivalent logarithmic form.

a. $8^3 = x$

b. $x^4 = 16$

c. $6^x = 216$

Solution

a. $\log_8 x = 3$

b. $\log_x 16 = 4$

c. $\log_6 216 = x$

Some of the common uses of logarithms make them an important part of daily life. For instance, logarithms with a base of 2 are used in communications engineering and information technology. Think about your computer, cell phone, or just about any other electronic device. All electronic devices that have a programmed component use a series of 0s and 1s to represent information and tell the device how to function. For instance, "Off" on a device is represented by a 0 and "On" is represented with a 1. There are many other applications for logarithms, as we will see shortly.

It should be noted at this point that a logarithm is nothing more than the inverse operation for an exponent. This implies that a logarithm interchanges the coordinates from an exponential function. It also means that the logarithmic graph is a reflection of the graph of the exponential function.

Consider the function $f(x) = 2^x$ and the function for its corresponding logarithm, $g(x) = \log_2 x$. Graphs for both functions are shown in Figures 1 and 2.

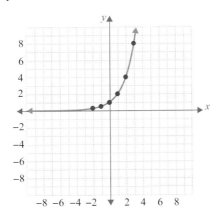

Figure 1: Graph of $f(x)$

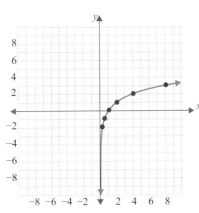

Figure 2: Graph of $g(x)$

Now consider Tables 1 and 2 of values for $f(x) = 2^x$ and $g(x) = \log_2 x$.

Table 1						
Values for $f(x) = 2^x$						
x	−2	−1	0	1	2	3
$f(x) = 2^x$	$\frac{1}{4}$	$\frac{1}{2}$	1	2	4	8

Table 2						
Values for $g(x) = \log_2 x$						
x	$\frac{1}{4}$	$\frac{1}{2}$	1	2	4	8
$g(x) = \log_2 x$	−2	−1	0	1	2	3

☞ **Helpful Hint**

A horizontal asymptote of a graph is a line that the graph approaches as the values of x tend to infinity or negative infinity.

The graphs of the two relationships illustrate an interesting connection between exponential functions and logarithms. Notice that the input values and the output values of $f(x)$ in Table 1 and $g(x)$ in Table 2 are the reverse of each other. Also recognize the asymptotic relationship between the functions. Function $f(x)$ has a horizontal asymptote of $y = 0$ and function $g(x)$ has a vertical asymptote of $x = 0$. This causes the graphs to be symmetric about the line $y = x$, as shown in Figure 3.

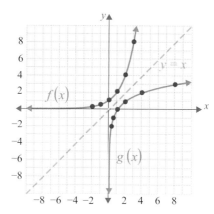

Figure 3: Relationship Between Graphs $f(x)$ and $g(x)$

Up to this point, we have discussed logarithms with many different bases. Let us now look at the logarithm with a base of 10. We call this the **common logarithm**.

> ## Common Logarithm
>
> A **common logarithm** is a logarithm with base 10. We usually express this function as $f(x) = \log_{10} x = \log x$, where $x > 0$.

The logarithmic model has a period of rapid growth, followed by a period where the growth slows, but the growth continues to increase without bound. This makes the model inappropriate where there needs to be an upper bound (or where there is a maximum number that cannot be exceeded). The main difference between this model and the exponential growth model is that the exponential growth model begins slowly and then increases very rapidly as time increases.

In general, we use logarithms to model growth patterns that deal with the intensity of natural phenomena. Since intensity can vary widely depending on situations such as earthquakes and sound, we must use a logarithm to allow us to discuss the magnitude of the number we are discussing.

When we are born, our body lengths are measured. As we continue to grow into adulthood, our bodies do not grow linearly or exponentially, they grow logarithmically. Do you know why? Consider this: if our bodies were to grow linearly or exponentially, then we would continue to grow larger and larger forever. Common sense tells us this is not the case. Our bodies grow rapidly when we are very young, and that growth slows down as we get older. This growth pattern does not match linear or exponential growth. Therefore, there must be some other growth taking place.

Example 3: Evaluating Logarithmic Functions

Past research tells us that our bodies grow at a given rate based on our DNA and the environmental factors in which we grow up. Assume that a baby girl is born and is 22 inches long. If we consider that the baby will grow at a rate proportional to her body size until adulthood, then the percentage of her adult height attained can be modeled by the logarithmic function

$$f(x) = 22 + 55\log(x+1),$$

where x represents her age in years between 0 and 25, and $f(x)$ represents the percentage of her adult height reached at age x. What percent of the girl's adult height is attained by age 12?

Solution

If we wish to know the percentage of her adult height that the girl will have attained at age 12, then we evaluate $f(12)$.

$$f(x) = 22 + 55\log(x+1) \qquad \text{Given function}$$
$$f(12) = 22 + 55\log(12+1) \qquad \text{Substitute for } x$$
$$= 22 + 55\log(13)$$
$$\approx 83.267$$

So, she will have attained about 83% of her adult height by the time she is 12 years old.

TECH TRAINING

Use the following keystrokes on a TI-30XIIS/B or TI-83/84 Plus calculator to evaluate the logarithmic function in Example 3.

22 [+] 55 [LOG] 13 [ )] [ENTER]

Skill Check #2

Using the same information from Example 3, what percentage of her adult height will a girl of age 18 have attained? Round your answer to the nearest percent.

Several physical applications have logarithmic models. From the beginning of this section, recall the discussion on measuring the intensity of earthquakes using logarithms. Other logarithmic models are the intensity of sound, the capacity of human memory, and the acidity of a solution.

Earthquakes

The Richter scale is used to measure the intensity of an earthquake. The actual model is a little more complex, but can be simplified to the equation

$$R = \log\left(\frac{I}{I_0}\right),$$

where R is the magnitude on the Richter scale of the earthquake, I_0 is the intensity of an earthquake that is barely felt, or a zero-level earthquake, and I is the intensity of the earthquake measured relative to a reference value. That reference value is the smallest seismic activity that can be measured and has the value $I_0 = 1$.

Example 4: Application of Logarithmic Functions Involving Earthquakes

Every increase of 1 in the Richter scale means the magnitude of the earthquake is 10 times greater. What is the magnitude of an earthquake that is 1000 times stronger than a zero-level earthquake?

Solution

Since $R = \log\left(\frac{I}{I_0}\right)$, an intensity of 1000 times that of a zero-level quake would mean the following.

$$R = \log\left(\frac{I}{I_0}\right) = \log\left(\frac{1000}{1}\right) = \log(1000) = 3$$

So, an earthquake that is 1000 times as intense as a zero-level earthquake has a Richter scale value of $R = 3$.

Example 5: Application of Logarithmic Functions Involving Earthquakes

Recall from the beginning of the section that the earthquake that created a huge tsunami off the coast of Sumatra in October 2010 was of magnitude 7.7. If I_0 is 1, what was the intensity level I of the Sumatran earthquake?

Solution

Every increase of 1 in the Richter scale means the magnitude of the earthquake is 10 times greater.

Since $R = \log\left(\frac{I}{I_0}\right)$ and the Sumatran earthquake had a magnitude of 7.7, we have the following.

$$R = \log\left(\frac{I}{I_0}\right)$$
$$7.7 = \log\left(\frac{I}{1}\right)$$
$$7.7 = \log(I)$$
$$10^{7.7} = I$$
$$I = 50,118,723.36$$

So the intensity level is approximately $I \approx 50,118,723$ and the Sumatran earthquake was over 50 million times stronger than a zero-level earthquake.

Example 6: Application of Logarithmic Functions Involving Earthquakes

An earthquake that hit Chile in 1960 was estimated to have been about 9.5 on the Richter scale. (This is the strongest earthquake ever recorded!) Find the intensity I of this earthquake if I_0 is 1.

Solution

Using our formula for the Richter scale, $R = \log\left(\frac{I}{I_0}\right)$, we can substitute our values for R and I_0 to obtain

$$9.5 = \log\left(\frac{I}{1}\right) = \log I.$$

Now, using the definition of logarithms, we can obtain

$$10^{9.5} = I$$

and the intensity level is approximately $I \approx 3{,}162{,}277{,}660$. That means that the Chilean earthquake was over 3.16 billion times stronger than a zero-level earthquake.

Skill Check #3

Determine how much stronger the Chilean earthquake was than the Sumatran earthquake.

Sound Intensity

The sound intensity level, measured in decibels and given by the formula

$$dB = 10\log\left(\frac{I}{I_0}\right),$$

measures the intensity of sound based on the reference value I_0. This value of I_0 is the threshold (minimum sound intensity) of hearing at 1 kHz, so we consider the reference value of I_0 to be equal to 1. The following chart compares the decibels for various daily exposures to hearing.

Table 3
Decibels for Daily Exposures

Decibels	Examples	Result
0–20	Silence to a quiet whisper	Very faint
40–60	Quiet conversation to regular conversation	Moderate
80–90	Yelling to the sound of a hair dryer	Loud
100–120	Car horns to rock concerts	Very loud with hearing loss
120–140	Fire crackers and gun shots	Very loud with hearing loss
150+	Jet engine from 25 feet	Eardrum rupture

There are 10 decibels to a bel. While the bel is the actual unit, like meter, liter, or gram, we use decibel for all practical purposes. An increase of 10 decibels is equivalent to a sound that is 10 times as strong in intensity. An increase of 20 decibels is equivalent to a sound intensity that is 100 times greater.

Example 7: Application of Logarithm Functions Involving Sound Intensity

If the decibel level of a certain sound is 115 dB, what is the level of intensity of the sound?

Solution

Substituting known values, we get the following.

$$dB = 10\log\left(\frac{I}{I_0}\right)$$

$$115 = 10\log\left(\frac{I}{1}\right)$$

$$115 = 10\log(I)$$

$$11.5 = \log(I)$$

$$10^{11.5} = I$$

So, the intensity of the sound is about $10^{11.5}$ times greater than silence, or about 316,227,766,017 times greater than silence (this represents the increase in intensity from our zero-decibel threshold). That's 316 billion times louder!

One more example of a logarithm with a special base is the natural logarithm.

Natural Logarithm

The **natural logarithm** is a logarithm with a base of e, where e is an irrational constant approximately equal to 2.718281828. The natural logarithm is generally written as $\ln x$ or $\log_e x$. Furthermore, the natural logarithm can be defined by the relationship between the common logarithm and the natural logarithm as $\ln x = \frac{\log x}{\log e}$.

Just like the common logarithm and logarithms with different bases, the natural logarithm of a number has an exponential equivalent. The value of a natural logarithm is the answer to the question, *to which power would e have to be raised to equal x?*

For example:

$$\ln(20.085536\ldots) \text{ is 3, since } e^3 = 20.085536\ldots$$

The natural log of e is equal to 1, or $\ln e = 1$ because $e^1 = e$. Also, $\ln 1 = 0$ since $e^0 = 1$.

Human Memory Capacity

The capacity of the human mind to process information may also be modeled with logarithmic growth. Let's assume that the percentage of information the average person can recall after a certain period of months have passed is modeled by

$$f(t) = 85 - 15\ln(t+1).$$

In this case, t is the number of months that have gone by after being presented the information (up to one year) and $f(t)$ is the percent of the information retained after t months.

Example 8: Human Memory Capacity and Logarithms

What percentage of learned material was retained 6 months after being presented the information?

Solution

Using the equation $f(t) = 85 - 15\ln(t+1)$, we can substitute 6 months for t and evaluate.

$$f(t) = 85 - 15\ln(t+1)$$
$$f(6) = 85 - 15\ln(7)$$
$$\approx 55.8113$$

So, this means that after 6 months, the amount of material that was retained was about 56%.

In this chapter, we have developed a library of mathematical growth that occurs around us in daily life. This chapter should give you a basis for understanding that mathematical growth occurs everywhere.

Skill Check Answers

1. $5^2 = 25$.

2. 92% of her adult height will be attained by the time she is 18 years old.

3. About 63 times stronger.

5.5 Exercises

Find the missing values for each logarithm using the definition.

1. $\log_b 64 = 6$

2. $\log_5 a = 2$

3. $\log_3 27 = x$

4. $\log_9 3 = x$

5. $\log_b 6 = \dfrac{1}{3}$

6. $\log_4 a = \dfrac{3}{2}$

7. $\ln 1 = x$

8. $\ln e = x$

9. $\ln e^2 = x$

10. $\ln x = 5$

Use the given information and logarithmic equation to answer each question. Round your answer to the nearest whole percentage when necessary.

The capacity of the human mind to process information may also be modeled with logarithmic growth. Let's assume that the percentage of information the average person can recall after t months have passed is modeled by:
$f(t) = 95 - 25\ln(t + 1)$ where $0 \le t \le 12$.

11. What percentage of information is retained after 1 month?

12. What percentage of information is retained after 7 months?

13. What percentage of information is lost after 8 months?

14. What percentage of information can the average person recall after a year has passed?

15.   Plot the information you found in Exercises 11, 12, 13, and 14 on a graph. Can you predict when a person will no longer be able to recall essentially any of the original information? You may plot more points if needed.

Use the given information and logarithmic equation to answer each question. Round your answer to the nearest hundredth when necessary.

pH = −log[H⁺] where [H⁺] is the hydrogen ion concentration, measured in moles per liter. Solutions with a pH-value of less than 7 are acidic; solutions with a pH-value of greater than 7 are basic; solutions with a pH-value of 7 (such as pure water) are neutral.

16.   Suppose that you test apple juice and find that the hydrogen ion concentration is $[H^+] = 0.0003$. Find the pH value and determine whether the juice is basic or acidic. Round your answer to the nearest hundredth.

17.   You test some ammonia and determine the hydrogen ion concentration to be $[H^+] = 1.3 \times 10^{-9}$. Find the pH value and determine whether the ammonia is basic or acidic. Round your answer to the nearest hundredth.

18.   An orange has a pH of about 3. Find the concentration of hydrogen ions in an orange.

19.   An antacid tablet has a pH of 11. Find the concentration of hydrogen ions in the tablet.

20.   Acid rain has a pH of 4 and normal rain has a pH of 6. Compare the pH of each and determine how much more acidic acid rain is than normal rain.

Use the given information and logarithmic equation to answer each question. Round your answer to the nearest whole number when necessary.

Recall that sound intensity is measured by the formula $dB = 10 \log\left(\dfrac{I}{I_0}\right)$.

21.   A dog bark is 400 times as intense as the threshold of sound. How many decibels is a dog bark? Round your answer to the nearest whole decibel.

22.   A jet engine is about 155 decibels. What is the intensity level of the jet engine? Round your answer to the nearest whole number.

23.   The threshold of hearing loss is about 125 decibels. If a gunshot has an intensity of 5.012×10^{12}, should you wear ear protection to prevent hearing loss?

Use the given information and logarithmic equation to answer each question. Round your answer to the nearest hundredth when necessary.

Earthquake intensity is measured by $R = \log\left(\dfrac{I}{I_0}\right)$.

24. What is the magnitude of an earthquake that is 5×10^8 times as intense as a zero-level quake?

25. A dump truck has an intensity of about 975. If a seismograph will register a quake of anything larger than a magnitude of 3, will the dump truck cause a small earthquake on the seismograph?

Use the given information and logarithmic equation to answer each question. Round your answer to the nearest whole percentage when necessary.

A child is 20 inches long at birth. If we consider that the baby will grow at a rate proportional to its body size until adulthood, then the percentage of her adult height attained can be modeled by the logarithmic function $f(x) = 20 + 47\log(x + 2)$, where x represents her age in years, and $f(x)$ represents the percentage of her adult height reached at age x.

26. What percent of the child's adult height is attained by age 4?

27. What percent of the child's adult height is attained by age 12?

28. What percent of the child's adult height is attained by age 19?

29. At what age will the child reach 95% of her adult height?

⑤

Chapter 5   Summary

Section 5.1   The Language of Functions

Definitions

Function

A function is a mathematical equation that describes the relationship between the dependent and independent variables for a given situation.

Dependent Variable

The value of the dependent variable changes with respect to the values of the independent variable.

Independent Variable

The value of the independent variable does not rely on the values of the other variables in an expression or function, and its value determines the values of the other variables.

Domain

The domain of a function is the set of input values (values of the independent variable) for which a function is defined. You can only evaluate a function at values that are in the domain.

Range

The range of a function is the set of output values (values of the dependent variable) that correspond to the domain values.

Section 5.2   Linear Growth

Definitions

Direct Variation

When two expressions are proportional, we say that there is a direct variation among the expressions.

Slope

The slope of a line is defined as the ratio of the change in the y-values ("rise") divided by the change in the x-values ("run") between two points.

Linear Function

The graph of a relationship that creates a straight line with constant change is a linear function.

Function Notation

Notation that labels the independent and dependent variables separately and allows for better association between variables.

x- and y-Intercepts

The x-intercept is the point where the graph intersects the x-axis (horizontal axis) and the y-intercept is the point where the graph intersects the y-axis (vertical axis).

Slope-Intercept Form of a Linear Equation

The slope-intercept form of a linear equation is $y = mx + b$, where m represents the slope of the line and b is the value where the graph crosses the y-axis. In other words, the point $(0, b)$ is the y-intercept.

Formula
Slope

The slope of a line is defined by the following formula.

$$\text{slope} = m = \frac{\text{rise}}{\text{run}} = \frac{\text{change in } y\text{-values}}{\text{change in } x\text{-values}} = \frac{y_2 - y_1}{x_2 - x_1}$$

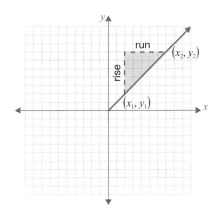

Section 5.3   Discovering Quadratics

Definitions
Quadratic Function

A quadratic function is a function of the form $f(x) = ax^2 + bx + c$, where $a \neq 0$. The common name for the graphical representation of a quadratic function is parabola.

Attributes of Parabolas

If $a > 0$, the parabola will open up.

If $a < 0$, the parabola will open down.

Every parabola has a maximum or minimum point called the vertex.

Parabolas are symmetrical about a vertical line passing through the vertex. This line is called the axis of symmetry.

Graphing Quadratic Functions

1. Establish the shape of the parabola as opening up ($a > 0$) or opening down ($a < 0$).

2. Find the vertex of the parabola: $\left(\dfrac{-b}{2a}, f\left(\dfrac{-b}{2a} \right) \right)$.

3. Determine the x-intercepts (if they exist) by substituting 0 for y or $f(x)$ and solving for x.

4. Determine the y-intercept by substituting 0 for x and solving for y.

5. Plot the vertex and the intercepts.

6. Connect the points to give a graph of the parabola.

Section 5.4   Exponential Growth

Definitions

Exponential Function

An exponential function is a function of the form $f(x) = b^x$, where $b > 0$, $b \neq 1$, and x is any real number.

General Form of an Exponential Function

The general form of an exponential function is

$$y = a \cdot b^x \text{ or } f(x) = a \cdot b^x$$

where b, the base, is a positive number not equal to 1, a is the initial value, and x is any real number.

Section 5.5   Logarithmic Growth

Definitions

Logarithm

If $b^x = a$, then the logarithm with base b of a is x. Symbolically, we can express this exponential as an equivalent logarithm $\log_b a = x$, where $b > 0$, $b \neq 1$, and $a > 0$.

Common Logarithm

A common logarithm is a logarithm with base 10. We usually express this function as $f(x) = \log_{10} x = \log x$, where $x > 0$.

Natural Logarithm

The natural logarithm is a logarithm with a base of e, where e is an irrational constant approximately equal to 2.718281828. The natural logarithm is generally written as $\ln x$ or $\log_e x$ Furthermore, the natural logarithm can be defined by the relationship between the common logarithm and the natural logarithm as

$$\ln x = \frac{\log x}{\log e}.$$

Formulas

Earthquakes

The magnitude R of an earthquake on the Richter scale is calculated by the following formula.

$$R = \log\left(\frac{I}{I_0}\right)$$

where I_0 is the intensity of the earthquake that is barely felt and I is the intensity of the earthquake measured relative to the reference value.

Sound Intensity

The sound intensity level dB, measured in decibels is calculated by the following formula.

$$dB = 10\log\left(\frac{I}{I_0}\right)$$

where I_0 is the threshold of hearing at 1 kHz and I is the intensity of the sound measured relative to the reference value.

Human Memory Capacity

The percentage of information the average person can recall after a certain period of months t is calculated with the following formula.

$$f(t) = 85 - 15\ln(t+1)$$

Chapter 5 Exercises

Graph each equation by evaluating integer values of *x* from –2 to 2, plotting the resulting points, and then connecting the points with a smooth, unbroken line.

1. $y = -2x$

2. $y = 3x + 1$

3. $y = \dfrac{2}{3}x - 1$

4. $y = x^2 - 1$

5. $y = -x^3 + 2x + 3$

Evaluate each function for the given values of *x*.

6. $f(x) = x^2 + 2x - 3$

x	$f(x) = x^2 + 2x - 3$
–3	
–2	
–1	
0	
1	
2	
3	

Determine the dependent and independent variables. Then write a function to represent each situation and answer each question.

7. A car purchased in 2010 had an initial value of $19,500. If the car depreciates at $1200 per year, write an equation that represents the depreciation. What is the value of the car after 5 years?

8. The cost to rent a car for a day is an initial cost of $35 and $0.25 per mile driven. Write an equation that represents to the cost of the car rental. What is the total cost if you drove 250 miles?

9. Jeff delivers pizza for Papa Jim's Pizza. Papa Jim's pays Jeff an hourly wage of $6.25. In addition to his hourly wage, Jeff earns $0.25 for every pizza he delivers. Write an equation that represents the wages earned by Jeff, where *x* represents the number of pizzas delivered when Jeff works a five-hour shift. If Jeff delivers 21 pizzas during his shift, how much money will he earn?

Solve each problem.

10. A basketball's height (in feet) is a function of time in flight (in seconds), modeled by an equation such as $h(t) = -16t^2 + 40t + 6$.

   a. Describe the graph of the function in terms of time in flight and height.

   b. Describe how you could use the given function relating time in flight to height to find when the shot will reach the height of the basket (10 feet)?

   c. Describe how you could find the time when the ball would hit the floor if it missed the basket entirely?

11. The figure contains the graph of a function, $f(x) = -0.00875x^2 + 0.775x + 13$, that measures the average fuel economy for a certain model of car based on the speed x of the car (in miles per hour).

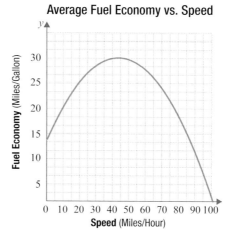

Average Fuel Economy vs. Speed

   a. What is the miles per gallon average if the vehicle is traveling 80 miles per hour?

   b. What is the miles per gallon average if the vehicle is traveling 40 miles per hour?

   c. At approximately what speed will the car have the greatest fuel economy? Round your answer to the nearest whole mile per gallon.

Find the slope and *y*-intercept of each linear equation.

12. $y = -3x + 4$

13. $y = \dfrac{3}{4}x - 3$

Write a function to represent each situation and answer each question.

14. Jenae works a job that pays $11.50 per hour. Find an equation that represents her pay based on the number of hours worked.

15. A movie company spent $12,000,000 making a feature film. If tickets to see the movie cost $8.00 each, write a function that represents the income from the movie.

16. A movie company spent $12,000,000 making a feature film. If tickets to see the movie cost $8.00 each, determine the number of tickets that must be sold in order for the company to break even.

17. The price of a computer is $1500 today. The price rises at a constant rate of $115.00 per year. How much will the computer cost in 5 years?

18. A population of mold decays at a rate of 30 mold spores per day. If the initial population of mold spores is 1560 spores, how many days will it take for the population to be less than 1000 mold spores?

Solve each problem.

19. The temperature outside is $15\,°C$; using the formula $F = \frac{9}{5}C + 32$, find the equivalent temperature in $°F$.

20. The function $C(x) = 0.0086x^2 + 1.11x - 1.37$ represents the stopping distance in feet while talking on a cell phone and driving at a speed of x mph. What distance will it take you to stop while talking on a cell phone if you are driving 50 mph? 80 mph?

21. Whispertown has a population growth model of $P(t) = at^2 + bt + P_0$ where P_0 is the initial population. Suppose that the projected future population of Whispertown t years after January 1, 2012, is described by the quadratic model $P(t) = 0.8t^2 + 6t + 24000$.

    a. What is the projected population of Whispertown on January 1, 2030?

    b. In what month and year will the population reach 60,000?

22. A rifle fires a projectile vertically upward from a height of 150 feet above the ground, with an initial velocity of 1250 ft/sec. Recall that projectiles are modeled by the function $h(t) = -16t^2 + v_0 t + y_0$.

    a. Write a quadratic equation to model the height in feet above the ground after t seconds.

    b. During what time interval will the projectile be more than 10,000 feet above the ground?

    c. After how many seconds will the projectile hit the ground?

23. A ball is dropped from the top of a 750-foot tall building. How long does it take the ball to hit the ground? Recall that projectiles are modeled by the function $h(t) = -16t^2 + v_0 t + y_0$. (Note that $v_0 = 0$.)

24. A polymerase chain reaction is a process in molecular biology to amplify DNA, meaning from very little DNA, scientists can generate millions of copies of a DNA sequence. The equation $f(x) = m \cdot 2^x$ models the mass (in grams) of DNA that is generated after x cycles from an initial mass of m grams.

   a. Find the mass of DNA that is generated from 0.0002 grams of DNA after 5 cycles.

   b. A forensic scientist needs about 0.5 grams of DNA to run an exhaustive set of tests to establish conclusive evidence for a criminal case. Use a spreadsheet or chart to find how many cycles it would take the scientist to have enough DNA if he begins with a human hair weighing 0.00015 g. Round your answer to the nearest whole cycle.

   c. If a cycle takes about 30 minutes to run, how long will it take the forensic lab to generate enough DNA for the criminal case in part **b.**?

25. For exponential equations of the form $y = a \cdot b^x$, if b is a number greater than 1, then it is called exponential growth. If b is a number between 0 and 1, then it is called exponential decay. If b is equal to 1, the function is constant. Let $a = 10$.

   a. Choose a number for b that is larger than 1 and graph at least 5 points from the equation on a graph.

   b. Choose a number for b that is between 0 and 1 and graph at least 5 points from the equation on a separate graph.

   c. Which of your two graphs could be a model for the height to which a ball bounces after x bounces?

26. As a birthday present, Aunt Tess bought you a share of stock in the newest technology gadget. When Aunt Tess purchased it, the price was $10 per share.

   a. To your surprise, the stock grew at a monthly rate of 8%. Find the number of months it took for the stock to at least double in value.

   b. When the value of the stock doubled, you decided to invest and purchase 9 more shares, giving you 10 shares in the stock. How much was the total value of your stock rounded to the nearest dollar at this point?

   c. Unfortunately, the stock began decreasing at a monthly rate of 7% right after your purchase. How much was each of your stocks worth after another 10 months?

   d. How much money did you personally lose at the end of the 20 months rounded to the nearest dollar? (**Hint:** Only consider the cost of the stocks you purchased, and not the gift from Aunt Tess.)

27. This world population chart shows how the world's most populous countries have grown over the two decades of 1990–2010. Assuming that the populations grow at the same rates over the next decade, use the formula for each country's growth to answer the questions.

$$\text{Population} = \left(2010 \text{ population}\right) \cdot e^{\frac{\text{growth \%}}{100}(\text{year} - 2010)}$$

		Population 2010	Population 1990	Growth (%) 1990–2010
Rank	Country			
	World	6,895,889,000	5,306,425,000	30.0%
1	China	1,341,335,000	1,145,195,000	17.1%
2	India	1,224,614,000	873,785,000	40.2%
3	United States	310,384,000	253,339,000	22.5%
4	Indonesia	239,871,000	184,346,000	30.1%
5	Brazil	194,946,000	149,650,000	30.3%
6	Pakistan	173,593,000	111,845,000	55.3%
7	Nigeria	158,423,000	97,552,000	62.4%
8	Bangladesh	148,692,000	105,256,000	41.3%
9	Russia	142,958,000	148,244,000	–3.6%
10	Japan	128,057,000	122,251,000	4.7%

Population Growth of the World's Most Populous Countries 1990–2010

Source: Wikipedia, s.v. "Population growth." Accessed March 2013. http://en.wikipedia.org/wiki/Population_growth

a. Estimate the population of Nigeria in 2015.

b. When will India overtake China as the most populous country?

c. When will Japan overtake Russia as the ninth most populous country?

d. Estimate the population of the United States this year. Compare your answer to the population currently given on Wikipedia's website. Explain why you do or do not believe that last decade's growth is a good indicator of this decade growth.

28. Radioactive substances lose their radioactivity over time in a way that is described by exponential decay. The half-life of a radioactive substance is the amount of time it takes for half of the radioactive material to be gone. The radioactive isotope carbon 15 decays at a rapid rate. If we begin with 350 grams, the mass of carbon after t seconds is modeled by $C(t) = 350e^{-0.283t}$. Round your answer to the nearest hundredth.

a. How much of the sample is left after 15 seconds?

b. What is the half-life of carbon 15?

Bibliography

5.3

1. Wolfram Alpha LLC. 2009. Wolfram|Alpha. http://www.wolframalpha.com/input/?i=2x%5E2%2B3x-1 (accessed July 8, 2014).

2. Wolfram Alpha LLC. 2009. Wolfram|Alpha. http://www.wolframalpha.com/input/?i=-16t%5E2%2B750t%2B400 (accessed July 2, 2014).

3. National Highway Traffic Safety Administration. http://www.nhtsa.gov/

Chapter 6
Geometry

Sections

Objectives

- Demonstrate an understanding of points, lines, and planes
- Apply the concepts of parallel and perpendicular
- Explore the properties of polygons
- Demonstrate an understanding of angle measure, angle sum, and applications of angles
- Apply the concepts of similar triangles
- Demonstrate an understanding of sine, cosine, and tangent functions
- Apply the concepts of perimeter and area
- Apply the concepts of surface area and volume

Geometry

Have you ever thought about how your cell phone gives you turn-by-turn directions to a location? What about when you lose your cell phone and you use an application on another device to locate your lost phone? These and *many* more concepts of location are based on a Global Positioning System (GPS). It may not be obvious, but geometry plays a large role in these scenarios.

What about your house or apartment? Have you ever stopped to consider how your house was actually constructed? The contractor building your house needs to have a firm grasp of plane geometry in order to make sure the walls are parallel to each other, while being perpendicular to the floor at the same time. In addition, the steepness of the roof on top of your house requires the contractor to have knowledge of slope, parallel lines, transversal lines, complementary angles, and supplementary angles, just to name a few!

In fact, geometry permeates most aspects of our lives every day. If you live in a hilly or mountainous area, the steepness of the road, usually given in the form "the grade of the road for the next 5 miles is 6%," is nothing more than the angle the road makes in relation to level ground. What about when you are driving your car and you look into your rearview mirror? The angle at which the mirror is placed allows you to see in several directions at once. The world around us is full of examples of geometry, and having an understanding of these concepts makes us better able to make decisions and solve problems. The purpose of this chapter is to use concepts of geometry as a means of introducing real-world applications of these concepts.

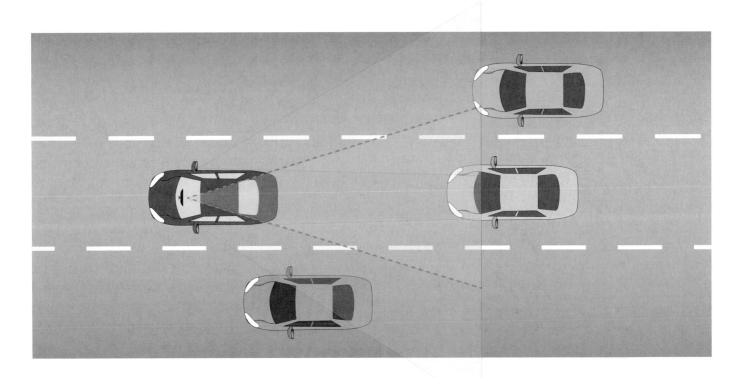

Figure 1

6.1 Everyday Geometry and Applications

MATH MILESTONE

Euclid was a Greek Mathematician and has often been called the "father of geometry." He wrote *Elements*, which was a collection of mathematics—some of which had been discovered by other mathematicians—that Euclid organized into a clear, concise presentation of ideas. *Elements* was used as a prominent math textbook until the early 20th century—over 2000 years after its publication!

Geometry is one of the most useful topics for real life that you will ever study. Its use in the construction of houses, buildings, and roads is almost limitless. The use of geometry in art, music, and other fields of humanities means that geometry is involved in most of the subjects you will study. Much of what we understand today regarding geometry can be traced to the Greek mathematician Euclid (325–270 BC). It is through his work in *Elements* that we understand the notions of plane geometry. We begin this chapter with an overview of basic definitions and ideas needed to apply the concepts of geometry.

Points, Lines, Planes, and More

The word geometry translates from Greek to mean "earth measure." In order to build or create anything requiring geometry, we need to begin with the basics of geometry. We start with three terms—point, line, and plane—that allow us to develop some basic ideas in geometry.

A **point** has no length, width, or shape.

A **line** is a collection of points that extends infinitely in opposite directions, and will always be assumed to be straight. A **line segment** is the portion of a line that lies between two points, called **endpoints**. Thus, a line segment is a part of a line consisting of two endpoints and all points between them. For example, a thread or piece of string pulled tight between two nails could be used to represent a line segment. A **ray** is defined as a half-line with an end point at a point A and directionality through a point B, continuing on infinitely in that direction. So a ray is a line that has a starting point and continues infinitely in only one direction.

A **plane** is a flat surface that has no thickness, depth, or boundaries. A table, desktop, or piece of paper can be used to represent planes, but a plane continues infinitely in all directions.

Table 1	
Summary of Points, Lines, and Planes	
Term/Example	**Definition**
Point • P	A **point** has no length, width, or shape. A point is denoted by an uppercase letter.
Line $\overleftrightarrow{A \quad B}$	A **line** is a collection of points that extends infinitely in opposite directions, and will always be assumed to be straight. A line is denoted by two points, such as $\overleftrightarrow{AB}$.
Line Segment $\overline{A \quad B}$	A **line segment** is the portion of a line that lies between two specified points called **endpoints**. A line segment is denoted by its end points, such as $\overline{AB}$.
Ray $\overrightarrow{A \quad B}$	A **ray** is a half-line with an endpoint A with directionality through point B. A ray is denoted by its endpoint and one other point on the line, such as $\overrightarrow{AB}$.
Plane $R \quad T$	A **plane** is a flat surface with no thickness, depth, or boundaries that extends indefinitely in all directions. A plane is denoted by an uppercase letter.

Example 1: Determining the Length of a Line Segment

Given $KB = 10$ inches, $BM = 12$ inches, and $MC = 13$ inches, find the length KC.

K   B   M   C

Solution

A line segment can be made up of other line segments. Since the measure of each smaller segment is given, we can find the measure of segment KC by adding the lengths of all segments together.

$$KB + BM + MC = KC$$

$$10 \text{ inches} + 12 \text{ inches} + 13 \text{ inches} = 35 \text{ inches}$$

This gives that $KC = 35$ inches.

Lines in the same plane can either intersect or be **parallel**. When two lines are parallel they never intersect and the distance between the two lines never changes. When two lines intersect at a point, angles are formed at the point of intersection. We need several definitions for angles before continuing.

An **angle** is formed when two rays have a common end point called a **vertex**. An angle is denoted by a point on each ray and the vertex, such as $\angle ABC$, where the vertex is the middle letter. Using the point of intersection as the vertex, four rays are formed when two lines in a plane intersect. The **measure of an angle** is the number of degrees that separates the two rays. We denote the measure of an $\angle ABC$ as $m\angle ABC$.

For our purposes, we would like to be able to classify angles based on their measure in an effort to make their use more manageable. An angle that has a measure greater than $0°$ but less than $90°$ is called an **acute angle**. An angle that has a measure greater than $90°$ but less than $180°$ is an **obtuse angle**. Two special angles that we need are a **right angle**, which is an angle with a measure of $90°$, and a **straight angle**, which is an angle with a measure of $180°$.

Table 2	
Summary of Parallel Lines and Angles	
Term/Example	**Definition**
Parallel Lines *C D* *A B*	Two lines, $\overleftrightarrow{AB}$ and $\overleftrightarrow{CD}$, in the same plane are **parallel** if they never intersect, no matter how far they extend, and can be denoted by $\overleftrightarrow{AB} \parallel \overleftrightarrow{CD}$.
Vertex	The **vertex** is the common endpoint of two line segments or rays.
Angle *A C* *B*	An **angle** is the union of two rays at a common vertex. The measure of an angle is the number of degrees (°) that separate the two rays. An angle is denoted by $\angle ABC$, where B is the vertex.

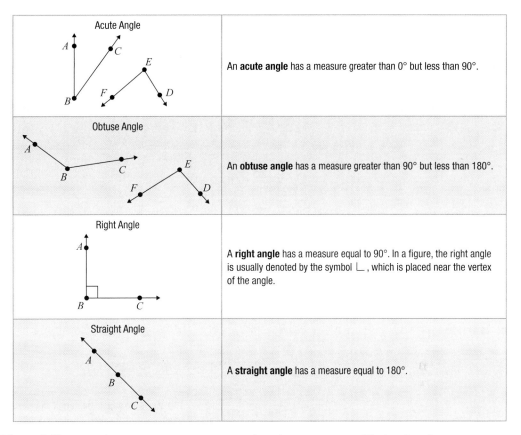

Figure 1 illustrates how we can measure an angle using a protractor. Notice that the protractor can measure an angle from 0° to 180°.

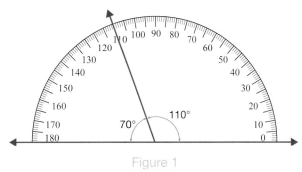

Figure 1

When two lines in the same plane intersect, the angles formed are **adjacent** if they have a common vertex and a common ray. The angles opposite of each other are called **opposite angles**. In Figure 2, ∠AEB and ∠BEC are adjacent angles while ∠AEB and ∠CED are opposite angles.

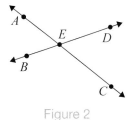

Figure 2

In addition, two angles are **complementary** if the sum of their measures is 90°, while two angles are **supplementary** if the sum of their measures is 180°.

We say that two lines in the same plane are **perpendicular** if the angles formed at their intersection are right angles.

Table 3	
Summary of the Properties of Angles and Perpendicular Lines	
Term/Example	**Definition**
Adjacent Angles	When two lines in the same plane intersect, four angles are formed. Two angles are adjacent if they have a common vertex and a common ray. In the example, two adjacent angles are $\angle BEA$ and $\angle AED$.
Opposite Angles	When two lines in the same plane intersect, four angles are formed. Two angles are opposite if they are opposite of each other, that is they share a vertex but do not have any rays in common. In the example, two opposite angles are $\angle AEB$ and $\angle CED$.
Complementary Angles $m\angle ABC = 90°$	Two angles are called **complementary** if the sum of their measures is 90°. In the example given, $\angle ABD$ and $\angle DBC$ are complementary.
Supplementary Angles $m\angle JKL = 180°$	Two angles are **supplementary** if the sum of their measures is 180°. In the example given, $\angle JKM$ and $\angle MKL$ are supplementary.
Perpendicular Lines	Two lines, $\overleftrightarrow{AB}$ and $\overleftrightarrow{CD}$, in the same plane are **perpendicular** if the angles formed at their intersection are right angles, and can be denoted by $\overleftrightarrow{AB} \perp \overleftrightarrow{CD}$.

Example 2: Complementary Angles

Determine if the following angles are complementary.

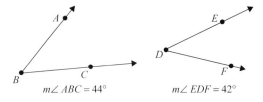

$m\angle ABC = 44°$ $m\angle EDF = 42°$

Solution

The definition of complementary angles says that two angles are complementary if the sum of the angle measures is equal to 90°. We have that $m\angle ABC = 44°$ and $m\angle EDF = 42°$, so the sum of the angles is $44° + 42° = 86°$. Since the sum is not equal to 90°, the angles are not complementary.

Example 3: Supplementary Angles

In the following figure, the road is represented by line $\overleftrightarrow{CD}$ while the trunk of the tree is represented by line $\overleftrightarrow{AB}$. The point of intersection of the two lines is labeled as E. Use the definitions given in Table 3 to show that $\angle BED$ is supplementary to $\angle DEA$.

Solution

By definition, two angles are said to be supplementary if the sum of their measures is $180°$. A straight angle has an angle measure of $180°$. Since the combination of the two angles forms the line $\overleftrightarrow{AB}$, the angles are supplementary.

Example 4: Complementary and Supplementary Angles

Given the figure, find $m\angle EBD$.

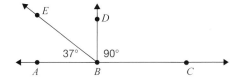

Solution

Since $\angle ABC$ is a straight angle and we are given that $\angle DBC$ is a right angle, we know that $\angle ABD$ also forms a right angle since the two angles are supplementary. This means that $\angle ABE$ and $\angle EBD$ are complementary angles, whose measures add to $90°$. So,

$$m\angle ABE + m\angle EBD = 90°$$
$$37° + m\angle EBD = 90°$$
$$m\angle EBD = 90° - 37° = 53°.$$

Skill Check #1

If two angles are supplementary and the measure of one of the angles is $113°$, find the measure of the missing angle.

We begin our discussion of fundamental geometry in the context of art. Consider an artist that wishes to paint a landscape scene, such as the figure in Example 3. In order for the artist to accurately portray the landscape, the artist must have an understanding of the basic components of geometry. From the intersection of the branches with the trunk of the tree to the angle that is created between the tree and the horizon, artists must understand the relationships between angles. Of course, artists are not the only ones that need to understand the relationship between angles and lines. Careers in many fields, including architecture, engineering, painting, plumbing, and carpentry, require knowledge of geometry.

Applications of Angles and Global Positioning

Do you know how your cellular phone receives information? Have you ever wondered how your smart phone knows where you are at any given time? Whenever you use your phone to check in with your friends using social media, to tag the location of where a picture is taken, or simply use your phone to find the location of a new restaurant, you are using the "location services" that are built into your phone. The ability to locate you or your phone (when not connected to a Wi-Fi signal) comes from a global positioning system (GPS). Developed by the United States Department of Defense as a means of electronic navigation, GPS has evolved into a very popular consumer device. GPS uses eighteen satellites (6 satellites orbiting the Earth in 3 separate

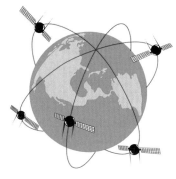

Figure 3

planes, 120° apart) so that any time you have a signal on your phone, there are at least three GPS satellites that are used to determine your position.

Think of your position as that of a single point on the *xy*-coordinate grid system. Much like that of the coordinate grid system, the Earth's surface can also be divided into single points. Instead of an *xy*-coordinate system, a single point on the surface of the Earth is a position given by latitude and longitude.

Longitude and the Prime Meridian

Longitude lines connect the north and south poles of the Earth and give us the positioning of east and west. With a longitudinal measure of 0°, the **prime meridian** is the starting point that divides the eastern and western hemispheres.

Latitude and the Equator

Latitude lines are used to measure the north and south locations on the surface of the Earth. With a latitudinal measure of 0°, the **equator** is the starting point that divides the northern and southern hemispheres.

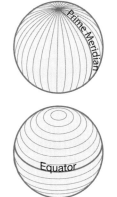

Latitude and longitude lines allow us to use the Earth's surface as a coordinate system. This means that two global reference points, latitude and longitude, determine your position on the Earth's surface.

As a beginning reference, we use the equator as our line of latitude. When we travel north or south of the equator, our relative position is based on the angles created by rays that extend from the center of the Earth to our location.

For lines of longitude, we use the Prime Meridian as a beginning reference. Often called the Greenwich Mean Time Line, this line serves as the reference line for angular position east and west.

Fun Fact

Did you know that ancient societies also had a marker for east versus west? The ancient Greeks used the longitude line that passed through the Canary Islands as a marker for many centuries. To read more about the development of the prime meridian, visit http://www. wwp.greenwichmeantime. com/info/prime-meridian.htm.

Lines of latitude and longitude are measured in degrees, minutes, and seconds. Degree, minute, and second are defined as follows.

1 degree (denoted by °) = 60 minutes

1 minute (denoted by ′) = 60 seconds

1 second (denoted by ″) = 61.6 feet

Most GPS units use a system of degrees/minutes instead of a degrees/minutes/seconds system, since the calculation of seconds can be done in decimals to the nearest foot.

Example 5: Converting Decimal Degrees to Degrees, Minutes, and Seconds

Convert 17.78° to degrees/minutes/seconds format.

Solution

First note that in the expression, 17 is already in degrees, so we need only need to convert the decimal portion of the expression. We know that $1° = 60$ min, so we can use dimensional analysis to write the decimal part of 17.78° in minute form as follows.

$$0.78° \cdot \frac{60'}{1°} = 46.8'$$

When we convert the portion of a degree 0.78°, our result does not yield a whole number of minutes. For this reason, we need to calculate the number of seconds represented by 0.8′. So,

$$0.8' \cdot \frac{60''}{1'} = 48''.$$

This means that $17.78° = 17° + 46' + 48'' = 17°46'48''$.

Example 6: Converting Degrees, Minutes, and Seconds to Decimal Degrees

Convert 56°26′23″ to decimal degrees.

Solution

We begin by writing 56°26′23″ in expanded form. So,

$$56°26'23'' = 56° + 26' + 23''.$$

Notice that 56° will be the whole number portion of the decimal degrees form. Our first step is to calculate the total number of seconds:

$$26'23'' = 26 \cdot 60 \text{ seconds} + 23 \text{ seconds} = 1583 \text{ seconds}.$$

The decimal part is the total number of seconds divided by 3600 since there are $60 \cdot 60 = 3600$ seconds in 1 degree.

$$\frac{1583}{3600} \approx 0.44.$$

Add fractional degrees to whole degrees to produce the final result.

$$56 + 0.44 = 56.44°$$

Therefore, the final answer is 56.44°.

Skill Check #2

Convert 125.345° to degrees/minutes/seconds format.

Example 7: Application of Angles

A rocket will be launched making an angle of θ with the moon that is positioned directly overhead. Given that the rocket travels at 18,000 mph and the moon completes a circular orbit of radius 240,000 miles approximately every 30 days, at what angle should the rocket be fired so that the rocket will intercept the moon?

Solution

We must employ several problem-solving strategies here in order to solve this problem efficiently. First, we need a picture to help us see the situation. In the figure, E represents the Earth and M represents the moon.

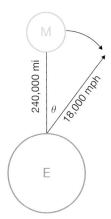

Recall that the relationship among the variables for distance, rate, and time is represented by the equation distance = rate · time, or $d = rt$.

Remember that we are looking for θ. To find this value, we can use our known values for the rocket speed (rate, $r = 18,000$ mph), the distance of the moon (distance, $d = 240,000$ miles) and the time it takes the moon to make one orbit (time, $t = 30$ days).

First, let's calculate the time required for the rocket to travel the required 240,000 miles to reach the moon's orbit.

$$d = rt \Rightarrow \quad t = \frac{d}{r}$$

$$t = \frac{240,000 \text{ miles}}{18,000 \text{ mph}}$$

$$\approx 13.33 \text{ hours}$$

So the rocket will reach a distance of 240,000 miles after traveling for approximately 13.33 hours.

Now, we need to consider that the moon is moving on an arc of 360° every 30 days. There are 720 hours in 30 days, so we can use this proportional relationship to determine the value for theta.

$$\frac{\theta°}{360°} = \frac{13.33 \text{ hours}}{720 \text{ hours}}$$

$$\overset{2}{\cancel{720}}\left(\frac{\theta}{\cancel{360}}\right) = \left(\frac{13.33}{\cancel{720}}\right)\cancel{720}$$

$$2\theta = 13.33$$

$$\theta = \frac{13.33}{2} = 6.665°$$

So, the rocket must be fired at an angle of 6.665°.

Now that we have an understanding of degrees, minutes, and seconds, we turn our attention back to the GPS coordinates. Triangulating is based on the prime meridian line and the equator. Moving east from the prime meridian, angles of longitude are positive values. However, if we were to move west from the prime meridian, the angles of longitude are negative values. This is quite helpful in that we can determine location going either direction. For instance, 60° east longitude is the same location as −300° west longitude. Similarly, as we move north from the equator, values of latitude are positive, while values south of the equator are negative. Therefore, we can determine the hemisphere in which we are located based on these simple facts.

Fun Fact

The GPS coordinates given in Example 8 are for Fort Worth, Texas.

Example 8: Determining Location

If your location is given in GPS coordinates of 32.75° north −79.98° west, determine whether you are located in the northern or southern hemisphere, and whether you are located in the western or eastern hemisphere.

Solution

Since the location has a positive value for latitude, the location is located in the northern hemisphere. Since the location has a negative value of −79.98° for longitude, we can conclude that the location must be in the western hemisphere. (Notice that if the negative value had an absolute value greater than 180°, then it would be located in the eastern hemisphere.)

MATH MILESTONE

Did you know that the launch of Sputnik by the Russians in 1957 was a catalyst to the development of the GPS navigational system or NAVSTAR GPS? When the Russian rocket was launched, US scientists discovered that they could track the rocket by the changes in frequency made by the rocket as it traveled in space. To learn more about the history of GPS navigation, go to http://www.time.com/time/magazine/article/0,9171,1901500,00.html.

One of the best applications of GPS devices tells us how far our current location is from a desired destination. In essence, that is what GPS devices were created for: relative location in terms of a desired location. It should be noted that at the equator, the distance between longitude lines is about 69 miles for every one degree. However, as the lines of longitude approach the poles, they get closer together and the distance between the lines of longitude eventually becomes 0 at the poles. For the purposes of this discussion, we will assume the standard that for every 10 degrees north (or south) of the equator, the lines of longitude get 4 miles closer together.

Longitudinal Adjustment Factor

The **longitudinal adjustment factor** (LAF) is used to find the distance between lines of longitude at a certain latitude measure. LAF is calculated by the formula

$$LAF = 69 - 4\left(\frac{\text{latitude measure}}{10}\right).$$

For example, consider the distance between lines of longitude at 30° north of the equator.

$$LAF = 69 - 4\left(\frac{30}{10}\right) = 69 - 4 \cdot 3 = 69 - 12 = 57$$

So, at 30° north or south, the distance between a degree of longitude is about 57 miles. Latitude is much simpler than longitude. Each degree of latitude represents approximately 69 miles. We explore this concept further in Example 9.

Example 9: Using GPS to Determine Location and Distance

If Amanda starts at GPS location 32.75° north −79.98° west, how far, in miles, is Amanda away from her destination of 35.15° north −90.06° west?

Solution

Our solution will begin with drawing the locations in a right triangle format. Recall that the Pythagorean Theorem states that the hypotenuse of a right triangle can be found if we know the length of the other two sides. We need to know the distance from the starting location to point P and the distance from P to the ending location. Once we know those, we can find the distance from Amanda's starting position to her destination.

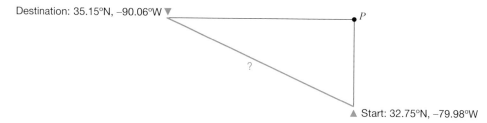

To find the legs of the triangle, we can see that the change in latitude between the starting position and P is

$$35.15° - 32.75° = 2.4°.$$

Since each degree represents 69 miles, the distance from the starting point to P would be

$$69 \cdot 2.4 = 165.6 \text{ miles.}$$

Similarly, the change in longitude is

$$90.06° - 79.98° = 10.08°.$$

To find the distance between longitude lines at 35.15 degrees above the equator, we apply the LAF as follows.

$$\text{LAF} = 69 - 4\left(\frac{35.15}{10}\right) = 69 - 4 \cdot 3.515 = 69 - 14.06 = 54.94$$

Each degree of longitude would represent approximately 55 miles. Therefore, the distance from P to the destination would be

$$55 \cdot 10.08 = 554.4 \text{ miles.}$$

Using the Pythagorean Theorem, we find the direct distance to be

$$165.6^2 + 554.4^2 = c^2$$
$$27423.36 + 307359.36 = c^2$$
$$334782.72 = c^2$$
$$\sqrt{334782.72} = \sqrt{c^2}$$
$$c \approx 578.60 \text{ miles}$$

Therefore the distance between the two cities is about 578.60 miles.

Trigonometry

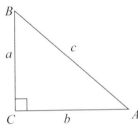

Figure 4

A right triangle consists of a triangle with one angle that measures 90° (also known as a right angle). One of the most useful applications of angles is based on the special relationships that exist between the lengths of the sides of a right triangle and the corresponding angles. This concept is the basis for courses such as trigonometry (which means "triangle measurement"). The basic idea of trigonometric relationships is that since all right triangles with the same angle measures are proportional to one another, we can use trigonometric relationships to determine any missing parts of a right triangle.

Consider the right triangle in Figure 4. The side opposite the right angle is always called the **hypotenuse**, and the other two sides are labeled based on the acute angle of interest, and are labeled as the **opposite side** and the **adjacent side** to a particular acute angle. For example, if angle A is the angle of interest, a is the length of the opposite side and b is the length of the adjacent side, as shown in Figure 5.

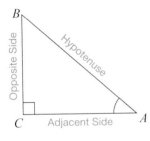

Figure 5

The three basic trigonometric functions that can be used with right triangles are **sine**, **cosine**, and **tangent**. Each function relates the measure of the angle with the ratio of two sides of the triangle.

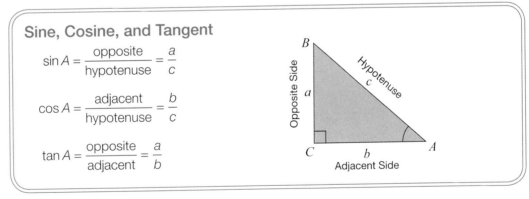

Sine, Cosine, and Tangent

$$\sin A = \frac{\text{opposite}}{\text{hypotenuse}} = \frac{a}{c}$$

$$\cos A = \frac{\text{adjacent}}{\text{hypotenuse}} = \frac{b}{c}$$

$$\tan A = \frac{\text{opposite}}{\text{adjacent}} = \frac{a}{b}$$

☞ **Helpful Hint**

Similar to the familiar acronym Please Excuse My Dear Aunt Sally used to indicate the order of mathematical operations, there are many mnemonic devices that can represent the trigonometric relationships. Can you make up your own?

Additionally, we can further describe these relationships using the mnemonic device SOHCAHTOA to help us to easily remember the ratios.

SOH
$$\text{Sine} = \frac{\text{Opposite}}{\text{Hypotenuse}}$$

CAH
$$\text{Cosine} = \frac{\text{Adjacent}}{\text{Hypotenuse}}$$

TOA
$$\text{Tangent} = \frac{\text{Opposite}}{\text{Adjacent}}$$

Example 10: Calculating Trigonometric Values

Sam is building a deck on the back of his house. The deck is 4 feet off the ground. According to the building code, the maximum angle for the incline of the steps for the deck is 30° and the maximum length of the step incline can be 8 feet. Using the given figure, determine the minimum distance the landing step (Point A) must be from the house.

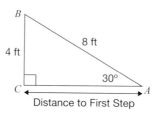

Distance to First Step

Solution

Using the information given and trigonometric ratios, we can determine the minimum distance from A to C. First, we verify that the angle of 30° is accurate for the height of the deck and the length of the incline. It is always true that $\sin 30° = \frac{1}{2}$, so we need to verify that the trigonometric ratio is also equal to $\frac{1}{2}$. Therefore,

$$\sin 30° = \frac{\text{opposite}}{\text{hypotenuse}} = \frac{4 \text{ ft}}{8 \text{ ft}} = \frac{1}{2}.$$

Now, to find the minimum distance AC, which we will denote by x, we can use the cosine function. The adjacent side to the angle is represented by x and the hypotenuse of the triangle is 8 feet. Substituting the values into the trigonometric ratio for cosine, we get the following.

$$\cos 30° = \frac{\text{adjacent}}{\text{hypotenuse}} = \frac{x}{8}$$
$$8 \cdot \cos 30° = x$$
$$x \approx 6.93 \text{ ft}$$

So, the distance from the deck (Point C) to Point A is about 6.93 feet. We can confirm this value by using the tangent function.

$$\tan 30° = \frac{\text{opposite}}{\text{adjacent}} = \frac{4}{x}$$
$$x \cdot \tan 30° = 4$$
$$x = \frac{4}{\tan 30°} \approx 6.93 \text{ ft}$$

☞ Helpful Hint

You may need to adjust the mode of your calculator to make sure that your calculator is set to work in degrees— there are other units used to measure angles, so if your calculator is in the wrong mode, you will not get correct answers. Your calculator may have a DRG button that converts between degrees, radians, and gradients. Press this button until you see that it says degrees. Alternatively, you may need to press a MODE button, where you can adjust the angle measurement mode. If you are unsure, try evaluating sin(30) in your calculator. If the answer is 0.5, then your calculator is set to degrees. If the answer is not 0.5, then you will need to adjust the mode.

TECH TRAINING

To evaluate the expression from Example 10 with a TI-30XIIS/B or TI-83/84 Plus calculator, type 8 × COS 3 0) = . The calculator will display 6.92820323...

Skill Check #3

Use the Pythagorean Theorem and the result from Example 10 to confirm that the side lengths form a right triangle.

Now we extend these ideas so that we can use a calculator to determine the measure of any angle in a right triangle, given the lengths of at least two sides, and to determine the length of any side given the measure of an angle and one side.

Example 11: Using Trigonometric Functions

A hiker knows that the distance from a point, A, on a map to another point, B, is 7 miles. She began at point A and headed west until she was due south of point B, and is now at point C. She needs to determine her distance from both point A and point B, given that the heading of her route from point A was at a $46°$ angle from the line between points A and B, as shown in the figure. When she arrives at point B, she also needs to know what angle she will need to make with her due-north route to return directly to point A.

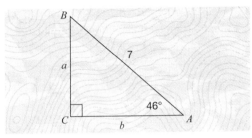

Solution

We need to find the length of sides a and b, and the measure of angle B in the figure. We know that $\sin A = \frac{a}{c}$, which means $\sin 46° = \frac{a}{7}$. Using a scientific calculator and a little algebra, we can solve for a to obtain the following.

$$7 \sin 46° = a$$
$$5.04 \approx a$$

So, $a \approx 5.04$.

Now that we know a, we can find the value of b using the Pythagorean Theorem.

$$a^2 + b^2 = c^2$$
$$5.04^2 + b^2 = 7^2$$
$$25.4016 + b^2 = 49$$
$$b^2 = 49 - 25.4016$$
$$b^2 = 23.5984$$
$$b \approx 4.86$$

Now we have a, b, and c, as well as two angles. To determine the third angle, recall that the sum of the interior angles of a triangle is $180°$. So, we subtract the two known angles from $180°$.

$$m\angle B = 180° - 90° - 46° = 44°$$

So, she has traveled 4.86 miles due west of point A and she will need to walk 5.04 miles due north to point B. After she faces south, she can turn $44°$ to walk the 7 miles back to point A.

TECH TRAINING

To evaluate the expression from Example 11 with a TI-30XIIS/B or TI-83/84 Plus calculator, type [7] [×] [SIN] [4] [6] [ )] [=]. The calculator will display 5.035378602...

Another interesting application of trigonometric functions is the use of **angles of elevation** and **angles of depression** to determine location. In general, if your eye level is horizontal, then looking down your line of sight toward the ground is considered an angle of depression. Conversely, if your eye level is horizontal, then looking up your line of sight toward the sky would be an angle of elevation. See Figure 6 for an illustration of this idea. When two lines, such as $\overrightarrow{BP}$ and $\overrightarrow{OA}$, are parallel, the angle of depression and the angle of elevation are equal.

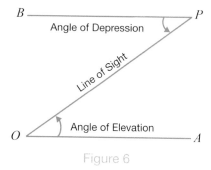

Figure 6

Example 12: Using Trigonometry

A ship's sonar locates a wrecked ship at a 17° angle of depression. A diver is lowered 105 meters to the ocean floor. How far does the diver need to travel along the ocean floor to get to the wreckage?

Solution

Make a sketch to illustrate the situation.

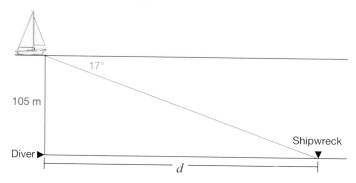

If we assume that the ocean floor is parallel to the surface of the water so that the angle of elevation from the wreckage to the ship is equal to the angle of depression from the ship to the wreckage.

The distance the diver is lowered (105 m) is the length of the leg *opposite* the 17° angle of elevation. The distance the diver must walk, d, is the length of the leg *adjacent* to the 17° angle. This means we need to use the tangent ratio to find the length the diver must travel.

$$\tan 17° = \frac{105 \text{ m}}{d}$$

$$d \tan 17° = 105$$

$$\frac{d \cancel{\tan 17°}}{\cancel{\tan 17°}} = \frac{105}{\tan 17°}$$

$$d = \frac{105}{\tan 17°}$$

$$d \approx 343.44 \text{ m}$$

Therefore, the diver must walk approximately 343 m to the wreckage.

TECH TRAINING

To evaluate the expression from Example 12 with a TI-30XIIS/B or TI-83/84 Plus calculator, type [1][0][5][÷][TAN][1][7][)][=]. The calculator will display 343.4395249.

Skill Check Answers

1. $67°$

2. $125°20'42''$

3. $4^2 + b^2 = 8^2$; $b^2 = 48$; $b \approx 6.928$

6.1   Exercises

Determine whether each angle is acute, right, or obtuse.

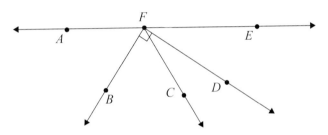

1. $\angle BFD$	**2.** $\angle DFE$	**3.** $\angle AFD$	
4. $\angle AFB$	**5.** $\angle EFC$	**6.** $\angle EFB$	

In the figure, $\overrightarrow{YU}$ bisects (divides into 2 equal parts) $\angle WYV$ and $\overrightarrow{YT}$ bisects $\angle XYV$. Use the figure to answer each question.

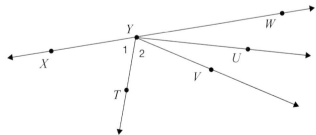

7.   If $m\angle WYV = 36°$, find $m\angle WYU$.

8.   If $m\angle 1 = 56°$, find $m\angle 2$.

9.   If $m\angle VYW = 82°$ and $m\angle WYU = (4r + 25)°$, find r.

10.   If $\angle XYV$ has a measure of $162°$, find $m\angle 1$.

Solve each problem.

11. The measures of two complementary angles are $(16z - 1)°$ and $(12z + 7)°$. Find the measures of the angles.

12. Find $m\angle T$ if $m\angle T$ is 20 more than four times its supplement.

13. Two angles are supplementary. One angle is 12° more than the other. Find the measures of the angles.

Use the figure to answer each question.

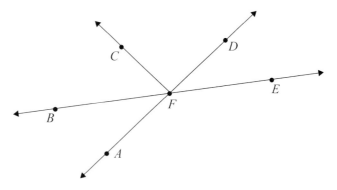

14. If $m\angle CFD = (15a + 45)°$, find a so that $\overrightarrow{FC} \perp \overrightarrow{FD}$.

15. If $m\angle AFB = (8x - 6)°$ and $m\angle BFC = (14x + 8)°$, find the value of x so that $\angle AFC$ is a right angle.

16. If $m\angle BFA = (3r + 12)°$ and $m\angle DFE = (-8r + 210)°$, find $m\angle AFE$. (**Hint:** Recall that lines that intersect at a common point share vertical angles that are congruent.)

Use the figure to find the measure of each angle.

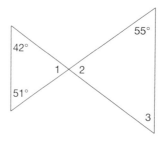

17. $m\angle 1$ **18.** $m\angle 2$ **19.** $m\angle 3$

Convert each measure to decimal degrees. Round your answer to the nearest thousandth when necessary.

20. $37°22'12''$ **21.** $121°6'32''$ **22.** $96°16'2''$

Convert each measure to degrees/minutes/seconds format.

23. 41.22°   **24.** 221.34°   **25.** 13.86°

Determine the distance between the GPS locations. Round your answer to the nearest mile.

26. 13.86°N, 46.25°E to 24.86°N, 52.31°E

27. 17.86°S, −52.31°W to 15.3°S, −52.1°W

28. 12.3°S, 56.15°E to 12.3°N, 45.15°E

Solve each problem.

29. Find the lengths of $\overline{AC}$ and $\overline{BC}$ in the figure and round to the nearest tenth.

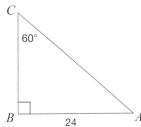

30. Use $\triangle ABC$ to find $\sin A$, $\cos A$, $\tan A$, $\sin B$, $\cos B$, and $\tan B$. Round your answers to the nearest hundredth.

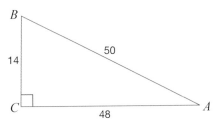

Find the length of *x* in each triangle. Round your answer to the nearest hundredth.

31.

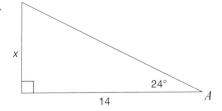

32.

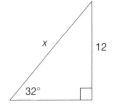

33.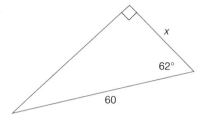

34.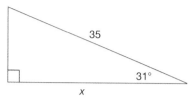

35.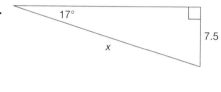

36.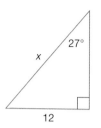

Solve each problem. Round your answer to the nearest hundredth when necessary.

37. A plane is two miles above the ocean when it begins to climb higher at a constant angle of 4° for the next 50 miles. About how far above the ocean is the plane after its climb?

38. A 16 ft ladder is leaning against a wall where it makes an angle of 68° with the ground and 22° with the building.

  a. How far is the ladder base from the building?

  b. How far does the ladder reach up the building?

39. A plane is flying at 20,000 feet and the pilot wishes to take the plane to 35,000 feet. The angle of elevation that should be used is 4°. How many miles will the plane need to fly to reach the correct height? (**Hint:** There are 5280 feet in 1 mile.)

40. The sun hits a building of unknown height so that the building casts a shadow of 517 feet. If the angle of elevation from the tip of the shadow to the sun is 16°, how tall is the building?

41. A snow ski slope has a run of 1.5 miles with an angle of elevation of 24°. What is the length of the vertical drop of the ski slope?

42. A wheelchair ramp is to be constructed with a maximum incline of 6°. If the height of the ramp is to be 5 feet, how long should the ramp be after construction?

43. A section of highway has an incline of 10°. If the length of the incline is 5 miles, what is the horizontal distance of the incline?

6.2   Circles, Polygons, Perimeter, and Area

We begin this section with a brief review of **plane geometry**. Plane geometry refers to objects like circles, polygons, and triangles, which have measurements such as perimeter and area. These ideas are quite useful in everyday life.

Table 1 is a summary of the basic information you should remember from earlier courses concerning circles and polygons.

Table 1		
Circles and Polygons		
Object/Example	**Definition**	**Basic Attributes**
Circle	A **circle** is the set of all points in a plane that are a certain distance from a fixed center point.	The distance from the center to any point on the circle is called the **radius**, r. The **diameter**, d, is the length of a line segment that passes through the center of the circle and intersects the circle at two points.
Polygons — Triangle, Quadrilateral, Pentagon, Hexagon	A **polygon** is a closed figure that is determined by three or more straight line segments.	Two line segments of a polygon intersect to form a **vertex**. A **diagonal** of a polygon is a line segment joining two nonconsecutive vertices.
Convex Polygons	A **convex polygon** is a polygon in which all of the interior angles measure less than 180°.	Every diagonal of a convex polygon remains inside or on the boundary of the polygon.
Concave Polygons	A **concave polygon** is a polygon with one or more interior angles that measure greater than 180°. Note that a triangle can never be concave.	The diagonals of a concave polygon might lie outside the polygon and might not intersect.

☞ **Helpful Hint**

A **regular polygon** is a polygon where all angles have the same measure and all sides have the same length. Examples include squares, and equilateral triangles.

Now that we have completed a quick review of basic concepts surrounding circles and polygons, we can use the measure of the angles inside a triangle to study other polygons. Recall that the sum of the three angles in a triangle is 180°. Consider the quadrilateral $ABCD$ in Figure 1.

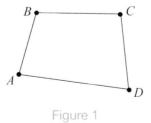

Figure 1

By inserting a diagonal line segment connecting vertices B and D, the quadrilateral can be cut into two triangles, as shown in Figure 2.

⑥

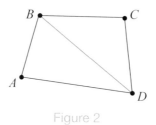

Figure 2

This implies that the angle sum of the quadrilateral will be two times that of a triangle, or $2 \cdot 180° = 360°$. What about polygons with even more sides? Let's try one more: a hexagon. A hexagon is six-sided polygon, as seen in Figure 3.

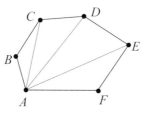

Figure 3

Notice that, if we choose to extend diagonals from vertex A to the remaining nonadjacent vertices (that is, the vertices that are not next to each other), then the hexagon is cut into four triangles as shown in Figure 3. Therefore, a hexagon has an angle sum of $4 \cdot 180° = 720°$.

Believe it or not, beginning with our knowledge that the interior angle sum of a triangle is 180°, we have just proved that the interior angle sum of a quadrilateral is 360°, and the interior angle sum of a hexagon is 720°. Furthermore, we can develop a formula for determining the interior angle sum for any polygon based on the number of sides.

Table 2		
Interior Angle Sum for Any Polygon Based on the Number of Sides		
Sides	**Number of Triangles**	**Sum of the Measures of the Interior Angles**
3	1	$1 \cdot 180 = 180°$
4	2	$2 \cdot 180 = 360°$
5	3	$3 \cdot 180 = 540°$
6	4	$4 \cdot 180 = 720°$

Sum of the Measures of the Interior Angles of a Polygon

The sum of the measures of the interior angles of an n-sided polygon is $(n - 2) \cdot 180°$.

Skill Check #1

What is the sum of the measures of the interior angles of an octagon?

Similar Triangles

In the previous section, we began a discussion of right triangles with trigonometry. In this section, we will continue our discussion of right triangles with applications that involve similar triangles.

> ## Similar Triangles
>
> Two triangles are **similar** if the corresponding angles are congruent and the ratio of the lengths of the corresponding sides are equal.

☞ Helpful Hint

When two angles have the same measure, they are called **congruent angles**. When two shapes have corresponding sides that are equal and corresponding angles that have the same measure, they are called **congruent shapes**.

A common use of similarity is for artists to create large works of art. Many artists use scale drawings to create a shape that is easy to manage. Then they scale the drawing by some factor to a larger figure in order to achieve the desired size.

Figure 4 shows an example of similar triangles. The figure consists of a triangle within a larger triangle. In this case, both triangles share a vertex angle at A and each triangle has a right angle. Because we now know that two of the three angles in both triangles are congruent, and that all the angles of a triangle must add to $180°$, the third angle must have the same measure in both triangles. Since the corresponding angles are congruent, the lengths of the sides must be proportional, and thus the triangles are **similar**.

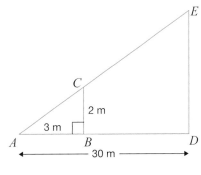

Figure 4

Example 1: Using Similar Triangles

☞ Helpful Hint

The method used here is also called *indirect measurement*. When you are not able to measure a length directly, you create a pair of similar triangles where one of the triangles can be measured.

Suppose a man who is 6 feet tall is standing 20 feet from the base of a tree. The sun shines on the tree and the man, and casts a 30-foot shadow, as seen in the figure. How tall is the tree?

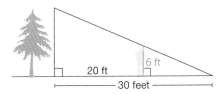

Solution

By drawing the figure, it is easier for us to see that the triangles created by this situation are indeed similar. The large triangle with the tree on the left and the small triangle with the man on the left both contain right angles. Since they also share a common angle, we know the triangles are similar and the lengths of corresponding sides are proportional. Begin by letting the height of the tree equal x. Since the distance from the tree to the man is 20 feet and the total distance is 30 feet, the length of the man's shadow is 10 feet. Then, we can set up a ratio, as we did in Chapter 4.

$$\frac{x}{30 \text{ ft}} = \frac{6 \text{ ft}}{10 \text{ ft}}$$

We can solve by multiplying both sides by the LCD to obtain the following.

$$30 \text{ ft} \cdot \left(\frac{x}{30 \text{ ft}} \right) = 30 \text{ ft} \cdot \left(\frac{6 \text{ ft}}{10 \text{ ft}} \right)$$

$$\cancel{30 \text{ ft}} \cdot \left(\frac{x}{\cancel{30 \text{ ft}}} \right) = \overset{3}{\cancel{30 \text{ ft}}} \cdot \left(\frac{6 \text{ ft}}{\cancel{10 \text{ ft}}} \right)$$

$$x = 18 \text{ ft}$$

So the height of the tree is 18 feet.

Perimeter

You may recall from previous courses in mathematics that the perimeter of a 2-dimensional figure is the distance around the boundary of the figure.

If you were to walk the bases of a professional baseball field, how far would you walk? Consider that the bases are 90 feet apart and they form a square as shown in Figure 5.

Notice that there are four edges, each of length 90 feet, so you will walk a total of $4 \cdot 90 = 360$ feet. Now consider a more general idea of perimeter that relates to *any* geometric figure.

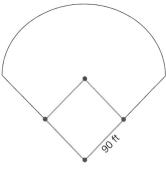

Figure 5

> **Perimeter**
>
> The **perimeter** P of a polygon is the sum of the lengths of the sides of the polygon.

Example 2: Finding Perimeter

Each of the figures represents a polygon of some kind. Assume that the distance between two grid lines, either horizontally or vertically, is one unit. What is the perimeter of each figure?

a. 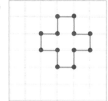   b. 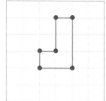   c.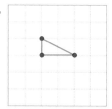

Solution

a. We can count the units and determine that the figure has a perimeter of 12 units.

b. Similarly, we can count the units and find that the figure has a perimeter of 10 units.

c. Since the figure is a right triangle, we can use the Pythagorean Theorem to find the length of the hypotenuse. We can see that the lengths of the two legs of the triangle are 1 unit and 2 units. Using these values, we find c as follows.

$$a^2 + b^2 = c^2$$
$$1^2 + 2^2 = c^2$$
$$1 + 4 = c^2$$
$$5 = c^2$$
$$\sqrt{5} = c$$

This means the perimeter of the triangle is $P = 1 + 2 + \sqrt{5} = 3 + \sqrt{5} \approx 5.24$ units.

Area

> ## Area
>
> The **area** A of a polygon is defined as the number of square units required to cover the polygon.

You might recall from earlier courses some of the formulas for area of polygons. Table 3 contains the basic formulas for perimeter and area with which you should be familiar.

Table 3			
Summary of Perimeter and Area Formulas			
Shape	**Definition**	**Formula for Perimeter**	**Formula for Area**
Rectangle	A **rectangle** is a quadrilateral with two pairs of parallel sides and four right angles. We denote the longer side as the length l and the shorter side as the width w.	$P = 2l + 2w$	$A = lw$
Square	A **square** is a quadrilateral with two pairs of parallel sides, all four sides are congruent, and four right angles. We denote the side length as s. Note that all squares are also rectangles.	$P = 4s$	$A = s^2$
Triangle	A **triangle** is a three-sided polygon that is identified by its base b and height h.	$P = $ sum of sides	$A = \dfrac{1}{2}bh$
Parallelogram	A **parallelogram** is a quadrilateral with two pairs of parallel sides. The base of the parallelogram is denoted by b and the height by h.	$P = $ sum of sides	$A = bh$
Trapezoid	A **trapezoid** is a quadrilateral with exactly one pair of parallel sides. The sides that are parallel are called bases and are denoted b_1 and b_2. The height is denoted by h.	$P = $ sum of sides	$A = \dfrac{1}{2}\left(b_1 + b_2\right)h$

Using the formulas and concepts in Table 3, we can now work a few examples involving area.

Example 3: Finding Area

Jim wants to build a garden in his back yard. Three different configurations that he is considering for the shape of the garden are given. Determine the area of each of the configurations, assuming the grid lines are each 1 foot apart.

a. 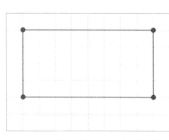   b.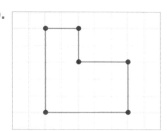

c.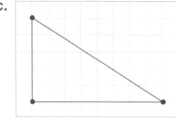

Solution

a. This figure is a rectangle with length 8 feet and width 4 feet. We can use the formula to find the area as follows.

$$A = lw$$
$$= (8 \text{ feet})(4 \text{ feet})$$
$$= 32 \text{ feet}^2$$

b. This figure can be broken into two pieces: a square with sides of length 2 feet and a rectangle with length 5 feet and width 3 feet. Therefore we have the following area.

$$A = s^2 + lw$$
$$= (2 \text{ feet})^2 + (5 \text{ feet})(3 \text{ feet})$$
$$= 4 \text{ feet}^2 + 15 \text{ feet}^2$$
$$= 19 \text{ feet}^2$$

c. This figure is a triangle with a base of length 8 feet and height of length 5 feet. We then use the formula to obtain the following area.

$$A = \frac{1}{2}bh$$
$$= \frac{1}{2}(8 \text{ feet})(5 \text{ feet})$$
$$= (4 \text{ feet})(5 \text{ feet})$$
$$= 20 \text{ feet}^2$$

Example 4: Solving Problems Involving Area

You wish to paint the floor shown in the figure.

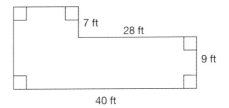

a. Determine the area to be painted.

b. If a gallon of paint will cover 350 ft², how many gallons of paint are needed?

Solution

a. We have established that the area of a rectangle can be found using the formula $A = lw$. Since we have a formula for rectangles, we need to divide our figure into rectangles, if possible. Using dashed lines, consider the following rectangles.

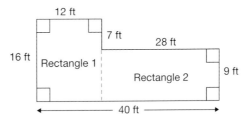

Now, we have two rectangles for which we can find the area. Once we have the area of each, we can find the sum of the areas to determine the total area to be painted.

$$\text{Rectangle 1:   Area}_1 = lw = (16 \text{ ft})(12 \text{ ft}) = 192 \text{ ft}^2$$
$$\text{Rectangle 2:   Area}_2 = lw = (28 \text{ ft})(9 \text{ ft}) = 252 \text{ ft}^2$$
$$\text{Area of floor } = \text{Area}_1 + \text{Area}_2 = 192 \text{ ft}^2 + 252 \text{ ft}^2 = 444 \text{ ft}^2$$

b. Since a gallon of paint will cover 350 ft², we can determine the total number of gallons needed by dividing the total area to be painted by the number of square feet that a gallon of paint will cover.

$$\text{gallons needed} = \frac{444 \text{ ft}^2}{350 \text{ ft}^2} \approx 1.27 \text{ gallons}$$

Since we are unable to purchase partial gallons at this paint store, we will need to purchase two gallons of paint to complete the project.

Skill Check #2

If carpeting costs $2 per square foot and we wish to carpet rooms of size 12 ft by 16 ft and 15 ft by 20 ft, how much would it cost to carpet these two rooms?

Example 5: Finding the Area of a Parallelogram

A new bridge consists of trusses that are parallelograms. The given figure shows an outline of one of the trusses. Find the area of each of the trusses.

Solution

From the figure, we can see that the base b of the parallelogram is 13 meters and the height h is 6 meters. This means the area would be

$$A = bh = (13 \text{ m})(6 \text{ m}) = 78 \text{ m}^2.$$

So, the area of each truss is 78 m².

Example 6: Finding the Area of a Triangle

A triangular plot of land has dimensions as shown in the figure. A farmer wishes to fertilize the plot for spring planting. If one bag of fertilizer will cover 15 m², how many bags of fertilizer are needed?

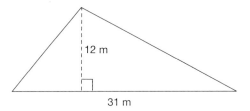

Solution

We begin by finding the area of the triangular plot.

$$A = \frac{1}{2}bh = \frac{1}{2}(31 \text{ m})(12 \text{ m}) = 186 \text{ m}^2$$

Now that we know the area is 186 m², we can determine the amount of fertilizer to purchase. Since one bag will cover 15 m², we know that to fertilize the whole plot will require $\frac{186 \text{ m}^2}{15 \text{ m}^2} = 12.4$ bags. Since the farmer can't purchase part of a bag, 13 bags of fertilizer will be purchased.

Example 7: Finding the Area of a Trapezoid

Find the area of the trapezoid in the figure.

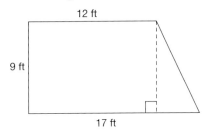

Solution

The area formula for the trapezoid is $\frac{1}{2}(b_1 + b_2)h$, where the bases measure $b_1 = 12$ ft and $b_2 = 17$ ft, and the height measures $h = 9$ ft. So,

$$
\begin{aligned}
A &= \frac{1}{2}(b_1 + b_2)h \\
&= \frac{1}{2}(12\text{ ft} + 17\text{ ft}) \cdot 9\text{ ft} \\
&= \frac{1}{2}(29\text{ ft}) \cdot 9\text{ ft} \\
&= (14.5\text{ ft}) \cdot 9\text{ ft} \\
&= 130.5\text{ ft}^2.
\end{aligned}
$$

So the area of the trapezoid is 130.5 ft^2.

Circles

Lastly, we discuss the area and circumference of circles. The distance around a circle (that is, the perimeter of a circle) is called its **circumference**.

The ratio of the circumference of a circle to its diameter yields the never-ending, never-repeating number we call π. We use 3.14 as an approximation of π.

Circumference of a Circle

The perimeter of a circle is called the **circumference** and is defined by $C = 2\pi r$, which can also be written as $C = \pi d$.

Area of a Circle

The **area of a circle** is defined by the number of square units inside the circle and is calculated using the formula $A = \pi r^2$.

Fun Fact

The number π has been part of mathematical calculations since about 2500 B.C. In fact, the Great Pyramid at Giza has a perimeter of 1760 cubits and a height of 280 cubits. If we compare the perimeter to the height we get

$$\frac{1760\text{ cubits}}{280\text{ cubits}} \approx 6.28 \approx 2\pi.$$

Example 8: Area and Circumference of Circles

A fish tank has a circular top with a radius of 15 cm. Jeff wants to cut a cover for the fish tank out of glass. Find the area and circumference of the cover of the tank.

15 cm

Solution

Using the formula for the area of a circle, we can calculate the area using the formula $A = \pi r^2$ with $r = 15$ cm as follows.

$$A = \pi r^2$$
$$= \pi (15 \text{ cm})^2$$
$$= \pi (225 \text{ cm}^2)$$
$$\approx (3.14)(225 \text{ cm}^2)$$
$$= 706.5 \text{ cm}^2$$

To find the circumference, we can use the formula $C = 2\pi r$ with $r = 15$ cm to obtain the solution as follows.

$$C = 2\pi r$$
$$= 2\pi (15 \text{ cm})$$
$$= \pi (30 \text{ cm})$$
$$\approx (3.14)(30 \text{ cm})$$
$$= 94.2 \text{ cm}$$

The glass cover will have an area of 706.5 cm² and a circumference of 94.2 cm.

Skill Check #3

Find the area and circumference of a circle with a diameter of 10 inches.

Example 9: Using Area

Consider the size of the lawn at a local park. The park consists of a grassy rectangular area determined by the dimensions 228 feet by 380 feet. We need to install circular sprinklers in the park in order to cover the grassy areas for the most complete coverage of the park grounds. A sprinkler has a maximum radius of 38 feet and the sprinklers will be placed in a grid-like pattern as seen in the figure. If we have 15 sprinklers, how much of the grass will not get water?

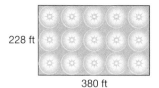

228 ft

380 ft

Solution

Using a drawing of the park, we can use what we know about the area of rectangles and circles to find the amount of grass that will not be watered. We will place our sprinklers in a grid-like pattern as shown in the figure. Now, to calculate how much of the park is being watered and how much is not being watered, we need to use a combination of the areas of rectangles and circles.

Using the given figure and the area formula for a rectangle, we calculate the total area of the park to be

$$A_{total} = lw = (380 \text{ ft})(228 \text{ ft}) = 86,640 \text{ ft}^2.$$

With the indicated position of the sprinklers, we can see that there are 15 sprinklers spread throughout the park. Each sprinkler makes a circular pattern with a radius of 38 feet. So the area watered by each sprinkler is as follows.

$r = 38$ ft

$$\begin{aligned}
\text{Area}_{sprinkler} &= \pi r^2 \\
&= \pi (38 \text{ ft})^2 \\
&= \pi (1444 \text{ ft}^2) \\
&\approx (3.14)(1444 \text{ ft}^2) \\
&= 4534.16 \text{ ft}^2
\end{aligned}$$

Then the total area watered is calculated as follows.

$$\text{Area}_{watered} = 15 \cdot (4534.16 \text{ ft}^2) = 68,012.4 \text{ ft}^2$$

This result indicates that the amount of the grass not watered will be

$$A_{total} - A_{watered} = 86,640 \text{ ft}^2 - 68,012.4 \text{ ft}^2 = 18,627.6 \text{ ft}^2.$$

With 15 sprinklers, 18,627.6 ft², or about 21.5% of the grass, is not watered.

Skill Check Answers

1. 1080°

2. $984

3. $A = 78.5$ in.²; $C = 31.4$ in.

6.2   Exercises

Solve each problem.

1.   Jeff was curious about the height of the Eiffel Tower. He used a 2-meter model of the tower and measured the shadow of the tower. The length of the shadow was 1.2 meters. Then he measured the length of the shadow of the Eiffel Tower and found it to be 192 meters. What is the height of the Eiffel Tower?

2.   The lengths of the sides of triangle *ABC* are 6 centimeters, 4 centimeters, and 9 centimeters. Triangle *DEF* is similar to triangle *ABC*. The length of one of the sides of triangle *DEF* is 36 centimeters. What is the greatest perimeter possible for triangle *DEF*?

3.   Miguel wants to enlarge the perimeter of an 18-by-24-inch picture by 30%. What will be the perimeter of the enlarged picture?

4.   Find the measure of each angle in the following figure.

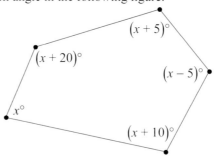

5.   The measure of an interior angle of a regular polygon is 108°. Find the number of sides in the polygon.

6.   Find the sum of the measures of the interior angles of each polygon.

   a. 12-gon

   b. 18-gon

   c. 32-gon

   d. 27-gon

7.   Find the measure of each interior angle.

Solve each problem involving perimeter and area. Use π = 3.14 and round your answer to the nearest hundredth when necessary.

8.   A triangular box is to be made from cardboard. The following figure represents the top of the box. How many centimeters of rope would be required to travel the perimeter of each box? How many square centimeters of material would be required to cover the top of the box?

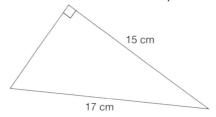

9.   A swimming pool is in the shape of the following figure. The homeowner wishes to build a fence around the perimeter of the pool and create a canvas cover. How many feet of fencing are required? How many square feet of canvas would be required to make a cover for the pool?

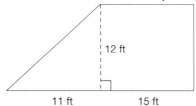

10. Find the perimeter and area of the floor tile represented by the parallelogram.

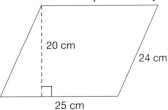

11. The side view of a wheel for a skate is shown. What is the circumference of the wheel? How many square inches of paint are required to paint the wheel?

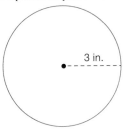

12. A piece of cake has the shape of the following figure. What is the area of the top of the cake slice?

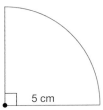

13. A window in the shape of the following figure is needed for a new home. If glass costs $0.85 per square foot, how much will the window cost?

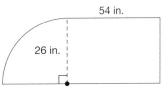

14. A rectangular piece of land has dimensions 1200 ft by 850 ft. Find the amount of fencing needed to construct a fence around the perimeter of this piece of land. Also find the area of the land to be enclosed.

15. A hot tub will have the footprint of a regular polygon, shown in the following figure. If your builder wishes to cover the area under the hot tub with plastic, how many square meters of plastic are needed?

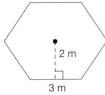

16. The roof of a house has a truss system like that in the following figure. The contractor wishes to know the area created by the truss so that mechanical workings such as electrical, plumbing, and HVAC may run through the opening. Determine the amount of material needed for the perimeter and the area of the opening.

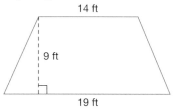

17. The gable end of a house is represented in the following figure. Find the amount of paint required to cover the gable.

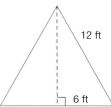

18. A piece of paper is in the shape of a parallelogram. Jack wants to know the perimeter and area of the parallelogram in order to determine how much paint he will need in order to cover the paper. Find the area and perimeter of the given parallelogram.

19. Find the perimeter and area of the top of the portion of the wheel of cheese given in the figure.

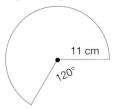

Solve each problem. Round your answer to the nearest tenth when necessary.

20. A rectangular pane of antique glass has dimensions of 25 inches by 48 inches. If glass sells for $0.50 per square inch, how much will each window cost?

21. Carpet Pro sells carpet for $1.50 per square foot. If you wish to carpet a room that measures 16 feet by 25 feet, how much will the carpet cost?

22. A 25-lb bag of grass seed will cover 1200 square feet. If your lawn is in the shape of a rectangle and measures 205 feet by 145 feet, how many bags of grass seed will you need to cover the lawn?

Find the area of the shaded region in each figure.

23.

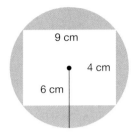

24.

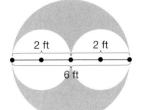

25.

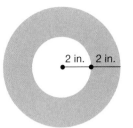

6.3   Volume and Surface Area

In the previous sections, we discussed 2-dimensional figures and their attributes. Now, we transition to figures in three dimensions by looking at their respective volumes and surface areas. **Volume** is a way to measure the amount of space occupied or consumed by an object in three dimensions. Whereas **surface area** is a measurement of the outside surface of the object. Both volume and surface area have many practical applications. We become aware of water conservation by keeping track of the volume of water used in households for baths, showers, laundry, or washing the car. We pay for the volume of space used to ship goods around the world or the volume of gas we use to heat a house in the winter. By knowing the surface area of 3-dimensional objects, we can calculate how many bricks are needed to construct the outside of a new home or how much wall paper is needed on the interior. Let's begin by looking at volumes first.

> ## Volume
>
> The **volume** of a 3-dimensional object is the amount of space occupied or consumed by a solid object, and is measured in cubic units.

The volume of a rectangular object is the number of cubes required to fill it completely, like blocks in a box, as Figure 1 illustrates.

Cubic Unit

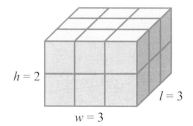

$h = 2$
$l = 3$
$w = 3$

Figure 1

The volume of shapes that are not rectangular are a little more involved. Table 1 gives the volume formulas for some common shapes.

	Table 1	
	Summary of Volume Formulas	
Shape	**Definition**	**Volume Formula**
Rectangular Solid	A **rectangular solid** is a 3-dimensional figure that has six faces, all of which are rectangles. We denote the longer side of the base as the length l and the shorter side as the width w. The height is h.	$V = lwh$
Cube	A **cube** is a rectangular solid that is made of six congruent squares. The side length is denoted by s.	$V = s^3$
Sphere	A sphere is a the set of all points in a 3-dimensional space that are a certain distance from a fixed center point. The radius is denoted by r.	$V = \dfrac{4}{3}\pi r^3$

Right Circular Cylinder	A **right circular cylinder** is a cylinder where the bases are circles with a radius of r and are perpendicular to the height h of the cylinder.	$V = \pi r^2 h$
Right Circular Cone	A **right circular cone** is a figure with a circle base with radius r tapering to a point. The height is denoted by h.	$V = \dfrac{1}{3}\pi r^2 h$
Square Pyramid	A **square pyramid** is a figure with a square base with sides s and four triangular sides that taper to a point. The height is denoted by h.	$V = \dfrac{1}{3}s^2 h$

Example 1: Determining Volume

Assume you want to build a concrete patio in the back of your house. The patio is to be 18 feet long, 12 feet wide, and 6 inches deep.

a. How many cubic yards of concrete are needed to build the patio?

b. If a concrete mixer truck holds 5 cubic yards of concrete, how many truckloads of concrete will be needed?

Solution

a. The volume of the concrete patio is determined by the formula $V = lwh$.

Before we substitute the values in the formula, notice that not all of our dimensions are in the same units. Because we should always work with dimensions that are in the same units, we need to first convert some of the measurements. The easiest option in this example is to convert the depth, which is given in inches, into feet. Therefore, the depth is 6 in. $= \dfrac{6 \text{ in.}}{12 \text{ in.}} = \dfrac{1}{2}$ ft or 0.5 ft. So, the total volume is

$$\begin{aligned} V &= lwh \\ &= (18 \text{ ft})(12 \text{ ft})(0.5 \text{ ft}) \\ &= 108 \text{ ft}^3. \end{aligned}$$

We can calculate the number of cubic yards required by recalling that 1 yd = 3 ft and cubing both sides to get $(1 \text{ yd})^3 = (3 \text{ ft})^3$ or 1 yd^3 = 27 ft^3. Multiplying by $\dfrac{1 \text{ yd}^3}{27 \text{ ft}^3}$ gives us the following.

$$108 \text{ ft}^3 \cdot \dfrac{1 \text{ yd}^3}{27 \text{ ft}^3} = 4 \text{ yd}^3$$

Therefore, our patio requires 4 cubic yards of concrete to construct.

b. Because we need less than 5 cubic yards of concrete, only one truckload of concrete would be needed to pour the patio.

In Example 1, we used volume to help us identify the amount of concrete we needed to purchase for our project. In the next example, we consider when restrictions are put in place on the volume of an object.

Example 2: Using Volume to Meet Restrictions

Suppose you want to build a swimming pool in your backyard. The city in which you live has a restriction on the volume of water that the pool can hold. The restriction is that pools may contain at most 4800 cubic feet of water. Your plans are to a build a pool that is 16 feet wide, 24 feet long, and has a depth that varies from 2 feet to 12 feet on a constant slope. Does the pool meet the restrictions required by the city?

Solution

To begin our solution, let's recall some of our problem-solving techniques from Chapter 1. By drawing a picture of the pool with views from both the top and side, we can better understand what is being asked. The following figure shows the top view of the swimming pool.

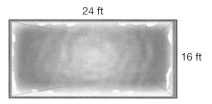

You can see that the top view of the pool shows that the pool is rectangular with a width of 16 ft and a length of 24 ft. Now let's take a look at a side view of the pool.

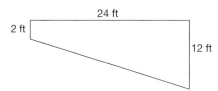

The side view of the pool is actually a trapezoid. The formula for the area of a trapezoid is

$$\text{Area}_{\text{trapezoid}} = \frac{1}{2}(b_1 + b_2)h,$$

where b_1 and b_2 are the lengths of the bases. In our case, the lengths of the bases are 2 ft and 12 ft. The height, or h, would be 24 ft. So, its area would be

$$\text{Area}_{\text{trapezoid}} = \frac{1}{2}(2 \text{ ft} + 12 \text{ ft})(24 \text{ ft}) = 168 \text{ ft}^2.$$

Another way—though unnecessary since we know the formula—to find this side view area would be to break the trapezoid into pieces.

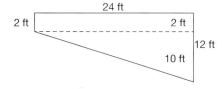

Here, we cut the side view of the trapezoid into a small rectangle and a triangle. We know the formula for the area of each shape. So, if we combine the two areas, we will have the area of the trapezoid.

$$\text{Area}_{\text{rectangle}} = (2 \text{ ft})(24 \text{ ft}) = 48 \text{ ft}^2$$

$$\text{Area}_{\text{triangle}} = \frac{1}{2}(24 \text{ ft})(10 \text{ ft}) = 120 \text{ ft}^2$$

$$\text{Area}_{\text{trapezoid}} = \text{Area}_{\text{rectangle}} + \text{Area}_{\text{triangle}}$$

$$= 48 \text{ ft}^2 + 120 \text{ ft}^2 = 168 \text{ ft}^2$$

You can see that either way we calculate it, the area of the trapezoid is 168 ft².

Finally, we can find the volume of the pool. The volume will be equal to the product of the area of the trapezoidal side view and the width of the pool. Take a moment to make sure you understand we have taken *all* three dimensions (length, width, and height) of the pool into account. The final volume of the pool will be

$$V_{\text{pool}} = \text{Area}_{\text{trapezoid}} \cdot \text{height} = (168 \text{ ft}^2)(16 \text{ ft}) = 2688 \text{ ft}^3.$$

Now to answer our initial question of whether or not we can build the pool. The answer is yes because we are under the maximum allowable volume for swimming pools.

Spheres

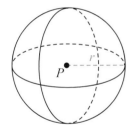

$$V = \frac{4}{3}\pi r^3$$

Figure 2:
Volume of a Sphere

In the next example, we consider the **sphere**. A sphere is a solid where all points on the surface of the solid are a certain distance from a fixed center point P, as seen in Figure 2. The distance from P to any point on the sphere is called the **radius** of the sphere, while the **diameter** of the sphere is defined as the length of any line segment with endpoints on the sphere that passes through the center P.

Great examples of spheres are a basketball, a baseball, or the Earth.

Example 3: Finding the Volume of a Sphere

Whitt is filling balloons with an air pump.

a. If after 2 pumps of air a balloon has a radius of 2 inches, find the volume of air contained in the balloon.

b. Whitt wants to make the balloon bigger, so he puts in 2 more pumps of air and the volume of air doubles. How much volume of air is in the balloon now?

c. How big is the balloon's radius after the 4 pumps of air?

d. By what percentage does the radius increase when Whitt goes from 2 pumps of air to 4 pumps?

Solution

a. We know that the volume can be determined by $V = \frac{4}{3}\pi r^3$, where $r = 2$ inches.

6

$$V = \frac{4}{3}\pi \left(2 \text{ in.}\right)^3$$

$$= \frac{4}{3}\pi \left(8 \text{ in.}^3\right)$$

$$= \frac{32\pi}{3} \text{ in.}^3$$

$$\approx \frac{32(3.14)}{3} \text{ in.}^3$$

$$\approx 33.49 \text{ in.}^3$$

b. The volume doubles after 2 more pumps of air, so the volume after 4 pumps is as follows.

$$33.49 \text{ in.}^3 \cdot 2 = 66.98 \text{ in.}^3$$

c. To find the new radius, we can substitute the known values into the volume formula and solve for r.

$$V = \frac{4}{3}\pi r^3$$

$$66.98 \text{ in.}^3 = \frac{4}{3}\pi r^3$$

$$\frac{3}{4} \cdot 66.98 \text{ in.}^3 = \frac{3}{4} \cdot \frac{4}{3}\pi r^3$$

$$50.235 \text{ in.}^3 = \pi r^3$$

$$\frac{1}{\pi} \cdot 50.235 \text{ in.}^3 = \frac{1}{\pi} \cdot \pi r^3$$

$$\frac{50.235 \text{ in.}^3}{\pi} = r^3$$

$$\frac{50.235 \text{ in.}^3}{(3.14)} \approx r^3$$

$$16.00 \text{ in.}^3 \approx r^3$$

Taking the cube root of each side gives $r \approx 2.52$ in. This means that the radius of the balloon after 4 pumps is approximately 2.52 in.

d. Finally, we can find the percentage increase of the radius as follows.

$$\text{percent increase} = \frac{\left(\text{new radius} - \text{old radius}\right)}{\left(\text{old radius}\right)} \cdot 100\%$$

$$= \frac{\left(2.52 \text{ in.} - 2.0 \text{ in.}\right)}{2.0 \text{ in.}} \cdot 100\%$$

$$= 0.26 \cdot 100\%$$

$$= 26\%$$

We can see that although the volume doubled (that is, it increased by 100%) between the two measurements, the radius only increased by 26 percent.

Right Circular Cylinders

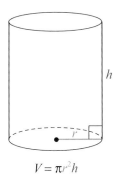

$V = \pi r^2 h$

Figure 3: Volume of a
Right Circular Cylinder

A **right circular cylinder** is the most common cylinder we encounter. A right circular cylinder is defined as a cylinder where the bases are circles with a radius of r and are perpendicular to the height h of the cylinder.

If we think intuitively about volume, the formula for calculating the volume of a cylinder is quite easy. The volume of any 3-dimensional object can be thought of as the product of the area of the "base" of the object times its height. So, for a right circular cylinder, the base is a circle and its height is given by h.

Example 4: Finding the Volume of a Right Circular Cylinder

The sound a drum makes varies according to the diameter and depth of the drum (assuming that the materials for the make of the drum remain the same).

a. Find the volume of air in a snare drum that has a diameter of 14 inches and a depth 5 inches.

b. A tom-tom drum has a radius of 5 inches and a depth of 8 inches. Find the volume of air in the tom-tom.

Solution

a. Using the formula for the volume of a right circular cylinder, $V = \pi r^2 h$, with $r = 14 \div 2 = 7$ inches and $h = 5$ inches, we compute the volume of the snare drum as follows.

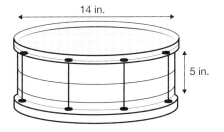

$$V = \pi r^2 h$$
$$= \pi (7 \text{ in.})^2 (5 \text{ in.})$$
$$= \pi (49 \text{ in.}^2)(5 \text{ in.})$$
$$= 245\pi \text{ in.}^3$$
$$\approx 245(3.14) \text{ in.}^3$$
$$= 769.3 \text{ in.}^3$$

So, the approximate volume of the snare drum is 769.3 in.^{3}.

b. To find the volume of air in the tom-tom drum, we use the same formula, with $r = 5$ and $h = 8$.

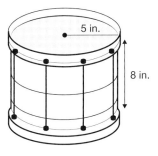

$$V = \pi r^2 h$$
$$= \pi (5 \text{ in.})^2 (8 \text{ in.})$$
$$= \pi (25 \text{ in.}^2)(8 \text{ in.})$$
$$= 200\pi \text{ in.}^3$$
$$\approx 200(3.14) \text{ in.}^3$$
$$= 628 \text{ in.}^3$$

So, the approximate volume of the tom-tom drum is 628 in.^{3}.

Right Circular Cones

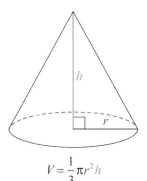

$$V = \frac{1}{3}\pi r^2 h$$

Figure 4: Volume of a
Right Circular Cone

A **right circular cone** has a volume formula very similar to that of a cylinder. Consider a cone of radius r that is formed out of a cylinder with the same radius. The volume is $\frac{1}{3}$ of the volume of the cylinder from which the cone was cut. The height of the cone is defined by h.

Example 5: Finding the Volume of a Right Circular Cone

E.D.'s Candy Shoppe is creating a new solid chocolate in the shape of a right circular cone. The height of the chocolate cone is 2 cm and the diameter of the base is 1.2 cm.

a. Find the volume of the new chocolate candy.

b. A rectangular block of chocolate measuring 22 cm by 10 cm by 10 cm is melted down and used to make a number of the new candies. Find the maximum number of candies that can be made from the block of chocolate.

Solution

a. The formula for the volume of a right circular cone is given as $V = \frac{1}{3}\pi r^2 h$. So, we can calculate the volume of the cone using this formula with $r = 1.2 \div 2 = 0.6$ cm and $h = 2$ cm.

$$V = \frac{1}{3}\pi r^2 h$$

$$= \frac{1}{3}\pi (0.6 \text{ cm})^2 (2 \text{ cm})$$

$$= \frac{1}{3}\pi (0.36 \text{ cm}^2)(2 \text{ cm})$$

$$= 0.24\pi \text{ cm}^3$$

$$\approx 0.24(3.14) \text{ cm}^3$$

$$\approx 0.75 \text{ cm}^3$$

The volume of the new chocolate is approximately 0.75 cm^3.

b. We begin by finding the volume of the chocolate block.

$$V = lhw$$

$$= (22 \text{ cm})(10 \text{ cm})(10 \text{ cm})$$

$$= 2200 \text{ cm}^3$$

Now we divide this total volume by the amount of chocolate it takes to make one of the new candies to find out how many candies can be made from one block.

$$\text{number of candies from one bar} = \frac{2200 \text{ cm}^3}{0.75 \text{ cm}^3} = 2933.33$$

Since we are asked to find the maximum number of chocolates that can be made from one bar, we need to round the number down to the nearest whole number. So, a maximum of 2933 candies can be made from the block of chocolate.

Square Pyramids

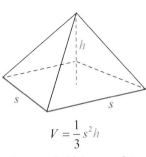

$$V = \frac{1}{3}s^2h$$

Figure 5: Volume of a Square Pyramid

The last 3-dimensional figure we will discuss is the **square (or regular) pyramid**. A square pyramid is defined as a 3-dimensional figure that has a square base and isosceles triangles as faces. It should be noted that a pyramid can have *any* polygon as its base. However, for the purposes of this course, we will only consider the case of the square pyramid.

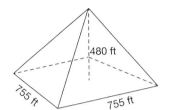

> ## Example 6: Finding the Volume of a Square Pyramid

The Great Pyramid of Giza was originally about 480 ft tall with a square base whose sides measured about 755 ft long.

a. Find the volume of the Great Pyramid of Giza.

b. A scale model of the Great Pyramid is made out of sand. If the model has a ratio of $1:100$, what is the volume of sand needed to build the model?

Solution

a. The given formula for the volume of a square pyramid is $V = \frac{1}{3}s^2h$. So, the volume of the Great Pyramid can be calculated as follows.

$$
\begin{aligned}
V &= \frac{1}{3}s^2h \\[2mm]
&= \frac{1}{3}\left(755 \text{ ft}\right)^2 \left(480 \text{ ft}\right) \\[2mm]
&= \frac{1}{3}\left(570{,}025 \text{ ft}^2\right)\left(480 \text{ ft}\right) \\[2mm]
&= 91{,}204{,}000 \text{ ft}^3
\end{aligned}
$$

Therefore, the volume of the Great Pyramid of Giza was about $91{,}204{,}000 \text{ ft}^3$.

b. Using our knowledge of ratios, we know that if the scale model has a ratio of $1:100$, then we need to divide by 100. We can do this in one of two ways. We can either divide each of the original length measurements by 100, and then multiply them together, or we can divide the volume by $100^3 = 1{,}000{,}000$. Either calculation will give the same result. The following shows the volume method. We leave the other method as an exercise. We take the volume of the Great Pyramid and divide by $100^3 = 1{,}000{,}000$ as follows.

$$\frac{91{,}204{,}000}{1{,}000{,}000} = 91.204$$

Therefore, the model pyramid will require 91.204 ft^3 of sand.

Skill Check #1

Show that dividing the original measurements in Example 6 by 100 produces the same volume as given in part **b.**

Surface Area

Whereas volume refers to the amount of material you can put *inside* a 3-dimensional object, surface area focuses on the total area on the *surface* of the object. We will focus on the same objects we used for the discussion involving volume. It should be noted that where volume measures the space taken up by a 3-dimensional object and is calculated as cubic units, surface area only measures a 2-dimensional aspect of an object, and is measured in square units.

Surface Area

The **surface area** of a 3-dimensional object is the area required to cover the outside surfaces of the object and is measured in square units.

Rectangular Solids

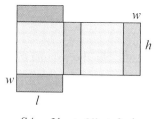

$$SA = 2lw + 2lh + 2wh$$

Figure 6: Surface Area of a Rectangular Solid

The surface area of a rectangular solid can best be described by considering its "footprint" as seen in Figure 6. When the rectangular solid is cut open, we can see that the dimensions of length, width, and height create six rectangles: two each for lw, wh, and lh. Therefore, the surface area will be the sum of all six rectangular areas.

Surface Area of a Rectangular Solid

The **surface area of a rectangular solid** with length l, width w, and height h can be computed by the formula $2lw + 2lh + 2wh$.

Example 7: Finding the Surface Area of a Rectangular Solid

Aimee is icing a cake for her sister's birthday. It's a half-sheet cake and measures 12 in. by 18 in. by 2 in.

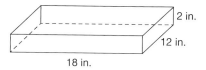

a. Find the surface area of the cake that will need icing.

b. If the cake is cut into pieces that are 2-inch squares, how many pieces of cake can be served?

Solution

a. We recognize that the rectangular solid has length $l = 18$ in., width $w = 12$ in., and height $h = 2$ in. Using our formula, we calculate the surface area as follows.

$$SA = 2lw + 2lh + 2wh$$
$$= 2(18 \text{ in.})(12 \text{ in.}) + 2(18 \text{ in.})(2 \text{ in.}) + 2(12 \text{ in.})(2 \text{ in.})$$
$$= 432 \text{ in.}^2 + 72 \text{ in.}^2 + 48 \text{ in.}^2$$
$$= 552 \text{ in.}^2$$

Therefore, the total surface area of the cake is 552 in.^{2} However, remember the bottom of the cake will not be iced, so we need to subtract off the area of the bottom of the cake. The area for the bottom of the cake is found by multiplying the length times the width. Hence, it is $12 \cdot 18 = 216$ in.^{2} So, the surface area of the cake that needs icing is

$$552 \text{ in.}^2 - 216 \text{ in.}^2 = 336 \text{ in.}^2$$

b. To find the number of 2-inch square servings in the cake, we only need to consider the top of the cake. We know that the top area of the cake is the same as the bottom, which we have already calculated as 216 in.^{2} So, we can divide the top area of the cake by the top area of a single slice, 2 in. $\cdot$ 2 in. $=$ 4 in.^{2}, and obtain the following.

$$\frac{216 \text{ in.}^2}{4 \text{ in.}^2} = 54 \text{ slices}$$

Skill Check #2

Find the volume and surface area for the given box.

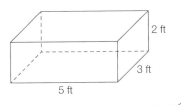

Cubes

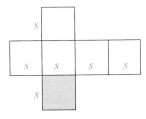

$SA = 6s^2$

Figure 7: Surface Area of a Cube

Similar to the rectangular solid, the surface area of a cube consists of the sum of the areas of the surfaces of the cube. Since a cube has six square surfaces, all with side s, and all with an area of s^2, the surface area follows from Figure 7.

Surface Area of a Cube

The **surface area of a cube** with side s can be computed by the formula $SA = 6s^2$.

Sphere

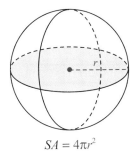

$SA = 4\pi r^2$

Figure 8: Surface Area of a Sphere

The surface area of a sphere can be found by considering the outside coverage of the sphere and the formula is $SA = 4\pi r^2$.

Surface Area of a Sphere

The **surface area of a sphere** with radius r can be computed by the formula $SA = 4\pi r^2$.

Example 8: Calculating Surface Area

Tyson's Athletic Gear, Inc. manufactures baseballs, among other things.

a. If a baseball has a diameter of 2.8 in., what is the surface area of the baseball?

b. If the quota for a day is to make 225 baseballs, what is the minimum amount of covering material needed to meet the quota in square feet?

Solution

a. Since the diameter of a circle is twice the radius, the radius will be 2.8 ÷ 2 = 1.4 in. Now using the formula, we can calculate the surface area of the baseball.

$$SA = 4\pi r^2$$
$$= 4\pi (1.4 \text{ in.})^2$$
$$= 4\pi (1.96 \text{ in.}^2)$$
$$\approx 4(3.14)(1.96 \text{ in.}^2)$$
$$\approx 24.62 \text{ in.}^2$$

So, the surface area of the baseball is approximately 24.62 in.^{2}

b. To find the amount of material needed to cover the baseballs, we can multiply the surface area by 225 to give

$$24.62 \text{ in.}^2 \cdot 225 = 5539.5 \text{ in.}^2$$

To find the amount of material in square feet, we use the fact that 1 ft = 12 in. and square both sides to get 1 ft^2 = 144 in.^{2} Therefore, we can divide the total material by 144 to give

$$\left(5539.5 \text{ in.}^2\right)\left(\frac{1 \text{ ft}^2}{144 \text{ in.}^2}\right) \approx 38.47 \text{ ft}^2.$$

So the amount needed to meet a day's quota is 38.47 ft^2 of material.

Right Circular Cylinder

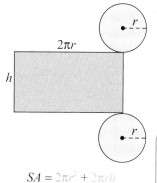

$SA = 2\pi r^2 + 2\pi rh$

Figure 9: Surface Area of a Right Circular Cylinder

A **right circular cylinder** has a top and bottom that are circles with radius r and a height of h. If we consider these attributes, then the surface area will be the sum of the areas of the circles and the lateral surface area. If we were to cut a cylinder along its height and lay it out flat, the lateral surface area would be a rectangle with width h (this is the same as the height h) and a length of the circumference of a circle, $2\pi r$.

This means the formula for a right circular cylinder is $SA = 2\pi r^2 + 2\pi rh$.

> ### Surface Area of a Right Circular Cylinder
>
> The **surface of a right circular cylinder** with radius r and height h can be computed by the formula $SA = 2\pi r^2 + 2\pi rh$.

Example 9: Finding the Surface Area of a Right Circular Cylinder

Find the surface area of a cylinder with a radius of 4 cm and a height of 10 cm.

Solution

The formula for the surface area of a cylinder is $SA = 2\pi r^2 + 2\pi rh$. So, the surface area of this cylinder can be calculated as follows.

$$SA = 2\pi r^2 + 2\pi rh$$
$$= 2\pi(4 \text{ cm})^2 + 2\pi(4 \text{ cm})(10 \text{ cm})$$
$$= 2\pi(16 \text{ cm}^2) + 2\pi(40 \text{ cm}^2)$$
$$= 32\pi \text{ cm}^2 + 80\pi \text{ cm}^2$$
$$= 112\pi \text{ cm}^2$$
$$\approx 112(3.14) \text{ cm}^2$$
$$= 351.68 \text{ cm}^2$$

Table 2 contains a summary of the volume and surface area formulas.

Table 2		
Summary of Volume and Surface Area Formulas		
Shape	**Volume Formula**	**Surface Area Formula**
Rectangular Solid	$V = lwh$	$SA = 2lw + 2lh + 2wh$
Cube	$V = s^3$	$SA = 6s^2$
Sphere	$V = \dfrac{4}{3}\pi r^3$	$SA = 4\pi r^2$
Right Circular Cylinder	$V = \pi r^2 h$	$SA = 2\pi r^2 + 2\pi rh$
Right Circular Cone	$V = \dfrac{1}{3}\pi r^2 h$	$SA = \pi r^2 + \pi r\sqrt{r^2 + h^2}$
Square Pyramid	$V = \dfrac{1}{3}s^2 h$	$SA = s^2 + 2s\sqrt{h^2 + \left(\dfrac{s}{2}\right)^2}$

Example 10: Using Volume

A farmer wishes to build a grain silo that can hold 400,000 cubic feet of grain. There is a limit to the height of the silo of 100 feet. Find an example of a silo that meets the necessary requirements for the farmer's needs.

Solution

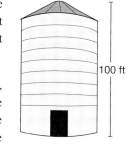

100 ft

A silo is in the shape of a right circular cylinder. Note that we are asked to find *an* example, not *the* example. As the radius and height of the silo can vary, there are many differently sized silos that meet the criteria.

Using what we know about the volume of right circular cylinders, the height of the silo, and the required volume, we can find the solution using $V = \pi r^2 h$, where r is the radius of the silo, h is the height of the silo, and V represents the volume. We are given that the farmer wants the silo to be able to hold 400,000 cubic feet of grain and that the height is to be no more than 100 feet. There are many possible answers to this question since the height can be any number between 0 and 100 feet. We are going to assume that the height is 100 feet. That means we only need to find the radius to determine the full dimensions of the silo. Using the formula for the volume of a cylinder, we can calculate the radius as follows.

$$V = \pi r^2 h$$

$$400{,}000 \text{ ft}^3 = \pi r^2 \left(100 \text{ ft}\right)$$

$$\frac{400{,}000 \text{ ft}^{3\ 2}}{\pi \left(100 \text{ ft}\right)} = \frac{\pi r^2 \left(100 \text{ ft}\right)}{\pi \left(100 \text{ ft}\right)}$$

$$\frac{4000}{\pi} \text{ ft}^2 = r^2$$

$$\sqrt{\frac{4000}{\pi} \text{ ft}^2} = \sqrt{r^2}$$

$$\sqrt{\frac{4000}{3.14} \text{ ft}^2} \approx \sqrt{r^2}$$

Using a calculator, we calculate that the radius would be $r \approx 35.69$ ft.

Thus, at 100 feet tall, in order for the silo to hold 400,000 cubic feet of grain, the radius should be approximately 35.69 feet.

Example 11: Using Surface Area

You have decided to paint the exterior of your house, and you need to determine the amount of paint required to complete the job. Assume your house is a one-story rectangular building with a length of 87 feet and a width of 45 feet. The height of each wall to be painted is 10 feet. There are 6 windows of size 30 inches by 36 inches and two doors, each measuring 30 inches by 84 inches. The gable end (triangular portion) of the house will not be painted. Assuming a gallon of paint will cover 400 square feet of surface area, and that you will apply two coats, how much paint do you need to paint your house?

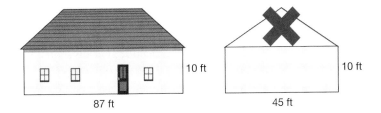

Solution

There are several aspects that need to be considered to determine the surface area to be painted and the amount of paint required to do so. Let's begin by finding the total surface area to be painted. There are two "long" sides to the house and two "short" sides to the house, all of which are rectangular. Therefore the total wall area can be calculated by

$$2 \cdot (\text{Area of the long side}) + 2 \cdot (\text{Area of the short side}).$$

The total area of the two long sides can be calculated as follows.

$$\text{Area}_{\text{long}} = 2lh = 2(87 \text{ ft})(10 \text{ ft}) = 1740 \text{ ft}^2$$

The total area of the two short sides is

$$\text{Area}_{\text{short}} = 2wh = 2(45 \text{ ft})(10 \text{ ft}) = 900 \text{ ft}^2.$$

Notice that we did not calculate the area for the top or bottom of the rectangular solid that is our house, since we will not be painting those sides.

Since we will not be painting the windows and doors, the amount of surface area they take up should be calculated and removed from our surface area to be painted. Each of these is a rectangle as well. For ease of computation, we need to convert the size of each from inches to feet. This means each window is 2.5 feet by 3 feet (30 in. = 2.5 ft and 36 in. = 3 ft), and each door is 2.5 feet by 7 feet (30 in. = 2.5 ft and 84 in. = 7 ft). Since there are six windows, the total window area is

$$\text{Area}_{\text{windows}} = 6lw = 6(3 \text{ ft})(2.5 \text{ ft}) = 45 \text{ ft}^2.$$

Since there are two doors, the total door area is

$$\text{Area}_{\text{doors}} = 2lw = 2(7 \text{ ft})(2.5 \text{ ft}) = 35 \text{ ft}^2.$$

We have now taken into account all sides to be painted and the spaces to be eliminated. The total area to be painted can now be found by finding the sum of the exterior walls and subtracting the areas that will not be painted (doors and windows).

$$\text{Wall Area} = \text{Area}_{\text{long}} + \text{Area}_{\text{short}} = 1740 \text{ ft}^2 + 900 \text{ ft}^2 = 2640 \text{ ft}^2$$

$$\text{No Paint Area} = \text{Area}_{\text{windows}} + \text{Area}_{\text{doors}} = 45 \text{ ft}^2 + 35 \text{ ft}^2 = 80 \text{ ft}^2$$

$$\text{Area}_{\text{painted}} = (\text{Wall Area}) - (\text{No Paint Area}) = 2640 \text{ ft}^2 - 80 \text{ ft}^2 = 2560 \text{ ft}^2$$

This indicates that there are 2560 ft^2 to be painted for one coat. Since we want to have two coats of paint, we will need to paint a total of 2560 ft$^2 \cdot 2 = 5120$ ft^2. Considering that each gallon will cover 400 ft^2, we must buy

$$\frac{5120 \text{ ft}^2}{400 \text{ ft}^2} = 12.8 \text{ gallons.}$$

Since we are unable to purchase partial gallons, we will need to purchase 13 gallons of paint to complete the job.

Example 12: Using Volume

The I Scream, U Scream Ice Cream shop is famous for their 0.5 L ice cream sundae. They had a special cup (instead of a cone) created just for them that enhances the flavor of their ice cream. The cup is in the shape of a conical frustum. (If we cut the top—or bottom depending on the orientation—off of a cone, we get a conical frustum, as shown in the figure where the top radius is R_1, the bottom radius is R_2, and the height is h.)

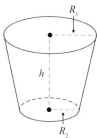

The original cup has dimensions of $R_1 = 4.5$ cm, $R_2 = 3$ cm, and $h = 11$ cm. Recently, the owner decided that he wanted to break the world record for the size of a sundae, which currently is approximately 37,500 L of ice cream. He needs to enlarge his special cup in order to achieve the goal of breaking the world record. The cup for breaking the world record has a constraint that $R_1 = 2.5$ meters and $R_2 = 1$ meter. The owner wants the final amount of ice cream that will fit in the cup to be 38,000 L. If the volume of a frustum can be found using the formula $V = \frac{1}{3}\pi h\left(R_1^2 + R_1 R_2 + R_2^2\right)$, what should the dimensions of the enlarged cup be?

Solution

We are given that the formula $V = \frac{1}{3}\pi h\left(R_1^2 + R_1 R_2 + R_2^2\right)$ can be used to find the volume of a frustum. Now, to find our missing value of h, we simply need to input values for V, R_1, and R_2 and solve for h. If we use the fact that 1000 L = 1 m^3, we obtain the following.

$$38 \text{ m}^3 = \frac{1}{3}\pi h\left((2.5 \text{ m})^2 + (2.5 \text{ m})(1 \text{ m}) + (1 \text{ m})^2\right)$$

$$= \frac{1}{3}\pi h\left(6.25 \text{ m}^2 + 2.5 \text{ m}^2 + 1 \text{ m}^2\right)$$

$$= \frac{1}{3}\pi h\left(9.75 \text{ m}^2\right)$$

$$38 \text{ m}^3 = \frac{9.75\pi h \text{ m}^2}{3}$$

$$\left(\frac{3}{9.75\pi \text{ m}^2}\right)38 \text{ m}^{\cancel{3}\,1} = \frac{9.75\pi h \text{ m}^2}{\cancel{3}}\left(\frac{\cancel{3}}{9.75\pi \text{ m}^2}\right)$$

$$\frac{114 \text{ m}}{9.75\pi} = h$$

$$\frac{114 \text{ m}}{(9.75)(3.14)} \approx h$$

$$3.72 \text{ m} \approx h$$

So, the new cup to be built to hold 38,000 L of ice cream will have a top radius of 2.5 meters, a bottom radius of 1 meter, and a height of about 3.72 meters.

Skill Check Answers

1. $\frac{1}{3}\cdot\left(\frac{755}{100}\right)^2\cdot\left(\frac{480}{100}\right) = \frac{1}{3}(7.55)^2(4.8)$

$$= \frac{1}{3}(57.0025)(4.8)$$

$$= 91.204$$

2. $V = 30$ ft^3; $SA = 62$ ft^2

6.3   Exercises

Find the volume and surface area of each figure. Use π = 3.14 and
round your answer to the nearest hundredth when necessary.

1.

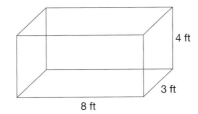

4 ft

3 ft

8 ft

2.

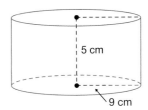

5 cm

9 cm

3.

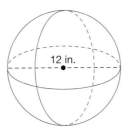

12 in.

4.

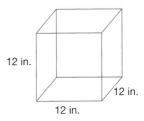

12 in.

12 in.

12 in.

5.

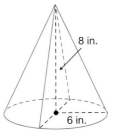

5 ft

5 ft   4 ft

6 ft

12 ft

Find the volume of each figure. Use π = 3.14 and round your answer
to the nearest hundredth when necessary.

6.

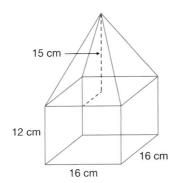

8 in.

6 in.

7.

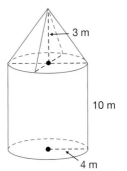

2.5 in.

4 in.

8.

15 cm

12 cm

16 cm

16 cm

9.

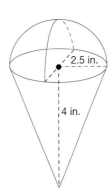

3 m

10 m

4 m

⑥

10.

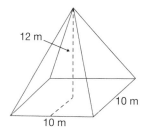

12 m

10 m

10 m

Solve each problem. Use π = 3.14 and round your answer to the nearest hundredth when necessary.

11. You want to make a fish tank in the shape of a rectangular solid that can hold 26,250 cubic centimeters of water. If the length is to be 50 cm and the height is to be 35 cm, then how wide is the tank? How many square centimeters of glass are needed to construct the tank?

12. The concrete foundation of a house requires 15 truckloads of concrete. The concrete was poured at a depth of 0.5 meters and covered an area of 540 square meters.

    a. How many cubic meters of concrete were used?

    b. If each truckload was the same size, how many cubic meters did each truck carry?

13. A grain silo in the shape of a circular cylinder is 75 feet tall and has a diameter of 24 feet.

    a. What is the volume of the silo?

    b. If a bushel of grain is 1.25 cubic feet, then how many bushels of grain will the silo hold?

    c. If a truck can unload 200 cubic feet of grain per minute into the silo, how long will it take to fill the entire silo?

14. A company is shipping a piece of equipment that requires a cubic box with a volume of 216 cubic feet. How many square feet of wood are required to construct the box?

15. The Earth has a diameter of about 7900 miles.

    a. What is the approximate surface area of the Earth?

    b. What is the approximate volume of the Earth?

    c. If $\frac{2}{3}$ of the Earth's surface is covered with water, about how many square miles of the surface are covered by land?

16. The surface of a house is covered with vinyl siding. The gable end (triangular portion) of the house and the roof will not need siding. Determine the amount of siding required.

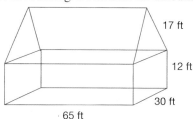

17 ft

12 ft

30 ft

65 ft

17. Refer to Example 12 of the section to solve the following problems.

    a. Find the surface area of the record-breaking cup, given that the surface area of a frustum can be calculated using the formula $SA = \pi(R_1 + R_2)\sqrt{(R_1 - R_2) + h^2} + \pi R_2^2 + \pi R_1^2$.

    b. Paper is not a viable option for the cup because it is not strong enough, so the owner has opted to use steel. If steel costs $7.25 per square meter, what is the cost of material to make the world record cup?

Chapter 6   Summary

Section 6.1   Everyday Geometry and Applications

Definitions

Longitude and the Prime Meridian

Longitude lines connect the north and south poles of the Earth and give us positioning of east and west. With a longitudinal measure of $0°$, the prime meridian is the starting point that divides the eastern and western hemispheres.

Latitude and the Equator

Latitude lines are used to measure the north and south locations on the surface of the Earth. With a latitudinal measure of $0°$, the equator is the starting point that divides the northern and southern hemispheres.

Formulas

Longitudinal Adjustment Factor

The longitudinal adjustment factor (LAF) is used to find the distance between lines of longitude at a certain latitude measure. LAF is calculated by the formula

$$\text{LAF} = 69 - 4\left(\frac{\text{latitude measure}}{10}\right).$$

Sine, Cosine, and Tangent

$$\sin A = \frac{\text{opposite}}{\text{hypotenuse}} = \frac{a}{c} \qquad \cos A = \frac{\text{adjacent}}{\text{hypotenuse}} = \frac{b}{c} \qquad \tan A = \frac{\text{opposite}}{\text{adjacent}} = \frac{a}{b}$$

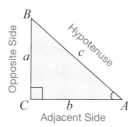

Section 6.2   Circles, Polygons, Perimeter, and Area

Definitions

Similar Triangles

We say two triangles are similar if the corresponding angles are congruent and the ratio of the lengths of the corresponding sides are equal.

Perimeter

The perimeter P of a polygon is the sum of the lengths of the sides of the polygon.

Area

The area A of a polygon is defined as the number of square units required to cover the polygon.

Circumference of a Circle

The perimeter of a circle is called the circumference.

Formulas

Sum of the Measures of the Interior Angles of a Polygon

The sum of the measures of the interior angles of an n-sided polygon is $(n-2)\cdot 180°$.

Summary of Perimeter and Area Formulas		
Shape	**Formula for Perimeter**	**Formula for Area**
Rectangle	$P = 2l + 2w$	$A = lw$
Square	$P = 4s$	$A = s^2$
Triangle	$P =$ sum of sides	$A = \dfrac{1}{2}bh$
Parallelogram	$P =$ sum of sides	$A = bh$
Trapezoid	$P =$ sum of sides	$A = \dfrac{1}{2}\left(b_1 + b_2\right)h$
Circle	$C = 2\pi r$ or $C = \pi d$	$A = \pi r^2$

Section 6.3   Volume and Surface Area

Definitions

Volume

The volume of a 3-dimensional object is the amount of space occupied or consumed by a solid object, and is measured in cubic units.

Surface Area

The surface area of a 3-dimensional object is the area required to cover the outside surfaces of the object and is measured in square units.

Formulas

Summary of Volume and Surface Area Formulas		
Shape	Volume Formula	Surface Area Formula
Rectangular Solid	$V = lwh$	$SA = 2lw + 2lh + 2wh$
Cube	$V = s^3$	$SA = 6s^2$
Sphere	$V = \dfrac{4}{3}\pi r^3$	$SA = 4\pi r^2$
Right Circular Cylinder	$V = \pi r^2 h$	$SA = 2\pi r^2 + 2\pi r h$
Right Circular Cone	$V = \dfrac{1}{3}\pi r^2 h$	$SA = \pi r^2 + \pi r \sqrt{r^2 + h^2}$
Square Pyramid	$V = \dfrac{1}{3}s^2 h$	$SA = s^2 + 2s\sqrt{h^2 + \left(\dfrac{s}{2}\right)^2}$

Chapter 6 Exercises

Solve each problem.

1.   The measures of two complementary angles are $(18z + 5)°$ and $(2z + 5)°$. Find the measures of the angles.

2.   Find $m\angle T$ if $m\angle T$ is 30 degrees larger than four times the measure of its supplement.

3.   Two angles are supplementary. One angle measure is 24° larger than the other. Find the measures of the angles.

Convert each measure to decimal degrees. Round your answer to the nearest thousandth when necessary.

4. $23°12'42''$ 5. $211°6'12''$ 6. $115°26'52''$

Convert each measure to degrees/minutes/seconds format.

7. $31.34°$ 8. $181.63°$ 9. $14.56°$

Determine the distance between the GPS locations. Round your answer to the nearest mile.

10.   **a.** 14.86°N, 66.25°E to 44.86°N, 32.31°E

   b. 21.56°S, −62.51°W to 18.3°S, −42.1°W

Solve each problem. Round your answer to the nearest hundredth when necessary.

11.   Find the lengths of $\overline{AC}$ and $\overline{BC}$ in the figure.

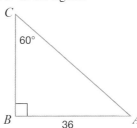

12. Use $\triangle ABC$ to find sin A, cos A, tan A, sin B, cos B, and tan B.

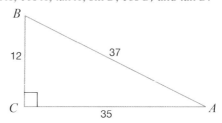

13. A plane is 1.5 miles above the ocean when it begins to climb higher at a constant angle of 4° for the next 60 miles. About how far above the ocean is the plane after its climb?

14. An 18-foot ladder is leaning against a wall where it makes an angle of 58° with the ground and 32° with the building.

    a. How far is the ladder base from the building?

    b. How far does the ladder reach up the building?

15. A plane is flying at an altitude of 15,000 feet and needs to fly an altitude of 25,000 feet. The angle of elevation that should be used is 4°. Over how many miles will it take the plane to reach the correct height? (**Hint:** There are 5280 feet in 1 mile.)

16. The sun hits a building of unknown height so that the building casts a shadow of 410 feet. If the angle of elevation of the sun is 16°, how tall is the building?

17. The lengths of three sides of a triangle ABC are 12 centimeters, 8 centimeters, and 15 centimeters. Triangle DEF is similar to triangle ABC. The length of one of the sides of triangle DEF is 120 centimeters. What is the greatest perimeter possible for triangle DEF?

18. Jessica wants to enlarge the dimensions of an 18-by-24-inch photo by 25%. What will be the perimeter of the enlarged photo?

19. Find the sum of the measures of the interior angles of each polygon.

    a. 15-gon

    b. 22-gon

    c. 30-gon

20. Find the measure of each interior angle.

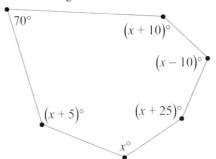

Find the perimeter and area of each figure. Use $\pi = 3.14$ and round your answer to the nearest hundredth when necessary.

21.

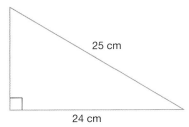

25 cm

24 cm

22.

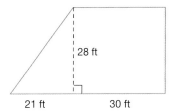

28 ft

21 ft   30 ft

23.

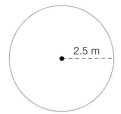

2.5 m

24.

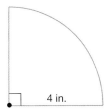

4 in.

25.

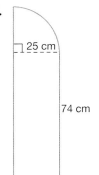

25 cm

74 cm

26.

5 m

6 m

27.

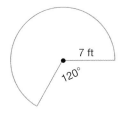

7 ft

120°

Find the area of the shaded region in each figure. Round your answer to the nearest hundredth when necessary.

28.

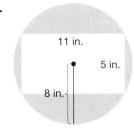

11 in.

5 in.

8 in.

29.

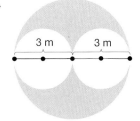

3 m   3 m

Solve each problem. Use $\pi = 3.14$ and round your answer to the nearest hundredth when necessary.

30. A window pane has dimensions of 30 inches by 22 inches. If glass sells for $0.80 per square inch, how much will each window cost?

31. A 50-pound bag of fertilizer will cover 5500 square feet. If your lawn measures 215 feet by 168 feet, how many pounds of fertilizer will you need to cover the lawn?

32. Civil War cannonballs were made out of cast iron. A collector found a set of three cannonballs for sale at an auction. The balls had diameters of 4.52 in., 4.346 in., and 3.908 in., respectively. Find the total volume of cast iron contained in the three Civil War cannonballs.

33. Gregory wanted a soccer-ball piñata for his birthday. Find the maximum amount of candy Gregory's mother could put in the piñata if the tag describes the ball as 12 inches in diameter.

34. A tube of lip balm has a height of 2.75 in. and a diameter of 0.75 in.

   a. If the cap on the tube is 0.20 inches high, find the maximum volume of lip balm that the tube can hold.

   b. What are the minimum dimensions of a box that can hold 30 tubes of lip balm, which are placed in six rows, with five tubes in each row?

35. A stick of butter measures 4.5 in. long, 1.25 in. wide, and 1.25 in. deep.

   a. How much butter is in each stick in cubic inches?

   b. What is the minimum surface area of a box containing 4 sticks of butter?

36. An 8-ounce tub of sour cream needs to contain 19.32 in.³ of sour cream. The tub must be in the shape of a right circular cylinder and leave a quarter inch between the top of the container and the sour cream. If the height of the container is 2 in., find the diameter of the tub of sour cream.

37. Oddie's Packaging Company always includes a serving size on its packages. The company has recently started making cheese balls and needs to estimate the number of servings in a ball with a diameter of three inches. Assuming the ball is a perfect sphere and that each serving is four fluid ounces, what should Oddie's estimate be? (**Hint:** There are 1.8 cubic inches in 1 fluid ounce.)

Chapter 7
Probability

Sections

Objectives

- Calculate basic probabilities
- Use the Fundamental Counting Principle to calculate probabilities
- Calculate permutations and combinations
- Use the addtion rule of probability and the multiplication rule of probability
- Calculate the expected value of an event

7   Probability

Consider how likely it would be for you to. . .

- Be struck by lightning
- Win the lottery
- Play football in the NFL
- Become President of the United States
- Be in a car accident

- Die of a spider bite
- Get married
- Get divorced
- Throw 2 sixes on a set of dice

We have a natural understanding of likelihood. Certainly, if you were asked to categorize these occurrences listed above as *likely to happen*, *unlikely to happen*, or *very unlikely to happen*, you would be able to. If instead the task was to order them in increasing likelihood, that would be much more of a challenge. Getting married is clearly more likely than being struck by lightning, but is winning the lottery more or less likely than being President of the United States? The study of **probability** is the mathematical approach to analyzing how likely things are to happen. In this chapter, we'll look at terminologies used in the study of probability, some different techniques to calculate certain probabilities, and how to use probabilities to help us make informed decisions.

Figure 1

7.1   Introduction to Probability

Classical Probability

To begin our discussion of probability, we need to define a few terms. First, a **trial**, or **probability experiment**, is any process that produces a random result, such as flipping a coin, drawing a number from one to ten out of a hat, or having a computer randomly generate a three-digit number. Each of these examples easily illustrates the possible individual results, called **outcomes**, in a trial. For instance, when you flip a coin, the outcomes possible are either a head or a tail. When you draw a number from the hat described above, four is a possible outcome, as is nine. In fact, all the numbers from one to ten are outcomes. The set of all possible outcomes from a given probability experiment is called the **sample space** of that experiment. For example, we've already seen that the sample space for a coin flip is {head, tail}. The sample space for the computer example is all possible three-digit numbers.

Example 1: Determining Sample Spaces

Identify the sample space for each of the following experiments. Note that the outcomes in a sample space are listed between brackets and separated by commas.

a. Rolling a single die.

b. Birth order gender for two children in a single family.

Solution

a. The sample space consists of all of the possible outcomes of rolling a die. It can land on any of the six sides of the die. Therefore, the sample space is the following.

$$\{\boxed{\cdot}, \boxed{\because}, \boxed{\therefore}, \boxed{::}, \boxed{\because\cdot}, \boxed{:::}\}$$

b. The sample space consists of all possible gender outcomes for a family with two children. Let B = boy and G = girl. The sample space is

$$\{BB, BG, GB, GG\}.$$

Skill Check # 1

Identify the sample space for tossing two coins together. You can list it inside { } separating the elements by commas.

Often, in probability we will be interested in grouping together outcomes in the sample space. A group, or subset, of outcomes in the sample space is called an **event**. It is possible for an event to include one, some, or all the members of the sample space.

Trial

A **trial**, or probability experiment, is any process that produces a random result.

Outcome

The **outcomes** of a trial are the possible individual results.

Sample Space

The **sample space S** of a trial is the set of all possible outcomes.

Event

An **event E** is a group, or subset, of outcomes in the sample space.

Informally, when we talk of the likelihood of an event, we are often referring to either a precise theoretical probability, called classical probability, or more of an experimental approach, called empirical probability. First let's consider classical probability.

Think Back

Recall that real numbers are not limited to fractions and whole numbers. Numbers like $\frac{1}{\sqrt{2}}$ and $\frac{3}{\pi}$ are viable values for probabilities since they are numbers between 0 and 1.

The scale used for the **classical probability** of an event is a real number between 0 and 1, where 0 means it will never happen and 1 means it is certain to happen. The notation $P(x)$ refers to the probability that x will happen. It is sometimes easier to think about classical probabilities in terms of fractions.

An event having a probability of $\frac{1}{2}$, that is, $P(x) = \frac{1}{2}$, means that there is an equal chance that the event will happen or will not happen. In other words, the event is one of two equally likely possibilities. Examples of events with probabilities of $\frac{1}{2}$ are a coin coming up heads or rolling an even number on a die. Similarly, $P(x) = \frac{1}{6}$ means that the event is one of six equally likely possibilities, such as rolling a 3 on a standard die. Whenever we can divide the situation into a known number of **equally likely** outcomes, the probability follows immediately. Be careful, the term "equally likely" here is important. There is an obvious flaw in the statement, *There's a 50% chance I'll win the lottery tomorrow.* Even though the sample space contains the two outcomes {win the lottery, don't win the lottery}, the probability of winning the lottery is *not* $\frac{1}{2}$ because the outcomes are definitely not equally likely!

☞ Helpful Hint

When an event includes the entire sample space, the probability of the event occurring is equal to 1.

Classical Probability

If all outcomes are equally likely, **classical probability** is calculated with the formula

$$P(\text{event}) = \frac{\text{the number of possible outcomes in the event}}{\text{the number of outcomes in the sample space}}.$$

$P(\text{event})$ will always be a real number between 0 and 1, inclusive.

Example 2: Calculating Classical Probability

Suppose you were asked to draw a card from a standard deck of 52 cards. A standard deck of cards contains the following cards.

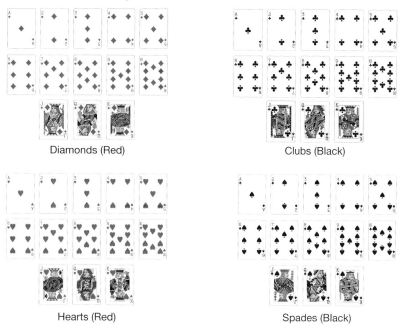

Diamonds (Red)   Clubs (Black)

Hearts (Red)   Spades (Black)

a. What is the probability that the card you draw is red?

b. What is the probability that the card you draw is a diamond?

c. What is the probability that the card you draw is a face card (king, queen, or jack)?

d. What is the probability of drawing a red spade?

Solution

a. Because this is a standard deck of cards we are drawing from, each card has the same probability of being chosen. We know that the sample space contains 52 cards. We also know that since there are two red suits (and two black suits) each with 13 cards, there are 26 possible red cards to choose. So, the probability that the card you draw is red is

$$P(\text{red card}) = \frac{\text{number of red cards}}{\text{total number of cards in deck}} = \frac{26}{52} = \frac{1}{2} = 0.5.$$

b. There are 13 cards in the diamond suit. So the probability that your card is a diamond is

$$P(\text{diamond}) = \frac{13}{52} = \frac{1}{4} = 0.25.$$

c. Each of the four suits contains three face cards (king, queen, and jack), so there are $4 \cdot 3 = 12$ face cards to choose from. So the probability that your card is a face card is

$$P(\text{face card}) = \frac{12}{52} = \frac{3}{13} \approx 0.230769.$$

d. Because all spades are black, it is impossible to draw a red spade. Therefore,

$$P(\text{red spade}) = 0.$$

Example 3: Calculating Classical Probability

Suppose that you grab a snack from a bag of chocolates that contains 4 caramel with milk chocolate, 4 peppermint with white chocolate, 6 dark chocolate with mint, and 2 raspberry with dark chocolate. What is the probability that you randomly grab a raspberry with dark chocolate for your snack?

Solution

Remember, to find the probability, we first need to know the number of outcomes in the sample space. We can add together all of the chocolates in the bag to find out the number of possible outcomes.

$$4 + 4 + 6 + 2 = 16 \text{ different chocolates in the bag}$$

Next, we know that there are 2 possible outcomes for the event of choosing a raspberry with dark chocolate. Therefore, the probability of the event is

$$P(\text{raspberry with dark chocolate}) = \frac{2}{16} = \frac{1}{8} = 0.125.$$

Empirical Probability

We have seen that the probability of rolling a 3 on a standard die is $\frac{1}{6}$, since it is 1 of 6 equally likely possibilities. Unfortunately, not every situation is so easily analyzed. For instance, what if we suspected that a die is loaded (meaning it is unfair). How would we know? Suppose we were to roll the die 600 times. If it were a fair die, we would expect to roll a 3 about 100 times, maybe not precisely 100, but close. In fact, doing an experiment like this is a way to estimate probability, called **empirical probability**. Empirical probability is built around the **law of large numbers**, which says that the greater the number of trials, the closer the experimental probability will be to the *true* probability.

> ### Empirical Probability
>
> If all outcomes are based on an experiment, **empirical probability** is calculated with the formula
>
> $$P(\text{event}) = \frac{\text{the number of times the event occurs}}{\text{total number of times the experiment is performed}}.$$
>
> $P(\text{event})$ will always be a real number between 0 and 1, inclusive.

Example 4: Empirical Probability with Multiple Trials

For her elementary school science fair project, Libby is conducting research on the accuracy of the weather prediction from her local news channel. She recorded the forecast and the actual weather for two weeks. The following table shows her results.

Table 1	
Accuracy of Weather Prediction	
Forecast	**Actual Weather**
Rain	Rain
Chance of snow	Rain
Snow	Snow
Cloudy	Clear
Cloudy	Cloudy
Rain	Rain
Clear	Drizzling rain
Clear	Clear
Cloudy	Clear
Chance of snow	Cloudy
Clear	Clear
Clear	Clear
Rain	Rain
Chance of rain	Cloudy

Using empirical probability, what is the probability that the news channel accurately predicts the next day's weather?

Solution

For Libby to calculate the probability, she needs to count the number of days the weatherman correctly predicted the weather and divide it by 14 (the total number of days she did the experiment).

$$P(\text{correct prediction}) = \frac{\text{number of times the forecast was correct}}{\text{total number of times the weather was recorded}} = \frac{8}{14} = \frac{4}{7} \approx 0.571429.$$

Because Libby can now estimate that the news channel correctly predicts the weather 57% of the time, she knows that this is also the probability that the prediction for the following day's weather will be correct.

Think Back

Remember, when changing from a decimal to a percent, you move the decimal 2 places to the right and add the % sign.

Example 5: Classical vs. Empirical Probability

Determine if the scenarios given are examples of classical or empirical probability techniques.

a. Katie is curious about her chances of winning an e-reader from the student government association. She polled her friends to find out how many of them filled out the survey to be entered in the contest.

b. Tristan is interested in his chances of winning at the black jack table. He determines the probability of what his next card will be by knowing the cards that have already been played.

c. Based on the recent United States Census, the local government estimates the amount of growth the community will experience in the coming years.

Solution

a. Because Katie is conducting an informal survey and not all students are included, the probability is empirical.

7

b. This is an example of classical probability since all cards have an equal chance of being dealt at the beginning, and Tristan adjusts his chances by accounting for those cards that have already been drawn.

c. Since the United States Census is actually an incomplete count, any probability calculated from it would be empirical.

As you might imagine, carrying out trials and surveys to estimate probabilities using an empirical approach can be very time consuming and potentially costly. Consequently, although using empirical probability may sometimes be unavoidable, it is certainly preferable to user classical probability whenever it can be obtained, So, in order for us to calculate classical probability where the sample space is not easily known, we need to be able to count the number of outcomes in a given sample space. We'll spend the next section looking at ways to count outcomes, and then in Section 7.3, we'll begin calculating classical probabilities.

Here's a recap of this section.

Probability Concepts

- A **trial**, or probability experiment, is any process in which the result is random in nature.

- An **outcome** is an individual result that is possible from a probability experiment.

- The **sample space** S is the set of all possible outcomes from a given probability experiment.

- An **event** E is a subset of outcomes from the sample space.

- If all outcomes are equally likely, **classical probability** is calculated with the formula

$$P(\text{event}) = \frac{\text{the number of possible outcomes in the event}}{\text{the number of outcomes in the sample space}}.$$

  $P(\text{event})$ will always be a real number between 0 and 1, inclusive.

- If all outcomes are based on an experiment, **empirical probability** is calculated with the formula

$$P(\text{event}) = \frac{\text{the number of times the event occurs}}{\text{the total number of times the experiment is performed}}.$$

- $P(\text{event})$ will always be a real number between 0 and 1, inclusive.

Skill Check Answer

1. {HH, HT, TT, TH}

7.1   Exercises

Identify the sample space for each experiment.

1.   A coin is flipped and then a single die is rolled.

2.   Choosing a number from all positive two-digit integers where the digits are repeated (for example, 11).

3.   Five marbles are in a bag, one of each of the following colors: blue (B), clear (C), green (G), yellow (Y), and red (R). Two marbles are drawn consecutively. Assume that the first marble is not put back in the bag before the second marble is drawn. Order of the selection matters. In other words, BG and GB are two different selections.

4.   When choosing an outfit, you have a choice of 3 shirts: white (W), black (B), or patterned (P); a choice of 2 types of jeans: faded (F) or dark wash (D); and a choice of 4 pairs of shoes: sandals (S), running shoes (R), climbing boots (C), or mules (M). List the sample space in regard to the outfits (combination of shirt, jeans, and shoes) you could pick from.

Determine whether each probability is empirical or classical.

5.   In order to find the percentage of bass that Troy had in his pond this spring, he decided to spend three days catching a total of 15 fish each day and counting how many of those were bass.

6.   Jason wants to know how likely he is to win a raffle if he bought 3 of the 1000 tickets that were sold.

7.   Virginia wants to know how likely it is for her to win a backgammon game if she only needs to roll a double six to win.

8.   Emre wants to see how many students own a smartphone. He surveys 100 college freshmen at a Winter Welcome event and asks what kind of phone they carry.

Calculate each empirical probability. Round your answer to the nearest millionth when necessary.

9.   A news organization asked a selection of voters exiting a polling place their age bracket in order to paint a picture of turn out on election day. Here's the record of the results collected so far.

Voting Age			
17-29	30-44	45-64	65 and Older
9	8	32	15

   a. What is the probability that the next voter to exit will be between 30 and 44?

   b. What is the probability that the next voter to exit will be in either of the youngest two age groups?

   c. What is the probability that the next voter to exit will be under 65?

7

10. The blood types of 200 people are collected at a doctor's office. The table shows the breakdown of patients per blood type. If a person from this group is selected at random, what is the probability that this person has type O blood?

Blood Type Survey Results	
Blood Type	Number of Patients
A	50
B	65
O	70
AB	15

11. As students were exiting the student center on campus, Vicki took note whether they were listening to headphones. The table shows results that she collected.

Based on Vicki's data, what is the probability that a randomly selected student will have headphones in their ears when exiting the student center?

Data for Students Exiting the Student Center	
	Number of Students
Headphones	33
No Headphones	51

12. A sample of 500 active-duty military showed that 82 of them suffered from Post Traumatic Stress Disorder (PTSD). However, 95 of those studied reported that they would be too embarrassed to seek mental health services.

   a. Based on this sample, if a soldier is chosen at random, what is the probability that he/she suffers from PTSD?

   b. Based on this sample, if a soldier is chosen at random, what is the probability that he/she would be willing to seek mental health services?

Use classical probability to calculate each probability. Assume individual outcomes are equally likely. Round your answer to the nearest millionth when necessary.

13. Find the probability of obtaining exactly one head when flipping four coins.

14. What is the probability that, out of 235 attendees (including yourself) at a conference, you are selected to win the door prize at the opening session?

15. A standard die is rolled.

   a. Find the probability that the roll produces a number less than 3.

   b. Find the probability that the number rolled is an even number.

   c. Find the probability that the number rolled is greater than 0.

16. On an American roulette wheel, there are 18 red pockets, 18 black pockets, and 2 green pockets. What is the probability of landing on a red pocket?

17. The table shows a breakdown for all employees on nonfarm payrolls in the United States during March 2014 (the values are not seasonally adjusted).

Employees on Nonfarm Payrolls (in Thousands), March 2014		
	Area of Employment	Number of Employees (in Thousands)
Private Sector	Goods-Producing	18,558.2
	Wholesale Trade	5803.7
	Retail Trade	15,004.0
	Transportation and Warehousing	4524.8
	Utilities	550.3
	Information	2653.0
	Financial Activities	7870.0
	Professional and Business Services	18,832.0
	Education and Health Services	21,481.0
	Leisure and Hospitality	14,143.0
	Other Private Service-Providing Services	5464.0
Public Sector	Federal Government	2705.0
	State Government	5217.0
	Local Government	14,341.0
	Total Nonfarm Employees	**137,147.0**
Source: Bureau of Labor Statistics. "Table B-1. Employees on nonfarm payrolls by industry sector and selected industry detail." Accessed June 2014. http://www.bls.gov/news.release/empsit.t17.htm		

    a. Find the probability that a random employee was employed in the leisure and hospitality sector during March 2014.

    b. Find the probability that a random employee was in the public sector during March 2014.

18. What is the probability that a card drawn randomly from a standard deck of cards will be an ace?

19. A book contains 321 pages numbered 1, 2, 3, . . . , 321. If a student randomly opens the book, what is the probability that the page's number has all of its digits the same (ignoring single-digit page numbers)?

20. Consider parents with four biological children. Find the probability that all four siblings are boys.

21. Mason has 213 songs on his iPod. He's categorized them in the following manner: 20 from sound tracks, 8 spiritual, 31 jazz, 16 Latin American, 27 R&B, 47 rock, and 64 pop. If Mason puts his iPod on shuffle, what is the probability that the next song played is a pop song?

22. Landon decided to play a joke on his friends at work. When he stocked the vending machine with colas, he randomly put the drinks into the different slots. If he put in 15 Diet Coke, 15 Coke, 12 Sprite, 17 Dr. Pepper, and 10 Fanta cans, what is the probability that the next person will get a Fanta drink when they put their money into the machine?

7.2 Counting Our Way to Probabilities

Recall that classical probability is calculated as follows.

$$P(\text{event}) = \frac{\text{the number of possible outcomes in the event}}{\text{the number of outcomes in the sample space}}$$

You can see that it is important to know how many events are in the sample space. That sounds easy enough, and in many cases it is, as we saw in Section 7.1.

For instance, in how many ways can you roll two dice such that the outcome of the second die is less than the outcome of the first die? An easy way to count these outcomes is by listing out all of the outcomes in an orderly way, as shown in Figure 1.

6, 1	5, 1	4, 1	3, 1	2, 1	1, 1
6, 2	5, 2	4, 2	3, 2	2, 2	1, 2
6, 3	5, 3	4, 3	3, 3	2, 3	1, 3
6, 4	5, 4	4, 4	3, 4	2, 4	1, 4
6, 5	5, 5	4, 5	3, 5	2, 5	1, 5
6, 6	5, 6	4, 6	3, 6	2, 6	1, 6

Figure 1: Rolling 2 Dice (1st Die, 2nd Die)

You can see that only the pairs of dice rolls in bold fit the criteria that the outcome of the second die is less than the outcome of the first die. That gives us 15 possible ways in which to roll two dice this way.

As another example, what if we were to tell you that you won a new iPod? All you have to do is go to the store and pick it out. You have a number of decisions to make before you claim your prize. How many different iPods can you choose from? Let's use a **tree diagram** to help us visualize all the possibilities. Just as its name suggests, a tree diagram uses branches to indicate possible choices at the next stage of outcomes. Here's what your choices would look like for an iPod.

> ### Tree Diagram
> A tree diagram uses branches to indicate possible choices at the next state of outcomes.

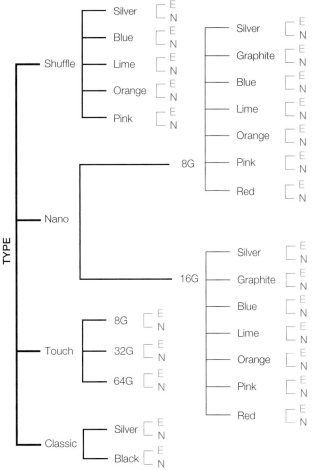

Figure 2

As you can see, there are many choices. In fact, if you count the iPods along the ends of each branch, you can see that there are 48 different iPods for you to choose from, which is the number of possible outcomes in our sample space.

The Fundamental Counting Principle

As we illustrated with the iPod example, even with just 48 outcomes a tree diagram can quickly become unmanageable. This is where we'll turn to a counting technique called the **Fundamental Counting Principle**. When there are several outcomes at each stage of an experiment, the Fundamental Counting Principle states that you can multiply together the number of possible outcomes at each stage of the experiment in order to obtain the total number of outcomes for the experiment.

Fundamental Counting Principle

For a sequence of n experiments where the first experiment has k_1 outcomes, the second experiment has k_2 outcomes, the third experiment has k_3 outcomes, and so forth, the total number of possible outcomes for the sequence of experiments is $(k_1)(k_2)(k_3)\cdots(k_n)$.

When counting outcomes in multistep experiments, we need to determine if repetition of outcomes is permitted (also known as an experiment with replacement) or if each outcome can only be used once (also known as an experiment without replacement).

Replacement

With replacement: When counting possible outcomes with replacement, objects are placed back into consideration for the following choice.

Without replacement: When counting possible outcomes without replacement, objects are *not* placed back into consideration for the following choice.

Example 1: Using the Fundamental Counting Principle with Replacement

In order to log in to your new e-mail account, you must create a password. The requirements are that the password needs to be 8 characters long consisting of 5 lowercase letters followed by 3 numbers. If you are allowed to use a character more than once, that is, **with replacement**, how many different possibilities are there for passwords?

Solution

If we think about each character in the password as a slot to fill, then we have 8 slots that need filling. The first 5 can be filled with letters and the last 3 with digits as the following figure shows.

a b c ... x y z	a b c ... x y z	a b c ... x y z	a b c ... x y z	a b c ... x y z	0 1 2 ... 8 9	0 1 2 ... 8 9	0 1 2 ... 8 9
Slot 1	Slot 2	Slot 3	Slot 4	Slot 5	Slot 6	Slot 7	Slot 8

Number of Choices: $26 \cdot 26 \cdot 26 \cdot 26 \cdot 26 \cdot 10 \cdot 10 \cdot 10$

The first 5 slots contain 26 possibilities each, one for each letter of the alphabet. The last 3 slots have 10 possible possibilities each, one for each digit 0 through 9. Using the Fundamental Counting Principle, we multiply each of the possibilities together to get $(26)(26)(26)(26)(26)(10)(10)(10) = 11,881,376,000$ possible passwords for the new e-mail account.

Example 2: Using the Fundamental Counting Principle without Replacement

Let's change the previous example slightly. The password still needs to be 8 characters long consisting of 5 lowercase letters followed by 3 numbers. However, now the characters may not be duplicated in the password, that is, we say we're counting **without replacement, or without repetition**.

Solution

We still have the first 5 slots being filled with letters and the last 3 with numbers. This time our picture changes slightly. The first slot still has a possibility of 26 letters, but the second slot now only has 25 choices since we used one letter for the first slot. Similarly, the third slot has 24 choices, and so forth. The same thing happens with the digits in the last 3 spaces.

a	a	a	a	a	0	0	0
b	b	b	b	b	1	1	1
c	c	c	c	c	2	2	2
.	.	.	.	.	.	.	.
.	.	.	.	.	.	.	.
.	.	.	.	.	8	7	6
x	x	x	x	x	9	8	7
y	y	y	y	y			
z	z	z	z	z			
Slot 1	Slot 2	Slot 3	Slot 4	Slot 5	Slot 6	Slot 7	Slot 8

Number of Choices: $26 \cdot 25 \cdot 24 \cdot 23 \cdot 22 \cdot 10 \cdot 9 \cdot 8$

So now we have $(26)(25)(24)(23)(22)(10)(9)(8) = 5,683,392,000$ possible passwords. That is almost half of the original amount of passwords possible if we allowed replacement!

Factorials

Notice that in Example 2, we multiplied successive decreasing numbers together to get our solution for the number of possible passwords. This is a common occurrence when solving this type of problem. Mathematically, we can represent something similar to this by using **factorials**.

$$5! = (5)(4)(3)(2)(1) = 120.$$

However, note in Example 2 we did not multiply all integers less than 26 because 26! was not the answer.

_n_ Factorial

In general, **_n!_** (read "**n factorial**") is the product of all the positive integers less than or equal to _n_, where _n_ is a positive integer.

$$n! = n(n-1)(n-2)(n-3)\cdots(2)(1)$$

Note that 0! is defined to be 1.

Example 3: Calculating Factorials

Calculate the value of the following factorial expressions.

a. $8!$

b. $\dfrac{3!}{0!}$

c. $\dfrac{89!}{87!}$

d. $\dfrac{7!}{(5-1)!}$

e. $\dfrac{5!}{3!(4-2)!}$

Solution

a. Multiply together all the positive integers less than or equal to 8.

$$8! = (8)(7)(6)(5)(4)(3)(2)(1) = 40,320$$

7

b. Calculate each factorial and then divide.

$$\frac{3!}{0!} = \frac{(3)(2)(1)}{1} = \frac{6}{1} = 6$$

c. Because the numbers are so large here, let's first look at taking a shortcut.

$$\frac{89!}{87!} = \frac{(89)(88)(87)(86)\cdots(2)(1)}{(87)(86)\cdots(2)(1)}$$

Many of the numbers being multiplied in the numerator and denominator will cancel, so let's do that first.

$$\frac{89!}{87!} = \frac{(89)(88)\cancel{(87)}\cancel{(86)}\cdots\cancel{(2)}\cancel{(1)}}{\cancel{(87)}\cancel{(86)}\cdots\cancel{(2)}\cancel{(1)}} = (89)(88) = 7832$$

d. Before we can start multiplying numbers, we need to do the subtraction in the denominator. Then we can cancel and multiply.

$$\frac{7!}{(5-1)!} = \frac{7!}{4!} = \frac{(7)(6)(5)\cancel{(4)}\cancel{(3)}\cancel{(2)}\cancel{(1)}}{\cancel{(4)}\cancel{(3)}\cancel{(2)}\cancel{(1)}} = (7)(6)(5) = 210$$

e. Before we can start multiplying numbers, we once again need to perform the subtraction in the denominator. Then, notice that $5! = (5)(4)3!$ allows us to cancel $3!$ in the numerator and the denominator before multiplying the remaining values.

$$\frac{5!}{3!(4-2)!} = \frac{5!}{3!2!} = \frac{(5)(4)\cancel{3!}}{\cancel{3!}(2)(1)} = \frac{(5)(4)}{(2)(1)} = \frac{20}{2} = 10$$

TECH TRAINING

To calculate $n!$ using a TI-83/84 Plus calculator, enter the number for n, then press MATH; scroll over to select PRB, then choose option 4: ! and press ENTER. The following screen shots illustrate 8!.

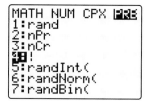 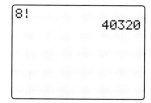

To calculate $n!$ using a TI-30XIIS/B calculator, enter the number for n, then press PRB, scroll to the !, and press ENTER.

Wolfram|Alpha can be used to compute factorials as well. Go to www.wolframalpha. com and type "8!" into the input line. Then, click the = button. Wolfram|Alpha will return the following. [1]

Permutations and Combinations

We often want to be able to count the number of ways that we can choose members from a group of objects. For instance, how many different sandwiches can be made with the ingredient choices at a sandwich shop? Or how many ways can the top three spots be filled at the end of a race of 140 people? Both of these are scenarios that can be calculated by using either a **permutation** or a **combination**. Let's define them both so you can see the difference.

☞ **Helpful Hint**

The following are all alternate notations for combinations and permutations.

$$_nC_r = C(n, r) = \binom{n}{r} = {}^nC_r = C_{n,r}$$

$$_nP_r = P(n, r) = {}^nP_r = P_{n,r}$$

Combinations

A **combination** involves choosing a specific number of objects from a particular group of objects, using each only once, when the order in which they are chosen is *not* important.

Permutations

A **permutation** involves choosing a specific number of objects from a particular group of objects, using each only once, when the order in which they are chosen *is* important.

You can see, just from their definitions, that the only thing that differentiates the two is whether the order of the objects is important. For the sandwich example, we'll contend that order is not important when deciding if one sandwich with turkey, mayonnaise, lettuce, and tomato is the same as another sandwich with the same ingredients. You might be particular in which order the tomato should be added, but it doesn't make a different sandwich. So, here we'll use a combination technique to count the number of possible sandwiches. However, if there are 140 people in a race of which Chloe, Blake, and Mary finish in the top three, then the order in which they finish makes a difference. Finishing first and winning the blue ribbon is certainly different than finishing third, so we need to use a permutation to count the possibilities.

Skill Check # 1

Decide whether you would use a permutation or combination to count the number of outcomes for each of the following scenarios.

a. In how many ways can 1st, 2nd, and 3rd place prizes be handed out to science fair winners if there are 30 students participating?

b. If each department needs two student representatives from each major on a campus committee, how many ways can the biology department chose the representatives from the 45 students who are majoring in biology?

Now that we know when to use a combination and when to use a permutation, how do we actually calculate the number of outcomes produced by each? Here are the formulas for combinations and permutations.

Combinations and Permutations of *n* Objects Taken *r* at a Time

The number of ways to select *r* objects from a total of *n* objects is found by the following two formulas. (Note that $r \le n$.)

When order is not important, use the following formula for a **combination**.

$$_nC_r = \frac{n!}{r!(n-r)!}$$

When order is important, use the following formula for a **permutation**.

$$_nP_r = \frac{n!}{(n-r)!}$$

Notice the difference in the formulas for combinations and permutations. When counting the number of possible permutations your result accounts for ALL possible arrangements. However, when counting the number of combinations you do not want to count groupings of the same *r* things more than once. So one has to divide by the number of arrangements for each group of *r*, which is the extra *r*! in the denominator of $_nC_r$.

Example 4: Using Combinations

Let's calculate the number of possibilities for our sandwich example. Suppose there are 18 toppings to choose from once you've decided on bread, meat, and cheese. How many different possible sandwiches are there if you choose 4 different toppings?

Solution

The order of sandwich toppings does not change the type of sandwich that is made. Therefore, this is a combination problem where we are choosing 4 toppings from a list of 18. Fill in the combination formula using $n = 18$ and $r = 4$.

$$_{18}C_4 = \frac{18!}{4!(18-4)!} = \frac{18!}{4!14!} = \frac{(18)(17)(16)(15)(14)(13)\cdots(2)(1)}{(4)(3)(2)(1)(14)(13)\cdots(2)(1)} = \frac{(18)(17)(16)(15)}{(4)(3)(2)(1)} = 3060$$

Therefore, there are 3060 different sandwich possibilities—far too many for you to say to a friend, "Just pick me up a turkey sandwich. It doesn't matter what kind. They're all alike!"

TECH TRAINING

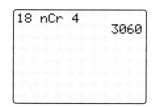

A TI-83/84 Plus can be used to find solutions involving combinations. To calculate $_{18}C_4$, type ⎡1⎤⎡8⎤, and then press ⎡MATH⎤. Scroll over to PRB and choose option 3: nCr. Then, type ⎡4⎤ and press ⎡ENTER⎤. The screen shot in the margin illustrates this.

A TI-30XIIS/B can also be used to find solutions involving combinations. To calculate $_{18}C_4$, type ⎡1⎤⎡8⎤, and then press ⎡PRB⎤. Scroll to nCr and press enter. Then, type ⎡4⎤ and press ⎡ENTER⎤.

Combinations can be calculated using Wolfram|Alpha. To calculate $_{18}C_4$, go to www.wolframalpha.com and type "Combination(18,4)" into the input bar. Then, click the = button. Wolfram|Alpha will return the following result. [2]

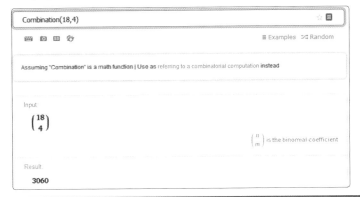

Example 5: Using Permutations

Consider a race with 140 participants. How many possible outcomes are there for the top three positions of gold, silver, and bronze?

Solution

Since the order of the winners matters in this example, we use a permutation to count the possibilities. We are choosing three runners from the original 140 that ran. Therefore, $n = 140$ and $r = 3$. Filling in the permutation formula with these values gives us the following work.

$$_{140}P_3 = \frac{140!}{(140-3)!} = \frac{140!}{137!} = \frac{(140)(139)(138)\cancel{(137)}\cancel{(136)}\cdots\cancel{(2)}\cancel{(1)}}{\cancel{(137)}\cancel{(136)}\cdots\cancel{(2)}\cancel{(1)}}$$

$$= (140)(139)(138) = 2,685,480$$

So, there are 2,685,480 possible ways the top three spots could be awarded.

TECH TRAINING

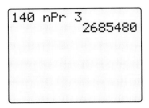

A TI-83/84 Plus can be used to find solutions involving permutations. To calculate $_{140}P_3$, type [1] [4] [0] and then press [MATH]. Scroll over to PRB and choose option 2: nPr. Then, type [3] and press [ENTER]. The screen shot in the margin illustrates this.

A TI-30XIIS/B can be used to find solutions involving permutations. To calculate $_{140}P_3$, type [1] [4] [0] and then press [PRB]. Scroll to nPr and press [ENTER]. Then, type [3] and press [ENTER].

Permutations can be calculated using Wolfram|Alpha. To calculate $_{140}P_3$, go to www.wolframalpha.com and type "Permutation(140,3)" into the input bar. Then, click the = button. Wolfram|Alpha will return the following result. [3]

---

2 Wolfram Alpha LLC, http://www.wolframalpha.com

3 Wolfram Alpha LLC, http://www.wolframalpha.com

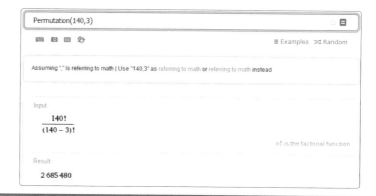

Example 6: Using Permutations

How many possible ways are there to arrange the order of appearance for the contestants in the local talent show, if there are 15 contestants all together?

Solution

Again, order is important here because being the first to perform is certainly not the same as performing last, or even second for that matter. So, this is a permutation situation with $n = 15$. However, for this problem, r is also 15 since all of the contestants are to be chosen for the talent show. Using these values in the permutation formula, we have the following.

$$_{15}P_{15} = \frac{15!}{(15-15)!} = \frac{15!}{0!} = \frac{(15)(14)(13)\cdots(2)(1)}{1} = 1,307,674,368,000$$

Having $1,307,674,368,000$ possible choices for the contestant lineup means that they will probably never choose the order by listing out all the possibilities and then randomly drawing one from a hat! Also, note that the result of $_{15}P_{15}$ is the same as $15!$.

So far, using permutations and combinations required us to have distinct objects from which to choose. In other words, none of the objects in the group were the same—no two runners, no two toppings, and no two contestants. However, suppose the objects we have to choose from contain some identical members, that is, objects are repeated within the group, and we'd like to count how many distinct ways we can arrange, or permute, the objects. For instance, how many ways are there to arrange the letters in the word MISSISSIPPI? In order to count this, we must use a slightly different permutation formula.

Permutations with Repeated Objects

The number of distinguishable **permutations** of n objects, of which k_1 are all alike, k_2 are all alike, and so forth is given by

$$\frac{n!}{(k_1!)(k_2!)(k_3!)\cdots(k_p!)},$$

where $k_1 + k_2 + \cdots + k_p = n$.

In general, to account for repetitions of objects when counting distinct permutations, we divide by the factorial representing the number of times each object is duplicated.

Example 7: Using Permutations with Repeated Objects

How many different ways can you arrange the letters in the word MISSISSIPPI?

Solution

Because there are repeated letters in the word, and no real distinction is made between each duplicated letter, we need to count the duplicate letters for our formula.

$$M = 1$$
$$I = 4$$
$$S = 4$$
$$P = 2$$

Note that since the letter M is not duplicated, $M = 1$ and its factorial is $1! = 1$, which will not change our fraction when we include it. This is always the case for unduplicated objects.

There are 11 letters in MISSISSIPPI, so $n = 11$. Substituting these values into the formula, we have

$$\frac{11!}{1!4!4!2!} = \frac{(11)(10)(9)(8)(7)(6)(5)\,4!}{4!4!2!} = \frac{(11)(10)(9)(8)(7)(6)(5)}{(4)(3)(2)(1)(2)(1)} = 34,650$$

Thus, there are 34,650 ways to arrange the letters in the word MISSISSIPPI.

TECH TRAINING

Permutations with repetitions can be calculated using Wolfram|Alpha. To calculate the number of rearrangements for the word MISSISSIPPI, we need to input the number of times each letter appears. Go to www.wolframalpha.com and type "Multinomial (1,4,4,2)" into the input bar. Then, click on the = button. Wolfram|Alpha will return the following result. [4]

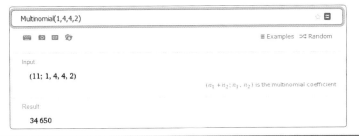

Skill Check Answers

1. **a.** permutation   **b.** combination

7.2   Exercises

Create a tree diagram to list the number of outcomes in each sample space.

1.   Find the sample space for the gender of each child in regard to birth order for a family with three children.

2.   In picking out a new car, there are several choices to make. The color can be red, white, or silver. The seats can be cloth or leather. Finally, it can have a sunroof, a moonroof, or neither. Find the sample space for the possible new car combinations using a tree diagram.

3.   Use a tree diagram to find the sample space for tossing a coin three times.

4.   Four students are randomly selected from an algebra class and asked whether they suffer from math anxiety. Find the sample space for the possible outcomes of the survey using a tree diagram.

Solve each problem.

5.   How many three-digit area codes can be made from the digits 0 through 9? Assume that the digits may repeat and that area codes beginning with 0 are allowed.

6.   How many three-digit area codes can be made from the digits 0 through 9 if the first digit is not allowed to be a 0 and the digits are allowed to repeat?

7.   How many six-character password codes are possible if you are allowed both lowercase letters and the digits 0 through 9, and repeating characters are allowed?

8.   When ordering a new e-reader, you have several choices to make. You can choose from five price ranges, decide between Wi-Fi and 3G, choose to have a one-year or three-year warranty, and pick a black, white, or silver casing. How many possible e-readers are there for you to choose from?

9.   The Youngs are planning their next family night. They always have dinner out somewhere and then do something fun together. There are two adults and four boys in the family. Each family member is allowed two meal suggestions, and each boy is allowed three activity suggestions. Assuming no family members choose the same thing, how many different family night possibilities are there?

10.   If there are seven children lining up for recess, how many different ways can they line up?

11.   The college's soccer team will play 11 games next fall. Each game can result in one of three outcomes: a win, a loss, or a tie. Find the total possible number of outcomes for the season record.

12. Harper is deciding on her schedule for next semester. She must take each of the following classes: English 102, College Algebra, History 102, and Biology 101. If there are 16 sections of English 102, 14 sections of College Algebra, 7 sections of History 102, and 12 sections of Biology 101, how many different possible schedules are there for Harper to choose from? Assume there are no time conflicts between the different classes.

13. If there were no restrictions on the use of the number zero, in theory, how many seven-digit telephone numbers are possible?

14. How many four-digit even whole numbers exist?

15. In the new ice creamery, several choices need to be made before tasting "a little bit of heaven," as the advertisement suggests. First, there are three cup sizes to choose from, then 39 different ice cream flavors to decide from, and finally several mix-ins to choose from: 10 candy bars, 8 fruits, 8 nuts, and 10 cookies or cakes. If you want a medium cup with one flavor of ice cream and two mix-ins, one candy bar and one nut, how many possible choices are there for you to have your "taste of heaven"?

Evaluate each factorial expression.

16. $5!$

17. $9!$

18. $\dfrac{8!}{6!}$

19. $1!$

20. $0!$

21. $\dfrac{10!}{2!4!}$

22. $\dfrac{5!}{3!2!}$

23. $\dfrac{12!}{8!(3-1)!}$

24. $\dfrac{23!}{11(25-4)!}$

Evaluate each permutation or combination.

25. $_8P_2$

26. $_7P_4$

27. $_5P_1$

28. $_4P_4$

29. $_3C_2$

30. $_{30}C_1$

31. $_5C_5$

32. $\dfrac{_3C_2}{_3P_2}$

33. $\dfrac{_5P_3}{_5C_3}$

34. $_7C_4 + {_7C_3} + {_7C_2} + {_7C_1}$

35. $_7P_4 + {_7P_3} + {_7P_2} + {_7P_1}$

Determine whether to use a permutation or combination to answer each question, and then determine the total number of outcomes.

36. There are 10 board members on the Community Arts Council. In how many ways can a president and treasurer be chosen? Assume that no member can hold both positions at the same time.

37. In how many ways can a committee of five people be chosen from a pool of 120 employees?

38. If there are 84 runners in a race, in how many ways can 1st, 2nd, and 3rd place ribbons be given out?

39. Elliot has to submit three photographs for the school art show. This semester he has taken 29 photographs that he thinks are show-worthy. In how many ways can he choose the photographs to submit?

40. Avery was born on 10/15/1995. How many eight-digit codes could she make using the digits in her birthday?

41. There are 18 tenured faculty in the biology department on campus. The department needs one tenured faculty member to facilitate undergraduate research, one member to supervise graduate advising, and one to coordinate grant proposals. In how many ways can these tasks be assigned, if a member may be appointed to only one duty?

42. A service organization on campus needs a group of six students from their organization's membership of 123 students to serve as program attendants at graduation. In how many ways can the attendants be chosen?

43. In how many ways can the letters in the word STATISTICS be arranged?

44. In how many ways can the letters in the word TENNESSEE be arranged?

Solve each problem.

45. Without calculating the permutations, decide which of the following words would produce the greatest number of four-letter arrangements.

    a. PASS

    b. TEST

    c. FAIR

    d. FREE

46. Determine how many ways a group of three students from your class could be selected.

47. Count the number of possible outfits you have in your closet based on the number of pants, shirts, and pairs of shoes you have. Assume that all match well together or that you would be making your own fashion statement some days.

48. Calculate the number of ways to arrange the letters in your:

    a. First name.

    b. Last name.

    c. The letters in both your first and last names.

7.3   Using Counting Methods to Find Probability

Our goal, as stated in Section 7.1, is to be able to calculate classical probability for certain events. Now that we've looked at several methods of counting the outcomes in a sample space in Section 7.2, we can begin to look at calculating probabilities. Recall that the definition of classical probability is

$$P(\text{event}) = \frac{\text{the number of possible outcomes in the event}}{\text{the number of outcomes in the sample space}}, \text{ and}$$

$P(\text{event})$ will always be a real number between 0 and 1, inclusive.

Let's reconsider the statement that *the probability will always be a number between 0 and 1, inclusive.* In other words, for any event E, $0 \leq P(E) \leq 1$. To put this in context, if an event will **not** occur, then its probability is 0 (or $P(E) = 0$). However, if an event is **certain** to happen, its probability is 1 (or $P(E) = 1$). All other probabilities fall somewhere between those two possibilities. The closer a probability is to 0, the less likely the event is to happen, and the closer the probability is to 1, the more likely the event is to happen. A probability of 0.5, or $\frac{1}{2}$, means an event is just as equally likely to happen as to not happen.

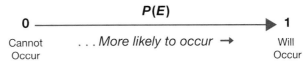

Figure 1: Range of Probability

Example 1: Calculating Classical Probability Using Combinations

Suppose that as one of the 20 graduate students in the physics department, you have a chance of being selected for one of the three student spots for a conference trip to Cancun. If the names of all the graduate students were put in a hat and three were drawn, what is the probability that you and your two friends, Leonard and Sheldon, end up being chosen?

Solution

The first thing we need to do is count the number of ways that the three student spots on the trip can be filled. Because the order in which the students are chosen is not important, we can count the outcomes using the combination formula. We have 20 students to choose from, so $n = 20$ and $r = 3$. That means there are

$$_{20}C_3 = \frac{20!}{3!(20-3)!} = \frac{(20)(19)(18)\,17!}{3!\,17!} = \frac{(20)(19)(18)}{(3)(2)(1)} = 1140$$

possible ways to choose 3 students from 20 for the trip. There is only one way in which to choose you and your two friends, so the probability of this event happening is

$$P(\text{You, Leonard, and Sheldon in Cancun}) = \frac{1}{1140} \approx 0.000877.$$

In other words, it is very unlikely that the three of you would randomly be chosen to go on the conference trip to Cancun together.

Let's change the previous example slightly. Suppose that as one of the 20 graduate students in the physics department, you have a chance of being selected for one of the three student spots for a conference in Cancun. The names of all the graduate students are put in a hat and three are drawn. However, if your name is drawn first, you get all expenses paid. If you are chosen second, everything is paid for except meals, and if you're chosen third, you must pay for your own meals and hotel. (This means that the department still picks up the tab for the flight and conference fees of the three lucky students, so it's not a bad deal!) What is the probability that you and your two friends, Leonard and Sheldon, all end up being chosen, and that your name is drawn first?

Solution

This time, the order in which the three students are chosen does make a difference when we are counting, so we'll use a permutation. Note that n is still 20 and r is still 3. Now there are

$$_{20}P_3 = \frac{20!}{(20-3)!} = \frac{(20)(19)(18)\,\cancel{17!}}{\cancel{17!}} = (20)(19)(18) = 6840$$

possible ways to choose the three lucky students for the trip. However, let's consider how many outcomes are in the event that you, Sheldon, and Leonard are chosen, and that your name is drawn first; we'll call this event E.

Let's list all the ways that the three of you could be chosen.

Table 1		
Combinations of You, Leonard, and Sheldon		
Possibility 1	**Possibility 2**	**Possibility 3**
1st pick: You	1st pick: You	1st pick: Sheldon
2nd pick: Leonard	2nd pick: Sheldon	2nd pick: You
3rd pick: Sheldon	3rd pick: Leonard	3rd pick: Leonard
Possibility 4	**Possibility 5**	**Possibility 6**
1st pick: Sheldon	1st pick: Leonard	1st pick: Leonard
2nd pick: Leonard	2nd pick: Sheldon	2nd pick: You
3rd pick: You	3rd pick: You	3rd pick: Sheldon

We can see that there are only two ways in which you are first in the list, and therefore get all expenses paid. So the probability that the event described occurs in this way is

$$P(E) = \frac{2}{6840} \approx 0.000292.$$

It seems that there is an even smaller chance of this happening, so it's better not to be greedy and wish for the top spot!

Complements

As we calculate probabilities, it will often be the case that the type of scenario we want to look at requires more than the basic classical probability formula. Let's look at some other vocabulary that will be helpful when solving probability questions. Recall from the first section in the chapter that the probability of choosing a diamond from a standard deck of cards is $P(\text{diamond}) = 0.25$.

Helpful Hint

Math Symbols:

The complement of an event E can be denoted in several different ways, such as E^c, $\overline{E}$, or E'.

Complement

The **complement** of event E, denoted by E^c, consists of all outcomes in the sample space that are *not* in event E.

Because of the definition of complement, we can determine the probability of not choosing a diamond. The probability of not choosing a diamond would be the probability of choosing any of the other cards; that is,

$$P(\text{not a diamond}) = \frac{39}{52} = \frac{3}{4} = 0.75.$$

Example 3: Finding the Complement of an Event

Describe the complement for each of the following events.

a. Rolling an even number on a die.

b. Choosing a number that doesn't end in 1, from all positive two-digit whole numbers.

c. From a class of 52 students, choosing a student who is over 21 years old.

Solution

a. The complement contains all the odd numbers on a die (that is, 1, 3, and 5).

b. The set of all positive two-digit whole numbers includes the numbers 10 through 99. The complement of our event would be all two-digit numbers that do end in 1 (that is, 11, 21, 31, 41, 51, 61, 71, 81, and 91).

c. The complement consists of the students in the class who are 21 years old or younger.

Skill Check #1

A pair of dice is rolled and the resulting sum is odd. Which of the following outcomes could be in the complement of this event?

a. A sum greater than 8

b. A sum that is an even number

c. A sum less than 5

d. A sum that is a multiple of 3

e. All of the above

By now, you're beginning to get the idea that between an event and its complement, the entire sample space is accounted for. It naturally follows that if you add the probability of event E to the probability of its complement E^c, you get 1. This gives us the following rules for complements.

Complement Rules of Probability

1. $P(E) + P(E^c) = 1$

2. $P(E) = 1 - P(E^c)$

3. $P(E^c) = 1 - P(E)$

Sometimes it is easier to find the probability of an event by calculating its complement rather than the probability of the event itself.

Example 4: Finding Probability Using Complements

Using the data given, find the following probabilities involving nuts imported into the United States between 2006–2011.

Table 2 US Import Destinations by Weight (in pounds) for Fresh or Dried Walnuts and Pistachios, 2006–2011		
Walnuts	India	4373
	Mexico	1268
	Spain	5239
	China	1533
	Austria	1938
	Other countries	4157
Pistachios	Iran	2012
	Turkey	2030
	Hong Kong	262
	Switzerland	64
	Italy	115
	Other countries	323

Source: USDA. "Fruit and Tree Nut Data." http://www.ers.usda.gov/data-products/fruit-and-tree-nut-data/data-by-commodity.aspx

a. The probability that the walnuts you consumed during this period were from Austria.

b. The probability that the walnuts you consumed during this period were from somewhere other than Austria.

c. Assume that you also purchased pistachios during this time period. What is the probability that the pistachios came from somewhere other than Italy and Switzerland?

Solution

a. The probability that the walnuts you consumed during this period were from Austria is found by dividing the weight of walnuts imported from Austria by the weight of total walnuts imported during this time period. The first number is given in the table as 1938 pounds. To find the total weight of walnuts imported, we need to add together all of the weights of walnuts imported.

$$\text{total walnut weight} = 4373 + 1268 + 5239 + 1533 + 1938 + 4157 = 18,508$$

The probability that the walnuts were from Austria is then found by

$$P(\text{walnuts from Austria}) = \frac{1938}{18,508} \approx 0.104711.$$

b. We could find the probability that the walnuts came from somewhere other than Austria by combining all the remaining places together. However, given that we just calculated the probability that the walnuts were from Austria, it is easier for us to just calculate the complement.

$$P(\text{not from Austria}) = 1 - P(\text{walnuts from Austria}) \approx 1 - 0.104711 = 0.895289$$

c. Again, it will be easier for us to calculate the complement here rather than all the other possibilities. First calculate the probability that the pistachios you consumed came from either Italy or Switzerland.

$$\text{total pistachio weight} = 2012 + 2030 + 262 + 64 + 115 + 323 = 4806$$

$$P(\text{from Italy or Switzerland}) = \frac{\text{Italy} + \text{Switzerland}}{\text{total pistachio imports}} = \frac{115 + 64}{4806} \approx 0.037245$$

Now, to calculate the probability that the pistachios came from somewhere other than Switzerland or Italy, we'll find the complement.

$$P(\text{not from Italy or Switzerland}) = 1 - P(\text{from Italy or Switzerland})$$
$$\approx 1 - 0.037245$$
$$= 0.962755$$

Skill Check Answer

1. e. all of the above

7.3 Exercises

Find the classical probability for each scenario. Round your answer to the nearest millionth when necessary.

1. There are two sets of balls numbered 1 through 5 placed in a bowl. If two balls are randomly chosen without replacement, find the probability that the balls have the same number.

2. William and Gavin are going to play video games after work. Together they have 48 games. If they decide to randomly choose two games to play, what is the probability that the two games they choose consist of William's favorite game and Gavin's favorite game? Assume they have different favorites.

3. A local pizza parlor has the following list of toppings available for selection. The parlor is running a special to encourage patrons to try new combinations of toppings. They list all possible three-topping pizzas (three distinct toppings) on individual cards and give away a free pizza every hour to a lucky winner.

Pizza Toppings			
Green Peppers	Onions	Pepperoni	Sausage
Baby Portabello Mushrooms	Black Olives	Ham	Spicy Italian Sausage
Roma Tomatoes	Pineapple	Beef	Grilled Chicken
Jalapeño Peppers	Banana Peppers	Bacon	Extra Cheese

   a. How many three-topping pizza cards are there?

   b. Find the probability that the next winner randomly selects the card with the pizza containing green peppers, ham, and bacon on it.

4. A combination padlock is a lock in which a sequence of numbers is used as the "key" to open the lock. Suppose a combination padlock has 10 digits to choose from for each of the four sections of the lock.

   a. Does the "key" for a combination padlock involve permutations or combinations?

   b. How many possible "keys" are there for the combination padlock?

   c. What is the probability that you randomly buy one of these locks whose "key" is made of 4 of the same digit?

5. Four students, three girls and a boy, have arranged to meet on the first day of class and sit in the front row. Suppose they agree to sit in the first four seats in the order that they arrive.

   a. How many possible seating arrangements are there for the four friends?

   b. What is the probability that all three girls end up sitting next to one another?

6. A committee of four is being formed randomly from the employees at a school: 5 administrators, 37 teachers, and 4 staff.

   a. How many ways can the committee be formed?

   b. What is the probability that all four members are staff?

   c. What is the probability that no member is an administrator?

7. A hand of poker is made up of five cards from a standard deck of cards.

   a. How many possible hands of poker are there in a standard deck of 52 cards?

   b. A royal flush consists of the cards Ace, King, Queen, Jack, and ten, all in the same suit. What is the probability of being dealt a royal flush?

8. Find the probability that three people randomly line up to buy tickets in order of their height (tallest, middle, shortest). Assume that no two people in the line are of the exact same height.

9. Matthew needs to set the pass code on his smartphone. It must be a four-digit number and repeated digits are allowed.

   a. How many possible pass codes are there for Matthew to choose from?

   b. How many possible pass codes are there for Matthew if he decides to choose four distinct numbers?

   c. A spy sneaks a look at Matthew's phone and sees his fingerprints on the screen over four numbers. What is the probability that the spy is able to unlock the phone on his first try?

   d. The spy knows the fingerprint trick and so on his phone he uses a repeated digit in his code. If you could see the three fingerprints on the spy's phone, what is the probability that you could unlock the phone on your first attempt?

   e. Based on parts **c.** and **d.**, is it better to repeat a digit or have four distinct digits in the code on your phone for security purposes?

10. A hand of blackjack consists of two cards. The dealer deals you a hand from a fresh deck.

   a. What is the probability that the two cards have the same face value, for instance, both cards are Kings or both cards are 5s ?

   b. If aces count 1 or 11, picture cards count 10, and card numbers 2 through 10 are equal to their face value, what is the probability that the two cards sum to 21?

11. One option to play the lottery is called "3-way any order." In order to play this method, you select three digits, from 0 to 9, such that precisely two of the digits are the same (for example, 1, 1, 2). You're a winner if your three digits show up in any order in the lottery's three randomly chosen digits. Digits may be repeated when the lottery chooses the winning number. Find the probability of winning with the "3-way any order" method.

12. Another option of playing the lottery is to choose three numbers (allowing repetition) in the exact order they will appear. Find the probability of winning the lottery with one ticket.

13. Ian is playing Scrabble. What is the probability that the next three letters he draws from the bag spell out his name in the order that he draws them? Assume there is one of each letter in the alphabet left in the bag.

14. For a pick-up game of basketball, jerseys are in a box and people start grabbing them. The box contains three extra-large, seven large, and four medium jerseys. If you are first to the box and grab two jerseys, what is the probability that you randomly grab two extra-large jerseys?

15. A junk drawer at home contains a half-dozen pens, two of which work. What is the probability that you randomly grab two pens from the drawer and don't end up with a pen that works?

Solve each problem.

16. Describe the complement of the set of odd numbers greater than 0 within the set of positive integers.

17. Let the event E be the sum of a pair of dice that is divisible by 3. List the events in E^c.

18. The following is a table of the ages of boys on a soccer team.

    Let $A = \{$soccer players older than 9$\}$. How many players are in the complement of A?

    | Ages of Boys on Soccer Team ||
Age	Number of Boys
8	3
9	6
10	7
11	2

19. Describe the complement of the set of face cards in a standard deck of cards.

20. In a company, all employees who have worked there for more than five years receive a gift. Describe the complement of this group of employees.

Find each probability using complements.

21. A bag contains each letter of the alphabet. Find the probability that a randomly selected letter from the bag will not be one of the five vowels.

22. Find the probability of randomly choosing a letter other than the letter O from a bag that contains the eighteen letters of the Italian city GUIDONIA MONTECELIO.

23. Using the table containing the breakdown of all employees on nonfarm payrolls in the United States during March 2014, find the probability that a randomly selected US worker was not in either retail trade or wholesale trade.

| Employees on Nonfarm Payrolls (in Thousands), March 2014 |||
	Area of Employment	Number of Employees (in Thousands)
Private Sector	Goods-Producing	18,558.2
	Wholesale Trade	5803.7
	Retail Trade	15,004.0
	Transportation and Warehousing	4524.8
	Utilities	550.3
	Information	2653.0
	Financial Activities	7870.0
	Professional and Business Services	18,832.0
	Education and Health Services	21,481.0
	Leisure and Hospitality	14,143.0
	Other Private Service-Providing Services	5464.0
Public Sector	Federal Government	2705.0
	State Government	5217.0
	Local Government	14,341.0
	Total Nonfarm Employees	**137,147.0**
	Source: Bureau of Labor Statistics. "Table B-1. Employees on nonfarm payrolls by industry sector and selected industry detail." Accessed June 2014. http://www.bls.gov/news.release/empsit.t17.htm	

24. In June 2011, the week of the final mission of the US space shuttle program, a Pew Research poll asked 1502 US adults whether the United States must continue to be a world leader in space exploration. The following table gives a breakdown of their opinions.

The United States Continuing to be a World Leader in Space Exploration is. . .		
Essential	**Not Essential**	**Don't Know**
871	571	60

Source: Pew Research Center. "Majority Sees U.S. Leadership in Space Essential." July 5, 2011. http://www.people-press.org/2011/07/05/majority-sees-u-s-leadership-in-space-as-essential/

   a. Find the probability that someone responded "essential."

   b. Find the probability that someone did not respond "essential."

25. Find the probability of rolling two dice and not getting the same number on both dice.

26. Suppose a family has five pets. Find the probability that at least one of the pets is male.

7.4 Addition and Multiplication Rules of Probability

We now turn our attention to those probabilities that involve two events, rather than just a singular event. There are really only two possibilities.

1. Event *A* happening *or* Event *B* happening
2. Event *A* happening *and* Event *B* happening

Of course, there are some subtleties of distinction that we will need to take note of as we go along.

Event *A* Happening OR Event *B* Happening

Let's start with the *or* events. Think about the probability of selecting a king *or* a spade from a standard deck of cards. The probability of selecting a king is $\frac{4}{52}$ and the probability of selecting a spade is $\frac{13}{52}$. It's tempting to want to just add the two probabilities together. However, think about the card that is both a king *and* a spade. The king of spades is in both events. We've counted it twice, so we need to take that into account. Therefore, the probability of choosing a king *or* a spade from a standard deck of cards is

$$P(\text{king or spade}) = P(\text{king}) + P(\text{spade}) - P(\text{king and spade})$$
$$= \frac{4}{52} + \frac{13}{52} - \frac{1}{52}$$
$$= \frac{16}{52}$$
$$\approx 0.307692.$$

P(king or spade) = P(king) + P(spade) − P(king and spade)

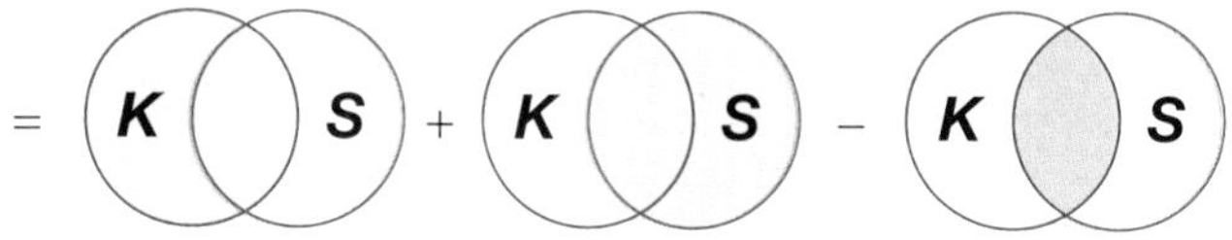

Figure 1: Probability of a King or a Spade

> **Helpful Hint**
>
> Venn diagrams, covered in Chapter 2 on Set Theory, are helpful in visualizing events.

If *A* and *B* are events that have some outcomes in common, then the probability that *A or B* will happen is calculated by adding the individual probability of each and then subtracting the probability that both events occur simultaneously.

Addition Rule for Probability

The probability of Event *A* happening *or* Event *B* happening is

$$P(A \text{ or } B) = P(A) + P(B) - P(A \text{ and } B).$$

Example 1: Applying the Addition Rule for Probability

Suppose that a student is chosen at random to receive a gift card for filling out a survey. The following table shows a breakdown of who filled out the survey.

Table 1			
Breakdown of Survey Takers			
Class	**Student Government Member**	**Non-Student Government Member**	**Total**
Freshman	3	15	18
Sophomore	1	11	12
Junior	2	7	9
Senior	4	3	7
Total	**10**	**36**	**46**

What is the probability that the winner was either a freshman or a member of student government?

Solution

We begin by finding the probability of choosing each of the individual criteria. We can see from the table that there were a total of 46 students who filled out the survey. The number of freshmen who filled it out was 18. So, the probability of choosing a freshman is

$$P(\text{freshman}) = \frac{18}{46}.$$

In previous sections, we converted our probability answers from fractions to decimals. However, since we are going to use the Addition Rule for Probability that requires us to add probabilities together, it's better to leave them as non-reduced fractions so that we avoid any error in rounding.

There were 10 members of the student government who filled out the survey, so the probability of choosing a member of the student governing board is

$$P(\text{student government}) = \frac{10}{46}.$$

There are 3 students who are both freshmen and members of the student government, so the probability of choosing a student who is in both groups is

$$P(\text{freshman and student government}) = \frac{3}{46}.$$

Using the Addition Rule for Probability, we have

$$P(\text{freshman or student government}) = \frac{18}{46} + \frac{10}{46} - \frac{3}{46}$$
$$= \frac{25}{46}$$
$$\approx 0.543478.$$

Example 2: Applying the Addition Rule for Probability

Recall the example from Section 7.2 where you won a new iPod. You can have any iPod you want, you just have to go to the store and pick it out. We can create a tree diagram to list the possible iPods you could choose from.

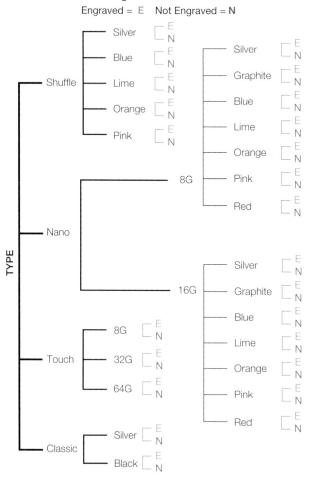

Tree Diagram of iPod Choices

Engraved = E   Not Engraved = N

Assuming that you are equally likely to choose any of the 48 iPods, what is the probability that the iPod you choose is orange or not engraved?

Solution

We begin again by finding the probabilities of both criteria individually. The tree diagram shows that of the 48 iPods, there are 6 orange possibilities, so the probability is

$$P(\text{orange}) = \frac{6}{48}.$$

There are 24 iPods that are not engraved, so

$$P(\text{not engraved}) = \frac{24}{48}.$$

There are three orange iPods that are also not engraved, so

$$P(\text{orange and not engraved}) = \frac{3}{48}.$$

Using the Addition Rule for Probability, we have

$$P(\text{orange or not engraved}) = \frac{6}{48} + \frac{24}{48} - \frac{3}{48}$$
$$= \frac{27}{48}$$
$$= 0.5625.$$

What about the case when the two events do not have any outcomes in common? Think about the following choices.

- Rolling a 1 or a 6 on a single roll of a die

- Living in the city or the country

- Buying a red car or a black truck as your first vehicle

- Going to Hawaii or Sweden for your one week of vacation

- Building a 1500 square-foot home or a 2100 square-foot home on your new single-home property

If we think about the formula for the Addition Rule for Probability, the formula adjusts for over-counting duplicates by subtracting off outcomes that are in both events. So, if two events have no outcomes in common, called **mutually exclusive events**, we end up subtracting 0. We will single out this type of probability by calling it the **Addition Rule for Mutually Exclusive Events**, but it is important to realize that this is simply a special case of the previous formula. There is just no overcounting to subtract.

Addition Rule for Mutually Exclusive Events

The probability of Event A happening *or* Event B happening when A and B have no outcomes in common is

$$P(A \text{ or } B) = P(A) + P(B).$$

Skill Check #1

For each pair of events, decide whether they are mutually exclusive or not.

a. Let event A consist of randomly selecting an adult from the mall who has shopped online at least once in the past 6 months. Let event B consist of randomly selecting an adult from the mall who has never shopped online.

b. Let event A consist of selecting an odd number and event B consist of selecting a prime number.

Example 3: Applying the Addition Rule for Mutually Exclusive Events

Suppose that you have decided it's time to get a pet. Your apartment complex allows you to have only one pet and you decide to go to the local animal shelter to adopt one of the available pets. Because you can't decide between a dog and a cat, you've left the choice up to chance. You're going to run your finger down the list of available animals without looking and let the lucky pet be the one you stop on. The following graph shows the available animals on the list at the shelter.

Available Animals for Adoption

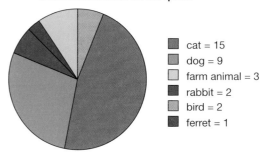

cat = 15
dog = 9
farm animal = 3
rabbit = 2
bird = 2
ferret = 1

Unfortunately, you didn't consider that other kinds of animals might be on the list. What is the probability that you choose either a cat or a dog to take home with you?

Solution

Once again, we'll begin by finding the probability of choosing a cat and the probability of choosing a dog individually. We can add the totals of each animal category to find out that there are currently 32 animals available at the shelter.

The probability of choosing a cat is

$$P(\text{cat}) = \frac{15}{32}.$$

The probability of choosing a dog is

$$P(\text{dog}) = \frac{9}{32}.$$

Because these are mutually exclusive events, that is, you cannot choose an animal that is both a cat and dog at the same time, we do not need to worry about duplicating the count of any animal. So, using the formula, we have

$$P(\text{cat or dog}) = \frac{15}{32} + \frac{9}{32}$$
$$= \frac{24}{32}$$
$$= 0.75.$$

Even though you forgot about the fact that other animals might be at the shelter, you still have a 75% chance of randomly choosing a cat or a dog. However, that means there is also a 25% chance you will select one of the other animals, some of which your landlord might not approve!

Example 4: Applying the Addition Rule for Mutually Exclusive Events

Choosing a college can be an exciting and nervous time in the life of a high school student. Emma has finally narrowed down her choices to the top 4. She's also given each school a probability based on certain characteristics.

Table 2		
University Probabilities		
University	**Characteristic**	**Probability**
A	Closest to home	$P(A) = 0.25$
B	Best sports	$P(B) = 0.10$
C	Her best friend's choice	$P(C) = 0.30$
D	Best academic program of her choice	$P(D) = 0.35$

What is the probability that Emma ends up at University B or University D?

Solution

Since Emma will choose one or the other, but not both at the same time, these events are mutually exclusive. So we just need to add the probability of her choosing University B to the probability of choosing University D.

$$P(\text{University B or University D}) = 0.10 + 0.35 = 0.45.$$

Event *A* Happening AND Event *B* Happening

Now we'll turn our attention to the possibility of two events both happening. Consider the following scenarios.

- Rolling a 6 on a die *and* drawing an ace from a deck of cards

- Rolling six 6s in a row

- You're dealt the queen of hearts *and* then a red face card in a blackjack hand

- Catching a cold *and* breaking your leg on the same day

- Being treated by an emergency room doctor who is female *and* over 35

- Choosing two boys from a group of eight girls and ten boys

- Winning the lottery *and* finding a ten dollar bill on the ground by your car

- Pulling three red Skittles in a row from the same bag

The key to calculating the probabilities for *and* scenarios is determining if one event influences the probability of the other event. Sometimes it does and sometimes it doesn't. Let's take the first scenario in the list. Does rolling a six on a die affect the probability of drawing an ace from a deck of cards? No, but how about being dealt the queen of hearts and then a red face card from a deck of cards? Here, the first event does influence the probability of the second event. If you were dealt the queen of hearts first, the probability of getting a red face card decreases for the second card because there is one less red face card in the deck to choose from.

Let's stop here and make a distinction between the two situations. We say two events are **independent** when the occurrence of one event *does not* influence the probability of the other event happening. If the result of one event *does* influence the probability of the second, we say that the two events are **dependent**.

Independent Events

Independent events are events where the result of one event does not influence the probability of the other.

Dependent Events

Dependent events are events where the result of one event influences the probability of the other.

Example 5: Independent vs. Dependent Events

Determine if the following pairs of events are independent.

a. **Event *A*:** Eating a red candy from a new bag of Skittles. **Event *B*:** Pulling a second Skittle from the same bag that is also red.

b. **Event *A*:** A woman giving birth to a daughter. **Event *B*:** The same woman's second child is also a girl.

c. **Event *A*:** Tina is the first woman to finish the 2020 Boston Marathon. **Event *B*:** Tina is the first woman to finish the 2021 New York Marathon.

Solution

a. These events are dependent. The chances of drawing a second red Skittle from the bag decreases after the first one is drawn, because it obviously was not replaced in the bag. It was eaten.

b. Although these events might appear dependent on one another, the probability that a child is a girl is the same for any given pregnancy. It is not affected by the gender of any previous pregnancies. Therefore, these events are independent.

c. At first glance these events might seem independent of one another. You might think that winning one race has no effect on winning a second race. In fact, the best starting positions for runners in large marathons are given to winners of previous races. Also, running a second marathon in consecutive years will have an effect on your body when training for and running the second one (whether that effect is positive or negative). Therefore, these events are dependent. This example illustrates determining dependence of events. Sometimes our own personal knowledge or experiences are not enough to rely on and we need to look to experts in other fields to help us determine whether events are dependent or not.

Let's focus on independent events first. If two events are independent, we can find the probability of both events occurring by multiplying the individual probabilities together. This is referred to as the **Multiplication Rule of Probability**. In fact, satisfying the multiplication rule defines independent events.

Multiplication Rule for Independent Events

A and *B* are **independent events** when the probability of Event *A* happening *and* Event *B* happening is given by $P(A \text{ and } B) = P(A) \cdot P(B)$.

Example 6: Multiplication Rule for Independent Events

Given a fair die and a standard deck of 52 cards, find the probability of rolling a 6 *and* drawing an ace.

Solution

Because the number rolled on the die does not affect the card drawn from the deck and vice versa, the events here are independent. Using the Multiplication Rule for Independent Events, we have

$$P(6 \text{ and ace}) = P(6) \cdot P(\text{ace}) = \frac{1}{6} \cdot \frac{4}{52} = \frac{4}{312} \approx 0.012821.$$

Example 7: Multiplication Rule for Independent Events

Suppose we know the following breakdown for internal medicine/pediatric hospitalists who work at Madison Regional Hospital and the ages of their patients on a given day.

Table 3

Hospitalists at Madison Regional Hospital

		Number
Hospitalist Gender	Male	6
	Female	7
Patient Age	< 35	16
	35 – 55	35
	> 55	21

What is the probability that the first patient treated is over 55 years old and treated by a male hospitalist at Madison Regional Hospital?

Solution

Because the age of the patient and the gender of the hospitalist have no effect on one another, these events are independent. So, we'll have to use the Multiplication Rule of Probability for Independent Events and multiply the individual probabilities together. Let's find the individual probabilities first.

$$P(\text{patient age} > 55) = \frac{21}{72}$$

$$P(\text{male hospitalist}) = \frac{6}{13}$$

Using the Multiplication Rule for Independent Events, we have the following.

$$P(\text{patient age} > 55 \text{ and male hospitalist}) = \frac{21}{72} \cdot \frac{6}{13}$$
$$= \frac{126}{936}$$
$$\approx 0.134615$$

7

Now, let's consider the case when the two events are dependent. Remember that dependence requires that one event influences the probability of the other event. A measure of that influence is **conditional probability**.

Conditional Probability

The **conditional probability** of Event B happening, given Event A, is the probability of Event B assuming that Event A has already, or will at some point, occur. The conditional probability is written $P(B \mid A)$, and read *the probability of B, given A*.

Once we know the conditional probability of an event, calculating the probability of two dependent events happening together is straightforward.

Multiplication Rule for Dependent Events

If A and B are **dependent events**, the probability of Event A happening *and* Event B happening is $P(A \text{ and } B) = P(A) \cdot P(B \mid A)$.

Example 8: Multiplication Rule for Dependent Events

Find the probability that, from a standard deck of cards, you're dealt two cards: the queen of hearts and then a face card.

Solution

As we've already seen, because being dealt the queen of hearts for the first card reduces the number of face cards left in the deck for the second card, these events are dependent. We begin by calculating the probability of first being dealt the queen of hearts. Since all cards are available the probability is

$$P(\text{queen of hearts}) = \frac{1}{52}.$$

When the second card is dealt, there are no longer 12 face cards in the deck. Only 11 face cards remain in a deck of 51 cards (remember that there is one less card). So, the probability is

$$P(\text{face card} \mid \text{queen of hearts}) = \frac{11}{51}.$$

We then use the Multiplication Rule for Dependent Events to get

$$P(\text{queen of hearts and face card}) = P(\text{queen of hearts}) \cdot P(\text{face card} \mid \text{queen of hearts})$$
$$= \frac{1}{52} \cdot \frac{11}{51}$$
$$= \frac{11}{2652}$$
$$\approx 0.004148.$$

Example 9: Conditional Probability

290 students were asked about their satisfaction with their interactions with the financial aid office on campus. Their responses are given in the following table.

Table 4			
Financial Aid Office Satisfaction			
Class	**Satisfied**	**Dissatisfied**	**Did Not Use**
Freshman	55	21	13
Sophomore	15	33	24
Junior	48	6	8
Senior	22	18	3
Graduate	4	1	19

If one response from the 290 students is selected at random, find the probability that the following occurred.

a. The student was satisfied with their experience.

b. The student was satisfied given that they were a senior.

c. The student was dissatisfied given that they were a freshman or sophomore.

d. The student was a graduate student given that they did not use the financial aid office.

Solution

a. To find the total number of students who were satisfied, we can find the sum of the first column.

$$\text{Students satisfied} = 55 + 15 + 48 + 22 + 4 = 144$$

To find the probability that the student we randomly selected is one of these, we divide the satisfied students by the total number of students:

$$P(\text{satisfied}) = \frac{144}{290}$$
$$\approx 0.496552.$$

b. To find the probability that the student was satisfied given that they were a senior, we need to limit our satisfied responses to those of senior students only. First, find the sum of the row of senior responses to find out how many of the 290 students were seniors.

$$\text{Senior students} = 22 + 18 + 3 = 43$$

Now, divide the number of seniors who responded satisfied by the total number of seniors.

$$P(\text{satisfied} \mid \text{senior}) = \frac{22}{43}$$
$$\approx 0.511628$$

c. Again, because we are looking at a conditional probability, we'll need to limit our student responses to only freshmen and sophomores. We can find the sum of their rows to calculate the number of students in these two classes.

$$\text{Freshman and sophomore students} = 55 + 21 + 13 + 15 + 33 + 24 = 161$$

This time we're looking at the number of dissatisfied students. Looking in the dissatisfied column, we see that there were $21 + 33 = 54$ dissatisfied freshmen and sophomores. We can then divide to find the probability that we randomly choose one of these students.

$$P(\text{dissatisfied} \mid \text{freshman or sophomore}) = \frac{54}{161}$$
$$\approx 0.335404$$

d. This conditional probability requires us to consider only the column of students who did not use the financial aid office.

Students who did not use the financial aid office $= 13 + 24 + 8 + 3 + 19 = 67$

Of these 67 students, 19 of them were graduate students. Thus, the probability that we randomly choose a graduate student given that they did not use the financial aid office is

$$P(\text{graduate student} \mid \text{did not use financial aid office}) = \frac{19}{67}$$
$$\approx 0.283582.$$

Example 10: Putting It All Together

Studies show that men have a 1 in 6 (or about 17%) chance of developing prostate cancer. A PSA test is used to detect prostate cancer and can give either a positive or negative result. Studies also show that of those men who have developed prostate cancer, their PSA test is negative 15% of the time. On the other hand, studies show that 58.5% of all men receive a positive PSA test. [1]

a. What is the probability that a man develops prostate cancer?

b. What is the probability that a man with cancer has a negative PSA test?

c. What is the probability that a man with cancer has a positive PSA test?

d. What is the probability that a man has cancer and a positive PSA test?

e. What is the probability that a man with a positive PSA test has cancer?

Solution

a. We are told this probability in the information given.

$$P(\text{cancer}) = \frac{1}{6} = 0.16\overline{6}$$

b. Again, we are told this information.

$$P(\text{negative} \mid \text{cancer}) = 15\% = \frac{15}{100} = 0.15$$

c. Here, we want to know the probability of a positive test, given that a man has cancer. Since the information we were given tells us that men with cancer have a negative PSA test 15% of the time, we can use the complement to find the probability of men with cancer having a positive test.

$$P(\text{positive} \mid \text{cancer}) = 1 - \frac{15}{100} = \frac{85}{100} = 0.85$$

1 National Cancer Institute, http://www.cancer.gov; Mayo Clinic, http://www.mayoclinic.com

d. Use the Multiplication Rule for Dependent Events to find the probability of having cancer and a positive PSA test using the probabilities we previously calculated.

$$P(\text{cancer and positive}) = P(\text{cancer}) \cdot P(\text{positive} \mid \text{cancer})$$
$$= \frac{1}{6} \cdot \frac{85}{100}$$
$$= \frac{85}{600}$$
$$\approx 0.141667$$

e. Be careful how you read this question. It's slightly different than the previous one we just answered. We want to know the probability of a man actually having cancer, given that he has a positive test. Since we weren't given this information, we'll need to rearrange the Multiplication Rule for Dependent Events to help us. Recall that the formula is

$$P(A \text{ and } B) = P(A) \cdot P(B \mid A).$$

Using algebra, we can divide both sides by $P(A)$ and get the following:

$$\frac{P(A \text{ and } B)}{P(A)} = P(B \mid A).$$

So,

$$P(\text{cancer} \mid \text{positive}) = \frac{P(\text{positive and cancer})}{P(\text{positive})}.$$

From part **d.**, we know that $P(\text{positive and cancer})$ is $\frac{85}{600}$. From the problem statement, we also know that 58.5% of all men receive a positive PSA test, which gives $P(\text{postive}) = \frac{58.5}{100}$. Therefore,

$$P(\text{cancer} \mid \text{positive}) = \frac{P(\text{positive and cancer})}{P(\text{positive})}$$
$$= \frac{\frac{85}{600}}{\frac{58.5}{100}}$$
$$= \frac{85}{600} \cdot \frac{100}{58.5}$$
$$= \frac{8500}{35100}$$
$$= \frac{85}{351}$$
$$\approx 0.242165.$$

This means that, given that a man has a positive PSA test, the probability that he actually has cancer is approximately 0.242, or 24.2%.

Parts **c.** and **d.** in Example 10 illustrate a common practice of probability. Very often we have the ability to find out $P(A \mid B)$, but what we really want to know is the $P(B \mid A)$. A mathematician named Thomas Bayes is credited with a theorem simplifying the relationship between the two. His theorem, known as Bayes' Theorem, shows how one relates to the other.

Bayes' Theorem

$$P(A\,|\,B) = \frac{P(B\,|\,A)\cdot P(A)}{P(B)},$$

when $P(B) > 0$.

Skill Check Answers

1. **a.** Mutually exclusive   **b.** Not mutually exclusive

7.4   Exercises

Calculate the probability of each set of mutually exclusive events. Round your answer to the nearest millionth when necessary.

1. Suppose that the probability of obtaining zero defective items in a sample of 50 items off the assembly line is 0.34 while the probability of obtaining 1 defective item in the sample is 0.46. What is the probability of the following?

   a. Obtaining no more than one defective item in a sample.

   b. Obtaining more than one defective item in a sample.

2. A pair of dice is rolled. What is the probability that the sum of the numbers is either 7 or 11?

3. A single letter from the word MISSISSIPPI is chosen. What is the probability of choosing an S or an I?

4. What is the probability that a card selected from a deck will be either an ace or a queen?

5. A reporter for an international newspaper is given an assignment that is randomly chosen from the following destinations worldwide: 13 continental United States assignments, 7 South American assignments, 21 European Union assignments, and 5 Asian assignments. Find the probability that he gets an assignment in Asia or South America.

6. The following table shows the breakdown of opinions for both faculty and students in a recent survey about the new restructuring of the campus to be a walking campus.

Survey Results on Restructuring Campus to a Walking Campus				
	Favor	**Oppose**	**Neutral**	**Total**
Faculty	12	4	3	**19**
Student	33	57	28	**118**
Total	**45**	**61**	**31**	**137**

   a. Find the probability that a randomly selected person is either a faculty member in favor of the change or a student who has an opinion either for or against.

   b. Find the probability that a randomly selected person is either neutral or in favor of the restructuring.

7.   The probability of the stoplight being green at the intersection of Meeting Street and Main Street is 0.55, while the probability of it being yellow is 0.15. Find the probability that the light is red when you get to the intersection of Meeting Street and Main Street. Assume that the light will be working and will be a solid color: red, yellow, or green

8.   In a box of pens and pencils, the probability of randomly choosing a sharpened pencil is 0.54 and the probability of randomly choosing a pen from the box is 0.39. Find the probability of randomly selecting either an unsharpened pencil or a pen from the box.

Calculate the probability of each set of events that are not mutually exclusive. Round your answer to the nearest millionth when necessary.

9.   A pair of dice is rolled. What is the probability that the sum of the numbers is an even number or a multiple of 3?

10.   A bag of eleven marbles contains five marbles with red on them, three with green on them, seven with black on them, and four with black and red on them. What is the probability that a randomly chosen marble has either black or red on it?

11.   What is the probability that a card selected from a deck will be either an ace or a spade?

12.   The following is a table showing the results of a poll taken on campus.

Will You Vote in the Upcoming Election?		
	Male	**Female**
Yes	16	24
No	19	11
Not decided	21	22

a.   What is the probability that a randomly selected student from this poll would be a male who has not decided whether he will vote in the upcoming election?

b.   What is the probability that a randomly selected student from this poll is female or will not vote in the upcoming election?

c.   What is the probability that a randomly selected student from this poll has decided to vote in the upcoming election?

13.   Out of a class of 30 students, there are 16 students who study Latin, 21 who study German, and 7 who study both. What is the probability that a randomly selected student from the class will study only Latin?

14.   Of the 11 instructors in the English department, four are new to the department and three are female. However, there is only one who fits all of the descriptions. Find the probability that if you randomly choose a course taught by these instructors, you get either a new instructor or a female instructor.

15. The following is a table representing the students who are on the Student Government Board.

Students on the Student Government Board		
	On-Campus Housing	**Off-Campus Housing**
Freshman	3	1
Sophomore	3	2
Junior	2	3
Senior	0	3
Graduate Student	0	2

Find the probability that a randomly chosen member of the Student Government Board is either a sophomore or lives in on-campus housing.

Determine whether each situation contains independent events.

16. The color of car driven by three randomly chosen classmates.

17. A password must be six characters long with no repeated characters. Are the choices of consecutive characters independent?

18. There are 15 board members, of which seven are men and eight are women. Two randomly chosen members will serve on the United Way campaign committee. If you wish to find the probability that both members chosen are the same sex, do you treat these selections as independent events?

19. Are receiving a bill in Monday's mail and receiving a letter from your grandparents in Monday's mail independent events?

20. Naomi and Amelia both put two business cards into the basket at a coffee shop. The shop owner selects three cards from the basket. Are the two events that Naomi's card is chosen and Amelia's card is chosen independent?

Calculate the probability of each set of independent events. Round your answer to the nearest millionth when necessary.

21. Suppose the probability that my pet will be alive in five years is 0.65 and the probability that my cousin's pet will be alive in five years is 0.48. Find the probability that both of these pets will be alive in five years assuming that they are independent events.

22. Two dice are thrown. Find the probability of getting an even number on the first die and an odd number on the second die.

23. The following table shows the student demographics for a sociology class.

Sociology 101 Student Demographics		
	Male	Female
Freshman	3	11
Sophomore	4	9
Junior	0	3
Senior	1	0

    a. Find the probability that a randomly selected student from the class is a male.

    b. Find the probability that two randomly selected students from the class are a female junior and a male sophomore.

24. Find the probability of choosing a heart and then an ace from a standard deck of cards with replacement.

25. On any given day at the beach, there is a 49% chance of precipitation. What is the probability that you will get precipitation for three days in a row on your beach vacation? Assume that the weather on a particular day at the beach is independent of the weather the day before.

Calculate each conditional probability. Round your answer to the nearest millionth when necessary.

26. A swim team consists of four boys and three girls. A relay team of four swimmers is chosen at random from the team members. What is the probability that there are two boys on the relay team given that there are two girls on the relay team?

27. Emma is playing Monopoly, a game played with two dice. What is the probability that the sum of the two dice she rolls is less than 4 given that she rolls an odd number?

28. Hunter bets his friend that he can draw two aces in a row from a standard deck of cards. What is the probability that Hunter draws a second ace given that his first card was an ace?

29. The probability that a student passes Intermediate Algebra is 0.55. The probability that a student passes College Algebra given that they pass Intermediate Algebra is 0.70. What is the probability that a student passes both College Algebra and Intermediate Algebra?

30. On each point in racquetball, a player is allowed two serves. Suppose while playing racquetball, Tim gets his first serve in about 75% of the time. He gets his first serve in and wins the point about 50% of the time. What is the probability that he wins the point, given that he gets his first serve in?

7

Calculate each probability. Round your answer to the nearest millionth when necessary.

31. Arianna likes chicken and apple sausage, but not chicken and asiago cheese sausage. There are 18 pieces of each kind of sausage on a sausage and cheese plate. What is the probability that Arianna randomly skewers three pieces of sausage that she likes given that the first two are to her liking?

32. A swim team consists of four boys and three girls. A relay team of four swimmers is chosen at random.

    a. What is the probability that two boys and two girls are chosen for the relay team?

    b. What is the probability that Jim is one of the two boys and Jane is one of the two girls?

33. James has 20 applications on the home screen of his smartphone. His nephew accidently deletes five of the apps on his home screen. What is the probability that the app originally in the top right corner and the app originally in the bottom left corner have not been deleted?

34. The probability that an e-mail is spam is 0.05, the probability that the word "offer" is in an e-mail is 0.02, and the probability that the word "bank" is in an e-mail is 0.1. The probability that the word "offer" appears given that the e-mail is spam is 0.2, and the probability that the word "bank" appears given that the e-mail is spam is 0.4.

    a. Find the probability that an e-mail contains the word "bank" and is spam.

    b. If the words are assumed to appear independently, find the probability that an e-mail that contains "offer" and "bank" is spam.

7.5   Expected Value

Now that we've explored ways to calculate probabilities, let's look at how we can use probabilities to predict the average outcome when an event takes on numerical values. **Expected value** is just that—what you would anticipate getting if you took the average of a sample of events.

For instance, suppose you throw a fair die. What would you "expect" to roll? Each of the numbers on the die has an equal chance of occurring, so you really shouldn't expect one number over another. However, if you threw it 60 times, what would you expect? Well, using the fact that each number has a $\frac{1}{6}$ chance of being rolled, you should expect each number to come up about 10 times, not exactly 10, but close. Suppose we took the average of all of those 60 rolls; in other words, we add them all up and divided by the number of rolls. Since the order of the rolls is irrelevant, we'll list them here in an organized manner so you can have the image in your head.

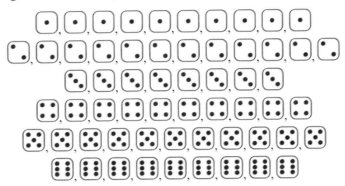

Figure 1: Sample of 60 Rolls of a Die

Adding these numbers up, we get 207. Now, divide by 60, and we get

$$\frac{207}{60} = 3.45.$$

So the average of these particular rolls of the die was 3.45. Of course if we did the experiment again, we would get a slightly different set of rolls and probably a slightly different average. If we perform the experiment multiple times, we would begin to build a picture of the *average* value. For instance, we might have the numbers 3.45, 3.51, 3.52, 3.47, 3.55, etc., all as average rolls of the die. We can see that the average of the rolls is hovering somewhere around 3.5 each time. Mathematically, the average is a number that describes the center of the data set. Although no member of the data set is actually 3.45, it does describe the central tendency of the data.

So how does finding averages link up with probabilities? Expected values are a way to find the anticipated value of a random event without having to go through this elaborate experimental process of listing out all the possibilities and manually calculating their average. To find the expected value, you can simply multiply each outcome by its probability and add them together.

> ## Expected Value
>
> The formula for **expected value** of an event X is
> $$E(X) = x_1 P(x_1) + x_2 P(x_2) + x_3 P(x_3) + \cdots + x_n P(x_n),$$
> where x_i is the ith outcome and $P(x_i)$ is the probability of x_i.

Let's use this formula to calculate the expected value for rolling a die. Remember that each number on the die has an equal probability of $\frac{1}{6}$. Constructing a table will make it easier to keep track of the calculations.

7

Table 1		
Outcomes and Probabilities of Rolling a Die		
Outcome x_i	Probability $P(x_i)$	$x_iP(x_i)$
1	$\dfrac{1}{6}$	$\dfrac{1}{6}$
2	$\dfrac{1}{6}$	$\dfrac{2}{6}$
3	$\dfrac{1}{6}$	$\dfrac{3}{6}$
4	$\dfrac{1}{6}$	$\dfrac{4}{6}$
5	$\dfrac{1}{6}$	$\dfrac{5}{6}$
6	$\dfrac{1}{6}$	$\dfrac{6}{6}$

Now, let's add them together.

$$E(X) = \frac{1}{6} + \frac{2}{6} + \frac{3}{6} + \frac{4}{6} + \frac{5}{6} + \frac{6}{6}$$
$$= \frac{21}{6}$$
$$= 3.5$$

So, the expected value for a roll of a single die is 3.5. Of course, this isn't the same value that we got when doing the 60 trials before. But, as we observed, our average could vary for each trial and these different values would all cluster around some central value, which is the expected value of 3.5. It is important to recognize that the expected value allows you to quickly calculate the average of a large number of trials without having to carry out any experiments. This is similar to the techniques used in classical probability as opposed to those used in empirical probability. Remember, the expected value gives us a *long term average*.

Skill Check # 1

Given the following table of outcomes and their probabilities, complete the table to help you calculate the expected value of the event.

Outcome, x_i	$P(x_i)$	$x_i \cdot P(x_i)$
3	0.45	
6	0.30	
9	0.25	

One of the nice properties of expectation is that it is *linear*. This means that, since we know that the average roll of one die is 3.5, then we immediately know that the average roll of two dice is $3.5 + 3.5 = 7.0$. In fact, the average roll of 20 dice is $20 \cdot 3.5 = 70$. This property is not just limited to dice.

The Sum of Expected Values

The expected value of the sum of two events, X and Y, is equal to the sum of the individual expected values. Formally, we can write this as

$$E(X+Y) = E(X) + E(Y).$$

You might wonder from this dice example whether this method is actually any quicker than just adding the trials up and dividing. To calculate the average of one die, there are six elements in the sample space to average. For two dice, there are 36 elements to average. For 20 dice, we would have 3,656,158,440,062,976 elements to average. Even if you have all the time in the world, you won't always be able to carry out the experiments you're interested in ahead of time. For instance, suppose you're considering investing in the stock market. You'd like to know the expected value of the return on your investment. Obviously, you'd like to calculate this *before* investing any money. Or, suppose you are a manager who needs to predict sales for the coming year to make hiring decisions. Knowing the average profit from each customer would be critical information.

Example 1: Calculating Expected Value

The campus dining service at State University is preparing for the upcoming year. Based on past years, they have observed the following data on the probability of selling different meal plans. Each plan consist of a set number of meals along with Plus Dollars, which can be used anywhere on campus but are restricted to food purchases. The university predicts an enrollment of 8421 students in the coming year.

Table 2

Meal Plans

Plan	Description	Price	Probability
1	19 meals per week and $100 Plus Dollars for the semester	$1245	$\frac{1}{10}$
2	14 meals per week and $250 Plus Dollars for the semester	$1245	$\frac{1}{15}$
3	10 meals per week and $350 Plus Dollars for the semester	$1245	$\frac{1}{30}$
4	$815 Plus Dollars to be used anytime during the semester (no meals)	$815	$\frac{1}{10}$
5	90 meals to be used anytime during the semester and $300 Plus Dollars	$870	$\frac{1}{12}$

Note: Students have the option of choosing not to buy a meal plan.

a. Suppose a random student from the university is chosen. What is the expected value for the amount that they have spent on their meal plan for that year?

b. Suppose that, for each meal plan sold, the university makes $510 from Plan 1, $550 from Plan 2, $600 from Plan 3, $220 from Plan 4, and $270 from Plan 5. What is the expected profit per student in the upcoming year?

c. What overall profit can dining services expect from student meal plans in the upcoming year?

Solution

a. To find the expected value, we need to multiply each meal plan cost by the probability of purchasing that meal plan and add them together. Remember that although they are not listed in the chart, many students will choose no meal plan at all. We've shown them in the following calculation although they spend $0 on a meal plan. This gives us

$$E(\text{meal plans}) = \frac{1}{10}(\$1245) + \frac{1}{15}(\$1245) + \frac{1}{30}(\$1245) + \frac{1}{10}(\$815) + \frac{1}{12}(\$870) + \frac{37}{60}(\$0)$$

$$= \$124.50 + \$83.00 + \$41.50 + 81.50 + 72.50$$

$$= \$403.$$

Therefore, a random student is expected to pay $403 for their meal plan. Of course, no

7

single student pays $403 for a meal plan. This expected value gives us an average per student across all students whether they bought a plan or not.

b. The profits per meal plan are summarized in Table 3.

Table 3	
Meal Plan Profits	
Plan	Profit per Meal Plan ($)
1	510
2	550
3	600
4	220
5	270

In order to calculate the expected profit per student, multiply the estimated profit for each meal plan by the probability of that meal plan being chosen.

$$E(\text{profit per student}) = \frac{1}{10}(\$510) + \frac{1}{15}(\$550) + \frac{1}{30}(\$600) + \frac{1}{10}(\$220) + \frac{1}{12}(\$270) + \frac{37}{60}(\$0)$$
$$= \$51.00 + \$36.67 + \$20 + \$22 + \$22.50$$
$$= \$152.17$$

Therefore, dining services can expect an average profit of $152.17 per student next year.

c Because expectation is linear, we can add together the expected profits for each of the 8421 students. Therefore, the expected profit for dining services is approximately $8421 \cdot \$152.17 = \$1,281,423.57$.

Casinos also rely heavily on expected values in order to turn a profit. Their entire business depends on creating games where the experience is appealing enough for you to play the game, but the probabilities end up in the casino's favor in the long run. Let's look at a simple example.

Example 2: Expected Winnings

In American roulette, the wheel contains the numbers 1 through 36, alternating between black and red. There are two green spaces numbered 0 and 00.

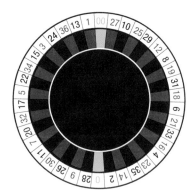

a. Calculate the probability of the roulette ball landing on a red pocket (or black for that matter, since the number of red and black pockets is the same).

b. Calculate the probability of the ball not landing on a red pocket.

c. A player bets $1.00 on red to play the game. Calculate his expected winnings.

Solution

a. In calculating the probability for landing on a red, note that there are 18 red pockets out of 38 possible pockets on the wheel. So the probability is

$$P(\text{red}) = \frac{18}{38}.$$

b. The probability of not landing on a red pocket is one minus the probability of landing on a red pocket, so

$$P(\text{not red}) = 1 - \frac{18}{38} = \frac{20}{38}.$$

Note: 20 is the sum of the 18 black pockets and the two green pockets.

c. If a player bets $1.00 on red, his chance of winning $1.00 is $\frac{18}{38}$ and his chance of losing $1.00 (or winning −$1.00) is $\frac{20}{38}$. So, the player's expected value is

$$E(X) = \left(\frac{18}{38}\right)(1) + \left(\frac{20}{38}\right)(-1)$$
$$= \frac{18}{38} - \frac{20}{38} = -\frac{2}{38}$$
$$\approx -0.05.$$

What this means is that the player can expect to lose about 5 cents on average for every dollar he bets throughout the night on red roulette pockets. Of course, he doesn't actually ever lose 5 cents on a single bet, but remember this is his expected average outcome for multiple trials. Suppose he played 10 games betting $1, he could expect to lose

$$(10)(0.05) = \$0.50 \text{ overall.}$$

That doesn't sound very profitable for the casino, but consider if they had 1000 people play that same bet throughout the day. Then their profits would be

$$(1000)(0.05) = \$50.00$$

Now consider that the casino has millions of roulette players per year. Because the casino has a very slight edge on the roulette wheel, the casino is expected to make money over time. Think about how this advantage changes if you place the same bet on a European roulette wheel; it is exactly the same as the American wheel except that it only has one green 0 pocket. Determining the difference in advantage is left as an exercise.

The last thing that we will mention is the concept of "odds," as you might hear at a casino or racetrack. Although it's a common mistake, the words odds and probability are not interchangeable. However, expected value helps us tie the two together nicely.

Let's take a look at the phrase "odds of 3 : 1 against." (Recall from Chapter 4 that we read 3 : 1 as "three to one.") Suppose you placed a $1.00 bet on a horse that had odds of 3 : 1. These odds do not mean that you have a $\frac{1}{3}$ chance of winning money. Instead, as we will see, you have a $\frac{1}{4}$ chance (or 25%) of winning. In real terms, it means that for every $1.00 you place on the bet, you win $3.00 if your horse wins. Let's use expected value to see why the probability works this way.

Suppose the odds of your horse winning are 3 : 1 against, and you place a bet of $1.00 on the horse. If your horse wins, you gain $3.00, but if it does not win, you lose $1.00 (or win −$1.00). Let the probability of your horse winning be p, which then means that the probability of your horse not winning is $1 - p$ as shown in the following table.

Table 4		
Odds of 3 : 1 Against		
Outcome	Gain	Probability
Winning	$3.00	p
Not Winning	−$1.00	$1 - p$

The following is the expected value formula with this information.

$$E(X) = (\$3.00)(p) + (-\$1.00)(1-p)$$

Both the bookie and yourself would think that the wager is a fair one if the expected gain and loss for both of you were $0.00. In other words, no money changes hand. So, we can say that the expected value should equal 0 for a fair bet, or

$$E(X) = (\$3.00)(p) + (-\$1.00)(1-p) = \$0.$$

Using a bit of algebra, we can solve the equation for p.

$$(3)(p) + (-1)(1-p) = 0$$
$$3p - 1 + p = 0$$
$$4p - 1 = 0$$
$$4p = 1$$
$$p = \frac{1}{4}$$

So, the probability of winning when the odds are $3:1$ against is $\frac{1}{4}$, and the probability of not winning is $1 - \frac{1}{4} = \frac{3}{4}$.

In fact, rather than being an expression for the probability of winning, "odds against" is the ratio of the chance of losing to the chance of winning. In our example, odds against of $3:1$ mean that 3 out of 4 times you're likely to lose. As we have already seen with ratios, you can also express odds as fractions.

$$\text{Odds against} = \frac{P(\text{losing})}{P(\text{winning})}$$

In this particular case, the fraction is

$$\text{Odds against winning} = \frac{\left(\frac{3}{4}\right)}{\left(\frac{1}{4}\right)} = \frac{3}{1} = 3.$$

Skill Check #2

If the odds on a bet are $4:1$ against, what is the probability of winning?

We should note that in the betting world, "odds against" are most often quoted because it is the most convenient way to understand the payout if the bet is a successful one for the player. So odds of $20:1$ mean that for every $1.00 you bet, the bookie will pay you $20.00 if you win. The bookie is out to make money, in reality, the odds represent how much the bookie is willing to pay out rather than the true ratio.

We might note that often when you are more likely to win than lose, it is referred to as "odds for" and the ratio is

$$\text{Odds for} = \frac{P(\text{winning})}{P(\text{losing})}.$$

Skill Check Answers

1.

$x_i \cdot P(x_i)$
1.35
1.80
2.25

$E(X) = 5.4$

2. $\frac{1}{5}$ or 20%

7.5   Exercises

Calculate the expected value of each scenario. Round your answer to the nearest hundredth when necessary.

1.

x_i	$P(x_i)$
1	0.21
2	0.58
3	0.06
4	0.15
5	0.0

2.

x_i	$P(x_i)$
−$1.50	0.3
$0.00	0.5
$2.75	0.1
$5.00	0.1

3.

x_i	$P(x_i)$
25	$\frac{1}{3}$
15	$\frac{2}{5}$
10	$\frac{1}{15}$
5	$\frac{1}{5}$

4.   The number of even numbers showing when a pair of standard six-sided dice are rolled.

x_i	$P(x_i)$
0	$\frac{9}{36} = \frac{1}{4} = 0.25$
1	$\frac{18}{36} = \frac{1}{2} = 0.5$
2	$\frac{9}{36} = \frac{1}{4} = 0.25$

Solve each problem. Round your answer to the nearest millionth when necessary.

5. Suppose Piper eats out twice a week 15% of the time, she eats out once a week 35% of the time, and she doesn't eat out anytime during the week 50% of the time. What is the expected value for the number of times Piper eats out during a week?

6. Suppose that you and a friend are playing cards and decide to make a bet. If you draw two aces in succession from a standard deck of cards without replacing the first card, you win $50.00. Otherwise, you pay your friend $10.00.

   a. What is the expected value of your bet?

   b. If the same bet was made 25 times, how much would you expect to win or lose?

7. A European roulette wheel has only one green slot instead of two. Using Example 2 from this section as a guide, calculate the expected winnings on a European roulette wheel if a player bets $1.00 on red to play the game.

8. Jim likes to day-trade on the Internet. On a good day, he averages a $1100 gain. On a bad day, he averages a $900 loss. Suppose that he has good days 25% of the time, bad days 35% of the time, and the rest of the time he breaks even.

   a. What is the expected value for one day of Jim's day-trading hobby?

   b. If Jim day-trades every weekday for three weeks, how much money should he expect to win or lose?

9. A university in town is raffling off $20,000 for student scholarships. You can buy one ticket for $10, three tickets for $25, or five tickets for $40. Assume that the university sells 10,000 tickets.

   a. Find the expected value for each of the three ticket options: purchasing just one ticket, purchasing three tickets, or purchasing five tickets.

   b. Should you buy one, three, or five tickets in order to maximize the money you expect to have at the end of the raffle?

10. You need to borrow money from your sister. She's feeling quirky on the day you ask and says she wants you to flip a coin. Heads, you get $15, tails you get $5. Thinking this is weird, you ask your mother for money instead. She says she'll let you roll a die and she'll give you $2 times the number that appears on the die. Before agreeing to either of these unique offers from the "mathy" folk in your family, you decide to see which is the better offer by calculating the expected value for each method (realizing that you too fit the bill of a "mathy" member of your family). Which offer should you take? Explain your reasoning.

11. Assume that stock in Degree Compass, a predictive analytics company in higher education, returns the percentages shown in the table.

Degree Compass Stock Returns	
Annual Return Rate	Probability
15%	0.17
30%	0.51
45%	0.32

Calculate the expected value of the return rate for stock in Degree Compass.

12. During the NCAA basketball tournament season, affectionately called *March Madness*, part of one team's strategy is to always foul their opponent's tall forward. Because he is so tall, he makes 57% of shots he takes close to the basket. However, when he is fouled, his free-throw shooting percentage is only 51.5%. The shots he makes close to the basket are worth two points and each of the two free throw shots after being fouled are worth one point.

   a. Calculate the expected value of the number of points the forward makes when he takes a shot close to the basket.

   b. Calculate the expected value of the number of points the forward makes when he shoots two foul shots.

   c. Based on these expected values, is fouling the tall forward a good strategy? Explain your answer.

13. On your next multiple-choice test, each question has four incorrect answers and one correct answer to choose from. Your professor tells you that each correct answer you make, you receive 1 point, but you lose $\frac{1}{4}$ point for each incorrect answer.

   a. What is your expected gain or loss on a question if you have no idea of the correct answer and end up simply guessing?

   b. What is your expected gain or loss if you guess on all 25 questions?

14. The stock prices of Web Movies on the 1st of the month for the first 6 months of 2014 were $177.41, $207.90, $214.63, $239.09, $242.19, and $259.99 respectively.

   a. Calculate the average change in the stock prices of Web Movies in a month.

   b. Use linearity of expectation to estimate the stock price on January 1st, 2015.

15. If the odds on a bet are $6:1$ against, what is the probability of winning?

16. Suppose the probability of a football team winning a playoff game is 0.25. What are the odds of winning?

17. The odds of a teenage male having an accident are $2:3$. What is the probability of a teenage male having an accident?

18. An insurance company claims the probability of surviving a certain type of cancer is 95%. What are the odds of surviving?

19. The UVest investment company publishes that the odds of increasing your wealth with their company is $5:2$. What is the probability of UVest increasing your investment?

20. Odds against being struck by lightning in one year are 1,000,000 to 1. [1]

   a. If you live to be 80, what are the odds against being struck by lightning over your lifetime? Assume each year has the same probability.

   b. The National Weather Service gives the odds against being struck by lightning over an 80-year lifetime as 10,000 to 1. Why do you think this is different from the answer you got in part **a.**?

1 NWS Lightning Safety, http://www.lightningsafety.noaa.gov

21. Overall odds in favor of winning in a state lottery game are $4.63 : 1$.

    a. Find the probability of winning in the lottery game.

    b. The prize for this lottery game is $100. If the cost to play the game is $2.00, what is the expected value for playing this game?

22. Suppose the odds for a bet are $10 : 1$. Your friend tells you he thinks the odds are too generous. Write down some less generous odds.

Chapter 7   Summary

Definitions

Probability

The study of probability is the mathematical approach to analyzing how likely things are to happen.

Trial

A trial, or probability experiment, is any process that produces a random result.

Outcome

The outcomes of a trial are the possible individual results.

Sample Space

The sample space S of a trial is the set of all possible outcomes.

Event

An event E is a group, or subset, of outcomes in the sample space.

Formulas

Classical Probability

If all outcomes are equally likely, classical probability is calculated with the formula

$$P(\text{event}) = \frac{\text{the number of possible outcomes in the event}}{\text{the number of outcomes in the sample space}}.$$

$P(\text{event})$ will always be a real number between 0 and 1, inclusive.

Empirical Probability

If all outcomes are based on an experiment, empirical probability is calculated with the formula

$$P(\text{event}) = \frac{\text{the number of times an event occurs}}{\text{total number of times the experiment is performed}}.$$

$P(\text{event})$ will always be a real number between 0 and 1, inclusive.

Definitions

Tree Diagram

A tree diagram uses branches to indicate possible choices at the next stage of outcomes.

Replacement

With replacement: When counting possible outcomes with replacement, objects are placed back into consideration for the following choice.

Without replacement: When counting possible outcomes without replacement, objects are *not* placed back into consideration for the following choice.

Combination

A combination involves choosing a specific number of objects from a particular group of objects, using each only once, when the order in which they are chosen is *not* important.

Permutation

A permutation involves choosing a specific number of objects from a particular group of objects, using each only once, when the order in which they are chosen *is* important.

Formulas

Fundamental Counting Principle

For a sequence of n experiments where the first experiment has k_1 outcomes, the second experiment has k_2 outcomes, the third experiment has k_3 outcomes, and so forth, the total number of possible outcomes for the sequence of experiments is $(k_1)(k_2)(k_3)\cdots(k_n)$.

n Factorial

In general, $n!$ is the product of all the positive integers less than or equal to n, where n is a positive integer.

$$n! = n(n-1)(n-2)(n-3)\cdots(2)(1)$$

Note that $0!$ is defined to be 1.

Combinations and Permutations of n Objects Taken r at a Time

The number of ways to select r objects from a total of n objects is found by the following two formulas. (Note that $r \leq n$.)

When order is not important, use the following formula for a combination.

$$_nC_r = \frac{n!}{r!(n-r)!}$$

When order is important, use the following formula for a permutation.

$$_nP_r = \frac{n!}{(n-r)!}$$

Permutations with Repeated Objects

The number of distinguishable permutations of n objects, of which k_1 are all alike, k_2 are all alike, and so forth is given by

$$\frac{n!}{(k_1!)(k_2!)(k_3!)\cdots(k_p!)},$$

where $k_1 + k_2 + \cdots + k_p = n$.

Section 7.3   Using Counting Methods to Find Probability

Definition

Complement

The complement of event E, denoted by E^c, consists of all outcomes in the sample space that are *not* in event E.

Formulas

Complement Rules of Probability

1. $P(E) + P(E^c) = 1$

2. $P(E) = 1 - P(E^c)$

3. $P(E^c) = 1 - P(E)$

Section 7.4   Addition and Multiplication Rules of Probability

Definitions

Independent Events

Independent events are events where the result of one event does not influence the probability of the other.

Dependent Events

Dependent events are events where the result of one event influences the probability of the other.

Conditional Probability

The conditional probability of Event B happening, given Event A, is the probability of Event B assuming that Event A has already, or will at some point, occur. The conditional probability is written $P(B \mid A)$, and read *the probability of B, given A*.

Formulas

Addition Rule for Probability

The probability of Event A happening *or* Event B happening is

$$P(A \text{ or } B) = P(A) + P(B) - P(A \text{ and } B).$$

Addition Rule for Mutually Exclusive Events

The probability of Event A happening *or* Event B happening when A and B have no outcomes in common is

$$P(A \text{ or } B) = P(A) + P(B).$$

Multiplication Rule for Independent Events

A and B are independent events when the probability of Event A happening *and* Event B happening is given by

$$P(A \text{ and } B) = P(A) \cdot P(B).$$

Multiplication Rule for Dependent Events

If A and B are dependent events, the probability of Event A happening *and* Event B happening is

$$P(A \text{ and } B) = P(A) \cdot P(B \mid A).$$

Bayes' Theorem

$$P(A \mid B) = \frac{P(B \mid A) \cdot P(A)}{P(B)},$$

when $P(B) > 0$.

Section 7.5   Expected Value

Formulas

Expected Value

The formula for expected value of an event X is

$$E(X) = x_1 P(x_1) + x_2 P(x_2) + x_3 P(x_3) + \cdots + x_n P(x_n),$$

where x_i is the ith outcome and $P(x_i)$ is the probability of x_i.

The Sum of Expected Values

The expected value of the sum of two events, X and Y, is equal to the sum of the individual expected values. Formally, we write this as

$$E(X + Y) = E(X) + E(Y).$$

Chapter 7 Exercises

Calculate the probability of each scenario. Round your answer to the nearest millionth when necessary.

Mrs. Okeela is a contestant on the game show *Wheel of Fortune*. The wheel that is used on the day she is on the show is given.

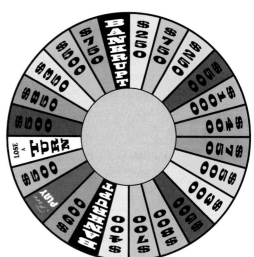

1. Identify the sample space for Mrs. Okeela spinning the wheel once.

2. Describe how you would create the sample space for spinning the wheel twice.

3. Find the classical probability of landing on

    a. $300

    b. Bankrupt

    c. $500

    d. A tropical vacation

4. Mrs. Okeela was an avid viewer who kept detailed records of the contestants' spins. From her data during the last season, of the 1237 spins she recorded, "Bankrupt" came up 150 times. Find the empirical probability of landing on Bankrupt from Mrs. Okeela's data. Does it match up with what you would expect? Explain your answer.

5. Dr. Gladden was interested in studying whether gender affected students' preferred social media outlet. A random sample of students was selected from a sociology class and asked to choose their top social media site out of Facebook, Twitter, or Pinterest. Their social media choice, gender, and education level (freshman, sophomore, junior, or senior) were recorded.

    a. Find the sample space for the possible outcomes of the survey using a tree diagram.

    b. Dr. Gladden decides to add Instagram and Tumblr to the choices for preferred social media sites. She also decides to include high school students as an additional education level. Use the Fundamental Counting Principle to establish how many possible outcomes are now in the sample space.

6. The students in the survey are asked to rank their top two social media sites from the five possibilities. How many ways are there for students to rank their top two favorites?

7. Billy got a new tablet computer for Christmas, but his envious sister Meg set a passcode on it without his knowing. The passcode is a sequence of five digits.

   a. What is the probability that Billy can guess the correct code the first time by entering five random digits?

   b. Frustrated, Billy looks at the screen much more carefully and sees Meg's sticky fingerprints on five of the number-pad keys. What is the probability that Billy can correctly guess the code the first time now?

   c. Under duress, Meg admits that she only actually used four of the five smudged digits, but that she did keep them in increasing order. Using this information, what is the probability that Billy can correctly guess the code on his first attempt?

8. Emre was delighted with his BB gun for Christmas and couldn't wait to get outside to try out his marksmanship. He grabbed six bottles and cans to shoot at as he ran from the house. Unfortunately, no one noticed that he grabbed one of his Nana's antique glass tumblers as one of the six targets to shoot at.

   a. Initially he decides to only set up four of the targets. What is the probability that the antique tumbler is not one of them?

   b. How many ways are there to arrange all six targets on the wall?

   c. What is the probability that he places the antique tumbler as the left-most target, if they are all on the wall?

   d. After placing all six targets and shooting from left to right, he had already downed four of the targets by the time an adult discovered the tumbler was missing from the cabinet. What is the probability that the antique survived?

9. A single letter from the word MISSISSIPPI is chosen. What is the probability of choosing an S or a P?

10. An experiment consists of tossing a coin and then rolling a die. List the sample space for this experiment.

An urn contains 100 marbles. Fifty are purple, ten are green, fifteen are red, and twenty-five are orange. Two marbles are drawn at random one after another *with* replacement. Calculate each probability.

11. $P(\text{purple and orange})$ **12.** $P(\text{green and green})$ **13.** $P(\text{same color})$

An urn contains 100 marbles. Fifty are purple, ten are green, fifteen are red, and twenty-five are orange. Two marbles are drawn at random one after another *without* replacement. Calculate each probability.

14. $P(\text{purple and orange})$ **15.** $P(\text{green and green})$ **16.** $P(\text{same color})$

Two dice are rolled. Calculate each probability. It may be helpful to refer to Figure 1 in Section 7.2.

17. $P(\text{sum of } 8)$

18. $P(\text{sum} > 7)$

19. $P(\text{sum is odd})$

20. $P(\text{sum is 6 or 8 or 10})$

One card is drawn from an ordinary deck of cards. Calculate each probability. Round your answer to the nearest millionth when necessary.

21. $P(\text{heart or a face card})$

22. $P(\text{ace and a spade})$

23. $P(\text{face card and an ace})$

Solve each problem.

24. A study of attendance at a football game, based on weather, shows the following pattern. Find the expected value of the attendance at a football game.

Football Attendance Based on Weather		
Weather	Attendance	Probability
Extremely Cold	10,000	0.05
Cold	15,000	0.30
Moderate	30,000	0.40
Warm	35,000	0.25

25. The odds in favor of Fast Enough winning the horse race are 6 to 5. If you place a bet of $2500 on Fast Enough and he wins the race, you win $6500.

   a. What is the expected value of your bet?

   b. Suppose you win. What are your winnings to take home, that is, how much ahead are you after the win?

26. On a roulette wheel, the slots are numbered 1 through 36, 0, and 00 for a total of 38 slots. Assume you bet on a single slot. The payout for landing on a single slot is 35 to 1. If a chip is worth $5, what is the expected value for a player that plays the number 12 repeatedly?

27. In a raffle, a ticket costs $50. If there are 5000 tickets sold in the raffle and the payout is $10,000, what is the expected value of purchasing one ticket?

Bibliography

7.2

1. Wolfram Alpha LLC. 2009. Wolfram|Alpha. http://www.wolframalpha.com/input/?i=8%21 (accessed July 2, 2014).

2. Wolfram Alpha LLC. 2009. Wolfram|Alpha. http://www.wolframalpha.com/input/?i=Combination%2818%2C4%29 (accessed July 2, 2014).

3. Wolfram Alpha LLC. 2009. Wolfram|Alpha. http://www.wolframalpha.com/input/?i=Permutation%28140%2C3%29 (accessed July 2, 2014).

4. Wolfram Alpha LLC. 2009. Wolfram|Alpha. http://www.wolframalpha.com/input/?i=Multinomial%281%2C4%2C4%2C2%29 (accessed July 2, 2014).

7.4

1. National Cancer Institute. "Prostate-Specific Antigen (PSA) Test." http://www.cancer.gov/cancertopics/factsheet/Detection/PSA

   Mayo Clinic. "Prostate cancer screening: Should you get a PSA test?" http://www.mayoclinic.com/health/prostate-cancer/HQ01273

7.5

1. Cooper, Mary Ann. "Medical Aspects of Lightning." National Weather Service Lightning Safety. http://www.lightningsafety.noaa.gov/medical.htm

Chapter 8
Statistics

Sections

Objectives

- Distinguish between sampling techniques
- Interpret different types of graphs
- Calculate numerical descriptors of data, such as measures of center, standard deviation, percentiles, and z-scores
- Use the normal distribution to determine probabilities using z-scores
- Calculate and interpret the correlation between two variables
- Calculate and appropriately use the linear regression line for a given set of data

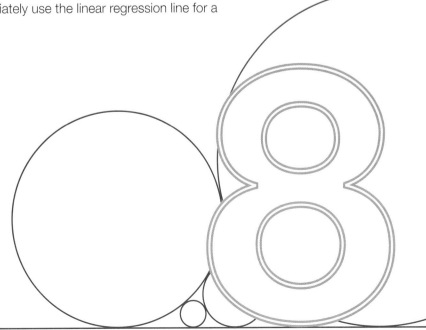

Statistics

As of July 2010, Facebook had more than 500,000,000 active users, with more than 50% of them logging in on any given day. In "Self-Presentation 2.0: Narcissism and Self-Esteem on Facebook," researchers at York University showed that students with comparatively lower self-esteem scores and higher narcissism scores not only spent more time on Facebook, but also tended to "self-promote" more than the students with higher self-esteem and lower narcissism scores.[1] With people averaging over 700 billion minutes per month on Facebook, are we becoming a more narcissistic society?

Or how about the influence of Twitter? In the study *Tweets in Action: Retail*,[2] researchers from Compete, a Boston based company, analyzed 5200 panelists across the United States. They looked at the online shopping habits of people exposed to retail tweets throughout the day. They concluded that Twitter users are heavy online retail shoppers and that the more tweets they see, the more likely they are to buy.

In order to deconstruct studies like these and determine the valuable conclusions that can be drawn, one must first understand the basic groundwork of how statistical studies are set up. This chapter focuses on collecting, displaying, and analyzing statistical data.

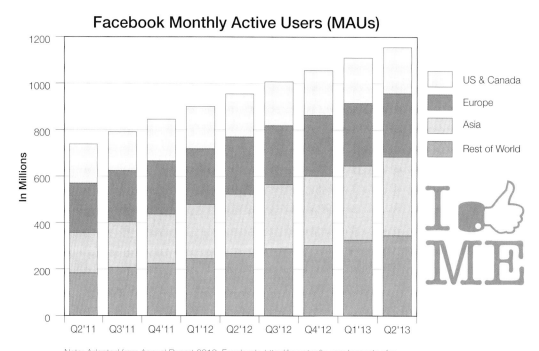

Note: Adapted from Annual Report 2013. Facebook. http://investor.fb.com/annuals.cfm

Figure 1

1 Soraya Mehdizadeh, *Cyberpsychology*

2 Compete and Twitter, *Tweets in Action: Retail*

8.1 Collecting Data

Populations vs. Samples

Most statistical studies are brought about by the desire to know the answer to a particular question. For instance, you might want to know which car gets the best gas mileage, which university is the best value for the money, or how often you might expect to be called for jury duty. The search for the answers to many questions just like these is helped along by collecting data and then drawing conclusions.

When we begin a statistical study, we must first define the **population**, or the particular group in which we are interested. Considering our desire to know which car gets the best gas mileage, we might first assume that the desired population for the study is "all cars." However, stop for a moment and think about the implications of trying to study all makes and models of all cars, in all parts of the world! This might take a while and might not be very useful. What if the car with the best gas mileage was only available in Romania 20 years ago? By being more precise about how we define the population of interest, we narrow our potential population to just those in which we are truly interested. In most cases, this also creates a more manageable population with which to work. When defining the statistical group of interest, it is impossible to over-clarify the population you wish to study. For our gas mileage example, let's narrow our attention to all passenger cars manufactured in the United States in 2013.

> ### Population
>
> A **population** is the particular group of interest in a study.

Helpful Hint

The singular form of data is datum.

Collecting **data**, or information, from each member of the population always results in the best analysis we can have of a particular group. When we are able to achieve this, it is called a **census**. A census gives us the ability to speak with 100% confidence about the data we gathered from the group as a whole. Data gathered from a census allows us to identify population parameters. A **parameter** is the numerical description of a particular population characteristic. For instance, suppose the average miles per gallon (mpg) for all passenger cars manufactured in the United States in 2013 was 23.6. Then 23.6 mpg is a population parameter. All populations have fixed parameters that numerically describe the structure of the population. The goal of a statistical study is to identify as best we can the population parameter we are interested in.

> ### Data
>
> **Data** is information collected for a study.
>
> ### Census
>
> A **census** involves collecting data from every member of the population.
>
> ### Parameter
>
> A **parameter** is the numerical description of a particular population characteristic.

Although conducting a census is the ideal way to identify population parameters, as you might imagine, this is often an unrealistic goal. Maybe the population is too big and hard to get to—think gas mileage from all cars around the globe. Even with a small population, data from every member may not be available to you—think obtaining the actual weight of all people in your class right now. When a census cannot be obtained, we gather information from a subset of the population, or a **sample**. The numeric descriptions of particular samples are called **sample statistics**. For instance, suppose you collected data on the gas mileage of the cars driven by half of the students in your class and found that the average mpg was 19.1. Then, 19.1 mpg is a sample statistic.

Sample

A **sample** is a subset of a population.

Sample Statistic

A **sample statistic** is the numerical description of a characteristic of a sample.

Table 1	
Population vs. Sample	
Population	**Sample**
Whole group	Part of a group
Group I want to know about	Group I do know about
Numerical descriptions of characteristics are called parameters	Numerical descriptions of characteristics are called statistics
Parameters are generally unknown	Statistics are always known
Parameters are fixed	Statistics vary with the sample chosen

Example 1: Identifying Parts of a Survey

Two shortened survey reports are given. In each report, identify the following: the population, the sample, the results, and whether the results represent a sample statistic or a population parameter.

a. A headline about the rising obesity among young people led a school board to survey local high school students. Out of 231 students surveyed, 58% reported eating a "high fat" snack at least 4 times a week.

b. A nonprofit organization interviewed 618 adult shoppers at malls across Louisiana about their views on obesity in youths. The resulting report stated that an estimated 48% of Louisiana adults are in favor of government regulation of "high fat" fast food options.

Solution

a. Population: local high school students

Sample: the 231 students who were surveyed

Results: 58% of students surveyed eat a "high fat" snack at least 4 times a week.

The result refers to only those students who were surveyed, thus the result is a sample statistic.

b. Population: Louisiana adults

Sample: the 618 adult Louisiana mall shoppers who were surveyed

Results: 48% of Louisiana adults are in favor of government regulation of "high fat" fast food options.

The results refer to all Louisiana adults, thus this is a population parameter. This population parameter is an estimate based on the sample statistics, which were not reported.

Since we've already noted that obtaining the values of population parameters by conducting a census is often unrealistic, we will focus our discussion on samples and how their characteristics help us know more about the population. It's important to stop here and note that the way we choose a sample from the population is important. Care must be taken in choosing your sample so that the population is well-represented and the results of the study are meaningful. For instance, consider our earlier example of a study wishing to know the average gas mileage for cars in the United States. If the only data you gathered were from half the students in your class, would this be an accurate picture of the population of cars driven in the United States? What about if we only chose cars of people who lived within the city limits of Nashville? How about in the city limits of San Diego? Let's look more closely at how to choose a sample that is representative of the entire population you are studying.

Choosing a Sample

When choosing a sample from the population, it's important that the sample is **representative** of the entire population.

> ### Representative Sample
>
> A **representative sample** is one that has the same relevant characteristics as the population and does not favor one group of the population over another.

Without a representative sample, it might not be possible to accurately generalize the results to the population as a whole. If the sample is poorly chosen, a form of bias, or favoring of a certain outcome, might occur. Consider the sample of cars driven by residents of the city of San Diego. Because of its dense population, hilly terrain, and popularity with a young affluent crowd, cars driven there tend to be either compact cars, which are more fuel efficient, or luxury vehicles. This sample would certainly favor a narrow sector of cars for our study on average miles per gallon.

There are several standard statistical methods for choosing samples that allow for the best possible representative samples. As you might imagine, some methods are better suited to particular situations than others. Let's look at some of the methods.

Random Sampling

> ### Random Sample
>
> A **random sample** is one in which every member of the population has an equal chance of being selected.

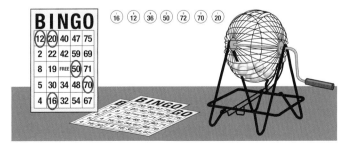

Figure 1: Random Sample

Drawing for door prizes by placing all of the ticket stubs in a basket and mixing them up is a very common example of random sampling. However, you should always be cautious when letting a human choose a "random" sample. In reality, it is against our human nature to choose members of the population entirely at random. Take a moment to choose 3 random digits between 0 and 9. Try to make sure they are truly random. Now, consider these questions.

- Are the numbers spaced out or grouped closely together—probably spaced out, right?

- Are they all even, odd, or some of both?

- Did you have a reason for choosing any of the numbers—maybe your favorite?

- Did you alter any of your original responses, and if so, why?

- Did you repeat any digits? Why not? It wasn't against the rules.

- Are your digits consecutive, such as 1, 2, 3?

If any reasons influenced your choice of digits, then your sample is not random. Because it is hard for humans to completely eliminate influences, it's best to take the human element out of it and allow technology to choose a random sample for us. For our fuel efficiency study we mentioned earlier, a random sample would occur if we allowed a computer to choose cars from a list of all possible passenger vehicles manufactured in the United States in 2013.

Stratified Sampling

Stratified Sample

A **stratified sample** is one in which members of the population are divided into two or more subgroups, called strata, that share similar characteristics like age, gender, or ethnicity. A random sample from *each* stratum is then drawn.

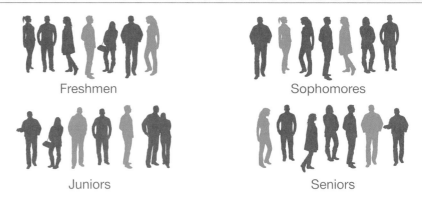

Freshmen   Sophomores

Juniors   Seniors

Figure 2: Stratified Sample

Stratified sampling is used when we want to be confident that certain subgroups of a population are represented in the sample. For instance, first we could divide our vehicles into strata based on seating capacity, and then randomly choose 10 vehicles from each group. This would produce a stratified sample of cars for us to collect data from while making certain that we include two-seater vehicles as well as those that carry as many as 15 people.

Skill Check # 1

Can you think of some other ways to divide the cars into strata that might represent a broader scope of vehicles on the market?

Cluster Sampling

Cluster Sample

A **cluster sample** is one chosen by dividing the population into groups, called clusters, that are each similar to the entire population. The researcher then randomly selects some of the clusters. The sample consists of the data collected from *every* member of each cluster selected.

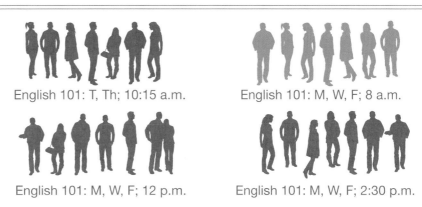

English 101: T, Th; 10:15 a.m.   English 101: M, W, F; 8 a.m.

English 101: M, W, F; 12 p.m.   English 101: M, W, F; 2:30 p.m.

Figure 3: Cluster Sample

This type of sampling resembles stratified sampling in that the population is first divided into groups. However, whereas stratified sampling chooses some members from each group, cluster sampling chooses all members from some of the groups. Often populations naturally lend themselves to subgroups from which each type of sampling might be done. For instance, if you wanted to know the average grade on the first test of all students in English 101, you could divide the students up by the section they are in. For cluster sampling, you would then randomly choose several sections and collect the grades from all students in each class.

Skill Check #2

For our mpg study, do you think this method would produce a good representative sample of vehicles if we allowed our clusters to be price ranges? Why or why not?

Systematic Sampling

Systematic Sample

A **systematic sample** is one chosen by selecting every n^{th} member of the population.

Every 10th Bottle

Figure 4: Systematic Sample

Assembly lines in a factory are often tested for quality control by systematic sampling. If every 10th product is checked for defects, then the manufacturer can feel confident that his plant is producing quality products. As a manager of the production line, you would want to ensure that there was no obvious bias in choosing every 10th product as opposed to every 9th one. If there was a unique characteristic about every 10th product, such as they came only from Machine A, you would not want to choose 10 as your counting interval.

Skill Check #3

Identifying every 8th car accessing the interstate on a particular entrance ramp during rush hour traffic is an example of systematic sampling for our fuel study. Can you identify any potential biases that we might need to be aware of when choosing the observation spot?

Convenience Sampling

Convenience Sample

A **convenience sample** is one in which the sample is "convenient" to select. It is so named because it is convenient for the *researcher.*

Figure 5: Convenience Sample

Although convenience sampling is the least reliable for obtaining a representative sample, in some instances it is all the data we can obtain, and the "some is better than none" factor comes into play. For instance, suppose you need 20 students to fill out a survey for your sociology project. Because of time constraints, you ask students in your English 101 class to participate. This sample of convenience is better than no sample at all. On the whole, however, caution should be used when results are obtained merely from convenience sampling.

Skill Check #4

Earlier we mentioned gathering data from half the students in your class for our comparison of fuel-efficient cars. Do you think this example of convenience sampling would be an accurate picture of the population of cars driven in the United States?

Example 2: Identifying Sampling Techniques

Identify the sampling technique used to obtain a sample in each of the following situations.

a. To conduct a survey on collegiate social life, you knock on every 5th dorm room door on campus.

b. Student ID numbers are randomly selected from a computer print out for free tickets to the championship game.

c. Fourth grade reading levels across the county were analyzed by the school board by randomly selecting 25 fourth graders from each school in the county district.

d. In order to determine what ice cream flavors would sell best, a grocery store polls shoppers that are in the frozen foods section.

e. To determine the average number of cars per household, each household in 4 of the 20 local counties were sent a survey regarding car ownership.

Solution

a. Because the sample is obtained by choosing every n^{th} dorm room, this is systematic sampling. This is a representative sample, as long as students were randomly assigned to dorm rooms and there are no hidden potential biases, like only males may live in every n^{th} room.

b. Since every member has an equal chance of being selected, this is random sampling.

c. The students were divided into strata based on their schools and then a random sample from each school was chosen. This is stratified sampling.

d. Because of the ease of choosing shoppers right in their own store, this is convenience sampling. In this case, convenience sampling is a viable method for gaining a representative sample since the store would be interested in knowing the thoughts of their customers.

e. Cluster sampling was used here because the counties are the natural clusters and all of the households in some of the counties received the surveys.

Skill Check Answers

1. Answers will vary. For example: size of engine, manufacturer, make, safety rating, number of doors.

2. Because cluster sampling is an "all from one group" method, comparing mpg's from cars in only certain price ranges would not produce a representative sample.

3. The location of the entrance ramp might lend itself to having cars only on one end of the price scale depending on the businesses located in the area.

4. It is unlikely that students (or any age group for that matter) will drive a wide range of cars. Newer, more expensive cars are less likely to be driven by students and would not be well represented in the student sample.

8.1   Exercises

Fill in each blank with the correct term.

1.   A _____ involves gathering data from every member of a population.

2.   A subset of a population is called a _____. _____ are numerical descriptors of characteristics of this subset.

3.   A _____ is a particular group of interest in a statistical study. _____ describe numerical characteristics of this entire group.

4.   A _____ does not favor one subgroup over another from a population.

Answer the question thoughtfully.

5.   Explain how cluster sampling is different from stratified sampling.

Decide if each numerical value from a statistical study is a population parameter or sample statistic.

6.   After the presentation by the United Way representative, 67 of the night's 300+ attendees were interviewed about their reaction. Of the 67 people interviewed, **79%** said they were motivated to donate more of their time and/or money to helping nonprofit organizations.

7.   The Nashville Country Music Marathon reported an average finish time of **4:46:28** for the 4082 finishers.

8.   A medical study showed that the heart-rate profile during exercise and recovery is a predictor of sudden death. Of the 5713 men studied, the mean maximum heart rate during exercise was found to be **96 bpm** in subjects who died suddenly from cardiac causes during the 23 years when the follow-up was conducted. [1]

9.   An estimated **65%** of drivers in rural or town areas view traffic conditions in a good light versus **39%** of drivers in the city or suburbs based on a recent news report that aired nationally.

In each scenario, identify the population, the sample, and any population parameters or sample statistics that are given.

10. A medical study followed 675 cancer patients who received either the new drug *Lafent* or a standard chemotherapy drug. 76% of those who received *Lafent* were still living after eight months, compared to only 63% of those who received the standard chemotherapy drug.

---

[1] Xavier Jouven et al., "Heart-Rate Profile during Exercise as a Predictor of Sudden Death."

11. US realtors across the country are encouraged by the latest reports on the real estate market. Taking into account seasonal factors, sales rose by nearly 9% in the West, 3.5% in the South, 3.4% in the Northeast, and 1% in the Midwest.

12. The online version of *Health* magazine, Health.com, reported the following: "Data from the Women's Health Study . . . examined the medical records of more than 36,000 women who had no history of depression at the start of the study. Roughly 18% of the women were experiencing some form of migraine or had suffered from the headaches in the past. Over the next 14 years, 11% of the study participants received a depression diagnosis." One of the conclusions of the study was that middle-aged women who experience migraines are 40% more likely to become depressed. [2]

13. In a recent study, US graduates of for-profit higher education institutions were between 4.8 and 6.7 percentage points more likely to be unemployed than those at nonprofit institutions and community colleges. [3]

Identify the sampling technique used in each scenario.

14. To ensure the quality of its product, a company tests every 15th item off the assembly line.

15. A student committee in the biology department was formed by randomly choosing three biology majors from every level of student, that is, 1st year, 2nd year, etc.

16. In order to choose the winners of free tickets to the campus concert, student housing printed out all 60 pages of student ID's and then chose three of the pages randomly.

17. A candidate for the local school board surveyed parents picking up their children from the public school on three school days in the last month.

18. Surveying movie-goers as they exited the late movie, researchers determined that only young adults under the age of 22 see movies in the theater anymore.

19. In order to sample the production yield of milk from the farm's herd of cows, random samples are taken from each of the five different types of cows.

20. Using Excel to generate an arbitrary list of customers, the marketing department sent promotional materials to the top 250 names on the list.

Determine an appropriate sampling method for each scenario. Give your reasons for choosing the method. Describe any potential biases that the researcher might need to consider.

21. A state senator wishes to survey his constituents on issues in the upcoming legislature session.

2 Matt McMillen, *Health*, http://www.health.com

3 Dan Berrett, *The Chronicle of Higher Education*, http://www.chronicle.com

22. A pharmaceutical company wishing to test the outcomes of a new drug for a particular skin disease finds 1200 patients willing to participate in the study. They can only choose 100 of them for the actual study.

23. Medical scientists wish to evaluate the effects of Vitamin E and low-dose aspirin in primary prevention of cardiovascular disease and cancer in apparently healthy women.

24. Researchers wish to study the effects of watching too much television at a young age on autism.

Answer each question thoughtfully.

25. *Self-selected* samples are a type of convenience sampling in which the participants volunteer to participate in the study rather than being chosen by the researcher. One issue with this type of sampling is that people who "self-select" to be in a survey often either have very strong opinions about an issue or desire monetary rewards for their participation. For instance, consider online polls regarding political views that are conducted by popular news outlets. Describe some of the potential responses from those who take the time to log on and respond. Is it reasonable to generalize the results from a study like this to include all of the American public? Why?

26. Consider how you would collect data for a study of illegal immigrants working in the United States. How would you go about collecting the data? Is it possible to conduct a census on this population of people? How reliable do you think your results would be? Discuss any issues you might encounter.

27. Often words like "best" and "worst" are used in reporting data. However, descriptive words like these are hard to measure in a concrete way. Consider how you could measure small towns to compile a list of "America's 50 Best Small Towns to Live In." Name at least five distinct measurements.

28. Researchers try to eliminate as many biases in studies as they can in order to get a true picture of the population they are studying. Describe as many potential sources of bias as you can if you were asked to study the effects of alcohol on college campuses.

8.2 Displaying Data

Once we have gathered data from the sample, our attention turns to how to communicate the results. As a society that has been inundated with many forms of media trying to get our attention, simply presenting the raw data in the format that it was gathered would be a fruitless act—it would neither elicit attention nor prompt those who do look at it to fully digest the information. As a result, it's important to display data so that it is organized clearly and it effectively conveys the intended message.

Organizing the data into a table format is an easy and natural way to begin. A **distribution** is a way to describe the structure of a particular data set or population.

Frequency Distribution

A **frequency distribution** is a count of every member of the data set and how often each value occurs, that is, the frequency of the data value.

In the simplest form of a frequency distribution, each observed data value is listed along with its corresponding frequency in a table. Sometimes other numbers are also included in a frequency distribution. One such number is that of the **relative frequency** of a data value. The relative frequency of a class is the percentage of all data that fall into that particular class. It can be displayed as a percentage or a fraction. Let's look at an example of this type of frequency distribution.

Example 1: Interpreting a Frequency Distribution

The following frequency distribution shows the number of crashes with roadside objects along State Route 3 in the state of Washington. After looking over the table, answer the questions that follow.

Table 1	
Crashes with Roadside Objects Along State Route 3, Washington	
Roadside Object	**Number of Crashes (% of total)**
Guardrail	57 (15.36)
Earth Bank	55 (14.82)
Ditch	42 (11.32)
Tree	42 (11.32)
Concrete Barrier	38 (10.24)
Over Embankment	31 (8.36)
Utility Pole	20 (5.39)
Wood Sign Support	19 (5.12)
Bridge Rail	17 (4.58)
Culvert	7 (1.89)
Boulder	6 (1.62)
Luminaire	6 (1.62)
Mailbox	5 (1.35)
Fence	5 (1.35)
Building	5 (1.35)
Other Object	16 (4.31)

Source: Jinsun Lee and Fred Manning. "Analysis of Roadside Accident Frequency and Severity and Roadside Safety Management." Washington State Transportation Commission. December 1999. http://www.wsdot.wa.gov/research/reports/fullreports/475.1.pdf

    a. What do the numbers in the parentheses represent?

    b. Which roadside object was involved in the most crashes?

    c. How many total crashes did the survey cover?

Solution

    a. The numbers in parentheses represent the relative frequency of each class; that is, the number of crashes in a particular category as a percentage of the total number of crashes. For example, there were 57 crashes involving a guardrail. Crashes involving a guardrail account for 15.36% of the total crashes along State Route 3.

    b. Because the "Guardrail" category has the highest frequency listed, guardrails were involved in the most crashes.

    c. By adding up the frequencies of all the different categories, we can see that there were 371 crashes involving roadside objects.

Often data will not be in categories that are names or labels, but categories that are numerical. Consider the data in the next example. People were asked to report the number of pets they had in their household. When making the frequency distribution, we'll still list each count of pets as a different category.

Example 2: Constructing a Frequency Distribution

The number of household pets for 52 families is recorded. Construct a frequency distribution of the data collected.

0 1 0 2 0 0 3 1 1 1 2 3 1 0 0 2 0 3 4 2 0 1 2 5 1 3

1 3 2 1 1 2 2 2 0 0 0 0 1 6 1 2 2 1 0 0 0 1 0 1 2 2

Solution

List each possible value for the number of household pets in the "Number of Pets Per Household" column and the number of times that particular data value occurs in the "Number of Households" column.

Table 2

Number of Household Pets

Number of Pets Per Household	Number of Households
0	16
1	15
2	13
3	5
4	1
5	1
6	1

Skill Check #1

What is the most popular number of household pets in the data given in Example 2?

You can imagine that if you had 500 unique possible data values, this frequency distribution would be tedious and not much more help than the original list. Consequently, sometimes it is more helpful to group the data into different categories, or **classes**. A frequency distribution of this type is called a **grouped frequency distribution**. When we group data into classes, we have to be mindful of certain important features of the classes.

First of all, the **class width** of a grouped frequency distribution is the number of distinct data points each class may contain. The class widths must be the same for all classes in a grouped frequency distribution. Secondly, the classes should never overlap. In other words, a particular piece of data must have only one possible category into which it could fall.

Suppose that the ages (in years) of children who attended church last week were recorded. Let's walk through creating a grouped frequency distribution of the ages, if they ranged from 0 to 15. We will arbitrarily let the first class contain the ages 0 through 5. This means that there are six possible distinct data points in the first class: 0, 1, 2, 3, 4, and 5. So that our classes do not overlap, but have the same width, the second class should contain the next six distinct points: 6, 7, 8, 9, 10, and 11. Then the final class contains the remaining possible ages: 12, 13, 14, and 15. Since there are only four ages to include in this last class, we must extended the range to include the values 16 and 17 as well. Although our original data didn't contain these last two ages, we must include them so that all of the class widths of the frequency distribution will all be the same.

Table 3	
Ages of Children Who Attended Church Last Week	
Age	Number of Children
0–5	10
6–11	14
12–17	7

The numbers 0, 6, and 12 (found on the left-hand side of the classes) are referred to as the **lower class limits** and the numbers 5, 11, and 17 are the **upper class limits**. The **class width** will always equal the difference between either consecutive lower class limits or consecutive upper class limits. For instance, the difference between consecutive lower class limits is the following.

Lower Class Limits

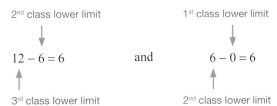

2nd class lower limit

$12 - 6 = 6$ and $6 - 0 = 6$

1st class lower limit

3rd class lower limit   2nd class lower limit

The same is true for the upper class limits.

Upper Class Limits

2nd class upper limit

$17 - 11 = 6$ and $11 - 5 = 6$

1st class upper limit

3rd class upper limit   2nd class upper limit

⑧

Notice that the class width is 6 for each calculation.

The arbitrary choice of our first class width in the example about the children's ages dictated the layout for the classes that followed. However, the width and the number of classes for any grouped frequency distribution can be adjusted as long as we adhere to the guidelines so the classes maintain equal widths and do not overlap with one another.

Let's try constructing another grouped frequency distribution. Consider the data given and think about how you would divide up the classes so that they make the data easier to digest and analyze. There is really no one correct answer, as long as you follow the guidelines we mentioned. Of course, there will always be cases where the data divides into classes more naturally than others.

Example 3: Constructing a Grouped Frequency Distribution

The number of text messages used per month for 52 individuals was recorded and is listed. Construct a grouped frequency distribution for the data.

357	239	379	381	298	377	130	312	333	287	389	357	328
345	301	278	222	388	391	295	342	327	276	289	367	399
265	298	326	301	333	344	313	389	372	369	328	337	317
352	189	355	318	272	361	340	322	371	358	266	345	207

Solution

☞ Helpful Hint

Although choosing the smallest data value as the first lower class limit is easy, it's not necessary to start there. Classes can begin with any number at or below the smallest data value.

The first thing we need to do is identify the smallest and largest pieces of data to ensure that our distribution will include all of the data values. The smallest number of text messages used was 130 and the largest was 399. When choosing the number of classes, we don't want too many or too few class divisions. Remember, if we have too many divisions, we might as well have listed out each individual piece of data, which is overwhelming. However, if we have too few classes, then the data are all lumped together and don't tell us much. A good rule of thumb is to aim for somewhere between 5 and 20 classes. Since our data set covers a spread of almost 270, a class width of 50 is a good place to start. If our classes each have a width of 50, where should the first class start? Let's try with the smallest data value. By adding the width to 130, we can find each of the lower class limits. Keep adding classes until all data values can be tallied in the distribution. We then have the following start to our table.

Table 4	
Number of Text Messages Used Per Month	
Number of Text Messages Used Per Month	**Frequency**
130–	
180–	
230–	
280–	
330–	
380–	
430–	

Since our largest data point is 399, it will safely fall into the class that has a lower limit of 380, since we can now see that the following class would start at 430. Therefore, we do not actually need to add the class beginning with 430, and we can stop with 6 classes. Next, in order to find upper class limits, we can use the lower limits. For the first upper limit, simply count backwards one unit from the lower limit of the second class, making sure the classes don't overlap with one another. For example, the first upper class limit will be $180 - 1 = 179$. We could continue this same process of subtraction to obtain the remaining upper limits. Alternatively, we can simply add the class width of 50 to the upper limit we just found. Either method produces identical upper limits. We then have the following table.

Table 5

Number of Text Messages Used Per Month

Number of Text Messages Used Per Month	Frequency
130–179	
180–229	
230–279	
280–329	
330–379	
380–429	

The only remaining thing to do is to tally the frequencies of each class as shown in Table 6.

Table 6

Number of Text Messages Used Per Month

Number of Text Messages Used Per Month	Frequency
130–179	1
180–229	3
230–279	6
280–329	16
330–379	20
380–429	6

When you complete your grouped frequency distribution, the sum of the frequencies should add up to the total number of data values. Check for yourself that it does in Example 3.

As consumers of information, we're likely to be asked to digest information in the form of a table more often than to construct one. Interpreting data correctly is a crucial step in making informed decisions. Let's look at reading and interpreting a frequency distribution.

Example 4: Interpreting a Frequency Distribution

The San Francisco Bay Area surveyed their residents about the nearly 17 million intraregional daily trips the population makes. Along with other data that they collected, they asked people to identify where their trips originated and how long they traveled. Table 7 organizes the travel data by trip purpose and travel time. The trip purposes recorded in the survey were work, shop, social/recreational, and school. Home-based trips were those that either started or ended at the residence of the trip maker. Non-home based trips were those that neither started nor ended at the residence.

Answer the following questions about the frequency chart and its data.

			Table 7			
			Number of Intraregional Trips			
Travel Time (in Minutes)	Home-Based Work	Home-Based Shop (Other)	Home-Based Social/Rec.	Home-Based School	Non-Home Based	Total Purposes
0–5.0	269,350	807,952	272,973	242,906	953,920	2,547,101
5.1–10.0	416,075	881,203	357,196	282,305	917,191	2,853,970
10.1–15.0	886,859	1,217,440	491,636	434,810	1,206,126	4,236,871
15.1–20.0	440,776	316,201	137,957	139,623	331,661	1,366,218
20.1–25.0	283,777	200,142	84,876	103,114	213,205	885,114
25.1–30.0	807,327	391,393	228,241	211,851	455,129	2,093,941
30.1–35.0	155,706	53,270	27,501	40,179	71,154	347,810
35.1–40.0	169,675	50,564	34,060	38,579	79,675	372,553
40.1–45.0	327,028	101,392	60,827	69,478	150,929	709,654
45.1–50.0	88,325	25,253	16,074	11,509	30,240	171,401
50.1–55.0	56,228	15,686	10,876	13,804	25,282	121,876
55.1–60.0	227,844	76,162	48,138	38,221	94,638	485,003
60.1–65.0	45,851	10,352	5287	3284	16,144	80,918
65.1–70.0	42,824	11,974	7697	3952	14,434	80,881
70.1–75.0	81,350	20,182	14,642	10,800	35,675	162,649
75.1–80.0	22,123	5876	3311	3321	9344	43,975
80.1–85.0	13,594	3852	3121	2495	4950	28,012
85.1–90.0	50,220	13,375	12,996	4947	27,684	109,222
90 +	68,457	29,496	26,402	9446	47,242	181,043
Total	4,453,389	4,231,765	1,843,811	1,664,624	4,684,623	16,878,212

Source: National Transportation Library. "San Francisco Bay Area 1990 Regional Travel Characteristics - WP #4 - MTC Travel Survey." http://ntl.bts.gov/DOCS/SF.html

a. How many classes are there in the frequency distribution?

b. Find the class width. (Exclude the first and last classes.)

c Which guideline is ignored in this published frequency distribution? What sorts of questions arise from this inconsistency?

d. Which class contains the largest number of school-related trips, and what is its frequency?

e. Is the statement "Most people take 10–15 minutes to travel to school in the San Francisco Bay Area" a true statement?

Solution

a. There are 19 classes in this frequency distribution. This can be found by counting the number of data rows in the frequency chart.

b. The class width is 5 minutes and can be found by finding the difference between the lower class limits (or the upper class limits) of two consecutive classes, excluding the first and last class.

c. The classes are not all of equal width. In fact, the last class might go on forever! Since this class contains more than 181,000 pieces of data (more than some of the other classes) it cannot be ignored. However, we don't know if the data lie within 5 minutes, 10 minutes, or 2 hours of the 90-minute mark. From the table, there is no way to know.

d. The class with the largest number of trips that are school related is the 10.1–15.0 minute class with a frequency of 434,810.

e. Although it is the largest class in that category, in order to say "most people" the class would need to contain more than half of all the people who travel to school. Since we can see that a total of 1,664,624 people travel to school each day, 434,810 is not more than half of these. Therefore, the statement is not true. Would you fall into the trap of saying that statement yourself?

Skill Check #2

Use the frequency distribution from Example 4 to determine how many people take between 10.1 and 25 minutes to travel.

Graphical Displays of Data

Although a frequency distribution is a reasonable place to start for displaying data, sometimes even a table can contain an overwhelming amount of information. That's when graphical displays such as pie charts, bar graphs, histograms, and line graphs help to convey the data more clearly. Because technology helps us make all of these types of graphs with relative ease, we'll focus our attention on interpreting the graphs.

Pie Charts

When we want to show a comparison between part of the data and the whole, we use a **pie chart**, also called a **circle graph**. A pie chart shows how large each category is in relation to the whole; that is, it uses the relative frequency distribution to divide the "pie" into different-sized wedges. The size of each wedge in the pie chart is determined by the measure of the central angle of the wedge. This central angle is calculated by multiplying the relative frequency of each class by 360° and rounding to the nearest whole degree.

Pie Chart

A **pie chart** shows how large each category is in relation to the whole.

Consider the following pie chart of religious affiliations of adults in the United States made from data gathered by the Pew Research Center. Notice that, although there is quite a bit of data being displayed, the information is visually clear and easy to read with the key given at the side of it. [1]

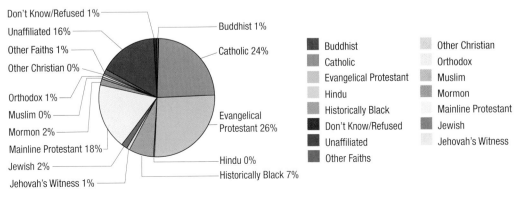

Figure 1

Using the graph, we can easily answer questions along the lines of which religion has the largest US adult following. Obviously it's between the two largest pie slices, Catholics and Evangelical Protestants. By looking at the percentages, we know that Evangelical Protestants have a slightly larger group with 26% of the adult population. However, as we critically look at the graph, one question you might consider is how large of a study was done? Luckily for us, the Pew Research Center gives an exact description of their survey methodology on their website.

"The US Religious Landscape Survey completed telephone interviews with a nationally representative sample of 35,556 adults living in continental United States telephone households. The survey was conducted by Princeton Survey Research Associates International (PSRAI). Interviews were done in English and Spanish by Princeton Data Source, LLC (PDS), and Schulman, Ronca, and Bucuvalas, Inc. (SRBI), from May 8 to Aug. 13, 2007. Statistical results are weighted to correct known demographic discrepancies." [2]

Together, the pie chart and survey methodology give us a better picture of the study that was done. We could have conveyed some of this information on the pie chart by stating how many adults were surveyed when creating the chart. There is a fine line between keeping the information displayed clearly and giving too little information. As a general rule, more information is always better than less.

Let's look at another graph.

Entry-Level Salaries

Total Number of Professions Surveyed = 541

Figure 2

In this pie chart, the percentages are not given, but instead, each wedge is labeled with the salary level. At a glance, is it possible to tell which salary level occurred most? Certainly we know that it's either the orange or red wedge that represents the most common salary. In order to determine the underlying frequencies of the original data, we either need the percentages of each category or the angles of the wedges. Figure 3 shows the same graph as Figure 2, but with more information. This enables us to answer more questions about the graph.

Entry-Level Salaries

Total Number of Professions Surveyed = 541

Figure 3

With this new information, it is clear that the $38,000 salary occurred the most.

Let's calculate how many of the entry-level salaries were $30,000. Since we know that there were 541 salaries in the study, we can multiply 541 by the appropriate percentage for $30,000 (21%).

Don't forget to change the percentage to a decimal in your calculation.

$$541(21\%) = 541(0.21) = 113.61$$

Consider this answer. We know that there couldn't have been a decimal number of salaries in this category, so we can conclude that the percentages were rounded off when constructing the pie chart. We'll do the same with our answer and estimate that there were 114 salaries that fell into the $30,000 category.

What percentage of salaries were above $25,000 in the study? To answer this question, we can simply add together the percentages for all categories that are above $25,000.

$$23\% + 21\% = 44\%$$

Therefore, 44% of salaries were above the $25,000-level in the study.

Is it then correct to say that most jobs in the study had entry-level salaries that were more than $25,000? Because the percentage is not above 50%, it is incorrect to say that "most salaries" were more than $25,000. However, it would be correct to say that "almost half" of the salaries were more than $25,000.

Bar Graphs

One of the most common ways to display categorical data is to use a **bar graph**. Like a pie chart, a bar graph represents the amount of data in each category. However, it uses either vertical or horizontal bars instead of parts of a circle.

Bar Graph

A **bar graph** is a chart of categorical data in which the height of the bar represents the amount of data in each category.

One advantage of bar graphs over pie charts is the ability to see small differences between individual categories. Thus, we might choose to use a bar graph instead of a pie chart if we wish to compare different categories that have similar amounts of data in them. It's also a useful tool if we are more interested in the frequencies in individual categories rather than how the categories compare to the whole.

<hr>

☞ **Helpful Hint**

Although there is no difference between using vertical bars versus horizontal bars in a graph, Microsoft Excel refers to a vertical bar graph as a "column graph" since the bars in the graph look like vertical columns. It uses horizontal bars to create what it calls a "bar graph."

When creating a bar graph, one axis displays the categories of data and the other axis displays the frequencies. Traditionally, for vertical bar graphs, the horizontal axis contains the categories and the vertical axis represents the frequencies. Because each bar represents a category, the width of the bar along the horizontal axis does not relay anything about the data itself. So that we don't cause misunderstandings about the data, the bars should be of uniform width and not touching one another. If the bars were different widths, readers might infer that one category held more significance than another. Similarly, bars touching or overlapping in some way might suggest that the categories were not completely separate from one another. The height of the bar should equal the frequency (or relative frequency) of the category.

Example 5: Interpreting Bar Graphs

The following horizontal bar graph shows the percentage of adults in each of four generational categories who think the United States is the greatest country in the world.[3] According to the bar graph, which generation has the largest percentage of approval amongst its age group?

---

[3] Pew Research Center, http://www.people-press.org

Which generation has the second largest approval?

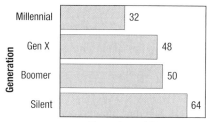

Generational Divide Over American Exceptionalism
% saying the United States is "the greatest country in the world"

Solution

We can easily scan the graph to see that the Silent generation has the largest percentage of approval amongst its age group. 64% of people in the Silent generation feel that the United States is the greatest country in the world.

Which generation has the second largest percentage of approval? To answer this question, we'll need to look a little closer at the labels on the ends of the bars to know the precise percentage of each category. The Boomer generation was second with 50% of its generation believing that the United States is the greatest country in the world.

Bar graphs are also commonly used to show a comparison of categories between multiple populations. We can either use a **side-by-side bar graph** or a **stacked bar graph** to show these comparisons. A side-by-side bar graph is just as the title suggests. The bars are placed next to one another to show the similarities and/or differences between populations. A stacked bar graph places the bars from each population on top of one another. This allows the reader to see a combined total for each category, yet at the same time, see the individual frequencies for each population.

Example 6: Interpreting Bar Graphs with Multiple Populations

The Surveillance, Epidemiology, and End Results (SEER) Program of the National Cancer Institute works to provide information on cancer statistics in an effort to reduce the burden of cancer among the US population.[4] In order for the results of their studies to have the best impact, SEER researchers need representative samples of the US population. The following side-by-side bar graph shows the demographic breakdown of participants in a SEER study versus the same demographic breakdown in the overall US population. Which two sub-populations are more represented in the SEER group?

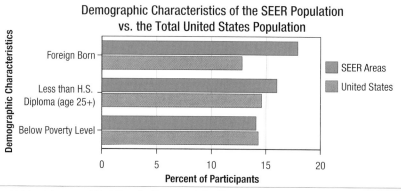

Demographic Characteristics of the SEER Population vs. the Total United States Population

4 Surveillance, Epidemiology, and End Results Program, http://seer.cancer.gov

Solution

The bars are obviously longer for SEER than for the United States in both the Foreign Born category and the High School Diploma category, meaning that these have more representation in the SEER group than the general US population. However, if we look carefully at the Poverty Level category, although the percentages are very close, the SEER group slightly underrepresents this demographic.

Is it possible to determine how many people are in the SEER group using the figure from Example 6? Unfortunately, it is not possible. However, by looking at their website we know that there were over 86,000,000 participants in SEER. When producing any type of graph, it's always a good idea to make sure that pertinent information is included for the reader; for example, the size of the sample.

A **stacked bar graph** not only compares different categories, but how the whole of each category is broken down. For instance, a college administrator might want to know how many spaces are available for the most popular classes on campus in the coming spring semester. The graph in Figure 4 shows not only how many total seats will be available for each class, but also how many are already filled. Notice the other categories in which each class is broken down.

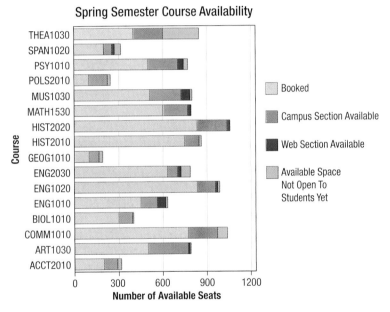

Figure 4

Let's take a closer look at the course Eng 1010. This is a freshman-level English course required by all liberal arts majors at the college. We can see that there are approximately 625 seats that could be made available for this class for the spring semester. Of those 625 seats, the blue part of the bar indicates that a little more than 450 are already booked. The orange portion tells us the available seats in physical classes that are offered on the campus; there are about 100 of these left. The purple shows the number of spaces available in web classes, which is approximately 50. Finally, we can see by the green portion that there are very few spaces to which students do not already have access.

A graph that contains this much information for each course would be good for an administrator who is trying to get an overall picture of the enrollment at his institution. However, if he required the exact numbers of spaces in the courses, this may not be the best way to convey the information.

Histograms

A special type of bar graph that displays the frequency distribution of numerical classes is called a **histogram**. There is a subtle difference between a standard bar graph and a histogram. In a histogram, the height of each bar represents the frequency of the corresponding class. Because the bars represent classes, and class limits are consecutive numbers, the bars touch in a histogram, as shown in Figure 5.

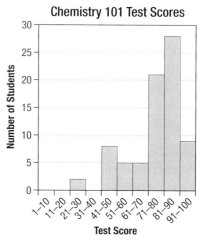

Figure 5

Example 7: Interpreting a Histogram

The following histogram displays data collected from delivery truck drivers. They were asked to record their wait times when an address had a closed entrance gate. Use the histogram to answer the questions that follow.

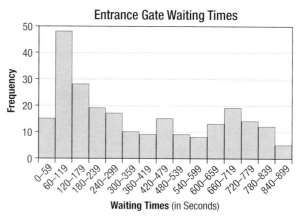

a. Determine approximately how many truck drivers were in the survey?

b. Approximately how many drivers waited between 60 seconds and 180 seconds?

c. Approximately how many drivers waited more than 7 minutes?

d. Is it accurate to say that about half the drivers waited more than 420 seconds? Why?

Solution

a. To determine the number of drivers surveyed, add together each of the frequencies for the different classes. We'll approximate the frequency of each class first and then add them together.

Table 8	
Entrance Gate Waiting Times	
Class	**Frequency**
0–59	15
60–119	48
120–179	28
180–239	19
240–299	17
300–359	10
360–419	9
420–479	15
480–539	9
540–599	8
600–659	13
660–719	19
720–779	14
780–839	12
840–899	5
Total	**241**

So, we estimate that there were approximately 241 truck drivers in the survey. We'll use these estimates in the following questions.

b. To determine how many drivers waited between 60 and 180 seconds, we need to add together the estimated frequencies of each of the bars between those times. Approximately 48 drivers waited between 60 and 120 seconds, and approximately 28 drivers waited between 120 and 180 seconds. Therefore, we can estimate that about $48 + 28 = 76$ drivers waited between 60 seconds and 180 seconds.

c. To approximate the number of drivers who waited more than 7 minutes, we need to add together all of the bars to the right of 7 minutes. We know that 7 minutes is 420 seconds, so we'll add the frequencies of the last eight bars together.

$$15 + 9 + 8 + 13 + 19 + 14 + 12 + 5 = 95$$

Therefore, there were an estimated 95 drivers who waited more than 7 minutes at the entrance gates.

d. Although 420 seconds represents the middle waiting time for the classes, it is inaccurate to say that half of the drivers recorded times less than this time and half recorded longer times. Instead, we need to think about the numbers of drivers. If there were approximately 241 drivers, then half would be around 120 drivers. We can add together the class frequencies until we reach approximately 120 and then note the waiting time.

The first two classes added together are $15 + 48 = 63$.

Adding the third class gives us $63 + 28 = 91$.

Continuing with the fourth class, we get $91 + 19 = 110$.

This is not quite the 120 that we're looking for, so we can include the fifth class as well: $110 + 17 = 127$. The fifth class has a lower limit of 240 seconds, so it is more accurate to say that about half of the drivers waited more than 240 seconds at entrance gates.

Line Graphs

When your data consists of measurements over time, it is best to display these data using a **line graph**. In a line graph, the horizontal axis represents time. The vertical axis represents the variable being measured. Each point on a line graph represents a data value and the appropriate time period it was observed. By joining the points together with line segments, changes over time are more easily observed. It is these line segments that give the graph its name.

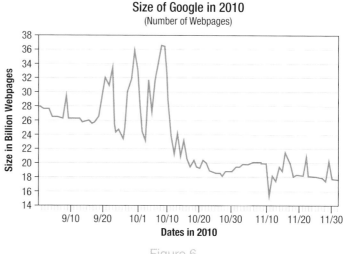

Figure 6

From this line graph of the size of Google, we can see that, overall, the number of web pages decreased over the time period shown from September 1, 2010 to December 2, 2010.[5] Approximately when did the number of pages peak for Google? Determining this date is not as easy as you might think. Although it's easy enough to see the peak, the way the horizontal tick marks are displayed on the graph makes it difficult to easily identify the date of the peak. A good guess would be somewhere around October 8th, 2010. Making every 7th tick mark (a week) bold or identifying the first day of the new month on the tick marks would help tremendously with reading and interpreting the graph. All of these small details are helpful to remember when you're the one constructing the graph.

The next line graph shows how you can have data from two different sources with different scales on the same graph.

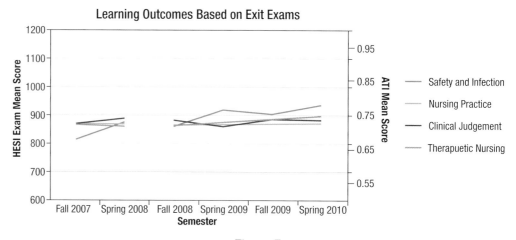

Figure 7

Each of the colored lines represents a content strand of knowledge that the nursing department is assessing in their students. However, in the summer of 2008, the test they used for the assessment was changed from the HESI to the ATI. The *y*-axis on the *left* shows the scale out of 1200 points on the HESI test that was given until Spring 2008. The *y*-axis on the *right* shows that the new test, ATI, was based on percentages. By carefully arranging the two scales so that points on the HESI test correspond to percentages on the ATI test, the sequences of outcomes before and after the change of tests can still be appreciated. By placing both test results on a single graph, the nursing department was able to visually show the progress of their students over time, even though the mode of assessment changed. Leaving a break in the lines helps to emphasize the fact that a change in tests was made.

The Power of Graphs

To show how powerful the visual impact of data can be, take a look at the 1861 map of the distribution of the slave population of the southern United States made from a census taken in 1860, shown in Figure 8.[6] It was one of the first maps to present statistical data, and not just topography and geography. It showed President Lincoln the layout of the Southern states by slave population and played a role in shaping his views about slavery and the framing of the Emancipation Proclamation. The map is even included in a portrait of President Lincoln with his cabinet.

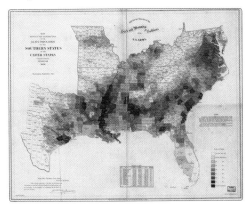

Figure 8: Distribution of the Slave Population of the United States, 1861

The data in the table on the map is difficult to read, so we include the data presented there as Table 9.

| | | **Table 9** | | | |
| | | **Census of 1860** | | | |
No.	**States**	**Free Population**	**Slave Population**	**Total Population**	**% of Slaves**
1	South Carolina	301,271	402,541	703,812	57.2
2	Mississippi	354,700	436,696	791,396	55.1
3	Louisiana	376,280	333,010	709,290	47.0
4	Alabama	529,164	435,132	964,296	45.1
5	Florida	78,686	61,753	140,439	43.9
6	Georgia	595,097	462,232	1,057,329	43.7
7	North Carolina	661,586	331,081	992,667	33.4
8	Virginia	1,105,192	490,887	1,596,079	30.7
9	Texas	421,750	180,682	602,432	30.0
10	Arkansas	324,323	111,104	435,427	25.5
11	Tennessee	834,063	275,784	1,109,847	24.8
12	Kentucky	930,223	225,490	1,155,713	19.5
13	Maryland	599,846	87,188	687,034	12.7
14	Missouri	1,067,352	114,965	1,182,317	9.7
15	Delaware	110,420	1,798	112,218	1.6
		8,289,953	**3,950,343**	**12,240,296**	**32.2**

Source: NOAA Office of Coast Survey, "Map Showing the Distribution of the Slave Population of the Southern States of the United States 1860," http://historicalcharts.noaa.gov/historicals/preview/image/CWSLAVE

The slavery map is in the lower right of the painting *President Lincoln Reading the Emancipation Proclamation to His Cabinet* by Francis Bicknell Carpenter, shown in Figure 9. [7]

Figure 9: *President Lincoln Reading the Emancipation Proclamation to His Cabinet* by Francis Bicknell Carpenter

Example 8: Reading Graphs

Considering the map of the distribution of the slave population of the Southern United States in Figure 8 and the data in Table 9, answer the following questions. The darker the shading is on the map, the higher the slave population was in that area.

a. Which states had a higher slave population than free population in the 1860 census?

b. Which state had the highest number of slaves in 1860?

c. One of the reasons that made the map so popular was the visual account of slavery by shading. Which areas had the heaviest concentration of slaves?

Solution

a. Using the frequency distribution of the 1860 census found on the bottom part of the map, we can see that South Carolina and Mississippi had a higher slave population than free population.

b. Although only 30.7% of its population was slaves, Virginia had the highest slave population at 490,887.

c. The darkest shadings occur along the Mississippi River, along the coast of South Carolina, in central Alabama, and in the middle of Georgia.

Misleading Graphs

As we said at the beginning of the section, it's important to display data so that it is organized clearly and it effectively conveys the intended message. Ideally, graphs should be able to stand alone without the need for additional information in order to be understood. However, sometimes graphs either intentionally or unintentionally convey the wrong message about data or are not quite clear enough to get their message across. It's important to be aware that there are visually misleading and/or ambiguous graphs out there. For example, making the bars different widths on a bar chart might imply that one category is somehow larger than another. Similarly, a distorted piece of a pie graph might inaccurately lead the reader to assume that one section of data is larger than another. An example of this is the graph in Figure 10.

Spirit Week Participation

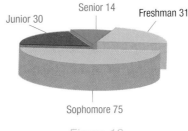

Figure 10

In their graph, the freshmen wanted to emphasize the fact that they came in second place for spirit week—although they just grabbed second place by a tiny margin. If the exact numbers were not on the graph, the emphasis on the freshmen piece of pie might visually imply that the green wedge is considerably larger than the purple wedge. Watch out for these visual manipulations when interpreting graphs.

Skill Check Answers

1. 0 pets

2. 6,488,203 people

8.2   Exercises

Fill in each blank with the correct term.

1.   A _____ is a graph that represents a frequency distribution.

2.   When comparing parts of data to the whole, a _____ visually shows this using sections of a circle.

3.   A ____ graph is best to use when showing data over a time period.

4.   When comparing multiple categories from different populations, a _____ graph or a _____ graph can be used.

5.   A _____ is a literal count of each member of a data set and how often it occurs.

Solve each problem.

6. The grades on the first statistics test for Ms. Seago's class are listed in the following table. Construct a frequency distribution for the grades.

Grades on Statistics Test 1						
A	C	F	C	C	D	F
B	D	F	B	A	A	F
B	C	C	A	B	F	D

7. The following table gives the grouped frequency distribution of weights of 194 babies in kilograms. Answer the questions that follow based on the distribution.

Weights of Babies						
Birth Weight (kg)	0.00–0.99	1.00–1.99	2.00–2.99	3.00–3.99	4.00–4.99	5.00–5.99
Frequency	2	17	39	89	46	1

   a. How many classes are in the grouped frequency distribution?

   b. What is the class width?

   c. What is the value of the lower class limit of the 3rd class?

   d. What is the value of the upper class limit of the 5th class?

   e. What is the relative frequency of the 4th class? Give your answer as a percentage rounded to the nearest tenth.

8. The following table represents a grouped frequency distribution of the number of hours spent on the computer per week for 55 students.

Hours	Number of Students
0.0–4.4	9
4.5–8.9	14
9.0–13.4	21
13.5–17.9	11

   a. Calculate the relative frequencies (as percentages rounded to the nearest tenth) for each class.

   b. What percentage of the students used the computer between 9 and 13.4 hours per week?

   c. What percentage of the students used the computer less than 9 hours per week?

9. Sisscon is a phone answering service. The following data are the numbers of calls per day reported by the company for the last month.

10	72	64	32	78	62	11
37	45	32	52	38	70	66
13	21	14	13	39	73	62
41	63	44	23	27	22	21
55	24	53	43	20	16	22

a. Create a grouped frequency distribution for the data using 8 classes and then use it to answer the following questions. Let the first lower class limit be 0 and the class width equal 10.

b. Calculate the relative frequencies for each class. Give your answer as a percentage rounded to the nearest tenth.

c. For what percentage of the days is the number of calls between 40 and 49?

d. For what percentage of the days is the number of calls in the single digits?

e. What is the most common range for the number of calls per day?

10. Consider the bar graph of the predicted fastest growing occupations between 2010 and 2020. [8]

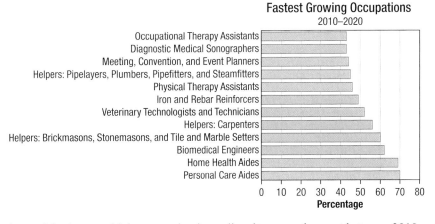

a. At a quick glance, which occupation is predicted to grow the most between 2010 and 2020? What is the predicted amount of growth?

b. How many new jobs will be available in the fastest-growing occupation in 2020?

11. Answer the following questions about the Job Growth graph. [9]

Job Growth Rates by State, 2000 to 2005

United States = 4.5%
- 7% or More (12 states)
- 4.5% to 6.9% (11 states)
- 2.5% to 4.4% (16 states)
- Less than 2.5% (10 states)
- Losing Jobs (2 states)

a. How many states lost jobs between 2000 and 2005?

b. In what part of the country are jobs growing the most in this time period?

12. The following pie chart shows destinations of recent University of Kent mathematics graduates, including business, financial math, and statistics. [10]

Destinations of Recent University of Kent Mathematics Graduates

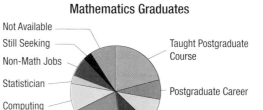

Not Available
Still Seeking
Non-Math Jobs
Statistician
Computing
Finance

Taught Postgraduate Course
Postgraduate Career
Teacher Training
Teaching Job

a. Does it appear that any destination category accounts for more than 25% of the graduates?

b. Which pairs of destinations appear to have similar percentages of graduates?

c. How many mathematics graduates from the University of Kent were surveyed?

d. Is the graph misleading in any way?

9 InContext, http://www.incontext.indiana.edu

10 University of Kent, http://www.kent.ac.uk

13. The following graph shows the US unemployment rate from February 2011 to February 2013.[11]

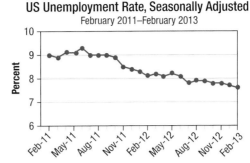

US Unemployment Rate, Seasonally Adjusted
February 2011–February 2013

a. Describe the trend of the unemployment percentage from February 2011 to February 2013.

b. Approximate the month and rate of the highest unemployment during this time period.

c. Approximate the month and rate of the lowest unemployment during this time period.

d. Is the graph misleading in any way?

14. The following is a portion of a table about state health facts. It lists 8 of the 50 states along with the percentage of women age 50 and older who report having had a mammogram between 2008 and 2010.

Percentage of Women Age 50 and Older Who Had a Mammogram Between 2008 and 2010	
Massachusetts	87.50%
Connecticut	83.80%
North Carolina	81.20%
Virginia	79.10%
Alabama	77.60%
Mississippi	70.90%
Nevada	69.90%
Idaho	68.30%

Source: Centers for Disease Control and Prevention (CDC).
Behavioral Risk Factor Surveillance System Survey Data. Atlanta,
Georgia: U.S. Department of Health and Human Services, Centers for
Disease Control and Prevention, 2010, available at http://apps.nccd.
cdc.gov/brfss/list.asp?cat=WH&yr=2010&qkey=4427&state=All

Is the following pie chart a good way to display this data? Explain why or why not.

Percentage of Women Age 50 and Older Who had a
Mammogram Between 2008 and 2010

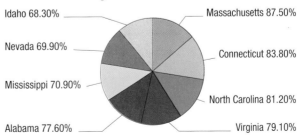

Idaho 68.30%
Nevada 69.90%
Mississippi 70.90%
Alabama 77.60%
Massachusetts 87.50%
Connecticut 83.80%
North Carolina 81.20%
Virginia 79.10%

15. The stacked bar graph shows the average number of hours that married people in Japan spend each day doing various activities. [12]

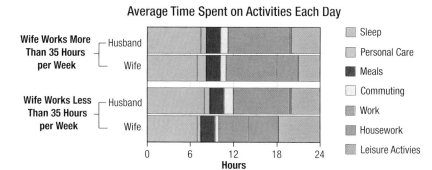

Average Time Spent on Activities Each Day

a. For wives who spend less than 35 hours per week working, how many hours on average are spent each day for leisure activities?

b. For husbands whose wives work more than 35 hours per week, approximately how many hours on average are spent on sleep?

c. Compare the number of hours spent sleeping for wives in each category.

d. What type of graph could be used to represent these data in a clearer fashion?

8.3 Describing and Analyzing Data

As you can tell from the last section, displaying data in a clear and informative way is certainly an important and necessary step in research. Just as important is describing the data numerically, so that they can be compared and analyzed. Imagine if the only way we had of comparing data sets that contain thousands of data points was to list each individual point, we would all throw our hands in the air and scream in frustration. The data would be cumbersome and useless. However, because of statistical tools like measures of central tendency and measures of dispersion (all of which we'll cover in this section) we can begin to get a better picture of what the data are able to tell us by using summary numbers.

Measures of Central Tendency

A number that best describes a typical value in a data set is referred to as a **measure of central tendency**. We'll begin our discussion of measures of central tendency with the most common ways to describe the "center" or "average" of the data: the mean, median, and mode.

Mean

Often we hear people refer to the "average" of a set of data, as in the average age of first-time parents or the average score on a test. Using the term average can refer to any of three measures of central tendency: the mean, the median, or the mode. Most commonly, the use of the word average is in reference to the **arithmetic mean**.

> ☞ **Helpful Hint**
>
> When calculating the mean, round to one more decimal place than the largest number of decimal places given in the data. Occasional exceptions to this rule can be made when the type of data lends itself to a more natural rounding method such as rounding values of currency to two decimal places.

Arithmetic Mean

The **mean** is the sum of all of the data values divided by the number of data points. Formally, the formula for the **population mean** is

$$\mu = \frac{x_1 + x_2 + \cdots + x_N}{N}.$$

The formula for the **sample mean** is

$$\overline{x} = \frac{x_1 + x_2 + \cdots + x_n}{n}.$$

x_i is the ith data value, N is the number of data values in the population, and n is the number of data values in the sample.

Example 1: Finding the Mean

A sample of the number of sick days employees at Witt's Insurance Agency took during last year is listed below. Calculate the mean of the sample data.

$$14, \ 5, \ 7, \ 11, \ 9, \ 7, \ 12, \ 6$$

Solution

There are 8 pieces of sample data, so in order to find the sample mean, add all the values together and divide by 8.

$$\bar{x} = \frac{14+5+7+11+9+7+12+6}{8} = \frac{71}{8} \approx 8.9$$

Therefore, the mean of this sample is 8.9.

TECH TRAINING

When calculating the mean using a TI-30XIIS/B calculator, always remember to clear the data list first. See the Helpful Hint in the margin for instructions.

After clearing the data list, enter the data from Example 1 into the calculator using the following commands.

1. Press 2ND DATA.
2. Choose 1-VAR and press ENTER.
3. DATA (X₁= should appear.
4. Enter a data value and press the down arrow key twice, since the frequency is 1 for each data point.
5. After the last data point is entered, press ENTER.

To calculate the mean, press STATVAR. The top of the screen will display a list of values that the calculator computed using the data entered. Use the left arrow key to move the cursor to $\bar{x}$. The mean of the data will be displayed as 8.875.

To calculate the mean using the list function on a TI-83/84 Plus calculator, clear the data list and then perform the following instructions.

1. Press STAT, then choose 1:Edit..., and enter your data in L1.
2. Press STAT again and now scroll to the right to CALC.
3. Choose option 1:1-Var Stats and press ENTER. If your data are in L1, press ENTER again since L1 is the default list. If you did not type your data in L1, enter the list where your data are located, such as L2 or L3. (These list names are in blue, above the numeric keys.)

The mean is labeled as $\bar{x}$ in the calculator output, as seen in the screen shot in the margin. Therefore, $\bar{x} = 8.875$.

Wolfram|Alpha can be used to calculate the mean of a set of numbers. Go to www.wolframalpha.com and type "mean $\{14, 5, 7, 11, 9, 7, 12, 6\}$" into the input line. Then, click the = button. Wolfram|Alpha will return the following. [1]

1 Wolfram Alpha LLC, http://www.wolframalpha.com

Notice that the mean number of sick days in Example 1 isn't actually a member of the sample set. No employee took 8.875 sick days last year. However, it is a description of all the data points.

A nice image to have of the mean is that of a seesaw or teeter-totter. If all the data points were placed on the seesaw with even weights and the pivot point was at the mean, the seesaw would balance evenly. We've illustrated this in Figure 1 with the data from the previous example.

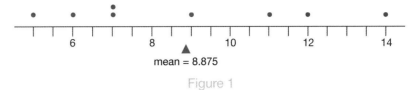

mean = 8.875

Figure 1

It's important to note that it is meaningless to find the mean of data that have no measurable values, such as the ratings of a hotel: very satisfactory, satisfactory, needs improvement, or don't plan to visit again. Because there is not a measurable difference between each of the ratings, even assigning a number value to each and calculating the mean, carries little meaning. However, you might often see such calculations being exhibited.

When we calculate the mean, it is often the case that the value we get is not actually a member of the data set, as shown in Example 1. The mean is simply a descriptive value of the entire set of data. It is an especially useful summary statistic when data sets are large and you may not be able to examine all the pieces of data individually.

Median

Another measure of the center of a data set is the **median**.

> ### Median
>
> The **median** of a data set is the middle value in an ordered array of the data.

In other words, after listing the data in either ascending or descending order, the median is the middle value of the list. If there is no single middle value, such as when there is an even number of values, then the median is the piece of data that lies exactly between the two middle data values, that is, the mean of the two middle data values. Note that when there is an even number of data points, the median may not be a member of the data set.

In circumstances where the data points are tightly grouped values, except for one or a few values, the median is a better choice for describing the "average" member of the data set. This is because extreme values do not affect the median in the way that they affect the mean. For example, the highest mean salary earned by University of North Carolina graduates is not earned by Accountancy, Law, or Medicine graduates, but by Geography majors. This is because Michael Jordan was a Geography major at the University of North Carolina. His salary alone is enough to outweigh the influence of all of the other salary points.

Example 2: Finding the Median

A VO_2 max score is the maximum amount of oxygen that one's body can transport and use during exercise. It is measured in liters of oxygen per minute (L/min). Given the following VO_2 max scores for 12 women, find the median score.

28.3, 27.7, 23.0, 25.5, 27.1, 26.94, 27.0, 27.52, 26.8, 27.2, 26.97, 27.53

Solution

First, put the data in ascending numerical order.

$$23.0, \ 25.5, \ 26.8, \ 26.94, \ 26.97, \ 27.0, \ 27.1, \ 27.2, \ 27.52, \ 27.53, \ 27.7, \ 28.3$$

Since there are 12 pieces of data, the median will be the value between the middle two data points, 27.0 and 27.1. To find this, add the two together and divide by two.

$$\text{Median} = \frac{27.0 + 27.1}{2} = 27.05.$$

Once again, the value of this "average" is not a member of the data set. However, it is a typical value in the sense that it is located in the middle of the data set when it is arranged numerically.

TECH TRAINING

Wolfram|Alpha can be used to calculate the median of a set of numbers. Go to www.wolframalpha.com and type "median $\{28.3, 27.7, 23.0, 25.5, 27.1, 26.94, 27.0, 27.52, 26.8, 27.2, 26.97, 27.53\}$" into the input bar. Then, click the = button. Wolfram|Alpha will return the following. [2]

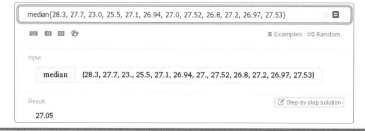

Mode

Sometimes a data set will not lend itself to numerical calculations. For instance, if you survey students on the color cell phone case they prefer, you would have a list of colors, which could not be added, subtracted, or put in ascending or descending order. This is also the case with the hotel ratings we mentioned earlier. It is impossible to add or subtract *very satisfactory, satisfactory, needs improvement,* or *don't plan to visit again.* However, having a reasonable description for this type of data is still a valuable thing. Even if data are numerical values, and not simply descriptions, sometimes we'd like to know the most common data value.

> ### Mode
>
> The **mode** is the value in the data set that occurs most frequently.

If all the data values occur only once, or they each occur an equal number of times, we say that there is **no mode**. If only one value occurs the most, then the data set is said to be **unimodal**. If exactly two values occur equally often and more than any other data value, the data set is said to be **bimodal**. If more than two values occur equally often and more than any other data value, the data set is **multimodal**. Note that, unlike the mean and the median, if there is a mode, it will always be a value in the data set.

2 Wolfram Alpha LLC, http://www.wolframalpha.com

Example 3: Finding the Mode

Find the mode of each of the following sets of data. State if the data set is unimodal, bimodal, multimodal, or has no mode.

a. Preferred color of cell phone cases among students

lemon, gunmetal, violet, turquoise, lime, violet, lemon, orange, red, lemon, pink, violet, lime, violet, lemon, pink, gunmetal, red, turquoise, violet, violet, gunmetal, turquoise, red, violet, turquoise, orange, pink, violet, violet, turquoise, violet, pink

b. Favorite football jersey number

32, 18, 99, 12, 7, 10, 28, 56, 13, 16, 19, 51, 23, 78

c. Ages of children at the community playground one afternoon

12, 4, 2, 7, 8, 4, 10, 6, 5, 7, 7, 4, 3

d. Number of ATM withdrawals per hour at the downtown branch of University Bank

10, 13, 9, 13, 9, 14, 10, 14

Solution

a. The color violet occurs more than any other color, so the mode is violet. This data set is unimodal.

b. Each value occurs only once, so there is no mode.

c. The values 4 and 7 both occur an equal number of times, which is more than any other value. Thus, the set is bimodal with the modes 4 and 7.

d. Be careful here. Since each value occurs the same number of times, there is no mode in this data set.

📖 **Helpful Hint**

Symmetric

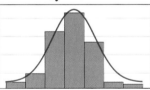

Skewed to the Left

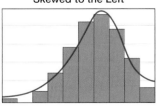

Skewed to the Right

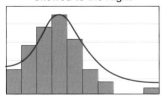

Of the three "averages" (mean, median, and mode), the mean should be used with data consisting of counts and measurements when the data set doesn't include any **outliers**.

Outlier

An **outlier** is a data value that is extreme compared with the rest of the data values in the set.

An outlier will influence the value of the mean of a data set, but will not affect the median or mode. Because it is an extreme value, the outlier drags the value of the mean toward itself. When this happens, the data are said to be skewed, or pulled in the direction of the outlier. The outliers cause the graph of the distribution to be skewed and not symmetrical.

Example 4: Finding the Mean, Median, and Mode

Given the following data set, find the mean, median, and mode, and decide which measure of center you think best describes the data set.

16, 44, 15, 48, 14, 77, 11, 84, 26, 61, 15

Solution

To find the mean, add up all of the data values and divide by 11 (the number of data values).

$$\text{Mean} = \frac{16+44+15+48+14+77+11+84+26+61+15}{11} = \frac{411}{11} \approx 37.4$$

To find the median, arrange the values in ascending order and find the middle value.

11, 14, 15, 15, 16, $\boxed{26}$, 44, 48, 61, 77, 84

Median = 26

The mode is the most commonly occurring value. Notice that 15 occurs twice, while all other values occur only once. Therefore, the mode is 15.

Although there is a mode, because it only occurs twice while all the other data points occur once, this is not the best descriptor of the "average" piece of data. A mode of 15 does not accurately reflect the middle of the data set since the data ranges from 11 to 84. That leaves the mean and the median. Since there are not any outliers it's appropriate to use the mean of the data as the measure of center for this data.

Skill Check #1

Find the mean, median, and mode of the following data.

8, 12, 10, 11, 13, 12, 15, 9, 11, 16

Measures of Dispersion

Measures of dispersion, like the range and standard deviation, describe the "spread" of the data. In other words, they tell us whether the data values are all very similar or if they cover a wide section of the number line. One of the simplest measures of dispersion is the **range** of the data.

Range

The **range** is the difference between the largest and smallest values in the data set, which tells you the distance covered on the number line between the two extremes.

range = maximum data value − minimum data value

Example 5: Finding the Range

Find the range of the following sets of data.

a. The number of students enrolled as computer science majors over the past 12 semesters

$$5,\ 21,\ 54,\ 33,\ 12,\ 14,\ 36,\ 40,\ 27,\ 29,\ 37,\ 22$$

b. The number of shoppers at a gas station downtown Monday through Sunday one week

$$1007,\ 1010,\ 1006,\ 1005,\ 1054,\ 1021,\ 1005$$

Solution

a. The maximum value is 54 and the minimum value is 5, so the range is

$$54 - 5 = 49.$$

b. The maximum value for the data set is 1054 and the minimum value is 1005, so the range is also

$$1054 - 1005 = 49.$$

Calculating the range is very easy. However, the range is not as descriptive as other measures of dispersion. Consider the two data sets in the previous example. Notice that both data sets have the same range. However, almost all of the values in the second data set are similar while the values in the first data set are more spread out. To distinguish between these two situations, we must use another measure of dispersion.

The **standard deviation** is a measure of how much we might expect a member of the data set to differ from the mean. The greater the standard deviation, the more the data values are spread out. Similarly, a smaller standard deviation indicates that the data values lie closer together. The standard deviation is always a number greater than or equal to 0. It is precisely the case when all of the data points are the same value that the standard deviation is equal to 0.

Recall that we presented two formulas for the mean: the population mean and the sample mean. We also have two formulas for the standard deviation: the population standard deviation and the sample standard deviation.

☞ Helpful Hint

When calculating the standard deviation, round to one more decimal place than the largest number of decimal places given in the data. Occasional exceptions to this rule can be made when the type of data lends itself to a more natural rounding method, such as rounding values of currency to two decimal places.

Standard Deviation

The **standard deviation** is a measure of how much we might expect a member of the data set to differ from the mean.

The formula for finding the population standard deviation is

$$\sigma = \sqrt{\frac{\sum\left(x_i - \mu\right)^2}{N}}$$

where x_i is the ith data value, μ is the population mean, and N is the size of the population.

For a sample, the standard deviation is

$$s = \sqrt{\frac{\sum\left(x_i - \bar{x}\right)^2}{n-1}}$$

where x_i is the ith data value, $\bar{x}$ is the sample mean, and n is the sample size.

Example 6: Calculating Standard Deviation by Hand

Calculate the sample standard deviation for a sample of nine ages of students working with a university theater production of *Macbeth*.

$$17, \ 21, \ 18, \ 18, \ 24, \ 19, \ 21, \ 20, \ 28$$

Solution

When calculating the standard deviation by hand, we need to first note the sample size n and find the sample mean $\bar{x}$. With $n = 9$, the mean is

$$\bar{x} = \frac{17+21+18+18+24+19+21+20+28}{9}$$
$$\approx 20.67.$$

Note that we will round the mean to the nearest hundredth in an effort to minimize any error introduced from rounding.

When calculating standard deviation by hand, it's helpful to use a table like Table 1 and build up to the formula.

Table 1

Sample Standard Deviation

x_i	$x_i - \bar{x}$	$(x_i - \bar{x})^2$
17	−3.67	13.47
21	0.33	0.11
18	−2.67	7.13
18	−2.67	7.13
24	3.33	11.09
19	−1.67	2.79
21	0.33	0.11
20	−0.67	0.45
28	7.33	53.73
		$\Sigma = 96.01$

We are now ready to substitute the values into the formula for the sample standard deviation.

$$s = \sqrt{\frac{\sum (x_i - \bar{x})^2}{n-1}}$$
$$= \sqrt{\frac{96.01}{9-1}}$$
$$= \sqrt{12.00125}$$
$$\approx 3.5$$

So, the sample standard deviation of ages is approximately 3.5. In other words, the age of the average student in the sample is about 3.5 years different (either younger or older) from the mean age of 20.67.

Since technology is so often used to find the standard deviation rather than the pencil and paper method, especially with large data sets, we will focus on showing you how to use technology for the calculations and how to interpret the results of what you find. The following example shows how to calculate standard deviation using a calculator.

Use your TI-30XIIS/B or TI-83/84 Plus calculator to find the standard deviation of the following data sets.

Helpful Hint

If you are using a TI-30XS/B Multiview calculator, the keystrokes will be slightly different for your calculator. See Appendix B for these instructions.

a. The following data represent the average number of Tweets per day posted on Twitter for a sample of 24 college students.

Table 2					
Tweets Per Day					
0.8	42.2	20.6	2.8	36.7	12.1
18.6	6.3	5.5	11.3	3.7	0.5
1.2	3.7	14.9	9.4	7.3	9.5
16.0	11.1	4.7	5.6	8.9	10.2

b. The SAT Critical Reading scores for the senior class at Richmond Prep High is given in Table 3.

Table 3					
SAT Critical Reading Scores					
520	640	750	620	470	520
630	600	590	660	700	580
460	600	640	690	530	490
500	560	630	760	650	760
580	610	710	610	590	570
590	550	610	490	630	550
590	620	610	600	570	690

Solution

a. In order to calculate the standard deviation using a TI-30XIIS/B calculator, begin by clearing the data lists in the calculator. Then enter the data points as before by using the following commands:

1. Press 2ND DATA.
2. Choose 1-VAR and press ENTER.
3. Press DATA. (X= should appear.
4. Enter a data value and press the down arrow key twice, since the frequency is one for each data point.
5. After the last data point is entered, press ENTER.

Because the values given are only a sample of students, we want the *sample standard deviation*. To calculate the sample standard deviation, press STATVAR. Scroll over to the sample standard deviation, which is denoted by sx in the list of calculated values. From the list we see that $s \approx 10.3$.

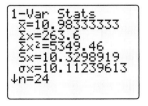

To calculate the standard deviation using the list function on a TI-83/84 Plus calculator,

1. Press STAT, then choose 1: Edit..., and enter your data in L1.
2. Press STAT again and now scroll to the right to CALC.
3. Choose option 1: 1-Var Stats and press ENTER. If your data are in L1, press ENTER again since L1 is the default list. If you did not type your data in L1, enter the list where your data are located, such as L2 or L3. (These list names are in blue, above the numeric keys.)

A list of numerical statistics will be generated for the data. The beginning of the list is shown in the margin. Because the values given are only a sample of college students, we want the *sample standard deviation*, which is denoted by s (on the calculator, this is displayed as Sx). From the list we see that $s \approx 10.3$.

Since the standard deviation tells us about the average distance away from the mean, we can conclude that student tweeting behavior usually varies from the mean by tens rather than hundreds of tweets.

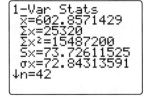

b. Begin by clearing the data lists in the calculator. Now enter the data as you did in part **a.** Since we are told that the values given represent an entire Senior class, we want the *population standard deviation*, which is denoted by σx. From the list we see that $\sigma \approx 72.8$.

Therefore, we know that SAT Critical Reading scores differ from the mean on average by 72.8 points. While we've got the calculator handy, we can see that the mean is actually approximately 602.9. Although there is not an actual score of 602.9, you can see that many of the students scores fall within about 70 points (or 1 standard deviation) of that mean, either larger or smaller.

TECH TRAINING

Wolfram|Alpha can be used to find the standard deviation of a set of numbers. To find the standard deviation of the data in Example 7a., go to www.wolframalpha.com and type "standard deviation $\{0.8, 42.2, 20.6, 2.8, 36.7, 12.1, 18.6, 6.3, 5.5, 11.3, 3.7, 0.5, 1.2, 3.7, 14.9, 9.4, 7.3, 9.5, 16.0, 11.1, 4.7, 5.6, 8.9, 10.2\}$" into the input bar. Then, press the = button. Wolfram|Alpha will return the following. [3]

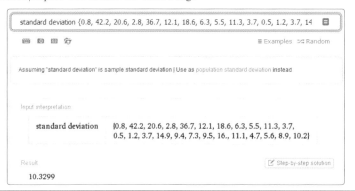

Skill Check #2

Find the population standard deviation for the following data.

8, 12, 10, 11, 13, 12, 15, 9, 11, 16

The standard deviation allows us to interpret the variation of the data with a sense of scale. Two standard deviations that are exactly the same number don't necessarily represent the same variation for different populations. For instance, consider Example 6. A standard deviation of 3.50 when referring to ages is not a large variation. However, if the standard deviation for gas prices at local gas stations had a value of $3.50, we would say that the data had a huge variation!

The standard deviation helps us to put into context one of the most useful estimation rules in statistics.

The Empirical Rule

When the distribution of a set of data is approximately bell-shaped, we can estimate the percentage of data values that fall within a few standard deviations of the mean in the following way:

Approximately 68% of all data points lie within 1 standard deviation above and below the mean.

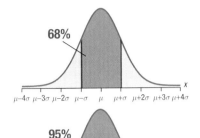

The Empirical Rule is also referred to as the Three Sigma Rule, or the 68–95–99.7 Rule.

Approximately 95% of all data points lie within 2 standard deviations above and below the mean.

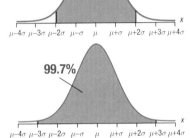

Approximately 99.7% of all data points lie within 3 standard deviations above and below the mean.

Example 8: Using the Empirical Rule

Suppose you suspect that emergency room waiting times for a local hospital have a bell-shaped distribution. The data have a reported mean of 116.9 minutes and a standard deviation of 56.1 minutes.

a. Identify the range of waiting times that 95% of patients are likely to experience.

b. Estimate the percentage of patients that will wait less than 2 hours and 53 minutes.

Solution

a. By the Empirical Rule, we know that 95% of the data set, which is approximately bell-shaped, will fall within two standard deviations of the mean. Therefore, we can double the standard deviation and then add and subtract it from the mean to find the range of times that 95% of patients are likely to experience.

$$(56.1)(2) = 112.2$$
$$116.9 - 112.2 = 4.7$$
$$116.9 + 112.2 = 229.1$$

So, we can estimate that 95% of patients will wait somewhere between 4.7 minutes and 229.1 minutes.

b. To answer this question, we'll need to first convert 2 hours and 53 minutes into minutes only, since our mean and standard deviation are both in minutes. Since 2 hours is 120 minutes, add 53 minutes to get a total time of 173 minutes. We know that this time falls above the mean, but by how much? If we subtract the mean, we have

$$173 - 116.9 = 56.1 \text{ minutes.}$$

In other words, the time we're interested in is one standard deviation above the mean. Don't jump to conclusions here and assume that 68% of the data falls below that. Let's draw a picture first.

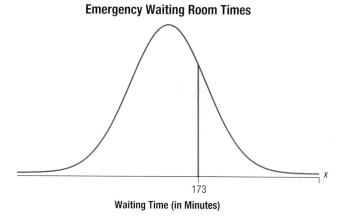

Emergency Waiting Room Times

173
Waiting Time (in Minutes)

Remember, we want to know what percentage of patients will wait less than 173 minutes. From the Empirical Rule, we know that 68% of data is within one standard deviation, so half of that, 34%, must be between the mean and one standard deviation.

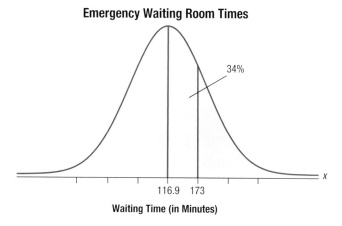

Emergency Waiting Room Times

34%

116.9 173
Waiting Time (in Minutes)

We also know that because of the symmetry of a bell-shaped curve, since the mean is in the middle, half of the data lies below it.

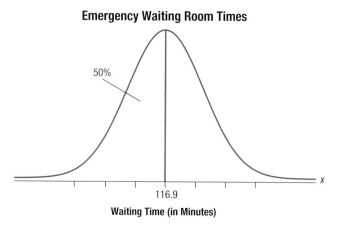

Emergency Waiting Room Times

50%

116.9
Waiting Time (in Minutes)

So, putting these two facts together, we have that 50% + 34% = 84% of the data lies below one standard deviation above the mean. In other words, approximately 84% of patients will wait less than 2 hours and 53 minutes.

Measures of Relative Position

When we want to describe the position of particular pieces of data in a set compared to the rest of the data, we use **measures of relative position** like percentiles and quartiles. They are each based on dividing the data into sections and then referencing the section in which the data value falls.

Percentile

Percentiles divide the data into 100 equal parts and tell you approximately what percentage of the data lies at or below a given value.

To have a data point at the 81st percentile means that 81% of the population is at or below the data value. It's important to note that percentiles don't convey anything about the data value itself, only its relative position to the data set as a whole. One of the most prominent places you might have heard of percentiles would be in connection with any standardized tests you've taken, such as the ACT or SAT college entry exams. Birth measurements are also commonly reported in terms of percentiles.

Example 9: Interpreting Percentiles

Sierra received her scores from taking a mathematics placement test for her chosen university. Choose the best explanation for what it means for her to be in the 61st percentile.

    a. She correctly answered 61% of the answers on the test.

    b. 61% of people taking the test scored the same as Sierra.

    c. Sierra's score was at least as good as 61% of the people taking the test.

    d. Sierra missed 39% of the test questions.

Solution

The correct interpretation of her score is **c.**: "Sierra's score was at least as good as 61% of the people taking the test." Both **a.** and **d.** are incorrect because they refer to how many questions she answered correctly on the test and not how she did in comparison to others taking the test. **b.** is not quite correct because percentiles tell you the percentage that scored at or below you. They are not all necessarily the same score as Sierra's.

The percentiles that divide the data into four even parts are called **quartiles**. Figure 2 illustrates that to divide something into 4 parts, you only need 3 dividers. We call these "dividers" the first quartile Q_1, the second quartile Q_2, and the third quartile Q_3.

Figure 2

Quartiles

Q_1 = **First Quartile** = 25th percentile, that is, 25% of the data is less than or equal to this value.

Q_2 = **Second Quartile** = 50th percentile, that is, 50% of the data is less than or equal to this value.

Q_3 = **Third Quartile** = 75th percentile, that is, 75% of the data is less than or equal to this value.

By definition, Q_2 will be the same as the median.

Example 10: Interpreting Quartiles

On Karl's recent standardized test results, the picture graph of his score showed he was above the third quartile in language arts. His classmate, Asher, said his score was at the 70th percentile, while Rylie said hers was at the 79th percentile. Which of the three had the best language arts test score?

Solution

We know the percentile ranks of both Asher and Rylie are the 70th and 79th respectively. What we know about Karl's score is that it was above the third quartile. Since the third quartile is the same as the 75th percentile, we know that his score was somewhere at or above the 75th percentile. We can conclude that he did better than Asher, whose score was at the 70th percentile, but can make no definite comparison with Rylie, whose score was at the 79th percentile, because we do not know for sure which one had the best language arts score.

As we've already seen, a calculator can be very helpful when calculating the various data descriptors that have been covered in this section. Often it is the case that several descriptors are desired when considering a set of data. Certainly, the more characteristics about the data we have, the better "picture" we can form of the data. One way to easily calculate several descriptors at once is to use Microsoft Excel or similar spreadsheet packages. The following example illustrates the commands for using Excel.

[TECH] Example 11: Using Excel to Find Data Descriptors

To receive federal financial aid for higher education, students must first complete a Free Application for Federal Student Aid form, commonly called FAFSA. Given the following data gathered from FAFSA on the number of applications by state, use Excel to find the minimum and maximum data values, mean, median, mode, range, and standard deviation for the data.

Table 4
2011–2012 Application Cycle

Quarter 4		Quarter 4	
State	**Number of FAFSA Applications**	**State**	**Number of FAFSA Applications**
Alberta	31	Nebraska	7742
American Samoa	177	Nevada	15,014
Arizona	39,134	New Brunswick	2
Arkansas	16,122	Newfoundland	3
Blank	28,459	New Hampshire	4511
British Columbia	36	New Jersey	34,906
California	197,594	New Mexico	11,335
Canada	104	New York	82,152
Colorado	28,373	North Carolina	57,614
Connecticut	14,757	North Dakota	2045
Delaware	4238	Northern Mariana Islands	238
District of Columbia	2958	Northwest Territories	2
Federated States of Micronesia	877	Nova Scotia	3
Florida	124,874	Nunavut	0
Foreign Country	3042	Ohio	59,066
Georgia	68,669	Oklahoma	18,377
Guam	695	Ontario	89
Hawaii	5211	Oregon	20,937
Idaho	9747	Palau	75
Illinois	59,081	Pennsylvania	45,316
Indiana	30,127	Prince Edward Island	6
Iowa	12,434	Puerto Rico	21,299
Kansas	13,366	Quebec	35
Kentucky	22,078	Rhode Island	4027
Labrador	3	Saskatchewan	7
Louisiana	23,576	South Carolina	27,203
Maine	4953	South Dakota	2941
Manitoba	5	Tennessee	32,411
Marshall Islands	125	Texas	134,463
Maryland	27,318	Utah	18,628
Massachusetts	23,675	Vermont	1748
Mexico	522	Virginia	41,507
Michigan	60,553	Virgin Islands	558
Minnesota	26,841	Washington	34,980
Mississippi	20,741	West Virginia	7091
Montana	4245	Wisconsin	25,411
Missouri	31,919	Wyoming	2452
		Yukon	7

Source: Federal Student Aid, "Application Volume Reports," http://federalstudentaid.ed.gov/datacenter/application.html

Solution

You may need to load the Analysis ToolPak in Microsoft Excel. If you do not have the Data Analysis option in the Data tab, follow these steps to load the Analysis ToolPak on a PC.

1. Under the File menu, click on Options.

2. Click Add-Ins (listed on the left side), and then in the Manage box, select Excel Add-ins.

3. Click Go.

4. In the Add-Ins available box, select the Analysis ToolPak check box, and then click OK.

   If Analysis ToolPak is not listed in the Add-Ins available box, click Browse to locate it.

   If you get prompted that the Analysis ToolPak is not currently installed on your computer, click Yes to install it.

5. After you load the Analysis ToolPak, the Data Analysis command is available in the Analysis group on the Data tab.

Begin by typing the data into Columns A and B. Go to the **Data** tab, then **Data Analysis**, then choose **Descriptive Statistics**, and click OK. In the *Input Range* box, enter the cells where your Number of FAFSA Applications data are located. Select **New Worksheet Ply** and type "Descriptive Statistics" in the box. Click on the box in front of **Summary Statistics**. The Descriptive Statistics menu should look similar to the following.

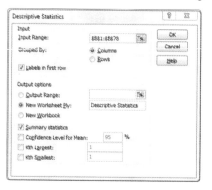

Clicking **OK** should produce the following list of descriptive statistics.

	A	B
1	*Number of FAFSA Applications*	
2		
3	Mean	21611.48
4	Standard Error	3877.837471
5	Median	9747
6	Mode	3
7	Standard Deviation	33583.05761
8	Sample Variance	1127821759
9	Kurtosis	11.48669404
10	Skewness	3.024608257
11	Range	197594
12	Minimum	0
13	Maximum	197594
14	Sum	1620861
15	Count	75

From the output, we can see that the values that we are looking for are as follows.

Minimum value = 0

Maximum value = 197,594

Mean = 21,611.5

Median = 9747

Mode = 3

Range = 197,594

Standard deviation = 33,583.1

Skill Check Answers

1. Mean: 11.7; Median: 11.5; Mode: 11, 12

2. 2.4

8.3   Exercises

Find the mean, median, mode, range, and standard deviation for each data set.

When applicable state whether the data set is: unimodal, bimodal, or multimodal.

Round answers to one more decimal place than the largest number of decimal places given in the data.

All data sets are samples unless stated otherwise.

1. 19, 32, 15, 21, 25, 22, 22, 28, 27, 27, 26

2. $11.40, $32.00, $22.50, $12.01, $10.08, $18.30, $18.40, $32.00

3. 45, 21, 26, 26, 45, 37, 22, 33, 26, 21, 42, 37, 41, 43, 46, 35, 31, 29, 46

4. 310, 310, 310, 310, 310, 310

5. 9, 3, −5, −3, −7, 3, 0, 6, −9, −7, −3, −8

6. The following data represent sample ACT scores from students at a local high school.

ACT Scores	
13	26
10	20
24	30
25	31
6	24
35	35
26	15

7. The following are lengths of each movie in the complete Harry Potter film series. Note that because these include all of the films in the series, this is a population.

Time Lengths for Harry Potter Film Series	
Movie Title	Time (in Minutes)
Harry Potter and the Philosopher's Stone (2001)	152 minutes
Harry Potter and the Chamber of Secrets (2002)	161 minutes
Harry Potter and the Prisoner of Azkaban (2004)	141 minutes
Harry Potter and the Goblet of Fire (2005)	157 minutes
Harry Potter and the Order of the Phoenix (2007)	138 minutes
Harry Potter and the Half-Blood Prince (2009)	153 minutes
Harry Potter and the Deathly Hallows - Part 1 (2010)	146 minutes
Harry Potter and the Deathly Hallows - Part 2 (2011)	130 minutes

8. The following table shows the top 15 busiest airports from January to November 2011.

Top 15 Busiest Airports from January to November 2011	
Airport	Total Passengers
Amsterdam Schiphol Airport	46,213,944
Beijing Capital International Airport	71,284,796
Dallas/Fort Worth International Airport	53,126,399
Denver International Airport	48,402,802
Dubai International Airport	46,287,234
Frankfurt Airport	52,191,355
Hartsfield-Jackson Atlanta International Airport	85,165,259
Hong Kong International Airport	48,587,000
John F. Kennedy International Airport	44,045,938
London Heathrow Airport	63,912,107
Los Angeles International Airport	56,819,805
Madrid Barajas Airport	46,019,110
O'Hare International Airport	61,370,268
Paris Charles de Gaulle Airport	56,254,938
Soekarno-Hatta International Airport	47,513,248
Tokyo International Airport	56,969,971
Source: Wikipedia, s.v. "World's busiest airports by passenger traffic," http://en.wikipedia.org/wiki/World%27s_busiest_airports_by_passenger_traffic	

Use the formula for the mean, $\bar{x} = \dfrac{x_1 + x_2 + \cdots + x_n}{n}$ to find the missing piece of data.

9. John knows that his first 4 tests grades were 84, 79, 82, and 88. Find John's grade on the fifth test if his average was 83.8.

10. A small boat that ferries visitors to a resort island has strict guidelines on the weight allowed for passenger luggage. Consequently the five vacationers are limited to a maximum average luggage weight of 40 pounds (lb). The following are the weights of three out of five pieces of luggage: 39 lb, 32 lb, and 43 lb. The two pieces of luggage that haven't been weighed will have to split the remaining weight allowance. Determine the maximum average possible weight allowance for each remaining bag.

For each data set, determine the most appropriate measure of center.

11. Styles of houses in a suburb: ranch, colonial, bungalow, etc.

12. Grades on the final in Biology 210 at State University.

13. The ratings on a customer satisfaction survey: strongly disagree, disagree, neither agree nor disagree, agree, and strongly agree.

14. Salaries for janitorial staff at the state governmental buildings that include the Director of Sanitation's salary.

Use the empirical rule to answer each question.

15. Although there is some controversy around the precise average body temperature of adults, new data suggest that the mean is 98.2° with a standard deviation of 0.6 and has a bell-shaped distribution.

   a. According to this distribution, approximately what percentage of body temperatures are between 97° and 99.4°?

   b. Approximately what percentage of temperatures are greater than 98.8°?

   c. Approximately what percentage of temperatures are no more than 98.2°?

16. 2011 high school seniors had the following mean and standard deviation on the mathematics portion of the SAT exam. $\mu = 514$ and $\sigma = 117$.

   a. Approximately what percentage of scores were greater than 397 but less than 631?

   b. What two scores have approximately 95% of the data between them?

   c. Approximately what percentage of high school seniors had mathematics scores in the second quartile?

   d. Describe where the best and worst 0.3% of scores lie.

Answer each question thoughtfully.

17. Given the following measures of center, decide the likely shape of the distribution: mean = 22.5, median = 17.0, mode = 17.0.

18. Accounting 101 has five class sections. All five classes took the same final. The mean scores on the final for each class were 72, 78, 76, 74, and 79. Can the mean final score for all students in Accounting 101 be found by averaging the mean scores in each class? Explain your answer.

19. Does the standard deviation of a data set equaling zero imply that all entries in the data set equal zero?

20. Is it possible for a data set to have a standard deviation of –2.5?

21. Explain the difference between Amelia making an 82 on her pre-calculus exam and scoring in the 82nd percentile in mathematics on the ACT test.

Solve each problem.

22. Marcel scored in the 91st percentile on the MCAT (Medical College Admissions Test). The medical school he is applying to only accepts students who score in the top 10% on the MCAT. Did Marcel score well enough to be considered for his school of choice?

23. The five-number summary is a numerical description of data which includes the minimum data point, the maximum data point, and the data points representing quartiles Q_1, Q_2, and Q_3. The following are house prices in one neighborhood.

$181,865 $119,442 $152,750 $100,960 $159,635

$150,963 $133,702 $149,788 $145,495 $182,500

$112,021 $120,900 $145,850 $164,590 $144,413

a. Find the five-number summary of the house prices.

b. What percentage of house prices is at or below $159,635?

c. What is the range of house prices for this neighborhood?

24. The following graph contains a box plot. A box plot is a graphic display of a five-number summary, which was introduced in Exercise 23. The endpoints represent the minimum and maximum data values, while the lines sectioning off the box in the middle represent each of the quartiles as shown in the graph.

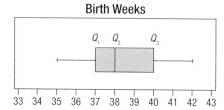

Birth Weeks

a. Based on the box plot, estimate each of the values in the five-number summary.

b. The mid-range is the average of the minimum and maximum data values. Estimate the mid-range from the box plot.

25. Using the information given in Exercise 24, calculate the values needed to construct a box plot for the following data as described in Exercise 24. Sketch a graph of the box plot.

310 320 450 460 470 500 520 540

580 600 650 700 710 840 870 900

1000 1200 1250 1300 1400 1720 2500 3700

26. Given the five-number summary for three data sets, sketch a box plot for each, side-by-side on the same graph. Then answer the following questions based on your box plots.

Committees and the Ages of Members			
	Membership	**Finance**	**Publicity**
Min	23	26	25
Q_1	27	32	26
Q_2	29	38	27
Q_3	33	44	29
Max	35	46	33

   a. Which committee has the largest range of ages?

   b. Which committee has the least variation in the ages?

   c. Which committee has the smallest median?

27. Describe two data sets, one that might have a large variation and one that might have a small variation.

28. Suppose that, in a list of data, 37% of the data are greater than 45. True or False: Q_1 must be greater than 45.

29. Lucas received an e-mail containing the five-number summary for the company sales data that he asked for. Unfortunately, the e-mail cut off the summary labels and scrambled their order. Can you still determine which number is the first quartile, Q_1? Explain your answer.

   five-number summary: 11, 17.5, 9, 13.5, 19

30. Suppose that 110 male students are surveyed and that 52% have a height less than 1.776 m.

   a. True or False: Of those surveyed, the mean height must be under 1.776 m.

   b. True or False: Of those surveyed, the median height must be under 1.776 m.

31. If we know that a salary of $65,300 was in the 67th percentile in a company survey, can we determine how many employees were in the sample? Why or why not?

8.4   The Normal Distribution

As we begin to create a better picture of the way data are dispersed, we realize that, in many instances, the "shape" of the data falls into one of a few recognizable forms. Suppose we looked at the histogram of data collected for the heights of 150 random men. It might look something like the following.

Table 1	
Heights of 150 Random Men	
Class	**Frequency**
4.5'–5'	10
5'–5.5'	44
5.5'–6'	49
6'–6.5'	35
6.5'–7'	12

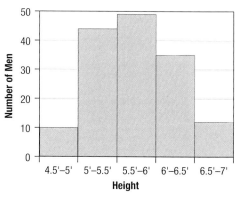

Figure 1

You can see that it's what we might describe as approximately bell-shaped and is roughly symmetrical about the middle of the graph. If we were to take bigger and bigger samples of men's heights, our histogram would begin to look even more symmetrical until, eventually, we wouldn't be able to notice where it wasn't exactly symmetrical anymore. It would become almost perfectly bell-shaped. If we were able to sample the heights of all men, we'd find that our histogram would become what we call the **normal distribution**, which is shown in Figure 2.

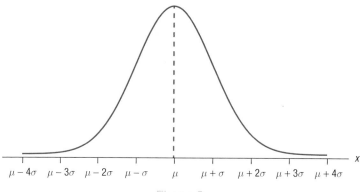

Figure 2

Characteristics of the Normal Distribution

a. It is bell-shaped, meaning it has only one mode that is at the center, and symmetrical.

b. The mean, median, and mode are all the same.

c. The total area under the curve is equal to 1.

d. It is completely defined by its mean and standard deviation.

The last characteristic of the normal distribution tells us is that if we know a normal distribution's mean and standard deviation, we know how the distribution is dispersed, or spread out. Recall the Empirical Rule we looked at in Section 8.3. By knowing the mean and the standard deviation of a given data set, we also know where the data lie in reference to the mean. Recall from the Empirical Rule that 68% of the data will lie within one standard deviation of the mean, 95% of the data will lie within two standard deviations, and 99.7% of the data will lie within three standard deviations of the mean. Remember, the standard deviation tells us how far on average a piece of data is away from the mean. So, the smaller the standard deviation, the closer the data are to one another. The larger the standard deviation, the more dispersed the data. Figure 3 is a picture of several normal distributions with equal means, but different standard deviations. Which distribution do you think has the largest standard deviation?

Various Normal Distributions

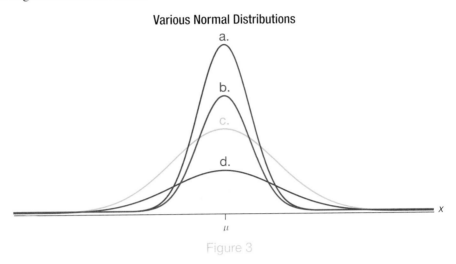

Figure 3

Curve **d.** has the largest standard deviation, because it is the curve that is most spread out. Note that it's not the one with the highest peak.

Because so many natural data sets follow a normal distribution, we are able to use its features to help estimate and analyze entire populations of data. The **z-score** tells us how many standard deviations a particular piece of data lies away from the mean in a normal distribution. So, if z is a positive number, the piece of data is greater than the mean μ; if z is a negative number, the piece of data is less than the mean μ; and if z is equal to 0, the data value equals the mean μ.

---

z-Score

The **z-score** tells how many standard deviations a particular piece of data lies away from the mean in a normal distribution. The formula for finding a z-score is

$$z = \frac{\text{data value} - \text{mean}}{\text{standard deviation}},$$

or more formally with symbols,

$$z = \frac{x - \mu}{\sigma} \text{ for populations and } z = \frac{x - \bar{x}}{s} \text{ for samples.}$$

Let's first try our hand at calculating some z-scores.

☞ Helpful Hint

z-scores are normally rounded to the nearest hundredth.

Example 1: Calculating z-scores

Given that the heights of Canadian women are normally distributed with a mean of 159.5 cm and a standard deviation of 7.1 cm, calculate the z-scores for the following pieces of data.

a. A height of 148.2 cm

b. A height of 160.3 cm

c. A height of 1.7 m

Solution

a. Using a height of 148.2 cm in the z-score formula, we have

$$z = \frac{\text{data value} - \text{mean}}{\text{standard deviation}} = \frac{148.2 - 159.5}{7.1} = \frac{-11.3}{7.1} \approx -1.59.$$

Because the height given is smaller than the mean, we expect the z-score to be negative, which it is. Data points below the mean will always have a negative z-score. What this tells us is that a height of 148.2 cm is 1.59 standard deviations below the mean.

b. The z-score for a height of 160.3 cm is calculated by

$$z = \frac{\text{data value} - \text{mean}}{\text{standard deviation}} = \frac{160.3 - 159.5}{7.1} = \frac{0.8}{7.1} \approx 0.11.$$

This tells us that a height of 160.3 cm, which we know to be only slightly larger than the mean, is 0.11 standard deviations above the mean. Remember, a data point greater than the mean will always have a positive z-score.

c. Before we can use the formula to compute the z-score, we must first convert the data into the same unit of measurement. That is, both should be in either meters or centimeters. Since the previous measurements were in centimeters, we'll change 1.7 m to centimeters. Note that you would get exactly the same z-score if you changed the mean and the standard deviation to meters instead. To change meters to centimeters, simply multiply by 100. So, we have 1.7 m = 170 cm. Substituting into the z-score formula we have

$$z = \frac{\text{data value} - \text{mean}}{\text{standard deviation}} = \frac{170 - 159.5}{7.1} = \frac{10.5}{7.1} \approx 1.48.$$

So, a height of 1.7 m is 1.48 standard deviations above the mean.

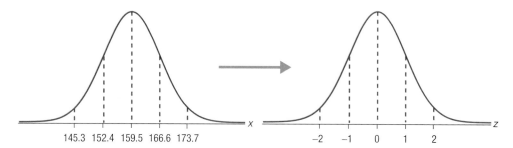

Skill Check #1

Given that the heights of women in Malaysia are normally distributed with a mean of 154.7 cm and a standard deviation of 6.46 cm, calculate the z-score of a Malaysian woman who is 159.1 cm tall.

Recall from Example 10 in Section 8.3 that we compared language arts scores from a standardized test amongst three classmates. Another way to compare their results would be to look at their z-scores. But to do that, we'd need to know their actual scores. Let's change it up a bit and suppose that they all took different standardized tests, but we wanted a way to compare their language arts scores. Suppose Karl's score was 18 on his test, where the mean was a 15 and the standard deviation was 1.4, Asher scored 18 on his test, where the mean was 17 and the standard deviation was 1.9, and Rylie scored 16 on her test, where the mean was 15.46 and the standard deviation was 0.8. To compare the scores, we need to calculate the *z*-score for each.

$$\text{Karl:} \quad z = \frac{\text{data value} - \text{mean}}{\text{standard deviation}} = \frac{18 - 15}{1.4} = \frac{3}{1.4} \approx 2.14$$

$$\text{Asher:} \quad z = \frac{\text{data value} - \text{mean}}{\text{standard deviation}} = \frac{18 - 17}{1.9} = \frac{1}{1.9} \approx 0.53$$

$$\text{Rylie:} \quad z = \frac{\text{data value} - \text{mean}}{\text{standard deviation}} = \frac{16 - 15.46}{0.8} = \frac{0.54}{0.8} \approx 0.68$$

Now, knowing each of the respective *z*-scores, we can see that Karl did the best on his test, even though his raw score was the same as Asher's. Karl's score of 18 was 2.14 standard deviations above the mean. However, Asher and Rylie's scores were fewer standard deviations above their respective means. The larger the *z*-score, the further above the mean a data point lies.

From *z*-Scores to Percentages TECH

Let's consider Karl's score again. He didn't just do a little better than Asher and Rylie, he actually did considerably better on his test. We were told in the last section that he was in the 4th quartile, which means at or above the 75th percentile. Now that we know his *z*-score, we can actually find out exactly what percentage of students he did better than.

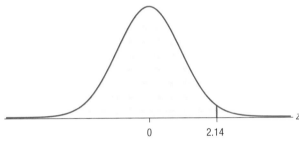

Figure 4

We can use a calculator to find what *percentage* of scores is below a *z*-score of 2.14. With a TI-83/84 Plus calculator, first find the correct distribution function by pressing 2ND and then VARS. Choose option 2: normalcdf(, which stands for the Normal Cumulative Density Function. The format for entering the statistics is normalcdf(lower bound, upper bound). The terms "*lower bound*" and "*upper bound*" define the boundaries of the area under the curve that we wish to find. In Figure 4, we only have an upper boundary of 2.14. There is essentially no lower boundary since we are interested in all data that fall below 2.14.

```
normalcdf(-1E99,
2.14)
        .9838226748
```

Since we want to know the percentage *below z*, we can think of the lower bound as $-\infty$. We cannot enter $-\infty$ into the calculator, so we will enter a very small value for the lower endpoint, such as -10^{99}. This number appears as -1E99 when entered correctly into the calculator. To enter -1E99, use the following keystrokes (-) 1 2ND , 9 9. Enter normalcdf(-1E99, 2.14). The percentage is 0.9838. In other words, 98.38% of the data falls to the left of a *z*-score of 2.14. We now know that Karl's score was better than 98.38% of students taking that test.

It is also possible to use Wolfram|Alpha to find the percentage of scores below a certain value. To find the percentage of scores below $z = 2.14$, go to www.wolframalpha.com and type "z-score calculator" into the input field and then click the = button. Wolfram|Alpha will return a worksheet with a field to enter the endpoint value. Type "2.14" into the endpoint field and then click the = button. Wolfram|Alpha will return the following. [1]

---

1 Wolfram Alpha LLC, http://www.wolframalpha.com

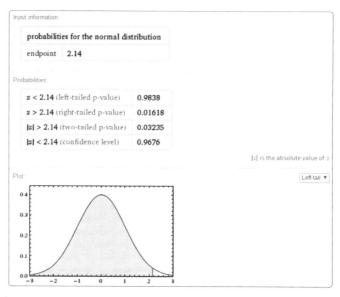

Notice that the left-tailed *p*-value, right-tailed *p*-value, and two-tailed *p*-value are provided. We are interested in finding the percentage of scores below $z = 2.14$, so we use the left-tailed *p*-value answer of 0.9838 or 98.38%.

Since one of the properties of the normal distribution is that the total area under the curve is equal to one, the percentage of data points that lie below a *z*-score together with the percentage of data points that lie above it always add up to 100%. Therefore, with Karl's *z*-score, we also know that $100\% - 98.38\% = 1.62\%$ of students did better than Karl.

```
normalcdf(1.34,1
E99)
      .0901227339
```

Let's look at an example where we find the percentage of data points above a *z*-score. Suppose the score we are interested in is $z = 1.34$. We know the lower bound of the interval is 1.34 and the upper bound is ∞. Once again, we cannot enter ∞ into the calculator, so we'll just use a very large number, say 10^{99}. This is the same entry in the calculator as before ($-1E99$) but without the negative sign. So, we have normalcdf(1.34, 1E99). The percentage of data points above $z = 1.34$ is 0.0901 or 9.01%. When using Wolfram|Alpha to find the percentage above a *z*-score, use the right-tailed *p*-value answer.

If we'd like to know the percentage of data points between two *z*-scores, we simply use both *z*-scores as our lower and upper bounds, making sure to enter them in the correct order. For example, to find the percentage of data between a *z*-score of −0.98 and a *z*-score of 2.01, we enter normalcdf(-0.98, 2.01) and find that 0.8142 or 81.42% of the data are between these *z*-scores.

Using Wolfram|Alpha, go to www.wolframalpha.com and type "*z*-score calculator" into the input field and then click the = button. Wolfram|Alpha will return a worksheet. Click "left end point and right end point" under the input field. This will return another worksheet with an input field for the left endpoint and the right endpoint. Type "−0.98" into the left endpoint input field and "2.01" into the right endpoint input field, then click the = button. Wolfram|Alpha will return the following. [2]

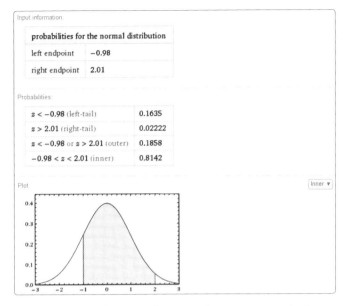

We are interested in the percentage between the two values, so we use the inner probability answer of 0.8142 or 81.42%.

Summary for Using the TI-83/84 Plus Calculator

Percentage of data points **below** z: `normalcdf (–1E99, z)`

Percentage of data points **above** z: `normalcdf (z, 1E99)`

Percentage of data points **between** two z-scores: `normalcdf (z1, z2)`

TECH Example 2: Putting It All Together

Suppose that the average caloric intake for women is 2050 calories per day, with a standard deviation of 175 calories. If we assume that caloric intake follows a normal distribution, find the percentage of females in your class that consume more than 2000 calories per day.

Solution

The first thing we need to do is find a z-score for the data point we are interested in of 2000 calories. Substituting into the formula, we have the following:

$$z = \frac{2000 - 2050}{175} \approx -0.29.$$

Now, because we're interested in knowing the percentage that consumes more than 2000 calories, we let −0.29 be the lower bound. Our upper bound in this case is ∞. That gives us `normalcdf(-0.29, 1E99)`. Calculating this using a TI-84 Plus gives a result of approximately `0.6141`. This means that 61.41% of females consume more than 2000 calories per day.

Skill Check Answer

1. $z \approx 0.68$

8.4   Exercises

Calculate the standard score for each given value. Round your answer to the nearest hundredth.

1. $\mu = 57, \sigma = 11$

    a. $x_1 = 63$

    b. $x_2 = 38$

    c. $x_3 = 58$

2. $\bar{x} = 1123, s = 241$

    a. $x_1 = 1284$

    b. $x_2 = 900$

    c. $x_3 = 1364$

    d. $x_4 = 1123$

3. $\bar{x} = 3.19, s = 0.06$

    a. $x_1 = 3.13$

    b. $x_2 = 3.22$

    c. $x_3 = 3.00$

4. $\mu = 178.15, \sigma = 49.3$

    a. $x_1 = 73.9$

    b. $x_2 = 267.3$

    c. $x_3 = 199.5$

5.   Scores on a test have a mean of 73 and a standard deviation of 11. Steve has a score of 68. Convert Steve's score to a z-score.

Answer each question thoughtfully.

6.   The annual rainfall in a town has a mean of 47.22 inches and a standard deviation of 10 inches. Last year there was 51 inches of rain. How many standard deviations from the mean is that?

7.   Mason's weekly poker winnings have a mean of $144 and a standard deviation of $51. Last week he won $165. How many standard deviations from the mean is that?

8.   The mean score for a set of data is marked by the dotted line on the following graph. Which value is a likely z-score for the indicated value? Choose from **a.** −2.1, **b.** 0, or **c.** 2.7.

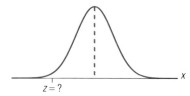

9.   Ava scored a 92 on a test with a mean of 71 and a standard deviation of 15. Charlotte had a score of 688 on a test with a mean of 493 and a standard deviation of 150. Which score was better with respect to their test?

10.   Avery started training to run a 5K. Her first race was a 5K for charity. She finished in 37.3 minutes. The average race time for the charity run was 36.42 with a standard deviation of 1.73 minutes. In her second race, Avery finished in 36.5 minutes. The race had a mean time of 33.02 minutes with a standard deviation of 2.45 minutes. In which race did Avery place higher in the list of finishers?

Use the *z*-score formula to complete each table.

11. Find the missing value in each row of the table. Round answers to the same number of decimal places given in the table.

	z	*x*	μ	σ
a.		82.1	74.0	6.3
b.	1.05	162.3		8.9
c.	3.04		34.5	5.02
d.	−2.73	379	634	

12. Find the missing value in each row of the table. Round answers to the same number of decimal places given in the table.

	z	*x*	μ	σ
a.		4.33	6.10	2.04
b.	−2.39	−57		139.8
c.	0.58		118	21.2
d.	2.78	68	43	

Find the percentage of data points that lie below each *z*-score.

13. $z = -0.19$

14. $z = 1.46$

15. $z = 3.07$

16. $z = -2.22$

17. $z = 0$

Find the percentage of data points that lie above each *z*-score.

18. $z = 1.03$

19. $z = -1.87$

20. $z = -3.10$

21. $z = 2.84$

22. $z = 0$

Find the percentage of data points that lie between each pair of *z*-scores.

23. $z_1 = -1.00$
$z_2 = 1.00$

24. $z_1 = -2.40$
$z_2 = 1.73$

25. $z_1 = 2.00$
$z_2 = 3.00$

26. $z_1 = -3.01$
$z_2 = -0.56$

27. $z_1 = 0$
$z_2 = 2.61$

Find the percentage of data points that lie below z_1 and above z_2.

28. $z_1 = -1.10$
$z_2 = 1.10$

29. $z_1 = -2.84$
$z_2 = 2.84$

30. $z_1 = -1.75$
$z_2 = 0.53$

31. $z_1 = 1.09$
$z_2 = 2.88$

32. $z_1 = -0.01$
$z_2 = 0.02$

Solve each problem.

33. The average IQ score for adults is 100 with a standard deviation of 15. Assume that the distribution of IQ scores is approximately normal.

   a. Find the percentage of adults who have an IQ score less than 90.

   b. Find the percentage of adults who have an IQ score which exceeds the mean by at least 15 points.

   c. Find the percentage of adults who have an IQ score between 100 and 120.

   d. Find the percentage of adults who have an IQ score less than 55 or more than 145.

34. Assume the average weights of offensive linemen in the NFL follow a normal distribution with a mean of 300 pounds and a standard deviation of 12.3 pounds.

   a. Find the percentage of linemen in the NFL who weigh more than 320 pounds.

   b. Find the percentage of linemen in the NFL who weigh between 275 and 325 pounds.

   c. Find the percentage of NFL linemen who weigh at least 260 pounds.

   d. Find the percentage of NFL linemen who weigh at most 315 pounds.

35. What is the minimum z-score that a piece of data would need to have in order to be in the top 10% of a set of data?

36. What is an "average" z-score? Explain your answer.

37. What z-score represents the 1st quartile? 2nd quartile? 3rd quartile?

8.5   Linear Regression

The last bit of data analysis we're going to look at involves evaluating the relationship between two variables. For example, we could compare the heights and weights of a sample of teenage boys, or the crime rate of small communities to the size of local police stations in those communities. Just as before, let's start by visually considering the data.

Scatter Plots

> ### Scatter Plot
>
> A **scatter plot** is a graphical display that is most commonly used to show two variables and how they might relate to one another. It is a graph on the coordinate plane that contains one point for each pair of data values.

Consider the scatter plots shown in Figure 1. Scatter plots **A**, **B**, and **C** have visible relationships while **D** and **E** don't display any obvious patterns.

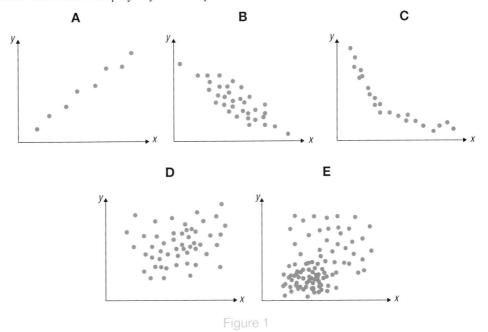

Figure 1

Scatter plots like the ones in Figure 1 help us to identify any trends in the data. If the trend seems to follow the pattern of a straight line, as in scatter plots **A** and **B**, there is said to be a **linear relationship** between the variables. Scatter plot **C** has a visible pattern; it's just not linear.

Consider the relationship between the number of children in a household and the number of bedrooms in the house. We can make a scatter plot that shows one point for each household, with the horizontal axis representing the number of children and the vertical axis the number of bedrooms, as shown in Figure 2.

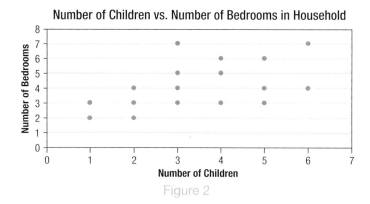

Figure 2

Although there seems to be a linear relationship between the number of children in a household and the number of bedrooms, it doesn't seem to be a very strong one. Both the direction and the position of the points, that is, how close the points are to lying in a straight line, tell us information about the relationship. When data are plotted in a scatter plot and the points seem to lie in a linear pattern, we say variables represented by the data are **correlated**.

Correlation

When data are plotted in a scatter plot and there appears to be an upward trend, that is, as one variable increases the other increases as well, we say there is a **positive correlation** between the variables. If the scatter plot trends downward, that is, as one variable increases the other decreases, we say there is a **negative correlation** between the variables.

Example 1: Identifying Correlations

Consider the relationship between the following variables and what kind of correlation might show up in a scatter plot of the data. Decide if the variables would likely have a positive correlation, negative correlation, or no linear correlation.

a. The number of cigarettes smoked and the probability of lung cancer

b. The number of minutes spent on social media sites by college students and their first semester grades

c. The amount of credit card debt incurred by college freshmen and their IQ score

Solution

a. As the number of cigarettes smoked increases, so does the chance of lung cancer. Thus, the scatter plot is likely to have upward-trending data points. The variables would be positively correlated.

b. As the number of minutes (or hours) spent on social media sites increases, your grades are likely to decrease. This would result in a downward-trending scatter plot and a negative correlation between the amount of time spent on social media sites and grade point average.

c. The scatter plot for these variables would likely contain a wide range of credit card debt and a wide range of IQ scores. It would be unlikely that there is a linear relationship between these two variables. Thus, they are neither positively or negatively correlated.

Correlation Coefficient [TECH]

As well as the direction, we can talk about the strength of a linear correlation by calculating what is called the **Pearson correlation coefficient**.

Pearson Correlation Coefficient

The **Pearson correlation coefficient**, rounded to the nearest thousandth, is a value between −1 and 1 that measures the strength of a linear correlation. For a sample, it is represented by the variable r.

The stronger the correlation, the closer the correlation coefficient is to either −1 or 1. If there is a very strong positive correlation, r will be close to 1. Conversely, if there is a strong negative correlation r will be close to −1. The closer r is to 0, the less correlation between the variables. A correlation coefficient of 0 means that there is no linear relationship between the variables at all.

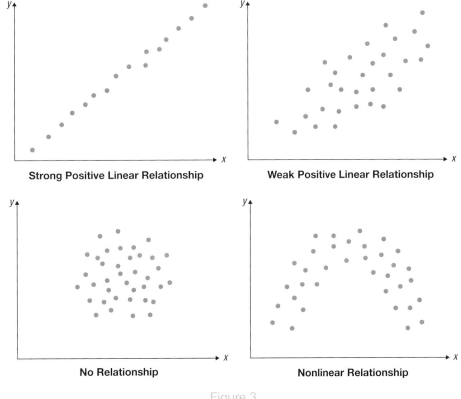

Figure 3

Using a TI-84 Plus calculator, we can easily calculate r, as well as some other variables we'll look at later in the section, to help us build a prediction model. Because the calculator does this simultaneously, we'll point these out as we go along. Begin by pressing [STAT] and then 1: Edit and enter the values for one variable in L1 and the values for the other variable in L2. Then press [STAT], scroll over to CALC, and select option 4: LinReg(ax+b). Press [ENTER] twice. The output will include the correlation coefficient r.

Let's try an example with actual data.

TECH Example 2: Using a TI-84 Calculator to Find the Pearson Correlation Coefficient

The following is a small sample of data collected from male participants in a 161 km trail ultramarathon. The survey collected the BMI (Body Mass Index) of each participant and their age. A table of the data, along with the scatter plot of the data, is given. Use your calculator to find the correlation coefficient r between the variables.

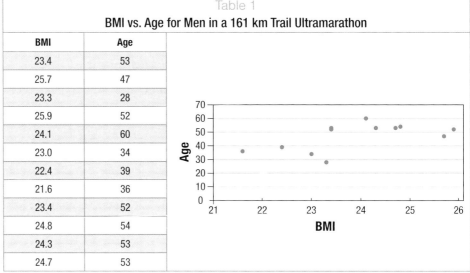

Table 1
BMI vs. Age for Men in a 161 km Trail Ultramarathon

BMI	Age
23.4	53
25.7	47
23.3	28
25.9	52
24.1	60
23.0	34
22.4	39
21.6	36
23.4	52
24.8	54
24.3	53
24.7	53

```
LinReg
  y=ax+b
  a=4.473199484
  b=-60.08491435
  r²=.3321307024
  r=.5763078191
```

Solution

Begin by pressing STAT and then 1: EDIT. Enter the values for the BMI's in L1 and the values for the ages in L2. Then press STAT and choose CALC and option 4: LinReg(ax+b). Press ENTER twice. The output should appear as shown in the screen shot in the margin.

Since $r = 0.5763078191$, we know that there is a weak positive correlation between the male participant's BMI and their age.

TECH TRAINING

A TI-30XIIS/B can be used to calculate the correlation coefficient from Example 2. Begin by pressing 2ND STAT, select 2-Var, then press ENTER. Press DATA. Enter the first x-value for X₁, press the down arrow key, and then enter the first y-value for Y₁. Continue to press the down arrow key and enter the values until all of the data are entered, then press ENTER.

Press STATVAR and scroll to the right until the value for r is displayed.

Significance   TECH

The correlation coefficient r can also help us determine if the relationship between two variables is statistically significant. In general, we say that if the correlation coefficient for a particular set of data is less than 0.5, then there is no significant correlation between the two variables. As we will see later, if the sample size is at least 30, we can actually be fairly confident of a linear correlation if $r = 0.5$.

So how do we know? By using critical values for the Pearson correlation coefficient, we can determine how large r needs to be for the relationship to be statistically significant. All we need is the sample size and a level of confidence we want to achieve. The **level of confidence** c is the probability that the assertions made about the data are in fact correct. The **level of significance** α determines the probability that we are wrong in our assertions made about the data. The level of confidence c and the level of significance α are related, in that together they sum to one ($c + \alpha = 1$).

Level of Confidence and Level of Significance

The **level of confidence c** is the probability that the assertions made about the data are correct.

The **level of significance α** is the probability that the assertions made about the data are incorrect.

$$c + \alpha = 1.$$

In other words, if $\alpha = 0.05$, then $c = 1 - \alpha = 0.95$. This means there is a 5% probability that although we think there is a correlation between the two variables, they are actually not related at all. It also implies that, there is a 95% probability that the variables are correlated. If $\alpha = 0.01$, then there is only a 1% probability that the results occurred by chance, and our level of confidence is $c = 0.99$. Table A in Appendix A gives us the critical values for r for both $\alpha = 0.05$ and $\alpha = 0.01$. A portion of Table A is shown in Table 2.

Table 2		
Critical Values of the Pearson Correlation Coefficient		
n	$\alpha = 0.05$	$\alpha = 0.01$
4	0.950	0.990
5	0.878	0.959
6	0.811	0.917
7	0.754	0.875
8	0.707	0.834
9	0.666	0.798
10	0.632	0.765
11	0.602	0.735
12	0.576	0.708

Statistically Significant

If $|r|$ is greater than the critical value listed in the table, then r is **statistically significant**, which means it is unlikely to have occurred by chance.

Example 3: Identifying Statistical Significance

Use the critical values in Table 2 to determine if the correlation between BMI and age in the previous example is statistically significant. Recall that $r = 0.5763078191$. Use a 0.05 level of significance.

Solution

There are twelve pairs of data in Example 2, so $n = 12$. By looking along the row where $n = 12$ and down the column where $\alpha = 0.05$, the critical value is 0.576. Comparing this critical value to the correlation coefficient we found for the data in the previous example, we have $|r| \approx 0.5763 > 0.576$. So the linear relationship between the variables is statistically significant at the 0.05 level of significance. Therefore, we have enough evidence to conclude that a linear relationship exists between BMI and age for male ultramarathoners.

When we do find data that prove to have a statistically significant r, even at the 0.01 level of significance, we need to be careful to not assign causation to the situation. When two variables are statistically correlated, we are often tempted to infer that one thing caused the other to happen. For instance, we might find that the number of watermelons consumed is positively correlated with the number of drownings that occur. This in no way means that eating a watermelon causes you to drown. Instead it might indicate that during the hotter months of the year, more people eat watermelons (which are in season) while at the same time the number of drownings increases simply because more people go swimming when it's hot outside.

Regression Lines $\boxed{TECH}$

Once we've established that there actually *is* a statistically significant correlation between two variables, we can use the **regression line**, or **line of best fit**, to help us make predictions.

> ## Regression Line
>
> The **regression line**, also known as the **line of best fit**, is a particular line that most closely "fits" the data points on the scatter plot. The regression line can be represented by
>
> $$\hat{y} = ax + b,$$
>
> where a is the slope of the line and b is the y-intercept.

Certainly, you have seen the equation of a line in slope-intercept form, such as $y = mx + b$, or $y = b + mx$, where m is the slope and b is the y-intercept. The regression line is similar in form. You can determine a regression line using a graphing calculator, such as a TI-83/84, or a statistical package, such as that in Microsoft Excel. The letters a and b are used to represent the slope and y-intercept, respectively.

Consider the following data for newborn male babies. The table shows each baby boy's birth weight (in grams) versus his gestational age (in completed weeks).

Table 3	
Birth Weight and Gestational Age for Baby Boys	
Gestational Age (Weeks)	**Birth Weight (Grams)**
22	401
26	908
26	686
31	1259
31	1698
31	2209
33	2127
37	2384
37	2552
37	3080
38	3665
39	2701
39	4049
39	3465
39	2942
40	3613
41	4328
41	3179
41	3733
42	3851

```
LinReg
 y=ax+b
 a=187.9457364
 b=-4030.573643
 r²=.8796609889
 r=.937902441
```

Be sure to enter the value of the gestational ages in the L1 column (inputs) and the birth weights in the L2 column (outputs) when you enter the data values in the calculator. Then, we will run the linear regression to find the line of best fit. In the calculator screen shot in the margin, we can see the values of r, a, and b that were calculated from the data.

Using the Pearson Correlation Coefficient Table of Critical Values in Appendix Table A, you can see that r is most definitely statistically significant. In fact, as long as the sample size is bigger than 5, r is significant at both the 95% and 99% confidence levels. Check for yourself that the r we calculated is larger than any critical r in the table, as long as the sample size is larger than 5.

Knowing that r is significant, we can write the regression line by simply substituting the calculated values for the slope, $a = 187.94573643$, and y-intercept, $b = -4030.573643$, to get

$$\hat{y} = 187.946x - 4030.574 \text{ (rounded to thousandths).}$$

So, given a value for x in this example, we can predict what the value for y will be by substituting the given value for the variable x into the regression line equation. Let's try it. Predict the weight of a baby boy if he is born after 40 complete weeks; that is, find the value of y when x is 40.

Evaluating the equation, we have

$$\hat{y} = 187.946x - 4030.574$$
$$= 187.946(40) - 4030.574$$
$$= 7517.84 - 4030.574$$
$$= 3487.266.$$

So, when x is 40, we can predict that y will be 3487.266. In other words, we can predict that a baby boy born at 40 weeks will weigh about 3487 grams.

As with all predictions, you should be careful not to get too carried away. We've already stated that you should only use the regression line for predictions if you find r to be statistically significant. You should also make sure that you are predicting for values that are within the range and population of the original sample data. In other words, don't try to predict something that is either way out of the scope of the original data or not from the same type of population the sample was drawn from. For instance, it would not be wise to try to predict a baby's weight for a gestational age of 19 weeks. Nor would it be reasonable to apply these results to baby girls, which are a different population.

[TECH] Example 4: Predictions Using the Regression Line

A recent study sought to find if any correlation existed between childhood weight and self-esteem. Children between the ages of 9 and 11 were surveyed. Weight was measured in pounds, while self-esteem was measured based on the average of the answers to 15 questions asking the participant to rate themselves on a scale of 1 to 5—where higher scores mean higher self-esteem. Here's the data for 12 cases.

Table 4	
Self-Esteem in Children	
Weight	**Self-Esteem**
54.3	2.6
62.0	4.6
88.1	1.0
61.8	3.7
69.0	4.7
55.8	2.9
77.2	1.5
66.4	4.8
63.3	3.9
75.2	1.1
79.0	1.9
68.3	2.9

a. Is r statistically significant at the 0.05 level? How about the 0.01 level?

b. Write down the linear regression line in the form $\hat{y} = ax + b$.

c. If appropriate, predict the value for the level of self-esteem given that the child weighs 64.1 pounds.

d. If appropriate, predict the value for the level of self-esteem given that the child weighs 45.0 pounds.

e. If appropriate, predict the value for the level of self-esteem given that the 13-year-old weighs 88.0 pounds.

Solution

```
LinReg
 y=ax+b
 a=-.0841824328
 b=8.721938986
 r²=.3661036892
 r=-.6050650289
```

a. Begin by putting both sets of data into your calculator as shown in the earlier examples. Input the weights in L1 and the self-esteem score in L2. After having the calculator compute the statistics for us, we can see that $r = -0.6051$. Looking at the Pearson correlation coefficient critical value table with $n = 12$ and $\alpha = 0.05$, we see that $|r| = 0.6051 > 0.576$, and is therefore statistically significant. However, with $\alpha = 0.01$, $|r| = 0.6051 < 0.708$, and hence is not statistically significant.

b. Using the values from the calculator which were calculated in the previous step, we know that the slope is $a = -0.084$ and the y-intercept is $b = 8.722$. So, the linear regression line, or line of best fit, is $\hat{y} = -0.084x + 8.722$.

c. Because the data are statistically significantly correlated at the 0.05 level, it is appropriate for us to consider the linear regression for predictions. Substituting $x = 64.1$ into the equation of the line written in the previous step, we have the following.

$$\hat{y} = -0.084x + 8.722$$
$$= -0.084(64.1) + 8.722$$
$$\approx -5.3844 + 8.722$$
$$= 3.3376$$

So, when a child's weight is 64.1 pounds, we can predict that his or her self-esteem score would be around 3.3.

d. The weight is outside of the range of the original data since it is smaller than any of the other data pieces, so it is not appropriate to use the regression line for prediction.

e. Once again it is not appropriate to use the regression line for predictions in this case. The study included only children between the ages of 9 and 11. A 13-year-old is not in the same population as the study and cannot be assumed to have the same characteristics.

Skill Check Answer

1. Answers will vary. Examples may include: The number of candy bars consumed daily and weight gain or the distance between two locations and the length of time it takes to drive between the two.

8.5   Exercises

In each scatter plot, determine whether there appears to be a positive linear correlation, a negative linear correlation, or no linear correlation.

1.
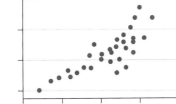

2.
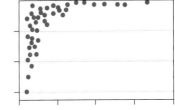

3.
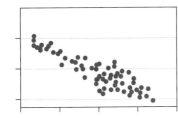

4.
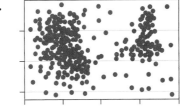

Consider each set of variables and predict whether the variables would have a weak negative relationship, a strong negative relationship, a weak positive relationship, a strong positive relationship, or no relationship at all.

5. Body weight and hours of exercise per week

6. A person's height and their self-esteem

7. Vision ability and IQ

8. Number of hours spent studying for a test and the grade on the test

Determine whether each correlation coefficient is statistically significant at the specified level of significance for the given sample size.

9. $r = 0.703$, $\alpha = 0.01$, $n = 12$

10. $r = 0.403$, $\alpha = 0.05$, $n = 25$

11. $r = 0.378$, $\alpha = 0.05$, $n = 29$

12. $r = 0.809$, $\alpha = 0.01$, $n = 8$

For each data set, find the following.

a. Estimate the correlation in words as positive, negative, or no correlation.

b. Calculate the correlation coefficient r. Round your answer to the nearest thousandth.

c. Determine whether r is statistically significant at the 0.01 level of significance.

13. The following table gives the number of hours a student watches TV per week and his or her overall GPA.

Hours of TV Per Week and Overall GPA								
TV Hours	20	10	25	15	14	13	21	9	5
GPA	2.0	2.46	2.3	2.9	3.0	3.2	3.5	3.3	3.7

14. The following table gives a sample of annual income and number of years of education.

Annual Income and Years of Education						
Annual Income	$21,000	$39,000	$40,000	$39,500	$42,000	$55,500
Years of Education	12	12	14	16	16	16
Annual Income	$61,000	$45,000	$100,000	$142,000	$240,000	$205,000
Years of Education	17	16	16	20	22	21

15. The following table shows the blood pressure reading and the stress test score for 20 adults.

| Blood Pressure Reading and Stress Test Score ||
Stress Test Score	Blood Pressure Reading
51	67
59	66
62	71
63	76
64	73
68	77
71	77
70	76
72	80
82	82
78	79
79	83
83	81
84	83
88	85
87	90
89	82
91	80
90	86
90	88

16. The following table shows the heights of identical twins in centimeters.

| Heights of identical twins ||
Sibling 1	Sibling 2
110.5	109.5
116.6	115.6
122.6	121.6
128.2	127.4
133.5	133.5
138.8	140.2
145.0	146.7
152.3	151.9
159.6	155.0
165.1	156.6
168.3	157.1
169.9	157.6
170.7	158.0

Use the linear regression model $\hat{y} = ax + b$, to predict the y-value for each value of x.

17. $\hat{y} = 28.01x + 17.83$

    a. $x = 21$

    b. $x = 31$

    c. $x = 40$

18. $\hat{y} = -16.5x + 230.55$

    a. $x = 5$

    b. $x = 13$

    c. $x = 35$

Solve each problem.

19. The following table gives the data for the number of cigarettes women smoked in their third trimester of pregnancy and the number of nonviolent crime arrests for their male babies.

Number of Cigarettes and Number of Arrests for Sons									
# of Cigarettes	0	5	3	10	22	19	30	15	8	12
# of Arrests	1	4	0	5	9	12	10	0	4	9

    a. Determine the regression line $\hat{y} = ax + b$. Round the slope and y-intercept to the nearest thousandth.

    b. Determine if the regression equation is appropriate, at the 0.05 level of significance, to use for making predictions. If so, answer part **c.**

    c. If a mother smokes eight cigarettes in her third trimester, make a prediction for the number of times her son will be arrested for a nonviolent crime, if appropriate.

20. The following table shows students' test grades on the first two tests in an introductory literature class.

Test Grades in Introductory Literature Class											
Test 1 (x)	61	45	71	81	89	55	84	91	95	59	77	88
Test 2 (y)	67	79	68	80	87	68	87	90	97	71	77	74

    a. Determine the regression line $\hat{y} = ax + b$. Round the slope and y-intercept to the nearest thousandth.

    b. Determine if the regression equation is appropriate, at the 0.05 level of significance, to use for making predictions. If so, answer part **c.**

    c. If a student scored a 70 on his first test, make a prediction for his score on the second test, if appropriate.

21. The following shows the results on evaluations measuring self-esteem and perceived family support from 10 adolescents.

Self-Esteem and Perceived Family Support Evaluation Results										
Self-Esteem	30	31	31	28	27	26	15	32	27	33
Family Support	13	13	19	21	8	4	10	12	7	17

a. Determine the regression line $\hat{y} = ax + b$. Round the slope and y-intercept to the nearest thousandth.

b. Determine if the regression equation is appropriate, at the 0.05 level of significance, to use for making predictions. If so, answer part **c.**

c. If an adolescent had a self-esteem score of 22, make a prediction for his perceived family support score, if appropriate.

22. A medical equipment company wishes to show that a new device works with the same degree of accuracy and precision as an earlier model to perform an electrocardiogram. One of the measurements tested was the change in radio electric waves during a cardiac cycle. The following results were collected from both healthy adults and those with cardiovascular problems.

Change in Radio Electric Waves During Cardiac Cycle	
# of 5 mm Squares Between R Waves	
Old	New
2	2
3	3
4	4.5
3	3
6	6
4	4.5
3	3
5	5
3	3.5
2	2
6	6
4	4
6	6
5	5
3	3
2	2

a. Determine the regression line $\hat{y} = ax + b$. Round the slope and y-intercept to the nearest thousandth.

b. Determine if the regression equation is appropriate, at the 0.01 level of significance, to use for making predictions. If so, answer part **c.**

c. If the old machine had a reading of 5.5, make a prediction for the new machine reading, if appropriate.

Chapter 8   Summary

Section 8.1   Collecting Data

Definitions

Population

A population is the particular group of interest in a study.

Data

Data is information collected for a study.

Census

A census involves collecting data from every member of the population.

Parameter

A parameter is the numerical description of a particular population characteristic.

Sample

A sample is a subset of a population.

Sample Statistic

A sample statistic is the numerical description of a characteristic of a sample.

Representative Sample

A representative sample is one that has the same relevant characteristics as the population and does not favor one group of the population over another.

Random Sample

A random sample is one in which every member of the population has an equal chance of being selected.

Stratified Sample

A stratified sample is one in which members of the population are divided into two or more subgroups, called strata, that share similar characteristics like age, gender, or ethnicity. A random sample from each stratum is then drawn.

Cluster Sample

A cluster sample is one chosen by dividing the population into groups, called clusters, that are each similar to the entire population. The researcher then randomly selects some of the clusters. The sample consists of the data collected from every member of each cluster selected.

Systematic Sample

A systematic sample is one chosen by selecting every n^{th} member of the population.

Convenience Sample

A convenience sample is one in which the sample is "convenient" to select. It is so named because it is convenient for the researcher.

Section 8.2   Displaying Data

Definitions

Frequency Distribution

A frequency distribution is a count of every member of the data set and how often each value occurs, that is, the frequency of the data value.

Relative Frequency

The relative frequency of a class is the percentage of all data that fall into that particular class. It can be displayed as a percentage or a fraction.

Grouped Frequency Distribution

In a group frequency distribution, the data is grouped into different categories, called classes. Each class must have the same width, called the class width, and the classes cannot overlap. The lowest value in a class is called the lower class limit. The highest value in a class is called the upper class limit.

Pie Chart

A pie chart shows how large each category is in relation to the whole; that is, it uses the relative frequency distribution to divide the "pie" into different-sized wedges.

Bar Graph

A bar graph is a chart of categorical data in which the height of the bar represents the amount of data in each category.

Side-by-Side Bar Graph

A side-by-side bar graph is a bar graph in which the bars are placed next to one another to show the similarities and/or difference between populations.

Stacked Bar Graph

A stacked bar graph is a bar graph in which bars from each population are stacked on top of one another. This bar graph not only compares different categories, but how the whole of each category is broken down.

Histogram

A histogram is a special type of bar graph that displays the frequency distribution of numerical classes. The height of each bar in a histogram represents the frequency of the corresponding class. Because the bars represent classes, and class limits are consecutive numbers, the bars in a histogram touch.

Line Graph

A line graph shows how data changes over time. The horizontal axis of a line graph represents time and the vertical axis represents the variable being measured. Each point on a line graph represents a data value, and then the points are joined together with line segments so changes over time are more easily observed.

Section 8.3   Describing and Analyzing Data

Definitions

Arithmetic Mean

The mean is the sum of all of the data values divided by the number of data points.

Median

The median of a data set is the middle value in an ordered array of the data.

Mode

The mode is the value in the data set that occurs most frequently.

Outlier

An outlier is a data value that is extreme compared with the rest of the data values in the set.

Range

The range is the difference between the largest and smallest values in the data set, which tells you the distance covered on the number line between the two extremes.

Standard Deviation

The standard deviation is a measure of how much we might expect a member of the data set to differ from the mean.

The Empirical Rule

When the distribution of a set of data is approximately bell-shaped, we can estimate the percentage of data values that fall within a few standard deviations of the mean in the following way:

Approximately 68% of all data points lie within 1 standard deviation above and below the mean.

Approximately 95% of all data points lie within 2 standard deviations above and below the mean.

Approximately 99.7% of all data points lie within 3 standard deviations above and below the mean.

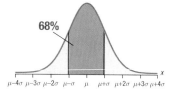 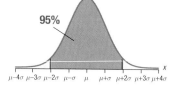 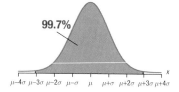

Percentile

Percentiles divide the data into 100 equal parts and tell you approximately what percentage of the data lies at or below a given value.

Quartiles

Q_1 = First Quartile = 25th percentile, that is, 25% of the data is less than or equal to this value.

Q_2 = Second Quartile = 50th percentile, that is, 50% of the data is less than or equal to this value.

Q_3 = Third Quartile = 75th percentile, that is, 75% of the data is less than or equal to this value.

By definition, Q_2 will be the same as the median.

Formulas

Arithmetic Mean

The formula for the population mean is

$$\mu = \frac{x_1 + x_2 + \cdots + x_N}{N}$$

and the formula for the sample mean is

$$\bar{x} = \frac{x_1 + x_2 + \cdots + x_n}{n}.$$

Standard Deviation

The formula for finding the population standard deviation is

$$\sigma = \sqrt{\frac{\sum (x_i - \mu)^2}{N}}.$$

For a sample, the standard deviation is

$$s = \sqrt{\frac{\sum (x_i - \bar{x})^2}{n-1}}.$$

Section 8.4   The Normal Distribution

Definitions

Characteristics of the Normal Distribution

a. It is bell-shaped, meaning it has only one mode that is at the center, and symmetrical.

b. The mean, median, and mode are all the same.

c. The total area under the curve is equal to 1.

d. It is completely defined by its mean and standard deviation.

z-Score

The z-score tells us how many standard deviations a particular piece of data lies away from the mean in a normal distribution.

Formula

z-Score

The formula for finding a z-score is $z = \dfrac{\text{data value} - \text{mean}}{\text{standard deviation}}$, or more formally with symbols,

$$z = \frac{x - \mu}{\sigma} \text{ for populations} \quad \text{and} \quad z = \frac{x - \bar{x}}{s} \text{ for samples.}$$

Section 8.5   Linear Regression

Definitions

Scatter Plot

A scatter plot is a graphical display that is most commonly used to show two variables and how they might relate to one another. It is a graph on the coordinate plane that contains one point for each pair of data values.

Correlation

When data are plotted in a scatter plot and there appears to be an upward trend, that is, as one variable increases the other increases as well, we say there is a positive correlation between the variables. If the scatter plot trends downward, that is, as one variable increases the other decreases, we say there is a negative correlation between the variables.

Pearson Correlation Coefficient

The Pearson Correlation Coefficient, rounded to the nearest thousandth, is a value between -1 and 1 that measures the strength of a linear correlation. For a sample, it is represented by the variable r.

Level of Confidence

The level of confidence c is the probability that the assertions made about the data are correct.

Level of Significance

The level of significance α is the probability that the assessment of the data is incorrect.

$$c + \alpha = 1$$

Statistical Significance for *r*

If $|r|$ is greater than the critical value listed in the Pearson Correlation Coefficient table, then *r* is statistically significant, which means it is unlikely to have occurred by chance.

Regression Line

The regression line, also known as the line of best fit, is a particular line that most closely "fits" the data points on the scatter plot. The regression line can be represented by

$$\hat{y} = ax + b,$$

where *a* is the slope of the line and *b* is the *y*-intercept.

Chapter 8 Exercises

In each scenario, identify the population, the sample, and any population parameters or sample statistics that are given.

1. The report titled *The American Freshman* looks at national norms of college freshman by analyzing the responses of 203,967 first-time, full-time freshmen entering 270 baccalaureate institutions across the United States. It found that over the past two years, the percentage of students who describe themselves as "liberal" (27.6%) or "conservative" (20.7%) has not changed significantly. The percentage of those identifying as "middle of the road" (47.4%) has risen slightly. [1]

2. *Social Metadata for Libraries, Archives, and Museums: Executive Summary* reported on work done in 2009–2010 by a 21-member RLG Partner Social Media Working Group from five countries. Their work reviewed 76 online sites that were relevant to libraries, archives, and museums. The sites supported social media features such as tagging, comments, reviews, etc. Among their findings, they reported that more than 70% of library sites have been offering social media features for 2 years or less and that 83% of respondents add new content at least monthly. Most of the respondents manage their own sites rather than use hosted services. [2]

Answer each question thoughtfully.

3. Suppose you want to know the proportion of students who wear glasses on your campus. Is simply surveying your current class a good sample to choose? Why or why not?

4. Suppose that in a race, male participants are given even entry numbers and females are given odd entry numbers. Would choosing every 10th runner to answer a survey give a representative sample of all racers? What about every 5th runner? Why?

5. You are assigned the task of determining the average age of people who shop at the local mall.

    a. Describe at least three different methods of sampling that could be used for the task. Which method would you say is the best one to use? Why?

    b. Describe any potential for bias in your method.

6. Determine which type of graph would most clearly depict the data described.

    a. The enrollment size at Austin Peay State University (APSU) over the past decade

    b. The size of each class—freshman, sophomore, junior, senior—for the current year at APSU

    c. The size of each class—freshman, sophomore, junior, senior—for the current year at APSU, specifically comparing the genders of students in each category

    d. The enrollment sizes of all state universities across the country

1 John H. Pryor et al., *The American Freshman*

2 Karen Smith-Yoshimura, *Social Metadata for Libraries, Archives, and Museums*

Solve each problem.

7. Use the stacked bar graph on US defense spending trends from 2000–2011 to answer the following questions. [3]

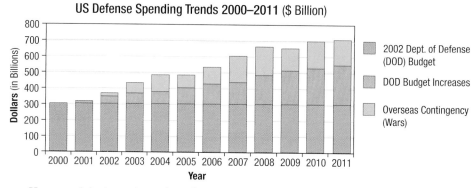

US Defense Spending Trends 2000–2011 ($ Billion)

a. How much is the estimated total spending for 2011?

b. How much has spending increased from 2000 to 2011 in the Overseas Contingency category?

c. By how much has the Department of Defense (DOD) budget increased from 2000–2011?

d. Would you say that the stacked bar graph is a good way to represent the data here? Why or why not?

8. The US Department of Agriculture estimates that between 2010 and 2015, the US economy will generate 54,000 annual openings requiring baccalaureate or higher degrees in food, renewable energy, and environmental specialties. Use the graph about employment opportunities to answer the following questions. [4]

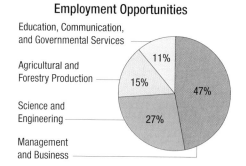

Employment Opportunities

a. What percentage of the jobs are expected to be in the business and science occupations?

b. Over the 5 years, how many total jobs openings are expected in education, communication, and governmental services?

c. How many annual openings are there expected to be in agricultural and forestry production?

d. Would you say that the pie chart is a good way to represent the data given? Why or why not?

3 Emily Skarbek, "How Would You Cut Defense Spending?" http://www.mygovcost.org

4 USDA, http://ag.purdue.edu

9. Consider the following graph showing the results of a personal entertainment survey.

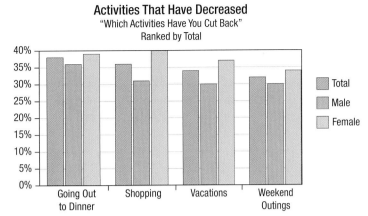

Activities That Have Decreased
"Which Activities Have You Cut Back"
Ranked by Total

a. Name at least two things that are missing from the graph for it to be an effective means of informing people about the data.

b. List ways in which a reader might be unclear or misled about what the graph shows.

For each data set, calculate the following numerical descriptors: mean, median, mode, range, sample standard deviation, and five number summary.

10. Area prices for unleaded gasoline for the week of 1/25/2013 are found to be the following.

$3.07 $3.49 $3.05 $3.35 $3.29 $3.06 $3.18 $3.29

$3.21 $3.29 $3.29 $3.31 $3.05 $3.21 $3.19

11. Whitt gathered the following data from students living in his dorm.

Daily Soda Consumption Number of Cans per Day		
2	0	1
0	1	0
6	1	1
0	0	2
0	3	0
4	0	1

Calculate the standard score for each given value. Round your answer to the nearest hundredth.

12. $\mu = 20, \sigma = 2$

 a. $x_1 = 22$

 b. $x_2 = 19$

 c. $x_3 = 25$

13. $\bar{x} = 16.9, s = 0.04$

 a. $x_1 = 17.15$

 b. $x_2 = 16.84$

 c. $x_3 = 17.4$

Solve each problem.

14. Scores on a test have a mean of 86 and a standard deviation of 13. Jerrica has a score of 94. Convert Jerrica's score to a z-score.

15. Hasef scored a 55 on a test with a mean of 47 and a standard deviation of 9. Kimberly had a score of 168 on a test with a mean of 145 and a standard deviation of 26. Which score was better with respect to the mean score on the given test?

Use the *z*-score formula to complete the table.

16. Find the missing value in each row of the table. Round your answers to the nearest hundredth when necessary.

	z	x	μ	σ
a.		26.2	35.0	4.3
b.	−1.24	152.2		12.2
c.	2.60		55.6	6.04
d.	−2.73	1250	1735	

Find the percentage of data points that lie below each *z*-score.

17. $z = -0.29$

18. $z = 1.16$

19. $z = 3.21$

20. $z = -2.75$

Find the percentage of data points that lie above each *z*-score.

21. $z = 1.17$

22. $z = -1.34$

23. $z = -3.25$

24. $z = 2.14$

Find the percentage of data points that lie between each pair of *z*-scores.

25. $z_1 = -2.00$, $z_2 = 2.00$

26. $z_1 = -1.40$, $z_2 = 2.56$

27. $z_1 = -2.00$, $z_2 = -1.00$

28. $z_1 = 0.62$, $z_2 = 3.1$

Solve each problem.

29. Suppose the length of Ethan's newborn son had a standard score of 0.48. True or false: Ethan's son is longer than the average newborn.

30. Salespeople for a car sales company have an annual sales average of $225,000, with a standard deviation of $18,000. What percentage of the salespeople will make the following sales? Assume the sales follow a normal distribution.

    a. More than $250,000

    b. Less than $190,000

    c. Between $200,000 and $240,000

31. A potato chip manufacturer processes bags of chips with an average weight of 6.5 ounces and a standard deviation of 0.8 ounces. What percentage of the bags of chips will have the following weights? Assume the weights follow a normal distribution.

    a. More than 6.7 ounces

    b. Less than 6.4 ounces

    c. Between 6.4 and 6.7 ounces

In each scatter plot, decide whether there appears to be a positive linear correlation, a negative linear correlation, or no linear correlation.

32.
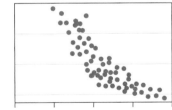

33.
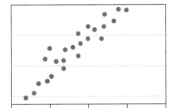

34.
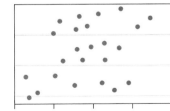

35.
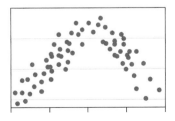

Consider each set of variables and predict whether the variables would have a weak negative relationship, a strong negative relationship, a weak positive relationship, a strong positive relationship, or no relationship at all.

36. Score on the ACT and first semester GPA in college

37. A person's eye color and their self-esteem

Determine whether each correlation coefficient is statistically significant at the specified level of significance for the given sample size.

38. $r = 0.913$, $\alpha = 0.01$, $n = 15$ **39.** $r = 0.525$, $\alpha = 0.05$, $n = 50$

For each data set, find the following.

a. Estimate the correlation in words as positive, negative, or no correlation.

b. Calculate the correlation coefficient r. Round your answer to the nearest thousandth.

c. Determine whether r is statistically significant at the 0.01 level of significance.

40. The following table gives the number of hours a student watches TV per week and his or her overall GPA.

Hours of TV per Week and Overall GPA								
TV Hours	30	10	28	12	11	13	21	9	5
GPA	1.8	2.46	2.4	3.1	3.4	3.2	3.5	3.3	3.7

41. The following table shows the annual salary of 10 adults and their corresponding years of education.

Adult Annual Salary and Education	
Salary ($)	**Years of Education**
125,000	19
100,000	20
40,000	16
35,000	16
41,000	18
29,000	12
35,000	14
24,000	12
50,000	16
60,000	17

Use the linear regression model $\hat{y} = ax + b$ to predict the y-value for each value of x.

42. $\hat{y} = 25.33x + 353.16$

    a. $x = 8$

    b. $x = 12$

    c. $x = 18$

Solve the problem.

43. The following table shows students' test grades on the first two tests in an introductory literature class.

Test Grades on the First Two Tests in an Introductory Literature Class											
Test 1 (x)	81	55	81	90	79	45	84	91	82	69	77	88
Test 2 (y)	79	68	83	87	77	55	87	90	87	71	77	74

    a. Determine the regression line $\hat{y} = ax + b$. Round the slope and y-intercept to the nearest thousandth.

    b. Determine if the regression equation is appropriate at the 0.05 level of significance to use for making predictions. If so, answer part **c.**

    c. If a student scored a 70 on his first test, make a prediction for his score on the second test, if appropriate.

Bibliography

Introduction

1. Mehdizadeh, Soraya. "Self-Presentation 2.0: Narcissism and Self-Esteem on Facebook." *Cyberpsychology, Behavior, and Social Networking*. August 2010, 13(4): 357–364. doi:10.1089/cyber.2009.0257

2. Compete and Twitter. *Tweets in Action: Retail*. https://g.twimg.com/business/pdfs/Tweets_In_Action_Retail_Study.pdf

8.1

1. Jouven, Xavier, Jean-Philippe Empana, Peter J. Schwartz, Michal Desnos, Dominique Courbon, and Pierre Ducimetiere. "Heart-Rate Profile during Exercise as a Predictor of Sudden Death." *New England Journal of Medicine* 2005; 352:1951–1958, May 12, 2005 doi: 10.1056/NEJMoa043012

2. McMillen, Matt. "Migraines May Raise Depression Risk in Women." *Health*. http://news.health.com/2012/02/22/migraines-depression-risk/

3. Berrett, Dan. "Graduates of For-Profits Lag Behind Their Peers in Earnings and Employment, Study Finds." *The Chronicle of Higher Education*. http://chronicle.com/article/Graduates-of-For-Profits-Lag/130900/

8.2

1. Pew Research Center's U.S. Religious Landscape Survey. "Religious Affiliation: Diverse and Dynamic." Accessed February 2008. http://religions.pewforum.org/affiliations

2. Pew Forum on Religion & Public Life. "U.S. Religious Landscape Survey. Appendix 4: Survey Methodology." Accessed January 2012. http://religions.pewforum.org/pdf/report-religious-landscape-study-appendix4.pdf

3. Pew Research Center. "The Generation Gap and the 2012 Election, Section 4: Views of the Nation." Accessed January 2012. http://www.people-press.org/2011/11/03/section-4-views-of-the-nation/

4. Surveillance, Epidemiology, and End Results Program. "Population Characteristics." Accessed April 2012. http://seer.cancer.gov/registries/characteristics.html

5. WorldWideWebSize.com. "The size of the World Wide Web (The Internet)." Accessed December, 2011. http://www.worldwidewebsize.com

6. NOAA Office of Coast Survey. "Map Showing the Distribution of the Slave Population of the Southern States of the United States 1860." http://historicalcharts.noaa.gov/historicals/preview/image/CWSLAVE

7. *President Lincoln Reading the Emancipation Proclamation to His Cabinet* by Francis Bicknell Carpenter. http://www.noaanews.noaa.gov/stories2011/images/cgs05195.jpg

8. Bureau of Labor Statistics Occupational Outlook Handbook. "Fastest Growing Occupations." http://www.bls.gov/ooh/fastest-growing.htm

9. InContext. "Earnings per Job Growing Better than Number of Jobs." http://www.incontext.indiana.edu/2007/june/2.asp

10. University of Kent. "Mathematics Careers." http://www.kent.ac.uk/careers/Maths.htm

11. Bureau of Labor Statistics. "The Employment Situation—February 2013." March 8, 2013. http://www.bls.gov/news.release/archives/empsit_03082013.pdf

12. Statistics Bureau (Japan). "2001 Survey on Time Use and Leisure Activities." September 30, 2002. http://www.stat.go.jp/english/data/shakai/2001/jikan/yoyakuj.htm

8.3

1. Wolfram Alpha LLC. 2009. Wolfram|Alpha. http://www.wolframalpha.com/input/?i=mean+%7B14%2C+5%2C+7%2c+11%2C+9%2C+7%2c+12%2C+6%7D (accessed July 3, 2014).

2. Wolfram Alpha LLC. 2009. Wolfram|Alpha. http://www.wolframalpha.com/input/?i=median%7B28.3%2C+27.7%2C+23.0%2C+25.5%2C+27.1%2C+26.94%2C+27.0%2C+27.52%2C+26.8%2C+27.2%2C+26.97%2C+27.53%7D (accessed July 3, 2014).

3. Wolfram Alpha LLC. 2009. Wolfram|Alpha. http://www.wolframalpha.com/input/?i=standard+deviation+%7B0.8%2C+42.2%2C+20.6%2C+2.8%2C+36.7%2C+12.1%2C+18.6%2C+6.3%2C+5.5%2C+11.3%2C+3.7%2C+0.5%2C+1.2%2C+3.7%2C+14.9%2C+9.4%2C+7.3%2C+9.5%2C+16.0%2C+11.1%2C+4.7%2C+5.6%2C+8.9%2C+10.2%7D (accessed July 22, 2014).

8.4

1. Wolfram Alpha LLC. 2009. Wolfram|Alpha. http://www.wolframalpha.com/input/?i=z-score+calculator&a=FSelect_**NormalProbabilities-.dflt-&f2=2.14&f=NormalProbabilities.z_2.14&a=*FVarOpt.1-_***NormalProbabilities.z--.***NormalProbabilities.pr-.**NormalProbabilities.l-.*NormalProbabilities.r---.*--&a=*FVarOpt.2-_**-.***NormalProbabilities.mu--.**NormalProbabilities.sigma---.**NormalProbabilities.z--- (accessed July 14, 2014).

2. Wolfram Alpha LLC. 2009. Wolfram|Alpha. http://www.wolframalpha.com/input/?i=z-score+calculator&a=FSelect_**NormalProbabilities-.dflt-&f2=-0.98&f=NormalProbabilities.l_-0.98&f3=2.01&f=NormalProbabilities.r_2.01&a=*FVarOpt.1-_***NormalProbabilities.l-.*NormalProbabilities.r--.***NormalProbabilities.z--.**NormalProbabilities.pr---.*--&a=*FVarOpt.2-_**-.***NormalProbabilities.mu--.**NormalProbabilities.sigma---.**NormalProbabilities.l-.*NormalProbabilities.r--- (accessed July 14, 2014).

Chapter Exercises

1. Pryor, J. H., L. DeAngelo, L. Palucki Blake, S. Hurtado, & S. Tran. (2011). *The American Freshman: National Norms Fall 2011*. Los Angeles: Higher Education Research Institute, UCLA. http://heri.ucla.edu/PDFs/pubs/TFS/Norms/Monographs/TheAmericanFreshman2011.pdf

2. Smith-Yoshimura, Karen. 2011. *Social Metadata for Libraries, Archives, and Museums: Executive Summary*. Dublin, Ohio: OCLC Research. http://www.oclc.org/research/publications/library/2012/2012-02.pdf

3. Skarbek, Emily. "How Would You Cut Defense Spending?" January 5, 2012. MyGovCost. http://www.mygovcost.org/2012/01/05/how-would-you-cut-defense-spending/

4. United States Department of Agriculture. "Employment Opportunities for College Graduates in Food, Renewable Energy, and the Environment, United States, 2010–2015." http://www3.ag.purdue.edu/USDA/employment/Pages

Chapter 9
Personal Finance

Sections

Objectives

- Create a budget
- Calculate sales prices and discounts
- Calculate percentage increase/decrease
- Calculate simple interest
- Understand present value
- Understand future value
- Calculate compound interest
- Understand savings plans
- Calculate annual percentage yield
- Calculate monthly payments
- Calculate credit card payments

Personal Finance

Making sound financial decisions is an important component of becoming an adult. What if you won the Powerball Lottery with a jackpot of $150 million? Would you know how to handle the money? Given an option of taking all of the money at once with a 40% penalty or acquiring the money in 25 equal payments over a 25-year period, which is the best option? Surprisingly, about 70% of lottery winners wind up bankrupt because of an inability to manage their money.

How about navigating your student loans, paying for a summer vacation, or purchasing a home or car? One of the first steps in understanding how to manage money is to create a budget and stick to that budget. When we begin making financial decisions on our own, understanding the concepts of money allows us to make better use of all of our funds in a manner that is best for us in the long term. For instance, where is it best to invest money for retirement? How much interest can be earned from an investment? What is the total cost of having a credit card? How much will you pay for a house over the total length of the mortgage loan?

This chapter focuses on how to create a budget using Microsoft Excel as well as understanding how compound interest works in our favor when investing money, but in the bank's favor when we borrow money. We will look at how to determine what expenses fall within the means of a budget and how to save for the future.

Figure 1

9.1 Understanding Personal Finance

As adults we all face financial challenges that affect our decisions in daily life. Determining if we have enough money to eat at a certain restaurant and go to a movie is just one example. What about if we wanted to start saving for home or a car? Learning how to manage the money earned from a job by making a budget and sticking to it is difficult for most people. This section will introduce you to some basic concepts of personal finance that can be used for a lifetime.

Many financial decisions we make, either in college or just after graduating, have long-term effects on our ability to save for retirement, put a down payment on a house, or buy a car. Learning a small number of financial concepts that can help you build good financial habits provides a powerful framework for life. Some of the concepts that will guide us throughout this chapter are listed below.

Sound Financial Concepts

- Learn how to make a budget.
- Determine what your long-term goals are for your money.
- Learn how to research major purchases that may need financing.
- Develop a credit history in a smart, positive manner.
- Always know your bank balance by keeping up with all transactions such as deposits, written checks, debit purchases, and credit card payments.
- Understand how interest works on credit cards.
- Understand how interest works on large purchases such as cars, houses, etc.

These concepts are not meant to be all-encompassing, but rather a foundation for learning to maintain a healthy financial livelihood. Let's begin our study of personal finance with budgeting.

Budgeting

A task that many people find challenging, including the federal government, is setting up a budget based on income. While budgeting may seem overwhelming when you first start out, with practice and dedication it can quickly become an important and useful part of your financial life.

Steps to Create a Budget

1. Calculate monthly and periodical income.
2. Calculate monthly and periodical expenses.
3. Subtract total expenses from income.
4. Allocate any remaining funds to savings, long-term goals, or an emergency fund.

A budget is essentially based on income. One of the biggest hurdles for young workers is to determine the **net income**, which is income after taxes, that they will receive from a given yearly salary. There are many factors that affect net income such as state taxes, federal taxes, social security/medicare taxes—usually listed as FICA (Federal Insurance Contributions Act) on a pay stub—and unemployment taxes. The amount a worker pays for each of these is based on their salary. The 2013 rates for each of these taxes for a worker earning between $18,000 per year and $72,000 per year are: income tax is 15%; social security tax is 6.2%; and 6.0% for unemployment taxes. State taxes vary by state with some states having no state income tax.

⑨

> ### Net Income
>
> **Net income** is equal to total, or gross, income minus all taxes.

Example 1: Determining Federal Income Tax

Jamie graduated from college in 2013 with a degree in accounting. Her first job pays an annual salary of $52,000. How much should Jamie expect to pay in federal income tax for this salary if the federal tax rate is 15%?

Solution

According to the 2013 tax rates, Jamie should expect to pay about 15% of her salary in federal income tax.

$$\$52,000 \cdot 15\% = \$52,000 \cdot 0.15 = \$7800$$

Therefore, Jamie can expect to pay about $7800 in federal income tax each year at her current salary.

Example 2: Determining Monthly Take Home Pay

Pria just graduated from a liberal arts college and acquired a job as a sociologist. Her yearly salary is $34,500. If the federal tax rate is 15%, the social security tax rate is 6.2%, and the unemployment tax rate is 6%, determine the following.

a. Pria's yearly taxes.

b. Pria's monthly take home pay.

Solution

a. Pria's salary of $34,500 means that her income tax rate is 15%, social security tax is 6.2%, and unemployment tax is 6%. Therefore, Pria's yearly taxes would be as follows.

$$
\begin{aligned}
\text{income tax:} \quad & \$34,500(0.15) = \$5175 \\
\text{social security:} \quad & \$34,500(0.062) = \$2139 \\
\text{unemployment:} \quad & \underline{\$34,500(0.06) = \$2070} \\
& \text{total} = \$9384
\end{aligned}
$$

So, Pria can expect to pay approximately $9384 in taxes for the year.

b. Pria's monthly take home pay would be equal to her yearly salary minus her yearly taxes divided over the 12 months of the year. So,

$$\frac{\$34,500 - \$9384}{12} = \frac{\$25,116}{12} = \$2093 \text{ per month.}$$

Therefore, Pria's monthly take home pay is $2093.

Example 3: Budgeting on a Given Income

In Example 2, Pria had a monthly net income of $2093. Use the steps to create a budget to allocate Pria's monthly income if her expenses consist of the following.

rent:	$600
gas:	$250
utilities:	$150
food/entertainment:	$600
student loans:	$210
	$1810

Solution

The steps to create a budget indicate that we need to do the following.

1. Determine monthly income.

2. Determine total monthly expenses.

3. Subtract the amount from Step **2.** from the amount from Step **1.**

4. Allocate additional funds to savings.

In Example 2, we calculated Pria's monthly income to be $2093. Given Pria's total monthly expenses of $1810, we can now find the difference between monthly income and total monthly expenses to obtain

$$\$2093 - \$1810 = \$283.$$

This means that Pria has $283 of remaining income each month that she can put toward savings, long-term goals, or an emergency fund. Making a sound financial decision would mean that Pria should understand that she may not always have an extra $283 each month. Therefore, instead of spending the funds, placing the money into a savings account would be a financially responsible decision.

TECH Example 4: Creating a Budget with a Spreadsheet

Gail works for an art gallery with an annual income of $44,000. Her net monthly pay is $2,669.33. Her basic monthly expenses are the following.

rent:	$550
utilities (water, gas, electricity, internet, cable):	$310
cell phone:	$60
insurance:	$40
transportation:	$65
food:	$300
loan repayment:	$290

In addition, Gail receives $30 a week on average from a side consulting job and gives $45 a month to charities. Use a spreadsheet, such as Microsoft Excel, to create a budget for Gail. How much does she have each month for expenses not listed in her budget?

Solution

Creating a budget using a spreadsheet allows us to let the computer do the arithmetic. This means that even though expenses might change month-to-month or new incomes are introduced, we can easily add them to the spreadsheet and the budget will be adjusted accordingly. Recall that to prepare a budget, we need to add both the income and expenses separately. Because spreadsheets can only perform calculations on numbers and not words, we need to list the numerical figures in one cell and their labels in another. Therefore, we'll create a column for expenses and one for incomes. Let's start with expenses. In cell A1 of the spreadsheet, type "Expense", and in the adjacent cell, B2, type "Amount". The spreadsheet should look like the following screenshot.

Under the column labeled "Expense" enter the name of each of Gail's expenses and under the column labeled "Amount" enter the amount associated with each expense. The new spreadsheet should now look like the following.

	A	B
1	Expense	Amount
2	rent	$550
3	utilities (water, gas, electricity, internet, cable)	$310
4	cell phone	$60
5	insurance	$40
6	transportation	$65
7	food	$300
8	loan repayment	$290
9	charities	$45

To obtain the total amount of expenses, we need a place for the total on row 10. Enter "Total" in A10. In B10, we'll put a formula that will sum the expenses. Enter "=sum(B2:B9)", which instructs the computer to add together the data from column B, row 2 through column B, row 9.

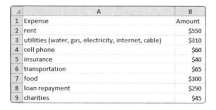

B10		f_x	=SUM(B2:B9)
	A		B
1	Expense		Amount
2	rent		$550
3	utilities (water, gas, electricity, internet, cable)		$310
4	cell phone		$60
5	insurance		$40
6	transportation		$65
7	food		$300
8	loan repayment		$290
9	charities		$45
10	Total		$1,660

The spreadsheet calculates that Gail has monthly expenses of $1,660.

Next, we'll follow the same process for Gail's income. Create a set of columns for Income and Income Amount in columns D and E. (We've intentionally left column C blank for clarity.)

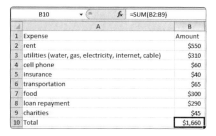

	A	B	C	D	E
1	Expense	Amount		Income	Income Amount
2	rent	$550			
3	utilities (water, gas, electricity, internet, cable)	$310			
4	cell phone	$60			
5	insurance	$40			
6	transportation	$65			
7	food	$300			
8	loan repayment	$290			
9	charities	$45			
10	Total	$1,660			

☞ Helpful Hint

All formulas in Excel must begin with an equal sign, =.

Enter Gail's income labels in column D and the amount in column E. Note that when we enter the amount for her consulting income, we'll need to write a formula for the monthly income, since we're given the weekly amount. The formula will multiply the weekly amount by 4; that is, "=30*4". (Note that during some months, Gail will be paid 5 times for her side consulting job.) Doing it this way means that if the amount Gail receives each week changes, she can simply replace the number 30 with the new weekly income, and the spreadsheet will adjust the monthly amount accordingly. To calculate Gail's total income for the month, enter "=sum(E2:E3)" into cell E10. The budget spreadsheet now looks like the following.

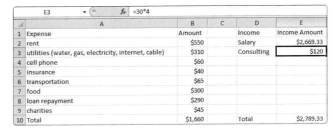

Finally, we need to subtract the Expense Total from the Income Total to find the remaining funds that Gail has available. We'll put this in column G; again, leaving a blank column F for the sake of clarity. We need to tell the spreadsheet to subtract the expense total from the income total by directing it to the cells that contain the total for expenses and incomes. The formula will be in the form income total − expense total, so type the formula "=E10–B10" into cell G10.

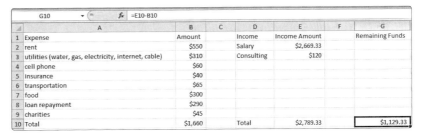

We can now see that Gail has $1,129.33 to put towards expenses not in her budget each month. One of the values of using a spreadsheet like this is that if any of the expenses or incomes change, the spreadsheet automatically adjusts the final calculation for us.

Skill Check #1

Using the spreadsheet from Example 4, find Gail's remaining income if her utilities increase by $50 a month.

Understanding Percentages for Consumer Purchases

In an attempt to be financially responsible, purchasing items on sale whenever possible is quite prudent. We have all been in a department store and noticed items that were on sale for an announced percentage off of the list price, or full-purchase price. Sometimes the reduced prices are indicated on the price tag, but often we are left calculating the discount on our own before reaching the cash register. Either way, knowing the amount that we save can help us stay within our budget. Similarly, having a sense of how much a store marks up an item from their original purchase price in order to determine the retail price often gives us an incentive to wait for those sales!

A few common concepts in consumer purchasing and financial mathematics are necessary for us to continue.

List Price

The price of an item as it is listed for public sale is the **list price**.

Discount

The **discount** is the reduction from the list price. This is usually given as a percentage of the list price.

Sale Price

The **sale price**, sometimes called the **net price**, is the actual cost of an item after a discount of some kind is applied.

$$\text{sale price} = \text{list price} - \text{discount}$$

Let's begin with a simple example regarding the price of a pair of shoes.

Example 5: Finding the Sale Price

Tennis Star Shoes is having a sale. A pair of tennis shoes has a list price of $129.99. The store is offering a discount of 25% off of the list price. Determine the sale price of the shoes before taxes.

Solution

We are to find the sale price of the shoes after a discount of 25% is taken. Since the list price is $129.99, the discount amounts to 25% of $129.99. Remembering our rules of problem-solving from Chapter 1, this indicates that

$$25\% \text{ of } \$129.99 = 0.25(\$129.99) = \$32.50.$$

So, the discount amounts to $32.50. Now, the final price before taxes is determined by subtracting the discount from the list price.

$$\begin{aligned} \text{sale price} &= \text{list price} - \text{discount} \\ &= \$129.99 - \$32.50 \\ &= \$97.49 \end{aligned}$$

The sale price of the pair of shoes is $97.49 before taxes.

Example 6: Determining a Discount

Andrea bought a stereo on sale for $198.45. The list price of the stereo was $330.75. What percentage is the discount?

Solution

We begin by recognizing that a percentage off the list price leads to our sale price. A discount of $x\%$ is subtracted from the original amount of $330.75. This means that the amount of the discount is $330.75 – $198.45 = $132.30. So, what we really need to know is what percentage $132.30 is of $330.75. Mathematically we can interpret the problem as the following.

$$x\% \text{ of } 330.75 = 132.30$$
$$330.75x = 132.30$$
$$x = \frac{132.30}{330.75}$$
$$= 0.4$$

So, the amount of the discount represents a 40% discount.

Skill Check #2

Max bought a computer on sale for $427. The original price of the computer was $1220. What percentage is the discount?

Percentage Change

The percentage change over a period of time can be calculated by the following formula.

$$\text{percentage change} = \frac{\text{new value} - \text{reference value}}{\text{reference value}} \cdot 100\%$$

Example 7: Computing Percentage Decrease

Freddie bought a tablet PC for $1200. One year later, the same tablet PC costs $900. Determine the percentage decrease for the price of the tablet PC.

Solution

To find the percentage change, we need to compare the difference between the values (that is, new price – old price) and divide by the reference value (old price). So, the percentage change is calculated as follows.

$$\text{percentage change} = \frac{\text{new value} - \text{reference value}}{\text{reference value}} \cdot 100\%$$
$$= \frac{900 - 1200}{1200} \cdot 100\%$$
$$= \frac{-300}{1200} \cdot 100\%$$
$$= -0.25 \cdot 100\%$$
$$= -25\%$$

This means the price of the tablet has decreased 25% over the last year. Note that if the percentage change in price was an increase of the price, then the percentage change would be positive.

Skill Check #3

Zola bought a camera for $1125. If the price of the camera six months later was $761, what is the percentage decrease?

Example 8: Computing Percentage Increase

Pear Bike Company uses a 40% profit margin to determine the list price of their products. If the company buys a bike at a wholesale price of $450, what would be the list price for the bike?

Solution

To determine a percentage increase, we need to recognize that the final price will be equal to the original price of $450 increased by 40%. So, the amount of the increase is

$$\$450\,(0.40) = \$180.$$

This means the total price for the bike would be

$$\$450 + \$180 = \$630.$$

Skill Check #4

Use the list price of $630 for the bike in Example 8 to find the sale price if the bike was reduced 40%. Compare the 40% reduction in price to the wholesale price in Example 8.

Skill Check Answers

1. $1079.33

2. 65%

3. −32.36%, or a 32.36% decrease

4. $378; It is less than the wholesale price in Example 8.

9.1 Exercises

Solve each problem.

1. Jeff was recently hired for a job with an annual income of $48,000. Using the federal income tax rate of 15%, what amount should Jeff expect to pay in federal taxes?

2. Megan has a job where she has a take-home salary each month of $1800. If Megan wants to spend no more than 25% of her income on rent, how much rent can Megan afford?

3. Erica earns $670 weekly at her job at a local newspaper. Calculate the weekly taxes for Erica's salary. If Erica wishes to purchase a car where her payments are no more than 15% of her take home pay for a month, what is the maximum monthly car payment she can afford? Assume that the tax rate is 27.2% and that a month has 4 weeks.

4. Jack rents an apartment for $650 per month, pays his car payment of $470 per month, has utilities that cost $420 per month, and spends $850 per month on food and entertainment. Determine Jack's monthly expenses.

5. Using the information from Exercise 4, if Jack has $1500 remaining after his monthly expenses, what is his take-home pay?

6. The given table represents the estimated cost for an on-campus student to attend the University of California-Los Angeles (UCLA) in the 2013–14 academic year (10 months). Create a monthly expense budget for attending UCLA and determine a monthly income that would allow a student to attend UCLA without incurring any debt.

Estimated Cost of Attending UCLA for the 2013–2014 Academic Year	
Budget Category	**On-Campus Student**
University Fees	$12,685
Room & Board	$14,454
Books & Supplies	$1536
Transportation	$807
Personal	$1395
Health Insurance	$1323
Loan Fees	$156
Total	$32,356

Source: Financial Aid Office, "2013-2014 Undergraduate and Graduate Budgets", UCLA. Accessed April 2014. http://www.fao.ucla.edu/publications/2013-2014/Budget_Figures.pdf

7. A car with a list price of $18,000 will be discounted 35% at the time of purchase. What is the sale price?

8. The discount on an LCD television amounts to $240. If the sale price is $1575, what was the list price of the television?

9. You purchase a pair of jeans with a list price of $175. If the sales tax is 8.5%, what is the total cost of the jeans? Round your answer to the nearest cent.

10. At a restaurant, your total bill is $48.55. You wish to give a tip of 18% of the total bill. What is the amount of the tip? Round your answer to the nearest cent.

11. Jamie found a receipt for an MP3 player for $127.59, tax included. If the sales tax rate was 9%, what was the selling price of the MP3 player? Round your answer to the nearest cent.

12. The average cost of a car in 1990 was $12,500. In 2010, the average price of a car was $21,800. What is the percentage increase in the average price of a car? Round your answer to the nearest tenth of a percent.

13. During the housing price decline of 2009, the value of a house decreased by 30% in one year. If the 2010 value of the house was $115,000, what was the house worth prior to the decline? Round your answer to the nearest cent.

14. The wholesale price of a coat is 50% less than the retail price. If the retail price is $199, what is the wholesale price?

15. A store is having a 60% off sale. The sale price of an item is $125. What is the list price?

16. The local sales tax is 8%. If a pair of shoes sells for $115, what is the total cost after tax?

17. The value of your house is $245,000. If your local property tax rate is 2.5% of the value, how much are your property taxes?

18. You receive a 10% decrease in your pay. What percentage increase in pay would you have to receive in order to gain your original pay rate again? Round your answer to the nearest whole percent.

9.2 Understanding Interest

Albert Einstein is reported to have once said, "the most powerful force in the universe is compound interest." Whether Einstein actually said this or not, it's a great financial principle to live by. Without attention, compound interest can rapidly become a force of ill intent and create financial burdens we never planned on. Compound interest can also become the means to our financial dreams through a prudent investment. Why? Because compound interest is an exponential function. As we've seen, exponential functions can quickly create extremely large numbers. This might explain why feelings of disappointment set in after years of paying a mortgage only to discover that most of the money went toward paying the lender for the privilege of borrowing the money, or feelings of elation when discovering that a small nest egg invested at the time of your birth is now available to help you pay for college!

We are often placed in a situation where we need to borrow money in order make a purchase and we must rely on credit to do so. Whether that is in the form of a personal loan, payday loan, or credit card, when we borrow money, we do so with the understanding that we will in turn reimburse the credit holder in the form of **interest**. In other words, we pay the lender a fee for allowing us to borrow money.

Interest

Interest is the amount charged by a lender for borrowing money.

Interest can be a positive thing if you are investing money; that is, if you are the lender. Consider an 18-year-old making a one-time $5000 contribution to an investment fund that had an average 6% annual return. If she never touches the money, $5000 will grow to over $77,000 by the time she retires at age 65. However, if she waits until she's 40 years old to make her investment, that $5000 would only grow to about $21,000. The key factor is in the length of time over which the compounding of interest takes place.

However, if you are the one needing to borrow money, interest is not as gratifying. For instance, if you place $1000 on a credit card with 15% interest rate and pay only the minimum payments required each month, it will take you almost nine years to pay off the debt and cost you $730 in interest.

Simple Interest

In order to calculate the amount of interest to be paid to a lender, you need to know a few things: the amount of the principal, the interest rate being charged, and the amount of time over which the interest will be calculated. The **principal** amount is the sum of money on which the interest is charged. It is also referred to as the initial value of an investment or the starting value of an investment. The **interest rate** is the amount charged expressed as a percentage of the principal. It is usually noted on an annual basis and called the **annual percentage rate (APR)**.

Principal

The **principal** is the sum of money on which interest is charged.

Interest Rate

The **interest rate** is the amount charged to the borrower expressed as a percentage of the principal.

Annual Percentage Rate (APR)

Annual percentage rate is the yearly interest rate that is charged for borrowing. APR is normally given as a percentage per year.

The most straightforward way to calculate interest is by using **simple interest**. This method of calculating interest only considers the principal amount.

Since interest is nothing more than a percentage, or rate, of the principal, for any given amount of principal P, we can see that the interest on that principal for a one-year loan can be found by multiplying the principal by the interest rate. When we combine this idea with the fact that we are introducing the variable of time, we can compute the amount of interest that accrues, or accumulates, for a given amount of time.

Simple Interest Formula

The amount of interest on a simple interest loan with principal P, annual interest rate r (written as a decimal), and loan term of t (usually in years) is calculated with the following formula.

$$I = Prt$$

Example 1: Calculating Simple Interest

Determine the interest that is accrued on $5500 for five years at a rate of 8.5%.

Solution

We have that the principal is $P = \$5500$ and the interest rate is 8.5%. Note that we need to change the interest rate to a decimal before substituting it in the formula, so $r = 0.085$. We also have that $t = 5$. Using our formula, we can determine the amount of interest owed on the $5500 as follows.

$$\begin{aligned} I &= Prt \\ &= (5500)(0.085)(5) \\ &= \$2337.50 \end{aligned}$$

So, if we borrowed $5500 at the given rate of 8.5% for five years, we would owe $2337.50 in interest along with the original principal of $5500 for a total purchase of $7837.50.

Retail stores often offer special financing through their store for purchases over a certain amount. They entice consumers by advertising a "no interest, same as cash" loan. But beware, these interest free loans are often short term and have very high interest rates after the free period is over. Example 2 examines such an offer.

Example 2: Calculating Simple Interest on Purchases

Ian is purchasing a new television with a "deal" from Big Screens R Us. The deal offers 90 days same-as-cash to make the purchase. This means that at the end of 90 days, if Ian has paid off the cost of the television, he owes no interest charge. However, if Ian does not pay off the amount, he owes simple interest for the original purchase amount calculated over the entire 90 days. If the price of the television is $2650 (tax included) with an annual interest rate of 21.99%, how much would Ian owe on the 91st day if he made no payments during the first 90 days?

Solution

First, we want to determine the entire amount due after the 90 days has expired assuming that Ian has not paid any money toward the amount he owes. This means the principal of $2650 is still due in addition to the amount of interest that accumulated over the 3 months that he borrowed the money. Since the formula for simple interest requires that time t must be given in years, we need to convert three months into a fraction of a year; that is, $\frac{3}{12} = \frac{1}{4}$ of a year. Remember that when using the formula, we need the interest rate to be in decimal form. Therefore, we have that $P = 2650$, $r = 0.2199$, and $t = \frac{1}{4}$.

The amount of interest gained over three months can then be calculated as follows.

$$I = Prt$$
$$= (\$2650)(0.2199)\left(\frac{1}{4}\right)$$
$$\approx \$145.68$$

So, the amount of interest due at the end of 90 days would be $145.68. Therefore, the total amount Ian must pay for the television on the 91st day is the sum of the principal and the interest.

$$\$2650 + \$145.68 = \$2795.68$$

Skill Check # I

Find the total cost of a loan for $4320 that has a simple interest rate of 15% for 18 months.

Compound Interest

Each of the previous examples was based on the use of simple interest. For certain types of loans, interest is calculated and applied more than one time per year. This is known as compound interest.

Compound interest is based on simple interest, except that there is a gain of interest added to the principal at each interval. In each compounding interval, interest is calculated on the principal as well as any interest accumulated during the previous interval. Thus, the principal plus the interest equals the new adjusted principal for the next interval.

Compound Interest

Compound interest is interest that is computed based on both the principal and the accrued interest as additional principal at each interval. The future value is the total amount of money A that has been accrued after compounding at an annual percentage rate r based on the initial principal P with n compounding intervals per year for t years. Future value of a compound interest account is calculated with the following formula.

$$A = P\left(1 + \frac{r}{n}\right)^{nt}$$

For example, if we let the principal be $500 and the rate be 4% per year, then the interest after the first year would be 4% of $500, or $0.04 \cdot 500 = \$20$. This amount is then added to the principal of $500 and the new adjusted principal will become $520. The second year's interest will be 4% of $520, or $20.80. In the same manner as before, the interest is added to the principal and the adjusted principal will be $540.80. After just two years, we have accrued $40.80 in interest.

Interest can be compounded more than once a year. Table 1 shows some common compounding intervals along with their terminology for reference.

Table 1
Compounding Intervals

Compounding	Number per Year
Annually	1
Semiannually	2
Quarterly	4
Monthly	12
Weekly	52
Daily	365

Example 3: Computing Compound Interest

Lilly deposits $12,000 into an account with an annual interest rate of 4.5% compounded monthly. If she leaves the money in the account for 10 years, what will the future value be at the end of this time period?

Solution

We can use the compound interest formula to compute the future value for Lilly after 10 years. In her case, $P = 12,000$, $r = 0.045$, and $t = 10$. Since interest is compounded monthly, $n = 12$. So we have the following.

$$A = P\left(1 + \frac{r}{n}\right)^{nt}$$
$$= 12,000\left(1 + \frac{0.045}{12}\right)^{12 \cdot 10}$$
$$= 12,000(1.00375)^{120}$$
$$\approx \$18,803.91$$

So, after 10 years, Lilly will have accumulated a total of $18,803.91, which includes her initial investment of $12,000.

TECH TRAINING

Wolfram|Alpha can be used to calculate compound interest. Go to www.wolframalpha.com and type "compound interest" into the input bar. Then, click the = button. Wolfram|Alpha will return a worksheet with fields to input the data, but it uses the names instead of variables to label the inputs. Note that we need to choose *future value* from the drop-down menu next to *Calculate*. We also need to click on *compounding frequency* at the bottom of the worksheet since the number of compounding intervals is more than one. Enter "$12000" for *present value*, "4.5%" for *interest rate*, and "10" for *interest periods*—this is asking for the number of years. Choose *monthly* from the drop-down menu next to "compounding frequency". Wolfram|Alpha displays the following. (Notice that Wolfram|Alpha rounds the future value to the nearest $10.) [1]

Skill Check #2

Use the information in Example 3 to find the future value of Lilly's account after 20 years.

The power of compound interest lies not only in the interest rate, but also in the number of times the interest is compounded per year. The following example compares the growth of an investment given different numbers of yearly compound intervals.

Example 4: Comparing Compound Interest for Different Compounding Intervals

Thomas uses $4500 to open an IRA (Individual Retirement Account) savings account that earns 3.8% APR. If he leaves the money alone, what is the future value after eight years for the following compounding intervals?

a. The interest is compounded yearly.

b. The interest is compounded quarterly.

c. The interest is compounded weekly.

Solution

a. We have that Thomas' principal is $P = 4500$. The interest rate is $r = 0.038$ and time is $t = 8$. If the interest is compounded yearly, then $n = 1$. Therefore, we have

$$A = P\left(1 + \frac{r}{n}\right)^{nt}$$

$$= 4500\left(1 + \frac{0.038}{1}\right)^{1 \cdot 8}$$

$$\approx \$6064.45.$$

So, if the interest is compounded yearly, Thomas will have approximately \$6064.45 at the end of eight years.

b. This time interest will be compounded quarterly. In other words, $n = 4$. We still have that $P = 4500$, $r = 0.038$, and $t = 8$. Substituting these into the compound interest formula, we have

$$A = P\left(1 + \frac{r}{n}\right)^{nt}$$

$$= 4500\left(1 + \frac{0.038}{4}\right)^{4 \cdot 8}$$

$$\approx \$6089.97.$$

This means that Thomas will have \$6089.97 after eight years if the interest is compounded quarterly.

c. Finally, if the interest is compounded weekly, $n = 52$. Substituting this into the formula, we have

$$A = P\left(1 + \frac{r}{n}\right)^{nt}$$

$$= 4500\left(1 + \frac{0.038}{52}\right)^{52 \cdot 8}$$

$$\approx 6098.03.$$

So, after eight years, Thomas' investment has a future value of \$6098.03 with interest compounded weekly.

You might notice that the more compounding intervals that occur in a given year, the larger the total accumulated amount of money. The other factor that has an impact on the accumulated amount is time. The longer an amount is invested, the larger the accumulated amount. Making sound financial decisions requires that we also understand the power of compounding interest over a long period of time.

Continuous Compound Interest

Sometimes there are instances where, instead of compounding interest a certain number of times per year, interest is compounding literally *continuously*—every moment of every day of every week of every year. We call this **continuous compound interest**. Interest that is continuously compounded uses a formula similar to that of interest compounded n times per year, except that the variation among the number of compounding intervals per year is accounted for using the irrational number e.

Continuous Compound Interest Formula

The future value A of a continuous compound interest account after t years at an annual interest rate of r and an initial amount, or principal, P is calculated with the following formula.

$$A = Pe^{rt}$$

Example 5: Calculating Continuous Compound Interest

Find the future value of $8900 invested at a rate of 2.05% that is compounded continuously over 15 years.

Solution

We know that $P = 8900$, $t = 15$, and $r = 0.0205$. Using the continuous compound interest formula, we have

$$A = Pe^{rt}$$
$$= 8900e^{0.0205 \cdot 15}$$
$$\approx 12{,}104.19.$$

So, the future value is $12,104.19 when interest is compounded continuously.

TECH TRAINING

Use the following keystrokes on a TI-30XIIS/B or TI-83/84 Plus calculator for the calculation of continuously compounded interest in Example 5.

8900 $\times$ 2ND LN 0.0205 $\times$ 15) =

Wolfram|Alpha can be used to calculate the result from Example 5. Go to www.wolframalpha.com and type "8900*e^(0.0205*15)" into the input bar. Then click the = button.

☞ Helpful Hint

In order to calculate the value of Pe^{rt}, notice that we must use the exponential function e^x. In order to enter this function on a calculator, we must press the buttons 2ND LN.

If you are using a TI-30XS/B Multiview calculator, the keystrokes will be slightly different for your calculator. See Appendix B for these instructions.

Although it is possible, you very seldom find banks offering interest that is compounded continuously. However, since continuous compounding represents the greatest possible number of compound intervals over a year, it can help us find the maximum amount of interest that is possible to earn in one year.

Example 6: Finding the Maximum Amount of Interest Possible in One Year

Assume you wish to deposit $2500 into an account bearing 6% interest for 10 years.

a. What is the maximum future value possible after 10 years?

b. What is the maximum amount of interest possible after 10 years?

Solution

a. Because we want to know the maximum future value, we need to use the continuous compound interest formula. We are given that the principal is $P = 2500$, $r = 0.06$, and $t = 10$. Substituting the given values into the formula we have the following.

$$A = Pe^{rt}$$
$$= 2500e^{0.06 \cdot 10}$$
$$\approx 4555.30$$

Therefore, the largest possible future value of $2500 invested at 6% over 10 years is $4555.30.

b. We can find the largest amount of interest the principal can earn over the 10-year period by subtracting the principal amount from the future value as shown below.

$$\text{interest earned} = \text{future value} - \text{principal}$$
$$I = A - P$$
$$I = 4555.30 - 2500.00$$
$$= 2055.30$$

So, the principal can earn at most $2055.30 in interest over the 10 years.

Annual Percentage Yield

An important component to managing your money and investments is determining the actual amount that an account will earn in a given year. In particular, the effect of the number of compounding intervals per year can make a significant difference in the total amount of money earned in a year. Consider investing $100 at 5% simple interest. At the end of the year, the future value will be $105 since the interest was not compounded during the year. However, if we had $100 invested at 5% compounded quarterly, the future value would be as follows.

$$A = P\left(1 + \frac{r}{n}\right)^{nt}$$
$$= 100\left(1 + \frac{0.05}{4}\right)^{4 \cdot 1}$$
$$= 105.09$$

After we take quarterly compounding into account, the practical interest rate is no longer 5%, it is effectively 5.09% over the year. It should make sense that this rate is higher, since at each compounding period, we earn interest on the original principal plus the interest earned in the previous periods. This practical interest rate is called the **annual percentage yield (APY)**, or **effective interest rate**, for a given period.

Annual Percentage Yield (APY)

The **annual percentage yield** (APY) is the effective annual interest rate earned in a given year that accounts for the effects of compounding. APY is calculated with the formula

$$\text{APY} = \left[\left(1 + \frac{r}{n}\right)^{n} - 1\right] \cdot 100\%,$$

where r is the annual percentage rate and n is the number of compounding intervals per year.

Example 7: Annual Percentage Yield

Samantha deposits $5000 in an account paying 5% interest per year.

a. Find the APY for Samantha's investment if the interest is compounded monthly.

b. Find the APY if the interest is compounded daily on Samantha's investment.

Solution

a. Notice, that to determine the APY, we only need the rate of interest and the number of compound intervals per year. The amount of the principal plays no role in finding the APY. So, we need $r = 0.05$ and $n = 12$ for monthly compounding intervals.

$$\text{APY} = \left[\left(1 + \frac{r}{n} \right)^n - 1 \right] \cdot 100\%$$

$$= \left[\left(1 + \frac{0.05}{12} \right)^{12} - 1 \right] \cdot 100\%$$

$$= (0.05116) \cdot 100\%$$

$$= 5.116\%$$

This means with monthly compounding, the APY for Samantha's investment is 5.116%.

b. Daily compounding means that $n = 365$. Therefore, our APY calculation is as follows.

$$\text{APY} = \left[\left(1 + \frac{r}{n} \right)^n - 1 \right] \cdot 100\%$$

$$= \left[\left(1 + \frac{0.05}{365} \right)^{365} - 1 \right] \cdot 100\%$$

$$= (0.05127) \cdot 100\%$$

$$= 5.127\%$$

With daily compounding, Samantha has an annual percentage yield of 5.127%.

Notice that although the advertised APR for Samantha's investment was 5%, when interest is compounded, none of the annual percentage yields were actually 5%.

TECH TRAINING

Wolfram|Alpha can be used to calculate the APY. Go to www.wolframalpha.com and type "((1+0.05/365)^365−1)*100" into the input bar. Then, click the = button.

Since the APY is a larger percentage than the annual percentage rate (APR) when compounding occurs, banks will often publicize the APY for investment accounts instead of the APR to entice investors to deposit money with their institution. Conversely, when advertising interest rates for loans, lenders will use the APR instead of the APY to entice borrowers.

Example 8: APR vs. APY

Suppose that the APD Bank of the South advertises the following rates for their personal loans.

Table 2

APD Bank of the South Personal Loan Rates

Loan Amount	APR*
< $20,000	10.49%
$20,000–$99,999	9.99%
≥ $100,000	7.50%

*interest rates are compounded quarterly

Find the APY, or effective interest rates, for each of the loan categories.

Solution

To find the APY for each loan category we need the published APR as well as the number of compounding intervals per year. In this case, each is compounded quarterly, so $n = 4$.

The APY for an APR of 10.49% compounded quarterly is calculated as follows.

$$\begin{aligned} \text{APY} &= \left[\left(1 + \frac{r}{n} \right)^n - 1 \right] \cdot 100\% \\ &= \left[\left(1 + \frac{0.1049}{4} \right)^4 - 1 \right] \cdot 100\% \\ &= (0.1091) \cdot 100\% \\ &= 10.91\% \end{aligned}$$

Therefore, the APY is 10.91%.

For an APR of 9.99%, we can calculate the APY as follows.

$$\begin{aligned} \text{APY} &= \left[\left(1 + \frac{r}{n} \right)^n - 1 \right] \cdot 100\% \\ &= \left[\left(1 + \frac{0.0999}{4} \right)^4 - 1 \right] \cdot 100\% \\ &= (0.1037) \cdot 100\% \\ &= 10.37\% \end{aligned}$$

That gives an APY of 10.37%.

Lastly, to calculate the APY for an APR of 7.50%, we have the following.

$$\begin{aligned} \text{APY} &= \left[\left(1 + \frac{r}{n} \right)^n - 1 \right] \cdot 100\% \\ &= \left[\left(1 + \frac{0.075}{4} \right)^4 - 1 \right] \cdot 100\% \\ &= (0.07714) \cdot 100\% \\ &= 7.714\% \end{aligned}$$

Therefore, the APY is 7.714%.

The following shows the comparison of the publicized APR versus the APY.

Table 3		
APR vs. APY		
Loan Amount	APR	APY
< $20,000	10.49%	10.91%
$20,000–$99,999	9.99%	10.37%
≥ $100,000	7.50%	7.714%

Therefore, when trying to determine which loan is best, look at the APY to get a clearer picture of the actual cost of the loan.

A great example of how the APR can differ substantially from the advertised interest rate is with a payday loan. A payday loan, sometimes called a cash advance, is a short-term loan that requires no collateral, just a history of having a paycheck. A typical payday loan is for two weeks, where the amount of interest owed per $100 borrowed varies from $15 to $25—for just two weeks! These loans became popular during the 1990s and have since been made illegal in many states because of the excessive interest rates charged. So, how expensive is it to get a payday loan?

Example 9: Calculating Interest on Payday Loans

Assume you wish to borrow $300 for two weeks in the form of a payday loan and the amount of interest you must pay is $25 per $100 borrowed. This means that at the end of two weeks, you owe $375. What is the APR?

Solution

Recall that APR is defined as the interest rate over a year. Since the loan was for two weeks, we need to convert this to a yearly rate. We can find the APR by using

$$\text{APR} = \text{short term interest rate} \cdot \frac{52 \text{ weeks}}{\text{length of loan}},$$

where the short term rate of interest is 25% for two weeks, and the loan is for two weeks.

Now, when we fill in the numbers, we obtain the following.

$$\text{APR} = 25\% \cdot \frac{52 \text{ weeks}}{2 \text{ weeks}}$$
$$= 650\%$$

Thus, the APR is 650%. That is a ridiculous rate of interest for one year. Although this rate is never paid because the loans are for a very short amount of time, the APR is the reason these types of loans have become illegal in many states.

Skill Check Answers

1. $5292

2. $29,465.60

9.2 Exercises

Calculate the simple interest for each situation. Round your answer to the nearest cent, if necessary.

1. Determine the interest owed on $800 for 5 years at a rate of 8.5%.

2. Determine the interest owed on $5000 for 2 years at a rate of 10%.

3. Determine the interest owed on $1200 for 30 months at a rate of 19.5%.

4. Determine the interest owed on $550 for 5 years at a rate of 8.5%.

5. You are purchasing a new computer using the store's "90 days same as cash" deal. If the cost of the computer is $1575 (tax included) with an annual interest rate of 19.99%, how much would you owe on the 91st day if you make no payments during the first 90 days?

Use the compound interest formula to find the following for each situation.

a. Calculate the total amount in the account after the given time period.

b. Determine the amount of interest earned for each.

6. $P = \$2500$, $r = 6.5\%$ compounded weekly, $t = 10$ years

7. $P = \$3500$, $r = 4.5\%$ compounded monthly, $t = 10$ years

8. $P = \$2500$, $r = 6.5\%$ compounded daily, $t = 10$ years

9. $P = \$5650$, $r = 8\%$ compounded biannually, $t = 15$ years

10. $P = \$2500$, $r = 6.5\%$ compounded yearly, $t = 10$ years

11. $P = \$15,000$, $r = 6\%$ compounded semiannually, $t = 25$ years

12. $P = \$12,500$, $r = 8\%$ compounded biweekly, $t = 15$ years

13. $P = \$7300$, $r = 19.9\%$ compounded weekly, $t = 20$ years

Solve each problem.

14. Assume you wish to borrow $500 for two weeks and the amount of interest you must pay is $20 per $100 borrowed. What is the APR at which you are borrowing money? Round your answer to the nearest hundredth.

15. A couple deposits $25,000 into an account earning 6% annual interest for 25 years.

    a. Calculate the future value of the investment if interest is compounded monthly.

    b. Calculate the future value if the interest on the investment is compounded weekly.

16. Suppose your salary in 2012 is $65,000. If the annual inflation rate is 4%, what salary do you need to make in 2020 in order for it to keep up with inflation?

17. Suppose that $15,000 is deposited for eight years at 5% APR.

    a. Calculate the simple interest earned.

    b. Calculate the interest earned if interest is compounded weekly.

    c. Calculate the interest earned if interest is compounded monthly.

18. A payday loan is made for six weeks, where the amount of interest owed per $100 borrowed is $20. If you borrow $500 for six weeks, answer the following questions.

    a. How much do you owe at the end of six weeks?

    b. What is the APR for this transaction?

19. Angela deposits $2500 into an account with an APR of 5.5% for 10 years. Find each of the following.

    a. Amount if interest is compounded annually

    b. Amount if interest is compounded monthly

    c. Amount if interest is compounded weekly

    d. Amount if interest is compounded daily

    e. Amount if interest is compounded continuously

20. David deposits $4000. Determine the APY for each of the following.

    a. APR of 5% compounded monthly

    b. APR of 5% compounded weekly

    c. APR of 5% compounded daily

    d. APR of 7.5% compounded monthly

    e. APR of 7.5% compounded weekly

    f. APR of 7.5% compounded daily

21. Determine the simple interest earned on $10,000 after 10 years if the APR is each of the following rates.

    a. 3%

    b. 6%

    c. 12%

    d. 24%

An account is compounded monthly for five years with an APR of 9%. For each principal amount, calculate the following.

    a. The total amount paid

    b. The amount of interest paid

22. $15,000   **23.** $30,000

24. $60,000   **25.** $120,000

Solve each problem.

26. Suppose the First Bank of Lending offers a CD (Certificate of Deposit) that has a 6.45% interest rate and is compounded quarterly for three years. You decide to invest $5500 into this CD.

    a. Determine how much money you will have at the end of the three years?

    b. Find the APY.

27. The First Bank of Lending lists the following APR for loans. Determine the APY, or effective interest rate for each category.

First Bank of Lending Loan APR	
Loan Amount	**APR**
< $20,000	11.25
$20,000–$99,999	8.99
≥ $100,000	5.75
*interest rates are compounded quarterly	

9.3 Saving Money

Interest is not always a bad thing. It can lend a helping hand when we are saving money. Whether it is for a particular short term goal or a long term retirement plan, we can use interest to our benefit.

Suppose we have a particular monetary goal in mind that we would like to save for. We can rearrange the interest formula we saw in Section 9.2 so that it indicates how much money needs to be invested *now* in order to meet that goal in the *future*. This amount needed to invest now is called the **present value**.

Present Value (*PV*)

Present value (*PV*) is the amount of principal needed now in order to reach a future value amount. Present value is calculated with the formula

$$PV = \frac{A}{\left(1 + \dfrac{r}{n}\right)^{nt}}$$

where *A* is the future value, *r* is the APR, *n* is the number of compound intervals per year, and *t* is time, or term, in years.

For example, many states have college savings plans that allow parents to deposit a lump sum amount of money into their child's college fund based on current college costs. The amount deposited will then earn enough interest to cover the predicted cost of college 18 years in the future. The only catch is you must have a lump sum payment to invest now.

Example 1: Computing Principal Needed for Future Value

State College predicts that in 18 years it will take $150,000 to attend the college for four years. Amber has a substantial amount of cash and wishes to save money for her newborn child's college fund. How much should Amber put aside in an account with an APR of 4%, compounded monthly, in order to have $150,000 in the account in 18 years?

Solution

We know that Amber wishes to have a future value of $150,000 for her child's college fund. Since she has found an investment opportunity that will give her monthly compounding at 4%, we know that $A = 150,000$, $r = 0.04$, $n = 12$, and $t = 18$. Substituting these values into the present value formula to find P, we have the following.

$$PV = \frac{150,000}{\left(1 + \dfrac{0.04}{12}\right)^{12 \cdot 18}}$$

$$\approx 73,100.31$$

This means that if Amber could deposit $73,100.31 into the account now, then after 18 years there will be $150,000 in the account.

TECH TRAINING

To perform the calculation from Example 1 on a TI-30XIIS/B or TI-83/84 Plus calculator, use the following keystrokes.

150000 [÷] [( ] 1 [+] [( ] 0.04 [÷] 12 [ )] [ )] [^] [( ] 12 [×] 18 [ )] [=]

To perform the calculation with Wolfram|Alpha, go to www.wolframalpha.com and type "150000/(1+(0.04/12))^(12*18)" into the input bar. Then, click the = button.

Skill Check #1

How much should you deposit now in an account with an APR of 7% compounded monthly if you wish for the account to have a balance of $200,000 in 25 years?

Often it is preferable to make regular payments into a savings program rather than depositing a lump sum of money at once. Although these payments might be considerably smaller than a lump sum amount, as we discussed in the previous example, compound interest can work in our favor in just the same manner. The significant point with this type of savings plan is that consistent payments are ongoing. For instance, suppose you deposit $50 each month into an account. The first month, interest is calculated on the initial $50 deposit. However, in the next period, after you've made your second deposit of $50, interest is calculated on both $50 payments as well as the interest you earned in the first period. So, although you personally are building up your savings by contributing $50 a month, interest continues to add to your savings as well. In broad terms, whenever fixed regular payments are made we refer to them as **annuity** payments.

Annuity

An **annuity** is a sequence of regular payments made into an account, or taken out of an account, over time.

We can calculate the future value of an annuity payment plan by using the following formula. It takes into account not only the additional compound interest each period, but also the regular payment added each time.

Annuity Formula for Finding Future Value

The future value (*FV*) of an annuity savings account is calculated with the formula

$$FV = PMT \cdot \frac{\left[\left(1+\dfrac{r}{n}\right)^{nt} - 1\right]}{\left(\dfrac{r}{n}\right)}$$

where *PMT* is the payment amount that is deposited on a regular basis, *r* is the APR, *n* is the number of regular payments made each year, and *FV* is the future value after *t* years.

Example 2: Calculating Future Value Using an Annuity Plan

Daigle begins placing $75 per month into an annuity savings plan when he begins receiving his first official paycheck in August 2014. The savings plan earns 2.5% APR, compounded monthly.

a. Determine the amount of money that Daigle will personally put in the savings plan over 10 years.

b. Calculate the future value of Daigle's savings in August 2024.

Solution

a. To find the amount of money that Daigle will personally add to the savings plan, we need to multiply the monthly payment amount by the number of payments each year and then by the number of years he plans to continue his payments. This gives us the following.

$$\text{payment amount} \cdot \text{payments per year} \cdot \text{number of years} = \$75 \cdot 12 \cdot 10$$
$$= \$9000$$

So, over the 10 years, Daigle will contribute $9000 to his savings plan.

b. In order to find the future value of Daigle's savings, we need several values. First, we are told that he will contribute $75.00 each month. So, we know that $PMT = \$75.00$ and the number of payments made each year gives us $n = 12$. The APR given is 2.5%, or $r = 0.025$. The time period is from August 2014 through August 2024, so $t = 10$ years. Now, we can substitute these values in the annuity formula to find the future value.

$$FV = PMT \cdot \frac{\left[\left(1 + \frac{r}{n}\right)^{nt} - 1\right]}{\left(\frac{r}{n}\right)}$$

$$= 75 \cdot \frac{\left[\left(1 + \frac{0.025}{12}\right)^{12 \cdot 10} - 1\right]}{\left(\frac{0.025}{12}\right)}$$

$$= 10,212.90$$

Therefore, after 10 years, Daigle's savings account will contain $10,212.90. Since we know that he contributed $9000, over 10 years Daigle's savings earned interest in the amount of: $10,212.90 - \$9000 = \$1212.90$.

TECH TRAINING

We can use Microsoft Excel to calculate the annuity formula by using the built-in formula "=FV(rate,nper,pmt,[pv],[type])". (Note that another spreadsheet program could be used, but the directions may be slightly different.) The built-in function uses the following notation.

rate: interest rate per period (note this is per period, so this value is $\frac{r}{n}$). Excel doesn't mind whether interest is in the form of a decimal or a percentage.

nper: total number of payment periods—because this is the total number of periods, this value is $n \cdot t$.

pmt: payment amount made each period—this must be the same for each period; the value you enter here must be negative if you are making a deposit into an account and positive if you are receiving a payment from an account.

fv: future value, or a cash balance you want to attain after the last payment is made.

[pv]: present value, or the lump-sum amount that a series of future payments is worth right now. If left blank, Excel will assume the value is $0.

[type]: a value of 0 if the payments are made at the beginning of the compounding periods or 1 if the payments are made at the end of the periods. If left blank, Excel will assume that payments are made at the beginning of the compounding periods.

To calculate the future value of an annuity, write the formula in any cell with the values filled in.

Therefore, for part **b.** of Example 2, to calculate the amount in the account after 10 years (or 120 months) using Microsoft Excel, we can type in "=FV(2.5%/12,120,−75)"—note that the payment is negative since we are making a payment into the account. The value given by Excel is $10,212.90.

Skill Check #2

Find the future value of the annuity plan in Example 2 after 15 years.

Example 3: Calculating Future Value for an Annuity (Savings Plan)

Use the annuity formula to calculate the future value after six months of monthly $50 payments for an annuity savings plan having an APR of 12%.

Solution

We have monthly payments of $PMT = \$50$, an annual interest rate of $r = 0.12$, $n = 12$ because the payments are made monthly, and $t = \frac{1}{2}$ because six months is a half year. Using the savings formula, we can find the future value after six months.

Replacing the variables in the formula with their values we obtain the following.

$$FV = PMT \cdot \frac{\left[\left(1+\dfrac{r}{n}\right)^{nt} - 1\right]}{\left(\dfrac{r}{n}\right)}$$

$$= 50 \cdot \frac{\left[\left(1+\dfrac{0.12}{12}\right)^{12 \cdot 0.5} - 1\right]}{\left(\dfrac{0.12}{12}\right)}$$

$$= 307.60$$

Therefore, after six months, the savings plan has accrued $307.60.

TECH TRAINING

To perform the calculation from Example 3 on a TI-83/84 Plus calculator, use the following keystrokes.

$\boxed{5\emptyset}\ \boxed{\times}\ \boxed{(}\ \boxed{(}\ \boxed{1}\ \boxed{+}\ \boxed{0.12}\ \boxed{\div}\ \boxed{12}\ \boxed{)}\ \boxed{\wedge}\ \boxed{(}\ \boxed{12}\ \boxed{\times}\ \boxed{0.5}\ \boxed{)}\ \boxed{-}\ \boxed{1}\ \boxed{)}\ \boxed{\div}$
$\boxed{(}\ \boxed{0.12}\ \boxed{\div}\ \boxed{12}\ \boxed{)}\ \boxed{=}$

To perform the calculation with Excel, input the following to obtain the future value. "=FV(12%/12,6,−50)"

To perform the calculation with Wolfram|Alpha, go to www.wolframalpha.com and type "50*((1+0.12/12)^(12*0.5)−1) / (0.12/12)" into the input bar. Then, click the = button.

☞ Helpful Hint

If you are using a TI-30XS/B Multiview calculator, the keystrokes will be slightly different for your calculator. See Appendix B for these instructions.

Example 4: Calculating Future Value of an IRA Using the Annuity Formula

At the age of 25, Angela starts an IRA (individual retirement account) to save for retirement. She deposits $200 into the account each month. Based on past performance of similar accounts, she expects to obtain an annual interest rate of 8%.

a. How much money will Angela have saved upon retirement at the age of 65?

b. Compare the future value to the total amount Angela deposited over the time period.

Solution

a. Because Angela starts making payments at age 25 and stops at 65, she is making payments over 40 years. Using the savings plan annuity formula, we need the following: payments of $PMT = \$200$, an interest rate of $r = 0.08$, and $n = 12$ for monthly deposits. The balance after $t = 40$ years is as follows.

$$FV = PMT \cdot \frac{\left[\left(1+\dfrac{r}{n}\right)^{nt} - 1\right]}{\dfrac{r}{n}}$$

$$= 200 \cdot \frac{\left[\left(1+\dfrac{0.08}{12}\right)^{12\cdot 40} - 1\right]}{\dfrac{0.08}{12}}$$

$$= 698,201.57$$

b. Now we wish to calculate the total deposit amount over the 40 years. Keep in mind that over the 40 years, Angela has made 480 monthly deposits of $200. That means she has deposited a total of

$$480 \cdot \$200 = \$96,000.$$

The rest is all interest. Or in other words, even though Angela only deposited $96,000, she made $602,201.57 in interest. Just think if she had started saving that same $200 per month when she was 18!

TECH TRAINING

To perform the calculation from Example 4 on a TI-83/84 Plus calculator, use the following keystrokes.

200 ✕ ((1 + 0.08 ÷ 12) ⌃ (12 ✕ 40) − 1) ÷ (0.08 ÷ 12) =

To perform the calculation with Excel, input the following to obtain the future value. "=FV(8%/12,(12*40),−200)"

To perform the calculation with Wolfram|Alpha, go to www.wolframalpha.com. In the input bar, type "200*((1+0.08/12)^(12*40)−1)/(0.08/12)". Then, click the = button.

Skill Check #3

Repeat Example 4 with Angela saving $200/month from age 18 to age 65.

To further show the power of compound interest, let's consider two different investment strategies. Suppose two friends, Kurt and Johnny, graduate from college on the same day. Each has a different view on saving for the future. Kurt decides to save for the future right away and begins depositing $200.00 per month into an account with an APR of 7.5%. After 15 years, Kurt decides to stop investing money into the account and NEVER make another deposit. Without withdrawing any of the money, the account will still continue to earn the same level of interest.

Johnny, however, waits to start saving. On the same day that Kurt stops adding monthly payments into his savings account, Johnny decides to begin depositing $200.00 per month into a savings account of his own, having the same APR of 7.5%. How many years will it take Johnny to catch up with Kurt's savings balance?

Kurt's savings plan after 15 years has a future value of $66,222.46. We find this just as we have before with the annuity formula, as shown.

$$FV = 200 \cdot \frac{\left[\left(1 + \dfrac{0.075}{12}\right)^{12 \cdot 15} - 1\right]}{\dfrac{0.075}{12}}$$

$$= 66,222.46$$

Since Kurt decides not to withdraw any money from the account, the savings will continue to grow at 7.5% interest, compounded monthly even though he has stopped depositing money each month. The following graph shows the growth of Kurt's account after this point.

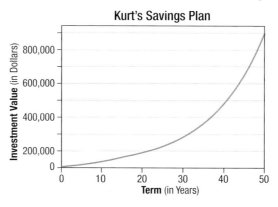

Figure 1

Johnny begins depositing money when he turns 37 years of age, 15 years after graduation. At this time, he begins to deposit $200.00 per month, just as Kurt did, into an account paying 7.5% interest. The future value of Johnny's savings account, beginning 15 years after graduation, is shown in comparison to Kurt's investment growth 15 years after graduation in Figure 1.

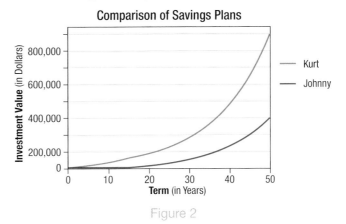

Comparison of Savings Plans

Figure 2

Remember that Kurt has a big head start on the amount of money in the account, even though he stopped his monthly payments. Will the amount of money in Johnny's account ever exceed the amount in Kurt's account?

When we consider the graphs of these two functions, we can see that Kurt has such a huge lead in money that, if the money continues to grow, Johnny will never catch up to Kurt even though he continues to deposit money long after the initial 15 years. Therein lies the power of compound interest over long periods of time! Lesson to learn: invest early, even if it's a small amount.

Working Toward a Goal

When saving money, there is often a goal amount we wish to reach for things such as college or retirement. In these cases, it is usually easier to make regular payments toward the goal, just as we have been looking at, rather than depositing a lump sum. We can use the same formula, but solve it for the payment amount instead of future value.

Annuity Formula for Finding Payment Amounts

The regular payment amount (*PMT*) needed to reach a future goal with an annuity savings account is calculated with the formula

$$PMT = FV \cdot \frac{\left(\dfrac{r}{n}\right)}{\left[\left(1+\dfrac{r}{n}\right)^{nt} - 1\right]}$$

where *FV* is the future value desired, *r* is the APR, *n* is the number of regular payments made each year for *t* years.

Example 5: Finding Monthly Payments in Order to Meet a Goal

Suppose you'd like to save enough money to pay cash for your next computer. The goal is to save an extra $3000 over the next three years. What amount of monthly payments must you make into an account that earns 1.25% interest in order to reach your goal?

Solution

We can find the monthly payment required to meet the goal by using the annuity formula for payments. The future value is $FV = 3000$, $r = 0.0125$, $n = 12$, and $t = 3$ years. Substituting these values into the formula, we have the following.

$$PMT = FV \cdot \frac{\left(\dfrac{r}{n}\right)}{\left[\left(1+\dfrac{r}{n}\right)^{nt} - 1\right]}$$

$$= 3000 \cdot \frac{\left(\dfrac{0.0125}{12}\right)}{\left[\left(1+\dfrac{0.0125}{12}\right)^{12 \cdot 3} - 1\right]}$$

$$\approx 81.82$$

Therefore, in order to reach $3000 in three years with a 1.25% APR, you need to make a monthly payment of $81.82.

TECH TRAINING

To perform the calculation from Example 5 on a TI-30XIIS/B or a TI-83/84 Plus calculator, use the following keystrokes.

3000 ⊠ ((0.0125 ÷ 12) ÷ (((1 + 0.0125 ÷ 12) ^ (
12 ⊠ 3) − 1) =

To perform the calculation with Excel, input the following to obtain the monthly payment. "=PMT(rate,nper,pv,fv,type) = −PMT(1.25%/12,(12*3),0,3000,0)"

To perform the calculation using Wolfram|Alpha, go to www.wolframalpha.com and type "3000*(0.0125/12)/((1+0.0125/12)^(12*3)−1)" into the input bar. Then, click the = button.

Let's return to the first example at the beginning of the section in which Amber wished to save up enough money for her newborn child to go to State College in 18 years. Recall that State College predicted that the cost of a degree would be $150,000 in 18 years. We calculated that Amber could put a lump sum of $73,100.31 into an account and leave it for 18 years in order to have enough money saved up. The next example determines how much Amber would need to deposit on a monthly basis in order to save the $150,000 for her child's college fund in 18 years. We will use the same APR of 4% as in Example 1.

Example 6: Calculating Monthly Payments

Amber wishes to deposit a fixed monthly amount into an annuity account for her child's college fund. She wishes to accumulate a future value of $150,000 in 18 years.

a. Assuming an APR of 4%, how much money should Amber deposit monthly in order to reach her goal?

b. How much of the $150,000 will Amber ultimately deposit in the account, and how much is interest earned?

Solution

a. Amber's goal is to have accumulated a balance of $150,000 after 18 years. She has an interest rate of 4% and is going to make monthly payments. Therefore, we know that $P = 150,000$, $r = 0.04$, $n = 12$, and $t = 18$. Now, substituting these into the formula, we can calculate the monthly payment needed.

$$PMT = FV \cdot \frac{\left(\dfrac{r}{n}\right)}{\left[\left(1+\dfrac{r}{n}\right)^{nt} - 1\right]}$$

$$= 150,000 \cdot \frac{\left(\dfrac{0.04}{12}\right)}{\left[\left(1+\dfrac{0.04}{12}\right)^{12 \cdot 18} - 1\right]}$$

$$= 475.30$$

So, Amber will need to deposit at least $475.30 per month in order to meet her goal of $150,000.

b. Amber will need to make 18 years worth of monthly payments of $475.30. We can find her contribution to the savings plan by multiplying these together.

$$18 \cdot 12 \cdot \$475.30 = \$102,664.80$$

The interest earned is then the difference between the goal and the contributed amount from Amber.

So, we can subtract and have the following.

$$\$150,000 - \$102,664.80 = \$47,335.20$$

So, the savings plan earned $47,335.20 in interest over the 18 years.

Retirement is such a long way off that it may seem silly to start thinking about it before you have even started your first job out of college. However, we know from our study on interest that time is on our side and interest works in our favor when saving money. The longer we can let money accrue interest, the better off we'll be. So, yes, now is the time to start saving for your retirement, even if it's just a small amount each month. Ideally, later on in life, our retirement funds will provide us with a consistent yearly pay without the worry of outliving our savings. In other words, the annuity payments would come solely from interest earned on the principal. That way the principal would not be affected at all and continue to earn us more money! Let's take a look at an example to help you imagine how you can create a nice nest egg for your own retirement.

Example 7: Calculating Monthly Payments for a Retirement Fund

Assume you have just graduated from college and have obtained your first job making $44,250 per year. Currently, the national average retirement income for people with a bachelor's degree is $60,000 per year. Assuming you are 22 years of age at graduation and you will retire at age 67, then you will be working—and investing—for 45 years. At current rates, a mutual fund has an APR of 8%. How much do you need to invest each month in order to live off of $60,000 per year when you retire?

Solution

You want to build an account large enough so that you can have an annual annuity payment of $60,000 without reducing the principal on your retirement fund. In other words, you will live on the interest the account earns. We can phrase this another way by asking, "What balance is needed so that it earns $60,000 in annual interest?" Since we are assuming there is an APR of 8%, the $60,000 must be equal to 8% of the total balance. We can write this as a mathematical equation, letting x equal the principal needed, and then solve for x as shown below.

$$0.08 \cdot x = 60,000$$

$$x = \frac{60,000}{0.08}$$

$$x = 750,000$$

With an 8% APR, a balance of $750,000 allows you to withdraw $60,000 per year without reducing the principal. Now that we know the total amount needed to generate our $60,000 per year income, we can use the monthly annuity formula to compute the monthly payment amount needed to meet this future value goal.

We have that $FV = \$750,000$ from our calculation, $r = 0.08$, $t = 45$ years, and $n = 12$.

$$PMT = FV \cdot \frac{\left(\dfrac{r}{n}\right)}{\left[\left(1 + \dfrac{r}{n}\right)^{nt} - 1\right]}$$

$$= 750,000 \cdot \frac{\left(\dfrac{0.08}{12}\right)}{\left[\left(1 + \dfrac{0.08}{12}\right)^{12 \cdot 45} - 1\right]}$$

$$\approx 142.19$$

So, if you wish to retire and make $60,000 per year without affecting the principal in your retirement fund, you need to start at age 22 depositing approximately $143 per month. With a fixed APR of 8% over the years of your retirement, you will never outlive your retirement savings.

Skill Check Answers

1. $34,931.95

2. $16,359.26

3. $1,242,475.48

9.3   Exercises

For each situation, use the formula for present value of money to calculate the amount you need to invest now in one lump sum. Round your answer to the nearest cent, if necessary.

1. $25,000 after 10 years with an APR of 8% compounded monthly

2. $25,000 after 10 years with an APR of 12% compounded monthly

3. $100,000 after 18 years with an APR of 6% compounded quarterly

4. $1,000,000 after 40 years with an APR of 10% compounded monthly

Solve each problem.

5.   Alexis and Will are purchasing a home. They wish to save money for 10 years and purchase a house that has a value of $180,000 with cash. If they deposit money into an account paying 12% interest, how much do they need to deposit each month in order to make the purchase?

6.   Marilyn wishes to retire at age 65 with $2,000,000 in the bank. At the age of 21, she decides to begin depositing money into an account with an APR of 11%. What is the monthly payment Marilyn must make in order to make this happen?

7.   Repeat Exercise 6 with an APR of 6%.

8.   Repeat Exercise 6 with a desired retirement amount of $1,500,000.

9.   Suppose you wish to retire at the age of 65 with $80,000 in savings. Determine your monthly payment into an IRA if the APR is 7.5% and you begin making payments at

   a. 20 years old.

   b. 30 years old.

   c. 40 years old.

10. Revere College predicts that in 18 years it will take $200,000 to attend the college for four years. Debbie wishes to save money for her child's college fund. How much should Debbie put aside in an account with an APR of 9% compounded monthly in order to have $200,000 in the account in 18 years?

11. Repeat Exercise 10 with an interest rate earned of 5%.

12. Repeat Exercise 10 with an interest rate earned of 3.5%.

13. Suppose you'd like to save enough money to pay cash for your next car. The goal is to save an extra $26,000 over the next 6 years. What amount of quarterly payments must you make into an account that earns 5.5% interest in order to reach your goal?

14. Repeat Exercise 13 if the interest rate earned is 6.5%.

15. Repeat Exercise 13 if the interest rate earned is 3.5%.

16. Willie deposits a fixed monthly amount into an annuity account for his child's college fund. He wishes to accumulate a future value of $75,000 in 15 years.

    a. Assuming an APR of 3.5%, how much money should Willie deposit monthly in order to reach his goal?

    b. How much of the $75,000 will Willie ultimately deposit in the account, and how much is interest earned?

17. Repeat Exercise 16 with an APR of 6%.

18. Repeat Exercise 16 with an accumulated amount of $125,000.

19. Blake starts an IRA (Individual Retirement Account) to save for retirement at the age of 22. He deposits $450 each month. The IRA has an average annual interest rate of 7%.

    a. How much money will he have saved upon retirement at the age of 65?

    b. Determine the amount of money Blake deposited over the length of the investment and how much he made in interest.

20. Jimmie has a job at an advertising agency earning $54,000 per year. Jimmie is currently 26 years old and wishes to retire at age 67 with a retirement income of $75,000. How much money would Jimmie need to invest each month into a growth stock mutual fund with an interest rate of 6.5% in order to withdraw $75,000 per year without reducing the principal?

9.4 Borrowing Money

Good financial habits, such as creating a budget and saving for the future, are key to staying financially healthy. However, even with the best budgeting skills, sometimes we must borrow money in order to make large purchases, such as a car or a house, or to help bridge the gap when emergencies arise. In this section, we will look at credit cards, car loans, and mortgages to learn best practices of borrowing money.

Credit Cards

☞ Helpful Hint

A creditor is any organization, such as a bank or credit card company, that lends money with the idea that when you pay the money back, you will also pay the interest charged as well.

Just as interest worked in our favor when saving money, borrowing money can make interest work against us. The financial institutions that loan the money are the ones earning interest, not the individual. Most savings accounts offer an APR of 10% or less. However, when borrowing money, institutions often charge a much higher rate of interest, such as 14.99%, 19.99%, or even 29.99%. In some cases the interest rate can soar even higher if payments are missed or are late.

We will begin by looking at credit cards and how they work. Before we do, one important fact we need to point out is that despite the fact that we're very used to seeing most plastic cards with a Visa or MasterCard emblem on them, not all of these cards are actual credit cards. Many are in fact debit cards, also known as *bank cards* or *check cards*. Debit cards are not cards for borrowing money, they are simply an electronic means by which consumers can access the money already in their bank accounts.

Credit cards work on the principle that money is lent out for a month. If the money is not repaid in full at the end of the month, and a balance remains, a charge of interest is incurred upon the balance. The month where no interest is accruing is called a **grace period**. However, if a balance is carried over to the next month, interest continues to accrue until the balance is 0.

> ### Grace Period
>
> A **grace period** is a period of time in which no interest accrues on a debt.

The interest rate on a credit card is often a *variable interest rate*. This means that the interest rate fluctuates over time based on a particular benchmark, often a country's prime rate of interest. Notice that Figure 1 shows the APR for a customer's credit card. The APR on the left hand side is for purchases only and is a variable rate, which means it is subject to change.

Annual Percentage Rates
Your annual interest rate as of August 22:

17.24% | **25.24%**
Purchases | Cash Advances
Variable Rate | Variable Rate

Figure 1: Credit Card Interest Rates

The APR on the right hand side is for cash advances. It also is a variable interest rate, but is quite a bit higher than the rate for purchases. This is standard practice for most credit cards. Unfortunately, most cardholders don't realize that other fees also exist for cash advances. Initially, an up front fee is charged, usually 2% of the amount. This is to cover the fee that is normally incurred by the merchant when a purchase is made. Along with the higher interest rate and the fee, there is also no grace period. In other words, the cash advance starts accruing interest the minute it is dispensed from the ATM. Lesson to learn: cash advances on credit cards should only be used in emergencies!

Using credit cards wisely means that you only charge what you can afford to pay off in one month. If it is the case that you cannot pay off the balance by the end of the month, then you must make a partial payment. The **minimum payment**, which is the minimum amount the lender requires you to pay each month, is usually equal to a percentage of your average balance due, say 2%. So, if your average daily balance for a certain month is $1000, your minimum payment for that month will be $20. Many people assume that it's a good deal to pay $20 per month in order to pay off a debt of $1000. However, what people tend to forget is that they are accruing interest on the balance every month it goes unpaid. Therefore, this is really not a good deal at all. If the interest rate is as high as 20% on the balance owed, the $20 payment each month goes almost completely towards interest, and has very little impact on reducing the total balance of the credit card. By requiring that you pay only small amounts each month, creditors know that it will take you much longer to pay off your debt, and that you'll end up paying a lot more interest over time. Of course, this is how lenders actually make money and stay in business.

Consider paying off a $1000 credit card balance by paying the minimum payment each month. Most of the payment at the beginning is simply interest, but as more and more payments are made, the balance is reduced. When the balance shrinks, so does the interest accrued. Since the minimum payment is a percentage of the balance, it shrinks as well. As you might imagine, this smaller and smaller payment method looks good to the consumer, but it is actually prolonging the debt.

Figure 2 is a graph showing the payment time line for a $1000 credit card balance with 14% interest where the minimum payment due is 4% of the balance each month—this is assuming no new additional purchases are made. Notice that after about six years (72 months) the balance is finally less than $120.00. This means that the minimum payment amount is less than $5.00 each month. Sounds good, but the payments will continue to go on and on. At this rate, the balance will never be completely $0. To prevent this from happening, credit card companies have a set minimum payment amount that will eventually force you to pay the balance in full.

Figure 2

Let's contrast that with a different picture of making payments. Suppose the same $1000 debt was paid off monthly with a consistent payment of $40 per month. Even though this is the starting minimum payment, it will soon become more than the required minimum payment and hence bring down the balance more quickly. Figure 3 shows the payment time line for this regular minimum payment. You can see that after about 30 payments, the debt is paid off.

Figure 3

Figure 4 shows just how pronounced the difference is by graphing both payoff schemes on one graph.

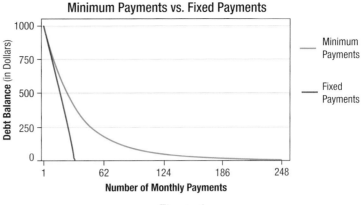

Figure 4

In May 2009, Congress concluded that certain practices in the credit card industry were neither fair nor transparent to consumers, and so they signed into law the Credit Card Accountability, Responsibility, and Disclosure Act (Credit CARD Act) with strong support in both the Senate and House of Representatives. Among other things, this Act protects consumers from excessive charges and wildly changing APRs. As a result of the Credit CARD Act, many credit card statements now provide information to their customers about paying off the debt. [1]

Figure 5 shows an example of the information given on a credit card statement. Notice also that there is a help line number for consumer debt. Although this number is fictitious as printed in this text, the Credit CARD Act requires that creditors provide a toll-free telephone number for the purposes of providing information about accessing credit counseling and debt management services.

Fun Fact

Section III of the Credit CARD Act pertains specifically to young consumers. It provides guidelines directed at helping consumers under 21 develop wise credit habits.

Late Payment Warning: If we do not receive your minimum payment due by the payment due date listed, you may have to pay a late fee of up to $35.00 and your purchase APR may be increased to the penalty APR of 27.24%.

Minimum Payment Warning: If you make only the minimum payment each period, you will pay more in interest and it will take you longer to pay off your balance. For example:

If you make no additional charges and each month you pay. . .	You will pay off the balance shown on this statement in about. . .	You will pay an estimated total of. . .
Only the minimum payment due	7 years	$2771
$60	3 years	$2152 (Savings = $619)
If you would like information about credit counseling services, call 1-800-555-0199.		

Figure 5: Credit Card Warnings

Although many consumers may have never taken the time to notice this type of warning on their monthly statement, the creditor is actually showing you how bad it is for you to only pay them the minimum payment. They even go as far as to tell you how much money you could save!

We can use the following formula to calculate the number of fixed payments required to pay down a debt that is accruing interest monthly, just like the credit card company did.

Number of Fixed Payments Required to Pay Off Credit Card Debt

The **number of fixed payments R required to pay off a credit card debt** is calculated with the formula

$$R = \frac{-\log\left[1 - \dfrac{r}{n}\left(\dfrac{A}{PMT}\right)\right]}{\log\left(1 + \dfrac{r}{n}\right)}$$

where n is the number of payments made per year for a loan of amount A, with interest rate r and monthly payment PMT.

Example 1: Paying Off Credit Card Debt with a Fixed Payment

Assume you want to buy a new computer that costs $2200 using a credit card that has an APR of 19.99%.

a. How long will it take you to pay off the computer if you make regular monthly payments of $40?

b. How much will you pay in the long run for the computer if you make monthly payments of $40?

c. How long will it take you to pay off the computer if you make regular monthly payments of $80?

d. How much will you pay in the long run for the computer if you make monthly payments of $80?

Solution

a. We wish to find the number of fixed payments required to pay off a credit card debt. We are told that the debt is $A = 2200$ and the APR = 19.99%, so $r = 0.1999$. The monthly payment is $PMT = 40$ and $n = 12$. Substituting these values into the formula, we have the following.

$$R = \frac{-\log\left[1 - \dfrac{r}{n}\left(\dfrac{A}{PMT}\right)\right]}{\log\left(1 + \dfrac{r}{n}\right)}$$

$$= \frac{-\log\left[1 - \dfrac{0.1999}{12}\left(\dfrac{2200}{40}\right)\right]}{\log\left(1 + \dfrac{0.1999}{12}\right)}$$

$$= 150.0760203$$

Therefore, it will take approximately 150 monthly payments, or $12\frac{1}{2}$ years, to payoff the debt.

b. If we make 150 payments at $40 each, then over $12\frac{1}{2}$ years we will have paid a total of

$$150 \cdot \$40 = \$6000.$$

As you might imagine, the computer would be considerably out of date in $12\frac{1}{2}$ years, and you paid nearly triple the original price of the computer. This is not a wise method for purchasing a computer.

c. This time, we will double the monthly fixed payment so the $PMT = 80$. The remaining variables stay the same; that is, $A = 2200$, $r = 0.1999$, and $n = 12$. Using the formula, we have the following.

$$R = \frac{-\log\left[1 - \frac{r}{n}\left(\frac{A}{PMT}\right)\right]}{\log\left(1 + \frac{r}{n}\right)}$$

$$= \frac{-\log\left[1 - \frac{0.1999}{12}\left(\frac{2200}{80}\right)\right]}{\log\left(1 + \frac{0.1999}{12}\right)}$$

$$= 37.084776$$

Therefore, it will take approximately 37 monthly payments, or just over three years, to pay off the debt.

d. If we make 37 payments at $80 each, then over three years we will have paid a total of

$$37 \cdot \$80 = \$2960.$$

This is a much more reasonable method for purchasing the computer. Although the computer still might be outdated in three years, it will probably still be functionally acceptable.

TECH TRAINING

Use the following keystrokes on a TI-30XIIS/B or TI-83/84 Plus calculator for the calculation in part **c.** of Example 1.

[(−)] [LOG] 1 [−] 0.1999 [÷] 12 [×] [( ] 2200 [÷] 80 [ )] [ )] [÷] [LOG] 1
[+] 0.1999 [÷] 12 [ )]

To perform this calculation with Wolfram|Alpha, go to www.wolframalpha.com. In the input bar, type "(−log(1−(0.1999/12)(2200/80))/log(1+0.1999/12)" then click the = button.

Car Loans and Mortgages

Credit cards lend money that is paid back over an unspecified period of time often with a variable interest rate. One of the dangers in using this type of loan is the risk of buying more than you can afford. Loans such as car loans or house mortgages don't use open-ended time frames like credit cards. Instead, they operate as fixed installment loans. For **fixed installment loans**, terms for the amount of the loan and the length of time it will take to pay off the loan are agreed upon up front so that both the lender and the borrower know exactly how much the loan will cost. Because everything is decided at the beginning of the loan period, fixed installment loans use *fixed interest rates*—rates where the APR never changes over the life of the loan. Normally, a **down payment** is required for a fixed installment loan. A down payment is a cash payment made up front towards the total purchase price. It usually equals a certain percentage, often 5% to 25%, of the total value of the purchase. Once the down payment is subtracted from the original price, the amount remaining is the principal that needs to be *financed*, or borrowed with interest.

> ### Fixed Installment Loans
>
> **Fixed installment loans** have a fixed interest rate and are paid off with monthly payments over a specified amount of time.
>
> ### Down Payment
>
> A **down payment** is a cash payment made up front towards the total purchase price of the goods or service.

When a fixed installment loan is taken out, several factors go into determining the monthly payment amount. The initial consideration is the size of the down payment. A larger down payment can often mean a better interest rate. For new car purchases, it is a wise rule of thumb to put down at least 20% of the car's price. Because vehicles *depreciate*, or decrease in value, very quickly, the value of your car depreciates by about 20% the minute you drive the new car off the lot. So, if you had an emergency and had to sell your new car, it would now only be worth 80% of its original value, and you don't want to owe more than it's worth!

Since the recession of 2007, down payment requirements have become stricter for the average house loan, or *mortgage*. In general, most mortgages require anywhere from 10% to 25% of the purchase price as a down payment. There are exceptions, however, that may decrease down payments to around 3% for first-time home buyers, or increase them to 35% for "credit-challenged" borrowers.

The other consideration is the length of time over which the loan will be paid off. For car loans, the time is usually between three and six years. Although a longer time period means lower payments, it also can mean higher interest rates. We already know from the earlier sections that higher interest rates over longer periods of time mean more interest accrues and hence you pay more for the car! For house mortgages, the length of a loan with a fixed interest rate is usually either 15 years or 30 years. Just as with car loans, a bigger down payment and shorter amount of time can mean a lower interest rate.

The following formula is used to calculate the monthly payment amount for fixed installment loans.

> ### Monthly Payment Formula for Fixed Installment Loans
>
> The amount of a **monthly payment** (*PMT*) **on a fixed installment loan** is calculated with the formula
>
> $$PMT = \frac{\left(P \cdot \dfrac{r}{n}\right)}{\left[1 - \left(1 + \dfrac{r}{n}\right)^{-nt}\right]}$$
>
> where *P* is the principal amount of money borrowed, *r* is the APR, *n* is the number of payments per year, and *t* is the number of years of the loan.

☞ Helpful Hint

Notice that the formula contains a negative exponent in the denominator. Be careful not to drop the negative sign during calculations.

Example 2: Purchasing a New Car

Kelly wishes to purchase a new car. The car she has chosen has a price of $34,000, including taxes and fees. She chooses to make a down payment of 20% of the price and wants to finance the remainder. If Kelly has acquired an APR of 3.99% for a 72-month loan, what is the amount of her monthly payment?

Solution

First we need to know the amount of the down payment Kelly will pay up front. We can find 20% of $34,000 by multiplying the two together. Remember to change the percentage to a decimal in order to multiply.

$$\text{down payment} = 0.20 \cdot \$34000 = \$6800$$

Therefore, Kelly will need to finance the remainder, found by subtracting the down payment from the original price of the car.

$$\text{principal to finance} = \$34,000 - \$6800 = \$27,200$$

Now we have that Kelly will borrow $P = 27,200$ for $t = 6$ years with an APR of 3.99%, so $r = 0.0399$. Substituting into the formula for fixed installment loans, we can calculate what Kelly's monthly payment will be.

$$PMT = \frac{\left(P \cdot \dfrac{r}{n}\right)}{\left[1 - \left(1 + \dfrac{r}{n}\right)^{-nt}\right]}$$

$$= \frac{\left(27,200 \cdot \dfrac{0.0399}{12}\right)}{\left[1 - \left(1 + \dfrac{0.0399}{12}\right)^{-12 \cdot 6}\right]}$$

$$= 425.425058$$

So Kelly's monthly payment for her new car would be $425.43.

TECH TRAINING

To find the payment amount using Excel, type in "= − PMT(rate,nper,pv,fv,type)" where

- $rate = APR/n$
- $nper =$ number of payment periods for the loan
- $pv =$ the principal of the loan
- $fv =$ the future value that is assumed to be 0 for a loan
- $type = 0$ for payments due at the end of the period and 1 for payments due at the beginning of the period

We type a negative sign in front of the payment formula so that the payment amount appears as a positive amount.

For our example, enter the following.

"=−PMT (3.99%/12,72,27200,0,0)"

The payment amount can be found using Wolfram|Alpha. Go to www.wolframalpha.com and type "$(27200*(0.0399/12))/(1-(1+0.0399/12)^{\wedge}(12*6))$" into the input bar. Then, click the = button.

Skill Check #1

Jeff wants to buy a new truck that costs $48,000. After a down payment of $10,000 he finances $38,000. If Jeff has an APR of 5% for a 48-month loan, what is Jeff's monthly payment?

Example 3: Determining Best New Car Incentive

Dean gets to choose from one of the new car incentives when he purchases his car next week. He can either choose 0.9% APR financing for 60 months or $1500 cash back with a 3.75% APR over 48 months. Compare the two incentives that Dean has to choose from if the new car he wishes to buy is $27,465 and he has saved a down payment of $5000.

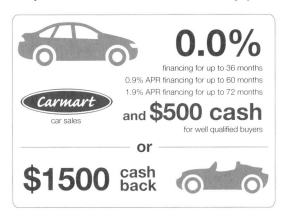

Solution

Let Option A be the 0.9% APR financing for 60 months and Option B will be $1500 cash back with a 3.75% APR over 48 months.

Option A: For this option, the principal to be financed is

$$\text{principal to be financed} = \$27{,}465 - \$5000 = \$22{,}465.$$

In addition, $r = 0.009$ and $t = 60 \div 12 = 5$. The payments each month are calculated as follows.

$$
\begin{aligned}
PMT &= \frac{\left(P \cdot \dfrac{r}{n}\right)}{\left[1 - \left(1 + \dfrac{r}{n}\right)^{-nt}\right]} \\[2em]
&= \frac{\left(22{,}465 \cdot \dfrac{0.009}{12}\right)}{\left[1 - \left(1 + \dfrac{0.009}{12}\right)^{-12 \cdot 5}\right]} \\[2em]
&= 383.0445874
\end{aligned}
$$

Dean would have a monthly payment of $383.04 for 60 months with Option A. His total cost for the car would be

$$\$383.04 \cdot 60 + \$5000 = \$27{,}982.40$$

Option B: For this option the APR is considerably more at 3.75%, but Dean can use the $1500 cash back to add to his down payment so that he has less to finance. Therefore, he will only need to finance

$$\$27{,}465 - \$5000 - \$1500 = \$20{,}965$$

For this option $r = 0.0375$ and the length of time is now $t = 48 \div 12 = 4$. So the payments each month will be the following.

$$PMT = \frac{\left(P \cdot \dfrac{r}{n}\right)}{\left[1 - \left(1 + \dfrac{r}{n}\right)^{-nt}\right]}$$

$$= \frac{\left(20{,}965 \cdot \dfrac{0.0375}{12}\right)}{\left[1 - \left(1 + \dfrac{0.0375}{12}\right)^{-12 \cdot 4}\right]}$$

$$= 471.0281082$$

Therefore, Dean's payments for Option B will be $471.03 for 48 months. His total cost for the car would be

$$\$471.03 \cdot 48 + \$5000 = \$27{,}609.44$$

Now that the calculations for both options have been done, you can see that Dean will pay less overall if he uses Option B. Note that the difference is about $400 between the two options. There are other factors that might sway Dean's decision. First, Option B only saves him money if he is disciplined and puts the cash back that he received towards the purchase of the car. However, the car is completely paid off an entire year earlier with this option. Dean will also need to consider which monthly payment option he can afford. Option A has the lower monthly payment. Now that Dean understands the outcomes of total cost based on length of the loan and interest rate, he can use his knowledge of budgeting and financial decision-making to choose the best option for his financial situation.

TECH TRAINING

Using the Excel payment formula, enter the following to obtain the monthly payment from Example 3.

For 48 months: "=−PMT(0.0375/12,48,20965,0,0)"

For 60 months: "=−PMT(0.009/12,60,22465,0,0)"

Now let's look at longer term fixed installment loans used for purchasing a house.

Example 4: Calculating Monthly Mortgage Payments

William and Helen are preparing to buy a new home in the Midwestern United States. They have saved $21,300 for a down payment. The total price of the house is $131,800, including taxes and fees. Find their monthly mortgage payment if the fixed interest rate is 3.52% for a 30-year loan.

Solution

Since William and Helen have a down payment, we can subtract that from the purchase price to find the amount they will need to finance to buy the home.

$$\text{principal amount to finance} = \$131,800 - \$21,300 = \$110,500$$

We are told that the APR = 3.52%, so $r = 0.0352$ and $t = 30$ years. Using the payment formula for fixed installment loans we can calculate their monthly mortgage payment.

$$PMT = \frac{\left(P \cdot \dfrac{r}{n}\right)}{\left[1 - \left(1 + \dfrac{r}{n}\right)^{-nt}\right]}$$

$$= \frac{\left(110,500 \cdot \dfrac{0.0352}{12}\right)}{\left[1 - \left(1 + \dfrac{0.0352}{12}\right)^{-12 \cdot 30}\right]}$$

$$= 497.4288455$$

Therefore, William and Helen will have a monthly mortgage of $497.43.

TECH TRAINING

Using the Excel payment formula, enter the following to obtain the monthly payment from Example 4.

"=−PMT(3.52%/12,12*30,110500,0,0)"

In 2007, the United States had a major financial crisis occur when variable interest rates on home loans escalated, causing many homeowners to lose their homes. As a result, financial education about mortgages became key for consumers. According to financial experts, monthly home mortgage payments should cost less than 25% of your monthly take-home salary to avoid being overburdened with home loan debt. This means that if your monthly take-home pay is $2000.00, your monthly mortgage payment should be no more than $500.00 per month.

If we want to stay within the recommended monthly mortgage payment—that is, 25% of our take-home pay—we can rearrange the payment formula to help us determine the maximum amount we should spend on buying a house. The following formula can be used to calculate the maximum purchase price attainable based on a set monthly payment and a predetermined interest rate and time period.

Maximum Purchase Price Formula

The **maximum purchase price** for a house can be found by

$$\text{maximum purchase price} = PMT \cdot \frac{\left[1 - \left(1 + \dfrac{r}{n}\right)^{-nt}\right]}{\left(\dfrac{r}{n}\right)}$$

where PMT is the monthly mortgage payment you wish to make, r is the APR, n is the number of payments per year, and t is the number of years of the loan.

Example 5: How Much House Can You Afford?

Suppose you have recently graduated from college and want to purchase a house. Your take-home pay is $3220 per month and you wish to stay within the recommended guidelines for mortgage amounts by only spending $\frac{1}{4}$ of your take-home pay on a house payment. You have $15,300 saved for a down payment. With your good credit and the down payment you can get an APR from your bank of 3.37% compounded monthly.

a. What is the total cost of a house you could afford with a 15-year mortgage?

b. What is the most that you could afford with a traditional 30-year mortgage instead of a 15-year?

Solution

a. The first thing to do is to calculate the size of the monthly mortgage payment you are willing to spend. Since you have $3220 per month in take-home pay, multiply this by 25% to find your maximum monthly payment.

$$\text{advised monthly payment} = \$3220 \cdot 0.25 = \$805$$

We know that $r = 0.0337$ and that because this is a 15-year mortgage, $n = 12$ and $t = 15$. Substituting these values in the formula, we have the following.

$$\text{maximum purchase price} = PMT \cdot \frac{\left[1 - \left(1 + \dfrac{r}{n}\right)^{-nt}\right]}{\left(\dfrac{r}{n}\right)}$$

$$= 805 \cdot \frac{\left[1 - \left(1 + \dfrac{0.0337}{12}\right)^{-12 \cdot 15}\right]}{\left(\dfrac{0.0337}{12}\right)}$$

$$= 113,617.8221$$

So, you could afford a 15-year mortgage of approximately $113,617.82.

Remember that the amount of down payment you have available will add to the maximum amount you can spend on a house. Therefore, with a 15-year mortgage you can afford to buy a house with a maximum price of

$$\$113,617.82 + \$15,300 = \$128,917.82.$$

b. For a 30-year mortgage, the only thing that changes is t. Now $t = 30$. The monthly payment you can afford stays the same, as well as the interest rate and n. Therefore, we have the following.

$$\text{maximum purchase price} = PMT \cdot \frac{\left[1 - \left(1 + \dfrac{r}{n} \right)^{-nt} \right]}{\left(\dfrac{r}{n} \right)}$$

$$= 805 \cdot \frac{\left[1 - \left(1 + \dfrac{0.0337}{12} \right)^{-12 \cdot 30} \right]}{\left(\dfrac{0.0337}{12} \right)}$$

$$= 182{,}201.1079$$

By this calculation, you should be able to afford a house that costs approximately $182,200 plus your down payment.

With the down payment added in, your total purchase price with a 30-year mortgage could be as much as

$$\$182{,}200 + \$15{,}300 = \$197{,}500.$$

This amount is considerably more than the 15-year mortgage, but remember that you will incur much more interest over the 30 years as well. In addition, we assumed that you could get the same interest rates for both time periods, but in reality, the 30-year mortgage would probably have a higher interest rate as well.

TECH TRAINING

Use the following keystrokes on a TI-30XIIS/B or TI-83/84 Plus calculator for the calculation of the maximum purchase price for part **a.** of Example 5.

805 $\times$ (1 $-$ (1 $+$ 0.0337 $\div$ 12) $\wedge$ ($\pm$ 12 $\times$ 15)) $\div$ (0.0337 $\div$ 12) $=$

Wolfram|Alpha can be used to find the maximum purchase price. Go to www.wolframalpha.com and type "805*((1−(1+0.0337/12)^(−12*15))/(0.0337/12)" into the input bar. Then, click the = button.

☞ Helpful Hint

If you are using a TI-30XS/B Multiview calculator, the keystrokes will be slightly different for your calculator. See Appendix B for these instructions.

Good Financial Habits

To end this chapter we want to point out some of the best practices for healthy financial living.

Make a budget. First and foremost, begin the habit of budget making early. Why not start today? It might surprise you how much money you are spending in certain areas, or how much money you could afford to be putting into an interest earning savings account for later. Always make room for savings. You can't afford not to. Remember to let time be on your side when saving for the big things.

Control impulse spending. One of the biggest budget breakers is impulse spending, or spending on a whim. Eating out, online purchases, or even picking up a small thing here and there when shopping can add up quickly. Evaluate your needs versus your wants. A budget will help you keep up with this.

Have an emergency fund. Saving for a goal is great, but don't forget that life unexpectedly throws you financial curve balls. It can happen to any of us.

Be a smart consumer. Look in the fine print or ask about extra fees or penalties when borrowing money. Some loans, especially mortgages, can carry a penalty for paying the loan off early.

Monitor your accounts and pay your bills on time. Use the online resources that most banks provide free of charge and check your account balance often. One of the best ways to get good interest rates on loans is to have a healthy credit score. Although we don't go into detail in the text about the components that make up a personal credit score, paying your bills on time, every time, will give you a good start to a good score.

Ask questions. Finally, don't assume you can't ask for help with financial matters. Ask questions and shop around to look for the best possibilities for you. Often, there isn't just one right answer that fits everyone's needs. There are plenty of resources available to the public, even on-line resources. Use them!

Skill Check Answer

1. $875.11

9.4   Exercises

Solve each problem. Round your answer to the nearest cent, if necessary.

1. Given the chart below, solve the following problems.

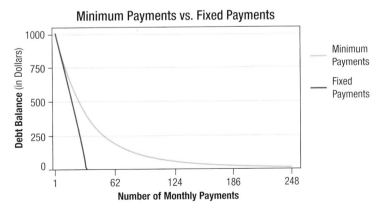

Minimum Payments vs. Fixed Payments

    a. Estimate the total amount paid when a debt balance was paid using a fixed monthly payment of $40.

    b. Estimate the total amount paid when a debt balance was paid using the minimum monthly payment of $18.

2.   Rachel is purchasing a new camera that costs $3800 for her photography business. Rachel uses a credit card that has an APR of 16.99%.

   a. How long will it take her to pay off the camera if she makes monthly payments of $75?

   b. How much will she pay in the long run for the camera if she makes monthly payments of $75?

   c. How long will it take her to pay off the camera if she makes monthly payments of $150?

   d. How much will she pay in the long run for the camera if she makes monthly payments of $150?

3.   Tommy gets to choose from one of the new car incentives when he purchases his car next week. He can either choose 0.9% APR financing for 48 months or $1000 cash back with a 4.75% APR over 48 months. Compare the two incentives that Tommy has to choose from if the new car he wishes to buy is $32,457 and he has saved a down payment of $3500.

4.   Mike bought a new car and financed $25,000 to make the purchase. He financed the car for 60 months with an APR of 6.5%. Determine each of the following.

   a. Mike's monthly payment

   b. Total cost of Mike's car

   c. Total interest Mike pays over the life of the loan

5.   Omar wants to purchase three vans for his delivery business. Each van costs $38,000. He wishes to finance the purchase for 48 months and has acquired an APR of 4.5%. Determine each of the following.

   a. Omar's monthly payment

   b. Total cost of Omar's vans

   c. Total interest paid by Omar over the life of the loan

6.   Jamal bought a new car for $32,000. He paid a 10% down payment and financed the remaining balance for 36 months with an APR of 4.5%. Determine each of the following.

   a. Jamal's monthly payment

   b. Total cost of Jamal's car

   c. Total interest Jamal pays over the life of the loan

7.   Susan wants to buy a new computer from Banana Computers. The company sells a laptop model for $2650. Susan decides to finance the computer for 24 months at an APR of 12.5%. Determine each of the following.

   a. Susan's monthly payment

   b. Total cost of the computer

   c. Total interest paid over the 24 months

8.   Amanda and Ferobee are buying a house on a 30-year mortgage. They can only pay $800 per month for a mortgage. If they have an APR of 3.75%, what is the maximum price of a mortgage that they can take out?

9. Brad decides to purchase a $250,000 house. He wants to finance the entire balance. He has received an APR of 4.5% for a 30-year mortgage.

   a. What is Brad's monthly payment?

   b. Over the course of the loan, how much interest will Brad pay?

   c. What is Brad's total cost if he takes all 30 years to pay off the house?

   d. If he changed the term to 15 years instead of 30 years, what would his monthly payment be?

   e. With a 15-year mortgage, how much interest will Brad pay?

   f. With a 15-year mortgage, what is the total cost of the house?

10. The city of Nettleton recently completed a new school building. The entire cost of the project was $19,000,000. The city has put the project on a 20-year loan with an APR of 2.4%. There are 15,000 families that will be responsible for paying the loan.

    a. Determine the amount of the monthly payment for the loan.

    b. Determine the amount that each family should be required to pay each year to cover the cost of the school.

    c. Determine the total cost of the school.

11. You want to buy a car and finance $20,000 to do so. You can afford a payment of up to $450 per month. The bank offers three choices for the loan: a four-year loan with an APR of 7%, a five-year loan with an APR of 7.5%, and a six-year loan with an APR of 8%. Which option best meets your needs, assuming you want to pay the least amount of interest?

Consider a credit card with a balance of $7000. You wish to pay off the credit card in each scenario. Calculate the following. Round your answer to the nearest cent, if necessary.

   a. The amount of a monthly payment within the time frame given

   b. The total amount paid over the time period

12. APR of 17.99% paid off within 1 year

13. APR of 12.5% paid off within 2 years

14. APR of 24% paid off within 3 years

Consider a credit card with a balance of $5560. You wish to pay off the credit card in each scenario. Calculate the following. Round your answer to the nearest cent, if necessary.

   a. The amount of a monthly payment within the time frame given

   b. The total amount paid over the time period

15. APR of 14.99% paid off within 1 year

16. APR of 11.99% paid off within 2 years

17. APR of 5.9% paid off within 3 years

Answer each question. Round your answer to the nearest cent, if necessary.

18. A credit card has a balance of $5000 at an APR of 9.99%. You plan to pay $500 each month in an effort to clear the debt quickly. How long will it take you to pay off the balance?

19. A credit card has a balance of $11,500 at an APR of 14.99%. You plan to pay $650 each month in an effort to clear the debt quickly. How long will it take you to pay off the balance?

20. Suppose you have a student loan of $80,000 with an APR of 4.5% for 25 years.

   a. What is your monthly payment?

   b. If you decide you want to pay off the loan in 15 years instead of 25, what is your monthly payment?

   c. What is your savings for paying the loan off in 15 years instead of 25?

21. Suppose you have graduated from college and want to purchase a house. Your take-home pay is $4560 per month and you wish to stay within the recommended guidelines for mortgage amounts by only spending $\frac{1}{4}$ of your take-home pay on a house payment. You have $18,500 saved for a down payment. With your good credit and the down payment you can get an APR from your bank of 4.35%, compounded monthly.

   a. What is the total cost of a house you could afford with a 15-year mortgage?

   b. What is the most that you could afford with a traditional 30-year mortgage instead of a 15-year?

22. What if you won the Powerball Lottery with a jackpot of $150 million? Calculate the amount of money you would receive over a 25 year period with each of the following two options. Which option gives you the most money over the 25 years?

   Option 1: Taking all the money at once with a 40% penalty, and pay the income tax of 38% on the lump sum, and investing the remaining amount into an account earning 6% interest for 25 years.

   Option 2: Acquire the money as part of an annuity to be paid out in 25 equal payments over a 25-year period, paying the income tax of 38% on the income from the winnings each year.

Chapter 9   Summary

Section 9.1   Understanding Personal Finance

Definitions

Steps to Create a Budget

1. Calculate monthly and periodical income.

2. Calculate monthly and periodical expenses.

3. Subtract total expenses from income.

4. Allocate any remaining funds to savings, long-term goals, or an emergency fund.

Net Income

Net income is equal to total, or gross, income minus all taxes.

List Price

The price of an item as it is listed for public sale is the list price.

Discount

The discount is the reduction from the list price. This is usually given as a percentage of the list price.

Formulas

Sale Price

The sale price, sometimes called the net price, is the actual cost of an item after a discount of some kind is applied.

$$\text{sale price} = \text{list price} - \text{discount}$$

Percentage Change

The percentage change over a period of time can be calculated by the following formula.

$$\text{percentage change} = \frac{\text{new value} - \text{reference value}}{\text{reference value}} \cdot 100\%$$

Section 9.2   Understanding Interest

Definitions

Interest

Interest is the amount charged by a lender for borrowing money.

Principal

The principal is the sum of money on which interest is charged.

Interest Rate

The interest rate is the amount charged to the borrower expressed as a percentage of the principal.

Annual Percentage Rate (APR)

Annual percentage rate is the yearly interest rate that is charged for borrowing. APR is normally given as a percentage per year.

Compound Interest

Compound interest is interest that is computed based on both the principal and the accrued interest as additional principal at each interval.

Formulas

Simple Interest Formula

The amount of interest on a simple interest loan with principal P, annual interest rate r (written as a decimal), and loan term of t (usually in years) is calculated with the following formula.

$$I = Prt$$

Compound Interest Formula

The future value is the total amount of money A that has been accrued after compounding at an annual percentage rate r based on the initial principal P with n compounding intervals per year for t years. Future value of a compound interest account is calculated with the following formula.

$$A = P\left(1 + \frac{r}{n}\right)^{nt}$$

Continuous Compound Interest Formula

The future value A of a continuous compound interest account after t years at an annual interest rate of r and an initial amount, or principal, P is calculated with the following formula.

$$A = Pe^{rt}$$

Annual Percentage Yield (APY)

The annual percentage yield is the effective annual interest rate earned in a given year that accounts for the effects of compounding. APY is calculated with the formula

$$APY = \left[\left(1 + \frac{r}{n}\right)^{n} - 1\right] \cdot 100\%,$$

where r is the APR and n is the number of compounding intervals per year.

Section 9.3 Saving Money

Definitions

Present Value

The present value of an investment is the amount of principal needed now in order to reach a future value amount.

Annuity

An annuity is a sequence of regular payments made into an account, or taken out of an account, over time.

Formulas

Present Value Formula

Present value (PV) is calculated with the formula

$$PV = \frac{A}{\left(1 + \frac{r}{n}\right)^{nt}}$$

where A is the future value, r is the annual percentage rate, n is the number of compound intervals per year, and t is time, or term, in years.

Annuity Formula for Finding Future Value

The future value (FV) of an annuity savings account is calculated with the formula

$$FV = PMT \cdot \frac{\left[\left(1+\dfrac{r}{n}\right)^{nt} - 1\right]}{\left(\dfrac{r}{n}\right)}$$

where PMT is the payment amount that is deposited on a regular basis, r is the APR, n is the number of regular payments made each year, and FV is the future value after t years.

Annuity Formula for Finding Payment Amounts

The regular payment amount (PMT) needed to reach a future goal with an annuity savings account is calculated with the formula

$$PMT = FV \cdot \frac{\left(\dfrac{r}{n}\right)}{\left[\left(1+\dfrac{r}{n}\right)^{nt} - 1\right]}$$

where FV is the future value desired, r is the APR, n is the number of regular payments made each year for t years.

Section 9.4   Borrowing Money

Definitions

Grace Period

A grace period is a period of time in which no interest accrues on a debt.

Fixed Installment Loan

Fixed installment loans have a fixed interest rate and are paid off with monthly payments over a specified amount of time.

Down Payment

A down payment is a cash payment made up front towards the total purchase price of the goods or service.

Formulas

Number of Fixed Payments Required to Pay Off Credit Card Debt

The number of fixed payments R required to pay off a credit card debt is calculated with the formula

$$R = \frac{-\log\left[1 - \dfrac{r}{n}\left(\dfrac{A}{PMT}\right)\right]}{\log\left(1 + \dfrac{r}{n}\right)}$$

where n is the number of payments made per year for a loan of amount A, with interest rate r and monthly payment PMT.

Monthly Payment Formula for Fixed Installment Loans

The amount of a monthly payment (PMT) on a fixed installment loan is calculated with the formula

$$PMT = \frac{\left(P \cdot \dfrac{r}{n}\right)}{\left[1 - \left(1 + \dfrac{r}{n}\right)^{-nt}\right]}$$

where P is the principal amount of money borrowed, r is the APR, n is the number of payments per year, and t is the number of years of the loan.

Maximum Purchase Price

The maximum purchase price for a house can be found by

$$\text{maximum purchase price} = PMT \cdot \frac{\left[1 - \left(1 + \dfrac{r}{n}\right)^{-nt}\right]}{\left(\dfrac{r}{n}\right)}$$

where PMT is the monthly mortgage payment you wish to make, r is the APR, n is the number of payments per year, and t is the number of years of the loan.

Chapter 9 Exercises

Solve each problem. Round your answer to the nearest cent, as necessary.

1. Timothy has an annual income of $38,750. Using the federal income tax rate of 15%, what amount should Timothy expect to pay in federal taxes?

2. William has a job where he has a take-home salary each month of $3375. If William wants to spend no more than 25% of his income on rent, how much rent can William afford?

3. Michael rents an apartment for $750 per month, pays his car payment of $360 per month, has utilities that cost $330 per month, and spends $476 per month on food and entertainment. Determine Jack's monthly expenses.

4. The given table represents the estimated cost of attending Harvard University in the 2013–14 academic year (10 months). Create a monthly expense budget for attending Harvard and determine a monthly income that would allow a student to attend Harvard while acquiring no debt.

Estimated Cost of Attending Harvard University for the 2013–14 Academic Year	
Budget Category	**On-Campus Student**
Tuition	$37,576
Room & Board	$13,630
Books & Supplies	$1350
Transportation	$5000
Personal	$3454
Health Insurance	$930
Total	**$61,940**

Source: Harvard College. "Tuition and Expenses." Accessed April 2013. http://www.admissions.college.harvard.edu/financial_aid/cost.html

5. A car with a list price of $32,547 will be discounted 35% at the time of purchase. What is the purchase price?

6. You purchase a pair of jeans with a list price of $110. If the sales tax is 8.5%, what is the total cost of the jeans?

7. At a restaurant, your total bill is $56.03. You wish to give a tip of 18% of the total bill. What is the amount of the tip?

8. During the housing price decline of 2009, the value of a house decreased by 25% in one year. If the 2010 value of the house was $126,750, what was the house worth prior to the decline?

9. The wholesale price of a coat is 65% less than the retail price. If the retail price is $129.99, what is the wholesale price?

10. A store is having a 75% off sale. The price of an item on sale is $176. What is the original cost?

11. You receive an 8% decrease in your pay. What percentage increase in pay would you have to receive in order to gain your original pay rate again? Round your answer to the nearest whole percent.

12. Determine the interest owed on $1200 for five years at a rate of 6.5%.

13. Assume you are purchasing a new computer with "90 days same as cash" to make the purchase. If the cost of the computer is $1255, tax included, with an annual interest rate of 16.99%, how much would you owe on the 91st day if you make no payments during the first 90 days?

14. Use the compound interest formulas to **a.** calculate the total amount in the account after the given time period and **b.** determine the amount of interest earned.

$P = \$5500$, $r = 2.5\%$ compounded weekly, $t = 10$ years

$P = \$4755$, $r = 4.5\%$ compounded monthly, $t = 10$ years

$P = \$7300$, $r = 19.9\%$ compounded continuously, $t = 20$ years

15. Assume you wish to borrow $750 for two weeks and the amount of interest you must pay is $17 per $100 borrowed. What is the APR at which you are borrowing money? Round your answer to the nearest whole percent.

16. A couple deposits $12,500 into an account earning 2.75% annual interest for 25 years.

a. Calculate the future value of the investment if interest is compounded monthly.

b. Calculate the future value of the investment if interest is compounded weekly.

17. Suppose your salary in 2012 is $46,500. If the annual inflation rate is 4%, what salary do you need to make in 2030 in order for it to keep up with inflation?

18. A payday loan is made for eight weeks, where the amount of interest owed per $100 borrowed is $15. Suppose you borrow $1000 for eight weeks.

a. How much do you owe at the end of eight weeks?

b. What is the APR for this transaction?

19. Miguel deposits $2850. Determine the annual percentage yield for each of the following. Round each answer to the nearest hundredth of a percent.

a. APR of 5% compounded monthly

b. APR of 5% compounded weekly

c. APR of 5% compounded daily

A savings account is compounded monthly for five years with an APR of 4.99%. For each principal amount, calculate the following. Round your answer to the nearest cent, as necessary.

   a. The future value of the investment

   b. The amount of interest earned

20. $15,000

21. $30,000

22. $60,000

23. $120,000

Solve each problem. Round your answer to the nearest cent, as necessary.

24. Suppose ABC Lending offers a CD (certificate of deposit) that has a 2.35% interest rate and is compounded quarterly for five years. You decide to invest $5500 into this CD.

   a. Determine how much money you will have at the end of the five years?

   b. Find the APY, rounded to the nearest hundredth of a percent.

25. ABC Lending lists the following APR rates for loans. Determine the APY for each category.

First Bank of Lending	
Loan Amount	**APR**
< $20,000	9.75
$20,000–$99,999	5.99
≥ $100,000	3.75
*interest rates are compounded quarterly	

26. Jay and Sybil are purchasing a home. They wish to save money for five years and purchase a house with a value of $195,000 with cash. If they deposit money into an account paying 12% interest, how much do they need to deposit each month in order to make the purchase?

27. Suppose you wish to retire at the age of 65 with $1,000,000 in savings. Determine your monthly payment into an IRA if the APR is 8.5% and you begin making payments at the following ages.

   a. 20 years old

   b. 30 years old

   c. 40 years old

28. Rueben would like to save enough money to pay cash for his next car. His goal is to save an extra $35,000 over the next six years. What is the quarterly payment that Reuben must make into an account that earns 4.5% interest in order to reach his goal?

29. Lacy deposits a fixed monthly amount into an annuity account for her child's college fund. She wishes to accumulate a future value of $135,000 in 18 years.

    a. Assuming an APR of 6.5%, how much money should Lacy deposit monthly in order to reach her goal?

    b. How much of the $135,000 will Lacy ultimately deposit in the account, and how much is interest earned?

30. Vanessa starts an IRA (individual retirement account) to save for retirement at the age of 22. She deposits $250 at the end of each month. The IRA has an average annual interest rate of 8.25%.

    a. How much money will she have saved upon retirement at the age of 65?

    b. Compare the future value to the total amount Vanessa deposited over the time period.

31. Max has a job at an advertising agency and earns $44,250 per year. If Max is 24 years old and wishes to retire at age 67, how much money would Max need to invest each month into a growth stock mutual fund with an interest rate of 7.5% in order to have a retirement income of $68,000 per year without reducing the principal?

32. Ozzie bought a new car and financed $14,950 of the purchase. He financed the car for 36 months with an APR of 5.75%. Determine each of the following.

    a. Ozzie's monthly payment

    b. Total cost of Ozzie's car

    c. Total interest Ozzie pays over the life of the loan

33. Chelsea and Bill are buying a house on a 30-year mortgage. They can pay $1200 per month for a mortgage. If they have an APR of 4.25%, what is the maximum mortgage that they can take out?

34. Amelia decides to purchase a $215,000 house. She wants to finance the entire balance. She has received an APR of 2.75% for a 15-year mortgage.

    a. What is Amelia's monthly payment?

    b. What is Amelia's total cost if she takes all 15 years to pay off the house?

    c. Over the course of the loan, how much interest will Amelia pay?

    d. If she changed the term to 30 years instead of 15 years, what would her monthly payment be?

    e. With a 30-year mortgage, what is the total cost of the house?

    f. With a 30-year mortgage, how much interest will Amelia pay?

35. The city of Wilsonville recently completed a new school building. The entire cost of the project was $37,500,000. The city has put the project on a 30-year loan with an APR of 1.4%. There are 19,450 families that will be responsible for paying the loan.

    a. Determine the amount of the monthly payment for the loan.

    b. Determine the amount that each family will be required to pay each year to cover the cost of the school.

    c. Determine the total cost of the school.

36. You want to buy a car and finance $27,450 to do so. You can afford a payment of up to $600 per month. The bank offers three choices for the loan: a four-year loan with an APR of 5.5%, a five-year loan with an APR of 6.5%, and a six-year loan with an APR of 7%. Which option best meets your needs, assuming you want to pay the least amount of interest?

Consider a credit card with a balance of $4875. You wish to pay off the credit card in each scenario. Calculate the following. Round your answer to the nearest cent, as necessary.

    a. The amount of a monthly payment within the time frame given

    b. The total amount paid over the time period

37. APR of 19.99% paid off within one year

38. APR of 21.5% paid off within two years

39. APR of 29.99% paid off within three years

Answer each question. Round your answer to the nearest cent, as necessary.

40. A credit card has a balance of $7250 at an annual percentage rate of 15.99%. You plan to pay $400 each month in an effort to clear the debt quickly. How long will it take you to pay off the balance?

41. Suppose you have a student loan of $76,500 with an APR of 2.5% for 25 years.

    a. What is your monthly payment?

    b. If you decide you want to pay off the loan in 15 years instead of 25, what is your monthly payment?

    c. What is your savings for paying the loan off in 15 years instead of 25?

Bibliography

9.2

1. Wolfram Alpha LLC. 2009. Wolfram|Alpha. http://www.wolframalpha.com/input/?i=compound+interest&a=*C.
compound+interest-_*Formula.dflt-&a=FSelect_**PresentValueFutureValue-.dflt-&a=*FS-
_**PresentValueFutureValue.FV-.*PresentValueFutureValue.PV-.*PresentValueFutureValue.i-.*PresentValueFutureValu
e.n--&f4=%2412000&f=PresentValueFutureValue.PV_%2412000&f5=4.5%25&f=PresentValueFutureValue.i_4.5&f6
=10&f=PresentValueFutureValue.n_10&a=*FP.PresentValueFutureValue.compoundingfreq-_Monthly (accessed July
8, 2014).

9.4

1. The Library of Congress THOMAS, s.v. "Bill Summary & Status 111th Congress (2009-2010) H.R.627."
Accessed February 2013. http://thomas.loc.gov/cgi-bin/bdquery/z?d111:H.R.627:

Chapter 10
Voting and Apportionment

Objectives

- Understand counting methods such as the majority, plurality, Borda count, plurality with elimination, and pairwise comparison methods

- Understand fairness criteria such as the Condorcet, majority, monotonicity, and irrelevant alternatives criteria

- Understand Arrow's impossibility theorem

- Use apportionment methods such as the Hamilton, Jefferson, Webster, and Huntington-Hill methods of apportionment

- Identify paradoxes such as the Alabama, population, and new states paradoxes

- Understand the Balinski & Young Theorem

- Understand weighted voting systems using coalitions and power indices such as the Banzhaf and Shapley-Shubik power indices

Voting and Apportionment

When it's time to elect the next President of the United States, the Constitution states that he or she must be chosen by electoral votes. Each state is assigned a number of electoral votes matching the number of Congressional members it has. (In addition, the District of Columbia has three electors in the Electoral College.) Out of the 538 possible electoral votes, a winner must receive at least 270 to be declared the winner. If no candidate receives the required votes, then each member of the House of Representatives casts one vote for one of the top three presidential candidates. However, the House of Representatives has only been called upon to elect a president two times in the history of the United States, once with Thomas Jefferson, in 1801, and then John Quincy Adams, in 1825.

Fun Fact

In most states, except Maine and Nebraska, the winner of the state's popular vote by the plurality method gets all of the state's electoral votes. In Maine and Nebraska, the winner of the state only gets two votes, one for each Senator. These states' other electoral votes are distributed according to the winner of each congressional district in that state.

The 2000 presidential election in the United States was quite a controversial affair. Republican candidate George W. Bush, then governor of Texas and son of former president George H. W. Bush, and Democratic candidate Al Gore, then Vice President, fought hard to win the seat of the highest elected official in America. The election will go down in history as a mathematical free-for-all. At the heart of the tension was Florida and its 25 electoral votes; questions about how the popular votes were counted and who would secure the state's electoral votes created quite a frenzy. To make matters worse, at some point during the evening and early morning hours of election night, each man was proclaimed the winner of Florida, and hence, President of the United States, by the news media. The morning after the election, a president was still not decided upon, but all hinged on the outcome in Florida. Because the votes were so close, the state had a mandatory recount, but even that was disputed. In the end, it was the fact that presidents are appointed based on electoral votes and not the popular vote that decided the election. Bush narrowly won the November 7th election with 271 electoral votes to Gore's 266—with one elector abstaining in the official tally—despite the fact that the official count gave Gore 50,999,897 votes from individual Americans to Bush's 50,456,002 votes. This made President Bush the fourth president in United States history to receive fewer popular votes than the runner-up candidate.

This chapter focuses not only on methods of counting votes used in elections around the world, but also ways in which those methods are deemed to be fair.

Electoral Map of the 2000 US Presidential Election

Democrat: 266
Al Gore

Republican: 271
George W. Bush

Figure 1

10.1   How to Determine a Winner

In elections, sometimes it's the case that we are simply concerned about an overall winner, but at other times, we desire a ranking of the candidates in terms of outcomes, that is, first place, second place, etc. In either case, there is still a choice to make about how to determine a winner.

We begin our discussion of how to determine a winner by looking at an election example. Let's consider a scenario in which four candidates are competing against one another in an election: Russo, Satou, Tremblay, and Williams. Voters have been asked to indicate their rankings of the candidates on what is called a **preference ballot**. This type of ballot allows the voter to give each candidate a ranking: first place, second place, third place, etc.

> ### Preference Ballot
>
> A **preference ballot** is a ballot that allows a voter to rank the items in order of preference from most preferred to least preferred.

Often we are asked to indicate only our top candidate on a ballot, but a preference ballot gives us the opportunity to voice how we rank all of the candidates. This information allows the comparison of different ways of tallying the votes.

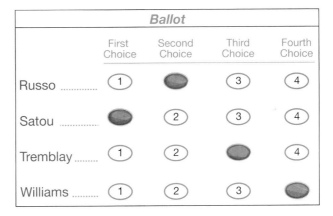

Figure 1: Preference Ballot

In order to compile the results of all of the voter rankings from the ballots, a **preference table** is used. A preference table summarizes how often each possible ranking occurred in the voting process. By convention, if a particular ranking receives no votes, it is not included in the preference table.

Think Back

The Fundamental Counting Principle states that for a sequence of n experiments where the first experiment has k_1 outcomes, the second experiment has k_2 outcomes, the third experiment has k_3 outcomes, and so on, the total number of possible outcomes for the sequence of experiments is $(k_1)(k_2)(k_3)\cdots(k_n)$.

> ### Preference Table
>
> A **preference table** summarizes all of the individual preference ballots in an election by tallying the number of ballots with the same order of ranking.

Since there are four candidates in our election example, let's consider how large the preference table for the election might be. From the probability discussion in Chapter 7, we know that if there are n distinct items to place in order, then there are $n!$ ways in which to order the items. So, there are $4! = 4 \cdot 3 \cdot 2 \cdot 1 = 24$ unique ways to order the four candidates on our preference ballot. Potentially, there could be 24 columns in the preference table. But in our particular election example, only four different rankings were chosen by the voters. Therefore, there are four columns in the preference table for the candidates. The following preference table shows how often each particular ranking occurred amongst the 776 voters in our election.

Table 1

Preference Table for Candidates

	Rankings			
1st	Williams	Tremblay	Satou	Russo
2nd	Satou	Satou	Williams	Williams
3rd	Tremblay	Williams	Tremblay	Satou
4th	Russo	Russo	Russo	Tremblay
Total Votes	**132**	**210**	**167**	**267**

Notice that the bottom row of the table indicates how many individual voters ranked the candidates in the particular order listed in each column. For example, according to the first column, 132 people ranked Williams first, Satou second, Tremblay third, and Russo fourth. In order to explore different methods of choosing a winner in an election, we'll refer back to this preference table of our four candidates throughout the section.

Example 1: Reading a Preference Table

Answer the following questions about the preference table shown for the election for Senior Class President at Clarkstown High School.

Table 2

Preference Table for Senior Class President

	Rankings				
1st	Sydney	Ava	Ava	Carley	Carley
2nd	Ryan	Carley	Carley	Zaire	Sydney
3rd	Ava	Ryan	Sydney	Ryan	Ava
4th	Carley	Zaire	Ryan	Ava	Zaire
5th	Zaire	Sydney	Zaire	Sydney	Ryan
Total Votes	**15**	**29**	**6**	**24**	**1**

a. How many different rankings were possible for the election?

b. How many students voted in the election?

c. Which student had the most first-place votes?

d. How many students thought that the order of candidates should be Ava, Carley, Sydney, Ryan, then Zaire?

Solution

a. Because there were five candidates on the ballot, we can use factorials to find the number of possible rankings.

$$5! = 5 \cdot 4 \cdot 3 \cdot 2 \cdot 1$$
$$= 120$$

Therefore, there were 120 possible rankings of the five candidates in the election.

b. To find the total number of students that voted in the election, we add the numbers in the row labeled "Total Votes."

$$15 + 29 + 6 + 24 + 1 = 75$$

Thus, there were 75 students who voted in the election.

c. To find the candidate with the most first-place votes, we need to count the number of votes each candidate received where they were ranked first place. By looking across the 1st row, we see that there was only one ranking order that placed Sydney first. At the bottom of that column, we know that this particular ranking received 15 votes. So, Sydney received 15 first-place votes overall.

	Table 3				
	Preference Table for Senior Class President				
	Rankings				
1st	Sydney	Ava	Ava	Carley	Carley
2nd	Ryan	Carley	Carley	Zaire	Sydney
3rd	Ava	Ryan	Sydney	Ryan	Ava
4th	Carley	Zaire	Ryan	Ava	Zaire
5th	Zaire	Sydney	Zaire	Sydney	Ryan
Total Votes	**15**	**29**	**6**	**24**	**1**

There were two rankings in which Ava received first-place votes: the second column of rankings and the third. Therefore, Ava had a total of $29 + 6 = 35$ first-place votes.

	Table 4				
	Preference Table for Senior Class President				
	Rankings				
1st	Sydney	Ava	Ava	Carley	Carley
2nd	Ryan	Carley	Carley	Zaire	Sydney
3rd	Ava	Ryan	Sydney	Ryan	Ava
4th	Carley	Zaire	Ryan	Ava	Zaire
5th	Zaire	Sydney	Zaire	Sydney	Ryan
Total Votes	**15**	**29**	**6**	**24**	**1**

Carley was also ranked first in two rankings, which are shown in the last two columns of the table. Her first place rankings total $24 + 1 = 25$.

	Table 5				
	Preference Table for Senior Class President				
	Rankings				
1st	Sydney	Ava	Ava	Carley	Carley
2nd	Ryan	Carley	Carley	Zaire	Sydney
3rd	Ava	Ryan	Sydney	Ryan	Ava
4th	Carley	Zaire	Ryan	Ava	Zaire
5th	Zaire	Sydney	Zaire	Sydney	Ryan
Total Votes	**15**	**29**	**6**	**24**	**1**

Notice that there are no rankings that place either Ryan or Zaire in first. Therefore, Ava received the most first place rankings with a total of 35 votes.

d. The number of students who prefer the order of ranking to be Ava, Carley, Sydney, Ryan, and then Zaire is found in the third column of rankings. The total votes for this column is 6, so there were 6 students who chose this ranking.

10

Skill Check #1

Which candidate in Example 1 had the most rankings for fifth place?

Summarizing the results from preference ballots is the first step towards determining a winner in an election. The remainder of the section focuses on five different methods that have historically been used to choose winners. In each case, we'll use the preference ballot for the four candidates we introduced earlier in order to highlight how each different method can produce varying outcomes, even while the vote count stays the same.

Majority Rule Decision

Using the **majority rule decision** to declare a winner means that the winner is supported by a majority of the voters, that is, more than 50% of the voters rank a single candidate in first place. When there are only two choices, the majority rule will almost always have a winner. The only possible time the majority rule will not produce a winner is in the case of a tie when both candidates receive exactly 50% of the votes. To avoid this stalemate, many voting organizations allow the chair of the committee to only have a vote in order to break a tie. However, when there are more than two candidates, there is no guarantee that a winner can be named with the majority rule decision. Let's return to our example of the four candidates Russo, Satou, Tremblay, and Williams. Suppose they were running for office in a voting system such as Sweden's, which uses a majority rule decision to choose its leaders. Who would win?

Example 2: Majority Rule Decision

Use a majority rule decision to determine a winner in the election results from Satou, Williams, Tremblay, and Russo. The preference table is reprinted as Table 6 for ease of reference.

Table 6				
Preference Table for Candidates				
	Rankings			
1st	Williams	Tremblay	Satou	Russo
2nd	Satou	Satou	Williams	Williams
3rd	Tremblay	Williams	Tremblay	Satou
4th	Russo	Russo	Russo	Tremblay
Total Votes	132	210	167	267

Solution

In order to be declared a winner with the majority rule decision, a candidate must have more than 50% of the first-place votes. The first thing to determine is the total number of voters so we can calculate the number needed for a majority. To find the total number of voters, we sum across the row labeled "Total Votes."

$$132 + 210 + 167 + 267 = 776$$

There were 776 votes cast in this election. Therefore, to have a majority, a candidate needs more than half of the votes, or more than

$$\frac{776}{2} = 388 \text{ votes.}$$

So a minimum of 389 votes are needed for a majority.

Looking across the 1st row, we can see that each candidate has only one ranking where they were placed first. The number of first place rankings are as follows.

Russo: 267 first-place votes

Satou: 167 first-place votes

Tremblay: 210 first-place votes

Williams: 132 first-place votes

This means that no candidate obtained the minimum of 389 first-place votes needed for a majority. Therefore, using the majority rule decision, we cannot declare an immediate winner. Instead, an alternative method for choosing a winner will be needed.

Plurality Method

What happens when the majority rule doesn't produce a winner, as we just saw in Example 2? A popular way to choose a winner is by the *plurality method*. The **plurality method** states that the candidate with the most first-place votes wins—majority or not. Often, this method is confused with the majority rule decision. When there are only two candidates, this method is equivalent to the majority rule decision. However, if there are more than two choices, a winner does not need to have a majority of votes to win. Suppose our same four candidates were running for office in India, where a plurality method is used to select a winner. Let's use our election preference table once again and determine the winner.

Example 3: Plurality Method

Use a plurality method to determine a winner with the four candidates, Satou, Williams, Tremblay, and Russo. The preference table is reprinted as Table 7.

Table 7				
Preference Table for Candidates				
	Rankings			
1st	Williams	Tremblay	Satou	Russo
2nd	Satou	Satou	Williams	Williams
3rd	Tremblay	Williams	Tremblay	Satou
4th	Russo	Russo	Russo	Tremblay
Total Votes	**132**	**210**	**167**	**267**

Solution

When using the plurality method, the only consideration is who has the most first-place votes. The other rankings play no role in the process here. Recall each candidate's first-place votes from Example 2.

Russo: 267 first-place votes

Satou: 167 first-place votes

Tremblay: 210 first-place votes

Williams: 132 first-place votes

Therefore, Russo wins the election with 267 first-place votes.

We know that in the plurality method, only the first-place votes count, and all other rankings are irrelevant. However, let's take a closer look at the last example. Russo was the winner with the most first-place votes, but look at the other rankings he received—all of the other sets of rankings placed him in fourth place. So, although he won with 267 votes, the rest of the voters $(132 + 210 + 167 = 509$ voters$)$ think he should be in last place. In other words, although he has more first-place votes than any other candidate, a majority of voters definitely think he should not be the winner. Do you think this poses any issues? Let's turn our attention to a method of vote counting that actually takes all of the rankings into account.

Borda Count Method

The **Borda count method** assigns each ranking a specific number of points based on how many candidates are in the election. A first-place ranking receives the most points, second place gets one fewer, and so on until the last place, which receives one point. For instance, in our election with four candidates, first-place votes receive four points, second-place votes receive three points, third-place votes receive two points, and fourth-place votes receive one point. Points are then totaled for each candidate. The winner is the candidate with the most points overall. One of the benefits in using the Borda count method is that all of the preferential rankings of each voter play a role in the process. This method is often used in polls that rank sports teams or players. In fact, the Heisman Trophy, given to the most outstanding college football player in the United States each year, uses the Borda count method to choose a winner. In the Republic of Slovenia, the Italian and Hungarian national communities each elect one deputy to the National Assembly by using the Borda count method. Suppose our four candidates were running to be a deputy for the Hungarian community. Who would be chosen to represent the Hungarian community in the Slovenia National Assembly?

Example 4: Borda Count Method

Use a Borda count method to determine a winner with our four candidates. The preference table is reprinted as Table 8.

Table 8			
Preference Table for Candidates			
	Rankings		
1st	Williams	Tremblay	Satou	Russo
2nd	Satou	Satou	Williams	Williams
3rd	Tremblay	Williams	Tremblay	Satou
4th	Russo	Russo	Russo	Tremblay
Total Votes	**132**	**210**	**167**	**267**

Solution

Using the Borda count method requires that we assign points for each ranking. Because there are four candidates in our election, each ranking will receive the following number of points.

$$1^{st} \text{ place } = 4 \text{ points}$$

$$2^{nd} \text{ place } = 3 \text{ points}$$

$$3^{rd} \text{ place } = 2 \text{ points}$$

$$4^{th} \text{ place } = 1 \text{ point}$$

To help keep the information straight, the calculations are organized in a table. Note, in the following table we will look at each individual's ranking from Table 8.

Table 9					
Borda Count Tally					
	Rankings				**Total Points**
Russo	4th 1(132) = 132	4th 1(210) = 210	4th 1(167) = 167	1st 4(267) =1068	1577
Satou	2nd 3(132) = 396	2nd 3(210) = 630	1st 4(167) = 668	3rd 2(267) = 534	2228
Tremblay	3rd 2(132) = 264	1st 4(210) = 840	3rd 2(167) = 334	4th 1(267) = 267	1705
Williams	1st 4(132) = 528	3rd 2(210) = 420	2nd 3(167) = 501	2nd 3(267) = 801	2250
Total Votes	**132**	**210**	**167**	**267**	

The final column in the table calculates each candidate's point total as follows.

Russo receives:	$132 + 210 + 167 + 1068$	$= 1577$ points
Satou receives:	$396 + 630 + 668 + 534$	$= 2228$ points
Tremblay receives:	$264 + 840 + 334 + 267$	$= 1705$ points
Williams receives:	$528 + 420 + 501 + 801$	$= 2250$ points

When we take into account the ranking preferences of all the voters, we have Williams as the overall winner with 2250 total points, although we know from the previous example that he received the fewest first-place votes.

You can now see that running elections with the Borda count method may produce a completely different result than using the plurality method. Russo, who won the election when using the plurality method, is actually in last place using the Borda count method. It might be surprising to you that precisely the same set of votes produces two different winners depending on the counting method used. This is not to say that one method is necessarily better than the other, but rather, it's important to choose a method that best fits the situation at hand and one that will represent the purpose of the election. It's also a wise decision to decide upon a method of choosing a winner *before* the elections occur and not after the votes have been cast.

Before you rush to decide on one of these methods for electing a president of your local club, let's consider two other variations.

Plurality with Elimination Method

The plurality method that we discussed earlier is a one-vote, winner-take-all method of naming a winner. However, a variation on this method, called **plurality with elimination**, uses a series of eliminations to choose a winner. This method can be done by running several consecutive elections, often called runoffs, or by using a preference ballot to begin with. After the first count of votes, if a candidate has a majority of votes, then he is named the winner. However, if no single candidate has a majority of votes, then the candidate with the least amount of first-place votes is removed from the ballot, and a runoff election is held with the remaining candidates.

If a preference ballot is used in the first election, there is no need for a second voting process. The losing candidate is simply removed from the preference ballots and then the rankings for each voter are adjusted accordingly. For instance, suppose the candidate who is removed from the ballot after the first count was originally ranked first by a voter. Once he is removed, the voter's second place choice is then moved up a ranking to first place, the third place moves to second place, and so on. Once this type of adjustment happens for each ballot, the votes are recounted. In each round, only first-place votes are counted to eliminate the candidate with the fewest first-place votes. This method of determining a winner is used to elect the leader of the Labour Party in the United Kingdom. Suppose our four contestants were running for Labour Party leader. Who would be appointed?

Example 5: Plurality with Elimination Method

Use a plurality with elimination to determine a winner with the four candidates. The preference table is reprinted as Table 10.

	Table 10			
	Preference Table for Candidates			
	Rankings			
1st	Williams	Tremblay	Satou	Russo
2nd	Satou	Satou	Williams	Williams
3rd	Tremblay	Williams	Tremblay	Satou
4th	Russo	Russo	Russo	Tremblay
Total Votes	**132**	**210**	**167**	**267**

Solution

In the plurality with elimination process, only first-place votes are considered in each counting cycle. By looking at the preference table, we know that each of the candidates has the following number of first-place votes.

Russo: 267 first-place votes

Satou: 167 first-place votes

Tremblay: 210 first-place votes

Williams: 132 first-place votes

We already know that Russo would win in a plurality contest with his 267 votes, but since he doesn't have the required 389 votes for a majority, an elimination must happen. Since Williams has the least amount of first-place votes with 132, he is eliminated as shown in Table 11.

	Table 11			
	Plurality with Elimination Cycle 1			
	Rankings			
1st	~~Williams~~	Tremblay	Satou	Russo
2nd	Satou	Satou	~~Williams~~	~~Williams~~
3rd	Tremblay	~~Williams~~	Tremblay	Satou
4th	Russo	Russo	Russo	Tremblay
Total Votes	**132**	**210**	**167**	**267**

Now, the preference table must be adjusted accordingly. For example, in the first rankings column, Satou will now be ranked first, Tremblay second, and Russo third. In the second rankings column, Tremblay and Satou remain in first and second place, but Russo will move into third place. In the third rankings column, both Tremblay and Russo move up one ranking. Finally, in the last rankings column, Satou and Tremblay move up one ranking. The new preference table for the second cycle is shown in Table 12. Note that the number of votes in the bottom row will stay the same in each cycle.

Table 12				
Plurality with Elimination Cycle 1 Simplified				
	Rankings			
1st	Satou	Tremblay	Satou	Russo
2nd	Tremblay	Satou	Tremblay	Satou
3rd	Russo	Russo	Russo	Tremblay
Total Votes	**132**	**210**	**167**	**267**

Now, we recount the number of first-place votes for each candidate and determine if anyone has the required 389 votes for a majority.

Satou now has two rankings that place him in first place, which gives him a total of

$$132 + 167 = 299 \text{ votes.}$$

Tremblay and Russo remain the same with 210 and 267 first-place votes, respectively.

Although Satou is now the leader with 299 votes, since he does not have a majority, we must remove a candidate and continue with another elimination. This time it is Tremblay who has the least amount of votes, so he is eliminated.

Table 13				
Plurality with Elimination Cycle 2				
	Rankings			
1st	Satou	~~Tremblay~~	Satou	Russo
2nd	~~Tremblay~~	Satou	~~Tremblay~~	Satou
3rd	Russo	Russo	Russo	~~Tremblay~~
Total Votes	**132**	**210**	**167**	**267**

Adjusting the table in the same manner as before, we have the following two remaining candidates, Satou and Russo.

Table 14				
Plurality with Elimination Cycle 2 Simplified				
	Rankings			
1st	Satou	Satou	Satou	Russo
2nd	Russo	Russo	Russo	Satou
Total Votes	**132**	**210**	**167**	**267**

Counting the votes, we see that Satou now has $132 + 210 + 167 = 509$ first-place votes to Russo's 267. Therefore, Satou now wins with a majority of the first-place votes.

Pairwise Comparison Method

The last type of counting we will consider is the **pairwise comparison method**. Each candidate is "paired" with every other candidate in a head-to-head vote count. The candidate with the most number of votes in each comparison receives one point. If there is a tie in the comparison count, each candidate receives $\frac{1}{2}$ of a point. After all pairs of comparisons are made, the candidate with the most points overall is the winner. The pairwise comparison method is helpful for making decisions involving fairly complicated criteria. It allows the comparison between each and every option available to play a role in the decision making process. This method of decision making is often used to choose leaders in organizations, assess risks about a certain decision, or even to select the most qualified candidate for a job.

So, how many head-to-head comparisons need to be made when using this method? Multiplying the total number of candidates, n, by the number of other candidates in the race he or she could be paired with, $n - 1$, is a good place to start. However, we'll end up with duplicates because a head-to-head matchup between Candidate A and Candidate B is the same as a matchup between Candidate B and Candidate A. So, we need to divide that product in half. We can also think of this as the combination of choosing two candidates from a pool of n candidates, or $_nC_2$. The following formula will give us the number of head-to-head comparisons needed for the **pairwise comparison method**.

Number of Pairwise Comparisons

The number of pairwise comparisons that must be made if there are n candidates is

$$\frac{n(n-1)}{2}$$

The 2008 Wikimedia Board Election Committee used the pairwise comparison method to elect a new board member for the coming year. Suppose our original four candidates were in the election for board member. Who would have been elected to the 2009 Wikimedia Board of Trustees? We will use the pairwise comparison method in Example 6 to examine this question.

Example 6: Pairwise Comparison Method

Use the pairwise method of comparison to determine a winner with our four candidates. The preference table is reprinted as Table 15.

Table 15				
Preference Table for Candidates				
	Rankings			
1st	Williams	Tremblay	Satou	Russo
2nd	Satou	Satou	Williams	Williams
3rd	Tremblay	Williams	Tremblay	Satou
4th	Russo	Russo	Russo	Tremblay
Total Votes	132	210	167	267

Solution

In the pairwise method of comparison, each candidate is pitted against every other candidate to see who received the highest ranking. We can use the formula to find out how many comparisons will need to be made in our tally. Since there are four candidates, $n = 4$. Substituting this into the formula, we have the following.

$$\text{number of comparisons } = \frac{4(4-1)}{2} = \frac{12}{2} = 6$$

In order to help us keep track of all six comparisons, we'll create a table where each candidate has a column and a row. There is no need for comparisons between a candidate and himself, so we've put an X in those cells that appear on the diagonal. Also note that each head-to-head comparison is listed twice in the chart, that is, a comparison between Russo and Satou (Row 1, Column 2) is the same as a comparison between Satou and Russo (Row 2, Column 1). Since there is no need to duplicate the comparisons, we've grayed out the cells that are duplicates. Check for yourself that this leaves us with the required six comparisons.

Table 16				
Pairwise Comparison Grid				
	Russo	**Satou**	**Tremblay**	**Williams**
Russo	X			
Satou		X		
Tremblay			X	
Williams				X

In each head-to-head cell, list the candidate with the higher ranking from each of the four original rankings, and the number of votes he received from that ranking. For the comparison in the Russo vs. Satou cell, we can create a table to compare the rankings for the two candidates.

Table 17				
Head-to-Head Comparison				
	Rankings			
Russo	4th	4th	4th	1st
Satou	2nd	2nd	1st	3rd
Total Votes	132	210	167	267

Using Table 17, we can see that Satou is ranked higher in the first three columns, so he receives all of the votes for those rankings. The last 267 votes go to Russo in this head-to-head comparison. The same is done for each of the head-to-head matchups, as shown in Table 18.

Table 18				
Completed Pairwise Comparison Grid				
	Russo	**Satou**	**Tremblay**	**Williams**
Russo	X	S = 132; S = 210; S = 167; R = 267	T = 132; T = 210; T = 167; R = 267	W = 132; W = 210; W = 167; R = 267
Satou		X	S = 132; T = 210; S = 167; S = 267	W = 132; S = 210; S = 167; W = 267
Tremblay			X	W = 132; T = 210; W = 167; W = 267
Williams				X

Once all of the votes have been awarded based on the different rankings, the overall winner in each head-to-head matchup is given 1 point. If there is a tie in the votes, each candidate gets $\frac{1}{2}$ of a point. Table 19 shows who received the point for each matchup.

Table 19				
Pairwise Comparison Grid of Winners				
	Russo	**Satou**	**Tremblay**	**Williams**
Russo	X	S = 509; R = 267 [Satou]	T = 509; R = 267 [Tremblay]	W = 509; R = 267 [Williams]
Satou		X	S = 566; T = 210; [Satou]	W = 399; S = 377; [Williams]
Tremblay			X	W = 566; T = 210; [Williams]
Williams				X

We can see that Russo received 0 points, Satou ended up with 2 points, Tremblay had 1 point, and Williams had 3 points. Therefore, Williams is the winner in the pairwise method of comparison.

Table 20 shows the winners for each of the different counting methods. As we've shown, each method is just as viable as the others. Since no method is inherently superior, this allows for the people doing the polling to choose the method that best fits their needs.

		Table 20			
Comparison Table of Winners for Different Counting Methods					
Candidate	**Majority Rule Decision**	**Plurality**	**Borda Count**	**Plurality with Elimination**	**Pairwise Comparison**
Russo		Winner			
Satou	No Winner			Winner	
Tremblay					
Williams			Winner		Winner

Table 21 gives a summary of all five election counting methods discussed in this section.

	Table 21
Election Counting Methods	
Majority Rule Decision	The winner must have more than 50% of the votes to win.
Plurality	The candidate with the most votes wins. No majority required.
Borda Count	Voting results are organized in a preference table and each ranking is assigned a specific number of points based on how many candidates are in the election. The candidate with the most ranking points is declared the winner.
Plurality with Elimination	A series of runoff elections, or eliminations, where the candidate with the least amount of first-place votes is removed from the ballot each round. A winner is declared when a candidate has a majority of the first-place votes.
Pairwise Comparison	Every candidate is compared head-to-head with the other candidates. In each pair of comparisons, the candidate with the greater number of higher rankings is given a point. The candidate with the most points after all head-to-head comparisons are made is the winner.

Skill Check Answer

1. Sydney, with 53 votes that gave her a fifth place ranking.

10.1   Exercises

Fill in each blank with the correct term.

1. The _____ method of choosing a winner in an election is a single vote with a winner-take-all approach. The winner need not have a majority of the votes to win; she just needs the most votes.

2. The _____ method for elections pits each candidate against all of the others individually in order to choose a winner.

3. Setting up an election so that voters are required to rank all the candidates requires use of a _____.

4. In order to give weight to each of the different rankings for candidates, the _____ attributes each ranking a certain number of points, and the candidate with the most points wins the election.

5. In a _____ election, the winner must receive more than 50% of the votes.

Create a preference table to show the results for each set of preference ballots.

6. Candidates in the election are: Maria, Daniel, Eisa.

Maria, Eisa, Daniel: 29

Eisa, Daniel, Maria: 43

Eisa, Maria, Daniel: 33

Daniel, Maria, Eisa: 17

Maria, Daniel, Eisa: 3

Daniel, Eisa, Maria: 21

7. The candidates are: Lars, Noah, Stephen, Oliver, Jack.

Stephen, Oliver, Jack, Lars, Noah: 212

Noah, Oliver, Jack, Lars, Stephen: 133

Oliver, Stephen, Jack, Noah, Lars: 543

Jack, Stephen, Lars, Noah, Oliver: 24

Oliver, Noah, Lars, Jack, Stephen: 179

Noah, Lars, Stephen, Jack, Oliver: 8

Stephen, Lars, Oliver, Noah, Jack: 201

Stephen, Jack, Lars, Noah, Oliver: 11

8. The candidates are: Jones (J), Brown (B), Wang (W), Cohen (C), Diaz (D).

JBDCW, CWJDB, WBDJC, BDCWJ, JBDCW, JBDCW, CWJDB, CWJDB, WBDJC,

JBDCW, BWJDC, WBDJC, CWJDB, BWJDC, BWJDC, BJWCD, WBDJC, BWJDC,

JBDCW, BJWCD, WBDJC, BWJDC, JBDCW, JBDCW, DBWCJ

9. Candidates in the election are: Nguyen (N), Dey (D), Smith (S), Mori (M), Abbadi (A).

MADNS, NDSMA, DSMAN, SMAND, SMAND, NDSMA, MANDS, DSMAN, SAMND,

NDSMA, SMAND, NDSMA, MANDS, SAMND, NDSMA, MANDS, MANSD, SDMAN,

MADNS, MANDS, MANDS, SMAND, NDSMA, DSMAN, SAMDN, MADNS, SAMDN,

SMAND, SMAND, NDSMA

Use the given preference table to answer each question.

10. Answer the following questions about the given preference table.

	Preference Table for Candidates				
	Rankings				
1st	A	C	C	E	E
2nd	B	E	E	D	A
3rd	C	B	A	B	C
4th	E	D	B	C	D
5th	D	A	D	A	B
Total Votes	153	129	63	404	111

a. How many possible unique rankings are there of the candidates in the election?

b. How many people voted in the election?

c. How many voters place Candidate C in first place?

d. Which candidate wins using the plurality method?

e. How many votes would a candidate need to have a majority?

f. Do any of the candidates have a majority of first-place votes?

11. A wedding planning website asked visitors to rank the following six all-time favorite wedding songs. The song choices are *Nothing Compares to You* by Sinead O'Connor, *Love Me Tender* by Elvis Presley, *Close to You* by Maxi Priest, *When You Say Nothing at All* by Ronan Keating, *Just the Way You Are* by Billy Joel, and *You Made Me Love You* by Al Jolson. Use the preference table to answer the questions.

	Preference Table for Favorite Wedding Songs			
	Rankings			
1st	Presley	Priest	O'Connor	Keating
2nd	Jolson	O'Connor	Presley	Joel
3rd	Keating	Joel	Priest	Presley
4th	Priest	Jolson	Keating	O'Connor
5th	Joel	Presley	Joel	Jolson
6th	O'Connor	Keating	Jolson	Priest
Total Votes	1320	2010	3167	2697

a. How many possible unique rankings are there of the six wedding songs?

b. How many people participated in the website questionnaire?

c. Which song was ranked number one by the most voters?

d. How many votes are required for a song to have a majority of first-place votes?

e. Do any of the songs have a majority of first-place votes?

Calculate the number of pairwise comparisons that must be made in each election.

12. In the history club on campus, five students are running for the office of president.

13. Running for president of the National Association of College Students are the following students: L. Mayer, R. Schartz, M. Gillmore, L. Bilbo, R. Rosevear, K. Colletta, T. Yawn, S. Oswald, and K. Smith.

Use the given preference table to solve each problem.

14. The following table shows the number of votes for first, second, and third places that the top five college football players received in 2011 for the Heisman Trophy—College Football's highest award. Determine the number of total points that each player would receive using the Borda count method if a first place vote receives 3 points, a second place receives 2 points, and third place receives 1 point.

Preference Table for 2011 Heisman Trophy Candidates				
Player	**School**	**1st**	**2nd**	**3rd**
Robert Griffin III	Baylor	405	168	136
Andrew Luck	Stanford	247	250	166
Trent Richardson	Alabama	138	207	150
Montee Ball	Wisconsin	22	83	116
Tyrann Mathieu	Louisiana State	34	63	99

Source: Heisman Trophy. "2011 - 77th Award Robert Griffin III Baylor University." http://www. heisman.com/winners/r-griffin11.php

15. In professional baseball, the Baseball Writers' Association of America hands out two Most Valuable Player (MVP) awards each year. One for the American League and one for the National League. The following table shows the voting breakdown for the top 10 players in the National League. Determine the number of total points that each player received using the Borda count method, given that a first place vote receives 14 points, a second place receives 9 points, third place receives 8 points, and on down to 1 point for tenth place.

National League MVP 2011											
Player	**Team**	**1st**	**2nd**	**3rd**	**4th**	**5th**	**6th**	**7th**	**8th**	**9th**	**10th**
Ryan Braun	Milwaukee Brewers	20	12								
Matt Kemp	Los Angeles Dodgers	10	16	6							
Prince Fielder	Milwaukee Brewers	1	4	11	9	1	3		2		1
Justin Upton	Arizona Diamondbacks	1		8	11	6	3	1	1		1
Albert Pujols	St. Louis Cardinals			1	6	11	6	4		2	
Joey Votto	Cincinnati Reds			4	3	2	8	3	3	4	1
Lance Berkman	St. Louis Cardinals			1	2	6	3	7	2	4	3
Troy Tulowitzki	Colorado Rockies						3	4	8	5	4
Roy Halladay	Philadelphia Phillies			1		1	1	6	2		3
Ryan Howard	Philadelphia Phillies				1	3	1	1		1	3

Source: Baseball Writers' Association of America. "2011 NL MVP: Ryan Braun Slugs His Way to Award." http://bbwaa. com/2011/11/2011-nl-mvp/

16. In an attempt to choose a list of the "Best Places to Live in America," readers of a magazine were asked to rank eight towns. The following preference table summarizes the results of the survey.

Preference Table for Best Places to Live in America					
	Rankings				
1st	Louisville, CO	Middleton, WI	Leesburg, VA	Leesburg, VA	Hanover, NH
2nd	Papillion, NE	Louisville, CO	Middleton, WI	Milton, MA	Leesburg, VA
3rd	Hanover, NH	Liberty, MO	Milton, MA	Louisville, CO	Milton, MA
4th	Milton, MA	Milton, MA	Liberty, MO	Papillion, NE	Solon, OH
5th	Liberty, MO	Papillion, NE	Hanover, NH	Solon, OH	Middleton, WI
6th	Solon, OH	Solon, OH	Papillion, NE	Hanover, NH	Papillion, NE
7th	Middleton, WI	Leesburg, VA	Solon, OH	Middleton, WI	Louisville, CO
8th	Leesburg, VA	Hanover, NH	Louisville, CO	Liberty, MO	Liberty, MO
Total Votes	**345**	**231**	**211**	**202**	**16**

a. Determine a winner using the plurality method.

b. Determine a winner using the plurality with elimination method.

c. Which method do you think chooses the winner that best reflects the votes, or would you choose a different method? Explain your answer.

17. The following preference table summarizes the election results for city mayor of Clamptonville.

Preference Table for Candidates			
	Rankings		
1st	Costa	Little	Little	Torres
2nd	Braugh	Allen	Costa	Braugh
3rd	Little	Torres	Braugh	Allen
4th	Torres	Braugh	Allen	Little
5th	Allen	Costa	Torres	Costa
Total Votes	**1231**	**542**	**1001**	**654**

a. Determine the winner for city mayor using the Borda count method.

b. Determine the winner for city mayor using the plurality with elimination method.

c. Determine the winner with the pairwise method of comparison.

d. Suppose that Braugh withdrew from the race before the votes were counted. Candidates ranked below him in any of the orders simply move up a ranking. Who would be declared the winner using the Borda count method with the new preference table? Is this a different winner than in part **a.**?

18. A university's Department of Business is electing a new chairman. As part of the process, each candidate gives a brief question and answer session with the business students. The students are then asked to rank the candidates in order of preference. The following table shows the results of the student votes.

Preference Table for Candidates						
	Rankings					
1st	B	A	C	C	D	B
2nd	A	B	D	B	A	A
3rd	D	C	B	D	B	C
4th	C	D	A	A	C	D
Total Votes	43	12	13	21	51	9

  a. Determine the student preference for chairman by using the pairwise method of comparison.

  b. Determine the student preference for chairman by using the plurality with elimination method.

  c. Determine the student preference for chairman by using the plurality method.

  d. If you were chosen to represent the students by announcing a preference for chairman, which candidate would you recommend? How would you defend your choice based on the answers to **a.–c.**?

19. The Tennessee Department of Transportation is considering a new bus route between its state capital, Nashville, and one of the other four largest cities: Knoxville, Memphis, Chattanooga, and Clarksville. Each of the 99 State Representatives was asked to rank the cities in order of preference for the new route. Their votes are summarized in the following table.

Preference Table for New Bus Route						
	Rankings					
1st	Chattanooga	Memphis	Chattanooga	Knoxville	Clarksville	Memphis
2nd	Memphis	Knoxville	Clarksville	Memphis	Knoxville	Knoxville
3rd	Clarksville	Clarksville	Knoxville	Chattanooga	Memphis	Chattanooga
4th	Knoxville	Chattanooga	Memphis	Clarksville	Chattanooga	Clarksville
Total Votes	23	13	9	33	12	9

  a. Which city would receive the new bus route using the pairwise method of comparison?

  b. Which city received the most first-place votes? Is this a majority?

  c. Which city would receive the new bus route using the plurality with elimination method?

  d. Which city would receive the new bus route using the Borda count method?

  e. Explain which method you think is most "fair" in choosing the new bus route.

20. In the 1968 United States Presidential Election, three candidates shared most of the November 5 popular vote, although there was a small portion of votes for other candidates. The following table displays the breakdown of the popular vote along with the electoral vote for the election.

1968 US Presidential Election Results			
Presidential Candidate	**Popular vote**		**Electoral Vote**
	Count	**Percentage**	
Richard Milhous Nixon	31,783,783	43.40%	301
Hubert Horatio Humphrey	31,271,839	42.70%	191
George Corley Wallace	9,901,118	13.50%	46
Eugene McCarthy	25,634	0.00%	0
Other	217,624	0.30%	0
Total Votes	**73,199,998**		**538**

Sources: 1968 Presidential General Election Results, http://www.uselectionatlas.org. Electoral Votes for President and Vice President, http://www.archives.gov.

a. Who won the popular vote by plurality? Was it a majority?

b. Who won the electoral vote by plurality? Was it a majority?

c. Compare the outcomes of the popular vote between Nixon and Humphrey to that of the electoral vote for the two candidates.

d. Suppose Wallace had dropped out of the election. Is it possible that Humphrey would have won the presidency? Explain your answer.

Complete each preference table.

21. The student programming board is deciding which band to bring to campus for a student concert. The possibilities are Matisyahu, Lady Antebellum, and The Black Keys. Complete the following preference table so that The Black Keys win the election using the majority rule.

Preference Table for Student Concert			
	Rankings		
1st	The Black Keys	Lady Antebellum	Matisyahu
2nd	Lady Antebellum	Matisyahu	The Black Keys
3rd	Matisyahu	The Black Keys	Lady Antebellum
Total Votes	**?**	**9**	**2**

22. A hospital needs a new doctor. The hiring committee wants to get feedback from the current staff. They have the nursing staff and resident doctors rank the candidates in order of their preference. Complete the following preference table so that Perez is the staff's top choice using the plurality method.

Preference Table for New Doctor			
	Rankings		
1st	Ross	Porter	Walker	Perez
2nd	Perez	Walker	Ross	Porter
3rd	Walker	Perez	Perez	Ross
4th	Porter	Ross	Porter	Walker
Total Votes	**?**	**21**	**22**	**36**

23. Phonetex is trying to decide which new feature to include in an advertising campaign: 12 megapixel camera, longer battery life, voice-activated commands, or reception quality. The company surveyed its current customers and asked them to rank the four features in order of preference. Complete the following preference table so that voice-activated commands is the favorite among current customers using the plurality with elimination method.

	Preference Table for New Features			
	Rankings			
1st	Camera	Battery Life	Voice-Activated	Voice-Activated
2nd	Reception Quality	Reception Quality	Reception Quality	Camera
3rd	Battery Life	Voice-Activated	Camera	Reception Quality
4th	Voice-Activated	Camera	Battery Life	Battery Life
Total Votes	44	11	?	56

24. Rhone County is electing a new county executive. The candidates are McMillian, Witt, and Longsdon. Complete the following preference table so that McMillian wins the election using the Borda count method.

	Preference Table for Candidates		
	Rankings		
1st	?	Witt	McMillian
2nd	?	McMillian	Longsdon
3rd	McMillian	Longsdon	Witt
Total Votes	601	632	650

25. *College Living Magazine* is compiling a list of the top five features of a college or university that students used when choosing an institution for higher education. They asked students to rank the importance of the following features of colleges and universities: size of the school, strength of the degree program, location, and number of student activities. Complete the following preference table so that location is the top influencer among students if the pairwise method of comparison is used.

	Preference Table for Most Important Features				
	Rankings				
1st	Size	Location	Activities	Size	Strength
2nd	Activities	Strength	Size	Location	Activities
3rd	Strength	Size	Location	Strength	Location
4th	Location	Activities	Strength	Activities	Size
Total Votes	100	?	55	105	89

10.2   What's Fair?

As you were reading through the different ways to choose a winner in the previous section, maybe one or more of the methods made you think to yourself, "That's not very fair." But someone hearing your thoughts might also respond, "Not fair to whom?" So, how do we begin to define "fairness"? For instance, is it fair that if a person has a majority of the votes, then that person should always win the election under any circumstances? Or, is it more fair that the candidate who is rejected by the fewest voters is declared the winner?

Historically, individual scholars have proposed a variety of conditions as standards or benchmarks for measures of fairness to help ensure electoral integrity. Five of the most natural conditions have become widely accepted yardsticks of fairness. They are the Condorcet criterion, the majority criterion, the monotonicity criterion, the irrelevant alternatives criterion, and the dictator criterion. These criteria have become a means by which we can evaluate election models. Although some seem more obvious than others, each strives to make the outcome of an election as fair as possible.

Despite the facts that these criteria describe fairness in a variety of different ways, and that there are so many different possible ways to tally electoral votes, it turns out that there is no single voting method that captures perfect fairness as defined by the five criteria. Kenneth Arrow, an American economist and social choice theorist, demonstrated that this is impossible. In his Ph.D. thesis, *A Difficulty in the Concept of Social Welfare* (1950), and later in his book, *Social Choice and Individual Values*, Arrow outlined and proved a theorem that states that if there are three or more choices on a ballot, there *cannot* be a voting method that will satisfy all five fairness criteria. His theory is called **Arrow's Impossibility Theorem**.

In this section, we will consider each of the five criteria for fairness. The examples of the election methods discussed in the previous section will be used to show how they each violate at least one of the fairness criteria and, in turn, support Arrow's Impossibility Theorem.

The first of the conditions is referred to as the **Condorcet criterion**. It is named for the French mathematician Nicolas de Condorcet. It simply states that if a candidate wins the head-to-head comparison—as in the pairwise comparison method—against *every other* candidate, then that candidate should also win the overall election in a fair voting system. By design, elections using the pairwise method of comparison adhere to the Condorcet criterion. However, the plurality method, the Borda count method, and the plurality with elimination method all have the possibility of violating this criterion. That doesn't mean that they will always violate it, but there is not a guarantee that they won't. Let's look at an example where the criterion is breached.

MATH MILESTONE

Kenneth J. Arrow is an American economist and the youngest winner of the Nobel Prize in Economics. In 1972, he was joint winner with British economist John Hicks. Amongst other titles, he is currently the Joan Kenney Professor of Economics and Professor of Operations Research, emeritus at Stanford University.

Arrow   Hicks

MATH MILESTONE

Marie Jean Antoine Nicolas Caritat, the Marquis de Condorcet, known as Nicolas de Condorcet, was an 18th-century French mathematician and philosopher. After leading a very productive scientific and political career, he died mysteriously in a Paris prison cell after being arrested during the French Revolution for his political views.

Example 1: Condorcet Criterion and the Plurality Method

Three students are in an election for president of the National Society of Collegiate Scholars on campus. Members were asked to rank the three candidates. The results of the membership votes are shown in the following preference table.

Table 1				
Preference Table for National Society of Collegiate Scholars				
	Rankings			
1st	Charles	Charles	Andrew	Bethany
2nd	Bethany	Andrew	Charles	Charles
3rd	Andrew	Bethany	Bethany	Andrew
Total Votes	30	31	65	21

a. Determine the winner if the society used the plurality method for determining the winner.

b. Determine the winner if the society used the pairwise method of comparison to determine the winner.

c. Does the plurality method adhere to the Condorcet criterion for this election? Explain your answer.

Solution

a. In order to find a winner using the plurality method, we need to see which candidate has the most first-place votes. We can read across the 1st row to find the total number of first-place votes each candidate received as shown.

Andrew: 65 first-place votes

Bethany: 21 first-place votes

Charles: $30 + 31 = 61$ first-place votes

Therefore, Andrew is the winner if the plurality method is used to elect the president of the National Society of Collegiate Scholars.

b. Using the pairwise method of comparison requires that we determine the winner of each head-to-head matchup between students. Using the formula we found earlier, we know that there are three comparisons required.

$$\text{number of comparisons } = \frac{3(3-1)}{2} = \frac{6}{2} = 3$$

Therefore, using a comparison table to keep track of the head-to-head comparison, we have the following.

Table 2

Completed Pairwise Comparison Grid

	Andrew	Bethany	Charles
Andrew	X	B = 30; A = 31 A = 65; B = 21	C = 30; C = 31 A = 65; C = 21
Bethany		X	C = 30; C = 31 C = 65; B = 21
Charles			X

Table 3

Pairwise Comparison Grid of Winners

	Andrew	Bethany	Charles
Andrew	X	B = 51; A = 96 [Andrew]	C = 82; A = 65 [Charles]
Bethany		X	C = 126; B = 21 [Charles]
Charles			X

Adding each candidate's votes together in the comparisons, we see that Andrew has 1 point while Charles has 2 points. Therefore, Charles is the winner using the pairwise method of comparison. Notice that Charles won the pairwise comparison against both of the other candidates, Andrew and Bethany.

c. In the pairwise comparison of the candidates, Charles wins the head-to-head comparisons against all other candidates. However, the plurality method chose Andrew as the winner. This means that for the election, the Condorcet criterion of fairness is violated by the plurality method.

The **majority criterion** is probably the most obviously named. It states that if a candidate receives a majority of votes in an election, that candidate should win. Although it seems that this should always be trivially true, this is not the case. Of the counting methods we've considered, it is possible that when using the Borda count method, the majority criterion is not satisfied. The Borda count method favors a candidate who is consistently ranked highly by all voters over a candidate who is simultaneously popular with a majority of voters and unpopular with the rest of the electorate. Example 2 illustrates this scenario.

Example 2: Majority Criterion and the Borda Count Method

In an election for chairman of the board of directors for a major company, shareholders were given the opportunity to rank the top five nominees. The results are posted in the preference table. Use it to answer the following questions.

Table 4				
Preference Table for Chairman Of The Board				
	Rankings			
1st	H. Beridze	M. Gruber	T. Taylor	T. Taylor
2nd	L. Wright	R. Jensen	H. Beridze	H. Beridze
3rd	M. Gruber	H. Beridze	M. Gruber	L. Wright
4th	T. Taylor	T. Taylor	R. Jensen	M. Gruber
5th	R. Jensen	L. Wright	L. Wright	R. Jensen
Total Votes	**2300**	**3100**	**4000**	**2200**

a. Determine the winner of the board elections using the majority rule.

b. Determine the winner of the board elections using the Borda count method.

c. Does the Borda count method satisfy the majority criterion for this election? Explain your answer.

Solution

a. To find a winner using the majority rule method, we simply count the number of first-place votes for each candidate. They are as follows.

H. Beridze: 2300 first-place votes

M. Gruber: 3100 first-place votes

R. Jensen: 0 first-place votes

T. Taylor: 4000 + 2200 = 6200 first-place votes

L. Wright: 0 first-place votes

In order to win an election with a majority rule decision, a candidate needs more than half of the votes. In this case, there are

$$2300 + 3100 + 4000 + 2200 = 11,600 \text{ votes available.}$$

$$\text{Half of the votes would be } \frac{11,600}{2} = 5800 \text{ votes.}$$

So, a minimum of 5801 votes are needed for a majority.

Therefore, T. Taylor is the winner in a majority rule election with 6200 votes.

b. To determine a winner using the Borda count method, each of the rankings receives a number of points. Since there are five candidates, each ranking will receive the following number of points.

$$1^{st} \text{ place } = 5 \text{ points}$$

$$2^{nd} \text{ place } = 4 \text{ points}$$

$$3^{rd} \text{ place } = 3 \text{ points}$$

$$4^{th} \text{ place } = 2 \text{ points}$$

$$5^{th} \text{ place } = 1 \text{ point}$$

The following table shows the point calculations. The final column is the sum of each candidate's points.

Table 5					
Borda Count Tally for Chairman of the Board					
	Rankings				**Total Points**
H. Beridze	1st $5(2300) = 11{,}500$	3rd $3(3100) = 9300$	2nd $4(4000) = 16{,}000$	2nd $4(2200) = 8800$	45,600
M. Gruber	3rd $3(2300) = 6900$	1st $5(3100) = 15{,}500$	3rd $3(4000) = 12{,}000$	4th $2(2200) = 4400$	38,800
R. Jensen	5th $1(2300) = 2300$	2nd $4(3100) = 12{,}400$	4th $2(4000) = 8000$	5th $1(2200) = 2200$	24,900
T. Taylor	4th $2(2300) = 4600$	4th $2(3100) = 6200$	1st $5(4000) = 20{,}000$	1st $5(2200) = 11{,}000$	41,800
L. Wright	2nd $4(2300) = 9200$	5th $1(3100) = 3100$	5th $1(4000) = 4000$	3rd $3(2200) = 6600$	22,900
Total Votes	**2300**	**3100**	**4000**	**2200**	

From the final column, it is clear that H. Beridze is the winner with 45,600 points using the Borda count method.

c. Using the Borda count method, H. Berdize would be declared the winner. However, T. Taylor received a majority of the votes and would win using the majority rule method. Therefore, the Borda count method does not satisfy the majority rule criterion in this election.

The third condition is often referred to as the **monotonicity criterion**. It requires that if Candidate X wins an election, then X would also win a second election if each voter were allowed to reorder the candidates in such a way that X only increases in ranking while the order of the other candidates remained the same. For example, if Jensen were ranked second in your list originally, then Jensen must be ranked in either first or second place in a reordering while the order of the other candidates is unchanged.

Example 3: Monotonicity Criterion and the Plurality with Elimination Method

The city council for the town of Whitman was electing a new vice president. All council members were asked to rank the four candidates in order of preference. Show that by using the plurality with elimination method, the monotonicity criterion is violated. The preference table shows the results of the voting.

Table 6
Preference Table for Whitman City Council Vice President

	Rankings			
1st	Clarke	Roberts	Green	Green
2nd	Green	Clarke	Roberts	White
3rd	Roberts	White	Clarke	Clarke
4th	White	Green	White	Roberts
Total Votes	**6**	**5**	**4**	**2**

Solution

In order to show that the monotonicity criterion is violated using the plurality with elimination method, we need to first find the winner of the election using the plurality with elimination method. Recall that when using this method, if no candidate has a majority of the first-place votes, the candidate with the least amount of first-place votes is eliminated in each round. A majority is more than half the number of votes. In this case, there were

$$6 + 5 + 4 + 2 = 17 \text{ votes.}$$

Half of 17 is 8.5, so a majority consists of at least 9 votes. The following is a list of the number of first-place votes received by each candidate.

Clarke: 6 first-place votes

Green: $4 + 2 = 6$ first-place votes

Roberts: 5 first-place votes

White: 0 first-place votes

Therefore, since no candidate had a majority of votes, White is eliminated with the least amount of first-place votes.

Table 7
Plurality with Elimination Cycle 1

	Rankings			
1st	Clarke	Roberts	Green	Green
2nd	Green	Clarke	Roberts	~~White~~
3rd	Roberts	~~White~~	Clarke	Clarke
4th	~~White~~	Green	~~White~~	Roberts
Total Votes	**6**	**5**	**4**	**2**

The candidate with the next-smallest number of first-place votes is Roberts with 5 first-place votes. Note that White did not have any first-place votes, so none of our vote counts change when he is eliminated. So the candidate with the next smallest number of first-place votes, Roberts, is eliminated next.

Table 8
Plurality with Elimination Cycle 2

	Rankings			
1st	Clarke	~~Roberts~~	Green	Green
2nd	Green	Clarke	~~Roberts~~	Clarke
3rd	~~Roberts~~	Green	Clarke	~~Roberts~~
Total Votes	**6**	**5**	**4**	**2**

That leaves Green and Clarke. The following table shows the standings without White and Roberts.

Table 9				
Plurality with Elimination Cycle 2 Simplified				
	Rankings			
1st	Clarke	Clarke	Green	Green
2nd	Green	Green	Clarke	Clarke
Total Votes	6	5	4	2

We can now compare Green and Clark and determine the winner. Clarke wins in the first two columns of rankings, but Green wins in the final two. Therefore, Clarke has 11 first-place votes and Green has 6. So, Clarke is the winner using the plurality with elimination method.

In order to show that the monotonicity criterion is violated using this method, we need to show that a different winner is produced if some of the council members change their votes in such a way that the winner, Clarke, increases his ranking while the order of the other candidates stay the same. Suppose that the 2 voters in the last column of the original preference table changed their rankings to place Clarke first. You can see that their ranking for the other candidates stay in the same preference order (Green, White, then Roberts).

Table 10				
Whitman City Council Vice President Original Election				
	Rankings			
1st	Clarke	Roberts	Green	Green
2nd	Green	Clarke	Roberts	White
3rd	Roberts	White	Clarke	Clarke
4th	White	Green	White	Roberts
Total Votes	6	5	4	2

Table 11				
Whitman City Council Vice President Second Election				
	Rankings			
1st	Clarke	Roberts	Green	Clarke
2nd	Green	Clarke	Roberts	Green
3rd	Roberts	White	Clarke	White
4th	White	Green	White	Roberts
Total Votes	6	5	4	2

Now, using the plurality with elimination method, we have the following first-place votes for each candidate.

Clarke: 6 + 2 = 8 first-place votes

Green: 4 first-place votes

Roberts: 5 first-place votes

White: 0 first-place votes

Once again, White is eliminated without affecting any of the first-place votes. Therefore, we look to find the candidate with the next lowest number of first-place votes. Green is eliminated with only 4 votes. The following preference table shows both White and Green eliminated.

Table 12

Plurality with Elimination Cycles 1 and 2 Simplified

	Rankings			
1st	Clarke	Roberts	Roberts	Clarke
2nd	Roberts	Clarke	Clarke	Roberts
Total Votes	**6**	**5**	**4**	**2**

Now we see that Clarke has 6 + 2 = 8 votes and Roberts has 5 + 4 = 9 votes. That gives Roberts a majority and therefore the title of Whitman City Council Vice President in this second voting process.

Notice that Clarke won in the first voting. He then gained 2 first-place votes in the second process, but ultimately lost the election to Roberts. This illustrates that by using the plurality with elimination method, it is possible to violate the monotonicity criterion.

Example 3 used the same method, plurality with elimination, to produce a different winner after voters were allowed to change their votes, illustrating that the plurality with elimination method can violate the monotonicity criterion. It is the only counting method we discuss that has a possibility of violating the monotonicity criterion.

The fourth condition considers cases where candidates drop out of the race. The **irrelevant alternatives criterion** states that if a candidate wins an election, then that same candidate would win the election even if one or more candidates pulls out. This criterion is the most stringent of all the criteria because it may be violated in four of the five voting methods, all except for the majority rule. Example 4 illustrates a situation in which the pairwise method of comparison does not adhere to the irrelevant alternative criterion.

Example 4: Irrelevant Alternatives Criterion and the Pairwise Method of Comparison

As part of the process of choosing the next chancellor of the university, faculty members were asked to rank the four finalists in order of preference. The following table summarizes their votes.

Table 13

Preference Table for Chancellor

	Rankings				
1st	Khan	Khan	Patel	Mason	Mason
2nd	Patel	Lewis	Lewis	Lewis	Khan
3rd	Mason	Patel	Mason	Khan	Lewis
4th	Lewis	Mason	Khan	Patel	Patel
Total Votes	**166**	**290**	**235**	**123**	**236**

a. Use the pairwise method of comparison to determine the favorite candidate amongst the faculty.

b. Before a chancellor could be named, Patel took a job elsewhere and Lewis decided the university was not the right fit for him after visiting the campus. Consequently, Patel and Lewis withdrew their names from the application pool. Determine the faculty favorite after Patel and Lewis are removed from the ballot.

c. Based on your solutions to parts **a.** and **b.**, does this election comply with the irrelevant alternative criterion?

Solution

a. Using the pairwise method of comparison requires that we determine the winner of each head-to-head matchup between the candidates. Using the formula, we know that there are six comparisons required.

$$\text{number of comparisons} = \frac{4(4-1)}{2} = \frac{12}{2} = 6$$

Therefore, using a comparison table to keep track of the head-to-head comparisons, we have the following.

Table 14
Completed Pairwise Comparison Grid for Chancellor

	Patel	Mason	Lewis	Khan
Patel	X	P = 166; P = 290; P = 235; M = 123; M =236	P = 166; L = 290; P = 235; L = 123; L = 236	K = 166; K = 290; P = 235; K = 123; K = 236
Mason		X	M = 166; L = 290; L = 235; M = 123; M = 236	K = 166; K = 290; M = 235; M = 123; M = 236
Lewis			X	K = 166; K = 290 ; L = 235; L = 123 ; K = 236
Khan				X

Table 15
Pairwise Comparison Grid of Winners for Chancellor

	Patel	Mason	Lewis	Khan
Patel	X	P = 691; M = 359 [Patel]	P = 401; L = 649 [Lewis]	K = 815; P = 235 [Khan]
Mason		X	M = 525; L = 525 [Tie]	K = 456; M = 594 [Mason]
Lewis			X	K = 692; L = 358 [Khan]
Khan				X

Helpful Hint

Remember that in a head-to-head comparison, a tie gives each candidate $\frac{1}{2}$ of a point.

Adding each candidate's votes together in the comparisons, we see that Patel has 1 point, Mason has $1\frac{1}{2}$ points, Lewis has $1\frac{1}{2}$ points, while Khan has 2 points. Therefore, Khan is the winner using the pairwise method of comparison.

b. When both Patel and Lewis withdraw from the race, only Mason and Khan remain. We know from the head-to-head comparison in Table 15, that when these two face one another, Mason is the winner. Therefore, Mason is the winner using the pairwise method of comparison after Patel and Lewis withdraw.

c. Using the pairwise comparison method both before and after Patel and Lewis withdrew furnished two different winners. Therefore, it does not satisfy the irrelevant alternative criterion in this election.

Finally, the fifth condition is more about the voter rather than the method of counting the votes. The **dictator criterion** states that no single voter is allowed to decide the outcome of an election. Simply put, according to Arrow's Impossibility Theorem, in order to be fair, an electoral system must be a democracy and not a dictatorship.

The following table summarizes each of the voting methods discussed in Section 10.1 versus the fairness criteria used in democratic elections.

Table 16 Voting Methods and Fairness Criteria				
	Condorcet Criterion	**Majority Criterion**	**Monotonicity Criterion**	**Irrelevant Alternatives Criterion**
Majority Rule Decision	Never violates	Never violates	Never violates	Never violates
Plurality	May violate	Never violates	Never violates	May violate
Borda Count	May violate	May violate	Never violates	May violate
Plurality with Elimination	May violate	Never violates	May violate	May violate
Pairwise Comparison	Never violates	Never violates	Never violates	May violate

As our table shows, it would appear that using a majority rule as an election method might in fact be the perfect voting system because it never violates any of the four fairness criteria in a democracy. However, the one "fly in the ointment," so to say, is exactly what we saw in Section 10.1—using the majority rule doesn't always produce a winner. This is why Arrow required that valid voting systems always produce a distinct winner. So, when developing a method of counting votes, it's important to think beyond a first round majority pick.

10.2   Exercises

Answer the question thoughtfully.

1. Briefly summarize each of the four fairness criteria.

Use the given preference table to answer each question.

2. A group of students were asked to listen to auditions by four bands for the homecoming dance and then rank the bands. The following preference table summarizes their selections.

Preference Table for Homecoming Bands					
	Rankings				
1st	Band B	Band A	Band C	Band D	Band A
2nd	Band D	Band D	Band B	Band A	Band C
3rd	Band C	Band B	Band A	Band B	Band B
4th	Band A	Band C	Band D	Band C	Band D
Total Votes	**23**	**15**	**16**	**11**	**5**

a. Which band is the plurality winner?

b. Which candidate wins the election using a pairwise method of comparison?

c. Does this election, using the plurality method, satisfy the Condorcet criterion? Explain your answer.

3.   Susan, Courtney, and Jade ran for Freshman SGA representative. The following preference table shows the results of the race.

Preference Table for SGA Representatives			
	Rankings		
1st	Susan	Courtney	Susan
2nd	Courtney	Susan	Jade
3rd	Jade	Jade	Courtney
Total Votes	32	55	16

a.   Which candidate is the winner using the Borda count method?

b.   Which candidate is the winner using the pairwise method of comparison?

c.   Does the Borda count method satisfy the Condorcet criterion in this situation? Why or why not?

4.   Work colleagues are trying to decide on which restaurant to use for their end-of-year party. Since they don't all agree on a favorite, they decide to rank the four possibilities and decide on a restaurant that way. The results are shown in the preference table.

Preference Table for Favorite Restaurants					
	Rankings				
1st	Tanvi's	O'Reilly's Pub	Ralphie's	Tanvi's	Salty's
2nd	Ralphie's	Salty's	Tanvi's	O'Reilly's Pub	Ralphie's
3rd	Salty's	Tanvi's	Salty's	Salty's	O'Reilly's Pub
4th	O'Reilly's Pub	Ralphie's	O'Reilly's Pub	Ralphie's	Tanvi's
Total Votes	8	3	4	11	9

a.   Which restaurant is the winner using the plurality with elimination method?

b.   Which restaurant is the winner using the pairwise method of comparison?

c.   Does the plurality with elimination method satisfy the Condorcet criterion if the colleagues chose a restaurant using plurality with elimination? Why or why not?

5.   Jack is scheduling classes for next semester and has three possible classes for a particular time slot. He surveys students to see which class has the most interest out of Electricity and Magnetism, Heat and Optics, and Mechanics.

Preference Table for Classes			
	Rankings		
1st	H & O	Mechanics	H & O	E & M
2nd	E & M	H & O	E & M	Mechanics
3rd	Mechanics	E & M	Mechanics	H & O
Total Votes	81	132	36	13

a.   Which class has the most interest using the majority rule?

b.   Which class has the most interest using the Borda count method?

c.   Does the Borda count method satisfy the majority criterion if Jack uses this method to fill the available time slot? Explain your answer.

6. The history department is adopting a new textbook for the Freshman Introduction class. Instructors in the history department were asked to rank the top four textbooks in order of usability for the class. The preference table shows the outcome of their rankings.

Preference Table for History Textbooks				
	Rankings			
1st	Excursions Through Time	Thinking Historically	A Look in Time	Through the Ages
2nd	Thinking Historically	A Look in Time	Thinking Historically	A Look in Time
3rd	A Look in Time	Excursions Through Time	Through the Ages	Thinking Historically
4th	Through the Ages	Through the Ages	Excursions Through Time	Excursions Through Time
Total Votes	**7**	**3**	**7**	**4**

a. Use the plurality with elimination method to choose a new textbook for adoption.

b. After some discussion in the department, the seven instructors whose rankings appear in the first column decide to change their votes so that the textbooks titled *A Look in Time* and *Thinking Historically* swap rankings. Create a new preference table to show this change and then use the table to decide on a textbook using the plurality with elimination method once again.

c. Is the monotonicity criterion satisfied? Explain your answer.

7. Charles Beauregard High School sponsors a senior class trip each year. This year, the students can choose between trips to Charleston, SC; Jacksonville, FL; New Orleans, LA; Chicago, IL; or Atlanta, GA. The students were asked to rank their choices in order to choose a destination.

Preference Table for Senior Class Trip				
	Rankings			
1st	New Orleans, LA	Charleston, SC	Atlanta, GA	Chicago, IL
2nd	Chicago, IL	New Orleans, LA	New Orleans, LA	Charleston, SC
3rd	Atlanta, GA	Atlanta, GA	Chicago, IL	Atlanta, GA
4th	Charleston, SC	Jacksonville, FL	Charleston, SC	New Orleans, LA
5th	Jacksonville, FL	Chicago, IL	Jacksonville, FL	Jacksonville, FL
Total Votes	**375**	**348**	**289**	**115**

a. Which city would be the senior trip destination using the plurality with elimination method?

b. After consideration of travel time, the senior class sponsors decided to eliminate Chicago, IL, from the options. Which city is now the winning destination using plurality with elimination?

c. Does the plurality with elimination method satisfy the irrelevant alternatives criterion? Explain your answer.

8. A nonprofit organization is looking to use social networking sites to increase interest. They poll donors to decide where to focus their resources. The donors are asked to rank the following sites by how often they use them.

	Preference Table for Social Networking Sites			
	Rankings			
1st	Facebook	Pinterest	Facebook	Twitter
2nd	Twitter	Facebook	Twitter	Facebook
3rd	Google+	Twitter	LinkedIn	LinkedIn
4th	LinkedIn	LinkedIn	Pinterest	Google+
5th	Pinterest	Google+	Google+	Pinterest
Total Votes	**23**	**40**	**44**	**35**

a. Which site is given the higher ranking using the pairwise method of comparison?

b. If the nonprofit decides to eliminate Pinterest and Google+ from the list and adjust the votes accordingly, which site would then be the top place for the nonprofit to focus on using the pairwise method?

c. Does the pairwise method satisfy the irrelevant alternatives criterion for this preference table? Why or why not?

9. The men of the fraternity Chi Rho on campus are electing a new president. The candidates are M. Jones, H. Kennedy, and T. Parchment. The summary of the rankings of the candidates from the members is given in the chart.

	Preference Table for Chi Rho Presidential Candidates			
	Rankings			
1st	H. Kennedy	M. Jones	H. Kennedy	T. Parchment
2nd	M. Jones	T. Parchment	T. Parchment	H. Kennedy
3rd	T. Parchment	H. Kennedy	M. Jones	M. Jones
Total Votes	**8**	**14**	**10**	**11**

a. Name the fraternity's next president using the Borda count method.

b. Because of unexpected family issues, M. Jones had to withdraw from the university on short notice. Create a new preference table to show this change and then use the table to determine the next president using the Borda count method.

c. Is the irrelevant alternatives criterion satisfied with this change? Explain your answer.

10. For the end of year banquet, the program committee needs to decide on the type of cuisine. They have four choices: Italian, French, Mexican, or Chinese. The following preference table summarizes the rankings of the members surveyed. Determine if the Condorcet criterion is satisfied if the committee uses the Borda count method to choose a cuisine.

	Preference Table for Banquet Cuisine				
	Rankings				
1st	French	Mexican	Italian	Italian	Mexican
2nd	Italian	Italian	Mexican	Chinese	French
3rd	Mexican	Chinese	French	Mexican	Chinese
4th	Chinese	French	Chinese	French	Italian
Total Votes	**4**	**21**	**14**	**11**	**9**

10

11. In preparation for their Oscar party, a sorority decides to have an election of their own. In the Best Actor category are George Clooney, Johnny Depp, Colin Firth, and Zac Efron. The rankings from the sorority members are shown in the following preference table. Determine if the majority criterion will be satisfied if the sorority uses the Borda count method to choose the winner for Best Actor.

Preference Table for Best Actor					
	Rankings				
1st	Johnny Depp	Zac Efron	Johnny Depp	George Clooney	Colin Firth
2nd	Colin Firth	Colin Firth	George Clooney	Colin Firth	Zac Efron
3rd	Zac Efron	George Clooney	Zac Efron	Johnny Depp	Johnny Depp
4th	George Clooney	Johnny Depp	Colin Firth	Zac Efron	George Clooney
Total Votes	26	16	16	15	9

12. *Pretty People* magazine wants to name the number one activity for a first date amongst singles. They enlist the help of an online dating site that asks members to rank five activities in order of their preference as part of their sign-up process. The five activities include: movie, nice dinner, picnic, sporting event, and concert. The following preference table displays the results of the rankings. Determine if the Condorcet criterion is satisfied if the magazine uses the plurality method to name the top activity.

Preference Table for First Date Activities						
	Rankings					
1st	Nice Dinner	Picnic	Picnic	Sporting Event	Sporting Event	Movie
2nd	Picnic	Sporting Event	Sporting Event	Concert	Nice Dinner	Nice Dinner
3rd	Concert	Concert	Nice Dinner	Picnic	Movie	Concert
4th	Movie	Nice Dinner	Concert	Nice Dinner	Concert	Picnic
5th	Sporting Event	Movie	Movie	Movie	Picnic	Sporting Event
Total Votes	40	34	21	22	19	20

13. United Way asks its volunteers to rank the nonprofit organizations it serves in order to give out the annual award for Local Volunteer of the Year. This year's organizations include the Breast Cancer Foundation, Community Garden, Big Brother/Big Sister, YMCA, and Manna Café. Complete the preference table so that Community Garden is the winner using the Borda count method, but the Condorcet criterion is violated.

Preference Table for Local Volunteer of the Year Award			
	Rankings		
1st	MC	CG	BB/BS
2nd	BB/BS	BCF	MC
3rd	CG	BB/BS	CG
4th	YMCA	YMCA	BCF
5th	BCF	MC	YMCA
Total Votes	?	16	10

14. An online dating site asks new users to rank certain traits in order of importance when they are matched with another user. The site uses the top choice as the first criterion for matching people together. The following preference table summarizes the rankings for appearance, personality, income, profession, and height. Complete the preference table so that personality is the top trait if the site uses the Borda count method to count the votes and the majority criterion is not violated.

	Preference Table for Important Traits			
	Rankings			
1st	?	Personality	Income	Appearance
2nd	?	Profession	Personality	Income
3rd	Profession	Appearance	Appearance	Personality
4th	Height	Height	Profession	Height
5th	Income	Income	Height	Profession
Total Votes	45	56	33	48

15. College dining is considering opening a national fast food franchise on campus. Their choices have been narrowed down to Subway, Taco Bell, McDonald's, KFC, or Pizza Hut. Students were asked to rank these establishments in order of preference, and the results are shown in the preference table. If college dining uses the plurality method to choose the students' top choice of Subway, complete the preference table so that the irrelevant alternatives criterion is violated when KFC is removed from the race.

	Preference Table for New Franchise on Campus			
	Rankings			
1st	Subway	Taco Bell	KFC	Subway
2nd	McDonald's	Subway	Taco Bell	Taco Bell
3rd	Taco Bell	Pizza Hut	Subway	Pizza Hut
4th	KFC	McDonald's	Pizza Hut	McDonald's
5th	Pizza Hut	KFC	McDonald's	KFC
Total Votes	56	60	?	45

16. Sigma Air is exploring the possibility of establishing a new hub in one of the following cities: Cheyenne, WY; Little Rock, AR; Amarillo, TX; and Portland, OR. They surveyed their Gold and Platinum members to ask their preference for the new hub. Complete the preference table of the rankings so that using the plurality with elimination method Amarillo is the winner, and the Condorcet criterion is satisfied.

	Preference Table for New Hub Location		
	Rankings		
1st	Amarillo, TX	Portland, OR	Cheyenne, WY
2nd	Portland, OR	Little Rock, AR	Amarillo, TX
3rd	Cheyenne, WY	?	Little Rock, AR
4th	Little Rock, AR	?	Portland, OR
Total Votes	70	34	37

17. The following preference table summarizes the outcome of an election.

Preference Table for Candidates				
	Rankings			
1st	Luke	Luke	Lauren	Blake
2nd	Hannah	Blake	Blake	Luke
3rd	Lauren	Lauren	Hannah	Lauren
4th	Blake	Hannah	Luke	Hannah
Total Votes	31	9	23	41

    a. Which candidate is the plurality winner?

    b. Which candidate wins the election using a pairwise method of comparison?

    c. Does this method satisfy the Condorcet criterion? Why or why not?

10.3 Apportionment

We most often think about elections in terms of electing a president, a senator, a governor, a mayor, etc. However, we also hold elections for things such as favorite ice cream flavor at the local ice cream shop, best movie of the year, and the best singer on television shows. In the United States, we cherish the concept of democracy and our individual rights to vote for the candidates we feel are the best fit for our beliefs. As we saw in Sections 10.1 and 10.2, how votes are counted and how to make methods of choice fair can be somewhat difficult. In this section, we extend the concepts of elections to encompass how we *apportion* resources equally.

As an example, in voting, **apportionment** is the method of distributing power in political systems based on population density, whereas in a public school district, apportionment may reflect the availability of buses for athletic events.

Apportionment

Apportionment is the method of dividing resources in such a way as to maximize the use of those resources.

Fun Fact

The US Census was mandated in Article I of the Constitution as a method to determine not only the total count of the population of the United States, but also each state. The population for each state then determines the number of representatives for that state. For more information on the US Census, visit http://www. census.gov/2010census/ about/.

In the United States, a method of apportionment is used to determine how many representatives each state is allowed. The US census is the determining factor in the apportionment of power. Each time the census is conducted, there is a new debate on how congressional seats are distributed throughout a state based on shifting populations. Recall a discussion from the chapter introduction concerning the election of the president of the United States. The census allocates a number of electors equal to the number of representatives for a given state. These electors then vote with the majority of their state on who the president of the United States is going to be for the next four years. The candidate with the majority of electoral votes wins the presidency. Currently the number required for the majority is 270 electoral votes.

If there are only a certain number of items to apportion, such as votes, a dilemma can arise. Apportionment methods often divide the "votes" of a representative into fractional pieces. This means one vote must be divided among several constituents. When talking about people as representatives, dividing them between two or more states is an impossibility. Some method of converting fractional parts of votes into whole numbers must be devised. Over the 235-year history of the voting system in the United States, there have been proposed compromises for how to divide portions of representatives, which have led to disputes on how votes should be apportioned.

We begin our discussion with the apportionment methods that have been used over the years, including the Hamilton method, the Jefferson method, the Webster method, and the Huntington-Hill method. It's important to keep in mind that, although we have introduced the concepts of apportionment based on election of officials, apportionment is used in many facets of decision making, such as class size at a school, budgeting, etc.

Consider the allotment of money for colleges in Texas. There are 104 public institutions in the state of Texas. If the state has $87 billion allocated to higher education, how should this money be apportioned? A few possible ways to apportion the money are:

1. The school with the largest population gets the most money.

2. To apportion the money evenly among each of the institutions regardless of population

3. To apportion the money based on credit hours per student

Of course there are more ways than these three to apportion the money.

Example 1: Apportioning Money Based on Number of Institutions

If Texas were to divide the available funds of $87 billion evenly among its 104 institutions of higher education, how much would each institution receive? Is this a fair apportionment method for all institutions on a per-student amount?

Solution

The allocation of money in this case is straightforward and can be represented by the following quotient.

$$\frac{\$87,000,000,000}{104 \text{ institutions}} = \$836,538,461.50/\text{institution}$$

Under this funding formula, each institution would receive about $836,538,462.

This method is not very fair for each institution. For instance, under this formula, Stephen F. Austin University has about 12,000 undergraduate students and would receive the same amount of funding as the University of Texas, which has an undergraduate enrollment of about 38,000 students. Thus, Stephen F. Austin would receive about $70,000 per student whereas the University of Texas would receive about $22,000 per student.

Skill Check # I

Assuming Texas has a total of 1,066,956 students in higher education, how much would colleges receive per student if the overall budget were $87 billion?

Determining a method that apportions resources in a manner that ensures all members within a population receive a fair share is not always straightforward. In order to accomplish this, we need a standard way in which to *divide* the available resources, such as money.

A **standard divisor** is the average number of members of the population that account for one apportioned item. For example, if the population we wish to use to apportion representatives is the United States population, then the standard divisor would be the average number of people per congressional seat.

Standard Divisor

A **standard divisor** (SD) is the average number of members of the population that will account for one apportioned item.

$$SD = \frac{\text{total population in a group}}{\text{total number of items to be apportioned}}$$

Whereas the standard divisor is an average of the number of members in the population that will account for one apportioned item, a **standard quota** represents the number of items that will be apportioned to each subgroup. In other words, in the United States, the standard quota for a congressional seat is the number of representatives assigned to each district.

Standard Quota

A **standard quota** (SQ) represents the number of items that will be apportioned to each subgroup and is calculated as follows.

$$SQ = \frac{\text{population of the subgroup}}{\text{standard divisor}}$$

Next, let's work an example of apportionment in a situation that is not related to politics.

Example 2: Distributing Money from the Super Bowl

In 2012, Super Bowl XLV had an allotment of about 100,000 tickets. The average price of a ticket was about $900, meaning that the total value of sales to the NFL was a whopping $90,000,000. The allotment of tickets for each of the groups that were allowed to purchase tickets is given in Table 1. Determine the standard divisor and standard quota for each constituency.

Table 1

Distribution of Super Bowl Tickets

	Super Bowl Teams (17.5% each)	Non-Super Bowl Teams	Host Team	NFL League Office	Public	Total Tickets
Percent	35%	34%	5%	25%	1%	100%
# Tickets	35,000	34,000	5000	25,000	1000	100,000

Solution

The standard divisor can be found using the formula, where the total population in the group is the number of tickets sold, or 100,000. The total to be apportioned would then be $90,000,000. So, the standard divisor would be

$$SD = \frac{\text{total population in a group}}{\text{total number of items to be apportioned}} = \frac{100{,}000 \text{ tickets}}{\$90{,}000{,}000} \approx 0.001 \text{ tickets/dollar}.$$

The standard quota is then computed for each constituency in Table 2.

Table 2

Standard Quota per Subgroup

Subgroup	standard quota $= \dfrac{\text{population of the subgroup}}{\text{standard divisor}}$
Super Bowl Teams	$\approx \dfrac{35{,}000}{0.001} = \$35{,}000{,}000$
Non-Super Bowl Teams	$\approx \dfrac{34{,}000}{0.001} = \$34{,}000{,}000$
Host Team	$\approx \dfrac{5000}{0.001} = \$5{,}000{,}000$
NFL	$\approx \dfrac{25{,}000}{0.001} = \$25{,}000{,}000$
General Public	$\approx \dfrac{1000}{0.001} = \$1{,}000{,}000$

For simplicity, we can abbreviate standard divisor as SD and standard quota as SQ.

The Hamilton Method of Apportionment

When the United States first gained independence, the US Constitution called for representation for each state by one representative for every 30,000 people. The difficulty in determining how many representatives each state should have arose when not all states had an even population divisible by 30,000. Therefore, a method was needed to apportion the seats of the House of Representatives. The **Hamilton method of apportionment** was one of the first methods of determining seats in the United States. The Hamilton method allocates delegates by assigning each group or state an appropriate percentage of the total number of representatives based on the population of the state.

Hamilton Method

1. Find SD: $SD = \dfrac{\text{total population in a group}}{\text{total number of items to be apportioned}}$. Round to the nearest thousandth.

2. Find SQ: $SQ = \dfrac{\text{population of the subgroup}}{\text{standard divisor}}$.

3. Round SQ down to the next lowest integer.

4. Award any remaining resources based on which group's SQ was rounded down the most. That is, whoever had the largest fractional part rounded down.

Example 3: Applying the Hamilton Method of Apportionment

Table 3 shows the population of the United States in 1810 by state. The total population of the states with representation in the United States in 1810 was 7,239,081, which includes the population of territories, such as Mississippi and Alabama, that were not official states at the time. Use the Hamilton method to apportion the 143 seats in the US House of Representatives.

Table 3	
1810 Census Results	
State	**Population**
Maine	228,705
New Hampshire	214,460
Vermont	217,895
Massachusetts	472,040
Rhode Island	76,931
Connecticut	261,941
New York	959,049
New Jersey	245,562
Pennsylvania	810,091
Tennessee	261,727
Kentucky	406,511
West Virginia	105,469
Virginia	877,683
North Carolina	555,500
South Carolina	415,115
Delaware	72,674
Maryland	380,546
Georgia	252,433

Solution

To apply the Hamilton method, we will follow the four steps previously outlined.

Step 1: Find SD and round to the nearest thousandth.

$$SD = \frac{\text{total population in a group}}{\text{total number of items to be apportioned}} = \frac{7,239,081}{143} \approx 50,622.944 \text{ people/seat}$$

Note that this value is much higher than the 30,000 that was set by the constitution.

Step 2: Find SQ for each subgroup, in this case, each state. This is done in Table 4.

Step 3: Round SQ to the next lowest integer.

	Table 4		
	SQ and Rounded SQ Based on 1810 Census		
State	**Population**	$SQ = \dfrac{\text{population of the subgroup}}{\text{standard divisor}}$	**Rounded SQ**
Maine	228,705	$\dfrac{228,705}{50,622.944} \approx 4.518$	4 seats
New Hampshire	214,460	$\dfrac{214,460}{50,622.944} \approx 4.236$	4 seats
Vermont	217,895	$\dfrac{217,895}{50,622.944} \approx 4.304$	4 seats
Massachusetts	472,040	$\dfrac{472,040}{50,622.944} \approx 9.325$	9 seats
Rhode Island	76,931	$\dfrac{76,931}{50,622.944} \approx 1.520$	1 seats
Connecticut	261,941	$\dfrac{261,941}{50,622.944} \approx 5.174$	5 seats
New York	959,049	$\dfrac{959,049}{50,622.944} \approx 18.945$	18 seats
New Jersey	245,562	$\dfrac{245,562}{50,622.944} \approx 4.851$	4 seats
Pennsylvania	810,091	$\dfrac{810,091}{50,622.944} \approx 16.002$	16 seats
Tennessee	261,727	$\dfrac{261,727}{50,622.944} \approx 5.170$	5 seats
Kentucky	406,511	$\dfrac{406,511}{50,622.944} \approx 8.030$	8 seats
West Virginia	105,469	$\dfrac{105,469}{50,622.944} \approx 2.083$	2 seats
Virginia	877,683	$\dfrac{877,683}{50,622.944} \approx 17.338$	17 seats
North Carolina	555,500	$\dfrac{555,500}{50,622.944} \approx 10.973$	10 seats
South Carolina	415,115	$\dfrac{415,115}{50,622.944} \approx 8.200$	8 seats
Delaware	72,674	$\dfrac{72,674}{50,622.944} \approx 1.436$	1 seats
Maryland	380,546	$\dfrac{380,546}{50,622.944} \approx 7.517$	7 seats
Georgia	252,433	$\dfrac{252,433}{50,622.944} \approx 4.987$	4 seats
Total Seats			**127 seats**

By using the rounded-down SQ, we have 127 seats taken and 143 to fill.

Step 4: The Hamilton method tells us that we should go back and add a seat to the states that were rounded down the most until all remaining seats are allocated. In other words, we must apportion the remaining 16 seats based on the fractional component of each SQ. The largest fractional component not awarded a seat in Step 3 takes priority when assigning additional seats. This means that Georgia will receive the first available seat out of the 16 remaining, followed by North Carolina, then New York, and so on.

Table 5				
Hamilton Apportionment of House Representatives Based on 1810 Census				
State	Population	SQ	Rounded SQ	Total Seats
Maine	228,705	$\frac{228,705}{50,622.944} \approx 4.518$	4	$4 + 1 =$ 5 seats
New Hampshire	214,460	$\frac{214,460}{50,622.944} \approx 4.236$	4	$4 + 1 =$ 5 seats
Vermont	217,895	$\frac{217,895}{50,622,944} \approx 4.304$	4	$4 + 1 =$ 5 seats
Massachusetts	472,040	$\frac{472,040}{50,622.944} \approx 9.325$	9	$9 + 1 =$ 10 seats
Rhode Island	76,931	$\frac{76,931}{50,622.944} \approx 1.520$	1	$1 + 1 =$ 2 seats
Connecticut	261,941	$\frac{261,941}{50,622.944} \approx 5.174$	5	$5 + 1 =$ 6 seats
New York	959,049	$\frac{959,049}{50,622.944} \approx 18.945$	18	$18 + 1 =$ 19 seats
New Jersey	245,562	$\frac{245,562}{50,622.944} \approx 4.851$	4	$4 + 1 =$ 5 seats
Pennsylvania	810,091	$\frac{810,091}{50,622.944} \approx 16.002$	16	16 seats
Tennessee	261,727	$\frac{261,727}{50,622,944} \approx 5.170$	5	$5 + 1 =$ 6 seats
Kentucky	406,511	$\frac{406,511}{50,622.944} \approx 8.030$	8	8 seats
West Virginia	105,469	$\frac{105,469}{50,622.944} \approx 2.083$	2	$2 + 1 =$ 3 seats
Virginia	877,683	$\frac{877,683}{50,622.944} \approx 17.338$	17	$17 + 1 =$ 18 seats
North Carolina	555,500	$\frac{555,500}{50,622.944} \approx 10.973$	10	$10 + 1 =$ 11 seats
South Carolina	415,115	$\frac{415,115}{50,622.944} \approx 8.200$	8	$8 + 1 =$ 9 seats
Delaware	72,674	$\frac{72,674}{50,622.944} \approx 1.436$	1	$1 + 1 =$ 2 seats
Maryland	380,546	$\frac{380,546}{50,622.944} \approx 7.517$	7	$7 + 1 =$ 8 seats
Georgia	252,433	$\frac{252,433}{50,622.944} \approx 4.987$	4	$4 + 1 =$ 5 seats
Total Seats			127	143

The Jefferson Method of Apportionment

The Hamilton method can cause some groups whose standard quota was rounded down to be awarded an additional seat, while at the same time other groups whose standard quota was also rounded down can be deprived a seat. As a result, George Washington vetoed the Hamilton method of apportionment in 1790 and another method of apportioning representatives was sought out. Thomas Jefferson then proposed a method of apportionment that was used from 1790 until 1840. The Jefferson method tries to remedy the rounding situation by using a *modified divisor* (MD). The

MD is used if the sum of the SQs is not exactly the number of items to distribute. The MD may be a lower number than the SD, and will require the introduction of a modified quota (MQ).

> ## Modified Divisor (MD)
>
> A **modified divisor** is a divisor, near the standard divisor (SD), that is chosen by guess-and-check in an attempt to make the sum of the modified quotas exactly equal to the number of items to be apportioned.

Jefferson Method

1. Find the SD and all of the SQs. Round all of the SQs down.

2. If there are leftover things to apportion, we need to find a new divisor—an MD.

3. Calculate new MQs with the MD. If we still have leftover things to distribute, try again.

4. Continue the guess-and-check process until we have apportioned all items.

Example 4: Applying the Jefferson Method of Apportionment

Suppose that a university has 18 scholarships to be apportioned among 225 math majors, 417 history majors, and 308 computer science majors. Use the Jefferson method to determine how the scholarships should be apportioned among the three major groups.

Solution

The total number of students vying for a scholarship is $225 + 417 + 308 = 950$. As with the Hamilton method, we begin by finding the standard divisor, SD.

$$\text{SD} = \frac{\text{total population in a group}}{\text{total number of items to be apportioned}} = \frac{950}{18} \approx 52.778 \text{ students/scholarship}$$

Next, we find the standard quota for each major, as shown in Table 6.

Table 6		
SQ and Rounded SQ for Scholarship Apportionment		
Major	$\text{SQ} = \dfrac{\text{population of the subgroup}}{\text{standard divisor}}$	**Rounded SQ**
Math	$\dfrac{225}{52.778} \approx 4.263$	4 scholarships
History	$\dfrac{417}{52.778} \approx 7.901$	7 scholarships
Computer Science	$\dfrac{308}{52.778} \approx 5.836$	5 scholarships
Total Scholarships		**16**

By this computation, there are only 16 scholarships given out—two short of the number needed. The Hamilton method would allocate the remaining two scholarships to history and computer science majors. However, the Jefferson method seeks to mathematically apportion the scholarships by modifying the standard divisor. Thus, the modified divisor will help eliminate surplus seats.

The Jefferson method tells us that we should modify the divisor $(\text{SD} = 52.778)$ either up or down and find the modified quota. Because we need larger quotas, we want a smaller divisor; if we needed smaller quotas, we would want larger divisors. Rounding the standard

Helpful Hint

Dividing by a smaller number yields a larger quotient while dividing by a larger number will yield a smaller quotient.

10

divisor $(SD = 52.778)$ down to MD = 52, we now calculate the modified quotas, as shown in Table 7.

Table 7		
First MQ and Rounded MQ for Scholarship Apportionment		
Major	**MQ**	**Rounded MQ**
Math	$\dfrac{225}{52} \approx 4.327$	4 scholarships
History	$\dfrac{417}{52} \approx 8.019$	8 scholarships
Computer Science	$\dfrac{308}{52} \approx 5.923$	5 scholarships
Total Scholarships		**17**

Again, we have not given out all of the scholarships. So, again we test a different MD. Let's try MD = 51 this time. The new modified quotas are calculated in Table 8.

Table 8		
Second MQ and Rounded MQ for Scholarship Apportionment		
Major	**MQ**	**Rounded MQ**
Math	$\dfrac{225}{51} \approx 4.412$	4 scholarships
History	$\dfrac{417}{51} \approx 8.176$	8 scholarships
Computer Science	$\dfrac{308}{51} \approx 6.039$	6 scholarships
Total Scholarships		**18**

Finally, we have the correct number of scholarships accounted for in the allocation.

It should be noted that whenever the sum of the quotas found is smaller than the number of items, then the MD should be made *smaller* to increase the sum. Conversely, if the sum of quotas is larger than the number of items, then MD should be made *larger* to decrease the number of items allocated to each subgroup.

The Webster Method of Apportionment

Proposed by Senator Daniel Webster of Massachusetts, the Webster method was intended to be a solution to apportionment and alleviate some issues with the Jefferson method. Webster's method of apportionment is essentially identical to the Jefferson method, with the exception that modified quotas are rounded in the usual mathematical manner and used in conjunction with a MD. If the fractional part of the modified quota is 0.5 or more, we round up. So, if the modified quota is smaller than 0.5, we round down.

The Jefferson method was thought to be the fairest method of apportionment until the 1820 census, when a flaw was discovered that could possibly award additional seats in congress to a state even though the state population could have decreased or vice versa. After a repeat of this anomaly in 1830, Jefferson's method was replaced with Webster's. The increase in the population of the country and the want for fair representation led to the implementation of Webster's method because of its mathematical simplicity.

Webster's Method

1. Find the SD and all of the SQs. Round all of the SQs down.

2. If the sum of the quotas is not equal to the number of items to be apportioned, then we modify the divisor.

3. Calculate the new MQs with the MD.

4. Round the MQs to the nearest whole number.

5. Continue to modify until the sum of the quotas is equal to the apportionment needed.

Example 5: Applying the Webster Method of Apportionment

A computer tech firm has three divisions with 135, 98, and 132 employees, respectively. A total of 12 administrative assistants must be allocated to the three divisions according to their size. Use Webster's method to determine how many administrative assistants should be allocated to each division.

Solution

The total number of employees in the divisions is 365 and the total number of assistants is 12. First, we calculate the standard divisor.

$$\text{SD} = \frac{\text{total population in a group}}{\text{total number of items to be apportioned}} = \frac{365}{12} \approx 30.417 \text{ employees/assistant}$$

We again use a table to determine the allocation.

Table 9		
SQ and Rounded SQ for Administrative Assistant Apportionment		
Division	$\text{SQ} = \dfrac{\text{population of the subgroup}}{\text{standard divisor}}$	**Rounded SQ**
1 : 135 employees	$\dfrac{135}{30.417} \approx 4.438$	4 assistants
2 : 98 employees	$\dfrac{98}{30.417} \approx 3.222$	3 assistants
3 : 132 employees	$\dfrac{132}{30.417} \approx 4.340$	4 assistants
Total Assistants		**11**

Helpful Hint

The lower quota of an allocation is the SQ rounded down, regardless of the decimal.

We can see from determining the *lower quotas* that we need one more administrative assistant to be allocated. This means that we need to find a modified divisor that will in turn give us the MQs. Remember that this time, we round each MQ to the nearest whole number. If we use a modified divisor of 30, then we get our required allotment.

Table 10		
MQ and Rounded MQ for Administrative Assistant Apportionment		
Division	**MQ**	**Rounded MQ**
1 : 135 employees	$\dfrac{135}{30} = 4.5$	5 assistants
2 : 98 employees	$\dfrac{98}{30} \approx 3.267$	3 assistants
3 : 132 employees	$\dfrac{132}{30} = 4.4$	4 assistants
Total Assistants		**12**

We have now allocated all of the administrative assistants to the respective divisions.

The methods of apportionment discussed so far are three of the four methods that have been used at some point for the apportionment of US Representatives. The Jefferson method was used until 1840 and was abandoned in lieu of the Webster method. The Webster method was replaced in 1850 with the Hamilton method, which was used until 1940 when the Huntington-Hill method was introduced.

The Huntington-Hill Method of Apportionment

The last method of apportionment that we will consider is the Huntington-Hill method. The Huntington-Hill method was introduced after the 1940 census and is still in use today. Like the other methods of apportionment, the Huntington-Hill method seeks to fairly apportion the number of representatives in the US House of Representatives based on population trends of individual states.

The Huntington-Hill method is similar to the Webster method, in that it uses the same procedure to determine the standard quota with one exception: the determining factor of whether quotas should be rounded up or down is not based on mathematically rounding the quota, but rather on the geometric mean of the integers on each side of the fractional part of the quota.

Geometric Mean

The **geometric mean** of any two numbers m and n is $\sqrt{m \cdot n}$.

If the geometric mean—which is found using the integers above and below the SQ—is less than the SQ, then the SQ gets rounded up. If the geometric mean is greater than the SQ, then the SQ gets rounded down. Suppose that a company has a standard quota of 345.46 administrators. It seems that either rounding up or down arithmetically would be dubious at best. Webster's method would have us round to 345. The Huntington-Hill method, however, uses a little more sophisticated mathematics to determine rounding based on the geometric mean. The geometric mean of 354.46 would be

$$\sqrt{m \cdot n} = \sqrt{345 \cdot 346} = \sqrt{119{,}370} \approx 345.500.$$

Notice that the geometric mean is greater than the SQ, or 345.500 > 345.46. So the Huntington-Hill method says that, since the geometric mean is greater than the SQ, we round the SQ down to 345.

Huntington-Hill Method

1. Find the SD and all of the SQs.

2. Find the geometric mean of each SQ. If the geometric mean is greater than the SQ, round the SQ down. If the geometric mean is less than the SQ, round the SQ up.

3. If the sum of the quotas is less than what we need, then we modify the divisor.

4. Find the MD and the MQs.

5. Find the geometric mean of each MQ. If the geometric mean is greater than the MQ, round the MQ down. If the geometric mean is less than the MQ, round the MQ up.

6. Continue to modify until the sum of the quotas is equal to the apportionment needed.

Example 6: Applying the Huntington-Hill Method of Apportionment

Suppose we are dividing 100 seats among five states with the following populations.

Table 11

Population per State

State	Population
A	35,589
B	17,425
C	3658
D	11,457
E	6871
Total	**75,000**

Use the Huntington-Hill method to apportion the 100 seats.

Solution

We begin our solution by finding the standard quota for each state based on the standard divisor.

$$\text{SD} = \frac{\text{total population in a group}}{\text{total number of items to be apportioned}} = \frac{75,000}{100} = 750 \text{ people/seat}$$

Now, we calculate the standard quotas.

Helpful Hint

Remember that if the geometric mean is less than SQ, you round the SQ up. If the geometric mean is greater than SQ, you round the SQ down.

Table 12

SQ and Rounded SQ for Seat Apportionment

State	Population	$\text{SQ} = \dfrac{\text{population of the subgroup}}{\text{standard divisor}}$	Geometric Mean	Rounded SQ
A	35,589	$\dfrac{35,589}{750} \approx 47.452$	$\sqrt{47 \cdot 48} = \sqrt{2256} \approx 47.497$	47
B	17,425	$\dfrac{17,425}{750} \approx 23.233$	$\sqrt{23 \cdot 24} = \sqrt{552} \approx 23.495$	23
C	3658	$\dfrac{3658}{750} \approx 4.877$	$\sqrt{4 \cdot 5} = \sqrt{20} \approx 4.472$	5
D	11,457	$\dfrac{11,457}{750} = 15.276$	$\sqrt{15 \cdot 16} = \sqrt{240} \approx 15.492$	15
E	6871	$\dfrac{6871}{750} \approx 9.161$	$\sqrt{9 \cdot 10} = \sqrt{90} \approx 9.487$	9
Totals	**75,000**			**99**

From the table, we can see that using the standard quota, we are one seat short. This means that we need to find a modified divisor. Recall that the concept of finding a modified divisor is done by trial and error and was introduced in the Jefferson method. Let MD = 749 and find the MQs. Then, we find the geometric mean of each MQ and determine which MQ we round down or up.

10

Table 13
MQ and Rounded MQ for Seat Apportionment

State	Population	MQ	Geometric Mean	Rounded MQ
A	35,589	$\dfrac{35,589}{749} \approx 47.515$	$\sqrt{47 \cdot 48} = \sqrt{2256} \approx 47.497$	48
B	17,425	$\dfrac{17,425}{749} \approx 23.264$	$\sqrt{23 \cdot 24} = \sqrt{552} \approx 23.495$	23
C	3658	$\dfrac{3658}{749} \approx 4.884$	$\sqrt{4 \cdot 5} = \sqrt{20} \approx 4.472$	5
D	11,457	$\dfrac{11,457}{749} \approx 15.296$	$\sqrt{15 \cdot 16} = \sqrt{240} \approx 15.492$	15
E	6871	$\dfrac{6871}{749} \approx 9.174$	$\sqrt{9 \cdot 10} = \sqrt{90} \approx 9.487$	9
Totals	75,000			100

It should be noted that State A received an additional seat. Comparing the results to previous methods of apportionment, we can see that State A received seats by this method.

Skill Check #2

Repeat Example 6 using the Jefferson method of apportionment.

Quotas and Paradoxes

Apportioning items fairly is a method that is inherently flawed in the sense that one group will receive an additional benefit when one group will lose the same benefit for no other reason than a mathematical elimination. We mentioned earlier that an anomaly was discovered after the 1830 census with the Jefferson method. Not much was thought of the issue until it happened again in 1840. So, what was that anomaly? The Jefferson method violated what we call the **quota rule**.

Quota Rule

The **quota rule** implies that any fair apportionment method should assign to every group either its lower or upper quota.

One of the primary flaws of the apportionment methods is that each violates the quota rule at some point. There are two ways in which the quota rule is violated: when a group ends up with an apportionment of resources smaller than its lower quota or when a group ends up with an apportionment larger than its upper quota.

The Hamilton method introduced an interesting set of paradoxes based on the number of items to be apportioned, the population size within a group, and the addition of new groups. What exactly is a paradox? Basically, a paradox is a situation that is opposed to common sense but seems to be true. The paradoxes in apportionment have become known as the **Alabama paradox**, the **population paradox**, and the **new states paradox**.

Alabama Paradox

The Alabama paradox occurs when an increase in the number of available items causes a group to lose an item—even though populations remain the same. With a growing United States population, the House of Representatives increased from 270 seats to 280 seats after the 1870 census. Before the census, Rhode Island had two seats. Even though the overall total number of seats increased, Rhode Island actually lost one of its seats, and ended up with only one representative. After the 1880 census, apportionments for all House sizes between 275 and 350 members were computed. It was determined that if the House of Representatives had 299 seats, Alabama would get eight seats but if the House of Representatives had 300 seats, Alabama would only get seven seats. Thus, the idea of increasing the number of seats and a state actually losing a seat is called the Alabama paradox.

Population Paradox

The population paradox states that Group A can lose an item to Group B even when the rate of growth of the population of Group A is greater than in Group B. The population paradox was discovered around 1900. At this time, the population of Virginia was growing much faster than the population of Maine. However, Virginia lost a seat in the House of Representatives while Maine gained a seat.

New States Paradox

The new states paradox happens when the addition of a new group, with a corresponding increase in the number of available items, can cause a change in the apportionment of items among the other groups.

When Oklahoma became a state in 1907, the House of Representatives had 386 seats. At the time, Oklahoma's population dictated it should have five seats in the House of Representatives. This meant that the number of total seats would have to increase from 386 to 391, leaving the number of seats unchanged for the other states. However, when the apportionment was recalculated under Webster's method, Maine gained a seat (four instead of three) and New York lost a seat (from 38 to 37).

Example 7: The Alabama Paradox

After the 1880 census, when the population of the United States was found to be 50,189,209, the chief clerk of the United States Census Office computed the number of representatives to either be 299 or 300 seats. According to the 1880 census, the population of Alabama was 1,262,505. Find the SD and the SQ when there are **a.** 299 seats, and **b.** 300 seats in the House of Representatives.

Solution

a. For an apportionment of 299 seats, the SD and the SQ are as follows.

$$SD = \frac{\text{US population}}{\text{total number of seats}} = \frac{50,189,209}{299} \approx 167,856.89 \text{ people/seat}$$

$$SQ = \frac{\text{Alabama population}}{SD} = \frac{1,262,505}{167,856.89} \approx 7.52 \text{ seats}$$

b. For an apportionment of 300 seats, the SD and the SQ are the following.

$$SD = \frac{US\ population}{total\ number\ of\ seats} = \frac{50,189,209}{300} \approx 167,297.36\ \ people/seat$$

$$SQ = \frac{Alabama\ population}{SD} = \frac{1,262,505}{167,297.36} \approx 7.55\ seats$$

The flaw that arose comes from Hamilton's Method of Apportionment. As we can see from the calculation of the SQ for each scenario in Example 7, it appears Alabama should get eight representatives. However, the chief of the census noted to congress that if the House of Representatives had 299 seats, Alabama would get eight seats but if the House of Representatives had 300 seats, Alabama would only get seven seats. It just so happens that in the same year, the population of Illinois and Texas increased as well. Although Alabama's quota increased to 7.55 with a 300-seat House, Illinois had an increase of quota to 18.702, and Texas had an increase of quota to 9.672. Since the fractional quotas for Illinois and Texas were larger, they were given their upper quotas of 19 and 10 respectively, while Alabama was left with its lower quota of seven. Thus, the paradox of increasing the number of seats in Congress and having a state that loses a seat in the process.

Does a method exist that fairly apportions resources and does not violate the quota rule? The short answer is no.

Balinski & Young Theorem

There cannot be a perfect apportionment method since any apportionment method that does not violate the quota rule must produce paradoxes. Any apportionment method that does not produce paradoxes must violate the quota rule.

Although there is no perfect apportionment method, each of the methods discussed in this section has merit. The important thing to note is that understanding the paradoxes that can be produced by each apportionment method allows us to make informed decisions about how to allocate resources.

Skill Check Answers

1. $81,540.38

2.

State	Population	SQ	Rounded SQ	MQ (MD = 730)	Rounded MQ
A	35,589	$\frac{35,589}{750} \approx 47.452$	47	$\frac{35,589}{730} \approx 48.752$	48
B	17,425	$\frac{17,425}{750} \approx 23.233$	23	$\frac{17,425}{730} \approx 23.870$	23
C	3658	$\frac{3658}{750} \approx 4.877$	4	$\frac{3658}{730} \approx 5.011$	5
D	11,457	$\frac{11,457}{750} = 15.276$	15	$\frac{11,457}{730} \approx 15.695$	15
E	6871	$\frac{6871}{750} \approx 9.161$	9	$\frac{6871}{730} \approx 9.412$	9
Totals	75,000		98		100

Seat Apportionment Using the Jefferson Method

10.3   Exercises

Use the given table to solve each problem.

Student Enrollment in the University of California System		
Campus	Enrollment	Full Time Equivalent (FTE)
Berkeley	33,558	14,161
Davis	32,290	20,883
Irvine	28,000	12,558
Los Angeles	37,221	28,292
Merced	2700	799
Riverside	20,956	4689
San Diego	25,938	18,274
San Francisco	18,140	4174
Santa Barbara	21,016	6081
Santa Cruz	15,012	4597

1. The University of California system consists of 10 campuses and has a budget of $2.6 billion in state funding. How much money in state funding would each campus receive if the state divided the money equally among the campuses (round to nearest dollar)?

2. If the state of California wanted to allocate the money equally based on total student enrollment, how much would be appropriated per student, rounded to the nearest cent?

3. Based on the result from Exercise 2, how much money would the campus of Los Angeles and Merced receive based on the number of students enrolled on each campus?

4. If the state apportions the funds based on FTE, how much money will the campuses of Davis and Santa Cruz receive (round to nearest dollar)?

5. As part of a "green" initiative, the state is wanting to apportion 500 new electric vehicles to their university system campuses. The state decides to apportion these vehicles based on the number of students at each university.

   a. Find the SD.

   b. Find the SQ for the San Francisco and Irvine campuses.

6. Supposed the state decides to apportion the 500 electric vehicles based on FTE.

   a. Find the SD.

   b. Find the SQ for the Riverside and Santa Barbara campuses.

7. Use the Hamilton method to apportion the 500 electric vehicles to all 10 campuses based on the number of students.

8. Use the Hamilton method to apportion the 500 electric vehicles to all ten campuses based on FTE. Compare your results with the apportionments from Exercise 7 and determine if the apportionments are different when the apportionment basis is different.

10

Solve each problem.

9. An English teacher at a high school can teach six classes. There are 35 students enrolled in English I, 43 in English II, and 48 in English III.

    a. Find the SD and SQ to determine how many sections of each course should be offered?

    b. Use the Jefferson method to determine the apportionment of students to the courses to determine the number of sections needed per course.determine the number of sections needed per course.

10. Repeat Exercise 9 **b.** using the Webster method.

11. Repeat Exercise 9 **b.** using the Hamilton method.

12. Repeat Exercise 9 **b.** using the Huntington-Hill method.

13. A county is divided into four districts with the populations of: Northern: 5500, Southern: 6350, Eastern: 3470, and Western: 1950. There are 16 seats on the county board to be apportioned.

    a. Use the Jefferson method to apportion the board seats.

    b. Use the Huntington-Hill method to apportion the board seats.

    c. Compare the apportionments for the Jefferson and Huntington-Hill methods and determine if there is a difference in how the seats are apportioned.

14. A biology department uses 25 graduate assistants in teaching its undergraduate courses. The enrollments for each of the courses that these students teach is as follows. How many graduate assistants should be assigned to each course using the Jefferson method?

Enrollment per Course	
Course	**Enrollment**
Survey of Biology	450
Zoology	200
Cell Biology	175
Plant Biology	280

15. Use the Hamilton method to round each of the following numbers to a whole number while preserving the total.

    $$12.65 + 3.48 + 2.57 + 4.39 + 1.91 = 25$$

16. Suppose there are 76 faculty members in the sciences, 86 in the humanities, and 16 in the professional and trade schools. An 11-person faculty committee is to be formed.

    a. Use Hamilton's method to determine the allocation of committee members based on department size.

    b. Use Jefferson's method to determine the allocation of committee members based on department size.

    c. Use the Huntington-Hill method to determine the allocation of committee members based on department size.

17. Suppose Learn-A-Lot University has enrollments on its three campuses as follows. There are 40 police officers to be distributed among these campuses based on enrollment. Use Hamilton's method to apportion the police officers.

Enrollment per Campus			
Campus	1	2	3
Enrollment	10,170	9150	680

18. If the number of police officers to be apportioned increases by 1 to 41 (see Exercise 17) determine the apportionment of officers using Hamilton's Method and show that the Alabama paradox occurs when the number of officers increases by 1.

19. Suppose a country has six states with populations as given in the table. There are 250 seats in the House of Representatives for this country. Use Webster's method to apportion the representatives.

Population by State	
State	Population
A	1646
B	6936
C	154
D	2091
E	685
F	988
Total	12,500

20. Suppose a college homecoming planning committee has 17 members. The makeup of the committed is to be based on the size of the classes: Freshman = 422, Sophomore = 356, Junior = 321, and Senior = 288.

    a. Find the number of members from each class to be apportioned to the committee using Webster's method.

    b. Find the number of members from each class to be apportioned to the committee using Jefferson's method.

    c. Find the number of members from each class to be apportioned to the committee using the Huntington-Hill method.

The given table shows the number of students enrolled in history, liberal arts math, and English during the fall and spring semesters. Use it to solve each problem.

Course Enrollment per Semester		
Subject	**Fall**	**Spring**
History	1902	1922
Liberal Arts Math	14,200	14,200
English	3898	3938

21. If there are 197 full time teaching positions available for apportionment among the three departments based on course enrollment, answer each of the following questions.

   a. Find the number of teaching positions that should be apportioned to each department in the fall using Jefferson's method.

   b. Find the number of teaching positions that should be apportioned to each department in the spring using Jefferson's method.

   c. Does Jefferson's method create an example of the population paradox?

   d. Find the number of teaching positions that should be apportioned to each department in the fall using Hamilton's method.

   e. Find the number of teaching positions that should be apportioned to each department in the spring using Hamilton's method.

   f. Does Hamilton's method create an example of the population paradox?

   g. Find the number of teaching positions that should be apportioned to each department in the fall using the Huntington-Hill method.

   h. Find the number of teaching positions that should be apportioned to each department in the spring using the Huntington-Hill method.

   i. Does the Huntington-Hill method create an example of the population paradox?

10.4 Weighted Voting Systems

Have you ever been told by a parent that decisions they make are not part of a democracy, but rather a dictatorship? Most likely, the answer is yes. In the previous sections of this chapter, the process of voting, counting votes, and apportioning items based on populations and resources were covered. We have previously discovered that apportioning votes based on population sizes is the primary way the United States House of Representatives and the Electoral College base voting power. This is an example of a weighted voting system. In this section, the concept of weighted voting will be discussed. **Weighted voting systems** are based on the concept that, in any given election that is not necessarily democratic, individual votes can carry additional power toward making a decision.

> ### Weighted Voting System
>
> A **weighted voting system** is a system where an individual voter may have more than one vote and, thus, more power than another voter.

As an example, if a company has 20,000 shares of stock and each share of stock has a vote at a stockholders meeting, then a person with 1500 shares would have 1500 votes. We call the person casting votes a **player**, denoted by $P_1, P_2, P_3, \ldots, P_N$, where N is the number of players. Each player is given a **weight**, denoted by $w_1, w_2, w_3, \ldots, w_N$, where N is the number of players and w_i is the number of votes that player i controls. A **quota** (q) is the minimum number of votes needed to pass a proposal, such as a bill in the US Congress.

Any weighted voting system can be represented using the notation $[q: w_1, w_2, \ldots, w_N]$. This notation allows us to quickly see the number of votes needed to pass a proposal, the number of players, and the weight that each player carries.

Example 1: Weighted Voting System

A voting system has four players: Jane, Marilyn, Kent, and Stan. Jane has 4 votes, Marilyn has 3 votes, Kent has 2 votes, and Stan has 1 vote. A majority of 6 votes is needed to pass a proposal. Use the notation for weighted voting systems to represent this voting system.

Solution

We can use our notation to represent the players in the system by Jane as P_1, Marilyn as P_2, Kent as P_3, and Stan as P_4. Also, the corresponding weights of each player are $w_1 = 4$, $w_2 = 3$, $w_3 = 2$ and $w_4 = 1$. Having a quota of 6 votes for a majority, the system can be represented as $[6: 4, 3, 2, 1]$.

Weights and Power

There are certain concepts that must be understood to comprehend how voting systems actually work. Not everyone in a given system has the same weight for voting. Consider for example the system $[12: 14, 5, 4, 3]$. Notice that Player 1 has more weight than is needed to actually pass the proposal. When one player in a system has the power to pass a proposal based on their vote alone, we say that Player 1 is a **dictator**. The primary factor that actually makes Player 1 a dictator is that, even with all the other players' votes combined, they do not outweigh the dictator's share of votes; that is, the dictator has more than enough votes to pass the proposal alone.

Dictator

A **dictator** is a player with power to pass a proposal single-handedly; that is, the dictator has at least as many votes as the quota.

Now consider the voting system $[10: 8, 5, 3, 1]$. You should notice that Player 1 has the ability to influence the vote, but with a quota of 10, needs a little help to pass a proposal. Also, even if Players 2, 3, and 4 combine their votes together, they do not have the quota either. In this case, in order for any proposal to pass, Player 1 must be in favor of its passage. When this is the case, we say that Player 1 has **veto power** in the system.

Veto Power

A player has **veto power** in a voting system when they can keep a proposal from passing with their vote, but do not have a majority of the votes.

One last term we need to define in a voting system is a **dummy player**. Consider the system $[12: 8, 7, 2]$. Notice that Player 3 has no effect on the outcome of the proposal. Player 1 and Player 2 must both be in support of the proposal for it to pass. We call Player 3 a dummy player.

Dummy Player

A **dummy player** is a player who does not have enough votes to have an effect on the outcome of a proposal.

Most systems of voting have one or more of these attributes—a dictator, veto power, or a dummy player. To that end, many systems of voting allow the development of *coalitions*. Think of the United States two-party political system of Democrats and Republicans. Each of these groups represents a **coalition** of individuals with similar political ideas and tend to vote the same way. This same concept can be applied to any voting system where groups of individuals can pool their power to create simple majorities and supermajorities.

Coalition

A **coalition** is formed when a group of players decide to vote together.

Simple Majority

A **simple majority** is when a proposal requires more than half of the total votes to pass.

Supermajority

A **supermajority** is the number of votes required to pass a proposal in a system that requires more votes than a simple majority. For instance, to override a presidential veto, the senate must have a $\frac{2}{3}$ majority or a supermajority.

MATH MILESTONE

In 1782, the Articles of Confederation required 9 of the 13 states to agree on all major issues to amend the articles. Since the founding fathers of the United States were so aware of the dictatorship of the British rule system, they wanted to ensure that decisions were based on more than a simple majority of free choice. This became known as a supermajority.

Example 2: Passing a Motion with a Coalition

Given the voting system $[18: 8, 7, 6, 4, 4, 4, 2]$, determine if a motion will pass if a coalition consisting of Players 3, 4, 5, and 6 vote in favor of the motion.

Solution

The weight of Players 3, 4, 5, and 6 is 6 votes, 4 votes, 4 votes, and 4 votes, respectively. The total number of votes for these four players is $6 + 4 + 4 + 4 = 18$. With a quota of 18, the four players will have enough for a simple majority; that is, there are 35 votes total from the players, so 18 is a simple majority since its more than half of the total votes.

Skill Check #1

With a quota of 18 being a simple majority, how many votes would be needed to pass in a supermajority of $\frac{2}{3}$ of the votes in the voting system from Example 2?

We have discussed the voting power that individual players can have on the outcomes of a given vote. We found that if one player can sway the entire vote, that player is a dictator. If one player has enough votes to not allow a proposal to pass, no matter how the other players vote, then that player has veto power. One more player that needs to be discussed is a **critical player**. A critical player is similar to the player with veto power, but in a coalition of votes rather than the entire system. It should be noted that a dummy player will never be a critical player in a voting system, and if the dummy player refuses to vote, the lack of votes has no effect on the outcome.

Critical Player

A **critical player** is a player with veto power within a coalition.

Example 3: Winning Coalitions

Consider the voting system $[10 : 7, 6, 3, 3, 1]$. Determine a winning coalition, and identify any critical players in each coalition.

Solution

There are several winning coalitions of players for this example. Recall that a winning coalition is made up of voters that can pass the proposal by voting in the same manner.

The number of votes for each player is as follows.

$$P_1 = 7 \quad P_2 = 6 \quad P_3 = 3 \quad P_4 = 3 \quad P_5 = 1$$

In this case, the winning coalitions are as follows. Notice that each winning coalition has at least 10 votes.

$$\{P_1, P_2\}, \{P_1, P_3\}, \{P_1, P_4\}, \{P_2, P_3, P_4\}, \{P_2, P_3, P_5\}, \{P_2, P_4, P_5\}, \{P_1, P_3, P_4\},$$
$$\{P_1, P_3, P_5\}, \{P_1, P_4, P_5\}, \{P_1, P_2, P_3\}, \{P_1, P_2, P_4\}, \{P_1, P_2, P_5\}, \{P_1, P_2, P_3, P_4\},$$
$$\{P_1, P_2, P_3, P_5\}, \{P_1, P_3, P_4, P_5\}, \{P_1, P_2, P_4, P_5\}, \{P_2, P_3, P_4, P_5\}, \{P_1, P_2, P_3, P_4, P_5\}$$

You might notice that there are two players where one or the other must appear in a coalition in order to create a majority, P_1 and P_2. These players are considered critical players. For, without their votes, the proposal would not pass for any winning coalition.

Banzhaf Power Index (BPI)

Fun Fact

Sometimes referred to as the Banzhaf-Coleman index, the Banzhaf Power Index became famous after John Banzhaf argued that the weighted voting system used by the Nassau County Board of Supervisors in New York was unfair because a few of the cities actually had 0% voting power.

The Banzhaf Power Index (BPI) of a voter is the number of winning coalitions in which the voter is critical. In other words, when a critical player is identified, the number of winning coalitions in which the critical player can belong determines the amount of power the voter has in the system.

Banzhaf Power Index for Player P (BPI(P))

1. Find all winning coalitions for the system.
2. Determine the critical voters for each winning coalition.
3. Determine how many times each player is critical—called Banzhaf power.
4. Determine the total number of times all players are critical players—total Banzhaf power.
5. Determine BPI for Player P using

$$\text{BPI}(P) = \frac{\text{number of times Player } P \text{ is a critical player}}{\text{total number of times all players are critical players}} \cdot 100\%.$$

Example 4: Banzhaf Power Index

Wilson, Clinton, & Niese law firm has the following voting system [16: 9, 6, 4, 3, 3, 2]. Determine each player's Banzhaf Power Index.

Solution

To determine the Banzhaf Power Index, we will follow the steps.

Step 1: Find all winning coalitions for the system.

Table 1	
Coalitions	
Winning Coalition	**Votes**
$\{P_1, P_3, P_4\}$	16
$\{P_1, P_3, P_5\}$	16
$\{P_2, P_3, P_4, P_5\}$	16
$\{P_1, P_2, P_6\}$	17
$\{P_1, P_4, P_5, P_6\}$	17
$\{P_1, P_2, P_4\}$	18
$\{P_1, P_2, P_5\}$	18
$\{P_1, P_3, P_4, P_6\}$	18
$\{P_1, P_3, P_5, P_6\}$	18
$\{P_2, P_3, P_4, P_5, P_6\}$	18
$\{P_1, P_2, P_3\}$	19

Coalition	Votes
$\{P_1, P_3, P_4, P_5\}$	19
$\{P_1, P_2, P_4, P_6\}$	20
$\{P_1, P_2, P_5, P_6\}$	20
$\{P_1, P_2, P_4, P_5\}$	21
$\{P_1, P_2, P_3, P_6\}$	21
$\{P_1, P_3, P_4, P_5, P_6\}$	21
$\{P_1, P_2, P_3, P_4\}$	22
$\{P_1, P_2, P_3, P_5\}$	22
$\{P_1, P_2, P_4, P_5, P_6\}$	23
$\{P_1, P_2, P_3, P_4, P_6\}$	24
$\{P_1, P_2, P_3, P_5, P_6\}$	24
$\{P_1, P_2, P_3, P_4, P_5\}$	25
$\{P_1, P_2, P_3, P_4, P_5, P_6\}$	27

Step 2: Determine the critical players for each winning coalition.

Table 2

Critical Players for Winning Coalitions

Coalition	Votes	Critical Players
$\{P_1, P_3, P_4\}$	16	P_1, P_3, P_4
$\{P_1, P_3, P_5\}$	16	P_1, P_3, P_5
$\{P_2, P_3, P_4, P_5\}$	16	P_2, P_3, P_4, P_5
$\{P_1, P_2, P_6\}$	17	P_1, P_2, P_6
$\{P_1, P_4, P_5, P_6\}$	17	P_1, P_4, P_5, P_6
$\{P_1, P_2, P_4\}$	18	P_1, P_2, P_4
$\{P_1, P_2, P_5\}$	18	P_1, P_2, P_5
$\{P_1, P_3, P_4, P_6\}$	18	P_1, P_3, P_4
$\{P_1, P_3, P_5, P_6\}$	18	P_1, P_3, P_5
$\{P_2, P_3, P_4, P_5, P_6\}$	18	P_2, P_3, P_4, P_5
$\{P_1, P_2, P_3\}$	19	P_1, P_2, P_3
$\{P_1, P_3, P_4, P_5\}$	19	P_1, P_3
$\{P_1, P_2, P_4, P_6\}$	20	P_1, P_2
$\{P_1, P_2, P_5, P_6\}$	20	P_1, P_2
$\{P_1, P_2, P_4, P_5\}$	21	P_1, P_2
$\{P_1, P_2, P_3, P_6\}$	21	P_1, P_2
$\{P_1, P_3, P_4, P_5, P_6\}$	21	P_1
$\{P_1, P_2, P_3, P_4\}$	22	P_1
$\{P_1, P_2, P_3, P_5\}$	22	P_1
$\{P_1, P_2, P_4, P_5, P_6\}$	23	P_1
$\{P_1, P_2, P_3, P_4, P_6\}$	24	P_1
$\{P_1, P_2, P_3, P_5, P_6\}$	24	P_1
$\{P_1, P_2, P_3, P_4, P_5\}$	25	None
$\{P_1, P_2, P_3, P_4, P_5, P_6\}$	27	None

10

Step 3: Determine how many times each player is critical—called Banzhaf power.

Table 3
Number of Times a Player is Critical

Player	Banzhaf
1	20
2	10
3	8
4	6
5	6
6	2

Step 4: Determine the total number of times all players are critical players—total Banzhaf power.

$$20 + 10 + 8 + 6 + 6 + 2 = 52$$

Step 5: Determine BPI for Player P using

$$\text{BPI}(P) = \frac{\text{number of times Player } P \text{ is a critical player}}{\text{total number of times all players are critical players}} \cdot 100\%.$$

Table 4
BPI for Player P

Player	Banzhaf	Percentage
1	20	$\frac{20}{52} \approx 0.38462 = 38.462\%$
2	10	$\frac{10}{52} \approx 0.19231 = 19.231\%$
3	8	$\frac{8}{52} \approx 0.15385 = 15.385\%$
4	6	$\frac{6}{52} \approx 0.11538 = 11.538\%$
5	6	$\frac{6}{52} \approx 0.11538 = 11.538\%$
6	2	$\frac{2}{52} \approx 0.03846 = 3.846\%$
Total	**52**	

The percentages obtained by the index indicate each player's ability to influence the vote. In other words, Player 1 controls about 38.46% of the power to sway the outcome of the vote.

The Shapley-Shubik Power Index (SSPI)

Developed in 1954 as a method to measure the voting power of the players in a voting system, the Shapley-Shubik Power Index (SSPI) has the ability to reveal the importance of a given player at a deeper level. When discussing the power of a coalition using the Banzhaf index, we did not care about the order in which players cast their votes. The Shapley-Shubik Power Index (SSPI), however, looks at how often a player is pivotal to the proposal passing rather than simply counting the number of times a player is a critical player. Just like in determining the probability of events and the permutations of the possible outcomes of a given event in Chapter 7, order of voting is sometimes important. So, we will call coalitions where the order in voting within a coalition is important as a **sequential coalition** and denote such coalitions as $\langle S_1, S_2, \ldots, S_N \rangle$ where S_1 was the first player to join, S_2 is the second player to join, and so on. Thus, in the SSPI, the coalition of $<P_1, P_2>$ and $<P_2, P_1>$ are not the same because the order in which a player enters the coalition matters.

> ### Sequential Coalition
>
> A **sequential coalition** is a coalition where the order in which players cast their vote is important.

Consider the United States House of Representatives, with its 435 members. Each member only has one unweighted vote. For a bill to pass the House, 218 members of the House—a simple majority—must vote "yes." If there are two coalitions controlling the vote, the 218th player to cast their vote is considered a pivotal player.

> ### Pivotal Player
>
> A **pivotal player** is the player in a coalition whose additional vote(s) causes the coalition to reach the quota, and makes their coalition win.

We denote the players in a coalition using subscripts and display players as $\langle P_1, P_2, \ldots, P_N \rangle$, where N is the number of players in the coalition.

Example 5: Determining a Pivotal Player

Recall the voting system from Example 4: $[16: 9, 6, 4, 3, 3, 2]$. Which player is the pivotal player in the sequential coalition $\langle P_2, P_4, P_3, P_6, P_5 \rangle$?

Solution

Recall that the order of the players listed in the sequential coalition is the order that the players vote in the coalition. So, for this coalition, $S_1 = P_2$, $S_2 = P_4$, $S_3 = P_3$, $S_4 = P_6$, and $S_5 = P_5$.

We will keep a running total of votes for when a player joins to determine which player casts the winning vote.

Table 5
Determining the Pivotal Player in a Sequential Coalition

Coalition	Weight of Votes	Enough to Win?
$\langle P_2 \rangle$	6 votes	Not enough to win
$\langle P_2, P_4 \rangle$	$6 + 3 = 9$ votes	Not enough to win
$\langle P_2, P_4, P_3 \rangle$	$6 + 3 + 4 = 13$ votes	Not enough to win
$\langle P_2, P_4, P_3, P_6 \rangle$	$6 + 3 + 4 + 2 = 15$ votes	Not enough to win
$\langle P_2, P_4, P_3, P_6, P_5 \rangle$	$6 + 3 + 4 + 2 + 3 = 18$ votes	Enough to win

So, Player 5 is the pivotal player in the coalition.

Shapley-Shubik Power Index for Player P (SSPI (P))

1. List all $N!$ sequential coalitions that contain all N players in the voting system.

2. Determine the pivotal player in each sequential coalition.

3. Count the number of times Player P is the pivotal player.

4. Divide the number of times a player is a pivotal player by the total number of sequential coalitions; that is, $\text{SSPI}(P) = \dfrac{\text{number of times Player } P \text{ is a pivotal player}}{\text{total number of sequential coalitions}}$.

Recall from Chapter 7 that N items can be arranged in $N!$ ways, where

$$N! = (N)(N-1)(N-2) \quad (3)(2)(1).$$

Therefore, the number of sequential coalitions in a voting system with N players is equal to $N!$. The formula can be rewritten as

$$\text{SSPI}(P) = \frac{\text{number of times Player } P \text{ is a pivotal player}}{N!},$$

where N is equal to the number of players in the voting system.

Example 6: Number of Sequential Coalitions

If there are 10 players in a voting system, how many sequential coalitions are there using all players in the voting system?

Solution

Since there are 10 players in the voting system and the number of sequential coalitions is $N!$, $10 \cdot 9 \cdot 8 \cdot 7 \cdot 6 \cdot 5 \cdot 4 \cdot 3 \cdot 2 \cdot 1 = 3{,}628{,}800$ sequential coalitions can be formed.

Now imagine the number of sequential coalitions that can be formed using the United States House of Representatives with 435 members; in other words, $435!$. That's a really big number—more than the number of stars in the universe. Obviously, we would not attempt to compute this number by hand, but there are many computer programs that can do so.

Example 7: Applying the SSPI

Consider the weighted voting system $[7:4,3,2]$. Find the SSPI for each player in this system.

Solution

Applying the four-step process for the SSPI, we have the following.

Step 1: List all $N!$ sequential coalitions which contain all N players in the voting system.

There are three players, so there are $3! = 3 \cdot 2 \cdot 1 = 6$ sequential coalitions containing all three players in the voting system. They are listed as follows.

$$\langle P_1, P_2, P_3 \rangle$$
$$\langle P_1, P_3, P_2 \rangle$$
$$\langle P_2, P_3, P_1 \rangle$$
$$\langle P_2, P_1, P_3 \rangle$$
$$\langle P_3, P_1, P_2 \rangle$$
$$\langle P_3, P_2, P_1 \rangle$$

Step 2: Determine the pivotal player in each sequential coalition.

$$\langle P_1, P_2, P_3 \rangle$$
$$\langle P_1, P_3, P_2 \rangle$$
$$\langle P_2, P_3, P_1 \rangle$$
$$\langle P_2, P_1, P_3 \rangle$$
$$\langle P_3, P_1, P_2 \rangle$$
$$\langle P_3, P_2, P_1 \rangle$$

Step 3: Count the number of times Player P is the pivotal player.

The number of times each players is pivotal is $P_1 = 3$ and $P_2 = 3$. P_3 is never the pivotal player. Thus, P_3 is a dummy player.

Step 4: Divide the number of times a player is a pivotal player by the total number of sequential coalitions; that is,

$$\text{SSPI}(P) = \frac{\text{number of times Player } P \text{ is a pivotal player}}{N!}.$$

Table 6		
Shapley-Shubik Power Index for Each Player		
Player	**Number of Times Pivotal**	**Percentage**
1	3	$\frac{3}{6} = 0.5 = 50\%$
2	3	$\frac{3}{6} = 0.5 = 50\%$
3	0	$\frac{0}{6} = 0 = 0\%$

So, Player 1 and Player 2 have equal weight in the outcome of the vote in this voting system, while Player 3 has no influence.

Skill Check #3

Determine the SSPI for each player in the voting system $[8: 4, 3, 3, 2]$.

Throughout this chapter, the reader has had the opportunity to discover different voting and apportionment methods. The ways in which these methods can be considered "unfair" and result in completely different outcomes has also been discussed. While it isn't possible to have a perfect voting system in reality, sometimes the knowledge gained from this chapter is the only thing that can be used to pick the best system for the situation.

Skill Check Answers

1. 24 votes

2. Yes, Player 4 is critical

3. $P_1 = P_2 = P_3 = P_4 = 25\%$

10.4   Exercises

Answer each question thoughtfully.

1.   What is a critical player?

2.   What is the weight of a voter?

3.   How many ways can five voters be ordered from first to last?

4.   What is veto power in a voting system?

5.   What is the quota for a voting system that has a total of 30 voters and uses a majority quota?

Solve each problem.

6.   Which players in the voting system $[20: 7, 7, 3, 3, 2]$ have veto power?

7.   Five partners start a business, with each owning the following number of shares: P_1 owns 9 shares, P_2 owns 6 shares, P_3 owns 5 shares, P_4 owns 3 shares, and P_5 owns 2 shares. If one share is equal to one vote and they use a simple majority quota, represent this weighted voting system.

8.   Using the information in Exercise 7, if the quota is a two-thirds majority, represent the weighted voting system.

9.   Consider the weighted voting system $[25: 10, 7, 4, 4, 2, 2, 1, 1, 1]$.
   a. What is the quota for this voting system?
   b. What is the weight of P_5?
   c. If only the last six voters vote for a motion, does the motion pass?
   d. If P_1 and P_2 vote against a motion, does the motion pass?

10.   Consider the weighted voting system $[42: 20, 16, 10, 6, 4]$.
   a. How many voters are in the system?
   b. What is the quota for the system?
   c. Is the coalition $\{P_2, P_3, P_4, P_5\}$ a winning coalition?
   d. What are all of the winning coalitions?
   e. Are there any critical players?
   f. Do any of the players have veto power?

11. Consider the weighted voting system $[42: 22, 16, 10, 6, 4]$. Does this voting system contain a dictator? Does this system contain a dummy player?

12. Consider the weighted voting system $[42: 20, 16, 10, 6, 4]$. Determine the Banzhaf Power Index for each player.

13. Consider the weighted voting system $[q: 10, 7, 4, 4, 2, 2, 1, 1, 1]$.

    a. What is the smallest quota that requires a majority for this voting system?

    b. What is the largest quota for this voting system?

14. In the weighted voting system $[q: 20, 19, 15, 8, 4, 2]$, if a two-thirds majority of votes is needed, what is q?

15. In the weighted voting system $[q: 18, 15, 12, 9, 6, 3]$, what is the largest value of q so that no voter has veto power?

16. In the weighted voting system $[q: 10, 7, 4, 4, 2, 2]$, if every voter has veto power, what is q?

17. Consider the weighted voting system $[7: 4, 2, 2, 2, 2]$. Find the Banzhaf Power Index for each voter. Are any of the voters dictators? Do any of the voters have veto power?

18. Consider the weighted voting system, $[100: 80, 60, 30, 20]$. Calculate the Shapley-Shubik Power Index for each player in this system. Are any of the voters dictators? Do any of the voters have veto power?

19. The United Nations Security Council consists of fifteen members of the United Nations; five permanent member countries and ten nonpermanent member countries. In order for a vote to pass, nine members, including all five of the permanent members, must be in agreement. An equivalent weighted voting system is $[39: 7, 7, 7, 7, 7, 1, 1, 1, 1, 1, 1, 1, 1, 1, 1]$. According to our previous calculations, this means there are 15! or about 1.3 trillion permutations for the members. [1]

    a. Without computing all possible coalitions, determine when a nonpermanent member would be a pivotal voter.

    b. What is the Shapley-Shubik Power Index for each nonpermanent member?

    c. What is the Shapley-Shubik Power Index for the combined nonpermanent members?

    d. What is the Shapley-Shubik Power Index for each permanent member?

    e. What is the Shapley-Shubik Power Index for the combined permanent members?

20. There are 435 members of the United States House of Representatives.

    a. If a simple majority is needed to pass a bill, what constitutes a winning coalition?

    b. If a two-thirds majority is needed to ratify an amendment, what constitutes a winning coalition?

---

1 United Nations, http://www.un.org

21. In the weighted voting system $[43: 25, 20, 19, 17]$, determine each of the following.

  a. List all permutations in which P_1 is a pivotal player.

  b. List all permutations in which P_2 is a pivotal player.

  c. Calculate the Shapley-Shubik Power Index for each player.

  d. Calculate the Banzhaf Power Index for each player.

  e. Is there a player with veto power?

22. If the quota in Exercise 21 is increased to 50, how would the Shapley-Shubik Power Index change?

23. A company has five shareholders and a total of 500 shares. The quota for passing a measure is the number of votes where shareholders own 251 or more shares. The number of shares owned by each shareholder is as follows: $S_1 = 200$, $S_2 = 123$, $S_3 = 120$, $S_4 = 40$, and $S_5 = 17$. Suppose there is an investor S_6 that wants to buy shares and currently owns no shares:

  a. What are the winning and losing coalitions? Compute the number of votes needed to make a losing coalition a winning coalition.

  b. How many shares can S_1 sell to S_2 without causing any of the winning coalitions listed in part **a.** to lose or any of the losing coalitions in part **a.** to win?

  c. How many shares can S_1 sell to S_5 without causing any of the winning coalitions listed in part **a.** to lose or any of the losing coalitions in part **a.** to win?

  d. How many shares can S_1 sell to S_6 without causing any of the winning coalitions listed in part **a.** to lose or any of the losing coalitions in part **a.** to win?

Chapter 10   Summary

Definitions

Preference Ballot

A preference ballot is a ballot that allows a voter to rank the items in order of preference from most preferred to least preferred.

Preference Table

A preference table summarizes all of the individual preference ballots in an election by tallying the number of ballots with the same order of ranking.

Election Counting Methods

Majority Rule Decision

The winner must have more than 50% of the first-place votes to win.

Plurality

The candidate with the most first-place votes wins. No majority is required.

Borda Count

Voting results are organized in a preference table and each ranking is assigned a specific number of points based on how many candidates are in the election. The candidate with the most ranking points is declared the winner.

Plurality with Elimination

A series of runoff elections where the candidate with the least amount of first-place votes is removed from the ballot each round if there is no winner. A winner is declared when a candidate has a majority of the first-place votes.

Pairwise Comparison

Every candidate is compared head-to-head with the other candidates. In each pair of comparisons, the candidate with the greater number of higher rankings is given a point. The candidate with the most points after all head-to-head comparisons are made is the winner.

Formula

Number of Pairwise Comparisons

The number of pairwise comparisons that must be made if there are n candidates is

$$\frac{n(n-1)}{2}.$$

Definitions

Arrow's Impossibility Theorem

If there are three or more choices on a ballot, there cannot be a voting method that will satisfy all five fairness criteria.

Condorcet Criterion

The Condorcet criterion states that if a candidate wins the head-to-head comparison—as in the pairwise comparison method—against every other candidate, then that candidate should also win the overall election in a fair voting system.

Majority Criterion

The majority criterion states that if a candidate receives a majority of votes in an election, that candidate should win.

Monotonicity Criterion

The monotonicity criterion requires that if a Candidate X wins an election, then X would also win a second election if each voter were allowed to reorder the candidates in such a way that X only increases in ranking while the order of the other candidates remained the same.

Irrelevant Alternatives Criterion

The irrelevant alternatives criterion states that if a candidate wins an election, then that same candidate would win the election even if one or more candidates pulls out.

Dictator Criterion

The dictator criterion states that no single voter is allowed to decide the outcome of an election.

Section 10.3   Apportionment

Definitions

Apportionment

Apportionment is the method of dividing resources in such a way as to maximize the use of those resources.

Modified Divisor

A modified divisor (MD) is a divisor, near the standard divisor (SD), that is chosen by guess-and-check in an attempt to make the sum of the modified quotas exactly equal to the number of items to be apportioned.

The Quota Rule

The quota rule implies that any fair apportionment method should assign to every group either its lower or upper quota.

The Alabama Paradox

The Alabama paradox occurs when an increase in the number of available items causes a group to lose an item—even though populations remain the same.

The Population Paradox

The population paradox states that Group A can lose an item to Group B even when the rate of growth of the population of Group A is greater than in Group B.

The New States Paradox

The new states paradox happens when the addition of a new group, with a corresponding increase in the number of available items, can cause a change in the apportionment of items of among the other groups.

Balinski & Young Theorem

There cannot be a perfect apportionment method since any apportionment method that does not violate the quota rule must produce paradoxes. Any apportionment method that does not produce paradoxes must violate the quota rule.

Methods of Apportionment

Hamilton Method

1. Find SD: $SD = \dfrac{\text{total population in a group}}{\text{total number of items to be apportioned}}$. Round to the nearest thousandth.

2. Find SQ: $SQ = \dfrac{\text{population of the subgroup}}{\text{standard divisor}}$.

3. Round SQ down to the next lowest integer.

4. Award any remaining resources based on which group's SQ was rounded down the most. That is, whoever had the largest fractional part rounded down.

Jefferson Method

1. Find the SD and all of the SQs. Round all of the SQs down.

2. If there are leftover things to apportion, we need to find a new divisor—an MD.

3. Calculate new MQs with the MD. If we still have leftover things to distribute, try again.

4. Continue the guess-and-check process until we have apportioned all items.

Webster Method

1. Find the SD and all of the SQs. Round all of the SQs down.

2. If the sum of the quotas is not equal to the number of items to be apportioned, then we modify the divisor.

3. Calculate the new MQs with the MD.

4. Round the MQs to the nearest whole number.

5. Continue to modify until the sum of the quotas is equal to the apportionment needed.

Huntington-Hill Method

1. Find the SD and all of the SQs.

2. Find the geometric mean of each SQ. If the geometric mean is greater than the SQ, round the SQ down. If the geometric mean is less than the SQ, round the SQ up.

3. If the sum of the quotas is less than what we need, then we modify the divisor.

4. Find the MD and the MQs.

5. Find the geometric mean of each MQ. If the geometric mean is greater than the MQ, round the MQ down. If the geometric mean is less than the MQ, round the MQ up.

6. Continue to modify until the sum of the quotas is equal to the apportionment needed.

Formulas

Standard Divisor

A standard divisor (SD) is the average number of members of the population that will account for one apportioned item.

$$SD = \frac{\text{total population in a group}}{\text{total number of items to be apportioned}}$$

Standard Quota

A standard quota (SQ) represents the number of items that will be apportioned to each subgroup and is calculated as follows.

$$SQ = \frac{\text{population of the subgroup}}{\text{standard divisor}}$$

Geometric Mean

The geometric mean of any two numbers m and n is $\sqrt{m \cdot n}$.

Section 10.4   Weighted Voting Systems

Definitions

Weighted Voting System

A weighted voting system is a system where an individual voter may have more than one vote and, thus, more power than another voter.

Dictator

A dictator is a player with power to pass a proposal single-handedly; that is, the dictator has at least as many votes as the quota.

Veto Power

A player has veto power in a voting system when they can keep a proposal from passing with their vote, but do not have a majority of the votes.

Dummy Player

A dummy player is a player who does not have enough votes to have an effect on the outcome of a proposal.

Coalition

A coalition is formed when a group of players decide to vote together.

Simple Majority

A simple majority is when a proposal requires more than half of the total votes to pass.

Supermajority

A supermajority is the number of votes required to pass a proposal in a system that requires more votes than half of the votes.

Critical Player

A critical player is a player with veto power within a coalition.

Sequential Coalition

A sequential coalition is a coalition where the order in which players join is important.

Pivotal Player

A pivotal player is the player in a coalition whose additional vote(s) causes the coalition to reach the quota, and makes the coalition win.

Power Indices

Banzhaf Power Index for Player P $\left(\text{BPI}(P)\right)$

1. Find all winning coalitions for the system.

2. Determine the critical voters for each winning coalition.

3. Determine how many times each player is critical—called Banzhaf power.

4. Determine the total number of times all players are critical players—total Banzhaf power.

5. Determine BPI for Player P using $\text{BPI}(P) = \dfrac{\text{number of times Player } P \text{ is a critical player}}{\text{total number of times all players are critical players}} \cdot 100\%.$

Shapley-Shubik Power Index for Player P $\left(\text{SSPI}(P)\right)$

1. List all $N!$ sequential coalitions containing all N players in the voting system.

2. Determine the pivotal player in each sequential coalition.

3. Count the number of times Player P is the pivotal player.

4. Divide the number of times a player is a pivotal player by the total number of sequential coalitions; that is,

$$\text{SSPI}(P) = \dfrac{\text{number of times Player } P \text{ is a pivotal player}}{\text{total number of sequential coalitions}}.$$

Chapter 10 Exercises

Create a preference table to show the results for each set of preference ballots.

1. The candidates are: Smith (S), Patel (P), Harvey (H), Knight (K), Jordan (J).

Preference Voting Results				
HKPJS	PJKSH	HKPJS	SHJPK	SHKPJ
SHJPK	KJPHS	PHSJK	PJSHK	PJSHK
JPHKS	SHJKP	KPJSH	KPJSH	SHJPK
PJKSH	SHJPK	PJSHK	HKPJS	SHJPK
SHKPJ	HKPJS	PJKSH	KJPHS	KJPHS
PJKSH	KJPHS	HKPJS	KJPHS	SHJPK
KJPHS	PJSHK	SHKPJ	SHJKP	KJPHS

2. Candidates in the election are: Lawson, Eric, Misty.

Misty, Eric, Lawson: 27

Lawson, Misty, Eric: 41

Misty, Lawson, Eric: 31

Eric, Misty, Lawson: 15

Eric, Lawson, Misty: 5

Lawson, Eric, Misty: 19

Calculate the number of pairwise comparisons that must be made in each election.

3. Seven students are running for senior class president.

4. Running for local councilman: J. Pitts, K. McMillian, J. Wallace, D. McLaughlin, K. Smith, and T. Knight.

Use the given preference table to answer each question.

5. Answer the following questions about the given preference table.

Preference Table for Candidates					
	Rankings				
1st	A	E	C	C	A
2nd	D	B	A	A	E
3rd	B	C	B	E	C
4th	C	A	D	B	D
5th	E	D	E	D	B
Total Votes	221	212	109	84	167

     a. How many possible unique rankings are there of the candidates in the election?

     b. How many people voted in the election?

     c. How many voters place Candidate A in first place?

     d. Which candidate wins the using plurality method?

     e. How many votes would a candidate need to have a majority?

     f. Do any of the candidates have a majority of first-place votes?

6. A photography club held a contest where the pictures had to be taken solely with a cell phone. The club members were asked to rank the following entries in order of preference. The preference table summarizes the results of the photo contest.

       1.   2.   3.   4.   5.

Preference Table for Best Photo				
	Rankings			
1st	3.	3.	2.	5.
2nd	1.	5.	4.	4.
3rd	2.	4.	1.	3.
4th	4.	1.	3.	2.
5th	5.	2.	5.	1.
Total Votes	54	83	65	92

     a. Determine the photo winner using the Borda count method.

     b. Determine the winner using the plurality with elimination method.

     c. Determine the winner with the pairwise method of comparison.

     d. Suppose that Picture 4 was thrown out for violating the rules. Pictures ranked below Picture 4 simply move up a ranking. Which picture would be declared the winner using the Borda count method with the new preference table? Is this a different winner than in part **a.**?

7.   In the Campus Sing-Off, each contestant is ranked by the general student body via the Internet, and by all student organizations sponsoring a contestant. The top 10 entries receive votes in each ballot. The votes are distributed in the following manner.

1st place = 12 points	2nd place = 10 points	3rd place = 8 points
4th place = 7 points	5th place = 6 points	6th place = 5 points
7th place = 4 points	8th place = 3 points	9th place = 2 points
10th place = 1 point		

The following table shows how the student organization ranked each contestant in the first semifinal round. The contestants are listed by their last names and sponsor organizations.

Voting Results for First Semifinal

Contestant																		
Harris; SGA				1st							3rd							
Icin; Social Work Club	6th		6th	6th			7th	6th	2nd	7th	8th	3rd	2nd	10th	7th	9th	10th	5th
Green; ΑΛΠ	2nd	6th			3rd	1st	8th	3rd		8th	4th	1st	7th	6th		10th	6th	8th
Lawson; FCA	9th													7th				
Albert; GSA	1st	8th	2nd	7th		7th	1st	2nd	6th	6th	2nd	2nd	4th	9th	2nd	1st	1st	7th
Roman; ΓΒΘ	4th	7th	3rd		6th		9th	7th		3rd	5th	5th	10th	3rd	8th	6th	2nd	1st
Switzer; Galois Club		9th		4th	8th	9th			10th	10th	9th			3rd	8th	3rd		
Belton; Hispanic Culture Center			7th		9th	10th		9th										10th
Finley; History Club		4th		5th	10th		10th	10th		9th			3rd		1st			
Issen; NAEA				10th		6th			8th			10th	8th	5th	6th	4th		9th
San Marie; NBS	7th		9th		2nd											8th		
Centus; NTSS	5th	1st	1st	8th	5th	4th		8th	10th	2nd				4th			4th	3rd
Dennis; ΩΨΦ		3rd	10th	3rd		8th	2nd		3rd		7th	7th		8th	10th		5th	
Russan; ΦA	3rd	5th	4th	1st		5th	3rd	1st	1st	1st	9th	4th	1st		4th	2nd	9th	4th
Hunter; ΦMA					4th	3rd	5th	5th	7th		6th	6th			7th			
Austin; ΣΓΡ		10th				6th	9th											
Molda; Student Design Group	8th		5th	9th	7th	2nd	4th		5th	5th	6th	8th	5th	1st	9th	5th	7th	2nd
Irene; Voices of Praise	10th	2nd	8th	2nd				4th	4th	4th	1st		9th	2nd	5th	3rd		6th

a.   Determine the number of total points each contestant received in the first semifinal round from the student organizations.

b.   Which top 10 contestants did the organizations think should move on to the second semifinal?

8. Online voting for the best student music video was available for two weeks. Students were asked to rank the videos in order of preference. The following table shows the results of the student votes.

Preference Table for Best Student Music Video						
	Rankings					
1st	Video 2	Video 1	Video 2	Video 3	Video 3	Video 4
2nd	Video 1	Video 2	Video 1	Video 4	Video 2	Video 1
3rd	Video 4	Video 3	Video 3	Video 2	Video 4	Video 2
4th	Video 3	Video 4	Video 4	Video 1	Video 1	Video 3
Total Votes	143	212	99	43	121	151

a. Determine the winning video by using the plurality method.

b. Determine the winning video by using the pairwise method of comparison.

c. Determine the winning by using the plurality with elimination method.

d. If you were determining a winning video, which video would you declare the winner based on the online votes? How would you defend your choice based on the answers to **a.–c.**?

9. In the 1992 United States Presidential Election, three candidates shared most of the November 3 popular vote, although there was a small portion of voters for other candidates. The following table displays the breakdown of the popular vote along with the electoral vote for the election.

1992 US Presidential Election Results			
Presidential Candidate	Popular Vote		Electoral Vote
	Count	Percentage	
George H. Bush	39,104,545	37.45%	168
Bill Clinton	44,909,889	43.01%	370
Ross Perot	19,742,267	18.91%	0
Other	669,958	0.63%	0
Totals	104,426,659	100%	538

Source: Federal Election Commission. "Federal Elections 92. Election Results for the US President, the US Senate, and the US House of Representatives." June 1993. http://www.fec.gov/pubrec/fe1992/federalelections92.pdf

a. Who won the popular vote by plurality? Was it a majority?

b. Who won the electoral vote by plurality? Was it a majority?

c. Suppose Perot had dropped out of the election. Is it possible that Clinton would have lost the presidency? Explain your answer.

10. The student government is deciding which author to bring to campus next year for the annual Books to Tables talk. The possibilities are Marci Shimoff, coauthor of *Chicken Soup for the Woman's Soul*, Peter Legge, author of *How to Soar with the Eagles*, and Tania Aebi, author of *Maiden Voyage*. Complete the following preference table so that Tania Aebi wins the election using the majority rule.

Preference Table for Author for Books to Tables Talk			
	Rankings		
1st	Tania Aebi	Peter Legge	Marci Shimoff
2nd	Peter Legge	Marci Shimoff	Tania Aebi
3rd	Marci Shimoff	Tania Aebi	Peter Legge
Total Votes	?	5	4

11. The Campus Housing Welcome Back committee is deciding what movie to show free to students the first week of school. The movie options are *Django Unchained*, *Zero Dark Thirty*, *Silver Linings Playbook*, and *Life of Pi*. Students were asked to rank the movies by preference. The following preference table summarizes their selections.

Preference Table for Movie Showing							
Movie	Rankings						
1st	Zero	Zero	Django	Silver	Life	Django	Django
2nd	Django	Life	Life	Zero	Django	Silver	Zero
3rd	Life	Silver	Zero	Django	Zero	Zero	Silver
4th	Silver	Django	Silver	Life	Silver	Life	Life
Total Votes	30	23	15	16	11	15	27

a. Which movie is the plurality winner?

b. Which movie wins using a pairwise method of comparison?

c. Does using the plurality method satisfy the Condorcet criterion? Explain your answer.

12. An online poll asked readers to rank the top three law schools in the country. The following preference table shows the results after two days of online voting.

Preference Table for Top Law Schools				
	Rankings			
1st	Stanford	Yale	Stanford	Harvard
2nd	Harvard	Stanford	Harvard	Yale
3rd	Yale	Harvard	Yale	Stanford
Total Votes	84	132	39	13

a. Which law school would win using the majority rule if voting closed today?

b. Which law school would win using the Borda count method if voting closed today?

c. Does the Borda count method satisfy the majority criterion for these poll results? Explain your answer.

13. The Association of Marketing Professionals is choosing a city for its next conference. The committee making the decision asks the association's members to rank the top five cities they would like to visit. The following preference table gives the results of the member survey. Would the plurality with elimination method satisfy the irrelevant alternatives criterion if the committee decided to eliminate Jacksonville, FL, from the options after the vote? Explain your answer.

Preference Table for Conference Location				
	Rankings			
1st	Boston, MA	Nashville, TN	San Diego, CA	New Orleans, LA
2nd	New Orleans, LA	Boston, MA	Boston, MA	Nashville, TN
3rd	San Diego, CA	San Diego, CA	New Orleans, LA	San Diego, CA
4th	Nashville, TN	Jacksonville, FL	Nashville, TN	Boston, MA
5th	Jacksonville, FL	New Orleans, LA	Jacksonville, FL	Jacksonville, FL
Total Votes	375	348	289	115

14. The Chamber of Commerce is electing a new president. The candidates are N. Pitts, A. Palmer, and J. Layne. The summary of the rankings of the candidates from the members is given in the table. Is the irrelevant alternatives criterion satisfied using the Borda count method if N. Pitts has to withdraw from the ballot after votes are cast because he moved? Explain your answer.

Preference Table for Chamber of Commerce President				
	Rankings			
1st	A. Palmer	N. Pitts	A. Palmer	J. Layne
2nd	N. Pitts	J. Layne	J. Layne	A. Palmer
3rd	J. Layne	A. Palmer	N. Pitts	N. Pitts
Total Votes	**12**	**15**	**13**	**16**

15. *Living Day-to-Day* magazine wants to name the number-one feature homebuyers want in a house. They ask their readers to rank eight home features in order of importance. The eight home features are open-concept homes, smaller homes, outdoor living spacings, neutral decor, modern kitchens, linen closest & smart storage options, energy-efficient fixtures & appliances, and two-car garage with organization. The following preference table displays the results of the rankings. Determine if the Condorcet criterion is satisfied if the magazine uses the plurality method to name the top feature.

Preference Table for Homebuyers' Number One Feature						
Feature	**Rankings**					
1st	Outdoor	Smaller	Open	Kitchen	Smaller	Outdoor
2nd	Open	Outdoor	Smaller	Outdoor	Efficient	Kitchen
3rd	Kitchen	Kitchen	Outdoor	Efficient	Garage	Storage
4th	Neutral	Efficient	Neutral	Open	Open	Efficient
5th	Efficient	Storage	Kitchen	Storage	Outdoor	Garage
6th	Garage	Garage	Storage	Garage	Kitchen	Neutral
7th	Storage	Neutral	Efficient	Smaller	Storage	Open
8th	Smaller	Open	Garage	Neutral	Neutral	Smaller
Total Votes	**421**	**234**	**331**	**472**	**253**	**119**

16. Driving Divas asks its readers to rank the following car features in order of "must haves" in a new car. The car features are remote keyless entry, OnStar system, antilock brakes (ABS), electronic stability/skid-control system, telescoping steering wheel/adjustable pedals, and rear-seat DVD player. Complete the preference table so that rear-seat DVD player is the winner using the Borda count method, but the Condorcet criterion is violated in a head-to-head matchup with antilock brakes.

Preference Table for Must-Have Car Feature				
Feature	**Rankings**			
1st	DVD	Keyless	Adjustable	Stability
2nd	Adjustable	?	DVD	ABS
3rd	ABS	?	OnStar	OnStar
4th	Keyless	Stability	ABS	Keyless
5th	OnStar	OnStar	Stability	Adjustable
6th	Stability	Adjustable	Keyless	DVD
Total Votes	**35**	**44**	**28**	**20**

17. In his next newspaper column, Tom plans to publish his results from a study on the top six qualities that make a great leader. The following preference table summarizes the rankings for the traits. Complete the preference table so that "sense of humor" is the top trait if the site uses the Borda count method to count the votes, and the majority criterion is not violated.

Preference Table for Top Leadership Qualities				
Quality	Rankings			
1st	?	Honesty	Honesty	Communication
2nd	?	Sense of Humor	Communication	Sense of Humor
3rd	Confidence	Ability to Delegate	Sense of Humor	Honesty
4th	Commitment	Commitment	Confidence	Ability to Delegate
5th	Ability to Delegate	Confidence	Ability to Delegate	Commitment
6th	Communication	Communication	Commitment	Confidence
Total Votes	159	66	42	50

Use the given table to solve each problem.

Population per County	
County	Population
A	8578
B	9878
C	10450
D	7565
E	4563
F	6347

18. Suppose a state decides to apportion 250 new highway patrol officers on the basis of number of residents in a county.

   a. Find the standard divisor, SD.

   b. Find the standard quota, SQ, for county A and county B. Round to three decimal places.

19. Use the Hamilton method to apportion the 250 patrol officers based on the population of the county.

Use the given table to solve each problem.

Student Enrollment in the University of Texas System		
Campus	Enrollment	Full Time Equivalent (FTE)
Arlington	33,439	20,594
Austin	51,112	46,402
Brownsville	13,836	9398
Dallas	18,684	12,537
El Paso	18,160	16,271
San Antonio	30,968	23,198
Tyler	5064	4810
Permian Basin	16,266	2696
Pan American	21,016	15,494

20. The University of Texas system consists of nine campuses and has a budget of $13.1 billion. How much money in funding would each campus receive if the system divided the money equally among the campuses (round to nearest dollar)?

21. If the state of Texas wanted to allocate the money equally based on total student enrollment, how much would be appropriated per student?

22. Based on the result from Exercise 21, how much money would the campuses of Arlington and Pan American receive based on the student enrollment on each campus.

23. If the state apportions the funds based on FTE, how much money will the campuses of Austin and Dallas receive (round to nearest dollar)?

24. The University of Texas system wants to apportion 825 new electric vehicles to their campuses. The state decides to apportion these vehicles based on the student enrollment at each university.

    a. Find the SD. Round to three decimal places.

    b. Find the SQ for the Brownsville and Tyler campuses. Round to three decimal places.

25. Suppose the state decides to apportion the 825 electric vehicles based on FTE.

    a. Find the SD. Round to three decimal places.

    b. Find the SQ for the San Antonio and Permian Basin campuses. Round to three decimal places.

26. Use the Jefferson method to apportion the 825 electric vehicles to all 10 campuses based on the number of students.

27. Use the Hamilton method to apportion the 825 electric vehicles to all 10 campuses based on FTE. Compare your results with the apportionments from Exercise 26 and determine if the apportionments are different when the apportionment basis is different.

Solve each problem.

28. A science teacher at a high school can teach five classes. There are 27 students enrolled in Physical Science, 24 in Biology, and 37 in Anatomy.

    a. Find the SD and SQ of each course. Round to three decimal places.

    b. Use the Jefferson method to determine the apportionment of students to the courses to determine the number of sections needed per course.

29. Repeat Exercise 28 **b.** using the Huntington-Hill method.

30. Repeat Exercise 28 **b.** using the Hamilton method.

31. An English department uses 19 graduate assistants in teaching its undergraduate courses. The enrollments for each of the courses these students teach is as follows. Using Webster's method, how many graduate assistants should be assigned to each course?

Students per Course	
Course	Students
Comp I	675
Comp II	455
Survey of Lit	197
Poetry	385

32. Suppose there are 112 faculty members in the sciences, 136 in the humanities, and 47 in the professional and trade schools. A 15-person faculty committee is to be formed.

    a. Use Hamilton's method to determine the allocation of committee members based on department size.

    b. Use Jefferson's method to determine the allocation of committee members based on department size.

    c. Use Huntington-Hill method to determine the allocation of committee members based on department size.

33. A state consists of six counties: A, B, C, D, E, and F.

The senate for the state is to have 30 members apportioned to the counties based on the old populations using Webster's method. The new populations are also given.

Population by County		
County	Old Population	New Population
A	8578	9244
B	9878	9166
C	10,450	10,580
D	7565	10,254
E	4563	6680
F	6347	5993

a. Apportion the members of the senate using the old populations.

b. Apportion the members of the senate using the new populations.

c. Apportion the members of the senate using the Huntington-Hill method and the old populations.

d. Apportion the members of the senate using the Huntington-Hill method and the new populations.

e. Does the population paradox occur when calculating the apportionments using the Huntington-Hill method and the old and new populations?

34. The city of Hillsboro wants to allocate $55 million to youth sports and activity programs. The current participation rates are listed in the table.

Participants per Activity	
Activity	Number of Participants
Swimming	245
Library	2595
Baseball	1150
Softball	978
Theater	753
Computer Hobbies	477
Music	1658

a. Find how much the city should allocate to each activity based on the number of participants using the Jefferson method.

b. Find how much the city should allocate to each activity based on the number of participants using Webster's method.

c. Find how much the city should allocate to each activity based on the number of participants using the Huntington-Hill method.

d. Find how much the city should allocate to each activity based on the number of participants using the Hamilton method.

35. Three groups, A, B, and C, have the number of members shown in the table. If Jefferson's method is used to apportion the representatives of the groups on a panel, does the Alabama paradox occur if the number of representatives is increased from 40 to 41?

Members per Group	
Group	**Members**
A	3340
B	4500
C	8875

36. Repeat Exercise 35 and increase the number of representatives from 60 to 61.

37. Using the table in Exercise 35, if Webster's method is used to apportion the representatives of the groups on a panel, does the Alabama paradox occur if the number of representatives is increased from 40 to 41?

38. Repeat Exercise 37 and increase the number of representatives from 60 to 61.

39. What is the quota for $[8,5,3,2]$ in a simple majority?

40. What is the quota for $[8,5,3,2]$ for a two-thirds majority?

41. Consider the weighted voting system $[25: 9, 7, 4, 3, 2, 2, 1]$.

    a. What is the quota for this voting system?

    b. What is the weight of P_5?

    c. If only the last six voters vote for a motion, does the motion pass?

    d. If P_1 and P_2 vote against a motion, does the motion pass?

    e. What are the possible winning coalitions for this voting system?

42. Six partners in a law firm have the following voting system $[96: 34, 25, 18, 18, 12, 4]$.

    a. How many voters are in the system?

    b. What is the quota for the system?

    c. What are all of the winning coalitions?

    d. Are there any critical players?

    e. Do any of the players have veto power?

    f. Is there a dummy player?

    g. Determine the Banzhaf Power Index for each player.

43. In the weighted voting system $[q: 10, 7, 4, 4, 2, 2]$, if every voter has veto power, what is q?

44. A voting system is represented by $[61: 35, 30, 24, 21]$.

    a. List all sequential coalitions in which the voter with weight 35 is pivotal.

    b. List all sequential coalitions in which the voter with weight 30 is pivotal.

    c. Calculate the Shapley-Shubik Power Index for each player in the system.

    d. Are any of the voters dictators?

    e. Do any of the voters have veto power?

45. A professional football team is getting ready for the amateur player draft. The head coach, general manager, head scout, and team physician will use the voting system $[8: 5, 4, 3, 2]$ to make decisions regarding players to draft.

    a. List all sequential coalitions in which P_1 is a pivotal player.

    b. List all sequential coalitions in which P_2 is a pivotal player.

    c. List all of the winning coalitions.

    d. Calculate the Shapley-Shubik Power Index for each player.

    e. Calculate the Banzhaf Power Index for each player.

    f. Is there a player with veto power?

Bibliography

10.4

1. United Nations. *Charter of the United Nations.* "Chapter V: The Security Council." Accessed February 2013. http://www.un.org/en/documents/charter/chapter5.shtml

Chapter 11
The Arts

Sections

Objectives

- Understand the relationship between mathematics and art/architecture
- Understand the use of geometry in art/architecture
- Understand the use of sequences and series in art and music
- Understand the use of the golden ratio, golden rectangles and triangles, and their use in art and architecture
- Understand triangular and square numbers
- Understand the use of regular polygons in creating tilings and tessellations
- Understand how rotations, translations, and reflections are used in art and architecture
- Understand how sound frequencies in music are used to tune a piano and their relationships in musical harmonies

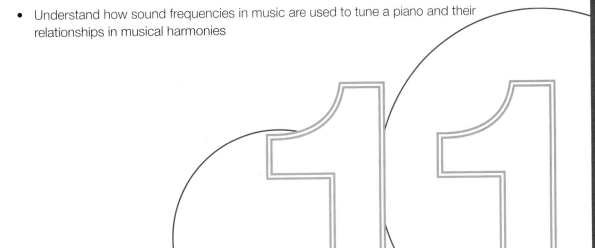

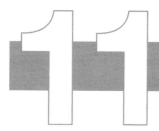

The Arts

Have you ever wondered how a piano is tuned or how the notes on the piano are arranged? Have you ever considered how much geometry is needed to build an office building? From ancient cave drawings to modern architecture, artists have used mathematical concepts such as ratios, proportions, and area for thousands of years to create works that are pleasing to look at as well as utilitarian in use.

One modern famous artist that employed the concepts of geometry into his work was M.C. Escher (1898–1972). Escher, a Dutch graphic artist, created magnificent works of art using rotations, reflections, and translations of common geometric figures, such as squares or rectangles. While Escher had no formal mathematical training or background, he understood math in a visual and intuitive sense, which is demonstrated in much of his art.

Leonardo da Vinci (1452–1519) was an Italian Renaissance painter, sculptor, architect, musician, scientist, mathematician, engineer, inventor, anatomist, geologist, cartographer, botanist, and writer. Most famous for his works of art that include the *Mona Lisa* and *Last Supper*, da Vinci was the master of bringing to life the Renaissance humanist ideal. Hence, he is the inspiration for the term "Renaissance man."

Da Vinci was a man well ahead of his time in many aspects. In art, he was one of the first artists to use a special rectangle called a golden rectangle, which we will discuss in this chapter. The golden rectangle is seen in many of his works, including his *Self-Portrait* and *Mona Lisa*.

For more information on da Vinci, visit http://www.biography.com/people/leonardo-da-vinci-40396.

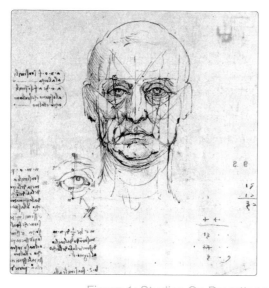
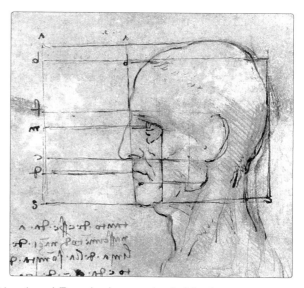

Figure 1: Studies On Proprtions of Head and Eyes by Leonardo da Vinci

11.1 Applications of Geometry to the Arts

What Is Art?

To begin our discussion of mathematics in art, we first need to define exactly what art is in order to discuss the mathematics needed to produce art. Do you have a definition for art? As human beings, we are surrounded by art and we often take its influence on our lives for granted. According to many art historians, art consists of two things: form and content. **Form** in art is the medium—that is, the material such as stone, marble, or canvas—used by an artist to deliver the content they wish to convey. **Content** of art simply refers to the actual piece of work. For instance, a given painting's content may focus on a building, a landscape, or a portrait. What about music? Is music art? In a short answer, yes! For example, songs often convey a deeper meaning and use literary concepts to enhance the performance aspect. Any of these examples help us understand the content contained in a piece of art.

> ### Art
>
> For the purposes of this book, **art** will be considered as any endeavor that requires creativity of the mind, whether it is a painting, a sculpture, architecture, a musical composition, etc.

Art can essentially be broken into two large categories: visual arts and performing arts. **Visual art** consists of works of art that are written or constructed. Examples include works of art created with a painter's paintbrush, a potter's wheel, a drafter's pencil, a weaver's cloth, a sculptor's chisel, a poet's pen, etc. **Performing arts** consist of art that is performed using voices and/or body movements. Examples include musical performances and theatrical arts. In this chapter, we study how geometry is used in the visual arts, how mathematics relates to music, and the use of numbers and sequences in visual art.

The relationship between mathematics and the visual and performing arts is sometimes not readily apparent to everyone. We found when discussing geometry in Chapter 6 that mathematics can be applied to a diverse range of subjects and the arts are no exception. For instance, the use of geometry is an integral component in having symmetrical pieces of artwork. Similarly, the presence of mathematical sequences in musical computation allows musicians to do what they do best: create music. We begin this chapter with the notions and ideas needed to apply the concept of mathematics to art. Geometry is one area of mathematical study that encapsulates pretty much every facet of everyday life—especially in art and architecture.

Let's start our exploration of geometry and art by introducing a few mathematical sequences and some basic number theory concepts that are frequently used. Consider the sequence 1, 1, 2, 3, 5, 8, 13, 21, 34, 55, 89, 144, . . . , which is known as the Fibonacci sequence. Can you determine the next term in the pattern? If so, can you use your problem-solving skills to develop a rule to continue the pattern?

MATH MILESTONE

The Fibonacci Sequence is named after mathematician Leonardo Pisano Bigollo, most commonly known as Fibonacci. He did not discover the sequence, but he did popularize it in his book *Liber Abaci*.

> ### Fibonacci Sequence
>
> The **Fibonacci sequence** is the sequence of numbers 1, 1, 2, 3, 5, 8, 13, 21, 34, 55, 89, 144, . . . that continues infinitely, where each successive number can be found by adding together the previous two numbers in the sequence.

The Fibonacci sequence is derived from a mathematical problem proposed in the book *Liber Abaci* in regards to rabbit breeding. The problem begins with a single pair of rabbits: one male and one female. Assuming that every month this rabbit pair produces a new pair of rabbits, and each new pair of rabbit themselves start producing an additional pair of rabbits when they are two months old, how many pairs of rabbits will be born in a year? The sequence is quite useful in defining and exploring phenomena that occur in nature and is used by artists to create character in the content and substance of works of art.

Example 1: The Fibonacci Sequence

Fibonacci's original version of the sequence goes something like the following.

Start with a pair of rabbits (one male and one female) born on January 1st. Assume that all months are of equal length and that: **1.** rabbits begin to produce offspring two months after their own birth; **2.** when the rabbits reach the age of two months, each pair produces a mixed pair (one male, one female), and then another mixed pair each month thereafter; and **3.** no rabbit dies. How many pairs of rabbits will there be after one year?

Solution

	Table 1	
	Fibonacci's Sequence: Original Version	
Month	**Pairs of Rabbits**	
1	1	=	original pair
2	1	=	original pair
3	2	=	original pair + new pair of offspring = 1 + 1
4	3	=	(original pair + offspring pair born in month 3) + new pair of offspring = 2 + 1
5	5	=	(original pair + pair from month 3 + pair from month 4) + (new pair of offspring from original pair + new pair from offspring born in month 3) = 3 + 2
6	8	=	5 + 3
7	13	=	8 + 5
8	21	=	13 + 8
9	34	=	21 + 13
10	55	=	34 + 21
11	89	=	55 + 34
12	144	=	89 + 55

So, we can see that at the end of the year, there will be 144 pairs of rabbits, all resulting from the one original pair born on January 1st of that year.

Example 2: More on the Fibonacci Sequence

What is the 25th Fibonacci number?

Solution

The most straightforward way to find the 25th number is to list the numbers in order. So, using the notation $F_1 = 1, F_2 = 1, F_3 = 1 + 1 = 2, \ldots$, where F_1 represents the first Fibonacci number, F_2 represents the second Fibonacci number, and so on, we find that the next 22 terms are as follows.

$$F_4 = 1 + 2 = 3$$
$$F_5 = 2 + 3 = 5$$
$$F_6 = 3 + 5 = 8$$
$$F_7 = 5 + 8 = 13$$
$$F_8 = 8 + 13 = 21$$
$$F_9 = 13 + 21 = 34$$
$$F_{10} = 21 + 34 = 55$$
$$F_{11} = 34 + 55 = 89$$
$$F_{12} = 55 + 89 = 144$$
$$F_{13} = 89 + 144 = 233$$
$$F_{14} = 144 + 233 = 377$$
$$F_{15} = 233 + 377 = 610$$
$$F_{16} = 377 + 610 = 987$$
$$F_{17} = 610 + 987 = 1597$$
$$F_{18} = 987 + 1597 = 2584$$
$$F_{19} = 1597 + 2584 = 4181$$
$$F_{20} = 2584 + 4181 = 6765$$
$$F_{21} = 4181 + 6765 = 10,946$$
$$F_{22} = 6765 + 10,946 = 17,711$$
$$F_{23} = 10,946 + 17,711 = 28,657$$
$$F_{24} = 17,711 + 28,657 = 46,368$$
$$F_{25} = 28,657 + 46,368 = 75,025$$

Therefore, the 25th Fibonacci number is 75,025.

The Fibonacci sequence has some interesting properties. In fact, mathematicians have been researching the applications of the sequence to the field of number theory since its introduction. One of the more interesting relationships that have been studied is the relationship of the Fibonacci numbers to nature. The numbers in the sequence occur in such instances as the number of spirals on a pine cone, the number of spines on a head of lettuce, and the number of petals on flowers, just to name a few, as shown in Figure 1.

Figure 1

Fun Fact

Pine cones also have counter-clockwise spirals. See if you can find them.

11

In the picture of the pine cone, the number of spirals in the pine cone is 13, a Fibonacci number. The head of lettuce has 5 "ribs," also a Fibonacci number. In the pictures of the flowers, the number of petals on each flower is 3 and 5, respectively—each of which is a Fibonacci number.

You might be asking yourself, "How is the Fibonacci sequence used in art?" This sequence is used as a means of developing aesthetic concepts in the items in a painting, the size or shape of a painting itself, and many features of architecture.

In the years since the time of Fibonacci, it has been discovered that the sequence of ratios of consecutive terms of the Fibonacci sequence, denoted as

$$\frac{1}{1}, \frac{2}{1}, \frac{3}{2}, \frac{5}{3}, \frac{8}{5}, \frac{13}{8}, \frac{21}{13}, \frac{34}{21}, \frac{55}{34}, \cdots$$

$$\frac{1}{1} = 1$$

$$\frac{2}{1} = 2$$

$$\frac{3}{2} = 1.5$$

$$\frac{5}{3} = 1.\overline{6}$$

$$\frac{8}{5} = 1.6$$

$$\frac{13}{8} = 1.625$$

$$\frac{21}{13} = 1.615384615$$

$$\frac{34}{21} = 1.619047619$$

$$\frac{55}{34} = 1.617647059$$

$$\cdots,$$

approaches the irrational number 1.618033988. . . as the ratio of larger numbers is considered. This number has been given the name of ϕ (read "phi").

We call the ratio ϕ the **golden ratio**. To better understand the relationship, construct a line segment $\overline{AB}$ of any given length. Recall that $\overline{AB}$ is the line segment with endpoints A and B, and AB is the length of $\overline{AB}$. The point C is placed on the line such that $AC = x$ and $CB = 1$, as shown in Figure 2.

Figure 2

The proportion $\frac{AB}{AC} = \frac{AC}{BC}$ can be solved algebraically as follows to determine the length of x.

$$\frac{AB}{AC} = \frac{AC}{BC}$$

$$\frac{x+1}{x} = \frac{x}{1}$$

$$x^2 = x + 1$$

$$x^2 - x - 1 = 0$$

Using the quadratic formula, we can obtain the value of x.

$$x = \frac{-(-1) \pm \sqrt{(-1)^2 - 4(1)(-1)}}{2(1)}$$

$$= \frac{1 \pm \sqrt{5}}{2}$$

Thus, $\phi = \frac{1+\sqrt{5}}{2}$. Note that since ϕ is a ratio of two lengths, we only consider the positive value for the equation.

Phi (ϕ)

The number ϕ is defined by the following mathematical representation.

$$\phi = \frac{1+\sqrt{5}}{2} \approx 1.618033988749894\ldots$$

☞ **Helpful Hint**

For the remainder of the text, we will use the ϕ approximation 1.618.

The ratio we have found, ϕ, has been used as a popular aesthetic for art and architecture since at least 3000 BC and has names such as the golden ratio or golden section.

Example 3: Approximating ϕ from the Fibonacci Sequence

Show that the ratio of the 22nd and 23rd Fibonacci numbers is approximately ϕ.

Solution

The 22nd Fibonacci number is 17,711 and the 23rd Fibonacci number is 28,657. The ratio of the 23rd and 22nd Fibonacci numbers is $\frac{28,657}{17,711} \approx 1.618$. So, the ratio of the 23rd and 22nd Fibonacci numbers is an approximation of ϕ accurate to the nearest thousandth.

Skill Check # I
Show that the ratio of the 25th and 24th Fibonacci numbers is approximately ϕ.

The golden ratio is quite common in nature and is a common theme in art. In fact, objects such as pinecones not only have a number of swirls that is equal to a Fibonacci number, as we saw earlier, but they also grow in a way that is called **gnomonic**. **Gnomonic growth** is growth that occurs based on a previous amount of growth. Think of a nautilus shell, as in Figure 3.

Figure 3

The shell of a nautilus consists of a spiral that begins at birth and continues to grow outwardly throughout its life. As it turns out, this spiral grows in a special manner that can be modeled by the golden ratio. We call this spiral the **golden spiral**. The golden spiral is a logarithmic spiral whose growth factor is ϕ, and the shape of the spiral can be approximated using Fibonacci numbers. Figure 4 shows how to construct a golden spiral. Every quarter turn of the spiral gets wider by ϕ and we have already seen that the ratios of consecutive terms in the Fibonacci series approach ϕ.

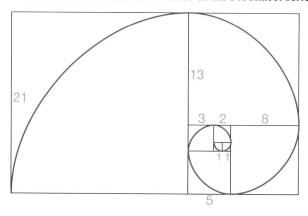

Figure 4

The golden ratio is also used to create a **golden rectangle**. This is a rectangle where the ratio of the length to the width is the golden ratio.

Golden Rectangle

A **golden rectangle** is a rectangle in which the ratio of the length to the width is the golden ratio.

Golden rectangles are often used in architecture and art as a means of "framing" aesthetics of a piece of art. The ancient Greeks were definitely familiar with this concept. Consider one of the most famous buildings of ancient Greece, the Parthenon, as illustrated in Figure 5. The width of the Parthenon compared to its height is very close to the golden ratio, indicating that the Greeks had a firm understanding of the naturally aesthetically pleasing nature of this ratio.

Figure 5

Fun Fact

Although we previously mention the Greeks' use of the golden ratio and golden rectangles in building, there is some speculation as to whether they actually used the ratio or if their buildings being constructed as "golden" is simply a coincidence. For a more intriguing discussion on this subject, you might be interested in NPR's *Math Guy*, Keith Devlin from Stanford University, and his Math Encounters talk entitled "Fibonacci and the Golden Ratio Exposed" at http://www.youtube.com/watch?v=JuGT1aZkPQ0.

To Construct a Golden Rectangle, Use a Compass and Straight Edge

1. Start with a segment of any length with endpoints A and B.

Figure 6

2. Next, construct a square with sides of the same length as $\overline{AB}$. Let's call the vertices of the square A, B, C, and D.

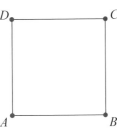

Figure 7

3. Then, find the midpoint of the base of the square and label this point M as shown in Figure 8.

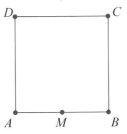

Figure 8

4. Open a compass with the needle on M and pencil tip on C. Draw a circle that contains the square. Extend $\overline{AB}$ so that it intersects the circle and label the intersection E as shown in Figure 9. Note that the radius of the circle we drew has length MC.

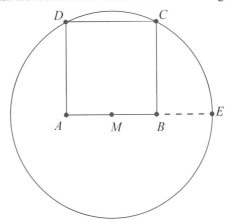

Figure 9

5. The segment $\overline{AE}$ is the length of the golden rectangle. Now we only need to complete the rectangle by extending $\overline{DC}$ to intersect with the line drawn through E parallel to AD. Label the point of intersection with F. The golden rectangle will have vertices A, E, F, and D, as in Figure 10.

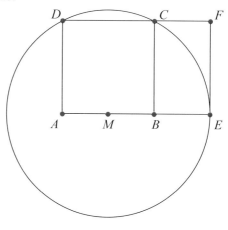

Figure 10

Now that the golden rectangle is complete, the ratio of the length to the width of the rectangle is $\frac{AE}{FE} = \phi$. In Figure 10, $AE = 14.042$ cm and $FE = 8.678$ cm. This gives that $\frac{AE}{FE} \approx 1.618$.

Example 4: Golden Rectangles

Verify that the rectangle in the given figure is a golden rectangle if $AB = 9.305$ and $BC = 5.75$.

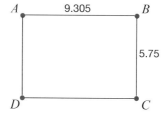

Solution

We know that golden rectangles have the attribute that the ratio of the length and width is ϕ. So, given length $AB = 9.305$ and width $BC = 5.75$, we can see that $\frac{AB}{BC} = \frac{9.305}{5.75} \approx 1.618261$. It is important to note that since ϕ is an irrational number, it is impossible to use numerical measurements to find an exact answer. Therefore, we must accept a level of accuracy that we feel comfortable with. We will consider rounding to the nearest thousandth as sufficient accuracy for our calculations.

Leonardo da Vinci, among his many other accomplishments, is considered by many to be a master at the use of the golden ratio and golden rectangles in his art. While many artists did not know of or use the golden ratio to design buildings and create art, there is evidence that da Vinci knew about this number and often utilized it in his art. In what is considered one of his most famous works, *Vitruvian Man*, shown in Figure 11, da Vinci used the golden ratio to illustrate how the human body itself is "golden" within a work of art that uses the golden ratio.

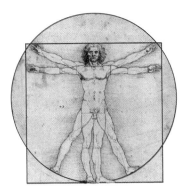

Figure 11: *Vitruvian Man* by Leonardo da Vinci

The golden ratio plays an important role in the human body, as rigorous mathematical analysis has shown that, on average, human beings and their skeletal structure have many golden relationships. Consider any of our fingers that have three knuckles. The bones of the fingers form a series of three terms, $1, \dfrac{1}{\phi}, \dfrac{1}{\phi^2}$, that correspond to lengths of the segments of the fingers. The ratio of the overall height of human beings compared to the height of their navels from the ground is also golden.

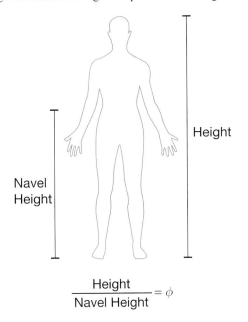

$$\frac{\text{Height}}{\text{Navel Height}} = \phi$$

$$\frac{1}{\dfrac{1}{\phi}} = \phi \qquad \frac{\dfrac{1}{\phi}}{\dfrac{1}{\phi^2}} = \phi$$

Figure 12

Skill Check #2

The ratio of the length of our arms, from the shoulder to the tips of our fingers, compared to the length from our elbow to the tips of our fingers is the golden ratio. Try it to see for yourself. It may be helpful, and more accurate, to accumulate multiple measurements from several individuals and find the average of the ratios.

Golden Triangles and Pentagons

In addition to the golden rectangle, we can also find a golden triangle. Golden triangles are used in art and architecture to create aesthetically pleasing objects. In photography, for instance, golden triangles are used as a "rule of thirds" to assist in aligning objects. Figure 13 consists of a golden rectangle with a diagonal drawn from opposite vertices. If perpendiculars to the diagonal line are drawn through the remaining vertices, then the ratio of DE to FG is the golden ratio. The triangles created within the figure, one of which is highlighted, have many golden relationships used by photographers to align the vertical (FG) and horizontal (DE) aspects of a potential photo.

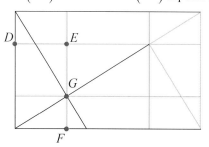

Figure 13

Think Back

An isosceles triangle has two congruent sides. In the triangle shown below, the two congruent sides have length a.

Golden Triangle

A **golden triangle** is defined as an isosceles triangle whose length of a side compared to the length of the base is the golden ratio.

To Construct a Golden Triangle, We Must Complete the Following Steps:

1. Construct a segment $\overline{AB}$ and its midpoint C.

Figure 14

2. Construct a segment perpendicular to $\overline{AB}$ through B. Then, construct a circle centered at B with radius $\overline{BC}$. Label the top point where this segment intersects the circle as D.

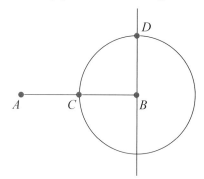

Figure 15

3. Construct segment *AD*. Then, construct a circle centered at *D* with radius *DB*. Now, label the intersection of the circle centered at *D* and segment *AD* as *E*.

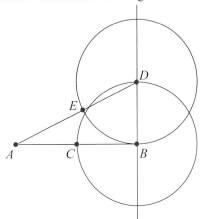

Figure 16

4. Construct a circle with radius $\overline{AE}$ centered at *A*. Next, label the intersection of this circle with segment $\overline{AB}$ as *G*.

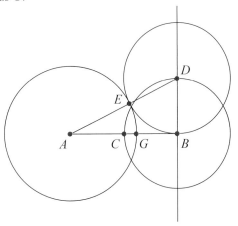

Figure 17

5. Focus only on segment AB. Point *G* divides line segment $\overline{AB}$ so that the ratio $\frac{AB}{AG}$ is the golden ratio.

Figure 18

6. Using Figure 18, reconstruct a circle centered at *A* with radius $\overline{GA}$ and construct another circle centered at *G* with radius $\overline{GB}$. Label one of the intersections of the circles as *H*.

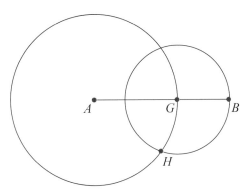

Figure 19

7. Construct line segments $\overline{AH}$ and $\overline{HG}$. The triangle HGA is a golden triangle. Notice that since $\overline{AH}$ and $\overline{AG}$ are both radii of the same circle, they have the same length. Thus, $\angle AHG$ is an isosceles triangle. In Figure 20, $AH = 6.80$ cm and $HG = 4.20$ cm. This gives that $\frac{AH}{HG} \approx 1.618$.

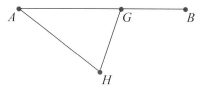

Figure 20

It should be noted that golden triangles are the basis of other geometric figures, such as pentagons and hexagons, that are used frequently in art and architecture. The pentagon is associated with the golden ratio more than any other geometric figure. For instance, when we connect the five vertices of a pentagon, in all possible ways, we end up with a multitude of golden triangles. In Figure 21, we can see that a golden triangle appears when we connect the vertices F, O, P.

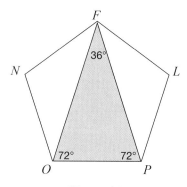

Figure 21

Fun Fact

When an isosceles triangle is inscribed into a regular pentagon as in Figure 21, the sum of one of the base angles of the golden triangle and the angle $\angle OFP$ is equal to the measure of one interior angle in a regular pentagon. So $72° + 36° = 108°$.

Recall from Chapter 6 that the sum of the measures of the interior angles of a polygon is $(n - 2) \cdot 180°$. We could easily verify this with the idea that the pentagon can be broken down into three triangles, and thus the sum would be $3 \cdot 180° = 540°$. Since the pentagon is regular, each of the vertex angles ($\angle FNO$, $\angle NOP$, $\angle OPL$, $\angle PLF$, and $\angle LFN$) will have a measure of $\frac{540°}{5} = 108°$.

As we mentioned previously, there are many examples of the relationship of pentagons and pentagrams in art and architecture. For example, you can see a golden triangle in both the US Pentagon building near Washington, D.C.,[1] and the chapel at the US Air Force Academy in Colorado Springs, Colorado.

Figure 22

---

[1] Photograph Courtesy of Patrick Neil

Golden Ratio in Art

Now that we have spent considerable time introducing the golden ratio, the golden triangle, and the golden rectangle, let's actually look at some works of art—ancient, renaissance, and modern—that contain aspects of the golden ratio.

Consider the artwork of Graham Sutherland (1903–1980) titled *Christ in Glory*.[2] A study of this tapestry, which measures 23.94 by 12.05 meters, indicates that many of the features are approximately golden. For ease in understanding, a line drawing of the tapestry is contained in Figure 23 and shows one example of a golden relationship. Verify for yourself that this indeed represents the golden ratio.

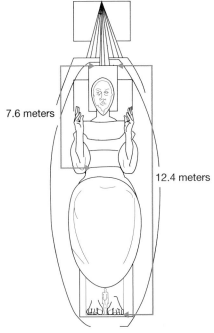

7.6 meters

12.4 meters

Figure 23

Golden Ratio in Architecture

We have spent a lot of time discussing the use of the golden ratio in art, but what about architecture? There are many instances where the golden ratio appears in architecture throughout history, from the ancient Egyptian pyramids to modern buildings. For instance, it is well documented from modern research that the Great Pyramid at Giza, shown in Figure 24, has a golden relationship. If we consider the base of the pyramid to be 2 units in length and we form a triangle as shown in Figure 24, then the triangle obtained by looking at its cross-section reveals that the height forms the square root of a golden ratio and the slant height of the pyramid forms a golden ratio.

Great Pyramid at Giza

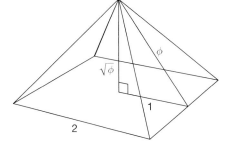

Figure 24: The Golden Relationship Demonstrated by the Great Pyramid at Giza

Example 5: Golden Rectangles in Modern Architecture

The United Nations headquarters is said to have been built using the golden ratio, where the width of the building compared to every 10 floors is the golden ratio. The height of the building is 516.75 feet and it contains 39 floors. That means that there are 13.25 feet per floor, or every 10 floors are approximately 132.5 feet. The width of the building is approximately 214.33 ft. Using the given information, show that this relationship indeed exists.

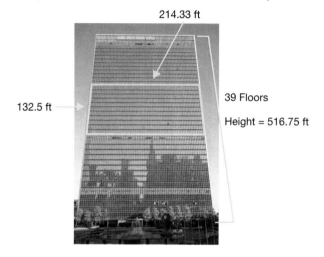

214.33 ft

132.5 ft

39 Floors

Height = 516.75 ft

Solution

We are given that the United Nations building is 214.33 feet wide and that every 10 floors measure approximately 132.5 feet, as indicated in the picture. Doing some arithmetic, we find that the ratio of the width to the height is

$$\frac{\text{width}}{\text{height}} = \frac{214.33 \text{ ft}}{132.5 \text{ ft}} \approx 1.618.$$

Therefore, the United Nations building does in fact display an approximate golden rectangle when the width of the building is compared to the height of 10 floors.

The golden ratio's naturally pleasing aesthetic to the human mind may be the reason it is used so often in architecture. It may be that the use of the golden ratio makes it easier to manufacture and build such a building. It may be as simple as the fact that the golden ratio is found so readily using basic construction tools such as a compass and straightedge. Whatever the case, the golden ratio, although quite mathematical in its computation, is very prevalent in works of art and architecture.

For more information on the golden ratio in art and architecture, please visit the following sites:

http://www.goldenmuseum.com/0805Painting_engl.html.

http://www.maths.surrey.ac.uk/hosted-sites/R.Knott/Fibonacci/fibInArt.html

Skill Check Answers

1. $F_{24} = 46,368; F_{25} = 75,025$

    $\frac{F_{25}}{F_{24}} = \frac{75,025}{46,368} \approx 1.618034$

2. Answers will vary.

11.1   Exercises

Find the next term of each sequence.

1. $4, 7, 10, 13, 16, \ldots$

2. $1, 7, 17, 31, 49, \ldots$

3. $3, 4, 7, 11, 18, \ldots$

4. $3, 5, 9, 17, 33, \ldots$

Use the Fibonacci sequence to evaluate each ratio. Recall that F_n represents the nth Fibonacci number. Round your answer to the nearest thousandth.

5.   Find $\dfrac{F_{12}}{F_{11}}$.

6.   Find $\dfrac{F_{15}}{F_{16}}$ and compare the result with $\dfrac{1}{\phi}$.

7.   Using the sequence, $3, 4, 7, 11, 18, \ldots$, find the ratio of the 11th and 10th terms and compare the result with the ratio of the 11th and 10th terms of the Fibonacci sequence.

8.   Determine if F_{3n} is an even number. Explain your reasoning.

Use the figure to answer each problem.

$EH = 12.94$ cm

$EA = 8$ cm

$BC = 3.06$ cm

$EC = 4.94$ cm

$GH = 8$ cm

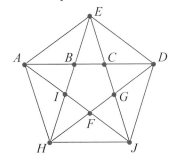

9.   Confirm that the ratio $\frac{EH}{EA} \approx \phi$.

10.   Confirm that $\frac{EC}{BC} \approx \phi$.

11.   Confirm that $\frac{EH}{EA} = \frac{EH + EA}{EH}$.

12.   What is the sum of $EC + BC$ and how does it relate to EA?

Solve each problem.

13. Confirm that $\left(\frac{1}{\phi}\right)^2 = 1 - \frac{1}{\phi}$. (**Hint:** Use the radical form of ϕ.)

14. Recall from Chapter 6 that we used the Pythagorean Theorem to illustrate the relationship between the three sides of a right triangle. Use the Pythagorean Theorem to show that the triangle given is a right triangle where $\phi = \frac{1+\sqrt{5}}{2}$.

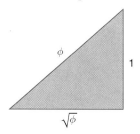

15. Suppose you want to buy a rectangular window whose sides have a proportion of the golden ratio and the width is less than the height. If the window is to be 12 feet tall, approximately how wide should the window be? Round your answer to the nearest tenth and use $\phi = 1.618$.

16. Given one side of a golden rectangle is 3.7 inches, find the possible lengths of the other side. Notice that the other side can be either longer or shorter than 3.7 inches. Round your answers to the nearest tenth and use $\phi = 1.618$.

17. Given one side of a golden rectangle is 12.6 miles, find the possible lengths of the other side. Notice that the other side can be either longer or shorter than 12.6 miles. Round your answers to the nearest tenth and use $\phi = 1.618$.

18. As an artist, you are commissioned to create a sculpture out of wire that is to be in the shape of a golden rectangle. If the total amount of wire available for the perimeter is 246 feet in length, what should the dimensions of the sculpture be? Round your answer to the nearest thousandth and use $\phi = 1.618$.

19. Show that the given parallelogram is golden.

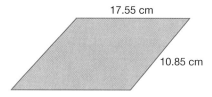

20. Determine if the following shapes are golden.

a.

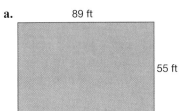

89 ft

55 ft

b.

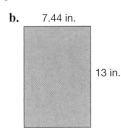

7.44 in.

13 in.

21. Create a piece of art featuring at least two instances where the golden ratio is used in the creating of the piece.

22. Given the following figure and measurements: $GK = 8.51$, $KL = 3.25$, $LD = 1.71$, $GL = 5.26$, and $EL = 2.76$, show all golden relationships for the labeled components.

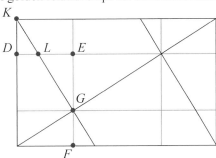

11.2   Tiling and Tessellations

There are several numerical sequences and number patterns that are used in art and architecture, such as triangular and square numbers. Once we develop an understanding of these basic notions, we will transition into how number theory is used to create works of art within a given architecture.

We will begin with a definition of **tiling**, or **tessellation**. We say that a tessellation is simply the mathematical idea of tiling a plane surface with a repeated pattern. Recall from the earlier chapter on geometry that a plane is defined as a two-dimensional surface that extends in each direction indefinitely. Planes are most often represented in an algebra course as the xy-coordinate plane. This, however, is only one representation of a plane. Consider the floor of a room, or the wall, or even the ceiling. Each of these is an example of a planar surface.

Triangular Numbers

Triangular numbers refer to numbers such as 1, 3, 6, 10, 15, . . . , where the number represents the number of objects that can be arranged in a *triangular manner*. Let's look at a few pictures of what triangular numbers look like when they are represented by a series of dots.

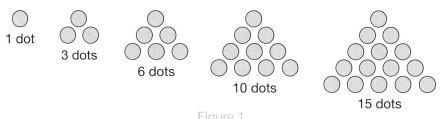

Figure 1

The pattern of triangular numbers is defined as the sum of the dots on successive rows that create equilateral triangles. We can represent each triangular number as the sum of integers as follows.

t_1: $1 = 1$

t_2: $3 = 1 + 2$

t_3: $6 = 1 + 2 + 3$

t_4: $10 = 1 + 2 + 3 + 4$

If we continue this pattern, then each successive term will be represented by the sum of the positive integers up to, and including, the index of the triangular number. In other words, the 5th triangular number t_5 can be found by adding the first five positive integers $(1 + 2 + 3 + 4 + 5)$ to get a sum of 15. So, $t_5 = 15$. Similarly, if we wanted to know what the 27th triangular number was, we simply need to add the first 27 positive integers.

It should be noted that this pattern continues indefinitely as long as the number of objects continue to form a triangle. Notice that the number 10 was identified as one of our triangular numbers. A popular usage of this triangular number is the game of bowling. The pins on a bowling lane are arranged in a *triangle* and there are 10 pins.

☞ Helpful Hint

Gauss's shortcut also works if there is an odd number of integers.

Find the 27th triangular number.

Solution

Recall from Chapter 1 that we discussed an example of how to find the sum of the first n

integers and discussed how Carl Gauss found this solution by considering the pairs of sums. Recall that to find the sum of, for example, the first 10 integers

$$1 + 2 + 3 + 4 + 5 + 6 + 7 + 8 + 9 + 10,$$

we can use the "shortcut." Instead of trying to add all 10 numbers, Gauss figured out that in any consecutive set of integers, it is possible to pair up the numbers to find the sums and, in effect, cut the work in half. For instance, $10 + 1 = 11$, $9 + 2 = 11$, $8 + 3 = 11$, and so on.

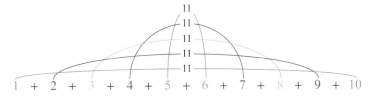

In this case, there are 5 pairs of 11. This means the sum of the numbers from 1 to 10 would be $5 \cdot 11 = 55$. We can rewrite this equation to get a general form of $\frac{10}{2} \cdot 11 = 55$.

Returning to our example, this means the 27th triangular number is then

$$\frac{27(27+1)}{2} = 378.$$

Formula to Find the nth Triangular Number

To find the nth triangular number t_n, use the following formula.

$$t_n = \frac{n(n+1)}{2}$$

Skill Check #1

Find the 12th triangular number.

Square Numbers

Similar to triangular numbers, square numbers can also be represented by a square. One of the many attributes of square numbers is their relation to triangular numbers. We begin our discussion of square numbers by trying to develop a formula for determining the nth square number. Consider the pattern of the first five square numbers in Figure 2.

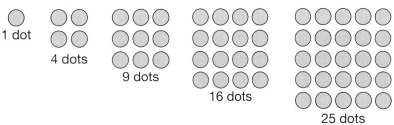

Figure 2

Notice that the number of dots inside each large square corresponds to the length of one side squared. So, square numbers are represented as 1, 4, 9, 16, 25, 36, . . . , where each term s_n is equal to n^2.

Formula to Find the nth Square Number

To find the nth square number s_n, use the following formula.

$$s_n = n^2$$

The relationship between triangular numbers and square numbers may not be obvious, so let's see what happens when we look at consecutive triangular numbers. Consider the first two triangular numbers, $t_1 = 1$ and $t_2 = 3$. The sum of these two triangular numbers is, of course, 4. But, moreover, the sum is a perfect square, $2^2 = 4$. This pattern actually continues for every pair of consecutive triangular numbers—their sum is a perfect square. Moreover, this should be the case, since two equilateral triangles of the same size geometrically form a square. Consider the pattern created in Figure 3.

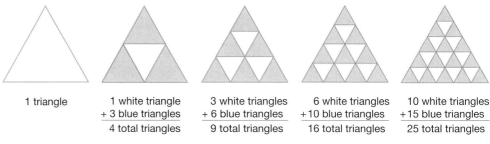

| 1 triangle | 1 white triangle + 3 blue triangles 4 total triangles | 3 white triangles + 6 blue triangles 9 total triangles | 6 white triangles +10 blue triangles 16 total triangles | 10 white triangles +15 blue triangles 25 total triangles |

Figure 3

The pattern represented can be illustrated as the relationship between square and triangular numbers. Recall that a square number is represented by the sum of consecutive triangular numbers, so algebraically, we have the following.

$$s_n = t_{n-1} + t_n = \frac{(n-1)n}{2} + \frac{n(n+1)}{2} = n^2$$

The relationship between triangular and square numbers can be further verified by considering that the number of blue triangles represents the nth triangular number, while the number of white triangles represents the $(n-1)^{th}$ triangular number. If we continue this pattern of understanding, we can see that the total number of triangles present in any stage is equal to the square number. So, in the 4th triangle, there are $4^2 = 16$ total triangles.

Example 2: Representing Square Numbers

How many total triangles are there in the following triangular number representation?

Solution

Since there are six blue triangles along the base, this triangle represents the 6th square number, and the total number of triangles present is $6^2 = 36$ triangles.

The purpose in discussing both triangular and square numbers in the context of art and architecture is to develop an understanding of how polygons are created mathematically in the process of designing a piece of artwork or building. For instance, consider the tile floor of a bathroom or kitchen in a house, as in Figure 4. The artistic pattern in which the tile is laid out uses triangular and square numbers of tiles to cover the floor so that mathematics is a necessary component and creates an artistic feature of the house.

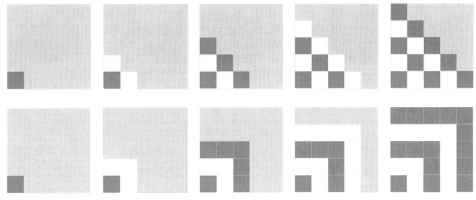

Figure 4

Tiling

We mentioned earlier in the section that tiling a geometric plane was similar to covering the space with geometric objects. Tiling is the process of covering a plane such as the floor, wall, or ceiling of a house with "tiles" that may be ceramic, clay, granite, etc. We could also be talking about the pattern that appears in wallpaper. In general, we say that a tiling or tessellation is any repeated pattern that covers a plane. For example, many floor tiles are in the shape of squares. Placing the squares next to each other, they form a pattern where the corners fit together so that there are no spaces left. This type of tiling is called a **regular tessellation** since the shapes used were regular polygons. Examples of regular tessellations can be seen in Figure 5.

 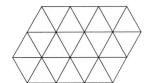 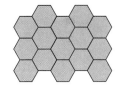

Figure 5

The process of tiling or tessellating a plane hinges on the key fact that the sum of the interior angles of the geometric figure at a common vertex must be 360°. In the case of a square, each vertex angle has a measure of 90°. Thus, when 4 squares come together at one point, the sum is $4 \cdot 90 = 360°$. This also means that the repetition of the pattern will completely cover a surface without any gaps.

These basic notions of tessellations allow us to develop more complex patterns of tiling and artistic creations used in covering a planar surface. Establishing an understanding of tessellations is only the beginning to more advanced concepts. In order to extend our ability to tile a plane, we need to have a deeper understanding of geometric transformations. A transformation in a given plane can be classified as a translation, a rotation, or a reflection.

Translation

A **translation** is a motion where an object is moved, or "translated," in a sliding manner vertically, horizontally, or diagonally. The shape and size of the object are not affected by the translation.

Rotation

A **rotation** of an object occurs when the object is "turned" about a center point. The amount of rotation is determined by a particular angle. The shape and size of the object are not affected by the rotation.

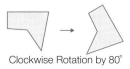

Clockwise Rotation by 80°

Reflection

A **reflection** of an object occurs when a mirror image of the object is reflected about some line called a line of reflection. Once again, the shape and size of the object are not affected by the reflection.

Rigid Motions

Each of the motions discussed above can be further classified as **rigid motions**. This simply means that the motion itself does not distort the object's size or shape.

Example 3: Rigid Motions

In order to create a perspective drawing, Troy wishes to translate the following figure to create a 3-dimensional drawing. Translate the figure 10 units to the right and 2 units up.

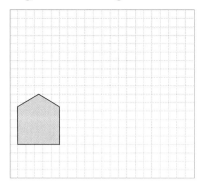

Solution

To complete the translation, we need to translate each point in the figure 10 units to the right and 2 units up. When done, the translation will be in the location of the green figure.

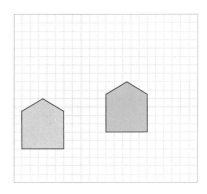

To finish the 3-dimensional figure, we only need to add line segments that connect the corresponding vertices of the figures, as follows.

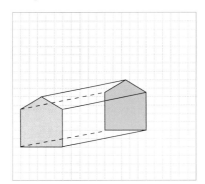

Although each transformation is unique in how it moves a geometric figure in a given plane, a combination of all three may be needed to properly tile a plane. Notice in Figure 6 that the surfaces, or planes, that have been "tiled" are a floor and wall. In the floor tiling, hexagons are used to cover the space. In the wall tiling, rectangles are used to cover the space. As we noted earlier, geometric figures that have a common vertex where the sum of the vertex angles adds to 360° will tessellate a plane. In our examples, three hexagons (each having a vertex angle measure of 120°) intersect at a common vertex where the sum of the angles is 360° and offset rectangles (each having a vertex angle measure of 90°) tessellate the planes. What is most important is to realize in each case is that the notion of translating the shapes is what creates the "pattern" that then becomes art. It is not hard to imagine how the black hexagons in the tiling are translated diagonally, horizontally, and vertically.

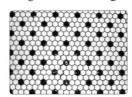 

Figure 6

While Figure 6 contains two types of tilings completed using ceramic tile, the same concepts also hold true for all tilings, including wallpaper.

We Will Discuss Only Two Types of Tiling and Their Relation to Art:

1. Regular tilings, including Archimedean tilings

2. Irregular tilings

Regular tilings are simply defined as tilings where the tiles are all the same and are made up of polygons such as equilateral triangles, squares, regular pentagons, regular hexagons, etc. An example of a regular tiling is the image on the left in Figure 7. **Archimedean tilings**, or semiregular

11

tilings, are a type of regular tiling composed of multiple regular polygons, but not necessarily the same regular polygon, to form a planar surface. There are only 12 of these tilings possible and they are named after the great Greek mathematician Archimedes. An example of an Archimedean tiling is the image on the right in Figure 7.

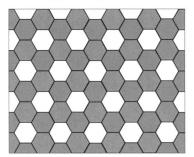 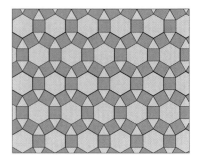

Figure 7: Regular and Archimedean Tilings

Regular Tilings

Regular tilings are tilings where the tiles are all the same and are made up of regular polygons such as equilateral triangles, squares, regular hexagons, etc.

Archimedean Tilings

Archimedean tilings are regular tilings that are composed of multiple regular polygons, but not necessarily the same regular polygon, to form a planar surface.

In Chapter 6, we learned that the sum of the measures of the interior angles of a regular polygon could be determined by the formula $(n - 2) \cdot 180°$, where n represents the number of sides of the polygon. Using this formula, we can determine that the measure of a single angle can be determined by the formula $180° - \frac{360°}{n}$. Archimedes determined the patterns that could be used to tile by considering k polygons with $n_1, n_2, n_3, \ldots, n_k$ sides, respectively, that meet to form a 360° angle and assuring that the pattern will tessellate the given plane. Recall that the most important thing in determining whether or not polygons will tessellate is that the sum of the angles meeting at the vertices is 360°. Using this knowledge, Archimedes found that if an unknown number of polygons, say k polygons, with the number of sides equal to n_1, n_2, and so on, meeting at a single vertex, the sum of the angles must satisfy the equation

$$\left(\frac{1}{2} - \frac{1}{n_1}\right) + \cdots + \left(\frac{1}{2} - \frac{1}{n_k}\right) = 1,$$

and thus,

$$\frac{k}{2} - \frac{1}{n_1} - \cdots - \frac{1}{n_k} = 1.$$

Finally,

$$\frac{1}{n_1} + \cdots + \frac{1}{n_k} = \frac{k}{2} - 1.$$

This last result indicates that, in order for us to have an Archimedean tiling, we need k whole numbers whose reciprocals add up to one less than $\frac{k}{2}$.

Example 4: Archimedean Tiling

Show that the given tiling is an Archimedean tiling.

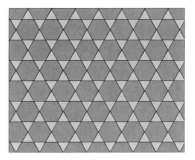

Solution

If we take a closer look at the intersection of polygons at any vertex, we can see that a given vertex consists of 4 polygons: 2 regular hexagons and 2 equilateral triangles.

So we have $k = 4$, $n_1 = 3$, $n_2 = 3$, $n_3 = 6$, and $n_4 = 6$. Now we need to make sure the formula is satisfied using these values.

$$\frac{1}{n_1} + \cdots + \frac{1}{n_k} = \frac{k}{2} - 1$$

$$\frac{1}{3} + \frac{1}{3} + \frac{1}{6} + \frac{1}{6} = \frac{4}{2} - 1$$

$$\frac{2}{3} + \frac{2}{6} = 2 - 1$$

$$\frac{2}{3} + \frac{1}{3} = 1$$

$$1 = 1$$

Since the equation is true, it is confirmed that two equilateral triangles in combination with two regular hexagons, does in fact tessellate the plane.

The second type of tiling is called an **irregular tiling**.

Irregular Tilings

Irregular, or **nonregular**, **tilings** are tilings in which there are no restrictions on the polygons used to tessellate, other than the requirement that the sum of the angles at meeting vertices must equal 360°. It should be noted that, since there are no restrictions on the polygons in terms of number of sides and interior angle measure, there are an infinite number of such tessellations.

MATH MILESTONE

M.C. Escher (1898–1972) was a Dutch graphic artist famous for his often mathematically-inspired works of art, which feature impossible constructions, as well as explorations of infinity, architecture, and tessellations. For examples of works by M.C. Escher, go to http://www.mcescher.com.

Popularized by the famous graphic artist M.C. Escher, irregular tilings take on an artistic look right away. In his etchings, Escher started with a regular polygon, such as a square, and cut out a portion of the square. Once the portion was removed from the square, Escher used the concepts of rotation and reflection to "move" the irregular piece of the square somewhere else. Since the square tessellates the plane, the irregular shape created from a square will also tessellate the plane.

We already know that squares, rectangles, equilateral triangles, and hexagons will tessellate a plane. Escher understood well that rotating a polygon can make a new, irregular shape which will also tessellate the plane. The following is an example of how to create an irregular tessellation out of a square.

Begin with a square.

Figure 8

Choose a section of the square and translate it to the opposite side.

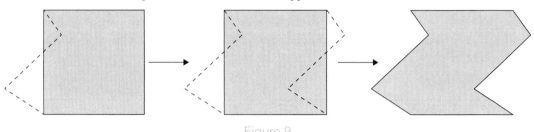

Figure 9

Choose a section of the bottom of the figure and translate it to the top of the figure. After completing the translations, a new irregular figure is formed.

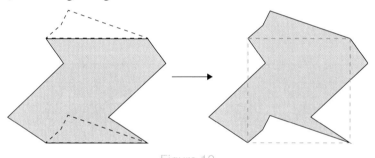

Figure 10

Now, we can create a tessellation of the plane by translating our figure in every direction.

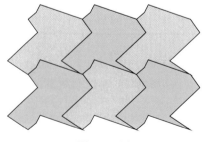

Figure 11

It should be noted that the same process for tessellating using a translation can be done using rotations. In other words, if we move a section of the original square by rotating it around a point instead of translating, we would still obtain an irregular figure that will tessellate the plane. Figure 12 demonstrates how rotations can used to create such a tessellation.

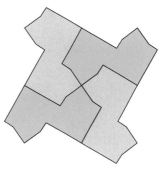

Figure 12

Skill Check #2

Using a square piece of paper, create an irregular tessellation that involves translations. Repeat the process and complete a tessellation that involves rotations.

As we can see from the previous discussion, the use of irregular tessellations allow for the artistic application of geometric figures on a planar surface. Thus, all flooring tile patterns, wallpaper patterns, and mosaic art pieces are based on polygons and the use of geometric transformations.

Skill Check Answers

1. 78

2. Answers will vary.

11.2   Exercises

Use the figure of triangular and square numbers to solve each problem.

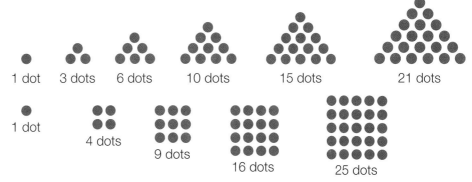

1. Find the next three triangular numbers that follow 21.

2. Find the next three square numbers that follow 25.

3. Is 66 a triangular number?

Use the figure to answer each question.

1 dot
3 dots
6 dots
9 dots
12 dots

4.   Using the pattern, how many dots would the 6th triangle have?

5.   Using the figure, can you find a formula for the nth triangle?

6.   How many dots would be in the 50th triangle?

Solve each problem.

7.   Rotate the figure clockwise by 90°.

8.   Translate the given figure 5 units to the left and 4 units down.

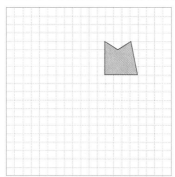

9.   Reflect the given figure about the line *l*.

10.   Is it possible to have a regular tessellation using only pentagons? Explain your answer.

11.   Is it possible to have an irregular tessellation using only rhombuses? Explain your answer.

12.   Is it possible to have a semiregular tessellation using 2 pentagons and 4 decagons?

13. Can you tessellate the plane with the following figure? Explain your answer.

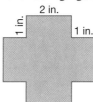

14. Quadrilateral *ABCD* has vertices on the coordinate plane at $A(-2, 1)$, $B(1, 3)$, $C(2, -1)$, and $D(-1, -2)$. Graph *ABCD* and reflect across the *x*-axis.

15. Triangle *DEF* has vertices $D(2, 2)$, $E(5, 3)$, and $F(7, 1)$. Draw the image of triangle *DEF* under a rotation of 90° clockwise about the origin.

16. Use the given figure to create a tessellation by rotating, reflecting, or translating. Describe which transformations you used to create the tessellation.

17. Using the given figure to create your own irregular tiling.

11.3 Mathematics and Music

Now, we turn our attention to performance art, specifically that of music. For centuries, music has been a medium that has allowed man to express the breadth of the human experience. Musical sound can evoke a wide range of emotions whether it is created vocally, or with a stringed instrument such as a violin or harp, with drums or other any other instruments.

Frequency and Exponential Functions

From experience, we know that different musical instruments have different sounds. For instance, a trumpet sounds drastically different than a flute, which sounds different from a harp. No matter the source of the sound, they all travel in waves of energy much like the waves of an ocean.

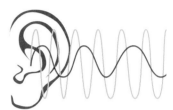

Figure 1

The **frequency** of a sound is the number of times a wave of sound energy completes a cycle in one second. We measure frequency in terms of cycles per seconds, or hertz (Hz). Frequency has an audible range of 16 Hz to 16,000 Hz. Sounds with a high pitch have higher frequencies and sounds with a low pitch have lower frequencies. **Pitch** is the tonal quality that is defined by how low or high an instrument sounds to our ear.

> ### Frequency
>
> The **frequency** of a sound is the number of energy wave cycles completed in one second and is measured in hertz (Hz).
>
> ### Pitch
>
> **Pitch** is the tonal quality that is defined by how low or high an instrument sounds to our ear.

We will explore the concept of frequency by looking at the keyboard arrangement of a piano. When a key is pressed on the piano, the sound that plays is the result of a small felt-covered hammer striking a string. The frequency of each sound, in turn, is used to arrange the strings of the piano in such a way that the notes progress from lower to higher pitches as you move up the keyboard. For instance, the notes on the left end of a piano keyboard have a very low pitch while the notes on the right end have a much higher pitch. Although frequency is objective and measurable, pitch is more subjective and might vary based on the individual listening. That is, two individuals might differ in the perceived pitch of a note, but the frequency will be exactly the same.

Fun Fact

Pythagoras and his group of Pythagoreans were the first to understand that the length of a string on a stringed musical instrument determines the pitch of the sound made when that particular string is plucked. The shorter the string, the higher the pitch.

Figure 2 shows a portion of a piano keyboard that will give us a basis for how musical notes are represented. The white keys represent the notes A, B, C, D, E, F, and G. These notes are repeated over the entire keyboard. To the left of any key, the pitch of the notes get lower and they have a lower frequency. To the right of any key, the pitch of the notes get higher and they have a higher frequency.

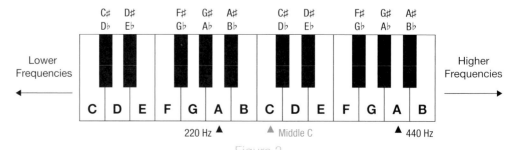

Figure 2

As Figure 2 indicates, a label is made for the note called middle C. This note gets its name from being the 4ᵗʰ C of eight on the piano; that is, it is the C located in the middle of the piano. Whether on a piano or on another instrument, the distance from one note to the next closest note is called a **half step**. Using Figure 2, if we start with middle C, a half step up is C♯ and a half step down is B.

Half steps will help us identify musical terms such as sharp, flat, octave, and chord. A sharp, represented by the symbol ♯, refers to the note that is one half step higher. For instance, C♯, read "C sharp," refers to the black key directly to the right of C on a piano. On other instruments, sharp still refers to a half step higher, but, for instance, it may be the next fret on a guitar.

Where sharps represent a half step higher, flats refer to a half step lower. Flats are represented by the symbol ♭. D♭ refers to the note one half step lower than D. Notice in the piano figure that both C♯ and D♭ refer to the same note on the keyboard.

We use this smallest measurement of distance, a half step, to also define larger distances on the piano. For instance, a **whole step** is defined as two half steps, while an **octave** covers 12 half steps.

Half Step

A **half step** is the distance between one note and the next nearest note on a piano.

Whole Step

A **whole step** is defined as the interval between two half steps on a piano.

Octave

An **octave** is the interval of notes between 12 half steps on a piano.

Skill Check #1

Starting at F, name the notes which are one half step up the piano and one half step down the piano.

Example 1: Determining Half Steps, Whole Steps, and Octaves on a Keyboard

Use the keyboard to help determine each of the following.

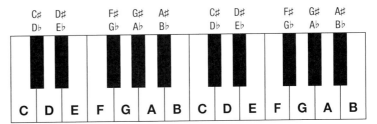

a. What note is one half step up from G?

b. What note is one whole step down from G♭?

c. What note is one octave up from G?

Solution

   a. If we start at G, one half step up the keyboard is G♯, also known as A♭.

   b. Beginning with G♭, to find one whole step down, we need to count 2 half steps. So, a whole step down from G♭ is the note E.

   c. Beginning at G, we need to count up 12 half steps to the next octave. Counting this, we see that another G is located an octave up.

As you probably noticed in part **c.** of Example 1, keys that are an octave apart have the same name. In fact, they sound alike. Try listening for yourself on a piano. If you don't have access to a piano, you can use the virtual piano on the website http://www.virtualpiano.net/. Take a moment and listen to notes that are an octave apart. For example, play different Cs on the piano and notice how they sound similar. In an octave, the highest note always has twice the frequency of the lowest note. So, as you play Cs up the piano, the frequency of the note is doubling each time. This is where the idea of pitch comes in. No matter where you play on the piano, an A is an A, is an A. This is true for every instrument. Even a person with a very low voice can sing in pitch with a person who has a high voice by singing in different octaves, but the notes they sing always have frequencies that are powers of 2 apart.

Fixed frequencies are used when tuning musical instruments. For instance, on a piano, the note of A—we will use the notation A_4—above middle C has a frequency of 440 Hz while the note of A_3 below middle C has a frequency of $440 \div 2 = 220$ Hz. In fact, when an orchestra begins to tune up before a concert, they tune based on a note played by the oboe, which is 440 Hz. Figure 3 shows this octave on a piano.

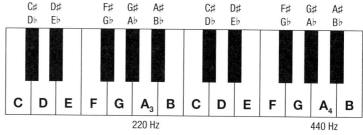

Figure 3

Example 2: Determining Frequencies of Octaves

What is the frequency of A_1 given that A_3 has a frequency of 220 Hz?

Solution

We know A_3 has a frequency of 220 Hz. We can use this information along with the fact that the frequency of a note is twice the frequency of the note an octave below it to find the frequency of A_2. Thus, A_2 has a frequency that is $\frac{1}{2}$ the frequency of A_3, or 110 Hz. Then, to find the frequency of A_1 from the frequency of A_2, we need to find $\frac{1}{2}$ of 110 Hz, since A_1 is one octave below A_2. Therefore, we arrive at our solution for the frequency of A_1, which is $\frac{1}{2}$ of 110, or 55 Hz.

As we've established, the transition over the 12 notes in an octave doubles the frequency as the scale moves from a lower frequency to a higher frequency. So how much of an increase is there between consecutive notes on a piano? We can determine exactly how much the frequency increases each time by the following. Let k be the amount of increase between each half step. Over the octave, there are 12 half steps, so we know that the following must be true.

$$k \cdot k \cdot k \cdot k \cdot k \cdot k \cdot k \cdot k \cdot k \cdot k \cdot k \cdot k = k^{12} = 2$$

Since $k^{12} = 2$, we use a calculator to find that

$$k = \sqrt[12]{2} \approx 1.059463.$$

Thus, the transition between notes on a piano differs by a factor of approximately 1.059463.

Therefore, if we start with A_3, which has a frequency of 220 Hz, we can determine the individual frequency for each note in the remainder of the octave leading up to A_4. Table 1 illustrates the frequency of each of the notes in this given octave. Notice that multiplying the frequency of the previous note by the factor of $\sqrt[12]{2} \approx 1.059463$ gives the frequency of the next higher note. Also notice that this progression means that the frequency of A_4, the next octave, is twice as large as the frequency for A_3.

Table 1	
Frequency of Notes from A_3 to A_4	
Note	**Frequency**
A_3	220
A♯	$220 \cdot \sqrt[12]{2} \approx 233.082$
B	$233.082 \cdot \sqrt[12]{2} \approx 246.942$
C	$246.942 \cdot \sqrt[12]{2} \approx 261.626$
C♯	$261.626 \cdot \sqrt[12]{2} \approx 277.183$
D	$277.183 \cdot \sqrt[12]{2} \approx 293.665$
D♯	$293.665 \cdot \sqrt[12]{2} \approx 311.127$
E	$311.127 \cdot \sqrt[12]{2} \approx 329.628$
F	$329.628 \cdot \sqrt[12]{2} \approx 349.229$
F♯	$349.229 \cdot \sqrt[12]{2} \approx 369.995$
G	$369.995 \cdot \sqrt[12]{2} \approx 391.996$
G♯	$391.996 \cdot \sqrt[12]{2} \approx 415.305$
A_4	$415.305 \cdot \sqrt[12]{2} \approx 440.000$

Recall from Chapter 5 that when a function grows in a manner where the relative rate of growth remains fixed while the units increase, we say that the function is growing exponentially. In this case, the base, or relative rate of growth, is $k \approx 1.059463$. This means that we can actually develop a function that represents the frequency F of the half steps x where F_0 is the reference frequency as follows.

Frequency of Notes on a Piano

$$F = F_0 \cdot 1.059463^x$$

where F_0 is the reference frequency and x is the number of half steps up from F_0.

Example 3: Musical Notes as Exponential Functions

Using the values from Table 1, find the frequency of the note that is five half steps above A_4.

Solution

Since A_4 is our reference frequency and we know that the frequency of $A_4 = 440$ Hz, we can use our formula to determine the frequency of the note five half steps above A_4.

$$F = F_0 \cdot 1.059463^x$$
$$F_0 = 440$$
$$x = 5$$
$$F = (440)(1.059463^5)$$
$$\approx 587 \text{ Hz}$$

So, the frequency of the note five half steps higher than A_4 is approximately 587 Hz. For reference, we call this note D_4. It should be noted that if the frequency of a note you wish to find is lower than the reference note, then the number of half steps x will be negative.

Skill Check #2

Find the frequency of the note that is 15 half steps below A_3.

Chords

We all know that a combination of notes must be used to actually compose a musical piece. However, the blending of the notes in such a way that they sound harmonious is the artistic component. Music notes are blended through the use of **chords**.

Chord

A **chord** is a set of two or more notes that produce one harmonious sound when played together.

In music, there are major chords and minor chords. Both major and minor chords consist of three notes played at the same time. We can use half steps to help us find these chords on a piano. For a major chord, the notes are spaced such that there are four half steps between the 1st and 2nd notes and three half steps between the 2nd and 3rd notes. No matter which note you start on—whether a white key or a black one—the spacing will always be the same. For instance, C major chord consists of the notes C, E, and G. Use the keyboard in Figure 4 to verify that there are four half steps between C and E and three half steps between E and G.

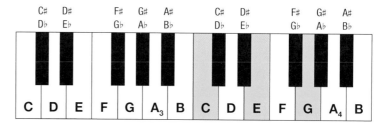

Figure 4

Use the virtual piano or a real piano to play the notes in C major. Notice how the sound is a "happy" or "sunny" sound. This is a feature of major chords. Chords that sound more "sad," or even "evil," are minor chords. For a minor chord, there are still seven half steps between the first note and the third note in the chord. However, the number of half steps is the opposite of a major chord. There are three half steps between the first two notes and four half steps between the last two notes. Essentially, the only difference is the middle note. For instance, a C minor chord consists of the notes C, D♯, and G. Figure 5 shows C minor.

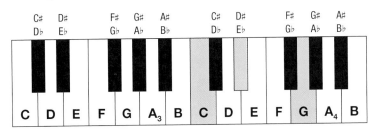

Figure 5

Example 4: Finding Major and Minor Chords on a Piano

Use the keyboard shown in Figure 5 to find the following chords.

a. F♯ major

b. A minor

Solution

a. To find F♯ major, we start with F♯ and count up four half steps to A♯, which gives us the middle note of the chord. Then, we count up another three half steps from there to reach the last note in the chord, C♯. Therefore, the chord F♯ major on a piano consists of the keys F♯, A♯, and C♯.

b. To find A minor on the keyboard, we start with the key A and count up three half steps to the middle note of the chord, which is C. Then, another four half steps give us the last note in the chord E. Thus, the chord A minor on a piano consists of the keys A, C, and E.

Another important musical harmony linked to the chord is called the **perfect fifth**. It consists of the top and bottom notes in either a major or minor chord. In other words, starting with one note and counting up seven half steps (the length of a major or minor chord) completes a perfect 5th. Just as the notes in an octave have a certain frequency ratio, notes that make up a perfect 5th also have a particular ratio of frequencies. For a perfect fifth, the ratio is 3:2. In other words, the top note vibrates three times in the same time it takes the bottom note to vibrate twice. Figure 6 shows a perfect fifth starting with D♯.

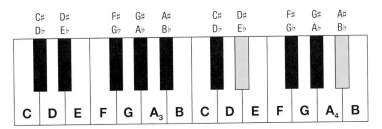

Figure 6

Fun Fact

The Museum of Mathematics—MoMath—opened its doors in December 2012 as the United States' first and only museum devoted completely to mathematics! It hopes to reveal the wonders of math through dynamic hands-on exhibits.

Example 5: Finding Frequency Ratios of Perfect Fifths

Show that the frequencies of the notes in the perfect fifth given in Figure 6 have a ratio of $\frac{3}{2}$. Use the approximate frequencies D♯ = 311.127 Hz and A♯ = 466.164 Hz.

Solution

To find the ratio of frequencies for A♯ and D♯, we need to divide the given values.

$$\frac{A\#}{D\#} = \frac{466.164}{311.127} \approx 1.49830777 \approx 1.50$$

Because we started with approximate frequencies, the ratio is approximate as well. However, you can see that the notes in the perfect fifth do have a frequency ratio of $\frac{3}{2}$.

Skill Check Answers

1. One half step up is F♯ and one half step down is E.

2. 92.499 Hz

11.3 Exercises

Answer each question.

1.   How many half steps are there between the A and G♯ on a keyboard?

2.   What note is one whole step down from B♭?

3.   If Eileen plays a note two octaves up from E, what note will it be?

Solve each problem. Round your answer to the nearest thousandth, when necessary.

4.   If a note has a frequency of 27.5 Hz, find the frequencies of the notes one, two, three, four, and five octaves higher.

5.   If a note has a frequency of 4176 Hz, how many octaves higher is the note than middle C (assume middle C has a frequency of 261 Hz)?

6.   If a note has a frequency of 220 Hz, find the frequencies of the notes one, two, and three octaves higher.

7.   If A_2 has a frequency of 110 Hz, find the frequencies of each note in the remainder of the octave leading up to A_3.

8. If A_5 has a frequency of 880 Hz, find the frequencies of each note in the remainder of the octave leading down to A_5.

9. A note on a piano has a frequency of 880 Hz. Find each of the following note frequencies.

   a. 10 half steps above 880 Hz

   b. 10 half steps below 880 Hz

   c. An octave higher than 880 Hz

   d. An octave lower than 880 Hz

   e. 15 half steps above 880 Hz

10. The note middle C on a piano has a frequency of approximately 261 Hz. Find each of the following note frequencies.

    a. 10 half steps above middle C

    b. 10 half steps below middle C

    c. An octave higher than middle C

    d. An octave lower than middle C

    e. 15 half steps above middle C

11. What is the frequency of the note that is nine half steps below middle C (assume middle C has a frequency of 261 Hz)?

12. What is the frequency of the note that is nine half steps above middle C (assume middle C has a frequency of 261 Hz)?

13. Where will Caitlin place her fingers to play a B♭ major chord?

14. Where will Aiden place his fingers to play a D minor chord?

15. Recall that a perfect fifth is represented by seven half steps (C → G).

    a. If C has a frequency of 261 Hz, what is the frequency of G?

    b. What are the frequencies of each note leading up to the fifth above middle C?

    c. What is the ratio of the frequencies of the top and bottom notes in a perfect fifth?

16. A major sixth is represented by nine half steps (A_4 → F♯).

    a. If A_4 has a frequency of 440 Hz, what is the frequency of F♯?

    b. What are the frequencies of each note leading up to the sixth above A_4?

    c. What is the ratio of the frequencies of the top and bottom notes in a major sixth?

17. Tristan wants to find the note that will complete a perfect fifth with F♯ on a piano. What is the note?

Chapter 11   Summary

Section 11.1   Applications of Geometry to the Arts

Definitions

Art

Art is considered as any endeavor that requires creativity of the mind, whether it is a painting, a sculpture, architecture, a musical composition, etc.

Fibonacci Sequence

The Fibonacci sequence is the sequence of numbers 1, 1, 2, 3, 5, 8, 13, 21, 34, 55, 89, 144, . . . that continues infinitely, where each successive number can be found by adding together the previous two numbers in the sequence.

Phi (ϕ)

The number ϕ, also known as the golden ratio, is defined by the following mathematical representation.

$$\phi = \frac{1+\sqrt{5}}{2} = 1.618033988749894\ldots$$

Golden Rectangle

A golden rectangle is a rectangle in which the ratio of the length to the width is the golden ratio.

Golden Triangle

A golden triangle is defined as an isosceles triangle whose length of a side compared to the length of the base is the golden ratio.

Section 11.2   Tiling and Tessellations

Definitions

Tessellation

A tessellation, or tiling, is the covering of a plane with a repeated geometric pattern with no gaps or overlaps.

Translation

A translation is a motion where an object is moved, or "translated," in a sliding manner vertically, horizontally, or diagonally. The shape and size of the object are not affected by the translation.

Rotation

A rotation of an object occurs when the object is "turned" about a center point. The amount of rotation is determined by a particular angle. The shape and size of the object are not affected by the rotation.

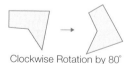

Clockwise Rotation by 80°

Reflection

A reflection of an object occurs when a mirror image of the object is reflected about some line called a line of reflection. Once again, the shape and size of the object are not affected by the reflection.

Rigid Motions

Translations, rotations, and reflections can be further classified as rigid motions. This simply means that the motion itself does not distort the object's size or shape.

Regular Tilings

Regular tilings are tilings where the tiles are all the same and are made up of polygons such as equilateral triangles, squares, regular pentagons, regular hexagons, etc.

Archimedean Tilings

Archimedean tilings are regular tilings that are composed of multiple regular polygons, but not necessarily the same regular polygon, to form a planar surface.

Irregular Tilings

Irregular, or nonregular, tilings are tilings in which there are no restrictions on the polygons used to tessellate, other than the requirement that the sum of the angles at meeting vertices must equal $360°$.

Formulas

Formula to Find the nth Triangular Number

To find the nth triangular number t_n, use the following formula.

$$t_n = \frac{n(n+1)}{2}$$

Formula to Find the nth Square Number

To find the nth square number s_n, use the following formula.

$$s_n = n^2$$

Section 11.3   Mathematics and Music

Definitions

Frequency

The frequency of a sound is the number of energy wave cycles completed in one second and is measured in hertz (Hz).

Pitch

Pitch is the tonal quality that is defined by how low or high an instrument sounds to our ear.

Half Step

A half step is the distance between one note and the next nearest note on a piano.

Whole Step

A whole step is defined as the interval between two half steps on a piano.

Octave

An octave is the interval of notes between 12 half steps on a piano.

Chord

A chord is a set of two or more notes that produce one harmonious sound when played together.

Formula

Frequency of Notes on a Piano

$F = F_0 \cdot 1.059483^x$, where F_0 is the reference frequency and x is the number of half steps up from F_0.

Chapter 11 Exercises

Find the next term of each sequence.

1. 2, 5, 7, 12, 19, . . .

2. 1, 10, 3, 12, 5, . . .

3. 2, 8, 27, 85, . . .

4. 2, 4, 6, 10, 16, . . .

Use the Fibonacci sequence to evaluate each ratio. Recall that F_n represents the nth Fibonacci number. Round your answer to the nearest thousandth.

5. Find F_{11} and $\dfrac{1}{F_{11}}$.

6. Find $\dfrac{F_{20}}{F_{19}}$ and compare the result with ϕ.

Solve each problem.

7. Using the sequence 2, 4, 6, 10, 16, . . . , find the ratio of the 13th and 14th terms. Determine if the ratio is comparable to ϕ.

8. True or False: F_{5n} is an odd number. Explain your answer.

Use the figure to solve each problem.

$AF = 3.29$ cm

$EA = 5.33$ cm

$AC = 8.62$ cm

$FG = 2.03$ cm

$EH = 5.33$ cm

$EG = 3.29$ cm

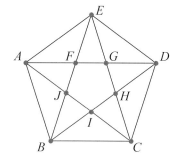

9. Confirm that the ratio $\frac{AC}{EA} \approx \phi$.

10. Confirm that $\frac{EG}{FG} \approx \phi$.

11. Confirm that $\frac{EA}{AC} = \frac{AC}{AC+EA}$.

12. What is the sum of $AF + EA$ and how does it relate to AC?

Solve each problem.

13. Suppose you want to buy a rectangular window that has a proportion of the golden ratio and the width is less than the height. If the window is to be 15 feet tall, approximately how wide should the window be? Round your answer to the nearest tenth and use $\phi = 1.618$.

14. Given a length of the golden rectangle to be 4.8 inches, find the possible lengths of the other side. Notice that the other side can be either longer or shorter than 4.8 inches. Round your answers to the nearest tenth and use $\phi = 1.618$.

15. Given a length of the golden rectangle to be 9.8 kilometers, find the possible lengths of the other side. Notice that the other side can be either longer or shorter than 9.8 kilometers. Round your answers to the nearest tenth and use $\phi = 1.618$.

16. As an artist, you are commissioned to create a sculpture out of wire that is to consist entirely of a golden rectangle. If the total amount of wire material available for the perimeter is 114 feet in length, what should the dimensions of the sculpture be? Round your answers to the nearest thousandth and use $\phi = 1.618$.

17. Show that the given parallelogram is golden.

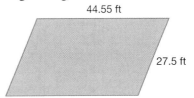

44.55 ft

27.5 ft

Use the figure to solve each problem.

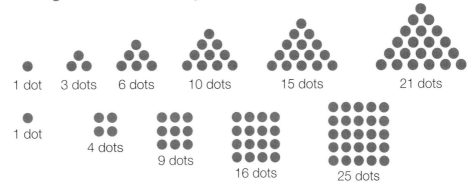

1 dot   3 dots   6 dots   10 dots   15 dots   21 dots

1 dot   4 dots   9 dots   16 dots   25 dots

18. Show that the triangular number 21 is the sum of the of numbers 1, 2, 3, . . ., 6.

19. Is 72 a triangular number?

20. What is the 30th triangular number?

21. What is the 30th square number?

Use the figure to answer each question.

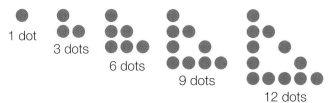

22. Using the pattern in the figure, how many dots would the 10th triangle have?

23. How many dots would be in the 30th triangle?

Solve each problem. Round your answer to the nearest thousandth, when necessary.

24. Write the square number 400 as the sum of two consecutive triangular numbers.

25. Write the square number 225 as the sum of two consecutive triangular numbers.

26. Is it possible to have a regular tessellation using only heptagons? Explain your answer.

27. Is it possible to have a regular tessellation using only hexagons? Explain your answer.

28. Is it possible to have a semiregular tessellation using two octagons and a square? Explain your answer.

29. Can you tessellate the plane with the following figure?

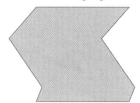

30. If a note has a frequency of 52 Hz, find the frequencies of the notes one, two, three, four, and five octaves higher.

31. What is the frequency of the note that is 17 half steps above middle C? Assume middle C has a frequency of 261 Hz.

32. If D_6 has a frequency of 1174 Hz, find the frequencies of each note in the remainder of the octave leading up to D_7.

33. A note on a piano has a frequency of 440 Hz. Find each of the following note frequencies.

    a. 10 half steps above 440 Hz

    b. 10 half steps below 440 Hz

    c. An octave higher than 440 Hz

    d. An octave lower than 440 Hz

    e. 15 half steps above 440 Hz

34. If A_3 has a frequency of 220 Hz, find the frequency of the note that is nine half steps below A_3?

35. Recall that a perfect fifth is represented by seven half steps (C → G). If C has a frequency of 522 Hz, what is the frequency of G?

11

Bibliography

11.1

1. Photograph by Patrick Neil, http://www.flickr.com/people/patrickneil/. This photograph is licensed under the Creative Commons Attribution 2.0 Generic license.

2. Graham Sutherland, *Christ in Glory*, tapestry in Coventry Cathedral, Coventry, England. Photograph by David Jones, http://www.flickr.com/photos/cloudsoup/. This file is licensed under the Creative Commons Attribution 2.0 Generic license.

Chapter 12
Sports

Objectives

- Calculate offensive statistics in baseball
- Calculate defensive statistics in baseball
- Calculate quarterback ratings in football
- Calculate scoreability and bendability in football
- Demonstrate an understanding of *a priori* probability and the NBA draft
- Calculate individual statistics in basketball
- Demonstrate an understanding of the probability of winning a tennis match
- Calculate reaction time to a tennis serve
- Demonstrate an understanding of the dimensional analysis of track and field
- Apply probability to putting on the PGA tour

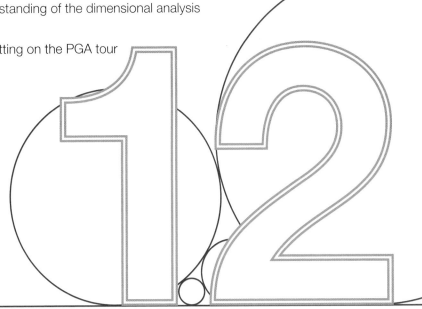

Sports

In an effort to improve winning outcomes, sports teams use mathematics as a way to make decisions. Consider a baseball player in Major League Baseball (MLB), such as Alex Rodriguez. In 2000, the Texas Rangers signed Rodriguez to a record contract of $252 million over 10 years. That is a tremendous amount of money to play baseball. So, how did the Rangers organization arrive at that amount of money to pay Rodriguez? The next highest contract at the time was about $60 million less than Rodriguez's contract. Was he really worth that much more than the next highest paid player? What players did he compare to at the time? Were his statistics that much better than every other player to warrant such a contract? The answers to these questions can be found quite easily with mathematics.

Professional sports teams, like those in Major League Baseball, take things a bit further in using mathematics to determine a player's value based on past performance. Most teams, if not all, keep detailed information of player performance from the number of at-bats to the number of hits to runs scored by each player. In combination with other factors of performance, a team can then determine the "value" of a player based on these comparisons to other players. This process of using mathematics to improve decision making allows teams that have smaller budgets to compete with teams that have larger budgets based on predicted outcomes.

Baseball, of course, is not the only sport where coaches and managers use mathematics to determine both strategy and value. Sports such as football, basketball, tennis, track, and golf also use mathematics to predict outcomes and make decisions regarding player value. Football teams use statistics that relate to quarterback reliability to strategize on defense, basketball teams use shot-making ability to create scoring options, track athletes use dimensional analysis to improve performance in running times, and golfers use probability to determine shot selection on a given course.

Major League Baseball teams make decisions on paying baseball players millions of dollars a year based on the expectation of how that player might perform in the coming years based on past experience. How about the strategy of National Basketball Association (NBA) teams regarding draft position in the draft lottery? All of these examples (and many more) involving sports and mathematics will be the focus of this chapter.

Rogers Hornsby Career Offensive Statistics

Career	G	AB	R	H	2B	3B	HR	RBI	AVG	OBP	SLG
23 Years	2,259	8,173	1,579	2,930	541	169	301	1,584	.358	.434	.577

Rogers Hornsby Career Defensive Statistics

Career	Games	Outs	Errors	FLD%
2B Totals	1,561	21,525	307	.965
SS Totals	356	9,120	142	.932
3B Totals	192	4,548	47	.924
1B Totals	35	903	4	.990
RF Totals	14	306	0	1.000
LF Totals	6	150	0	1.000
CF Totals	1	0	0	.000
23 Years	2,165	36,552	500	.958

Source: Baseball Almanac, "Rogers Hornsby Stats," accessed September 2014. http://www.baseball-almanac.com/players/player.php?p=hornsro01

Rogers Hornsby 1928

Figure 1

"I don't like to sound egotistical, but every time I stepped up to the plate with a bat in my hands, I couldn't help but feel sorry for the pitcher."

~ Rogers Hornsby

12.1   Baseball and Softball

Baseball and softball make use of numbers and statistics that help separate players by their individual ability. Baseball and softball team decisions are built on the statistics that account for a variety of feats. For example, in Major League Baseball (MLB), these statistics include the most consecutive games with a hit (Joe Dimaggio had a 56-game hitting streak in 1941), the last time a batter had a season batting average over 0.400 (Ted Williams hit 0.406 in 1941), or the player with the most home runs for their career (Barry Bonds with 762). With so many individual player statistics, it is easy for baseball organizations to use those statistics when creating a strategy for success.

Sabermetrics is the analysis of the statistics that are derived from the game of baseball through past performances of players' in-game activity. The book *Moneyball: The Art of Winning an Unfair Game* (Michael Lewis, 2003) was written as an homage to the use of sabermetrics by the Oakland A's baseball team and its General Manager, Billy Beane, and was made into a movie starring Brad Pitt in 2011. Sabermetrics takes into account the basic statistics of individual players, such as batting average, on-base percentage, and slugging percentage. For pitchers, sabermetrics includes earned run average and uses a combination of strikeouts, hits allowed, and walks to predict a pitcher's value.

Offensive Statistics in Baseball

Batting Average (BA)

We will begin our journey by talking about the easiest of these to compute: batting average. **Batting average** is a measure of an individual baseball player's offensive ability. First we will discuss a few definitions that will allow us to develop the ideas of the scoring side of baseball.

Figure 1 shows the basic layout of a baseball field where fielding positions are defined by their location on the field. When the ball is put into play by a batter, one of the position players attempts to catch, or field, the ball.

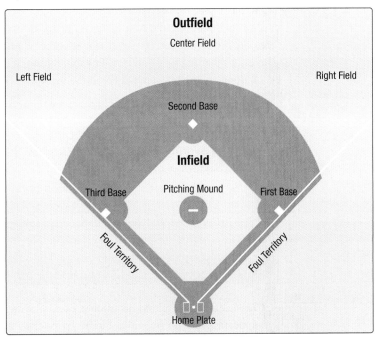

Figure 1: Basic Layout of a Baseball Field

At Bat (AB)

An **at bat** occurs when a batter has any outcome from a plate appearance except when the catcher interferes with the play or the player: **1.** is hit by the ball; **2.** makes a sacrifice hit; or **3.** is walked.

Sacrifice At Bat (SAC)

A **sacrifice at bat**, or sacrifice hit, is when a player intentionally gives up an out to advance a base runner to either 3rd base or home plate.

Hit By Pitch (HBP)

A batter earns a **hit-by-pitch** designation if the pitcher hits any part of the batter or his uniform with the ball. In this case, the batter is awarded first base.

Walk (BB)

A **walk**, or **base on balls**, occurs when a batter receives four pitches that are outside the strike zone, known as a **ball**, before being called out or getting a hit. A walk is an intentional walk if the pitcher makes no attempt at throwing a strike.

Hit (H)

A **hit** occurs when a batter successfully puts the ball into play and reaches a base safely without the defensive team making an error in attempting to catch the ball. A hit is a single if the runner reaches 1st base, a double if the runner reaches 2nd base, a triple if the runner reaches 3rd base, or a home run if the runner reaches home plate.

The most common batting statistic, or offensive statistic, in baseball is batting average. Batting average is computed by dividing the number of times a batter gets a hit (H) by the total number of times at bat (AB). When a player is issued a walk, reaches base on a hit by pitch, or has a sacrifice at bat, those plate appearances are not counted in the batting average. Batting average is a way for coaches to determine the best fit for a lineup. Consider that a team bats nine players and there are three outs in an inning. A player might come to plate three to five times per game depending on where their batting position lies. For this reason, a coach wants to ensure that the player with the best batting average is somewhere at the beginning of the batting order.

Batting Average (BA)

Batting average, rounded to the nearest thousandth, can be calculated using the following formula.

$$BA = \frac{\text{number of hits}}{\text{number of at bats}} = \frac{H}{AB}$$

Example 1: Calculating Batting Average

Jessica is a high school fastpitch softball player. After 5 games, Jessica had 6 hits in 12 plate appearances with no walks, no hit by pitches, and no sacrifice at bats. What is her batting average over those 5 games?

Solution

Using our formula, we compute her batting average over the 5 games as follows.

$$\mathrm{BA} = \frac{6 \text{ hits}}{12 \text{ at bats}} = \frac{6}{12} = 0.500$$

This means that over the 5-game span, Jessica got a hit during 50% of her plate appearances.

☞ **Helpful Hint**

Batting averages in baseball are rounded to the nearest thousandth and are read as hundreds. For instance, in Example 1, Jessica had a batting average of 0.500, which is read "five hundred." The lowest batting average would be 0.000 and the highest would be 1.000, or one thousand.

Example 2: Calculating Batting Average

During a baseball season, Henry had 247 hits in 720 plate appearances and Thomas had 148 hits in 401 plate appearances. Both players had no walks, no sacrifice hits, and no hit-by-pitches. Compare the batting averages of the two players and determine which player had the better batting average for the season.

Solution

We use our formula for batting average to get the season batting average for Henry as follows.

$$\mathrm{BA} = \frac{247 \text{ hits}}{720 \text{ at bats}} = \frac{247}{720} \approx 0.343$$

This is a batting average of 0.343, which means that Henry got a hit approximately 34% of the time he came to the plate.

Similarly, we find the batting average for Thomas as follows.

$$\mathrm{BA} = \frac{148 \text{ hits}}{401 \text{ at bats}} = \frac{148}{401} \approx 0.369$$

This means that Thomas got a hit almost 37% of the time he came to the plate.

In comparing the two batters, it is clear that Thomas has a batting average that is a little higher than Henry. Is the difference large enough to say that Thomas is a better batter than Henry? The answer is not as clear as it might seem. Many factors in baseball and softball influence batting, such as situations when a batter might sacrifice themselves to move a runner from first base to 2nd base. Events such as this count as at bats and, while they reduce batting average, are still helpful to the team.

Therefore, there is not a discernible difference in batting ability when comparing two batters such as Henry and Thomas, as their batting averages are relatively similar.

On-Base Percentage (OBP)

The next offensive statistic we will look at is **on-base percentage** (OBP). While OBP is a measure of the likelihood that a runner will reach base safely, it can be used for much more in terms of strategy by coaches of baseball or softball teams. Knowing a player's OBP gives a coach the ability to manage a hitting lineup in such a way that the players with the highest probability of getting on base are the players that most often get to bat. This, as a result, improves the team's chances at getting a player on base more often. The primary difference between OBP and batting average is that OBP takes into account the times a batter reaches base from walks and sacrifice at bats.

Fun Fact

If a batter reaches base 15 out of 15 times, or 100% of the time, we say the batter has an OBP of 1000.

On-Base Percentage (OBP)

On-base percentage, rounded to the nearest thousandth, can be calculated using the following formula.

$$OBP = \frac{H + BB + HBP}{AB + BB + HBP + SAC}$$

Here H is the number of hits, BB is the number of walks, HBP is the number of hit by pitches, AB is the number of at bats, and SAC is the number of sacrifice at-bats.

Example 3: Calculating On-Base Percentage

In 2011, the Boston Red Sox had 5710 at bats as a team. The Red Sox had a total of 1600 hits, 578 walks, 50 hit-by-pitch designations, and 50 sacrifice at bats. What was the team's OBP for 2011?

Solution

First, we identify the following statistics from the given information.

$$H = 1600 \qquad BB = 578 \qquad HBP = 50 \qquad SAC = 50 \qquad AB = 5710$$

Using our formula, we can calculate the OBP as follows.

$$OBP = \frac{H + BB + HBP}{AB + BB + HBP + SAC} = \frac{1600 + 578 + 50}{5710 + 578 + 50 + 50} = \frac{2228}{6388} \approx 0.349$$

So, as a team, the Boston Red Sox had an OBP of 0.349. That means that a Red Sox batter reached base safely in about 35% of their plate appearances. For a comparison of how good the Red Sox were at reaching base in 2011, the overall MLB OBP was 0.320. This means the Red Sox reached base about 3% more often than their competitors. This may not seem like a large percentage, but over the course of an entire season, they reached base over 100 more times than the average of all MLB teams.

Skill Check # 1

In 2011, the Seattle Mariners were last in on-base percentage in Major League Baseball. If the Mariners had 5421 at bats with 1261 hits, 435 walks, 37 hit-by-pitch designations, and 41 sacrifice at bats, what was the Mariners team OBP for 2011?

Slugging Percentage (SLG)

The next batting statistic we will look at is **slugging percentage**. Like on-base percentage, slugging percentage is used by decision makers in baseball as a method of determining offensive proficiency. Slugging percentage can be defined as the total number of bases a batter obtains divided by the total number of at bats. The total number of bases that a player obtains is calculated using one base for a single (1B), two bases for a double (2B), three bases for a triple (3B), and four bases for a home run (HR).

Slugging Percentage (SLG)

Slugging percentage, rounded to the nearest thousandth, can be calculated by the following formula.

$$SLG = \frac{1B + (2 \cdot 2B) + (3 \cdot 3B) + (4 \cdot HR)}{AB}$$

Here 1B is the number of singles, 2B is the number of doubles, 3B is the number of triples, HR is the number of home runs, and AB is the number of at bats.

A perfect slugging percentage can be calculated as 4.000. For a player to have a slugging percentage of 4.000, the player would have to hit a home run every single time they are at bat. Though this is the mathematically perfect slugging percentage, it is not very practical to assume that a batter could have a perfect 4.000 for more than a game or two in their entire career.

For example, let's say a batter has 60 singles (1B), 40 doubles (2B), 2 triples (3B), and 15 home runs (HR) in 375 at bats (AB) in a season. Recall that when we computed batting average, a single had the same weight as a home run, so this player's batting average is

$$BA = \frac{H}{AB} = \frac{60 + 40 + 2 + 15}{375} = \frac{117}{375} = 0.312.$$

Alternatively, for slugging percentage, we multiply each single by 1, each double by 2, each triple by 3, and each home run by 4. Then you add the products together and divide by 375 for a slugging percentage of

$$SLG = \frac{1B + (2 \cdot 2B) + (3 \cdot 3B) + (4 \cdot HR)}{AB} = \frac{60 + 2 \cdot 40 + 3 \cdot 2 + 4 \cdot 15}{375} = \frac{206}{375} \approx 0.549.$$

To fully interpret the value of this number, we need to understand what this number means. The average obtained is a ratio of the total bases a batter obtains to the total number of at bats of the batter. We already know that a perfect slugging percentage is 4.000. So, the closer a batter's SLG is to 4.000, the more total bases that batter obtained by putting the ball into play. Therefore, an SLG of 0.549 means that the batter obtains an average of about $\frac{1}{2}$ of a base per at bat. The MLB average for SLG is usually around 0.409. In MLB, the highest career SLG of all time belongs to Babe Ruth who had an SLG of 0.690.

Example 4: Calculating Slugging Percentage

During a softball season, Jessica had the following batting performances: 3 home runs, 5 triples, 12 doubles, 20 singles, 6 walks, and 35 outs. What is Jessica's slugging percentage?

Solution

First we need to determine the total number of at bats Jessica had. We see that she had 3 home runs, 5 triples, 12 doubles, 20 singles, and 35 outs. That means the total number of at bats was AB = 3 + 5 + 12 + 20 + 35 = 75. Now, using our formula, we can see that Jessica's slugging percentage for the season is calculated as follows.

$$SLG = \frac{1B + (2 \cdot 2B) + (3 \cdot 3B) + (4 \cdot HR)}{AB} = \frac{20 + 2 \cdot 12 + 3 \cdot 5 + 4 \cdot 3}{75} = \frac{71}{75} \approx 0.947$$

So, Jessica had a slugging percentage of 0.947. Thus, when Jessica was at bat, she averaged getting about 1 base each time. This is an extraordinary SLG—even higher than Babe Ruth. It is important to note that the more at bats a player has, the more opportunity there is for failure and success. Therefore, over the long term, it is more difficult to maintain a high

SLG. This is what makes Babe Ruth one of the best batters in the history of MLB—he had the ability to maintain a high SLG over his entire career and not just a few at bats.

Slugging percentage (SLG) is an effective tool in determining a player's ability to reach base successfully. However, it should be noted that SLG has one major flaw in determining a batter's true effectiveness—SLG does not take into account a batter's walks. What this means is that if we want an accurate portrait of the effectiveness of a batter, we must take into account each time the batter gets on base, including walks.

On-Base Plus Slugging Percentage (OPS)

The next, and last, statistic we will consider for batters is called **on-base plus slugging percentage**, or OPS. This measure of a hitter's ability to reach base is relatively new to the world of baseball. The book and movie *Moneyball* popularized the use of OPS as a way for the coach or general manager of a baseball or softball team to understand a hitter's ability to reach base. After all, the goal of the game is to reach base in order to score runs.

The first component of OPS is on-base percentage, OBP. Recall that OBP is calculated as follows.

$$OBP = \frac{H + BB + HBP}{AB + BB + HBP + SAC}$$

The second component of OPS is slugging percentage (SLG), which as discussed previously, measures the total number of bases a batter actually reaches through putting the ball in play. Recall that the formula for SLG is as follows.

$$SLG = \frac{1B + \left(2 \cdot 2B\right) + \left(3 \cdot 3B\right) + \left(4 \cdot HR\right)}{AB}$$

We calculate the formula for OPS by adding the formulas for OBP and SLG and simplifying.

> ## On-Base Plus Slugging Percentage (OPS)
>
> **On-base plus slugging percentage**, rounded to the nearest thousandth, can be calculated by the following formula.
>
> $$OPS = \frac{AB \cdot \left(H + BB + HBP\right) + \text{total bases} \cdot \left(AB + BB + HBP + SAC\right)}{AB \cdot \left(AB + BB + HBP + SAC\right)}$$
>
> Here, total bases is computed as $1B + \left(2 \cdot 2B\right) + \left(3 \cdot 3B\right) + \left(4 \cdot HR\right)$.
>
> Alternatively, we could simply write OPS = OBP + SLG.

Recall that OBP and SLG actually have values that indicate the effectiveness of a batter. Unlike OBP and SLG, OPS has no real value other than to be able to recognize an OPS that represents a good batter. Bill James, often called the father of sabermetrics, used the information in Table 1 to compare batters.

Table 1	
OPS Rating Values	
Classification	**OPS Range**
Excellent	0.900 and Higher
Good	0.833 to 0.899
Above Average	0.767 to 0.833
Average	0.700 to 0.767
Below Average	0.633 to 0.699
Poor	0.567 to 0.633
Horrible	0.567 and Lower

In general, an OPS of 1.000 is the standard for an exceptional player. It should also be noted that since OBP can be as large as 1 and SLG can be as large as 4, OPS has a range of $0 \leq OPS \leq 5$. As an example, Babe Ruth, with a career OPS of 1.1636, has the highest career OPS of all time of any major league player. Next on the list are Ted Williams with an OPS of 1.112 and Lou Gehrig with an OPS of 1.079, two of the greatest offensive players to ever play the game of baseball. An interesting facet of OPS is its ability to compare players at all levels. With comparable competition at all ages and all leagues, it is possible to determine the batting prowess of any player in any league (whether it be little league, college softball, or MLB) and compare them to other players in the same league over the course of a given season or even their entire career.

Example 5: Calculating OPS

Over the course of a ten-game span, Suzanne had 18 plate appearances. The results of each at bat are listed.

BB	1B	1B	1B	BB	Out	Out	2B	HBP
SAC	SAC	3B	Out	HR	Out	BB	Out	HR

Determine the OPS for Suzanne over the ten-game span.

Solution

Recall that the formula for OPS is as follows.

$$\text{OPS} = \frac{AB \cdot (H + BB + HBP) + \text{total bases} \cdot (AB + BB + HBP + SAC)}{AB \cdot (AB + BB + HBP + SAC)}$$

To make calculations easier, we tally each of the items needed.

AB = 12   **H** = 7 hits: 3-1B, 1-2B, 1-3B, 2-HR   **BB** = 3   **HBP** = 1   **SAC** = 2

total bases = $1B + (2 \cdot 2B) + (3 \cdot 3B) + (4 \cdot HR) = 3 + 2 \cdot 1 + 3 \cdot 1 + 4 \cdot 2 = 16$

Now we calculate OPS by substituting the known values into our formula.

$$OPS = \frac{AB \cdot (H + BB + HBP) + \text{total bases} \cdot (AB + BB + HBP + SAC)}{AB \cdot (AB + BB + HBP + SAC)}$$

$$= \frac{12 \cdot (7 + 3 + 1) + 16 \cdot (12 + 3 + 1 + 2)}{12 \cdot (12 + 3 + 1 + 2)}$$

$$= \frac{12 \cdot 11 + 16 \cdot 18}{12 \cdot 18}$$

$$= \frac{132 + 288}{216}$$

$$= \frac{420}{216} \approx 1.944$$

This means Suzanne has an OPS of 1.944 over the course of this ten-game span. According to Table 1, Suzanne would be considered an exceptional hitter over this time period.

Skill Check #2

Sam has 20 plate appearances with the following results: 4-1B, 3-2B, 0-3B, 2-HR, 2-HBP, 3-BB, 6 outs, and 0-SAC. Calculate Sam's OPS.

Defensive Statistics in Baseball

Earned Run Average (ERA)

Earned run average (ERA) is one of two important numbers used to evaluate the effectiveness of a pitcher. ERA represents the average number of earned runs (runs that are actually batted in by a batter who reaches base not on an error) given up by the pitcher per nine innings. For instance, if a pitcher gives up three solo home runs (a home run with no one on base), and then an error causes another runner to score, the pitcher is only credited with those first three runs that were the pitcher's fault.

Earned Run Average (ERA)

Earned run average, rounded to the nearest hundredth, is the average number of runs a pitcher allows per nine innings and can be calculated using the following formula.

$$ERA = \frac{\text{earned runs}}{\text{innings pitched}} \cdot 9$$

In terms of what is a good ERA and what is a bad ERA, the range for ERA is from 0.00 to infinity, where 0.00 is the best possible value. The larger a pitcher's ERA, the less effective the pitcher is for their team. Therefore, if a pitcher allows 18 earned runs in his first 75 innings pitched, his ERA would be

$$ERA = \frac{\text{earned runs}}{\text{innings pitched}} \cdot 9 = \frac{18}{75} \cdot 9 = 2.16.$$

In order to calculate the number of innings pitched, we add up the number of innings a pitcher was on the mound, including parts of an inning (that is, an inning in which the pitcher started to pitch, but did not finish or the part of an inning when the pitcher takes over). We use the value of $\frac{1}{3}$ of an inning if one out was recorded while the pitcher was on the mound and $\frac{2}{3}$ of an inning if two outs were recorded while the pitcher was on the mound. As an example, $6\frac{1}{3}$ or 6.33 innings means that the pitcher pitched six full innings and then pitched for only one out in the 7th inning that he was on the mound. Similarly, $6\frac{2}{3}$ or 6.67 innings means that the pitcher pitched six full innings and then pitched for only two outs in the 7th inning that he was on the mound.

Example 6: Calculating Earned Run Average (ERA)

In a single game, a pitcher pitches 5 innings and gives up 2 earned runs. Determine the pitchers earned run average (ERA).

Solution

Using the formula, we calculate the ERA as follows.

$$\text{ERA} = \frac{\text{earned runs}}{\text{innings pitched}} \cdot 9$$
$$= \frac{2}{5} \cdot 9$$
$$= 0.4 \cdot 9$$
$$= 3.6$$

So, over the course of a nine-inning game, the pitcher would be expected to allow 3.6 runs.

Skill Check #3

A pitcher gives up 9 earned runs over 21.33 innings. Determine the pitcher's ERA.

When trying to determine what a "good" ERA would be, we need to consider the statistics of the best pitchers in Major League Baseball. In 2011, Justin Verlander of the Detroit Tigers led the MLB with an ERA of 2.40. Similarly, Clayton Kershaw of the Los Angeles Dodgers led the MLB with an ERA of 2.53. Over the past 10 years, the lowest ERA of any MLB pitcher for a season was 1.87, by Roger Clemens of the Houston Astros. So, an ERA of 2.16 would be considered a very good number. Table 2 summarizes the effectiveness of a pitcher's ERA.

Table 2 ERA Rating Values	
Classification	**ERA Range**
Excellent	2.00 and Lower
Very Good	2.00 to 3.00
Good	3.00 to 4.00
Average	4.00 to 5.00
Poor	Above 5.00

As with batting average and on-base percentage, a player's age and league do not matter in terms of earned run average. No matter the age or league, ERA measures a pitcher's ability to get batters out and not allow runs based on giving up hits and walks. The use of ERA by coaches allows them to make decisions on pitching based on an understanding of which pitchers are most effective.

Walks Plus Hits per Inning Pitched (WHIP)

One last mathematical concept we will study for baseball is **walks plus hits per inning pitched** (WHIP). WHIP is the pitching statistic determined by the ratio of walks plus hits to innings pitched.

Walks Plus Hits per Inning Pitched (WHIP)

Walks plus hits per inning pitched (WHIP), rounded to the nearest hundredth, can be calculated as follows.

$$\text{WHIP} = \frac{\text{walks} + \text{hits}}{\text{innings pitched}}$$

While a pitcher's ERA gives us an accurate picture of the number of runs a pitcher allows, a pitcher's WHIP gives a much more accurate portrayal of a pitcher's effectiveness at getting batters out, not giving up walks, and completing innings pitched. This happens because WHIP is purely determined by things the pitcher controls (the number of walks and hits) whereas ERA has more variability based on runners reaching base. WHIP essentially gives us a number of how many **base runners** per inning a pitcher allows.

Base Runner

A **base runner** is any batter that successfully reaches base on either a hit or a walk.

Example 7: Calculating WHIP

A pitcher has pitched for 100 innings this season and has given up 85 hits and 35 walks. Determine the WHIP for this pitcher.

Solution

Using the formula, we can calculate WHIP as follows.

$$\text{WHIP} = \frac{\text{walks} + \text{hits}}{\text{innings pitched}} = \frac{85 + 35}{100} = \frac{120}{100} = 1.20$$

This means the pitcher has a WHIP of 1.20. Therefore, the pitcher allows about 1.2 base runners per inning.

The only difficulty of understanding WHIP is how to interpret the actual number derived from the calculation. In general, baseball/softball pitchers are considered *very good* if they have a WHIP less than 1.00. In terms of recent major league pitchers, Justin Verlander, who won the 2011 Cy Young Award for best pitcher in the American League, had a 2011 WHIP of 0.92.

Skill Check #4

In 2011, Justin Verlander of the Detroit Tigers had the lowest WHIP among major league baseball. If he allowed 174 hits and 57 BB in 251 innings, show that his WHIP = 0.92.

In 2011, there were 99 pitchers in Major League Baseball that had a recorded WHIP. Table 3 lists the possible WHIP ranges and the discussion that follows explains how to interpret WHIP for each range using the 2011 MLB pitchers.

Table 3	
WHIP Rating Values	
Classification	**WHIP Range**
Excellent	1.0 or Less
Very Good	1.01 to 1.20
Good/Average	1.21 to 1.40
Below Average	1.41 to 1.50
Well Below Average	Above 1.50

1.0 or Less: Best

Pitchers with a WHIP of less than 1.00 are considered the top pitchers in the league. In fact, in 2011, Major League Baseball only had three pitchers with a WHIP less than 1.00: Justin Verlander, Clayton Kershaw, and Cole Hamels.

1.01–1.20: Very good

A pitcher with a WHIP in this range is considered successful and effective. In 2011, Major League Baseball had 28 pitchers in this range.

1.21–1.40: Good/Average

These numbers are decent to fairly average. In 2011, Major League Baseball had 45 pitchers in this range.

1.41–1.50: Below Average

These are not very desirable numbers for a pitcher. In fact, a WHIP in this range is below average. In 2011, Major League Baseball only had 12 pitchers in this range.

1.50 and Above: Well Below Average

This number is reserved for pitchers who constantly have runners on base, which is a losing situation in Major League Baseball. In 2011, there were only four pitchers with a WHIP greater than 1.5.

Combinations and Basic Counting Principles

Up to this point, we have spent a considerable amount of time on the basic performance numbers for baseball and softball, such as on-base percentage, batting average, and ERA. Now we turn to the basic counting principles and how they might be used in baseball.

Managing a lineup for a team is a critical component for the strategy of winning a baseball or softball game. Therefore, the coach or manager of a team must have a deep understanding of how to create a lineup of players that best benefits the team goal of winning. The first thing we will look at is how many ways a coach or manager can organize a team's lineup.

Example 8: Determining the Number of Combinations of Baseball Lineups

On any Major League Baseball team, there are 25 players. Typically, about 11 of those players are pitchers. That means there are 14 players for the 8 remaining starting positions. Determine the number of ways in which a coach could choose and arrange a given 9-player lineup.

Solution

In order to determine the answer here, we need to recall the formula for a combination, where n is the number of players to choose from and r is the number of players chosen.

$$_nC_r = \frac{n!}{r!(n-r)!}$$

A combination tells us that we have the following number of possibilities when choosing 8 of 14 players.

$$_{14}C_8 = \frac{n!}{r!(n-r)!}$$
$$= \frac{14!}{8!(14-8)!} = 3003$$

> ☞ **Helpful Hint**
>
> Recall that a permutation is used when order of selection matters and a combination is used when order of selection does not matter.

This gives that there are 3003 different ways to choose the eight remaining players, other than a pitcher. Since there are 11 pitchers to choose from, there are $11 \cdot 3003 = 33{,}033$ ways to choose the 9 members of the starting lineup.

What about the order of the hitting lineup? Since the hitters are arranged from 1 to 9, a coach must use every available bit of information so as to maximize the probability of scoring runs. We learned from Chapter 7 that nine players can be arranged in 9! ways, or 362,880 ways. Then using the fundamental counting principle, the coach can determine that there are $33{,}033 \cdot 362{,}880 = 11{,}987{,}015{,}040$ ways to arrange a simple batting lineup. With so many ways to determine a lineup, coaches must use the information they have regarding batting average, slugging percentage, etc. to arrange the players in a way that maximizes the team's chance of success.

Skill Check #5

A baseball team has 12 players. How many different 9-player lineups can the coach make?

In this section, we have spent considerable time discussing many of the statistical numbers that are represented in baseball and softball, such as batting average, on-base percentage, slugging percentage, on-base plus slugging percentage, earned run average, and walks plus hits per inning pitched. It should be noted that our coverage of some the statistics in baseball is a mere introduction to the metrics used by baseball statisticians to assist in making decisions. In the rest of the chapter we will focus on how mathematics relates to additional sports such as football, basketball, and tennis.

Skill Check Answers

1. 0.292

2. 1.900

3. 3.797

4. $\dfrac{57 + 174}{251} = \dfrac{231}{251} \approx 0.920$

5. 220 lineups

12.1   Exercises

Solve each problem.

1. A batter had 15 at bats and 8 hits. What is the batting average of the player?

2. Jim has the highest batting average on his softball team with an average of 0.651. If Jim has had 175 at bats, how many hits has Jim had?

3. During a season, Larry had 257 hits in 755 at bats. What is Larry's batting average for the season?

4. During a softball season, Mikah had the following batting performances: 10 home runs, 7 triples, 6 doubles, 35 singles, 8 walks, and 2 outs. What is Mikah's slugging percentage for the season?

5. Over the course of a 12-game span, Shanda had 22 plate appearances. Over the same 12-game span, Ali had 27 plate appearances.

   Shanda's plate appearances over the 12-game span are as follows.

BB	1B	1B	1B	BB	Out	Out	2B	HBP	1B	HR
SAC	SAC	3B	Out	HR	Out	BB	Out	HR	2B	1B

   Ali's plate appearances over the 12-game span are as follows.

1B	1B	2B	HR	Out	Out	1B	BB	Out
Sac	2B	HR	1B	1B	Out	Sac	1B	HR
3B	Out	Out	1B	BB	BB	Sac	BB	2B

   Calculate the OPS for both Shanda and Ali and determine which player was more effective over the 12-game span.

6. A pitcher pitches 7 innings and gives up 3 earned runs. Determine the pitchers ERA.

7. A pitcher has thrown for 250 innings this season and has given up 178 hits and 27 walks. Determine the WHIP for this pitcher.

8. A youth baseball team has 14 players. Determine the number of ways in which a coach could choose 9 players for a lineup.

Use the table to answer each question.

2011 MLB Individual Batting Statistics														
Rk	Player	Team	G	AB	R	H	2B	3B	HR	RBI	BB	SO	HBP	SAC
1	Cabrera	DET	161	572	111	197	48	0	30	105	108	89	3	5
2	Gonzalez	BOS	159	630	108	213	45	3	27	117	74	119	6	5
3	Young	TEX	159	631	88	213	41	6	11	106	47	78	2	9
4	Reyes	NYM	126	537	101	181	31	16	7	44	43	41	0	6
5	Braun	MIL	150	563	109	187	38	6	33	111	58	93	5	3
6	Martinez	DET	145	540	76	178	40	0	12	103	46	51	2	7
7	Kemp	LAD	161	602	115	195	33	4	39	126	74	159	6	7
8	Ellsbury	BOS	158	660	119	212	46	5	32	105	52	98	9	8
9	Pence	PHI	154	606	84	190	38	5	22	97	56	124	0	5
10	Votto	CIN	161	599	101	185	40	3	29	103	110	129	4	6

Source: ESPN MLB. "MLB Player Batting Statistics - 2011." http://espn.go.com/mlb/stats/batting/_/year/2011
Note: H: sum of all 1B, 2B, 3B, and HR;, G: games; AB: at bats; R: runs

9. What was Cabrera's batting average?

10. What was Gonzalez's OBP?

11. How many singles did Pence have?

12. What was Ellsbury's SLG?

13. What was Cabrera's SLG?

14. Determine whether Cabrera or Braun has the higher OPS.

15. Determine whether Cabrera or Braun has the higher BA.

16. Determine whether Cabrera or Braun has the higher SLG.

17. What was Kemp's SO to hit ratio?

18. What was Votto's HR to at bat ratio?

Use the table to answer each question.

2011 MLB Team Batting Statistics													
Team	**GP**	**AB**	**R**	**H**	**2B**	**3B**	**HR**	**TB**	**RBI**	**SO**	**BB**	**SAC**	**HBP**
Boston	162	5710	875	1600	352	35	203	2631	842	110	578	72	50
NY Yankees	162	5518	867	1452	267	33	222	2451	836	1130	627	87	74
Texas	162	5659	855	1599	310	32	210	2603	807	930	475	88	39
Detroit	162	5563	787	1540	297	34	169	2412	750	1143	521	108	39
St. Louis	162	5532	762	1513	308	22	162	2351	726	978	542	124	44
Toronto	162	5559	743	1384	285	34	186	2295	704	1	525	78	48
Cincinnati	162	5612	735	1438	264	19	183	2289	697	1250	535	108	63
Colorado	162	5544	735	1429	274	40	163	2272	697	12	555	109	57
Arizona	162	5421	731	1357	293	37	172	2240	702	1	531	88	61
Kansas City	162	5672	730	1560	325	41	129	2354	705	1	442	112	39

Source: ESPN MLB. "MLB Team States - 2011." http://espn.go.com/mlb/stats/team/_/stat/batting/year/2011
Note: GP: games played; AB: at bats; R: runs scored; H: hits; 2B: doubles; 3B: triples; HR: home runs; TB: total bases; RBI: runs batted in; SO: strikeouts; BB: base on balls; SAC: sacrifice flies; HBP: hit by pitch

19. What was the team batting average of the St. Louis Cardinals?

20. What was the SLG of the Boston Red Sox?

21. What was the OBP for the Kansas City Royals?

22. What was the OPS for the NY Yankees? Calculate the OPS for the Colorado Rockies as well, and compare to the NY Yankees. Determine which team was more effective.

23. What was the average runs scored per game by the Arizona Diamondbacks?

Use the table to answer each question.

2011 MLB Individual Pitching Statistics												
Rk	**Pitcher**	**Team**	**W**	**L**	**G**	**IP**	**Hits**	**R**	**ER**	**HR**	**BB**	**SO**
1	Kershaw	LAD	21	5	33	233.33	174	66	59	15	54	248
2	Halladay	PHI	19	6	32	233.670	208	65	61	10	35	220
3	Lee	PHI	17	8	32	232.67	197	66	62	18	42	238
4	Verlander	DET	24	5	34	251	174	73	67	24	57	250
5	Weaver	LAA	18	8	33	235.67	182	65	63	20	56	198
6	Vogelsong	SF	13	7	30	179.67	164	62	54	15	61	139
7	Lincecum	SF	13	14	33	217	176	74	66	15	86	220
8	Hamels	PHI	14	9	32	216	169	68	67	19	44	194
9	Shields	TB	16	12	33	249.33	195	83	78	26	65	225
10	Fister	DET	11	13	32	216.33	193	76	68	11	37	146

Source: ESPN MLB. "MLB Player Pitching Stats - 2011." http://espn.go.com/mlb/stats/pitching/_/year/2011/order/false
Note: W: games won by the pitcher; L: games lost by the pitcher; G: games the pitcher pitched in; IP: innings pitched; Hits: hits allowed; R: runs scored by the other team; ER: earned runs scored by the opposing team; HR: home runs allowed; BB: bases on balls allowed; SO: strikeouts by batters faced

24. What was Kershaw's WHIP for the season? Calculate the WHIP for Fister and compare to Kershaw's. Make a determination of which pitcher was more effective.

12

25. What was Shields' ERA?

26. What was Lee's ratio of strikeouts to innings pitched?

27. What was Verlander's ratio of walks per innings pitched?

28. What was Weaver's average innings per game?

29. Which pitcher had the lowest ERA for the season?

30. Which pitcher had the lowest walks per inning ratio?

12.2   Football

As with baseball and softball, football collects statistics on just about everything each player does in the course of a game. From the number of rushing yards a running back obtains in a game to the number of turnovers committed by an offense, mathematical information gained during a football game and season allows coaches and players to make decisions on future strategy and to compare one performance to another. In this section, we will focus primarily on quarterback rating, scoreability index, and bendability index.

Quarterback Rating (NFL and NCAA)

In the game of football, every player is rated based on their effectiveness at their position. In no other position are players scrutinized more intensely than that of quarterback. **Quarterback rating**, also known as passing efficiency or passer rating, measures the efficiency and effectiveness of a quarterback's ability to pass the ball to other players. When another player on the same team catches the football, it is called a completion. Otherwise the pass is either incomplete, meaning no one caught the ball, or intercepted, which means it was caught by the opposing team. Thus, when Woody Hayes mentioned, "There are three things that can happen when you pass, and two of them ain't good," he was referring to these three outcomes. While it is possible for other players, such as a running back, to pass the football, only the quarterback position receives this rating.

The ratings of quarterbacks are calculated using each quarterback's completion percentage, passing yardage, touchdowns, and interceptions. It should be noted that different calculation methods are used for the National Football League (NFL) and National Collegiate Athletics Association (NCAA) ratings. Each of these calculations will be discussed in detail. The quarterback rating is essentially an evaluation of a quarterback's ability to successfully complete passes, accumulate yards, and score touchdowns.

First, we'll turn our attention to the National Football League (NFL). Quarterback ratings in the NFL have a range from 0 to 158.3. For an NFL quarterback to obtain a perfect rating of 158.3, the number of mistakes such as incomplete passes, interceptions, etc. must be very small. We will find each of the necessary components used to determine quarterback rating before finding the rating itself.

Completion Percentage (CP)

The first component for determining the quarterback rating is the **completion percentage**. If nothing else, completion percentage indicates the accuracy of a quarterback's throws to his receivers. The completion percentage of a quarterback is based on the number of passes completed to a receiver on his team compared to the number of passes attempted.

> ### Completion Percentage (CP)
>
> **Completion percentage**, rounded to the nearest tenth, can be calculated with the following formula.
>
> $$CP = \frac{\text{number of passes completed}}{\text{number of passes attempted}} \times 100\%$$

Example 1: Completion Percentage

A quarterback passes 145 times during a season and completes 84 of those passes.

a. What is the quarterback's completion percentage?

b. How would the completion percentage change with five more completions and the same number of pass attempts for the season?

Solution

a. We are given all of the necessary information to calculate completion percentage using the formula, so we substitute the known values and perform the calculation as follows.

$$CP = \frac{\text{number of passes completed}}{\text{number of passes attempted}} = \frac{84}{145} \approx 0.579$$

Therefore, the quarterback completed 57.9% of the passes he attempted.

Of course, the higher the completion percentage the better for a quarterback. In comparison, many NFL quarterbacks strive to have a completion percentage of 60% for the season.

b. If we compute the completion percentage with the additional five completions, we get

$$CP = \frac{84 + 5}{145} = \frac{89}{145} \approx 0.614.$$

Comparing the accuracy of the quarterback's throws with five extra completions and the same number of attempts to the original CP, we can see that the completion percentage improves to greater than 60%.

Yards per Pass Attempt

The second quarterback statistic we need for determining quarterback rating is yards per pass attempt. This value is calculated by dividing the total number of passing yards that a quarterback has for a game, season, or career (that is, the number of yards derived from a completed pass) by the number of pass attempts for the same period.

Yards per Pass Attempt (YPA)

Yards per pass attempt (YPA), rounded to the nearest hundredth, can be calculated with the following formula.

$$YPA = \frac{\text{total number of yards passing}}{\text{number of passes attempted}}$$

Example 2: Calculating Yards per Pass Attempt

In 2011, Eli Manning of the New York Giants passed for 4933 yards on 589 attempts. What was Manning's YPA for the 2011 season?

Solution

We are given all of the necessary information to calculate the YPA using the formula, so we substitute the known values and perform the calculation as follows.

$$\text{YPA} = \frac{\text{total number of yards passing}}{\text{number of passes attempted}} = \frac{4933}{589} \approx 8.38 \text{ YPA}$$

This means that each pass Manning made during the season gained an average of 8.38 yards. YPA factors into a quarterback's effectiveness at converting first downs, that is, moving the ball 10 yards or more, while retaining possession of the ball. A YPA greater than 7 is considered to be very good.

Touchdowns per Pass Attempt (TDPA)

The ultimate goal of any offensive possession in football is to score touchdowns. The next statistic we need to calculate the quarterback rating is the ratio of touchdowns thrown to passes attempted, or **touchdowns per pass attempt (TDPA)**.

Touchdowns per Pass Attempt (TDPA)

Touchdowns per pass attempt, rounded to the nearest hundredth, can be calculated with the following formula.

$$\text{TDPA} = \frac{\text{number of passing touchdowns}}{\text{number of passes attempted}}$$

Example 3: Calculating Touchdowns per Pass Attempt

In 2011, Aaron Rodgers of the Green Bay Packers had 45 touchdown passes and 502 attempts. What was Rodgers' touchdowns per pass attempt for 2011?

Solution

We are given all of the necessary information to calculate touchdowns per pass attempt using the formula, so we substitute the known values and perform the calculation as follows.

$$\text{TDPA} = \frac{\text{number of touchdowns}}{\text{number of passes attempted}} = \frac{45}{502} \approx 0.09$$

Thus, Rodgers scores a touchdown 9% of the time he passes the ball. This number, which is quite small, led the NFL in 2011 as the best among quarterbacks with more than 100 pass attempts.

Interceptions per Pass Attempt (IPA)

The last number we need in order to calculate the quarterback rating is the rate at which a quarterback throws interceptions. It is the ratio of the number of interceptions to the number of passes attempted.

Interceptions per Pass Attempt (IPA)

Interceptions per pass attempt, rounded to the nearest thousandth, can be calculated as follows.

$$IPA = \frac{\text{number of interceptions}}{\text{number of passes attempted}}$$

This number evaluates how often a quarterback throws an interception (that is, when the opposing team catches the ball) rather than a completion to his own team. The lower the IPA number is for a quarterback, the better.

Example 4: Interceptions per Pass Attempt

In 2011, Ryan Fitzpatrick of the Buffalo Bills had 23 interceptions in 569 attempts. What was Fitzpatrick's IPA for 2011?

Solution

Since Fitzpatrick had 23 interceptions in 569 attempts, his interceptions per pass attempt can be calculated as follows.

$$IPA = \frac{\text{number of interceptions}}{\text{number of passes attempted}} = \frac{23}{569} \approx 0.040$$

This means that the ball was intercepted about 4% of the time when Fitzpatrick threw a pass. To put this number in perspective, Fitzpatrick threw 1 interception for approximately every 25 passes he attempted. In general, NFL quarterbacks wish to limit their interceptions as much as possible. Based on an entire season with more than 400 passing attempts, most quarterbacks have a goal of less than 10 interceptions in a given year.

Skill Check #1

Drew Brees of the New Orleans Saints had 657 pass attempts in 2011. Brees completed 468 of those passes while throwing for 5467 yards, making 46 touchdowns, and had 14 interceptions. Calculate Brees' completion percentage, yards per pass attempt, touchdowns per pass attempt, and interceptions per pass attempt.

Fun Fact

The formula for the NFL quarterback rating was created by the Pro Football Hall of Fame as a way to compare quarterbacks of different eras with the same measures of performance.

Quarterback Rating Formula

Now that we have found the necessary components for determining the quarterback rating for an NFL quarterback, we can compile all of these into one number. Recall that the NFL quarterback rating ranges from 0 to 158.3.

NFL Quarterback Rating

The formula to calculate the **quarterback rating** for a player in the NFL, rounded to the nearest tenth, is as follows.

$$\text{NFL quarterback rating} = \left(\frac{5(CP - 0.3) + 0.25(YPA - 3) + 20(TDPA) + 25(0.095 - IPA)}{6} \right) \cdot 100$$

where CP is completion percentage (**Note:** CP is left as a decimal), YPA is yards per pass attempt, TDPA is touchdowns per pass attempt, and IPA is interceptions per pass attempt.

Don't let the formula intimidate you with its size. The formula can be written as a step-by-step procedure, as follows.

Step-By-Step Formula for the NFL Quarterback Rating

1. Divide a quarterback's completed passes by pass attempts.

2. Subtract 0.3 from the value obtained in Step 1.

3. Multiply the value from Step 2 by 5 and record the total. The product cannot be greater than 2.375 or less than zero.

4. Divide passing yards by pass attempts.

5. Subtract 3 from the value obtained in Step 4.

6. Multiply the value from Step 5 by 0.25 and record the total. The product cannot be greater than 2.375 or less than zero.

7. Divide touchdown passes by pass attempts.

8. Multiply the value from Step 7 by 20 and record the total. The product cannot be greater than 2.375 or less than zero.

9. Divide interceptions by pass attempts.

10. Subtract the value from Step 9 from 0.095.

11. Multiply the value from Step 10 by 25 and record the total. The product cannot be greater than 2.375 or less than zero.

12. Find the sum of the values from Steps 3, 6, 8, and 11.

13. Divide the value obtained in Step 12 by 6.

14. Multiply the value obtained in Step 13 by 100. This is the quarterback rating.

Example 5: NFL Quarterback Rating

Tim Tebow attempted 271 passes in 2011 while playing for the Denver Broncos. He completed 126 passes while throwing for 1729 yards, 12 touchdowns, and 6 interceptions. Calculate Tebow's NFL quarterback rating for 2011.

Solution

We know the following formula for the NFL quarterback rating.

$$\text{NFL quarterback rating} = \left(\frac{5(CP - 0.3) + 0.25(YPA - 3) + 20(TDPA) + 25(0.095 - IPA)}{6} \right) \cdot 100$$

We can use the numbers Tebow accumulated through the year to determine each component of the quarterback rating.

1. Divide a quarterback's completed passes by pass attempts.

$$CP = \frac{\text{number of passes completed}}{\text{number of passes attempted}} = \frac{126}{271} \approx 0.465$$

2. Subtract 0.3 from the value obtained in Step 1.

$$0.465 - 0.3 = 0.165$$

3. Multiply the value obtained in Step 2 by 5 and record the total. The product cannot be greater than 2.375 or less than zero.

$$0.165 \cdot 5 = 0.825$$

4. Divide passing yards by pass attempts.

$$YPA = \frac{\text{total number of yards passing}}{\text{number of passes attempted}} = \frac{1729}{271} \approx 6.38 \text{ YPA}$$

5. Subtract 3 from value obtained in Step 4.

$$6.38 - 3 = 3.38$$

6. Multiply the value obtained in Step 5 by 0.25 and record the total. The product cannot be greater than 2.375 or less than zero.

$$3.38 \cdot 0.25 = 0.845$$

7. Divide touchdown passes by pass attempts.

$$TDPA = \frac{\text{number of touchdowns}}{\text{number of passes attempted}} = \frac{12}{271} \approx 0.044$$

8. Multiply Step 7 by 20 and record the answer. The product cannot be greater than 2.375 or less than zero.

$$0.044 \cdot 20 = 0.88$$

9. Divide interceptions by pass attempts.

$$IPA = \frac{\text{number of interceptions}}{\text{number of passes attempted}} = \frac{6}{271} \approx 0.022$$

10. Subtract the value obtained in Step 9 from 0.095.

$$0.095 - 0.022 = 0.073$$

11. Multiply the difference obtained in Step 10 by 25 and record the total. The product cannot be greater than 2.375 or less than zero.

$$0.073 \cdot 25 = 1.825$$

12. Find the sum of the four components of quarterback rating (sum of Steps 3, 6, 8, and 11).

$$0.825 + 0.845 + 0.88 + 1.825 = 4.375$$

13. Divide the value obtained in Step 12 by 6.

$$\frac{4.375}{6} = 0.729$$

14. Multiply the value obtained in Step 13 by 100.

$$0.729 \cdot 100 = 72.9$$

The final number is your quarterback rating.

Alternatively, we can substitute these values into our formula for quarterback rating.

$$\text{NFL quarterback rating} = \left(\frac{5(CP - 0.3) + 0.25(YPA - 3) + 20(TDPA) + 25(0.095 - IPA)}{6}\right) \cdot 100$$

$$= \left(\frac{5(0.465 - 0.3) + 0.25(6.38 - 3) + 20(0.044) + 25(0.095 - 0.022)}{6}\right) \cdot 100$$

$$= \left(\frac{5(0.165) + 0.25(3.38) + 20(0.044) + 25(0.073)}{6}\right) \cdot 100$$

$$= \left(\frac{0.825 + 0.845 + 0.88 + 1.825}{6}\right) \cdot 100$$

$$= \left(\frac{4.375}{6}\right) \cdot 100$$

$$\approx 0.729 \cdot 100 = 72.9$$

We have calculated that Tim Tebow's NFL quarterback rating for the 2011 season was 72.9. The average quarterback rating in the NFL each year is around 75. As we can see, Tebow's rating is consistent with the average NFL quarterback.

Skill Check #2

Aaron Rodgers had the highest passer rating among NFL quarterbacks for the 2011 season. If Rodgers passed for 4643 yards with 343 completions in 502 attempts, 45 touchdowns, and 6 interceptions, what was Rodgers' NFL quarterback rating?

NCAA Quarterback Rating

Similar to the NFL quarterback rating formula, the NCAA quarterback rating formula allows us to compare quarterbacks and their individual performances. Luckily, there are no new calculations for the NCAA quarterback rating, but rather a slightly different formula using the same basic statistics. Additionally, the range of values is very different. The maximum rating for an NCAA quarterback is 1261.6 and the minimum rating is −731.6, where the average quarterback rating in the NCAA is approximately 120. Even though the maximum and minimum numbers are very unlikely, they are still possible for a quarterback to obtain. You might also notice that the NCAA quarterback rating can actually be a negative number. This is because the formula places a high negative value on throwing an interception, whereas the NFL rating formula is not affected as much by interceptions. Neither rating is better than the other; they are simply different methods used to compare quarterbacks to one another in each league.

NCAA Quarterback Rating

The formula for calculating the **NCAA quarterback rating**, rounded to the nearest tenth, is as follows.

$$\text{NCAA quarterback rating} = \frac{(8.4 \cdot \text{yards}) + (330 \cdot \text{TDs}) + (100 \cdot \text{comp}) - (200 \cdot \text{int})}{\text{number of pass attempts}}$$

Here, yards = passing yards, TDs = number of touchdowns, comp = number of completions, and int = number of interceptions.

In 2011, Robert Griffin III, or RG3 (as he has been nicknamed), won the Heisman Trophy, which is awarded annually to the nation's top college football player. He threw for 4293 yards, 37 touchdowns, and 6 interceptions with 291 completions in 402 attempts. What was RG3's NCAA quarterback rating for the 2011 season?

Solution

Using the formula, we can determine the quarterback rating for RG3. Since RG3 passed for 4293 yards, 37 touchdowns, 291 completions, and 6 interceptions in 402 attempts, we can calculate the quarterback rating as follows.

$$\text{NCAA quarterback rating} = \frac{(8.4 \cdot \text{yards}) + (330 \cdot \text{TDs}) + (100 \cdot \text{comp}) - (200 \cdot \text{int})}{\text{number of pass attempts}}$$

$$= \frac{(8.4 \cdot 4293) + (330 \cdot 37) + (100 \cdot 291) - (200 \cdot 6)}{402}$$

$$= \frac{36,061.2 + 12,210 + 29,100 - 1200}{402}$$

$$= \frac{76,171.2}{402}$$

$$\approx 189.5$$

So, RG3 had a NCAA quarterback rating of 189.5 for the 2011 season.

Skill Check #3

The best NCAA quarterback rating ever for a single season belongs to Russell Wilson of the Wisconsin Badgers during the 2011 season. Wilson passed for 3175 yards while throwing 33 touchdowns and only 4 interceptions on 309 pass attempts, where he completed 225 of those attempts. Determine Wilson's NCAA quarterback rating for the 2011 season.

Scoreability (YPS) and Bendability (YPPA)

Scoreability and bendability are measures in the game of football that evaluate the efficiency of a team's offense to score points, and a team's defense to stop the opposing teams from scoring points, respectively. In particular, each of these are measures of a football team's ability to convert the yards gained into points either given up by the defense or scored by the offense. One unique aspect is that they take into account the total yards gained by a team, including special team plays such as punt returns and kick-off returns.

The **scoreability index** measures the ability of an offense to score points and is found by calculating the ratio of total yards gained to total points scored by the offense. Since the scoreability index is a measure of the efficiency with which a team scores points, the lower the number, the more efficiently a team scores points. The result of this calculation is yards per point scored (YPS).

Fun Fact

A punt return occurs when the offensive team punts the ball to the defensive team, thus interchanging offense and defense for both teams.

A kick-off return occurs only when a team scores or at the beginning of the two halves of the game. The ball is kicked to one team who takes possession of the ball and begins an offensive series of plays.

Scoreability Index (YPS)

The **scoreability index**, rounded to the nearest hundredth, of an offense can be calculated with the following formula.

$$YPS = \frac{\text{total number of yards gained}}{\text{total number of points scored}}$$

Example 7: Scoreability

During the 2011 NFL season, the Green Bay Packers led the NFL in scoreability. Determine the YPS for the 2011 Green Bay Packers, where the Packers gained 6482 yards and scored 560 points.

Solution

We substitute the known values into the formula to calculate scoreability as follows.

$$YPS = \frac{\text{total number of yards gained}}{\text{total number of points scored}} = \frac{6482}{560} = 11.58$$

So, for the 2011 season the Packers had a scoreability index of YPS = 11.58. In theory, the smaller a team's YPS is, the more effective they are at scoring points per yard gained.

Like the scoreability index, the **bendability index** is calculated by dividing the total yards allowed by a defense (that is, total yards gained by opposing offenses) by the total points allowed. Since the bendability index is a measure of the ability of a team to not allow points to be scored on them, the higher the number, the more difficult it is to score points against at a team. Yards per point allowed (YPPA) is the official statistic for the bendability of a team.

Bendability Index (YPPA)

The **bendability index**, rounded to the nearest hundredth, of a defense can be calculated with the following formula.

$$YPPA = \frac{\text{total number of yards allowed}}{\text{total number of points allowed}}$$

Example 8: Bendability

One of the best defensive teams in the history of the NFL was the Baltimore Ravens defense in 2000. During the 2000 season, the Ravens defense gave up only 4037 total yards while allowing 165 points. Determine the YPPA for the 2000 Baltimore Ravens.

Solution

The calculation of the bendability index is the ratio of the total yards allowed to the total points allowed.

$$\text{YPPA} = \frac{\text{total number of yards allowed}}{\text{total number of points allowed}}$$
$$= \frac{4037}{165}$$
$$= 24.47$$

So, the Ravens had a bendability index of 24.47. This value represents one of the largest YPPA values in the history of the NFL.

To compare scorability to bendability, we only need to compute the values and determine which value is larger. Recall that for scorability, a lower number is better, while for bendability, a higher number is better. When comparing an offense to a defense, if the offensive number is larger than the defensive number, the offense has the advantage and vice versa. When the numbers are nearly the same, both are equally matched.

Skill Check #4

The New York Giants won Super Bowl XLVI (46). If the Giants scored 391 points after gaining a total of 6161 yards and allowed only 400 points and 6022 yards. What were the Giants' scoreability and bendability indices for the 2011 season?

Skill Check Answers

1. CP = 0.712, YPA = 8.32, TDPA = 0.07, IPA = 0.021

2. 122.5

3. 191.8

4. YPS = 15.76, YPPA = 15.06

12.2   Exercises

Answer each question.

1. A quarterback passes 225 times during a season and completes 142 of those passes. What is the quarterback's completion percentage?

2. A quarterback has a completion percentage of approximately 63.7% after completing 86 passes. How many passes did the quarterback attempt? (Round to the nearest whole number).

Use the table to answer each question.

2011 NFL Quarterback Statistics							
Rk	**Player**	**Team**	**Comp**	**Att**	**Yds**	**TD**	**Int**
1	Rodgers	GB	343	502	4643	45	6
2	Brees	NO	468	657	5476	46	14
3	Brady	NE	401	611	5235	39	12
4	Romo	DAL	346	522	4184	31	10
5	Stafford	DET	421	663	5038	41	16
6	Schaub	HOU	178	292	2479	15	6
7	Manning	NYG	359	589	4933	29	16
8	Ryan	ATL	347	566	4177	29	12
9	Smith	SF	273	445	3144	17	5
10	Roethlisberger	PIT	324	513	4077	21	14

Source: ESPN NFL. "NFL Player Passing Statistics - 2011." http://espn.go.com/nfl/statistics/player/_/stat/passing/sort/quarterbackRating/year/2011

3. What was Brady's touchdowns per pass attempt ratio?

4. What was Smith's completion percentage?

5. What was Manning's yards per pass attempt?

6. What was Ryan's quarterback rating?

7. What was Brees' interceptions per pass attempt ratio?

8. What was Romo's quarterback rating?

9. Which quarterback completed the highest percentage of passes?

10. Which quarterback had the highest touchdowns per pass attempt ratio?

11. Which quarterback had the highest interceptions per pass attempt ratio?

12. What was Brees' yards per completion average?

13. What was Manning's yards per completion average?

Use the table to answer each question.

2011 NCAA Quarterback Statistics							
Rk	**Player**	**Team**	**Att**	**Comp**	**Yds**	**TD**	**Int**
1	Keenum	HOU	603	428	5631	48	5
2	Weeden	OKST	564	408	4727	37	13
3	Jones	OKLA	562	355	4463	29	15
4	Smith	WVU	526	346	4385	31	7
5	Foles	ARIZ	560	387	4334	28	14
6	Griffin	BAY	402	291	4293	37	6
7	Osweiler	ASU	516	326	4036	26	13
8	Doege	TTU	581	398	4004	28	10
9	Carder	WMU	502	330	3873	31	14
10	Boyd	CLEM	499	298	3828	33	12

Source: ESPN College Football. "FBS (I-A) Player Passing Statistics - 2011." http://espn.go.com/college-football/statistics/player/_/stat/passing/sort/passingYards/year/2011

14. What was Griffin's average yards per attempt?

15. What was Keenum's NCAA quarterback rating?

16. What was Jones' completion percentage?

17. What was Weeden's yards per pass attempt ratio?

18. What was Boyd's touchdowns per pass attempt ratio?

19. What was Carder's NCAA quarterback rating?

20. Which quarterback had the highest completion percentage?

21. Which quarterback had the highest yards per attempt average?

22. Which quarterback had the lowest interceptions per pass attempt average?

Use the tables to solve each problem.

2011 NFL Team Defensive Statistics

Rk	Team	G	TotPts	Comp	Att	Yds	TD	Int
1	Pittsburgh Steelers	16	227	289	530	2751	15	11
2	Cleveland Browns	16	307	265	469	2959	16	9
3	Houston Texans	16	278	279	538	3035	18	17
4	Baltimore Ravens	16	266	288	535	3140	11	15
5	New York Jets	16	363	275	507	3216	15	19
6	Kansas City Chiefs	16	338	257	454	3221	23	20
7	St. Louis Rams	16	407	293	484	3301	21	12
8	Jacksonville Jaguars	16	329	326	513	3341	21	17
9	Cincinnati Bengals	16	323	319	539	3385	21	10
10	Philadelphia Eagles	16	328	301	518	3397	27	15

Source: ESPN NFL. "NFL Statistics - 2011." http://espn.go.com/nfl/statistics/team/_/stat/passing/position/defense/year/2011
Note: TotPts: points allowed; Comp: completions allowed; Att: pass attempts by opposing offenses; Yds: yards allowed (same as yards gained by opposing offense); TD: touchdowns allowed (same as TD's scored by opposing offense); Int: interceptions

2011 NFL Team Offensive Statistics

Rk	Team	Games	Total Yds	Pass	Rush Yds	Pts
1	Pittsburgh Steelers	16	4348	2751	1597	227
2	Houston Texans	16	4571	3035	1536	278
3	Baltimore Ravens	16	4622	3140	1482	266
4	San Francisco 49ers	16	4931	3695	1236	229
5	New York Jets	16	4993	3216	1777	363
6	Jacksonville Jaguars	16	5008	3341	1667	329
7	Cincinnati Bengals	16	5060	3385	1675	323
8	Philadelphia Eagles	16	5198	3397	1801	328
9	Seattle Seahawks	16	5315	3518	1797	315
10	Cleveland Browns	16	5318	2959	2359	307

Source: ESPN NFL. "NFL Statistics - 2011." http://espn.go.com/nfl/statistics/team/_/stat/passing/year/2011
Note: Total Yds: net total yards; Pass: net passing yards; Rush Yds: rushing yards; Pts: total points

23. Determine the scoreability index for the 2011 Baltimore Ravens and the bendability of the 2011 Pittsburgh Steelers. Based on the values, determine which team should win the game.

24. Determine the scoreability index for the 2011 New York Jets and the bendability of the 2011 Jacksonville Jaguars. Based on the values, determine which team should win the game.

25. What is the average yards allowed per game for the 2011 Kansas City Chiefs defense?

26. What is the average touchdowns allowed per game by the Philadelphia Eagles defense?

27. What is the average points per game allowed by the Cleveland Browns defense?

28. What is the bendability index of the St. Louis Rams?

29. Which team's defense gave up the most yards per pass attempt?

30. Which team's defense gave up the most yards per game?

12.3 Basketball

Winning the Lottery in the NBA

Did you know that NBA basketball teams participate in a lottery every year? Each year the NBA conducts a lottery among the 14 teams that do not make the playoffs to determine draft position. Baseball and football have many positions to fill, so their drafts for professional players consist of 40 rounds and 7 rounds, respectively. The larger number of players on the team reduces the impact of a single player on the outcome of a team's overall performances. Basketball, however, has only 2 rounds and a single player can be the difference between a mediocre team and a championship caliber team. This is because a single player represents 20%, or 1 out of 5, of the players on the court at one time. One of the most interesting aspects of the NBA lottery is that teams with the worst record each year are given the best chances to obtain the number one pick in the draft, and, as a result, dramatically improve their team. This player is generally considered the best player available for the draft in a given year. So, how do they stack the deck in favor of the worst team? How does this affect other teams in the lottery and their chances of winning in the lottery?

A team's probability of winning the draft lottery is determined by ***a priori*** **probability**. *A priori* probability assigns a probability to an event or outcome based on reason or some other information, thus altering the probability of that outcome actually occurring. For example, there are 6 sides on a standard die numbered 1 through 6, so the probability of obtaining a given number is $\frac{1}{6}$. However, if we were to paint the sides of the die where 3 sides are red, 1 side is blue, 1 side is yellow, and 1 side is green, then the probability of landing on a given color would be altered based on the number of sides that are that color. Since the sides are no longer unique, the probability of landing on a given side is changed to $\frac{3}{6}$ for landing on red and the probability of landing on blue, yellow, or green is each $\frac{1}{6}$.

To gain a better understanding of this concept, consider the idea that the probability of an event can change based on information you already have. For instance, in a professional basketball game where the Cleveland Cavaliers are playing, the probability of the Cavaliers winning will change based on whether LeBron James, their star forward, is playing. When James plays, the probability of the Cavaliers winning will be higher than when James does not play. So James playing gives the Cavaliers a higher priority in terms of the probability of a win. *A priori* probability is also used by the NBA to determine which team gets the first pick, second pick, and so on in the draft.

Example 1: Using *A Priori* Probability

A standard deck of playing cards has been modified. All of the 2s and the 3s in the deck are now aces, making the deck consist of twelve aces instead of four. What is the probability of drawing an ace on a single draw from this modified deck of cards?

Solution

Since the number of cards in the deck remains the same, the number of total possible outcomes remains 52. To answer the question, we simply need to calculate the probability of obtaining an ace as follows.

$$P(\text{ace}) = \frac{12}{52} = \frac{3}{13} \approx 0.231$$

Recall that in a standard deck of cards, $P(\text{ace}) = \frac{4}{52} = \frac{1}{13}$. By changing the number of aces in the deck from 4 to 12 (that is, three times more than the standard amount of aces) the probability of obtaining an ace has been tripled so that $P(\text{ace}) = 3\left(\frac{4}{52}\right) = \frac{3}{13} \approx 0.231$.

Skill Check #1

If a deck of cards has been modified to contain seven kings instead of four, and only one jack. What is the probability of drawing a king?

The NBA consists of 30 teams divided into two conferences (East and West) of 15 teams each. Only the top 8 teams in each conference actually make the playoffs. That means that there are 14 teams that do not make the playoffs. These 14 teams make up the teams participating in the "lottery."

Over the past 30 years, the NBA draft lottery has evolved into the system that is currently used. The lottery essentially works as follows. The 14 teams that did not make the playoffs are assigned combinations of the numbers 1 through 14. For example, a team may be assigned the combination of numbers 3-7-8-10, in any order. The total number of combinations that a team receives is based on the need to give the team with the worst record an advantage in obtaining the first pick in the draft. Statisticians wanted to ensure that the 14th worst team (the team particiapting in the lottery with the worst record) had some opportunity for selection. To that end, it was determined that the 14th worst team should have about $\frac{1}{2}$ of a 1% chance, or a probability of 0.005, of winning the lottery. NBA statisticians also decided that 1000 combinations of choosing a winner would be used.

Example 2: Combinations and the NBA Lottery

Think Back

We again revisit the idea of combinations, which was introduced in Chapter 7. Recall that a combination of n items chosen r at a time, without regard to order, is defined as follows.

$$_nC_r = \frac{n!}{r!(n-r)!}$$

Show that the number of 4-digit combinations of the 14 numbers used in the NBA lottery is 1001.

Solution

Using the formula for combinations, we can show the following when $n = 14$ and $r = 4$.

$$_{14}C_4 = \frac{n!}{r!(n-r)!} = \frac{14!}{4!(14-4)!} = 1001$$

There are 1001 possible combinations of 14 items chosen 4 at a time.

Fun Fact

Since the NBA only uses 1000 combinations to determine lottery probabilities, but there are 1001 possible combinations using the numbers 1 through 14, the combination 11-12-13-14 is never assigned to a team. If this combination is drawn, it is simply ignored. So far in the history of the NBA lottery, this combination has never been drawn.

The NBA assigns each team a number of combinations based on their overall record. The worst team is assigned 250 combinations, the 2nd worst team is assigned 199 combinations, and so on until all 14 teams are assigned a certain number of combinations. It should be noted that after the first three team selections in the draft are determined by the lottery, the remainder of the lottery picks continue in ranking order, from the worst record to the best. See Table 1 for the 2012 NBA lottery draft order.

Table 1			
2012 NBA Lottery Draft Order—Based on Overall Record			
Team	2011/12 Record	# of Combinations	Probability
Charlotte	7–59	250	$\frac{250}{1000} = 0.25$
Washington	20–46	199	$\frac{199}{1000} = 0.199$
New Orleans	21–45	156	$\frac{156}{1000} = 0.156$
Cleveland	21–45	119	0.119
Sacramento	22–44	88	0.088
New Jersey	22–44	63	0.063
Toronto	23–43	43	0.043
Golden State	23–43	28	0.028
Detroit	25–41	17	0.017
Minnesota	26–40	11	0.011
Portland	28–38	8	0.008
Milwaukee	31–35	7	0.007
Phoenix	33-33	6	0.006
Houston	32-32	5	0.005

Source: NBA. "2012 NBA Draft Lottery." http://www.nba.com/wizards/2012-nba-draft-lottery

Over the course of the last 30 years, the NBA has had several versions of the draft that have each striven to find the best way to give an advantage to each team that did not make the playoffs. At one time, the NBA had concerns that teams were losing on purpose at the end of a season in an effort to secure the 250 possible combinations in the lottery. In an effort to stop teams from intentionally losing games, the NBA instituted a rule that guaranteed the worst team in the league, by record, a pick in the draft of no worse than 4th. So, if the team with the worst record has not drawn a ball after the 3rd draw, the team automatically obtains the 4th draft pick.

Example 3: NBA Draft Lottery Probabilities

Given that Cleveland and New Orleans (as shown in Table 1) were the first two teams whose numbers were chosen in the 2012 lottery, what is the probability that Charlotte is the third chosen?

Solution

Since we know that the Cleveland and New Orleans were the first two chosen, their combinations can be eliminated from the total number of available eligible combinations to be drawn. This means that now there are $1000 - 156 - 119 = 725$ combinations left. With 725 possible combinations remaining, the probability that Charlotte is the 3rd draft lottery winner is as follows.

$$P(\text{Charlotte}) = \frac{250}{725} = 0.345$$

Charlotte has a 34.5% chance of obtaining the third pick, given that Cleveland and New Orleans were the first two chosen.

Basketball Player Statistics

We have already seen how the NBA uses the draft process to level the playing field among teams in its league, but what about other basketball leagues and how they use mathematics? Statistical data such as the scoring/shooting percentage, minutes played, length of a successful shot, points per game, free throws per game, etc. are basic ratios that help determine the effectiveness of a player.

Points per Game (PPG)

A basketball player's points per game average is just that, the average of the number of points a player scores in a game.

Points per Game (PPG)

Points per game, rounded to the nearest tenth, can be calculated with the following formula.

$$PPG = \frac{\text{total number of points scored}}{\text{number of games played}}$$

Example 4: Calculating Points per Game

Suppose Candace Parker, a member of the Los Angeles Sparks in the WNBA (Women's National Basketball Association), scored the following number of points over six games.

$$21 \quad 13 \quad 15 \quad 17 \quad 24 \quad 19$$

What is her PPG average over the six games?

Solution

We can substitute the known values into the formula to calculate the PPG as follows.

$$PPG = \frac{\text{total number of points scored}}{\text{number of games played}} = \frac{21+13+15+17+24+19}{6} = \frac{109}{6} \approx 18.2$$

So, Parker averaged about 18.2 PPG over the six games.

Skill Check #2

Over a ten-game span, Jack scored the following points in each game. What was Jack's PPG average over these ten games?

12 10 17 15 14 18 12 14 16 19

☞ **Helpful Hint**

A **free throw**, or foul shot, is an unopposed shot made from a designated area on the basketball court. These shots are generally awarded to a player after a foul is called on the opposing team.

Shooting Percentage

When a basketball player attempts a shot for either two or three points and makes the basket, it is called a **field goal**. To determine a basketball player's offensive prowess, we can consider the number of field goals made by the player compared to the number of field goals attempted. The higher the percentage, the more effective a player is at making shots that actually go in the basket.

Field Goal Percentage (FG%)

Field goal percentage, rounded to the nearest thousandth, can be calculated with the following formula.

$$FG\% = \frac{\text{field goals made}}{\text{field goals attempted}} \cdot 100$$

Example 5: Calculating Field Goal Percentage

In the 2011–2012 NBA regular season, LeBron James attempted 1169 field goals and made 621. What was LeBron's FG% for the season?

Solution

Field goal percentage is easily calculated as the ratio of field goals made to field goals attempted. So LeBron James' shooting percentage would be calculated as follows.

$$FG\% = \frac{\text{field goals made}}{\text{field goals attempted}} \cdot 100 = \frac{621}{1169} \cdot 100 \approx 53.1$$

This means that LeBron James had a FG% of 53.1%. In order to ascertain whether or not James' FG% is high, low, or average, we need only to look at the FG% of other players in the NBA. The 2011–2012 leader in FG% for a starting player was Tyson Chandler of the New York Knicks with a FG% of 69.7%. Out of all starters in the NBA, LeBron James was 13th on the list of highest FG% for the season. Considering there are 150 starters in the NBA, that's a pretty high ranking.

Fun Fact

The NBA has been keeping up with statistics since the beginning of the league. Do you know who is the all time NBA leader in field goal percentage? It's Artis Gilmore. For more information on statistical leaders of the NBA, check out http://www.nba.com/statistics.

Skill Check Answers

1. $\dfrac{7}{52} \approx 0.135$

2. 14.7

12.3 Exercises

Answer the question.

1. A player on a high school basketball team scored the following number of points over six games, respectively: 26, 17, 13, 18, 21, 22. What is her point-per-game average over the six games?

Use the table to answer each question.

2012 NBA Lottery Draft Order—Based on Overall Record			
Team	**2011–12 Record**	**# of Balls**	**Probability**
Charlotte	7–59	250	250/1000 = 0.25
Washington	20–46	199	199/1000 = 0.199
New Orleans	21–45	156	156/1000 = 0.156
Cleveland	21–45	119	0.119
Sacramento	22–44	88	0.088
New Jersey	22–44	63	0.063
Toronto	23–43	43	0.043
Golden State	23–43	28	0.028
Detroit	25–41	17	0.017
Minnesota	26–40	11	0.011
Portland	28–38	8	0.008
Milwaukee	31–35	7	0.007
Phoenix	33–33	6	0.006
Houston	32–32	5	0.005
Source: NBA. "2012 NBA Draft Lottery." http://www.nba.com/wizards/2012-nba-draft-lottery			

2. Given that Cleveland and Charlotte were the first two teams chosen in the lottery, what is the probability that New Orleans is the third chosen?

3. Given that New Orleans and Charlotte were the first two teams chosen in the lottery, what is the probability that Cleveland is the third chosen?

4. What is the probability that either Minnesota, Portland, Milwaukee, Phoenix, or Houston is chosen first?

Use the table to answer each question. In the table, FT stands for "free throw."

RK	Player	Team	GP	PTS	FG Made	FG Attempts	3-Pointers Made	3-Pointers Attempted	FT Made	FT Attempted
1	Durant	OKC	66	1850	643	1297	133	344	431	501
2	James	MIA	62	1683	621	1169	54	149	387	502
3	Bryant	LAL	58	1616	574	1336	87	287	381	451
4	Westbrook	OKC	66	1558	578	1266	62	196	340	413
5	Love	MIN	55	1432	474	1059	105	282	379	460
6	Griffin	LAC	66	1368	561	1022	2	16	244	468
7	Nowitzki	DAL	62	1342	473	1034	78	212	318	355
8	Jennings	MIL	66	1260	469	1121	129	388	193	239
9	Anthony	NY	55	1245	441	1025	68	203	295	367
10	Smith	ATL	66	1239	504	1101	28	109	203	322
11	Gay	MEM	65	1232	485	1067	54	173	208	263
12	Aldridge	POR	55	1191	483	943	2	11	223	274
13	Paul	LAC	60	1189	425	890	79	213	260	302
14	Pierce	BOS	61	1181	394	890	100	273	293	344
15	Ellis	GS/MIL	58	1181	450	1040	62	201	219	275
16	Jefferson	UTAH	61	1170	516	1048	1	4	137	177
17	Cousins	SAC	64	1160	448	999	2	14	262	373
18	Granger	IND	62	1159	391	941	123	323	254	291
19	Deron	NJ	55	1154	391	961	115	342	257	305
20	Lee	GS	57	1147	464	922	0	5	219	280

2012 NBA Offensive Player Statistics

Source: ESPN NBA. "NBA Statistics - 2012." http://espn.go.com/nba/statistics/player/_/stat/scoring/year/2012

5. What is Durant's free throw average?

6. What is Durant's PPG average?

7. What was Love's PPG average?

8. What was Lee's FG attempts per game average?

9. What was Gay's 3-pointers made per game average?

10. Which player had the highest PPG average?

11. Which player had the most FG attempts per game?

12. What was Paul's FT attempts per game?

13. What was Paul's FG attempts per game?

14. What was Anthony's average points per game?

15. Which player had the highest 3-pointer average?

16. What was the FG attempts average per game for Cousins?

17. Which player had the lowest FT average?

18. What was James' FG%?

19. What was James' 3-pointer percentage?

12.4 Additional Sports: Tennis, Golf, and Track & Field

In the previous sections, we looked at statistics measuring player and team performances in team sports. Now we turn our attention to individual sports such as tennis, golf, and track & field with some unique mathematical applications.

Scoring in Tennis

The game of tennis has one of the most interesting ways to keep score of any sport. Whether playing singles or doubles, tennis is scored as follows.

A point begins when a player serves the ball with the racquet. That player serves for an entire game that continues until either player earns four points, where each point is scored in multiples of 15, or almost 15. Scoring is then 0 (often called "love"), 15 for one point, 30 for two points, 40 for three points, and 60 (or "game") for four points. It should be noted that for a player to win a game, that player must win by at least two points. Also, if the two players are tied at 40, that is, the score is 40–40, we say that is a score of *deuce*, and play continues until a player wins two consecutive points. The first player to win six games (winning by two games) wins a set. However, if there is a 6–6 tie, a tie breaker game is played and the winner wins by one game. The winner of two out of three sets, or three out of five sets, wins the match.

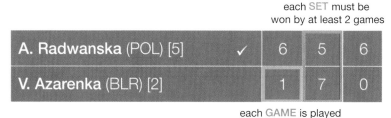

each SET must be won by at least 2 games

| A. Radwanska (POL) [5] | ✓ | 6 | 5 | 6 |
| V. Azarenka (BLR) [2] | | 1 | 7 | 0 |

each GAME is played for a point on the scoreboard

Figure 1: Scores from a Tennis Match

Fun Fact

The origins of scoring in tennis seem to be varied. The scores for a point of 15, 30, and 40 might have been derived from the scoring of a game played by British officers in India called sphairistike in the 19th century. Sphairistike's scoring system was based on the different gun calibers of the British naval ships of 15-pound guns on the main deck, followed by the 30-pound guns of the middle deck, and finally by the 40-pound lower gun deck.

Another explanation of the scoring system comes from a clock face that was used on the court in France, with a quarter move of the hand to indicate a score of 15, 30, and 45. When the hand moved to 60, the game was over.

Every time two players or two doubles pairs are pitted against each other in a tennis match, a probability can be attached to the likelihood that a given player wins any point. Consider two players, Player A and Player B. If we allow Player A to be the player *favored* to win a given point and Player B to be the *underdog* to win a given point, then we can assign probabilities based on these facts to determine the probability that a given player wins a game. If we let the probability that Player A wins a point be $P(\text{Player } A \text{ wins a point}) = f$ and let the probability that Player B wins a point be $P(\text{Player } B \text{ wins a point}) = u$, then we can use these values to calculate the probabilities of each player winning a game. Recall from Chapter 7 that

$$P(\text{Player } B \text{ wins a point}) = 1 - P(\text{player } A \text{ wins a point}).$$

Also recall that when multi-stage, independent events occur, the probabilities associated with the occurrences are also independent. For instance, if a coin is tossed three times, the probability of obtaining three heads is calculated as follows.

$$P(\text{HHH}) = P(\text{H}) \cdot P(\text{H}) \cdot P(\text{H}) = (P(\text{H}))^3 = (0.5)^3 = 0.125$$

We can use this same process to determine the probability of a tennis player winning a point, a game, a set, and ultimately a match.

Example 1: Probability of Winning in Tennis

If Susan is favored to win a tennis match and the probability that Susan wins a given point is $P(\text{Susan wins a point}) = f$, what is the probability that Susan wins four points in a row?

Solution

We are given that $P(\text{Susan wins a point}) = f$. We also know that the probability of winning a given point is an independent event. So, if we wish to determine the probability that Susan wins 4 points in a row, then $P(\text{Susan wins 4 points in a row}) = f \cdot f \cdot f \cdot f = f^4$.

Let's assume that the probability that Player A wins a point in a tennis match is

$$P(A \text{ wins a point}) = 0.7.$$

That means that the probability that Player B wins a point is

$$P(B \text{ wins a point}) = 1 - P(A \text{ wins a point}) = 1 - 0.7 = 0.3.$$

The possible outcomes of the match are given in Figure 2.

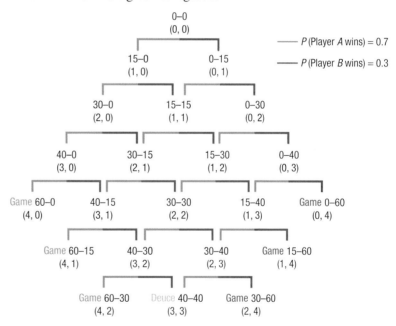

Figure 2: Score of Player A, Score of Player B
(Points Won by Player A, Points Won by Player B)

MATH MILESTONE

The Russian mathematician A. A. Markov (1856–1922) is known for his work in number theory, analysis, and probability theory. His work with the law of large numbers and the central limit theorem is the central component to understanding special classes of independent events that are now known as Markov chains.

As we progress through the possible scores for a given game of tennis, the probability of the transition is based on the present score and is independent of the other scores. The movement from point to point is known in probability theory as a Markov chain.

If Player A is serving and the score is 0–0, then after one point, the probability of the score being 0–15 is 0.3. Also, the probability of the score being 15–0 is 0.7. Symbolically, we can write the probabilities for the first "branch" of outcomes as follows.

$$P(15\text{–}0) = 0.7$$

$$P(0\text{–}15) = 0.3$$

It should be noted that this computation only represents the probability of a player winning the first point. Knowing that scoring each point in a tennis match is not affected by previous points scored

12

allows us to compute the remainder of the probabilities for each player to actually win the set. If we follow the tree diagram of the probabilities for each scoring scenario, we see that the probability of a score of Player A being 60 and Player B being 0 would be $0.7 \cdot 0.7 \cdot 0.7 \cdot 0.7 = 0.2401$. In the same manner, we could calculate the probability for each possible outcome in the game.

A formula can be developed to compute the probability of a particular game score for players A and B without using the tree. First, know that the probability to get the final game score through a single path through the tree is $P(\text{given path}) = 0.7^n \cdot 0.3^m$, and the probability is the same for all paths to that score. We need to find the number of possible paths from $(0, 0)$ to the desired score and then multiply that number by the $P(\text{given path})$ to obtain the total probability of a given score. So, we'll need to multiply the formula for the probability of a given path of the tree diagram by the number of ways to permute n–A's and m–B's. Thus, we have the following formula to compute the probability that a given player will win a game.

Think Back

In Chapter 7, we discussed probability and learned how to compute a permutation of n items and m items at the same time as $\dfrac{(n+m)!}{n!m!}$.

> ### Probability of a Game Score in Tennis
>
> $$P(\text{game score}) = \frac{\left(P_A\right)^n \left(P_B\right)^m (n+m)!}{n!m!}$$
>
> where n is the number of points Player A wins, m is the number of points Player B wins, P_A is the probability that Player A wins a point, and P_B is the probability that Player B wins a point.

Example 2: Probability of Winning a Point

Assume that two players, Jack and Glinda, are playing tennis. The probability that Glinda wins a given point is $P(\text{Glinda wins a point}) = 0.6$. What is the probability that after three points, the score will be 30–15, in favor of Jack?

Solution

The probability that Glinda wins any given point is $P(\text{Glinda wins a point}) = 0.6$, or $P(G) = 0.6$. This means $P(\text{Jack wins a point}) = 0.4$, or $P(J) = 0.4$. Our goal is to determine the probability that the score will be 30–15, in favor of Jack, after three games. So, we are looking for $P(\text{JJG})$ or $P(\text{JGJ})$ or $P(\text{GJJ})$. The paths JJG, JGJ, and GJJ are shown in blue in the figure.

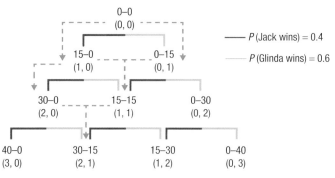

The order of obtaining a score of 30–15, in favor of Jack, can happen in three different, independent ways. So we have

$$P(\text{JJG}) = (0.4)(0.4)(0.6) = 0.096,$$

$$P(\text{JGJ}) = (0.4)(0.6)(0.4) = 0.096, \text{ and}$$

$$P(\text{GJJ}) = (0.6)(0.4)(0.4) = 0.096.$$

Therefore, the probability of Jack winning two points and Glinda winning one point is $0.096 + 0.096 + 0.096 = 0.288$.

Using the formula we developed, the solution can be calculated as follows.

$$P(\text{Jack 30 and Glinda 15}) = \frac{\left(P_A^n\right)\left(P_B^m\right)(n+m)!}{n!m!} = \frac{\left(0.4^2\right)\left(0.6^1\right)(2+1)!}{2!1!} = \frac{(0.16)(0.6)(3!)}{2!} = 0.288$$

So, the probability that Jack wins two points and Glinda wins one is 0.288.

Skill Check # 1

Assume that Jack and Glinda (from Example 2) are playing another game of tennis and the probability that Glinda wins a given point is still P(Glinda wins a point)=0.6. What is the probability that after 5 points, the score will be 40-30 in favor of Glinda?

Proportional Relationships in Sports

We now turn our attention to the use of proportional relationships to solve problems such as analyzing how fast a person runs and calculating reaction times for returning a tennis serve.

How Fast Is That Tennis Serve Anyway?

Eye-hand coordination is one of the most important skills for professional tennis players. Their ability to react to fast serves and place their return shots in the court are key to being successful tennis players. To understand just how little time professional tennis players have to react to a serve, we will use our understanding of proportions and dimensional analysis to determine a player's reaction time based on the speed of a tennis serve.

Fun Fact

In 2011, in a Davis Cup tennis match, Ivo Karlovic had a serve that was clocked at 156 mph. This is the fastest serve ever recorded in tournament play.

Figure 3 contains the dimensions of a tennis court. For singles play, the court is 78 feet long and is 27 feet wide. The speed of a serve off of the tennis racket is commonly measured in miles per hour (mi/hr or mph) or kilometers per hour (km/hr). However, since the size of the court is in feet, it might be easier to determine a reaction time based on feet per second (ft/s or fps). To do this calculation, the only information needed is how each of these dimensions relates to the others. Recall that there are 5280 feet in 1 mile and that there are 3600 seconds in one hour. Using these, we can change miles per hour to feet per second with ease.

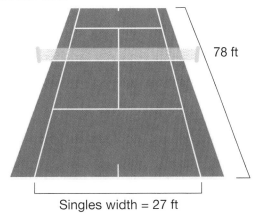

78 ft

Singles width = 27 ft

Figure 3: Tennis Court Dimensions

Example 3: Miles per Hour to Feet per Second

If a car is traveling at 70 miles per hour, what is the car's speed in feet per second?

Solution

We find the following based on our knowledge of proportions of time and distance.

$$\frac{70 \text{ mi}}{1 \text{ hr}} = \frac{70 \text{ mi}}{1 \text{ hr}} \cdot \frac{1 \text{ hr}}{3600 \text{ sec}} \cdot \frac{5280 \text{ ft}}{1 \text{ mi}}$$

$$= \frac{70 \text{ mi}}{1 \text{ hr}} \cdot \frac{1 \text{ hr}}{3600 \text{ sec}} \cdot \frac{5280 \text{ ft}}{1 \text{ mi}}$$

$$= \frac{(70 \cdot 5280) \text{ ft}}{3600 \text{ sec}} = \frac{369,600 \text{ ft}}{3600 \text{ sec}} \approx 102.667 \text{ ft/sec}$$

So, a car traveling at 70 mi/hr is traveling at 102.667 ft/sec.

To determine a tennis player's reaction time, we must take some things for granted, such as the velocity of the ball off of the racket versus the velocity of the ball when it reaches the returning player. The ball will lose velocity as it travels across the court due to air resistance as it travels through the air and friction when it bounces on the ground. But how much does the ball actually slow down? Resistance in the air and friction between the ball and the court are just two of the factors that determine how much the ball slows before reaching the returning player. For simplicity, we are going to assume (as many professionals training for a big match do) that the average velocity of the ball over the entire distance the ball travels is approximately 75% of its initial velocity, that is, the ball speed off of the racket.

Example 4: Reaction Time to a Tennis Serve

Venus Williams holds the record for the fastest women's tennis serve at 129 miles per hour. Using the standard dimensions of a tennis court, how long does the player returning the serve have to react to the ball?

Solution

We must take into account the information we have regarding the velocity of the ball reaching the player. If we assume that the ball will reach the returning player at 75% of its initial velocity, then the ball has an average velocity of $0.75 \cdot 129 = 96.75$ mi/hr. From Figure 3, we can see that the length of the court is 78 feet. Converting 96.75 mi/hr to ft/sec, we get

$$\frac{96.75 \text{ mi}}{1 \text{ hr}} = \frac{96.75 \text{ mi}}{1 \text{ hr}} \cdot \frac{1 \text{ hr}}{3600 \text{ sec}} \cdot \frac{5280 \text{ ft}}{1 \text{ mi}}$$

$$= \frac{96.75 \text{ mi}}{1 \text{ hr}} \cdot \frac{1 \text{ hr}}{3600 \text{ sec}} \cdot \frac{5280 \text{ ft}}{1 \text{ mi}}$$

$$= \frac{(96.75 \cdot 5280) \text{ ft}}{3600 \text{ sec}} = \frac{510,840 \text{ ft}}{3600 \text{ sec}} = 141.9 \text{ ft/sec}.$$

Therefore, the average velocity of the serve would be 141.9 ft/sec. In terms of reaction time, since the serve only travels 78 feet, we can determine that the reaction time would be

$$\frac{78 \text{ ft}}{\frac{141.9 \text{ ft}}{\text{sec}}} = \frac{78}{141.9} \text{ sec} \approx 0.5497 \text{ sec}.$$

The player moving to where the ball is located and returning the 129 mi/hr serve would have a little more than $\frac{1}{2}$ of a second to return the ball!

Skill Check #2

Andy Roddick once served a tennis ball 155 miles per hour. Using the same method as Example 4, determine the reaction time for the player returning the serve.

Track and Field

How Fast Are the World's Fastest Man and Woman?

Keeping with the theme of dimensional analysis, we turn our attention to determining exactly how fast the world's fastest man and woman can run. Since track and field events are measured in meters, we need a simple conversion from meters per second to kilometers per hour. Also, we will convert each race time to feet per second and miles per hour.

We can use the benchmark that 1 inch = 2.54 centimeters (cm) to determine that 1 foot = 30.48 cm. We also need to recall that 1 meter (m) = 100 cm and that 1 kilometer (km) = 1000 m. This means that 1 ft = 0.3048 m.

Example 5: How Fast Is Usain Bolt?

Usain Bolt is the 2008, 2009, 2010, 2011, 2012, and 2013 world champion in the 100-meter dash and also holds the world record with a time of 9.58 seconds. Determine how fast Bolt runs in

a. meters per second.

b. kilometers per hour.

c. feet per second

d. miles per hour.

Solution

Using the given conversions and knowledge of dimensional analysis, we can find each of these measurements. We know that Bolt runs 100 meters in 9.58 seconds, so we can calculate the following.

a. $\dfrac{100 \text{ m}}{9.58 \text{ sec}} = 10.438 \text{ m/sec}$

b.
$$\frac{10.438 \text{ m}}{\text{sec}} = \frac{10.438 \text{ m}}{\text{sec}} \cdot \frac{3600 \text{ sec}}{1 \text{ hr}} \cdot \frac{1 \text{ km}}{1000 \text{ m}}$$

$$= \frac{10.438 \text{ m}}{\text{sec}} \cdot \frac{3600 \text{ sec}}{1 \text{ hr}} \cdot \frac{1 \text{ km}}{1000 \text{ m}}$$

$$= \frac{(10.438 \cdot 3600) \text{ km}}{1000 \text{ hr}} = \frac{37{,}576.8 \text{ km}}{1000 \text{ hr}} \approx 37.577 \text{ km/hr}$$

c.
$$\frac{10.438 \text{ m}}{\text{sec}} = \frac{10.438 \text{ m}}{\text{sec}} \cdot \frac{1 \text{ ft}}{0.3048 \text{ m}}$$

$$= \frac{10.438 \text{ m}}{\text{sec}} \cdot \frac{1 \text{ ft}}{0.3048 \text{ m}} = \frac{10.438 \text{ ft}}{0.3048 \text{ sec}} \approx 34.245 \text{ ft/sec}$$

d.
$$\frac{34.245 \text{ ft}}{\text{sec}} = \frac{34.245 \text{ ft}}{\text{sec}} \cdot \frac{3600 \text{ sec}}{1 \text{ hr}} \cdot \frac{1 \text{ mi}}{5280 \text{ ft}}$$

$$= \frac{34.245 \text{ ft}}{\text{sec}} \cdot \frac{3600 \text{ sec}}{1 \text{ hr}} \cdot \frac{1 \text{ mi}}{5280 \text{ ft}}$$

$$= \frac{(24.245 \cdot 3600) \text{ mi}}{5280 \text{ hr}} = \frac{123{,}282 \text{ mi}}{5280 \text{ hr}} \approx 23.349 \text{ mi/hr}$$

Skill Check #3

Florence Griffith Joyner (December 21, 1959–September 21, 1998), nicknamed Flo Jo, holds the women's world record in the 100-meter dash with a time of 10.40 seconds. Determine how fast Joyner ran in

a. meters per second.

b. kilometers per hour.

c. feet per second.

d. miles per hour.

Relative Closeness of Scores in Golf

What's the Chance of Sinking That Putt and How Much Does It Matter?

In 2012, the top-ranked golfer on the money list, Rory McIlroy, had a scoring average of 68.87, while the 100th ranked player on the money list, Bob Estes, had a scoring average of 70.42. That is slightly more than $1\frac{1}{2}$ strokes per round, or 18 holes, of golf. In a standard tournament that lasts four days, that turns out to be about six strokes.

Professional golfers recognize that spending much of their practice time putting is necessary to be an excellent professional golfer. In terms of distance, putting is generally the shortest shot in golf. However, this shot can make drastic differences in scores. Consider the distance that a player putts the ball and the likelihood of them making the putt. The farther the distance of the putt, the less likely the putt will be made. Reducing the number of missed putts is a priority for professional golfers who wish to lower their golf scores and win more golf tournaments. Table 1 lists the distances of putts and the percentage of those putts made on the PGA tour for the 2013 season.

Table 1												
Percentage of Putts Made Based on Distance, PGA Tour 2013												
Distance (in feet)	1	2	3	4	5	6	7	8	9	10	11	12	13
Percent Made	100	96	90	85	75	65	56	49	43	38	34	30	30
Distance (in feet)	14	15	16	17	18	19	20	21	22	23	24	25	
Percent Made	25	22	20	19	17	16	14	13	12	11	11	10	

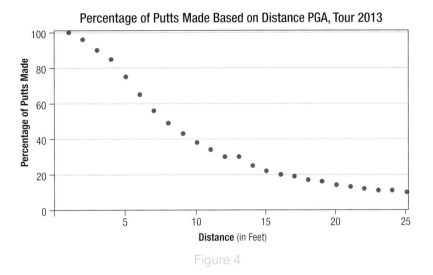

Figure 4

When we plot the data for putting distance compared to percentage made, we get a very interesting picture. Figure 4 illustrates that as the distance of the putt increases, the likelihood of making the putt decreases. Does the shape of the graph look familiar? It has the appearance of the graph of an exponential function. From our earlier study on growth and functions, we know that if we have the equation for the function, then we can determine the value of the function at any given point. In this situation, that means we can approximate the percent chance of making a putt from a given distance.

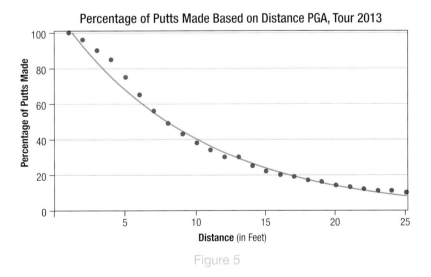

Figure 5

The function that closely models the data for the 2013 PGA Season is $f(x) = 113.374(0.902111)^x$, where x represents the distance of the putt and $f(x)$ represents the percentage of putts made. Figure 5 illustrates $f(x)$ graphed along with the data.

Example 6: Chances of Making a Putt

Fun Fact

Using the function for a 2013 PGA player making a putt, $f(x) = 113.374(0.902111)^x$, determine the percent chance of a player making a 19.5-foot putt during the 2013 PGA season.

Solution

We are given the function for percentage of putts made as $f(x) = 113.374(0.902111)^x$. If the distance is 19.5 feet, then the chance of making the putt is as follows.

$$f(19.5) = 113.374(0.902111)^{19.5} \approx 15.21\%$$

So, statistically speaking, a 2013 PGA player had approximately a 15.21% chance of actually making a 19.5-foot putt.

Skill Check Answers

1. $P(\text{Glinda 40, Jack 30}) = \dfrac{\left(P_A^{\,n}\right)\left(P_B^{\,m}\right)(n+m)!}{n!m!} = \dfrac{\left(0.6^3\right)\left(0.4^2\right)5!}{3!2!} = 0.3456$

2. 0.457 seconds

3. **a.** 9.615 m/s; **b.** 34.614 km/hr; **c.** 31.545 ft/sec; **d.** 21.508 mi/hr

12.4   Exercises

Solve each problem.

1. If Cydney is favored to win a tennis match and the probability that Cydney wins a given point is $P(\text{Cydney wins a point}) = f$, what is the probability that Cydney wins six points in a row?

2. Two players, Connor and Cade, are playing tennis. The probability that Connor wins a point is $P(\text{Connor wins a point}) = 0.8$ and the probability that Cade wins a point is $P(\text{Cade wins a point}) = 0.2$. Find each of the following probabilities.

    a. What is the probability that after three points, the score will be Cade-30 and Connor-15?

    b. What is the probability that after three points, the score will be Cade-15 and Connor-30?

    c. What is the probability that after two points, the score will be Cade-30 and Connor-0?

    d. What is the probability that after two points, the score will be Cade-0 and Connor-30?

    e. What is the probability that after four points, the score will be Cade-30 and Connor-30?

    f. What is the probability that after four points, the score will be Cade-40 and Connor-15?

    g. What is the probability that after four points, the score will be Cade-15 and Connor-40?

    h. What is the probability that after four points, Cade will have won the game?

    i. What is the probability that after four points, Connor will have won the game?

3.   Liz and Christie are playing tennis. The probabilities that each wins a given point are $P(\text{Liz wins a point}) = 0.55$ and $P(\text{Christie wins a point}) = 0.45$. Find each of the following probabilities.

   a. What is the probability that after three points, the score will be Liz-30 and Christie-15?

   b. What is the probability that after three points, the score will be Liz-15 and Christie-30?

   c. What is the probability that after two points, the score will be Liz-30 and Christie-0?

   d. What is the probability that after two points, the score will be Liz-0 and Christie-30?

   e. What is the probability that after four points, the score will be Liz-30 and Christie-30?

   f. What is the probability that after four points, the score will be Liz-40 and Christie-15?

   g. What is the probability that after four points, the score will be Liz-15 and Christie-40?

   h. What is the probability that after four points, Liz will have won the game?

   i. What is the probability that after four points, Christie will have won the game?

4.   A player serves a tennis ball at 85 miles per hour. How long does their opponent have to react to the serve?

5.   A player serves a tennis ball at 135 miles per hour. How long does their opponent have to react to the serve?

6.   Jeff runs the 100-meter dash in 10.55 seconds. Determine how fast Jeff runs in each of the following units of measurement.

   a. Meters per second

   b. Kilometers per hour

   c. Feet per second

   d. Miles per hour

7.   Blair runs marathons of length 26.2 miles. If Blair runs 26.2 miles in 4 hours and 15 minutes, determine how fast Blair runs in each of the following units of measurement?

   a. Meters per second

   b. Kilometers per hour

   c. Feet per second

   d. Miles per hour

8.   The world record in the 400-meter dash is held by Michael Johnson with a time of 43.18 seconds. Determine how fast Johnson ran in each of the following units of measurement.

   a. Meters per second

   b. Kilometers per hour

   c. Feet per second

   d. Miles per hour

9.   If a runner is running at 9.5 miles per hour, how long will it take the runner to run 26.2 miles?

10. How fast must a marathoner run to complete a 26.2 mile run in 3 hours or less?

11. How fast must a marathoner run to complete a 26.2 mile run in 4 hours or less?

12. Most marathons must be completed within 16 hours. How fast must a runner run to finish a 26.2 mile run in exactly 16 hours?

13. Using the formula for finding the percentage of putts made by 2013 PGA players, $f(x) = 113.374(0.902111)^x$, determine the chance of a player making the following putts.

   a. 10 feet   b. 15 feet

   c. 20 feet   d. 11 feet

   e. 7 feet   f. 9 feet

   g. 21 feet   h. 21.5 feet

   i. 22 feet

Chapter 12   Summary

Section 12.1   Baseball and Softball

Definitions

At Bat (AB)

An at bat occurs when a batter has any outcome from a plate appearance except when the catcher interferes with the play or unless the player: **1.** is hit by the ball; **2.** makes a sacrifice hit; or **3.** is walked.

Sacrifice At Bat (SAC)

A sacrifice at bat, or sacrifice hit, is when a player intentionally gives up an out to advance a base runner to either 3rd base or home plate.

Hit By Pitch (HBP)

A batter earns a hit-by-pitch designation if the pitcher hits any part of the batter or his uniform with the ball. In this case, the batter is awarded first base.

Walk (BB)

A walk, or base on balls, occurs when a batter receives four pitches that are outside the strike zone, known as a ball, before being called out or getting a hit. A walk is an intentional walk if the pitcher makes no attempt at throwing a strike.

Hit (H)

A hit occurs when a batter successfully puts the ball into play and reaches a base safely without the defensive team making an error in attempting to catch the ball. A hit is a single if the runner reaches 1st base, a double if the runner reaches 2nd base, a triple if the runner reaches 3rd base, or a home run if the runner reaches home plate.

Base Runner

A base runner is any batter that successfully reaches base on either a hit or a walk.

Formulas

Batting Average (BA)

Batting average, rounded to the nearest thousandth, can be calculated using the following formula.

$$ BA = \frac{\text{number of hits}}{\text{number of at bats}} = \frac{H}{AB} $$

On-Base Percentage (OBP)

On-base percentage, rounded to the nearest thousandth, can be calculated using the following formula.

$$ OBP = \frac{H + BB + HBP}{AB + BB + HBP + SAC} $$

Slugging Percentage (SLG)

Slugging percentage, rounded to the nearest thousandth, can be calculated by the following formula.

$$ SLG = \frac{1B + (2 \cdot 2B) + (3 \cdot 3B) + (4 \cdot HR)}{AB} $$

Here 1B is the number of singles, 2B is the number of doubles, 3B is the number of triples, HR is the number of home runs, and AB is the number of at bats.

On-Base Plus Slugging Percentage (OPS)

On-base plus slugging percentage, rounded to the nearest thousandth, can be calculated by the following formula.

$$OPS = \frac{AB \cdot (H + BB + HBP) + \text{total bases} \cdot (AB + BB + HBP + SAC)}{AB \cdot (AB + BB + HBP + SAC)}$$

Here, total bases is computed as $1B + (2 \cdot 2B) + (3 \cdot 3B) + (4 \cdot HR)$.

Alternatively, we could simply write $OPS = OBP + SLG$.

Earned Run Average (ERA)

Earned run average, rounded to the nearest hundredth, is the average number of runs a pitcher allows per nine innings and can be calculated using the following formula.

$$ERA = \frac{\text{earned runs}}{\text{innings pitched}} \cdot 9$$

Walks Plus Hits per Inning Pitched (WHIP)

Walks plus hits per inning pitched, rounded to the nearest hundredth, can be calculated as follows.

$$WHIP = \frac{\text{walks} + \text{hits}}{\text{innings pitched}}$$

Section 12.2   Football

Formulas

Completion Percentage (CP)

Completion percentage, rounded to the nearest thousandth, can be calculated with the following formula.

$$CP = \frac{\text{number of passes completed}}{\text{number of passes attempted}}$$

Yards per Pass Attempt (YPA)

Yards per pass attempt, rounded to the nearest hundredth, can be calculated with the following formula.

$$YPA = \frac{\text{total number of yards passing}}{\text{number of passes attempted}}$$

Touchdowns per Pass Attempt (TDPA)

Touchdowns per pass attempt, rounded to the nearest hundredth, can be calculated with the following formula.

$$TDPA = \frac{\text{number of touchdowns}}{\text{number of passes attempted}}$$

Interceptions per Pass Attempt (IPA)

Interceptions per pass attempt, rounded to the nearest thousandth, can be calculated as follows.

$$IPA = \frac{\text{number of interceptions}}{\text{number of passes attempted}}$$

NFL Quarterback Rating

The formula to calculate the quarterback rating for a player in the NFL, rounded to the nearest tenth, is as follows.

$$\text{NFL quarterback rating} = \left(\frac{5(CP - 0.3) + 0.25(YPA - 3) + 20(TDPA) + 25(0.095 - IPA)}{6} \right) \cdot 100$$

Note: CP is left as a decimal.

NCAA Quarterback Rating

The formula for computing the NCAA quarterback rating, rounded to the nearest tenth, is as follows.

$$\text{NCAA quarterback rating} = \frac{(8.4 \cdot \text{yards}) + (330 \cdot \text{TDs}) + (100 \cdot \text{comp}) - (200 \cdot \text{int})}{\text{number of pass attempts}}$$

Here, yards = passing yards, TDs = number of touchdowns, comp = number of completions, and int = number of interceptions.

Scoreability Index (YPS)

The scoreability index, rounded to the nearest hundredth, of an offense can be calculated with the following formula.

$$\text{YPS} = \frac{\text{total number of yards gained}}{\text{total number of points scored}}$$

Bendability Index (YPPA)

The bendability index, rounded to the nearest hundredth, of a defense can be calculated with the following formula.

$$\text{YPPA} = \frac{\text{total number of yards allowed}}{\text{total number of points allowed}}$$

Section 12.3   Basketball

Definition

A Priori Probability

A priori probability assigns a probability to an event or outcome based on reason or some other information, thus altering the probability of that outcome actually occurring.

Formulas

Points per Game (PPG)

Points per game, rounded to the nearest tenth, can be calculated with the following formula.

$$\text{PPG} = \frac{\text{total number of points scored}}{\text{number of games played}}$$

Field Goal Percentage (FG%)

Field goal percentage, rounded to the nearest thousandth, can be calculated with the following formula.

$$\text{FG\%} = \frac{\text{field goals made}}{\text{field goals attempted}}$$

Section 12.4   Additional Sports: Tennis, Golf, and Track & Field

Formula

Probability of a Game Score in Tennis

$$P(\text{game score}) = \frac{\left(P_A{}^n P_B{}^m\right)(n+m)!}{n!m!}$$

when n is the number of points Player A wins, m is the number of points Player B wins, P_A is the probability that Player A wins a point, and P_B is the probability that Player B wins a point.

Chapter 12 Exercises

Solve each problem.

1. A batter had 73 at bats and had 41 hits. What is the batting average of the player?

2. Les has the highest batting average on his softball team with an average of 0.721. If Les has had 215 at bats, how many hits has he had?

3. During a season, Chelsea had 112 hits in 335 at bats. What is Chelsea's season batting average?

4. Jacob has 3 home runs, 4 triples, 16 doubles, 24 singles, 3 walks, and 5 outs. What is Jacob's slugging percentage?

5. During a season in the National Pro Fastpitch softball league, a player had the following batting results after 58 plate appearances. Determine the player's OPS over the season.

BB	1B	1B	1B	BB	Out	Out	2B	HBP	1B	HR
SAC	SAC	3B	Out	HR	Out	BB	Out	HR	2B	1B
BB	1B	1B	1B	BB	Out	Out	2B	HBP	1B	HR
SAC	SAC	3B	Out	HR	Out	BB	Out	HR	2B	1B
BB	1B	1B	1B	BB	Out	Out	2B	HBP	1B	HR
SAC	SAC	3B								

6. A pitcher pitches 16 innings and gives up 5 earned runs. Determine the pitchers ERA.

7. A pitcher has thrown for 142 innings this season and has given up 29 hits and 15 walks. Determine the WHIP for this pitcher.

8. A youth baseball team has 12 players. Determine the number of ways in which a coach could choose 9 players for the starting lineup.

Use the table to answer each question.

2012 MLB Individual Batting Statistics													
Rk	Player	Team	Pos	G	AB	R	H	2B	3B	HR	RBI	BB	SO
1	Posey	SF	C	148	530	78	178	39	1	24	103	69	96
2	Cabrera	DET	3B	161	622	109	205	40	0	44	139	66	98
3	McCutchen	PIT	CF	157	593	107	194	29	6	31	96	70	132
4	Trout	LAA	CF	139	559	129	182	27	8	30	83	67	139
5	Beltre	TEX	3B	156	604	95	194	33	2	36	102	36	82
6	Braun	MIL	LF	154	598	108	191	36	3	41	112	63	128
7	Mauer	MIN	C	147	545	81	174	31	4	10	85	90	88
8	Jeter	NYY	SS	159	683	99	216	32	0	15	58	45	90
9	Molina	STL	C	138	505	65	159	28	0	22	76	45	55
10	Fielder	DET	1B	162	581	83	182	33	1	30	108	85	84
11	Hunter	LAA	CF	140	534	81	167	24	1	16	92	38	133
12	Butler	KC	DH	161	614	72	192	32	1	29	107	54	111
13	Cano	NYY	2B	161	627	105	196	48	1	33	94	61	96
14	Pacheco	COL	3B	132	475	51	147	32	3	5	54	22	61
15	Craig	STL	1B	119	469	76	144	35	0	22	92	37	89

Source: ESPN MLB. "MLB Player Batting Statistics - 2012." http://espn.go.com/mlb/stats/batting/_/year/2012
Note: H: the sum of all 1B, 2B, 3B, and HR; G: games; AB: at bats; R: runs

9. What was Posey's batting average for the 2012 season?

10. How many singles did Trout have in the 2012 season?

11. What was Mauer's SLG for the 2012 season?

12. What was Braun's SLG for the 2012 season?

13. Determine the player with the highest BA for the 2012 season.

14. What was Molina's strikeout to hit ratio?

12

Use the table to answer each question.

2012 MLB Individual Pitching Statistics											
Rk	Player	Team	W	L	G	IP	H	R	ER	HR	BB	SO
1	Kershaw	LAD	14	9	33	227.2	170	70	64	16	63	229
2	Price	TB	20	5	31	211.0	173	63	60	16	59	205
3	Verlander	DET	17	8	33	238.1	192	81	70	19	60	239
4	Dickey	NYM	20	6	34	233.2	192	78	71	24	54	230
5	Cueto	CIN	19	9	33	217.0	205	73	67	15	49	170
6	Cain	SF	16	5	32	219.1	177	73	68	21	51	193
7	Weaver	LAA	20	5	30	188.2	147	63	59	20	45	142
8	Lohse	STL	16	3	33	211.0	192	74	67	19	38	143
9	Gonzalez	WSH	21	8	32	199.1	149	69	64	9	76	207
10	Zimmermann	WSH	12	8	32	195.2	186	69	64	18	43	153
11	Sale	CWS	17	8	30	192.0	167	66	65	19	51	192
12	Hamels	PHI	17	6	31	215.1	190	80	73	24	52	216
13	Hernandez	SEA	13	9	33	232.0	209	84	79	14	56	223
14	Hellickson	TB	10	11	31	177.0	163	68	61	25	59	124
15	Lee	PHI	6	9	30	211.0	207	79	74	26	28	207

Source: ESPN MLB. "MLB Player Pitching Stats - 2012." http://espn.go.com/mlb/stats/pitching/_/year/2012/order/false
Note: W: games won by the pitcher; L: games lost by the pitcher; G: games the pitcher pitched in; IP: innings pitched; H: hits allowed; R: runs scored by the other team; ER: earned runs scored by the opposing team; HR: home runs allowed; BB: bases on balls allowed; SO: strikeouts by batters faced

15. What was Lee's WHIP for 2012?

16. What was Cueto's ERA for 2012?

17. What was Hamels' ratio of strikeouts to innings pitched?

18. What was Verlander's ratio of walks per inning pitched?

19. What was Weaver's average innings per game?

20. Which pitcher had the lowest ERA in 2012?

21. Which pitcher had the lowest walks per inning ratio?

Answer each question.

22. A quarterback passes 155 times during a season and completes 98 of those passes. What is the quarterback's completion percentage?

23. A quarterback has a completion percentage of 53.6% after completing 134 passes. How many passes did the quarterback attempt? (Round to the nearest whole number.)

Use the table to answer each question.

Rk	Player	Team	Pass Completions	Pass Attempts	Yards	TD	Int
			2012 NFL Quarterback Statistics				
1	Rodgers	GB	371	552	4,295	39	8
2	P. Manning	DEN	400	583	4,659	37	11
3	Griffin III	WSH	258	393	3,200	20	5
4	Wilson	SEA	252	393	3,118	26	10
5	Ryan	ATL	422	615	4,719	32	14
6	Brady	NE	401	637	4,827	34	8
7	Roethlisberger	PIT	284	449	3,265	26	8
8	Brees	NO	422	670	5,177	43	19
9	Schaub	HOU	350	544	4,008	22	12
10	Romo	DAL	425	648	4,903	28	19
11	Rivers	SD	338	527	3,606	26	15
12	Flacco	BAL	317	531	3,817	22	10
13	Dalton	CIN	329	528	3,669	27	16
14	E. Manning	NYG	321	536	3,948	26	15
15	Newton	CAR	280	485	3,869	19	12

Source: ESPN NFL. "NFL Player Passing Statistics - 2012." http://espn.go.com/nfl/statistics/player/_/stat/passing/sort/quarterbackRating/year/2012

24. What was Manning's touchdowns per pass attempt ratio?

25. What was Roger's completion percentage?

26. What was Griffin III's YPA?

27. What was Schaub's NFL quarterback rating?

28. Which quarterback completed the highest percentage of passes?

29. Which quarterback had the highest TDPA?

30. Which quarterback had the highest IPA?

Use the table to answer each question.

Rk	Player	Team	Comp	Att	Yds	TD	Int	Sack
\multicolumn{9}{c}{2012 NCAA Quarterback Statistics}
1	Nick Florence	BAY	286	464	4309	33	13	18
2	Landry Jones	OKLA	367	555	4267	30	11	13
3	Seth Doege	TTU	380	541	4205	39	16	17
	Geno Smith	WVU	369	518	4205	42	6	19
5	Rakeem Cato	MRSH	406	584	4201	37	11	26
6	David Fales	SJSU	327	451	4193	33	9	26
7	Colby Cameron	LT	359	522	4147	31	5	10
8	Derek Carr	FRES	344	511	4104	37	7	27
9	Mike Glennon	NCST	330	564	4031	31	17	36
10	Tajh Boyd	CLEM	287	427	3896	36	13	31
11	Aaron Murray	UGA	249	386	3893	36	10	26
12	Ryan Nassib	SYR	294	471	3749	26	10	16
13	Brett Hundley	UCLA	319	479	3745	29	11	52
14	Teddy Bridgewater	LOU	287	419	3718	27	8	28
15	Johnny Manziel	TA&M	295	434	3706	26	9	22

Source: ESPN College Football. "FBS (I-A) Player Passing Statistics - 2012." http://espn.go.com/college-football/statistics/player/_/stat/passing/sort/passingYards/year/2012

31. What was Johnny Manziel's average yards per pass attempt ratio?

32. What was Landry Jones' NCAA quarterback rating?

33. What was Landry Jones' completion percentage?

34. What was Geno Smith's yards per pass attempt ratio?

35. What was Aaron Murray's touchdowns per pass attempt ratio?

36. Which QB had the highest completion percentage?

37. Which QB had the highest YPA?

Use the tables to solve each problem.

2011 NFL Team Defensive Statistics

Team	Games	Total Pts	Comp	Att	Yds	TD	Int
Pittsburgh Steelers	16	227	289	530	2,751	15	11
Cleveland Browns	16	307	265	469	2,959	16	9
Houston Texans	16	278	279	538	3,035	18	17
Baltimore Ravens	16	266	288	535	3,140	11	15
New York Jets	16	363	275	507	3,216	15	19
Kansas City Chiefs	16	338	257	454	3,221	23	20
St. Louis Rams	16	407	293	484	3,301	21	12
Jacksonville Jaguars	16	329	326	513	3,341	21	17
Cincinnati Bengals	16	323	319	539	3,385	21	10
Philadelphia Eagles	16	328	301	518	3,397	27	15

Source: ESPN NFL. "NFL Statistics - 2011." http://espn.go.com/nfl/statistics/team/_/stat/passing/position/defense/year/2011
Note: Total Pts: points allowed; Comp: completions allowed; Att: pass attempts by opposing offenses; Yds: yards allowed (same as yards gained by opposing offense); TD: touchdowns allowed (same as TDs scored by opposing offense); Int: interceptions

2011 NFL Team Offensive Statistics

Team	Games	Total Yds	Pass	Rush Yds	Pts
Pittsburgh Steelers	16	4348	2751	1597	227
Houston Texans	16	4571	3035	1536	278
Baltimore Ravens	16	4622	3140	1482	266
San Francisco 49ers	16	4931	3695	1236	229
New York Jets	16	4993	3216	1777	363
Cincinnati Bengals	16	5060	3385	1675	323
Philadelphia Eagles	16	5198	3397	1801	328
Seattle Seahawks	16	5315	3518	1797	315
Cleveland Browns	16	5318	2959	2359	307
Kansas City Chiefs	16	5333	3221	2112	338
Atlanta Falcons	16	5338	3786	1552	350
Washington Redskins	16	5437	3553	1884	367
New York Giants	16	6022	4082	1940	400
New England Patriots	16	6577	4703	1874	342
Green Bay Packers	16	6585	4796	1789	359

Source: ESPN NFL. "NFL Statistics - 2011." http://espn.go.com/nfl/statistics/team/_/stat/passing/year/2011
Note: Total Yds: net total yards; Pass: net passing yards; Rush Yds: rushing yards; Pts: total points

38. Determine the YPS for the 2011 New York Giants.

39. Determine the YPS for the 2011 San Francisco 49ers.

40. What is the average number of yards allowed per game for the 2011 Baltimore Ravens defense?

41. What is the average number of touchdowns allowed per game by the Jacksonville defense?

42. What is the average number of points per game allowed by the Cleveland Browns defense?

43. What is the bendability index of the Pittsburgh Steelers for 2011?

44. Which team's defense gave up the most yards per game?

Answer the question.

45. A player on a high school basketball team scored the following number of points in six games, respectively.

15 14 18 12 19 23

What is her PPG average over the six games?

Use the table to answer each question.

2012 NBA Lottery Draft Order—Based on Overall Record			
Team	2011–12 Record	# of Chances to Win	Probability
Charlotte	7–59	250	0.250
Washington	20–46	199	0.199
New Orleans	21–45	156	0.156
Cleveland	21–45	119	0.119
Sacramento	22–44	88	0.088
New Jersey	22–44	63	0.063
Toronto	23–43	43	0.043
Golden State	23–43	28	0.028
Detroit	25–41	17	0.017
Minnesota	26–40	11	0.011
Portland	28–38	8	0.008
Milwaukee	31–35	7	0.007
Phoenix	33–33	6	0.006
Houston	32–32	5	0.005

46. Given that Washington and Charlotte were the first two teams chosen in the lottery, what is the probability that Toronto is the third chosen?

47. Given that New Orleans and Charlotte were the first two teams chosen in the lottery, what is the probability that Houston is the third chosen?

48. What is the probability that either New Jersey, Golden State, Detroit, Portland, or Houston is chosen first?

Solve each problem.

49. If Jackson is favored to win a tennis match and the probability that Jackson wins a given point is $P(\text{Jackson wins a point}) = f$, what is the probability that Jackson wins 12 points in a row?

50. Melissa and Heidi are playing tennis. The probability that Heidi wins a point is 0.75 and the probability that Melissa wins a point is 0.25. Find each of the following probabilities.

    a. What is the probability that after three points, the score will be Melissa–30 and Heidi–15?

    b. What is the probability that after three points, the score will be Melissa–15 and Heidi–30?

    c. What is the probability that after two points, the score will be Melissa–30 and Heidi–0?

    d. What is the probability that after two points, the score will be Melissa–0 and Heidi–30?

51. A player serves a tennis ball at 98 miles per hour. How long does a player have to react to the serve?

52. A player serves a tennis ball at 55 miles per hour. How long does a player have to react to the serve?

53. Jackie runs the 100-meter dash in 11.1 seconds. Determine how fast Jackie runs in each of the following measurements of speed.

    a. Meters per second

    b. Kilometers per hour

    c. Feet per second

    d. Miles per hour

54. If a runner is running at 6.8 miles per hour, how long will it take the runner to run 26.2 miles?

55. What average pace must a marathoner run to complete a 26.2-mile run in 4 hours?

Chapter 13
Graph Theory

Sections

Objectives

- Identify parts of a graph
- Determine if graphs are connected or disconnected
- Determine the chromatic number of a graph
- Know the properties of trees
- Find a spanning tree in a graph
- Determine if a graph is bipartite
- Find a matching in a bipartite graph
- Know the properties of planar graphs
- Use Euler's formula for planar graphs

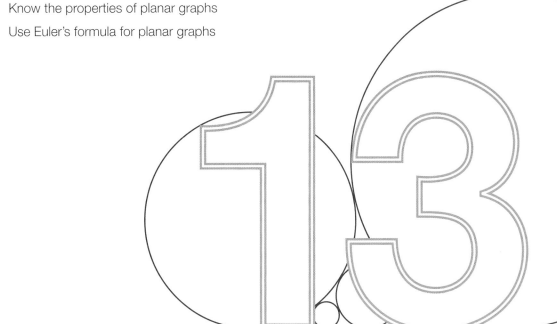

13 Graph Theory

The railway system in the Netherlands is one of the most intensively used networks within Europe. On any average workday, approximately 1.1 million passengers travel on 5500 trains across the country. However, making this operation run smoothly didn't just happen overnight. In 2002, those involved in the Dutch railway published a paper acknowledging the fact that the growth of passenger and freight transport was at a point of overwhelming the existing timetable that was created in 1970. There was no way to add more trains to the timetable and it was cost prohibitive to invest huge amounts of money into building more rail infrastructure.

In response to the paper, a team of specialists in railway logistics along with mathematicians, including graph theorists, set out to create a completely new timetable that would facilitate all of the current growth in travel, result in less train delays overall, and manage future anticipated growth across the country. By using graph theory to analyze and optimize the use of the existing train network, they were able to create a new railway schedule that achieved all of these goals without building a single extra mile of track.

According to "The New Dutch Timetable: The OR Revolution," the Netherlands Railways introduced this new timetable in December 2006. Because of the efficiency of the schedules, passenger travel increased, resulting in an additional $60 million in revenue. The authors of the paper, however, see more than just monetary benefits to their new time table.

". . . the benefits of the new timetable for the Dutch society as a whole are much greater: more trains are transporting more passengers on the same railway infrastructure, and these trains are arriving and departing on schedule more than they ever have in the past. In addition, the rail transport system will be able to handle future transportation demand growth and thus allow cities to remain accessible. Therefore, people can switch from car transport to rail transport, which will reduce the emission of greenhouse gases." [1]

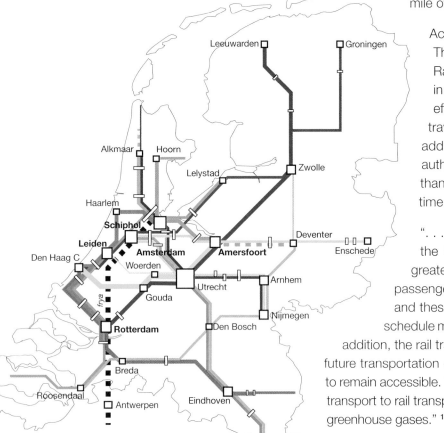

Figure 1

1 Leo Kroon et al., "The New Dutch Timetable"

13.1 Introduction to Graph Theory

In the world of mathematics, the field of graph theory is a relative newcomer. However, as one of the fastest-growing and most applicable branches of mathematics, graph theory permeates throughout academic disciplines such as computer science, linguistics, biology, chemistry, physics, cartography, and operations research. Even social media, such as Facebook, now utilizes graph theory to help users find potential friends by identifying the activities and demographics that they have in common.

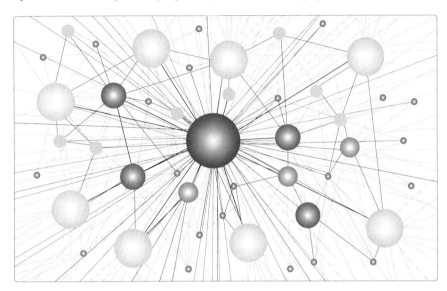

Figure 1: Social Media Friends

So, what is graph theory? Graph theory is the study of graphs—not the type we looked at in Chapter 5 on the Mathematics of Growth, however, but an abstract way to model different kinds of networks, systems, or structures like the one seen in Figure 1. Unlike Cartesian graphs of functions, in graph theory the actual position of the points on a graph is unimportant. What is important is whether the points are interconnected and if so, how they are connected. Graphs remove all extraneous information and only record what is important for the given problems—the connections. To study graphs of this type, we first need vocabulary that describes their different features.

In the field of graph theory, a **graph** consists of a set of points called *vertices* and a set of lines called *edges* that associate particular pairs of the vertices together. We can illustrate a graph by representing the vertices as dots on a page and the edges as lines that join two vertices together, just as in Figure 1. For instance, if you think of your Facebook friends as the set of vertices (dots), then the edges (lines) represent their connection to you and their connections to one another.

☞ **Helpful Hint**

The edges in a graph can be lines or curves. However, we will mostly use lines in this chapter.

> ### Graph
>
> A **graph** consists of a set of vertices and a set of edges that join pairs of vertices together. Graphs are generally represented with a capital letter, such as *G*.

Example 1: Identifying Parts of a Graph

In graph A, identify the vertices and edges.

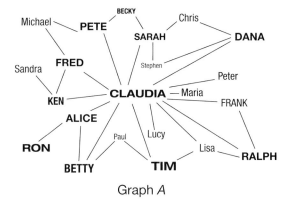

Graph A

Solution

In graph A, the vertices are represented by the names and the edges are the lines joining the names to one another.

As you can imagine, when working with graphs, we need a way to distinguish each vertex. A distinct lower case letter or number is commonly assigned to each vertex for identification. Figure 2 shows the five vertices in graph G labeled with the letters u, v, w, x, and y.

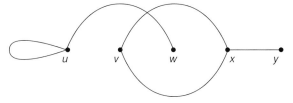

Figure 2: Graph G

Notice in graph G, that even though the edge joining vertices u and w crosses an edge joining v and x, the point where they cross is not a vertex. In graph theory, not every place where edges meet is a vertex.

Similarly, when we refer to the edges, a common practice is to denote them by the two vertices at each end of the edge. For instance, in graph G, the edge between vertices x and y is labeled xy as shown in Figure 3. Note that when labeling edges using vertices, the order in which the vertices are labeled is irrelevant. In other words, the edge between the vertices x and y can be labeled either xy or yx.

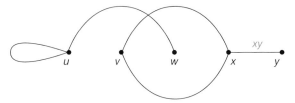

Figure 3: Graph G

How do we distinguish between the two edges joining vertices v and x in G? We cannot refer to them both as vx. The notation using the ends of the edges is no longer sufficient to distinguish between edges. When multiple edges exist between a pair of vertices, we name the edges with unique identifiers to set them apart. For instance, one could label the edges in G as illustrated in Figure 4.

$$e_1 = uu$$
$$e_2 = uw$$
$$e_3 = vx \text{ (drawn on top in Figure 4)}$$
$$e_4 = vx \text{ (drawn on bottom in Figure 4)}$$
$$e_5 = xy$$

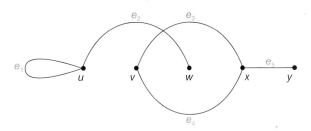

Figure 4: Graph G

Notice that both ends of edge e_1 are the same vertex, u. We call an edge like this a loop.

Loop

A **loop** is an edge that has the same vertex for both of its ends.

Example 2: Identifying Parts of a Graph

For the given graphs, label each vertex and edge. Identify any loops.

a.
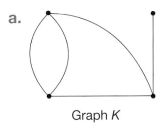
Graph K

b.
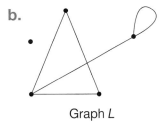
Graph L

Solution

a. First, label the four vertices in graph K with lowercase letters. We'll use a, b, c, and d. Note that the assignment of the letters is not critical; they can be in any order.

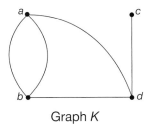
Graph K

Second, since the graph has two edges between the vertices a and b, it is not sufficient to use the ends as labels. Instead, use distinct identifiers in order to distinguish between the edges. We will use $f_1, f_2, f_3, f_4,$ and f_5.

Although graph K might appear to have a loop on the left-hand side of it, there is not an edge that begins and ends at the same vertex. Therefore, there are no loops in graph K.

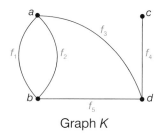

Graph *K*

b. In graph *L*, label the five vertices with lowercase letters. We use *v*, *w*, *x*, *y*, and *z*. Since there is at most one edge between any pair of vertices, we can label the edges using either format. This time, we will use the pairing method to label the edges. Note that there are only five vertices and not six, remembering that there is not a vertex every time edges intersect.

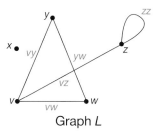

Graph *L*

Notice that edge *zz* begins and ends at vertex *z*, and is, therefore, a loop.

In 1735, Leonhard Euler used the ideas of vertices and edges to settle a long-standing dispute in the city of Königsberg, Prussia. The city was set on the Pregel River and included two islands that were connected to the mainland and each other by seven bridges. The towns people of Königsberg disagreed on whether it was possible to visit all parts of the city and cross each of its seven bridges once and only once. They were not allowed to cross half a bridge and then turn around and cross the other half of the bridge from the other side, however they did not have to begin and end the journey in the same spot. Figure 5 shows a map of the city with the bridges highlighted. [1]

Figure 5

Founding an entire new branch of mathematics, Euler's genius idea was to represent the islands and bridges as a graph. He used a different vertex to represent each of the four land masses and an edge to represent each of the seven bridges, as shown in Figure 6.

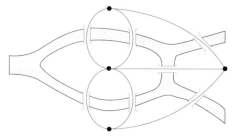

Figure 6: Graph of the Königsberg Bridge Problem

Routes around the city then became sequences of vertices and the edges that connect them. In order to settle the argument, Euler proved that a *walk* could not be found that visited every land mass and crossed every bridge precisely once. A **walk** in graph theory is a way to describe a particular sequence of vertices and edges in a given order.

Walk

A **walk** in a graph is a finite list of alternating vertices and connecting edges that begins and ends with a vertex.

For instance, in Figure 7, the walk v, x, y, x is highlighted in graph G. Notice that a walk is not a "jump" between vertices. Each vertex must be connected by an edge. For instance, u, v, w is not a walk.

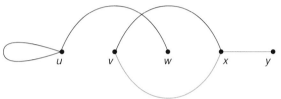

Figure 7: A Walk in Graph G

We say that the walk v, x, y, x is of length 3. The *length of a walk* is the number of edges in the walk. If a walk starts and ends at the same vertex, we say the walk is **closed**. We could make our walk closed in Figure 7 if we continue the walk back to the starting point v. The closed walk would then be v, x, y, x, v. Notice that there is no restriction on the number of times a vertex or edge is crossed in a walk.

Example 3: Identifying a Walk

Use graph B to answer the following questions.

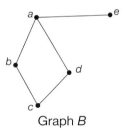

Graph B

a. Identify a walk of length 2 that starts at vertex a and ends at vertex c.

b. Is the walk you just found a closed walk?

Solution

a. There are two possible walks that start at vertex a, end at vertex c, and have a length of 2. The walk a, b, c is highlighted in blue and the walk a, d, c is highlighted in red. Either is an acceptable answer.

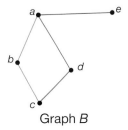

Graph B

b. Because the walks do not start and end at the same vertex, neither walk is closed.

Sometimes restrictions are placed on walks. For instance, in the Königsberg bridge problem, every land mass (vertex) had to be visited while crossing every bridge (edge) once. When we restrict a walk such that no edges or vertices can be repeated, we call it a **path**. The walk *v*, *x*, *y* is a path in graph *G*, as is shown in Figure 8.

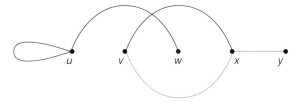

Figure 8: Graph *G*

Path

A **path** is a walk with no repeated edges or vertices.

Example 4: Identifying a Path

Identify a path of length 3 starting at vertex *a* and ending at vertex *d* in graph *C*.

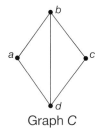

Graph *C*

Solution

Since the path must be of length 3 and start at vertex *a* and end at *d*, we cannot go straight to *d*. We need to move from *a* to *b*. From there, if we move to *d* next, we still only have a path of length 2. Instead, we need to move from *b* to *c* and then to *d*.

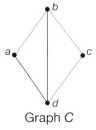

Graph *C*

This walk *a*, *b*, *c*, *d*, is a walk of length 3 and only crosses each vertex once and never duplicates an edge. Therefore, it is the required path.

Remember that the definition of a graph has no specifications as to where the vertices lie on the page nor the shape of the edges joining them. The edges can be drawn as curves or lines, short or long, etc. The definition of a graph simply describes which pairs of vertices are joined.

For instance, in Figure 9, we've drawn our original graph G on the left, but the right side is a different view of graph G, this time without any crossings. Notice that it has exactly the same vertices with the same edges joining them. If it did not, then they would not be different views of the same graph.

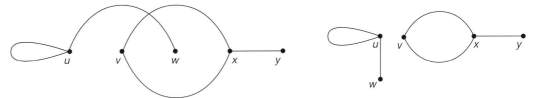

Figure 9: Two Different Views of Graph G

You might be wondering if this second view of G is actually one graph or two. It is certainly in two pieces, called *components*, however, it is still only one graph. Notice that, although you can move across edges between vertices u and w, and also between vertices v, x, and y, there is no way to reach vertex v from vertex w. A graph in which it is possible to move along a sequence of edges between any two vertices is called **connected**. As we stated, G is in fact one graph, but the graph is **disconnected**.

Connected and Disconnected Graphs

A graph is **connected** if there is at least one path connecting each pair of distinct vertices. A graph is **disconnected** if a pair of vertices exists so that there is no path between them.

Example 5: Identifying Connected and Disconnected Graphs

Label each graph as connected or disconnected.

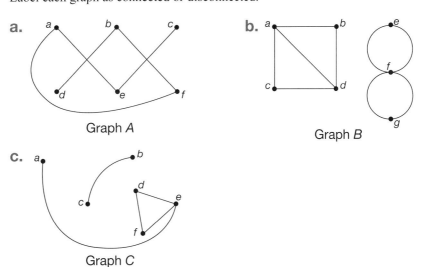

a. Graph A

b. Graph B

c. Graph C

Solution

a. Even though each vertex is not directly connected with every other vertex by single edges, this graph is connected because there are paths between each pair of vertices.

b. Because there is no path connecting vertex b to vertex e, the graph is disconnected. Only one example is needed to show that a graph is disconnected.

c. This graph is disconnected because there is no path between vertices a and c.

Example 6: Identifying Different Views of the Same Graph

Decide whether the pairs of graphs shown could be different views of the same graph or not.

a.

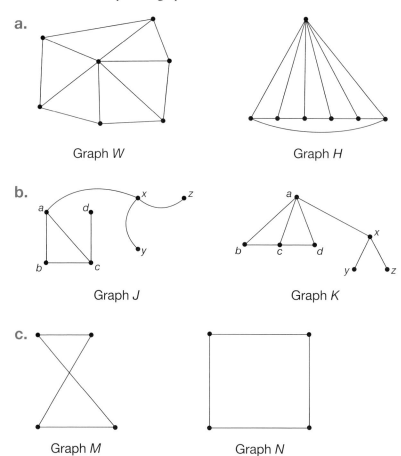

Graph W Graph H

b.

Graph J Graph K

c.

Graph M Graph N

Solution

a. In both graphs W and H, there are seven vertices. If you think of numbering the vertices 1 through 7, you can check that the edges join the same vertices in both diagrams. Check this for yourself in the following figure. Therefore, these are different views of the same graph.

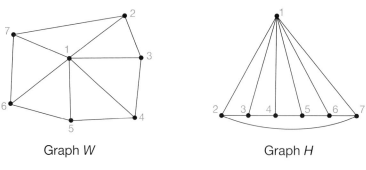

Graph W Graph H

b. Graphs J and K are not the same. Graph J does not include an edge between the vertices a and d, however, graph K does. Therefore, graphs J and K cannot be different views of the same graph.

c. Both graphs M and N contain four vertices. If you label each graph, you can see that the edges join the same vertices. Therefore, these are different views of the same graph.

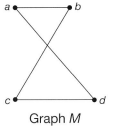   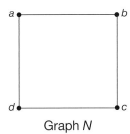

Graph M Graph N

Figure 10 shows a diagram of the graph G that we have been examining throughout the section. We can describe the placement of vertices that are connected together by an edge. When two vertices are joined by an edge we say the vertices are **adjacent**. For instance, vertices x and y are adjacent because they are joined by the edge e_5.

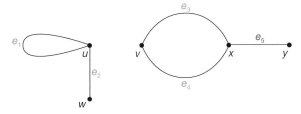

Figure 10: Graph G

We can also describe edges that are connected together. When two edges share a vertex, we say that they are **incident**. For instance, in Figure 10, edges e_1 and e_2 are incident because they share the vertex u.

Skill Check #2

Identify all pairs of adjacent vertices and all pairs of incident edges in Figure 10.

The degree of a vertex is the number of edges that are incident to that vertex. We denote the degree of the vertex by $d(\text{vertex})$. For instance, in Figure 10, the degree of vertex v is $d(v) = 2$ and the degree of vertex y is $d(y) = 1$. A loop contributes 2 to the degree count. In Figure 10, vertex u has two edges that are incident to it, and one of them is a loop. Therefore, the degree of u is three; that is $d(u) = 3$.

Adjacent

Two vertices that share an edge are **adjacent**.

Incident

Two edges that share a vertex are **incident** to one another. Also, we say that a vertex is **incident** to the edges that have that vertex as an endpoint.

Degree

The **degree** of a vertex u is the number of edges that are incident to u and is notated by $d(u)$.

Fun Fact

In the field of computer science, it is particularly useful for some graphs to be described by using an **adjacency matrix**. This matrix is an organized list to show which pairs of vertices are connected together and which are not.

Example 7: Identifying Parts of a Graph

Use Graph A to solve the following problems.

a. Determine the degree of each of the vertices.

b. Name the pairs of vertices that are adjacent.

c. Name the pairs of edges that are incident.

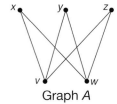

Graph A

Solution

a. To determine the degree of each vertex, we simply count the number of edges incident to it. Therefore,

$$d(x) = 2; \, d(y) = 2; \, d(z) = 2; \, d(v) = 3; \text{ and } d(w) = 3.$$

b. Two vertices are adjacent if they are ends of the same edge. Therefore, all the following pairs of vertices are adjacent.

x, v

x, w

y, v

y, w

z, v

z, w

c. Two edges are incident if they share a common vertex. Therefore, all of the following sets of edges are incident.

xv, xw

yv, yw

zv, zw

xv, yv, zv

xw, yw, zw

As Euler demonstrated with the bridges of Königsberg, it is often helpful to visually represent a graph given certain features or constraints. Sometimes this will involve finding the right visual representation, or drawing, of the given vertices and edges. Sometimes this will involve creating a graph that has a required structure. The next two examples demonstrate both of these possibilities.

Example 8: Drawing a Graph Given the Edges

a. Draw a five point star; that is, a graph with five vertices, v_1, v_2, v_3, v_4, and v_5, whose edges are v_1v_3, v_3v_5, v_5v_2, v_2v_4, and v_4v_1.

b. Determine the degree of each vertex.

c. Redraw the graph without edges crossing.

d. Determine the degree of each vertex in the new drawing.

Solution

a. Begin by placing the five vertices on the page and labeling them v_1 through v_5. Remember, the placement of the vertices is not important.

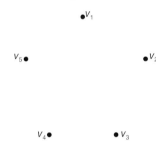

Now, connect the vertices with the given edges. In other words, edge v_1v_3 connects vertex v_1 with vertex v_3. The following graph shows the completed graph.

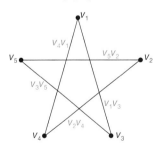

b. To determine the degree of each vertex, count the number of edges incident to each vertex. Therefore we have

$$d(v_1) = 2;\ d(v_2) = 2;\ d(v_3) = 2;\ d(v_4) = 2;\ \text{and}\ d(v_5) = 2.$$

c. To redraw the graph without edges crossing, think of "pulling" edge v_1v_3 to the outside of the graph as shown in the following figure.

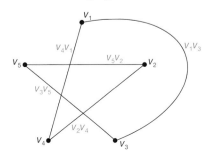

Next, do the same with edge $v_3 v_5$.

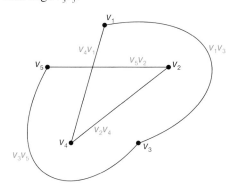

The last thing to do is to pull vertices v_2 and v_4 out of the middle. When you do, think of flipping their position, as shown in the following graph. Check for yourself that the edges are all still the same in this new layout.

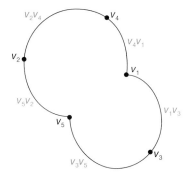

The graph is now drawn without any edges crossing. You can see that the "star" now looks more round.

d. The degree of each vertex does not change when the graph is redrawn. Each vertex still has the same edges connected to it and, hence, the same degree. (This will always be the case with any graph.) Therefore, all vertices in this graph still have degree 2.

Example 9: Drawing a Graph without Edges Predetermined

A research chemist is analyzing the structure of compounds of aluminum, iron, and oxygen. Each aluminum atom makes three chemical bonds, each oxygen atom makes two chemical bonds, and each iron atom makes six chemical bonds. Create a possible form for a molecule of two aluminum atoms, two oxygen atoms, and one iron atom where no atom makes a chemical bond to itself.

Solution

To create the structure of the molecule, we will represent the atoms and their bonds as a graph. Each vertex will represent an atom and each edge will represent a chemical bond that joins one atom with another. So, we have to draw a graph with no loops that has two vertices of degree 3, two vertices of degree 2, and one of degree 6.

Unlike the previous example, there are many graphs that meet the requirements given for the graph since there are no restrictions on which vertices must be joined together by edges. We will show one possible form for a molecule of this type.

First notice that there must be a total of five vertices, one for each atom. Place the vertices on the graph.

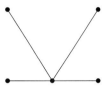

Let's begin with the vertex with the highest degree of degree 6. Recall that the degree of a vertex is the number of edges that are connected to it, not the number of vertices. Because there are no loops allowed, the vertex must have six edges joined to it. We'll join each of the other four vertices to it, giving us a vertex of degree 4.

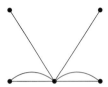

To get to degree 6, we'll need to draw two more edges somewhere. We'll choose to give the bottom two vertices on either side another edges with our central vertex, as shown.

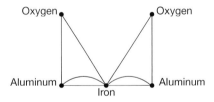

Let's stop and determine what the graph looks like so far. We have a graph with five vertices: one of degree 6, two with degree 2, and two with degree 1.

Notice that we have several possible ways to complete our graph here. One way is to join each vertex of degree 1 to a vertex of degree 2, which increases each of their degrees by one, giving us the desired degree count. Our final graph for the molecule's form is as follows.

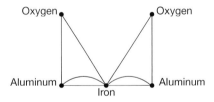

Skill Check #3

Draw a different graph with the same restrictions as in Example 9: no loops, having two vertices of degree 3, two vertices of degree 2, and one of degree 6.

Graphs can also be used to optimize resources in real-life situations. For instance, suppose you need to hire security guards to protect precious art work in a museum. You'd like to have enough guards so that every room in your museum can be seen by at least one guard at all times. At the same time, you want to be prudent with your money and not have an excess of guards. You need to establish how many security guards you need to hire and where they should be deployed. Alternatively, suppose you need to place Wi-Fi hotspots all over a building. Once again you would like to minimize the cost of covering the building with wireless Internet and, hence, both their number and placement are critical.

Both of the scenarios described can be modeled with graphs to help find optimal solutions. To do this we need to introduce a new graph structure—a **minimum vertex cover**. First, a **vertex cover** is a group of vertices with the property that every vertex in the graph is either in the group or adjacent to at least one of the vertices in the group. We can optimize by choosing the smallest group of vertices that will cover the graph. This we call a minimum vertex cover.

Vertex Cover

A **vertex cover** is a set of vertices *A*, so that every vertex in the graph is either in *A* or adjacent to a vertex in *A*.

Minimum Vertex Cover

When a vertex cover *A* is as small as possible, *A* is called a **minimum vertex cover**.

To see how a vertex cover can be useful, let's consider the problem involving the museum guards. Remember, that we want to hire the fewest (minimum) number of guards that will be able to guard all the rooms at once. To begin, we need to know the layout of the museum. Suppose we have a blueprint for the museum building, as shown in Figure 11.

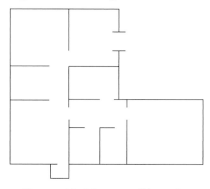

Figure 11: Museum Blueprint

The first step is to begin by representing the blueprint with a graph. Let each room be represented with a vertex. If the rooms share a door, then the vertices are adjacent. In other words, they will have an edge connecting them. Since there are ten rooms, the graph will have ten vertices. Remember that the position of the vertices makes no difference, but to make things easier, we will layout the vertices in the same manner as the blueprint.

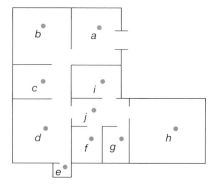

Figure 12: Museum Blueprint Graph with Vertices

Using the blueprint as a guide, join the vertices together with an edge if they share a doorway, as shown in Figure 13.

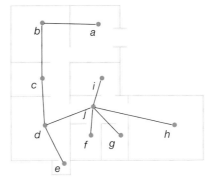

Figure 13: Museum Blueprint Graph with Vertices and Edges

Now that the building is represented by a graph, we can find a solution to the guard problem.

Example 10: Finding a Minimum Vertex Cover

Using the graph we created in Figure 13, determine the minimum number of security guards that would be needed if a guard can secure their room and all the rooms adjacent to it.

Solution

We can begin by placing a guard in room j, represented by a blue G. That way, rooms d, f, g, h, and i are all covered since they are adjacent to j.

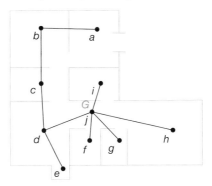

Then, we have the remaining four rooms a, b, c, and e that need a guard. Since these four rooms are connected in a row, this will require a minimum of two guards. One guard in room b who can cover room b as well as a and c, and one guard in either room d or e.

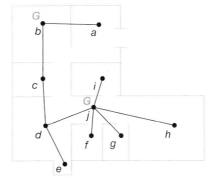

Here's where we can use our best judgment for the scenario. Although room d is secured by the guard in room j, the guard in room j might be a little overworked with all the rooms assigned to his position. We could technically place a second guard in room e and accomplish our objective. Placing the guard in room d might be a more sensible solution, helping to ease

13

the job of the guard in room *j*. Unless, of course, room *e* contains an incredibly valuable jewel. Either way, the minimum number of guards needed to cover every room in the museum is three.

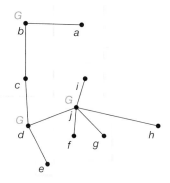

The set of vertices that form the guards' post assignments, $A = \{b, d, j\}$, is a minimum vertex cover for the museum graph.

Skill Check #4

Find another minimum vertex covering for the museum graph from Example 10.

Another way to use graphs to optimize resources is by categorizing the vertices. Suppose that we want to give cellular towers a frequency band so that nearby towers don't interfere with one another. How many different frequencies would be needed to cover an entire area? Or, suppose we want to paint a geometric mural such that no two bordering shapes have the same paint color. How many different paints would need to be purchased?

In both of these cases, a vertex coloring can be used to optimize the solutions. A **vertex coloring** of a graph is an assignment of colors to the vertices in such a way that no two adjacent vertices have the same color. In mathematics, the "colors" for vertex coloring might be colors, letters, or numbers, depending on which is most convenient.

Vertex Coloring

A **vertex coloring** of a graph is an assignment of colors to the vertices of the graph so that adjacent vertices receive different colors.

Chromatic Number

When the number of colors used in a vertex coloring is as small as possible, this number is called the **chromatic number** of a graph and is denoted $\chi(G)$, read "chi of G."

The field of designing maps, called cartography, illustrates how vertex colorings can be utilized. Consider Figure 14, which shows a map of the eleven counties in a state.

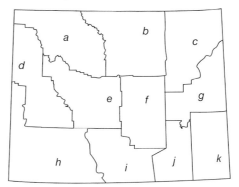

Figure 14: State Counties

Let's represent the map using a graph where the counties are vertices and an edge exists if counties share a border. Once again, begin by representing each of the eleven counties with a vertex, as shown in Figure 15.

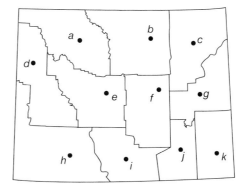

Figure 15: State Counties Graph with Vertices

Joining each neighboring county with an edge if they share a border gives us the following graph. Check this for yourself.

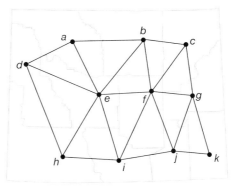

Figure 16: State Counties Graph with Vertices and Edges

Example 11: Finding a Vertex Coloring

Suppose a cell phone company needs to place cell towers in all counties of the state shown in Figure 16 in such a way that the frequencies for each tower do not interfere with each other. A vertex in the graph represents a county and an edge represents a shared border between counties.

When two cell towers are placed in adjacent counties, they interfere with each other's signal unless they are on different frequencies. Of course the cell phone company could place a different frequency in each county and have 11 frequencies for the state. However, to minimize their costs, the cell phone company needs to use as few frequencies as possible. Find the chromatic number of the graph and determine the best arrangement of the frequencies of the cell towers for the cell phone company.

Solution

We can represent the frequencies of a cell tower by using a number. Let's begin by putting the first frequency in county a, as shown in blue on the graph.

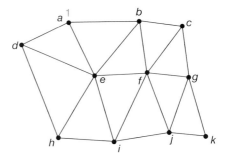

Every county (vertex) adjacent to county a is not allowed to use frequency 1. So, the counties b, d, and e all need new frequencies. Start by giving county b frequency 2. Notice, that county e can't be on 1 or 2 since it connects to both a and b, so we need to give it frequency 3. Here's how the graph looks so far.

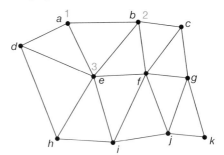

Returning to county d, we know that it cannot be the same as either of the adjoining counties a and e, which have frequencies 1 and 3 respectively. Since frequency 2 is available, we'll use it again as we are trying to use as few frequencies as possible.

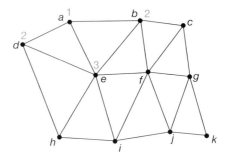

Continue in the same manner until all vertices are numbered.

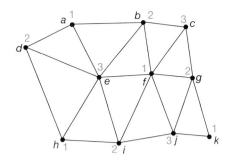

You can see that we needed three different frequencies to cover all the counties without interference. Therefore, the county graph has a chromatic number of 3; that is, $\chi(G) = 3$.

Think about the graph created in Figure 13 for the placement of museum guards and the graph created in Figure 16 for the placement of cell phone towers. Both graphs are shown again in Figure 17.

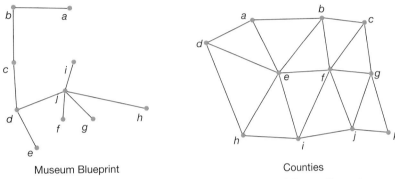

Museum Blueprint   Counties

Figure 17: Graphs for Museum Blueprint and State Counties

Although both graphs describe the structure of their underlying scenarios, the blueprint graph on the left is a little simpler than the county graph on the right. The difference is that the county graph has lots of interlocking **cycles**, which are walks that start and end at the same vertex. For instance, among the many cycles in the state counties graph are a, d, e, shown in red and c, f, e, h, i, j, k, g shown in blue in Figure 18.

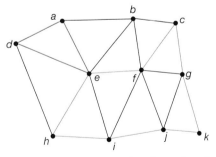

Figure 18: Two Cycles in the Counties Graph

Notice that although a cycle starts and end at the same vertex, we do not list it twice when referring to the cycle.

Cycle

A **cycle** is a closed walk; that is, a cycle starts and ends at the same vertex and has no edges or vertices repeated except for the starting vertex.

The length of a cycle is the number of edges or vertices in the cycle. For instance, the red cycle in Figure 18 is a cycle of length 3 and the blue cycle is a cycle of length 8. Notice that in a cycle, the number of edges is the same as the number of vertices.

Example 12: Identifying a Cycle in a Graph

Identify a cycle in graph G and determine its length.

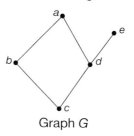

Graph G

Solution

The only cycle in graph G is the cycle among the vertices a, b, c, and d. To refer to the cycle we can begin at any of the vertices and list them in order as we trace the cycle. For instance, beginning with vertex b, the cycle is b, c, d, a. Because the cycle has four vertices and four edges, the cycle has length 4.

Skill Check #5

Identify a cycle of length 6 in the following graph.

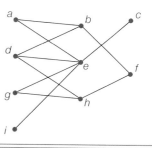

Skill Check Answers

1. Any of the following: c, b, a, e; c, b, a, d; c, b, a, b; c, b, c, b; c, d, a, e; c, d, a, b; c, d, a, d; c, d, c, d

2. Adjacent vertices: x and y; u and w; v and x; Incident edges: e_1 and e_2; e_3 and e_4; e_3 and e_5; e_4 and e_5

3. Answers will vary. For example,

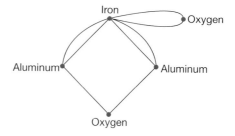

4. $B = \{b, e, j\}$

5. Answers will vary. For example, a, b, d, h, g, e; g, e, a, b, f, h; b, f, h, d, e, a; d, e, g, h, f, b.

13.1   Exercises

Fill in each blank with the correct term.

1. The _____ of a vertex is the number of edges that are incident to that vertex.

2. A _____ consists of a set of vertices and a set of edges that connect pairs of vertices.

3. A/an ____ associates a pair of vertices in a graph.

4. The end of an edge in a graph is called a _____.

5. A ____ is an edge whose ends are the same vertex.

6. Two edges are said to be _____ if they share a vertex.

7. The _____ of a graph is the minimum number of colors needed in a vertex coloring of a graph.

8. Two vertices are said to be _____ if they share an edge.

9. A sequence of vertices such that each vertex is adjacent to the next in the list, the last vertex is adjacent to the first, and there are no repeated edges or vertices, except for the starting vertex, is called a ____.

List the vertices and edges of each graph.

10.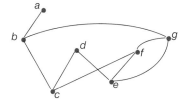

11.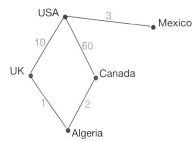

12.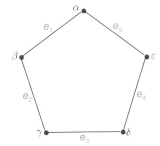

Draw a graph that satisfies each set of conditions.

13.   Draw a graph that has five vertices r, s, t, u, v, and the edges rt, ru, rv, st, su, and sv.

14.   Draw a graph with vertices u, v, w, x, y, z, and edges ux, uv, uy, vx, vy, vw, wy, wz, wx, with no crossings.

Solve each problem.

15.   Use the graph to answer the following.

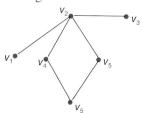

   a. Identify a walk of length two that starts at vertex v_1 and ends at vertex v_3.

   b. Is the walk in part **a.** closed?

   c. Identify a closed walk.

16.   Identify two paths in the following graph. What is the length of each path?

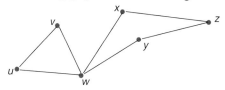

17.   Draw two different connected graphs, each with four vertices and four edges.

18.   Are the two given graphs the same graph? Explain.

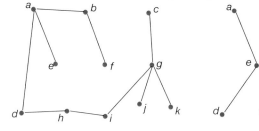

19.   Which of the given graphs is connected?

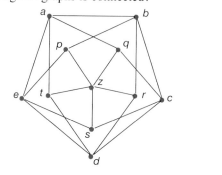
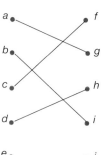

20. The length of a path is the number of edges in the path. Identify a path of the specified length in the graph.

    a. Path of length 2

    b. Path of length 3

    c. Path of length 6

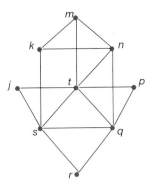

21. An **Euler trail** in a graph G is a walk that contains every edge in G precisely once. Each of the following graphs contains an Euler trail, identify it.

    a.

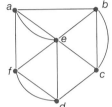

    b.

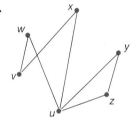

22. Identify a cycle in the following graph and determine its length.

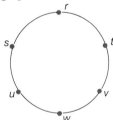

23. Identify a cycle of the required length in the graph.

    a. Cycle of length 4

    b. Cycle of length 6

    c. Cycle of length 8

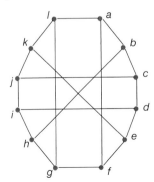

24. Identify the following features in the graph. In some cases, more than one answer is correct.

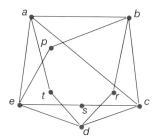

    a. An edge

    b. Two adjacent vertices

    c. Two incident edges

    d. Two vertices that are not adjacent

    e. A vertex of degree 4

    f. A cycle of length 5

25. A **Hamilton cycle** in a graph G is a cycle that contains every vertex in G. Decide if the following graphs contain Hamilton cycles.

    a.    **b.** 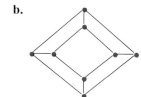   **c.**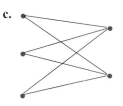

26. Find a vertex coloring of the graph. What is the chromatic number of that graph?

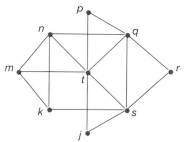

27. The graph shows a blueprint of rooms in a museum. Find a way to place guards in the rooms so that every room either has a guard or shares a doorway with a room that has a guard and the minimum number of guards is used.

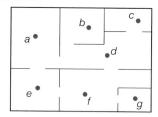

28. Use a vertex coloring to assign colors to the counties of Vermont so that adjacent counties have different colors. What is the minimum number of colors required? Find a vertex coloring using the minimum number of colors.

13.2 Trees

Some of the simplest graphs that we can consider have no cycles. Graphs without cycles prove to be convenient models for many real-life scenarios: family trees, finding shortest routes between cities, maximizing flow of pipes in a network, or minimizing cost for laying cable between states. All of these instances can be modeled using a type of graph called a **tree**.

> ### Tree
>
> A connected graph with no cycles is called a **tree**.

Remember that a connected graph is one in which there is a path between any two given vertices. In a tree, however, because there are no cycles, there is only one distinct path between any two vertices.

Trees vs. Non-Trees

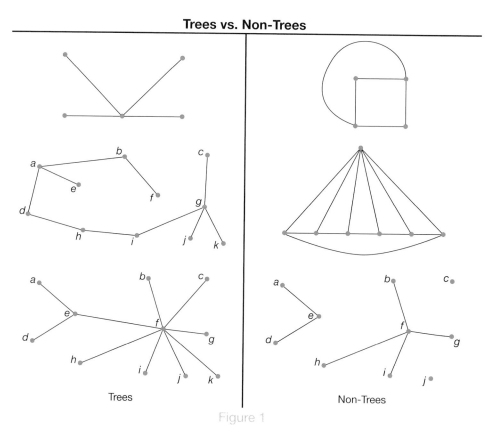

Figure 1

Example 1: Identifying Trees

Determine if each of the following graphs is a tree.

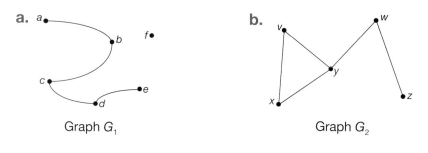

a.   Graph G_1

b.   Graph G_2

c.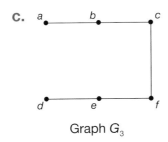

Graph G_3

Solution

a. Although G_1 does not have a cycle in it, it is not a connected graph because there is not a path between vertex f and other vertices. Therefore, it is not a tree.

b. G_2 is not a tree because there is a cycle among the vertices v, x, and y.

c. G_3 is a tree because it is a connected graph without any cycles.

Consider the set of vertices in Figure 2.

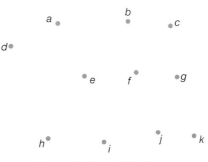

Figure 2: Set of Vertices

Suppose you are told to create a connected graph using the vertices from Figure 2 with no restrictions on the number of edges in the graph. How would you join the vertices? As you can imagine, there are an infinite number of possibilities. Figure 3 shows diagrams of two graphs, both of which are connected graphs using the vertices. They are just two of the many possible graphs you could create.

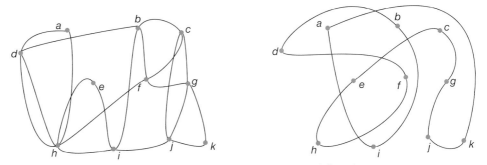

Figure 3: Two Possible Connected Graphs

Now, consider creating a graph with the same vertices, still having no restrictions on the number of edges, but with the knowledge that each edge you draw costs you $10. Would you draw the same graph? If the goal is to spend as little money as possible, you certainly would never join a pair of vertices with more than one edge. Most of us would begin to think of a way to find a connected graph of the vertices that uses the least number of edges. In other words, we would look to find a graph that is a tree.

Figure 4 shows two of the many connected graphs that could be drawn with the fewest number of edges; that is, they are trees. Notice that each of these would cost you the same amount of money to draw—$100.

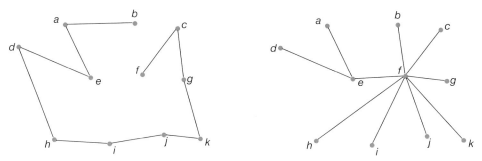

Figure 4: Two Possible Trees

Because both graphs are trees, neither one contains a cycle. If there were a cycle, any one of the edges of the cycle could be removed without disconnecting the graph. Therefore, a tree connects vertices with the fewest possible edges. In fact, the number of edges in a tree is always one fewer than the number of vertices. The following theorem states this principle.

> ### Tree Theorem
>
> Let graph T be a tree on v vertices. Then, graph T has $v - 1$ edges.

Example 2: Determining the Number of Edges Needed for a Tree

Determine the number of edges needed to connect the given vertices so that the resulting graph is a tree.

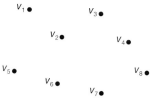

Solution

To determine the number of edges needed to construct a tree from the given vertices, we simply need to subtract 1 from the number of vertices. Since there are 8 vertices, the number of edges needed is $8 - 1 = 7$.

> ### Skill Check #1
>
> Construct two different trees with the vertices in Example 2.

Spanning Trees

By modeling situations with graphs, we can use trees to find ways to maximize or minimize efficiency in a network. We can create a tree within an existing graph by pruning away edges until only a tree remains. This tree is a *subgraph*, that is, a subset of the vertices and edges of the original graph. When a subgraph is a tree that contains all of the vertices, it is called a **spanning tree**. For many graphs, it is possible to have more than one spanning tree.

> ## Spanning Tree
>
> A **spanning tree** is a subgraph of a connected graph, which is itself connected and contains all the vertices of the original graph, but has no cycles.

The key to a tree is that it doesn't contain any cycles. So, to find a spanning tree in a graph, one needs to identify the cycles in the graph and delete enough edges to break the cycles while still keeping the graph connected. The following process gives a method for finding a spanning tree.

Steps for Constructing a Spanning Tree

1. In a connected graph G, identify a cycle. If more than one cycle exists, choose one at random.

2. Choose an edge from the cycle selected and delete it from the graph.

3. While the graph contains a cycle, repeat Steps 1 and 2.

Example 3: Using Spanning Trees

In an interview in *Premier Magazine* in 1994, Kevin Bacon commented that he had acted together with almost everyone in Hollywood. The game "Six Degrees of Kevin Bacon" was invented based on this comment. The goal is to link any named actor to Kevin Bacon by beginning with the named actor and naming no more than six actors, ending with Kevin Bacon, where each successive pair of actors have appeared in a movie together.

Find a spanning tree for the following graph, in which the vertices are actors, and two actors are joined by an edge if they appeared together in a movie where the edges are labeled with the movie.

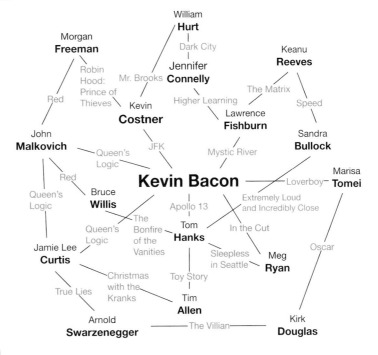

Solution

The algorithm says that we have to identify cycles one by one, and eliminate edges until no more cycles remain. There are lots of cycles to choose from here, but let's begin with the following cycle.

Kevin Bacon, Lawrence Fishburne, Keanu Reeves, Sandra Bullock, Tom Hanks

Do you see this cycle on the graph? We are free to delete any of the edges of this cycle to construct the spanning tree. Let's remove *Speed*.

The next cycle we'll consider is the following.

Kevin Bacon, Lawrence Fishburne, Jennifer Connelly, William Hurt, Kevin Costner

This time let's remove the *Mr. Brooks* edge. While we're at it, let's also remove the edge that joins Kevin Costner to Morgan Freeman. That eliminates the cycle

Kevin Bacon, Kevin Costner, Morgan Freeman, John Malkovich.

We can eliminate the cycle,

Kevin Bacon, Marisa Tomei, Kirk Douglas, Arnold Schwarzenegger, Jamie Lee Curtis

by removing the *The Villain* edge that joins Kirk Douglas to Arnold Schwarzenegger.

Although Jamie Lee Curtis and John Malkovich were both in *Queen's Logic* with Kevin Bacon, let's remove the edge that joins Jamie Lee Curtis and John Malkovich to break that cycle of length three. Now, there are three cycles that remain. Can you see them?

Kevin Bacon, John Malkovich, Bruce Willis, Tom Hanks

forms a cycle, so let's delete the *The Bonfire of the Vanities* edge.

Kevin Bacon, Tom Hanks, Tim Allen, Jamie Lee Curtis

also forms a cycle, so we'll delete the *Toy Story* edge. Finally, there is one cycle remaining:

Kevin Bacon, Meg Ryan, Tom Hanks.

We will remove this cycle by deleting the *In the Cut* edge. Now, we have the following spanning tree.

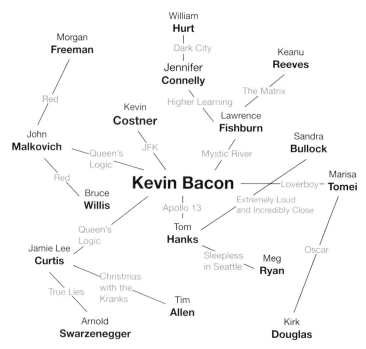

Skill Check #2

Find another spanning tree for the Kevin Bacon graph in Example 3 by deleting a different set of edges.

How does the spanning tree help to play the Six Degrees of Kevin Bacon Game? By using the algorithm to find a spanning tree where Kevin Bacon is always at most six vertices away from any other actor, you can play the Six Degrees of Kevin Bacon Game with all the actors in the graph. For example, according to the spanning tree in Example 3, Kirk Douglas is 2 degrees from Kevin Bacon.

We often assign values, or **weights**, to edges of a graph. We thought about this earlier when it cost us $10 an edge to connect vertices. We can assign weights to the edges to represent not only costs, but distances, time, numbers of people, volumes, or any other measurement. There are occasions when we want to minimize the spanning tree that we choose in order to find the most efficient route. One way to do this would be to list out all the possible weighted spanning trees and then choose the one with the smallest values. You can imagine that this would take a great deal of time and resources and it is not practical with a large graph, such as a graph of the US interstate system. A GPS (Global Positioning System) navigation system uses weighted graphs to determine shortest routes and does this quickly. The following algorithm can be used to find a minimum-weight spanning tree.

Steps for Constructing a Minimum-Weight Spanning Tree

1. Consider each edge in the graph in order of descending weight.

2. If the edge being considered is part of a cycle, remove it. If not, it must remain in the graph, and you can move on to the next edge.

3. Repeat the Steps 1 and 2 until all edges have been considered.

Example 4: Constructing a Minimum-Weight Spanning Tree

SpeedFirst Telecommunications is going to run its fiber optic cable in a new development. The edges in the given graph represent the possible ways that the cable can be run. The weight of each edge is the length of the cable needed, in meters. Obviously, SpeedFirst would like to find the most cost efficient way to lay the cable. Find the minimum spanning tree using the algorithm to obtain the most efficient process for SpeedFirst. What is the minimum amount of cable that SpeedFirst will need to run in the new development?

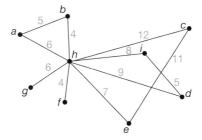

Solution

We begin with the edge with the largest weight, which is the edge $hc = 12$. Since it has the highest weight and is part of cycle h, c, e, remove it as shown.

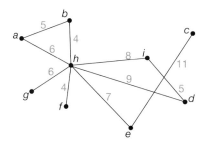

Next, look at edge $ec = 11$. This edge is no longer part of a cycle since the edge hc was removed, so this edge must remain in order to connect vertex c to the graph.

The next highest weighted edge is $hd = 9$. This edge is part of the cycle h, d, i, so we remove this edge. The figure shows the graph as it stands now.

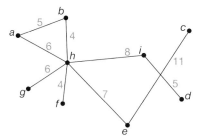

Edge $hi = 8$ has the next highest weight. This is not part of a cycle since the edge hd was removed, so this edge remains to connect vertex i.

Next, we consider $he = 7$. This edge is not part of a cycle since the edge hc was removed, so it must remain to connect the vertex e.

Look at edges $ah = 6$ and $gh = 6$. Edge ah is part of cycle a, h, b, so it will be removed. Edge gh is not part of a cycle, so it must remain to connect the vertex g.

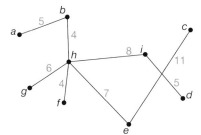

Next, we consider edges $ab = 5$ and $id = 5$. Both of these edges must remain to connect vertices a and d since they are not part of any cycles once the other edges were removed.

Finally, we look at edges $bh = 4$ and $fh = 4$. Again, neither of these edges are part of any cycles, so they must remain to connect vertices b and f.

We have considered all edges in G, so we are done. Thus, our minimum-weight spanning tree is the following.

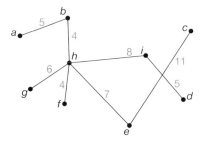

We can determine the minimum length of cable that SpeedFirst will need to run by adding together all of the weights in the minimum spanning tree.

Minimum length of cable = 4 + 4 + 5 + 5 + 6 + 7 + 8 + 11 = 50 meters.

Leaves on a Tree

No discussion of trees is complete without talking about its leaves. In graph theory, a vertex of degree one is called a **leaf**. We can determine the number of leaves on a tree by inspecting each vertex.

> **Leaf**
>
> A vertex of degree 1 in a tree is called a **leaf**.

Example 5: Determining the Number of Leaves on a Tree

Determine the number of leaves on the tree T.

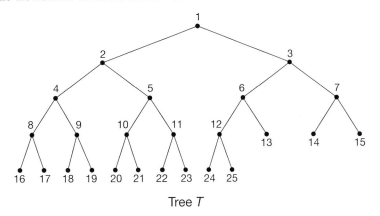

Tree T

Solution

To find the leaves in T, we need to look for vertices of degree 1. These will be vertices with only one edge connected to them. We've circled the leaves in the following diagram.

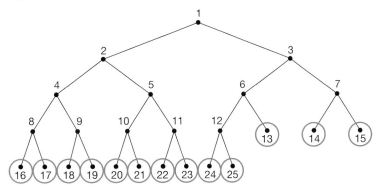

Tree T

There are 13 leaves on tree T.

As you can imagine, if the graph is very large, inspecting each vertex one at a time would get tedious. Instead, we can determine the number of leaves in a tree by knowing the degrees of all the non-leaf vertices, that is, the vertices with degree 2 or more.

☞ **Helpful Hint**

Σ is the Greek letter Sigma and is used to indicate "the sum of."

Number of Leaves on a Tree

If a tree has k vertices with degrees $d_1, d_2, \ldots, d_k$, each greater than 1, then the **number of leaves on the tree** is $\sum_i d_i - 2k + 2$.

Example 6: Determining the Number of Leaves on a Tree

Governor Airlines has airport hubs in seven cities. Those seven cities have connecting flights to 2, 3, 6, 8, 7, 4, and 5 cities respectively. If the graph formed by joining each pair of available flights is a tree, how many different destinations does Governor Airlines fly to?

Solution

The total number of destinations is the number of vertices in the tree. That number of vertices consists of the number of leaves on the tree as well as the internal vertices. We know that there are seven internal vertices representing the seven airport hubs. To complete the number of destinations, we need to know the number of leaves.

The formula for the number of leaves is

$$\sum_i d_i - 2k + 2,$$

where k is the number of vertices with degree greater than 1.

So, the number of leaves will be calculated as follows.

$$(2+3+6+8+7+4+5) - 2(7) + 2 = 35 - 14 + 2$$
$$= 23$$

Adding this to the number of airport hubs, the total number of destinations the airline serves is $23 + 7 = 30$.

Skill Check Answers

1.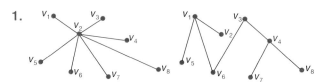

2. Answers will vary.

13.2   Exercises

Determine if each graph is a tree.

1.

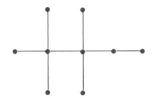

2.

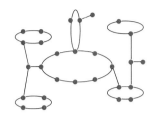

3.

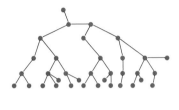

4.

Determine the number of edges needed to form a tree in each graph.

5.

6.

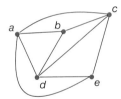

7.

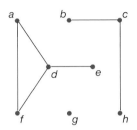

8.

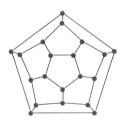

Find a spanning tree in each graph.

9.

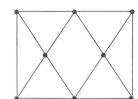

10.

11.

12.

13

Solve each problem.

13. The sidewalks at a university are laid out according to the following graph. Each corresponding edge is labeled with the length of that section of sidewalk. If all of the sidewalks are covered with snow, which sections of sidewalks should be cleared to connect all sections of the university while clearing as little snow as possible?

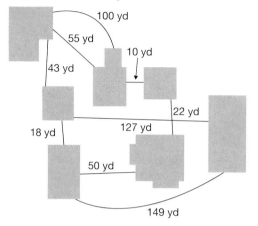

14. What is the chromatic number of a tree? **Hint:** Recall from Section 13.1 that the chromatic number of a graph is the number of colors used in a minimum vertex coloring.

15. A traveling salesman has to visit a number of destinations by a route that is as short as possible. Let G be a graph in which the vertices are the destinations, the edges represent roads that join the destinations, and the weights represent the lengths of the roads.

    a. If the minimum spanning tree has a weight of 100 miles, show that the salesman must drive at least 100 miles to visit all of the destinations.

    b. Show that he needs to drive no more than 200 miles to complete his route.

Find the number of leaves in each tree.

16.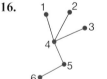

17.

18.

19.

20.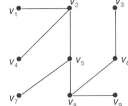

21.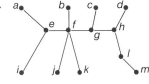

Solve each problem.

22. Let T be a tree with internal vertices of degree 4, 5, 6, 2, and 3. How many leaves does T have?

23. Eurojet Hovercrafts has five hubs connecting European destinations on both sides of the English Channel. If the routes form a tree, with the five hubs having routes to 4, 5, 7, 3, and 6 ports respectively, how many cities are served by Eurojet Hovercrafts?

13.3 Matchings

Each year over 30,000 medical students graduate from US medical schools and seek to move onto the next step in their education—*residency*. They dream of the perfect program for their needs and begin searching for their match. Each student has a number of residency programs that they aspire to be a part of and each applies to some of those programs. In a perfect world, each student would be matched up with a program that they applied to. The question is, can it be done?

This is an example of a situation that appears in a surprisingly wide number of settings. Whether it is high school students pairing up with prom dates that they know, students enrolling in classes, or telephone connections on a switchboard, these are all examples of things that have to be paired, or *matched* up.

To begin to analyze this problem, we'll use the medical residency problem and translate it into the language of graph theory. These types of problems require a graph with two groups of vertices. One group, the left-hand vertices, will represent the medical students with one vertex for each student. The other group, the right-hand vertices, will represent the residency places with one vertex for each place. We call a graph in two parts like this a **bipartite graph**.

> ## Bipartite Graph
>
> A **bipartite graph** is a graph in which the vertices can be partitioned into precisely two subsets so that every edge joins a vertex in one subset to a vertex in the other subset.

Notice that no vertices on the same side of the graph are joined together in a bipartite graph. In the case where every vertex on the left is joined with every vertex on the right, we call the graph a *complete bipartite* graph.

☞ **Helpful Hint**

A graph is a bipartite graph if and only if its chromatic number is 2.

Example 1: Identifying Bipartite Graphs

Determine if the following graphs are bipartite.

a.
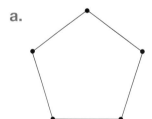

b.
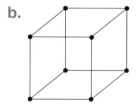

Solution

a. To determine if the graph is bipartite, we need to show that the vertices can be divided into two groups. Let's assume there are two groups, the left-side group and the right-side group. Begin by labeling the vertices either L for the left-hand side or R for the right-hand side. We will start at the top and label the vertices alternately as we go clock-wise around the graph. If vertices are adjacent, they cannot have the same label.

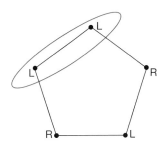

Notice that we end up with two adjacent vertices both labeled L. Therefore, this graph is not bipartite.

b. We can carry out the same process of labeling vertices with either L or R for the cube. We've also numbered each vertex to help keep track of them.

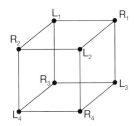

Notice that no adjacent vertices have the same label. Therefore this graph is bipartite. The next is an alternate drawing showing the vertices on their left and right sides.

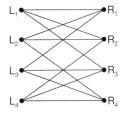

☞ Helpful Hint

Naming a graph by $K_{n, m}$ indicates that the graph is a complete bipartite graph where n identifies the numbers of vertices on the left-hand side of the graph, which are connected to all the other m vertices on the right-hand side of the graph. For instance, $K_{3, 2}$ identifies a graph with three vertices on the left-hand side and two vertices on the right-hand side, all of which are joined to everyone on the other side.

Skill Check #1

Determine if the following graph is bipartite.

Let's go back to our residency example. To draw the bipartite graph for the students, we place an edge joining each *student* vertex to each *residency* vertex for which the student is an applicant. Figure 1 shows a portion of this.

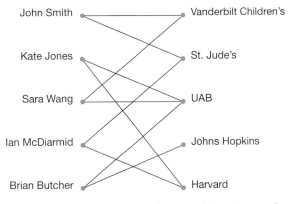

Figure 1: Diagram of a Bipartite Graph of Residency Students

In order to pair each student with one of their residency program choices so that every student has a spot in a particular residency, we need to find a *matching* in the graph. A **matching** is a subset of edges in the graph so that each vertex involved is incident with one matching edge, just like the red edges in Figure 2. In other words, a vertex never has two edges incident to it in a matching.

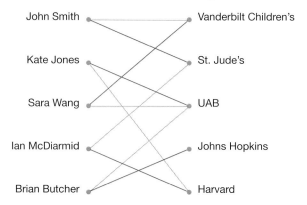

Figure 2: A Matching for Residency Students

Matching

A **matching** is a subset of edges in a graph so that each vertex is incident with only one edge.

A *matching from the left into the right* in a bipartite graph pairs each vertex on the left with exactly one vertex on the right. This does not imply that every vertex on the right is paired with a vertex on the left, but it does mean that every vertex on the left is paired.

In our matching, the edges pair the students with the places of residency. Since every vertex is incident with an edge in the matching, every student has a residency place. Moreover, every student is assigned to precisely one place and every place assigned has only one student.

This all sounds fine in principle, but you can imagine that finding a matching in a bipartite graph that is even remotely large can be very laborious to do by trial and error. So, how in general might we go about creating a matching or even deciding if one exists? First let us think about existence. What kind of situations would indicate that it would be impossible for a matching to exist?

Let's imagine that in our residency example, Vanderbilt Children's is not taking applications this year after all. Now these five students have only applied to four different residency places. In graph theory terms, the four residencies are the **neighborhood** of the five students.

Neighborhood

If we have a set of vertices A, then the **neighborhood** of A, denoted $N(A)$, is the set of all vertices connected to a vertex in A.

Example 2: Identifying a Neighborhood

Let A be the set of vertices $\{d, e\}$. Identify the neighborhood of A.

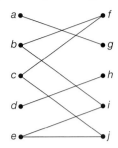

Solution

To find the neighborhood of A, we need to find all the vertices on the right-hand side that are adjacent to the vertices in A. Vertex d is adjacent to vertex h on the right-side and vertex e is adjacent to both vertices i and j.

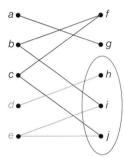

So, $N(A) = \{h, i, j\}$.

Skill Check #2

Let B be the set of vertices $\{a, b\}$ from the graph in Example 2. Identify the neighborhood of B.

It is clear that there is a problem trying to find a matching with unequal numbers of vertices in the two sets of vertices. In our residency example, we can certainly begin to match students up, but there is no way to match them all—there just aren't enough chosen residency places in the neighborhood for each residency applicant to have a place all to their own. For example, consider the situation in Figure 3.

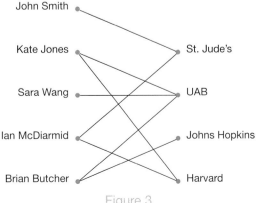

Figure 3

13

Of course, there is nothing magical about the numbers five and four. Any situation where the number of residency places is less than the number of applicants is going to create the same problem. In fact, no matter what the bipartite graph represents, we will not be able to find a matching if a set of vertices has a neighborhood smaller than itself. What is surprising, perhaps, is that this is the only possible barrier to finding a matching. This fact is described in a theorem by Philip Hall with the unusual name of the *Marriage Theorem*.

> ## Hall's Marriage Theorem
>
> Let *G* be a bipartite graph. Then there is a matching of the left-hand vertices into the right-hand vertices if and only if for every subset of left-hand vertices *A*, the number of vertices in $N(A)$ is at least as large as the number of vertices in *A*.

Example 3: Determining if a Matching is Possible

Use Hall's Marriage Theorem to determine if a matching is possible for the following graphs.

a.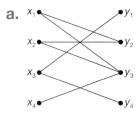

b. Bill, Bob, and Benny are looking for dates for the prom. They have Gill, Grace, and Gabi in mind. Bill feels that he is friendly enough with Gill and Grace to ask either of them. However, Bob only feels comfortable asking Gabi, and Benny only feels comfortable asking Gill. Is there a way for each boy to ask a different girl to the prom?

Solution

a. Hall's Marriage Theorem says that for there to be a matching, whichever subset of vertices we choose on the left, their neighborhood on the right must be at least as large. For several subsets of vertices on the left this property is satisfied. For instance, consider $A = \{x_3\}$, which has $N(A) = \{y_1, y_4\}$.

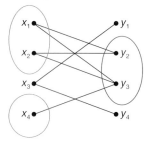

But if we look at $A = \{x_1, x_2, x_4\}$, the neighborhood is $N(A) = \{y_2, y_3\}$, which violates Hall's Marriage Theorem and shows that a matching is not possible.

b. To begin represent the boys and girls as a graph with the boys on the left and the girls on the right.

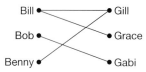

A matching from the left into the right is a way for each boy to ask a different girl to the prom. You can probably see that this can be done by looking at the graph, but let's see how Hall's Marriage Theorem also establishes this.

To use Hall's Marriage Theorem we need to consider each possible subset of boy vertices and check the size of the neighborhood. We have seven subsets to work through. We will deal with them one at a time.

1. $A = \{\text{Bill}\}$ $N(A) = \{\text{Gill, Grace}\}$
2. $A = \{\text{Bob}\}$ $N(A) = \{\text{Gabi}\}$
3. $A = \{\text{Benny}\}$ $N(A) = \{\text{Gill}\}$
4. $A = \{\text{Bill, Bob}\}$ $N(A) = \{\text{Gill, Grace, Gabi}\}$
5. $A = \{\text{Bill, Benny}\}$ $N(A) = \{\text{Gill, Grace}\}$
6. $A = \{\text{Bob, Benny}\}$ $N(A) = \{\text{Gill, Gabi}\}$
7. $A = \{\text{Bill, Bob, Benny}\}$ $N(A) = \{\text{Gill, Grace, Gabi}\}$

As we can see, for each subset the number of elements in $N(A)$ is greater than or equal to the number of elements in A. Therefore, Hall's Marriage Theorem ensures that there is a matching, and a way for each boy to ask a different girl to the prom.

Since Hall's Marriage Theorem works for every bipartite graph, it allows us to guarantee that a matching exists for an important special case of bipartite graphs. We say that a graph is a regular bipartite graph if there are the same number of vertices in both the left-hand side and right-hand side of the graph, and every vertex has the same degree.

Regular Graph

A **regular graph** is a graph where every vertex has the same degree.

Regular Bipartite Graph

A **regular bipartite graph** is a regular graph with the same number of vertices on both the left-hand side as the right-hand side.

Example 4: Identifying a Regular Bipartite Graph

Determine whether each graph is a regular bipartite graph.

a.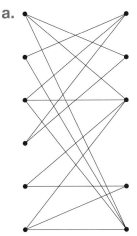

b.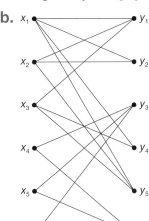

c.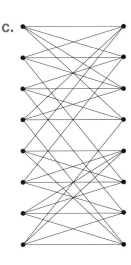

13

Solution

a. A regular bipartite graph always has the same number of vertices on the left and right-hand sides. This graph has six vertices on the left-hand side and five on the right. Therefore, this graph is not a regular bipartite graph.

b. This graph has six vertices on the left and right-hand sides, but in a regular bipartite graph every vertex also has the same degree. Among lots of other differences, vertex x_1 has degree 4, but vertex x_2 has degree 3. Therefore, this graph is not a regular bipartite graph.

c. This graph has eight vertices on each side, and every vertex has degree 4. Therefore, it is a regular bipartite graph.

Let's see how applying Hall's Marriage Theorem shows that every regular bipartite graph has a matching. Suppose that in a regular bipartite graph G, every vertex has degree 5. Choose a set A of left-hand vertices; say k of them. Each of those k vertices is adjacent with five right-hand vertices, but of course, they can overlap in all kinds of ways. We need to show that the neighborhood of these k vertices has at least k vertices itself.

Let's say that there are l right-hand vertices in the neighborhood of A. Now, let's count the edges between A and its neighborhood. There are $5k$ edges "coming out" of the k vertices in A on the left and going into the l vertices on the right. Because G is a regular bipartite graph, we know the l vertices also only have five edges each as well, some of which must come from A. However, not all of the edges have to come from A. (Remember we chose a subset of vertices from the graph, so l could have edges coming from vertices outside of A.) We know that,

$$5k \leq 5l, \text{ which means } k \leq l.$$

Hence, the condition of Hall's Theorem is satisfied for a regular graph; no matter how big k is, we must have a matching.

> **Regular Bipartite Graph Theorem**
>
> A regular bipartite graph has a matching.

Example 5: Deciding if a Matching is Possible

Let's consider the following scenario. At the British Museum in London, multimedia tours are offered in English, Korean, Arabic, French, German, Italian, Japanese, Mandarin, Russian, and Spanish. A group of ten students go to the museum. The graph shows which languages each student can understand. Determine if the graph has a matching.

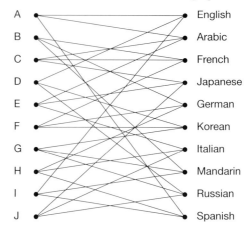

Solution

To determine if the graph has a matching, we need to determine if it is a regular bipartite graph. To do this, we need to check for two things:

1. All vertices must have the same degree.

2. The number of vertices on the left-hand side must be the same as the number of vertices on the right-hand side.

First to determine the degree of each vertex, we need to count the number of edges incident to each vertex on both sides of the graph.

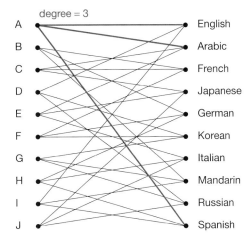

We see that all vertices have degree 3, so the first criterion is met.

We can count the number of vertices on each side of the graph to make sure there are equal numbers in each.

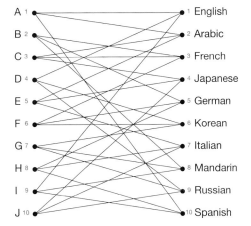

Since this graph is a regular bipartite graph, we know from the theorem that it has a matching.

Of course, knowing that a matching exists and finding it are two different things. To find a matching in a regular graph, we can use Schriver's Algorithm.

Schriver's Algorithm

Steps for Finding a Matching in a Regular Bipartite Graph

1. Given a regular bipartite graph G, give every edge in G a weight of 1.

2. Let C be a cycle in the edges of positive weight.

   • Number the edges of C successively in turn.

   • Increase the weight of the even-numbered edges by 1.

   • Decrease the weight of the odd-numbered edges by 1.

3. Repeat Step 2 until the edges with positive weight contain no cycle.

4. The positively weighted edges form a matching.

Example 6: Finding a Matching

Find a matching for the following regular bipartite graph.

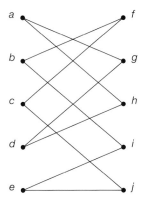

Solution

We'll use Schriver's Algorithm to find a matching. Let every edge have a weight of 1. Now, find a cycle in the graph. Take a moment and see if you can find one first before looking at the next figure that shows the cycle. Sometimes the hardest part is finding the first cycle.

After finding a cycle, label the edges consecutively. The next figure shows the cycle a, h, d, g with the edges labeled in order, not by weight.

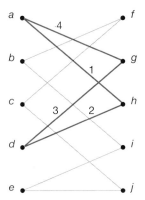

The next step is to increase the weights of the even numbered edges 2 and 4 by 1, which means they will now each have a weight of 2. We then decrease the odd numbered edges 1 and 3 by 1, which means they have a weight of 0 and can no longer be considered in the algorithm because only positive weighted edges are considered. In the next graph, the edges with positive weights are shown and the edges with zero weights are removed since they are not part of the matching.

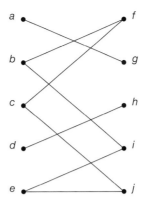

We begin again at Step 2 by finding a cycle from the edges with positive weights. The only remaining cycle is f, b, i, e, j, c. Label the cycle with consecutive numbers.

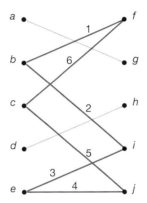

Once again, increase the weight of the even numbered edges by 1 and decrease the odd edges by 1. The only remaining positive weighted edges give us the following graph.

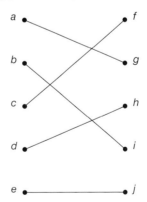

Because there are no more cycles, we can stop. The remaining edges form the matching we required.

Skill Check #3

Find a different matching for the graph in Example 6.

Skill Check Answers

1. Yes, the graph is bipartite, since *a* and *b* can be the left-hand side and *c and d* can be the right-hand side.

2. $N(B) = \{f, g, i\}$

3. Answers will vary. For example,

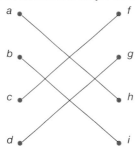

13.3   Exercises

Fill in each blank with the correct term.

1. A _____ is one in which the vertices can be broken into two distinct groups where each vertex in the first group is joined to a vertex in the second group.

2. By Hall's Marriage Theorem, we know that a _____ graph has a matching.

3. A _____ is a subset of edges in a graph so that each vertex is incident with only one edge.

4. In a _____, every vertex has the same degree.

Determine if each graph is bipartite. Justify your answer.

5.

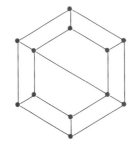

6.

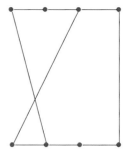

Use Graph *Q* to identify the neighborhood of each set.

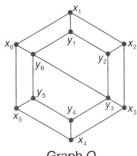

Graph Q

7. $A = \{x_1, y_1\}$

8. $B = \{\text{the set of odd numbered } x\text{'s}\}$

9. $C = \{y \text{ vertices}\}$

10. $D = \{\text{all vertices in } Q\}$

Determine if each graph is regular bipartite.

11.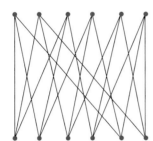

12.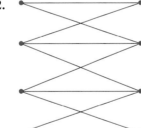

13.

14.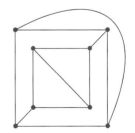

Determine if each graph has a matching. Justify your answer.

15.

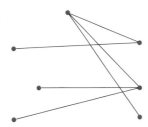

16.

17.

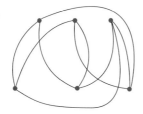

18.

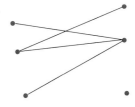

19.

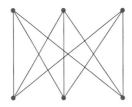

Solve each problem.

20. Use Hall's Marriage Theorem to determine if there is a matching in the following graph.

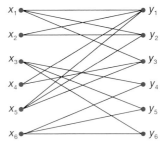

21. Is Grötzsch's graph bipartite?

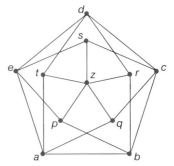

Find a matching in each regular bipartite graph.

22.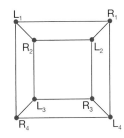

23.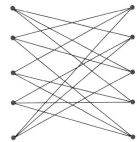

Form a bipartite graph from the information in each table. Find a matching for each scenario, if possible.

24.

Teacher	Courses
Mr. Hall	Political Science, American History
Ms. Cutlidge	English I, English II, English III, Advanced Writing, Computer Science
Mrs. Roseview	English I, English II, English III, Advanced Writing
Mr. Burden	Algebra II, Precalculus, Calculus, PE
Mr. Smith	American History, US History, Economics, Political Science
Mrs. Jones	Political Science, US History, PE, Computer Science
Ms. Rodriguez	Algebra I, Algebra II, Geometry, Calculus

25.

Volunteers	Positions Willing to Serve In
Kerri	greeter, usher, refreshments
Luke	parking attendant, usher
Ron	sound board, video, lights
Jose	lights, parking attendant, greeter
Cho	nursery, sound board, video
Mia	refreshments, nursery, greeter
Margaret	parking attendant, refreshments, lights
Lenton	video, nursery, greeter

Solve each problem.

26. Use Schriver's Algorithm to find a matching of the British Museum example given in the text. We know from the text that it is a regular bipartite graph, and hence does have a matching.

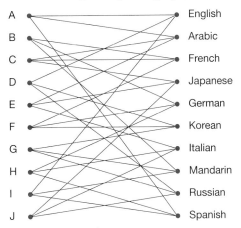

27. Let graph G be a bipartite graph with the same number of vertices on each side. Suppose that for every set of vertices B on the right-hand side, the number of vertices in $N(B)$ on the left-hand side is at least as large as the number of vertices in B. Explain why graph G has a matching from the left-hand side to the right-hand side.

28. An edge coloring of a graph G is an assignment of colors to the edges, so that incident edges have different colors. Explain why edges of the same color form a matching.

13.4   Planar Graphs

Recall from the definition of a graph that not every place where two edges cross is a vertex. In such graphs, edges that cross have no consequences. However, it's not always the case that we want graphs to allow such crossings. For instance, consider a circuit board, such as in Figure 1.

At first glance, it might seem that circuit boards can be arbitrarily complicated, having any design that you want, but that turns out not to be true. If the connections that join the components on a circuit board cross, it could cause the circuit board to malfunction. Although they can be very complicated, the design is limited to those graphs that can be drawn on a page without crossings. Chances are that you didn't think that when you first saw a circuit board, did you?

Figure 1: Circuit Board

This restriction of edge crossings can sometimes cause exasperation. For instance, no matter how hard we try, we cannot have five vertices all mutually linked together (that is, all possible edges drawn between them) without having at least two of the edges crossing. Try it for yourself. We will explore why you are doomed to failure later. This section begins the examination of graphs that have no edge crossings when drawn on a plane, known as **planar graphs**.

Think Back

Recall that a plane is a flat surface without thickness, depth, or boundaries; like an infinite piece of paper.

Planar Graphs

Graphs that can be drawn on a plane without edge crossings are called **planar graphs**.

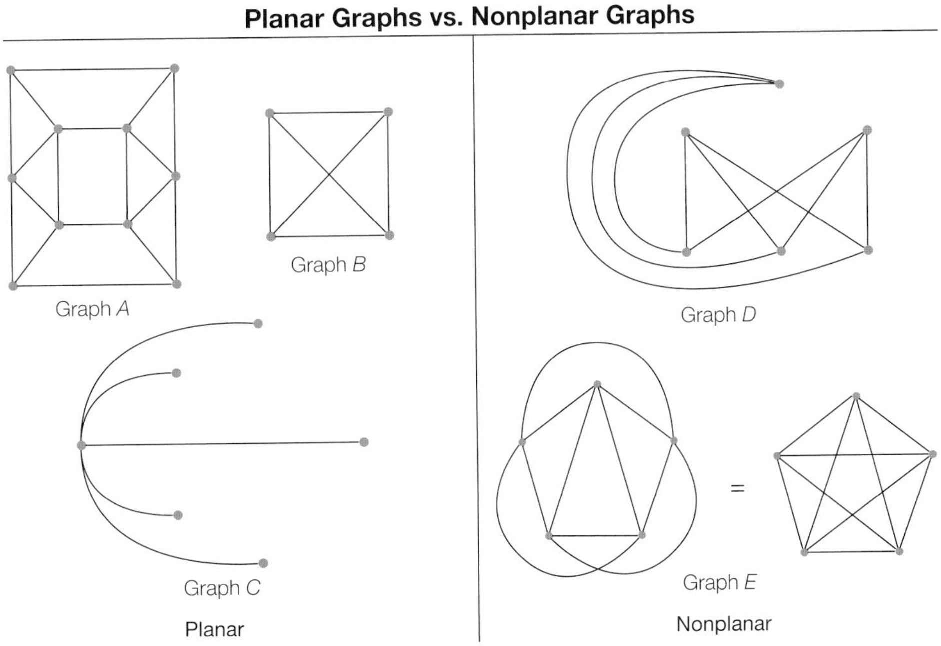

Planar Graphs vs. Nonplanar Graphs

Graph A

Graph B

Graph D

Graph C

Planar

Graph E

=

Nonplanar

Figure 2

Figure 2 shows both planar graphs and nonplanar graphs. At first glance, the planar graph *B* might appear to be nonplanar. However, if we redraw the graph, we see that it can be drawn without any edges crossing, as in Figure 3.

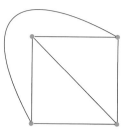

Figure 3: Alternate View of Graph B

Skill Check #1

Draw the following planar graph without any edge crossings.

What this tells us is that we cannot merely depend on drawings to determine if a graph is planar. However, we can use some algorithms and theorems to help us classify a graph as planar or not. We will begin with *Euler's formula*, which captures the structure of a planar graph, To explain the formula, we first need some new vocabulary.

Fun Fact

The website http://www. planarity.net/ has a browser-based and smartphone game related to planar graphs!

If you have a planar graph, the graph has another feature that we can describe in addition to the vertices and edges. The graph breaks the plane into certain areas, which we call **faces**.

Face

In a planar graph, a **face** is a region inside a cycle of edges or the infinite exterior region on the outside of the graph. We denote the number of faces of a graph by *f*.

A planar graph always has at least one face, which is the exterior face. For instance, in Figure 4, the exterior face on the tree *T* is shaded. Notice that an exterior face extends indefinitely.

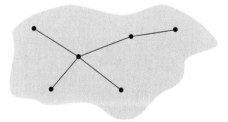

Figure 4: Faces of Tree *T*

If there is an interior face in a planar graph, it is bounded by a cycle. Figure 5 shows the graph *H*. There are three faces in graph *H*: one surrounded by the cycle *A, B, C, D*, which we've shaded orange, one surrounded by the cycle *C, D, E,* which we've shaded blue, and don't forget the exterior face on the outside the graph, which we've shaded in light green. Therefore, in graph *H*, $f = 3$.

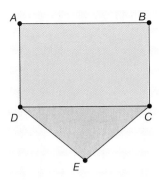

Figure 5: Faces of Graph *H*

You can think about faces in a "crafty" manner. If you were to cut the graph *H* out of this textbook by cutting along it's edges, the sections of paper you get (the orange piece, the blue piece, and the larger light green piece) would represent the three faces of the graph. Thinking in this way, although it has no cycles, even a tree has one face, as we saw in Figure 4.

Example 1: Determining the Number of Faces in a Planar Graph

Two drawings of the same graph *J* are shown. Verify that both drawings have the same number of faces.

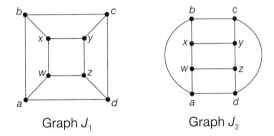

Graph *J*₁   Graph *J*₂

Solution

We can count each of the faces by counting the contained spaces in the graph as well as the exterior face. We've numbered the spaces for each graph in the following figure.

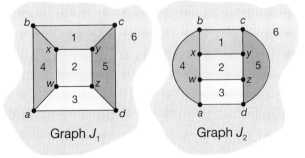

Graph *J*₁   Graph *J*₂

You can see that $f = 6$ in both diagrams. In fact, graph *J* will always have six faces no matter how we draw it.

Notice that if we were to remove any edge from a cycle in a graph, we decrease the number of faces by one. For instance, if we remove the edge *CD* from graph *H* in Figure 6, we create one large interior face.

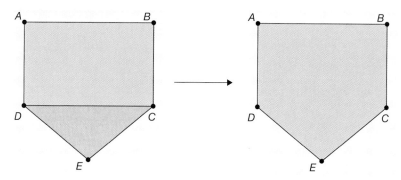

On the other hand, if we add any edge to a graph joining two of the existing vertices, we also increase the number of faces by one since that new edge will split one of the existing faces into two parts. For instance, if we add the edge *AC* to the graph, it will have four faces because it splits the orange face into two parts and creates a new cycle *A*, *B*, *C*, as Figure 7 shows.

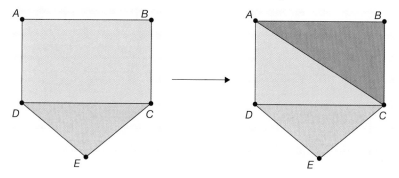

Skill Check #2

a. What would happen if we removed the edge *CE* from the original graph *H* in Figure 5?

b. What would happen if we add the edge *BD* to the original graph *H* in Figure 5?

Let's think for a minute about trees, which we looked at in Section 13.2. For a tree, we know the relationship between the numbers of edges, vertices, and faces. We've already seen in Figure 4 that every tree, since it has no cycles, has precisely one face. Because a tree has no cycles, it has no way to "surround" a specific area of the plane in order to make a face. Therefore, it has only the exterior face. So, $f = 1$.

On the other hand, we also know from Section 13.2 that the number of edges in a tree is one fewer than the number of vertices. If we denote the number of edges in a graph as e and the number of vertices in a graph as v, then we can write this as $e = v - 1$. Putting both v and e on the same side of the equation, we have $v - e = 1$. So, for a tree, we can add these together to see the following.

$$(f) + (v - e) = 1 + 1$$
$$f + v - e = 2$$

We now know from the previous discussion, that each time we add an edge without adding any extra vertices, we also add a face. So, this equation relating f, v, and e holds true for any planar graph. The mathematician Euler first proved this and it is now referred to as Euler's formula in graph theory.

Euler's Formula

If G is a connected planar graph with v vertices, e edges, and f faces, then

$$v + f - e = 2.$$

Let's use Euler's formula in an example.

Example 2: Verifying Euler's Formula for a Graph

Confirm Euler's formula for the planar graph G.

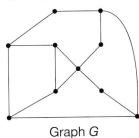

Graph G

Solution

We can see that G has 10 vertices and 14 edges. All that remains is to find the number of faces. It has 5 internal faces, as indicated in the graph.

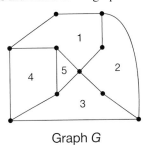

Graph G

However, we must also remember to include the external face. That makes a total of 6 faces.

So, now we have that $v = 10$, $e = 14$, and $f = 6$. Substituting into Euler's formula, we can confirm that the number of vertices plus the number of faces minus the number of edges is equal to 2.

$$v + f - e = 10 + 6 - 14 = 2.$$

Think Back

Recall that a pentagon is a
five-sided figure and that a
hexagon is a six-sided figure.

Example 3: Applying Euler's Formula

A soccer ball traditionally consists of hexagonal, white faces and pentagonal, black faces. Suppose that we use a box cutter and cut out one of the white, hexagonal faces from the ball to make a hole. Now, imagine that we could stretch the ball covering outward at the hole we made, so that it lies completely flat in a plane as shown. How many black pentagonal faces are there?

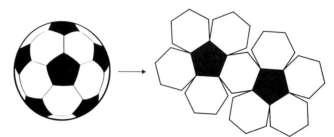

Solution

Remember that Euler's formula says that the number of vertices plus the number of faces minus the number of edges in a planar graph must equal 2. To use Euler's formula, we need to find a way to express the number of vertices, edges, and faces in a soccer ball, and then substitute them into the formula.

Let's tackle the vertices first. On the soccer ball, each black pentagon has five vertices, and of course, each white hexagon has six vertices. We can count the total number of vertices on the entire ball one face at a time by counting 5 for every black pentagon and 6 for every white hexagon. So, if we let P represent the number of pentagonal faces, we can multiply it by 5. Similarly, we'll let H represent the number of hexagonal faces and multiply it by 6. Then, the total number of vertices can be represented by the sum $5P + 6H$.

However, notice that every vertex on the soccer ball is a vertex of three adjacent faces. The following diagram shows how vertex v is adjacent to the three faces colored yellow, green, and blue.

So, if we claim the number of vertices on the soccer ball is $5P + 6H$, we end up counting each vertex multiple times. Look at vertex v in the previous figure again. It will be counted as part of each of the three faces in which it lies. In fact, every vertex lies in three faces, which means every vertex will be counted three times as we count around the faces. So, we must divide the number of vertices by 3 to give us

$$v = \frac{5P + 6H}{3}.$$

We can do the same to count the number of edges on the soccer ball. Every black pentagon has five edges, and every white hexagon has six edges. Again, we must be careful not to over count the edges. Notice that every edge is on the boundary of two faces. For instance, the edge highlighted is on the boundary of the yellow face as well as the green face.

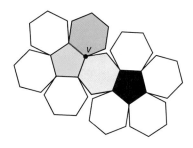

So, the number of edges is given by the following.

$$e = \frac{5P + 6H}{2}$$

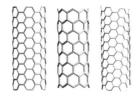
Now that we've found expressions for the number of vertices and edges, we just need to express the number of faces. The total number of faces will be the number of black pentagons P plus the number of white hexagons H. (Recall that we cut out one of the white hexagonal faces, which counts as the exterior face. As a result, we do not need to count an exterior face.) Substituting these expressions into the formula we have the following.

$$v + f - e = 2$$

$$\frac{5P + 6H}{3} + (P + H) - \frac{5P + 6H}{2} = 2$$

Next, simplify this equation by removing the fractions and combining like terms together.

$$\frac{5P + 6H}{3} + (P + H) - \frac{5P + 6H}{2} = 2$$

$$6\left(\frac{5P + 6H}{3} + (P + H) - \frac{5P + 6H}{2}\right) = 6(2)$$

$$10P + 12H + 6P + 6H - (15P + 18H) = 12$$

$$10P + 12H + 6P + 6H - 15P - 18H = 12$$

$$(10P + 6P - 15P) + (12H + 6H - 18H) = 12$$

$$P = 12$$

This means that no matter how many hexagons there are, there must always be precisely 12 black pentagons on the soccer ball.

The answer to Example 3 may be somewhat surprising to you. It is actually confirming that, no matter what size soccer ball you want to make, you will always need precisely 12 pentagons. Since hexagons are tessellating shapes, meaning that they can cover a surface without any gaps or overlaps, we can add more hexagons to the ball cover, but can't ever add more pentagons. If we tried to, a consequence of Euler's formula means that the configuration would no longer fit as a ball cover no matter how hard you try. This applies to any ball made of pentagons and hexagons.

One consequence of Euler's formula is that, given a fixed number of vertices, if we restrict ourselves to having at most one edge between each pair of vertices, there is a limit to how many edges can be added while keeping the graph planar. Euler's formula puts an upper limit on the number of total edges in a planar graph. In order for a graph to be planar, the number of edges it can have is at most $3v - 6$, where v is the number of vertices in the graph.

Corollary of Euler's Formula

A planar graph G with v vertices has at most $3v - 6$ edges. That is, $e \le 3v - 6$.

Example 4: Number of Edges in a Planar Graph

If graph G has 13 vertices, what is the greatest number of edges that graph G can have and be a planar graph?

Solution

If graph G is to be a planar graph, then the number of edges can be at most $3v - 6$. Since we know that graph G has 13 vertices, $v = 13$. Therefore, in graph G the number of edges, e, can be at most

$$e \le 3(13) - 6$$
$$e \le 39 - 6$$
$$e \le 33.$$

Skill Check #3

What is the maximum number of edges that a five-vertex graph can have and still be planar? (**Hint:** Draw a graph to help you find the answer.)

We've now looked at several characteristics of planar graphs. If a graph is planar, it will meet all of the following characteristics.

- It can be drawn with no edges crossing.

- It has definable faces.

- Euler's formula relates the number of vertices, faces, and edges by $f + v - e = 2$.

- The number of edges is always at most $3v - 6$.

However, not all graphs meet these criteria and, hence, are not planar. Consider the following graph, K. Graph K has five vertices that all have degree 4. In other words, every vertex is joined to every other vertex as shown in Figure 8.

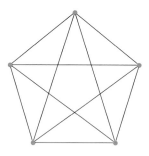

Figure 8: Graph K

If graph K is planar, it will meet all the characteristics. However, if we can find one characteristic it doesn't satisfy, we know it's not planar. For graph K, let's see if it satisfies the corollary of Euler's formula. From the diagram, we know that for K, $v = 5$ and $e = 10$. Count them yourself. If we substitute these values into the inequality we have,

$$e \overset{?}{\le} 3(v) - 6$$
$$10 \overset{?}{\le} 3(5) - 6$$
$$10 \overset{?}{\le} 15 - 6$$
$$10 \nleq 9$$

which is not possible. Therefore, it is impossible for graph K to be planar.

The graph K has a formal graph theory name of K_5. It is the graph with five vertices all of which are connected to each other. Formally, K_5 is the complete graph on five vertices. A complete graph is defined to be a graph in which all vertices are connected to all other vertices. We denote a complete graph by K_n, where n is the number of vertices in the graph.

Fun Fact

Complete graphs are represented by the capital letter K. Some sources claim that the K stands for the German word for complete, *komplett*, while others say that the notation gives honor to the graph theorist Kazimierz Kuratowski.

Complete Graphs

A **complete graph** is a defined to be a graph in which all vertices are connected to all other vertices. We denote a complete graph by K_n, where n is the number of vertices in the graph.

Example 5: Establishing That a Graph Is Not Planar Using Euler's Corollary

Use the corollary of Euler's formula to establish that graph G is not planar.

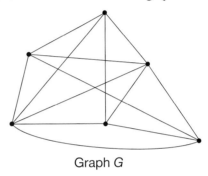

Graph G

Solution

To use the corollary of Euler's formula, we need to count the number of edges and the number of vertices in graph G. There are six vertices (numbered in blue) and 13 edges (numbered in red) as labeled in the following graph.

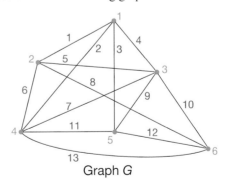

Graph G

So $v = 6$ and $e = 13$. Now we can substitute these values into the corollary to see if graph G meets the restriction to be planar.

$$e \overset{?}{\leq} 3(v) - 6$$
$$13 \overset{?}{\leq} 3(6) - 6$$
$$13 \overset{?}{\leq} 18 - 6$$
$$13 \nleq 12$$

The number of edges is too large, therefore it is not possible for G to be planar.

Let's consider another graph. This graph, U, is a regular bipartite graph where each vertex has degree 3. It has six vertices, three on the top, and three on the bottom.

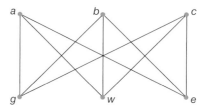

Figure 9: Graph U

Let's check to see if graph U satisfies $e \leq 3v - 6$ with $v = 6$ and $e = 9$.

$$e \overset{?}{\leq} 3(v) - 6$$
$$9 \overset{?}{\leq} 3(6) - 6$$
$$9 \overset{?}{\leq} 18 - 6$$
$$9 \leq 12$$

It does satisfy the inequality. However, that doesn't guarantee that graph U is planar. In Figure 9, graph U certainly has edges crossing. However, can we draw graph U without the crossings? Notice that the vertices a, g, b, w, c, e form a cycle of length 6. We've highlighted it in Figure 10.

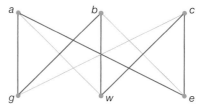

Figure 10: Cycle in Graph U

In any planar drawing of the graph, this cycle needs to be disentangled. We show this in Figure 11.

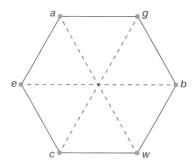

Figure 11: Alternate View of Graph U

That leaves us to uncross the edges shown by the dotted lines in Figure 11: aw, be, and gc. Notice that we can't possibly draw more than one of those edges "inside" the cycle without creating a crossing; and similarly we can only draw one edge "outside" the cycle. Since there are three edges remaining, that still leaves one edge left to draw that has to cross something either inside the cycle or outside. So, it is impossible to draw graph U in the plane without crossings. Therefore, U is not a planar graph.

The graph U has a formal graph theory name. It is called $K_{3,3}$. This graph is the regular bipartite graph on six (three on the left-hand side and three on the right-hand side) vertices of which all have degree 3.

Example 6: Classifying Graphs as Planar or Not

Decide whether graph F is planar or not.

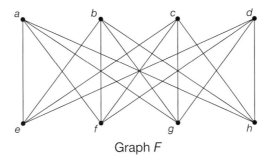

Graph F

Solution

Graph F has $v = 8$ vertices and $e = 16$ edges.

Let's first check that this satisfies the corollary of Euler's formula.

$$e \overset{?}{\le} 3(v) - 6$$
$$16 \overset{?}{\le} 3(8) - 6$$
$$16 \overset{?}{\le} 24 - 6$$
$$16 \le 18$$

It does satisfy the inequality. However, that doesn't guarantee that graph F is planar. Instead, notice that graph F has a subgraph on a, b, c, e, f, g that is the same as $K_{3,3}$, which we saw is not planar.

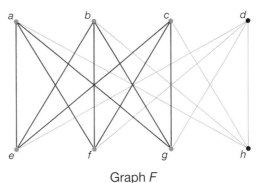

Graph F

So, since we can't draw $K_{3,3}$ without edges crossing, then graph F cannot be planar.

The logic that we employed to determine that F was not a planar graph is logic that can be applied much more broadly. In fact there is a theorem that states if a graph contains a subgraph that is not planar, then the whole graph is not planar. Consequently since we have shown that neither K_5 nor $K_{3,3}$ is planar, any graph that contains either one cannot itself be planar.

Planar Subgraph Theorem

Graph G is not planar if any subgraph of G is not planar.

Example 7: Establishing that a Graph is Not Planar Using Subgraphs

Use the theorem to show that the graph L is not planar by finding a suitable subgraph.

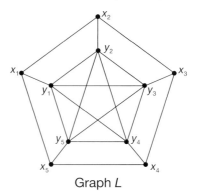

Graph L

Solution

In order to show that graph L is not planar, it is enough to find either a K_5 or $K_{3,3}$ subgraph. Let's consider the five vertices in the middle of the graph labeled y_1, y_2, y_3, y_4, and y_5. If they form the graph K_5, meaning each vertex is connected to all other vertices, then the graph is not planar.

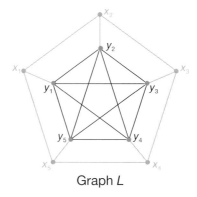

Graph L

Notice that each of the vertices y_1 through y_5 have degree 4 and are all connected to each other. Check this for yourself. The existence of this subgraph is sufficient to say that graph L is not planar.

Graph Minors

To close this section, we will improve on the previous theorem in a way that allows us to completely characterize planar graphs. Suppose that, given a graph G, we delete any vertex or edge to create a new graph H. If graph G is planar, then so is graph H. To see this, we simply draw graph G in the plane and then erase the edges or vertices that we want to delete, and we have a planar drawing of graph H.

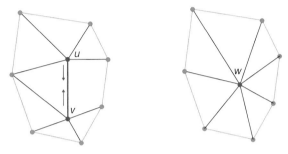

Figure 12: Graphs *G* and *H*

In fact, we can allow ourselves yet more flexibility by introducing one more graph operation called *contracting*. Given a graph *G* and an edge $e = uv$, the graph $H = G/e$ is obtained by shrinking *e* until the vertices *u* and *v* are the same vertex. In other words, contracting an edge means that it disappears in such a way that its end vertices become one. We illustrate this in Figure 13.

Figure 13: Contracting an Edge in a Graph

Happily, once again, if *G* is planar then *G/e* is also planar. We say that a graph that is obtained from *G* by a sequence of edge or vertex deletions and edge contractions is a *minor* of *G*. We can now finally characterize planar graphs completely. Since we know that neither $K_{3,3}$ nor K_5 are planar, no planar graph can have one of these as a minor. In fact, that is the only barrier to planarity.

Planar Graph Theorem

G is planar if and only if *G* has neither $K_{3,3}$ nor K_5 as a minor.

Example 8: Classifying Graphs as Planar or Not

MATH MILESTONE

The Grötzsch graph is the smallest triangle-free graph with chromatic number 4. It is named after German mathematician Herbert Grötzsch, born in 1902 in Halle, Germany. He died in 1993 a week before his 91st birthday.

The following is a famous graph called the Grötzsch graph. Show that it is not planar by finding a K_5 minor of the graph.

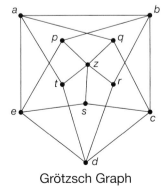

Grötzsch Graph

Solution

In order to show that the Grötzsch graph is not planar, we need to show that it contains a K_5 minor. Without stronger algorithms that are beyond the scope of this text, we need to simply "find" a minor by a process of deleting vertices, deleting edges, or using edge contractions. Remember that looking for a K_5 minor means that we are looking for five vertices that are all connected to each other. Obviously, there are too many vertices in the original graph, so we'll begin by deleting the middle vertex z. When we delete a vertex from a graph, all edges connected to it also get deleted because they no longer have a vertex at both of their ends. The following graph shows the Grötzsch graph after deleting z.

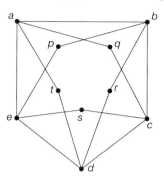

Can you begin to see where the five vertices in the minor might come from? If we can eliminate the inner vertices, we might just have the graph we are seeking. Begin by contracting edge aq so that the vertices a and q become one. We now have one edge between aq and c as the following graph shows.

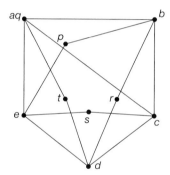

In the same manner, we can contract the edge pb as well as edge sc. Here's the contracted graph so far.

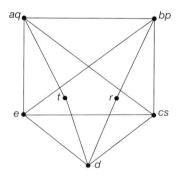

What remains is to contract edges td and rd. Once this is done, we have the following graph of five vertices of degree 4 all connected together. In other words, a K_5 minor of the Grötzsch graph.

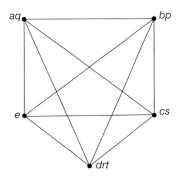

Hence, we have shown that the Grötzsch graph is not planar since it contains a K_5 minor.

Skill Check Answers

1. Answers will vary. Two possibilities are as follows.

2. **a.** There would then only be two faces.   **b.** We'd split the orange face in two, creating four faces in total.

3. Nine edges

13.4 Exercises

Draw each planar graph without any crossing edges.

1.

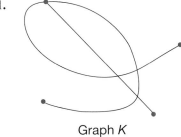

Graph *K*

2.

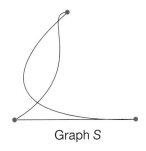

Graph *S*

Determine the number of faces in each planar graph.

3.

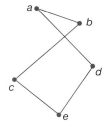

Graph *N*

4.

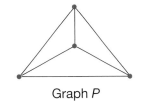

Graph *P*

5.

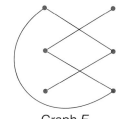

Graph *E*

Verify Euler's formula for each graph.

6.

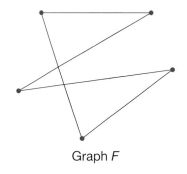

Graph *M*

7.

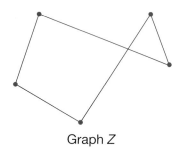

Graph *Z*

8.

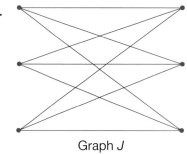

Graph *F*

9.

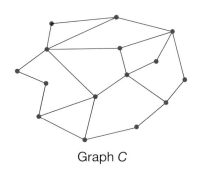

Graph *C*

Determine if each graph is planar.

10.

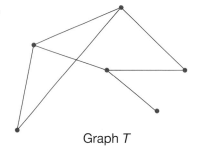

Graph *T*

11.

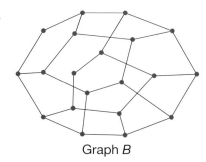

Graph *B*

12.

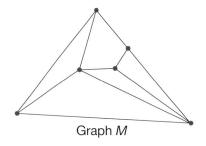

Graph *J*

13.

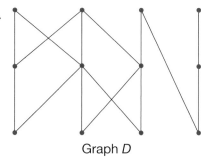

Graph *D*

Solve each problem.

14. A new housing development wants to connect three houses to their three utilities (gas, water and electric) without their cables/pipes crossing. Can it be done?

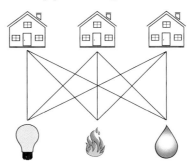

15. Show that a planar graph must have a vertex of degree at most 5.

16. Find the chromatic number of K_3, K_5, and K_n.

17. Find the chromatic number of $K_{5,5}$ and $K_{k,k}$.

18. Show that any planar graph can also be drawn on a sphere.

19. What happens if you draw the graph of a cube on a sphere and then flatten the faces so that they are the same size?

13

Chapter 13   Summary

Section 13.1   Introduction to Graph Theory

Definitions

Graph

A graph consists of a set of vertices (singular vertex) and a set of edges that join pairs of vertices together. Graphs are generally represented with a capital letter, such as G.

Loop

A loop is an edge that has the same vertex for both of its ends.

Walk

A walk in a graph G is a finite list of alternating vertices and connecting edges that begins and ends with a vertex.

Path

A path is a walk with no repeated edges or vertices.

Connected Graph

A graph is connected if there is at least one path connecting each pair of distinct vertices.

Disconnected Graph

A graph is disconnected if a pair of vertices exists so that there is no path between them.

Adjacent

Two vertices that share an edge are adjacent.

Incident

Two edges that share a vertex are incident to one another. Also, we say that a vertex is incident to the edges that have that vertex as an endpoint.

Degree

The degree of a vertex u is the number of edges that are incident to u and is notated by $d(u)$.

Vertex Cover

A vertex cover is a set of vertices A, so that every vertex in the graph is either in A or adjacent to a vertex in A.

Minimum Vertex Cover

When a vertex cover A is as small as possible, A is called a minimum vertex cover.

Vertex Coloring

A vertex coloring of a graph is an assignment of colors to the vertices of the graph so that adjacent vertices receive different colors.

Chromatic Number

When the number of colors used in a vertex coloring is as small as possible, this number is called the chromatic number of a graph and is denoted $\chi(G)$.

Cycle

A cycle is a closed walk; that is, a cycle starts and ends at the same vertex and has no edges or vertices repeated except for the starting vertex.

Section 13.2   Trees

Definitions

Tree

A connected graph with no cycles is called a tree.

Tree Theorem

Let graph T be a tree on v vertices. Then, graph T has $v - 1$ edges.

Spanning Tree

A spanning tree is a subgraph of a connected graph, which is itself connected and contains all the vertices of the original graph, but has no cycles.

Leaf

A vertex of degree 1 in a tree is called a leaf.

Formula

Number of Leaves on a Tree

If a tree has k vertices with degrees $d_1, d_2, \ldots, d_k$, each greater than 1, then the number of leaves is $\sum_i d_i - 2k + 2$.

Processes

Constructing a Spanning Tree

1. In a connected graph G, identify a cycle. If more than one cycle exists, choose one at random.

2. Choose an edge from the cycle selected and delete it from the graph.

3. While the graph contains a cycle, repeat Steps 1 and 2.

Constructing a Minimum-Weight Spanning Tree

1. Consider each edge in the graph in order of descending weight.

2. If the edge being considered is part of a cycle, remove it. If not, it must remain in the graph, and you can move on to the next edge.

3. Repeat Steps 1 and 2 until all edges have been considered.

Section 13.3   Matchings

Definitions

Bipartite Graph

A bipartite graph is a graph in which the vertices can be partitioned into precisely two subsets so that every edge joins a vertex in one subset to a vertex in the other subset.

Matching

A matching is a subset of edges in a graph so that each vertex is incident with only one edge.

Neighborhood

If we have a set of vertices A, then the neighborhood of A, denoted $N(A)$, is the set of all vertices connected to a vertex in A.

Hall's Marriage Theorem

Let G be a bipartite graph. Then there is a matching of the left-hand vertices into the right-hand vertices if and only if for every subset of left-hand vertices A, the number of vertices in $N(A)$ is at least as large as the number of vertices in A.

Regular Graph

A regular graph is a graph where every vertex has the same degree.

Regular Bipartite Graph

A regular bipartite graph is a regular graph with the same number of vertices on the left-hand side as the right-hand side.

Regular Bipartite Graph Theorem

A regular bipartite graph has a matching.

Process

Schriver's Algorithm

1. Given a regular bipartite graph G, give every edge in G a weight of 1.

2. Let C be a cycle in the edges of positive weight.

   - Number the edges of C successively in turn.

   - Increase the weight of the even–numbered edges by 1

   - Decrease the weight of the odd–numbered edges by 1.

3. Repeat Step 2 until the edges with positive weight contain no cycle.

4. The positively weighted edges form a matching.

Section 13.4   Planar Graphs

Definitions

Planar Graph

Graphs that can be drawn on a plane without edge crossings are called planar graphs.

Face

In a planar graph, a face is a region inside a cycle of edges or the infinite exterior region on the outside of the graph. The number of faces of a graph is denoted by f.

Complete Graphs

A complete graph is defined to be a graph in which all vertices are connected to all other vertices. We denote a complete graph by K_n, where n is the number of vertices in the graph.

Planar Subgraph Theorem

Graph G is not planar if any subgraph of G is not planar.

Planar Graph Theorem

Graph G is planar if and only if G has neither $K_{3,3}$ nor K_5 as a minor.

Formulas

Euler's Formula

If G is a connected planar graph with v vertices, e edges, and f faces, then $v + f - e = 2$.

Corollary of Euler's Formula

A planar graph G with v vertices has at most $3v - 6$ edges. That is, $e \leq 3v - 6$.

Chapter 13 Exercises

Solve each problem.

1. Draw a graph that has six vertices (v_1, v_2, v_3, v_4, v_5, v_6) and the edges v_1v_4, v_1v_5, v_1v_6, and v_2v_3. Is the graph connected?

2. Determine if the three graphs represent the same graph.

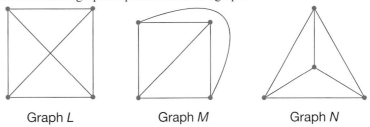

Graph *L*   Graph *M*   Graph *N*

3. Determine if the two graphs represent the same graph.

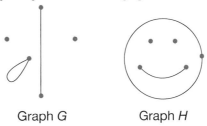

Graph *G*   Graph *H*

4. Use the following graph to answer the questions below.

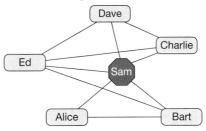

   a. What is the degree of "Sam"?

   b. Identify a path of length 5. If no such path exists explain why.

   c. Identify a path of length 6. If no such path exists explain why.

   d. Identify a cycle of length 5.

5. Find a vertex coloring of graph *W*. What is the chromatic number of that graph?

Graph *W*

6. Use a vertex coloring to assign colors to the regional electricity companies in England and Wales so that companies whose regions share a border have different colors. What is the minimum number of colors required? Find a vertex coloring using the minimum number of colors.

Regional Electricity Companies in England and Wales

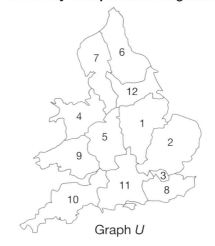

Graph *U*

7. Determine if the graph formed by the subway map is a tree.

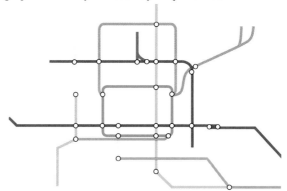

8. Determine whether the graph is a tree.

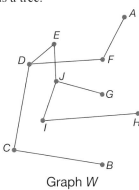

Graph *W*

Determine the number of edges needed to form a tree in each graph.

9.

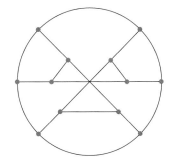

10.

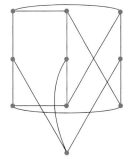

Solve each problem.

11. Find a spanning tree for the following graph.

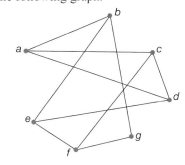

12. Find a minimum-weight spanning tree in the following graph.

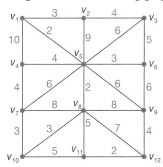

13

13. A national Internet grid wants to connect the cities in the following graph with optical fiber in the most cost-effective way possible. The weights on the edges represent the lengths of cable that would have to be used to connect each pair of cities, when that can be achieved.

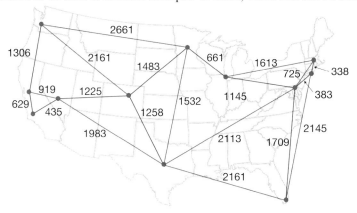

a. The proposed connections are represented by the red edges. Check that this graph is a spanning tree. Explain why a spanning tree is a suitable way to lay the optical fiber.

b. Is the proposed set of connections a minimum-weight spanning tree?

14. Find the number of leaves in the given trees.

a.

b.
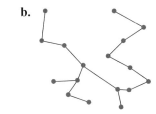

Use the given graph to identify the neighborhood of each set.

15. $A = \{\text{Bob, Evan}\}$

16. $A = \{\text{Sorelle, Peg, Evan}\}$

Determine if each graph has a matching. Justify your answer.

17.

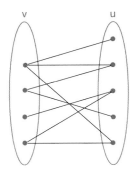

18.

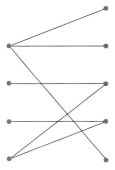

Find a matching in each regular bipartite graph.

19.

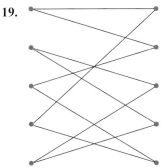

20.

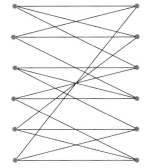

Determine the number of faces in each planar graph.

21.

22.

Solve the problem.

23.   Verify Euler's formula for the graphs in questions 21 and 22.

Determine if each graph is planar.

24.

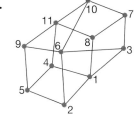

25.

Bibliography

Introduction

1. Kroon, Leo, Dennis Huisman, Erwin Abbink, Pieter-Jan Fioole, Matteo Fischetti, Gábor Marióti, Alexander Schrijver, Adri Steenbeek, and Roelof Ybema. "The New Dutch Timetable: The OR Revolution." *Interfaces*. Vol. 39, No. 1, January-February 2009, pp. 6–17. http://homepages.cwi.nl/~lex/files/Interfaces.pdf

13.1

1. Image courtesy of Bogdan Giuşcă. http://commons.wikimedia.org/wiki/File:Konigsberg_bridges.png. This file is licensed under the Creative Commons Attribution-ShareAlike 3.0 Unported license.

Chapter 14
Number Theory

Sections

Objectives

- Identify prime numbers
- Find the greatest common divisor of numbers
- Use modular arithmetic
- Use Fermat's Little Theorem to test for prime numbers
- Verify possible public-key encryption values

Number Theory

Security of information plays a large part in our everyday lives, although many of us never notice. Retail websites claim that credit card information is so secure from detection during transmission that we need never worry. Banks encourage us to do our monetary transactions via our mobile devices as well as online. But how can we be sure that our valuable information is safe? Confidence in the security of websites or mobile apps relies on cryptography. Cryptography is the technique of transferring information through secure codes in such a way that only those intended to understand the message will be able to.

Here is an example of a security code.

A = 1	B = 29	C = 9	D = 16	E = 14	F = 30	G = 28	H = 2	I = 15
J = 10	K = 11	L = 12	M = 7	N = 20	O = 27	P = 25	Q = 8	R = 6
S = 13	T = 26	U = 21	V = 22	W = 23	X = 18	Y = 31	Z = 5	

This code can be used to send secret messages, like the following.

15 2   27 25 14 31 27 21 12 15 11 14 26 2 15 13 29 27 27 11
— —   — — — — — — — — — — — — — — — — — —

At first sight, if you did not have the means to decipher the message, there would seem to be no pattern in the way the numbers have been arranged. Their apparent randomness is a good thing for security purposes if we do not want the code to be broken. However, these numbers were carefully chosen using the security code. A security code like this might be familiar to you. You just match the number to its letter to decipher the hidden message.

However, for the Internet or banking applications, codes like this are simply not secure enough. The aim of this chapter is to develop a sense of numbers and their properties by studying a branch of mathematics called number theory, which is the study of just what you might think: numbers. We'll look at prime numbers, modular arithmetic, and their uses. Then we can approach the most commonly used cryptographic technique for secure codes (public-key encryption) and do some coding of our own.

14.1 Prime Numbers

今有物不知其數三三數之剩二五五
數之剩三七七數之剩二問物幾何答
曰二十三

術曰三三數之剩二置一百四十五五
數之剩三置六十三七七數之剩二置
三十并之得二百三十三以二百一十
減之即得凡三三數之剩一則置七十
五五數之剩一則置二十一七七數之
剩一則置十五一百六以上以一百五
減之即得

Figure 1: Chinese Puzzle

The text in Figure 1 is from a classical Chinese text of Sun Tzu Suan-Ching dating from the third or fourth century. Here is a translation.

We have things of which we do not know the number; if we count by threes the remainder is two; if we count by fives the remainder is three; if we count by sevens the remainder is two. How many things are there?

Sun Tzu goes on to give his method to solve this puzzle.

If you count by threes and have the remainder two, put 140. If you count by fives and have a remainder three, put 63. If you count by sevens and have the remainder two, put 30. Add these numbers and you get 233. From this subtract 210 and you have the result. For each unity as a remainder when counting by three, put 70. For each unity as a remainder when counting by fives, put 21. For each unity as a remainder when counting by sevens, put 15. If the sum is more than 106 subtract 105 from this and you get the result.

Perhaps this makes things even more puzzling! It is certainly not clear where all these numbers come from. Did Sun Tzu actually solve the puzzle? If he did, it is not clear how this method could be generalized to solve a similar problem that does not involve counting by threes, fives, and sevens.

Think Back

The set of integers includes all whole numbers, their negatives, and 0.

Think Back

Remember that a *divisor*, or *factor*, of a number x is a number that divides x with a remainder of 0.

We're all familiar with the concept of odd and even numbers, but understanding Sun Tzu's solution requires understanding how numbers can be characterized into more complex groups. This is where number theory comes in. Number theory looks at the properties of families of numbers and how they interconnect. One of the fundamental building blocks of this interconnection is the set of prime numbers. Just as all matter is built up from the elements of the periodic table, we will see that every number can be built up from the "mathematical periodic table" of prime numbers.

We say that a positive integer is **prime** if it has precisely two divisors: 1 and itself. We should point out how the numbers 0 and 1 relate to the definition of a prime number. Since 0 is neither positive nor negative, it cannot be prime. The number 1 is also not considered prime because it does not have precisely two divisors. It has precisely one—itself. Therefore, the first prime number is 2. If a positive integer greater than 1 is not prime, it is **composite**. That is to say, a composite number has more than 2 divisors (or factors). Again, notice that 0 and 1 are eliminated from being composite as well. Therefore, the first composite number is 4.

14

Prime Numbers

A **prime number** is a positive integer that has precisely two divisors: 1 and itself.

Composite Numbers

A **composite number** is a positive integer that has more than two divisors.

Note: The numbers 0 and 1 are neither prime nor composite.

Fun Fact

The only even prime number is 2.

MATH MILESTONES

Eratosthenes (276-194 B.C.) was a Greek librarian of Alexandria who is remembered as a scholar, poet, and inventor. He is known for his measurement of the circumference of the Earth as well as his estimates of the distances from the Earth to the sun and the moon. He is also credited with the discipline of geography as we know it, inventing the system of longitude and latitude.

As we will show at the end of this section, there are an infinite number of both prime and composite numbers. The following is the beginning of the list of prime numbers.

$$2, 3, 5, 7, 11, 13, 17, 19, 23, \ldots$$

The following is the beginning of the list of composite numbers.

$$4, 6, 8, 9, 10, 12, 14, 15, 16, 18, 20, 21, 22, 24, \ldots$$

How do we find or determine prime numbers? A very early technique for finding primes is credited to Eratosthenes in the third century B.C. His idea was to organize the computations in order to find all the prime numbers smaller than a particular number N.

Begin by writing out all the positive integers up to N. We'll start with the number 2 since we've already established that 1 and 0 are neither prime nor composite. Highlight the number 2 and cross out all the other multiples of 2 since a multiple of 2 cannot be a prime number. The next smallest number remaining is 3. It must also be prime since it has no other divisors. Highlight the 3 and cross out all the remaining multiples of 3, and so on. At each stage, we cross out all the multiples of the next smallest remaining number that has not yet been crossed out. Once we reach N, we can stop. Now all the numbers that were not crossed out must be primes. Figure 2 shows the end result of this process for numbers up to 101. This process is commonly called the **sieve of Eratosthenes**, and may be likened to sifting the integers through a series of sieves.

	2	**3**	~~4~~	**5**	~~6~~	**7**	~~8~~	~~9~~	~~10~~
11	~~12~~	**13**	~~14~~	~~15~~	~~16~~	**17**	~~18~~	**19**	~~20~~
~~21~~	~~22~~	**23**	~~24~~	~~25~~	~~26~~	~~27~~	~~28~~	**29**	~~30~~
31	~~32~~	~~33~~	~~34~~	~~35~~	~~36~~	**37**	~~38~~	~~39~~	~~40~~
41	~~42~~	**43**	~~44~~	~~45~~	~~46~~	**47**	~~48~~	~~49~~	~~50~~
~~51~~	~~52~~	**53**	~~54~~	~~55~~	~~56~~	~~57~~	~~58~~	**59**	~~60~~
61	~~62~~	~~63~~	~~64~~	~~65~~	~~66~~	**67**	~~68~~	~~69~~	~~70~~
71	~~72~~	**73**	~~74~~	~~75~~	~~76~~	~~77~~	~~78~~	**79**	~~80~~
~~81~~	~~82~~	**83**	~~84~~	~~85~~	~~86~~	~~87~~	~~88~~	**89**	~~90~~
~~91~~	~~92~~	~~93~~	~~94~~	~~95~~	~~96~~	**97**	~~98~~	~~99~~	~~100~~
101									

Figure 2: Sieve of Eratosthenes

One of the first questions that you might ask yourself when trying to decide if a number is prime or composite is, "How do I know what the divisors of a number are?" Using the sieve of Eratosthenes is not a convenient or efficient method for finding primes once the numbers become large. Fortunately, there is a clever way to check if any of the numbers 2, 3, 4, 5, 6, 7, 8, 9, or 10 is a divisor of a number. We all know that if a number is even, then it's divisible by 2, but here's a list to help you know if a number if divisible by any of the others.

Table 1

Divisibility Rules

Divisor	Test	Example
2	The number is even.	The number 391,574 is divisible by 2 because it is even.
3	When the digits of the number are added together, the resulting number is divisible by 3.	87,408 is divisible by 3 because $8 + 7 + 4 + 0 + 8 = 27$ and 27 is divisible by 3.
4	The last 2 digits of the number form a number divisible by 4.	316 is divisible by 4 because 16 is divisible by 4.
5	The number ends in a 0 or a 5.	29,345 is divisible by 5 because it ends in a 5.
6	The number is divisible by both 2 and 3.	628,116 is divisible by 6 because it's divisible by 2, since it's even, and 3, since $6 + 2 + 8 + 1 + 1 + 6 = 24$ and 24 is divisible by 3.
7	Double the last digit, then subtract it from the remaining digits of the number. If the answer is divisible by 7, then so is the original number.	819 is divisible by 7 since $2 \cdot 9 = 18$ and $81 - 18 = 63$. 63 is divisible by 7, so 819 is also.
8	The last 3 digits of the number form a number divisible by 8.	2160 is divisible by 8 because 160 is divisible by 8.
9	When the digits of the number are added together, the resulting number is divisible by 9.	189 is divisible by 9 because $1 + 8 + 9 = 18$ and 18 is divisible by 9.
10	The number ends in a 0.	9,145,830 is divisible by 10 because it ends in a 0.

This list of divisibility rules will help us when determining if a number has more than two divisors. While it isn't helpful in finding divisors bigger than 10, it's a good start.

Example 1: Classifying a Number as Prime or Composite

Determine if the following numbers are prime or composite using the divisibility rules.

a. 312

b. 101

c. 2,344,017

Solution

a. Since the number 312 is even, we know that it is divisible by 2. It is enough to stop here and know that the number is composite, but we'll continue on to determine if there are other small divisors of 312.

If we add the digits together, we get $3 + 1 + 2 = 6$. Because 6 is divisible by 3, 312 is also divisible by 3.

In addition, the last 2 digits form the number 12, which is divisible by 4, so we know that 312 is also divisible by 4.

Although we have plenty of examples to show us the number is composite, one last point to notice is that because 312 is divisible by 2 and by 3, it must also be divisible by 6.

b. Since 101 is odd, it is not divisible by any of the even divisors: 2, 4, 6, 8, or 10. $1 + 0 + 1 = 2$, which is not divisible by 3, so 101 is not divisible by 3, 6, or 9. That leaves 5 and 7 to check. 101 is not divisible by 5 because it does not end in 0 or 5. Lastly, $2 \cdot 1 = 2$ and $10 - 2 = 8$. 8 is not divisible by 7, so 101 is not either.

That eliminates all of the small divisors, but to establish that 101 is prime, we have to deal with all divisors smaller than 101. Notice that if both a and b are integers bigger than 10, then $a \cdot b \geq 121$. So if 101 had divisors a and b, one of them must be smaller than 10. Since we've already established that there are no divisors smaller than 10, 101 is prime.

While 101 was shown to be prime earlier in the text, testing for factors is a method for checking without making a sieve or memorizing the list of primes.

C. We know that the number 2,344,017 is not divisible by 2, 4, 6, 8, or 10 since it is not even. Adding the digits together gives us $2 + 3 + 4 + 4 + 0 + 1 + 7 = 21$, which is divisible by 3, so the number is divisible by 3. Again, although we now know the number is composite, we'll continue to show how to check for other small divisors.

2,344,017 is not divisible by 5 since it does not end in a 5 or 0.

Doubling the last digit we have $7 \cdot 2 = 14$. Then, $234,401 - 14 = 234,387$. This is still a large number, so we can apply the rule again to check if it's divisible by 7. Continue repeating this process until we are able to definitively say if a number is divisible by 7.

$$234,387$$
$$7 \cdot 2 = 14 \Rightarrow 23,438 - 14 = 23,424$$
$$4 \cdot 2 = 8 \Rightarrow 2,342 - 8 = 2334$$
$$4 \cdot 2 = 8 \Rightarrow 233 - 8 = 225$$
$$5 \cdot 2 = 10 \Rightarrow 22 - 10 = 12$$

Since 12 is not divisible by 7, our original number is not divisible by 7. (Although we used the divisibility rule for 7 here, it may have been quicker to use long division and see if the number has a remainder when divided by 7.)

The only other number left in our list to check is 9. Adding the digits together, as we did earlier, we have 21. Since 21 is not divisible by 9, the number 2,344,017 is not divisible by 9.

However, we know that 2,344,017 is composite because it divisible by the number 3.

Skill Check #1

Determine if 3,743,216 is prime or composite.

Now that we have a working idea of the definition of prime numbers, we can more carefully examine how prime numbers are the building blocks of all numbers. The following theorem proves that every number can be built from prime numbers in a unique way. This particular theorem was proven by Euclid around 300 B.C.

☞ **Helpful Hint**

A **mathematical theorem** is a statement or rule that can be, and has been, proven to be (logically) true.

Fundamental Theorem of Arithmetic

Every positive integer greater than 1 is either a prime number or can be written as a unique product of prime numbers. This unique product of prime numbers is called its prime factorization.

One method to find the prime factorization of a number is to use a *factor tree* like the one shown in Figure 3.

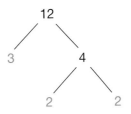

Figure 3: Factor Tree

To find the prime factorization of a number N using this method, begin by choosing any pair of factors, other than 1 and N, that multiply together to give you N. In Figure 3, we started factoring 12 with the numbers 3 and 4.

Then, find a pair of factors for each of these numbers. Continue until all factors are prime. Notice that in Figure 3, we continued to find factors for 4 but not for 3 since it is already prime.

Once all factors are prime, we have found the prime factorization of the number. We call this method a factor tree because of the way we organize the pairs of factors. The original number N is placed at the top of the "tree" and then the factors "branch off" below.

Example 2: Using a Factor Tree to Find a Prime Factorization

Use a factor tree to determine the prime factorization of the number 84.

Solution

Start by choosing any pair of factors you like, other than 1 and 84. We'll use the pair 4 and 21, as shown here.

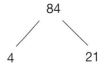

Next, notice that both of the numbers 4 and 21 also have factors. Branching each off into its own factors, we have the following.

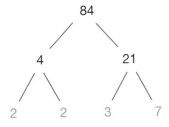

This time, as we look across the bottom row of factors, we can see that they are all prime. Thus, the prime factorization of 84 is $2 \cdot 2 \cdot 3 \cdot 7$.

Skill Check #2

Use a factor tree to find the prime factorization of 84 by starting with the pair of factors 2 and 42. Although your factor tree may not look identical to a tree drawn by someone else, the prime factorization of the number 84 will be identical.

14

One useful consequence of prime factorization is that we only need to check for prime factors when we are checking whether a number is prime or composite. In fact, we need only check the prime numbers that are less than the square root of the number. For large numbers, that cuts down the work considerably.

☞ Helpful Hint

When looking for the prime factors of a number n, you only need to check the prime numbers that are less than or equal to $\sqrt{n}$.

Example 3: Classifying a Number as Prime or Composite

Determine whether 197 is prime or composite.

Solution

If 197 is composite, it must have a prime divisor that is less than $\sqrt{197} \approx 14.0357$. The prime numbers less than 14 are 2, 3, 5, 7, 11, and 13.

197 is not even, so it is not divisible by 2. $1 + 9 + 7 = 17$ and 17 is not divisible by 3, which eliminates 3. 197 doesn't end in 0 or 5, which eliminates 5, and $19 - (7 \cdot 2) = 5$ and 5 is not divisible by 7, which eliminates 7. All that remains is to check 11 and 13.

$$\frac{197}{11} = 17.\overline{90} \quad \text{and} \quad \frac{197}{13} \approx 15.1538$$

This gives that 197 is prime.

Using prime factorizations, we can find the **greatest common divisor (GCD)** of any collection of numbers. The greatest common divisor of two numbers is the largest integer that divides both numbers without a remainder.

> ### Greatest Common Divisor (GCD)
>
> The largest integer that divides two numbers without a remainder is called the **greatest common divisor (GCD)**. If the GCD = 1 for a pair of numbers, the numbers are said to be **relatively prime**.

For example, the GCD of 10 and 15 is 5 because 5 is the largest number that divides both 10 and 15. When the GCD of two numbers, such as 24 and 16, is less obvious, it can be determined using factor trees, among other methods.

To find the GCD of two numbers using factor trees, begin by creating the factor tree of each number. The factor trees of 24 and 16 are shown in Figure 4.

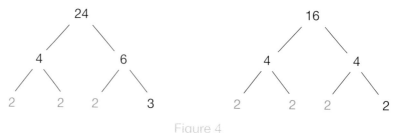

Figure 4

Once the trees are complete, the GCD of the two numbers is the product of all the prime factors that the two numbers have in common. In the case of 24 and 16, you can see that the prime number 2 appears three times in both of the prime factorizations. Therefore, the greatest common divisor for these numbers is $2 \cdot 2 \cdot 2 = 8$.

As we stated in the definition of the greatest common divisor, if there is no prime factor in common between two numbers, then the GCD is 1, and the numbers are said to be **relatively prime**, or *co-prime*. For instance, consider the numbers 10 and 21. Their factor trees are shown in Figure 5.

Figure 5

Since 10 and 21 have no common factors, they are relatively prime.

Example 4: GCD Using Factor Trees

Use factor trees to find the greatest common divisor of 40 and 60.

Solution

Begin by constructing the factor trees of 40 and 60 as shown.

Notice that both 40 and 60 have the factors 2 and 5 in common. In fact, both numbers have two 2s in common as well as one 5. So, the GCD of 40 and 60 is $2 \cdot 2 \cdot 5 = 20$.

Example 5: Using the GCD

One way that the Manna Café Pantry serves the hungry of Clarksville, TN, is by distributing boxes of food each week to the local shelters. As part of the stipulation for receiving local governmental funds, all boxes must contain the same number of items from each of the following categories: pasta, canned vegetable, and canned meat. This week, the pantry has the following in supply: 360 pasta items, 540 canned vegetables, and 240 canned meats.

a. Using all of the food, what is the maximum number of food boxes that Manna Café can distribute this week?

b. How many of each item will be in each box?

Solution

a. In order to distribute all of the food items evenly among the food boxes, the number of boxes must be a divisor of 360, 540, and 240. To find the *maximum* number of food boxes that can be made from all of the pantry supplies, we need to find the GCD of the three numbers. First, find the prime factorization of each number. We'll do this by constructing factor trees of each number.

14

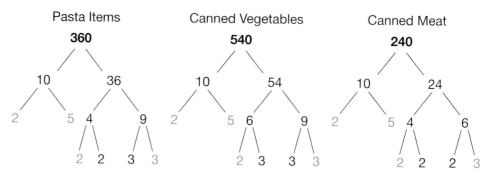

The GCD of 360, 540, and 240 is $2 \cdot 2 \cdot 3 \cdot 5 = 60$. Therefore, the pantry can give out 60 food boxes to the local shelters this week.

b. In order to find how many of each item will be in the boxes, we can use the prime factorizations that we found in part **a.** For each food category, if we remove the GCD from the prime factorization, the remaining factors will tell us how many of each item needs to go in a box. Each prime factorization with the GCD marked out is as follows.

$$\text{Pasta items: } \cancel{2} \cdot \cancel{2} \cdot 2 \cdot \cancel{3} \cdot 3 \cdot \cancel{5}$$
$$\text{Canned vegetables: } \cancel{2} \cdot \cancel{2} \cdot \cancel{3} \cdot 3 \cdot 3 \cdot \cancel{5}$$
$$\text{Canned meat: } \cancel{2} \cdot \cancel{2} \cdot 2 \cdot 2 \cdot \cancel{3} \cdot \cancel{5}$$

Therefore, we can see that there will be $2 \cdot 3 = 6$ pasta items, $3 \cdot 3 = 9$ canned vegetables, and $2 \cdot 2 = 4$ canned meats in each box.

Although we have used factor trees to help us find the GCD of numbers, there are other methods that lead to the same conclusion. The main point here is that the GCD can be found by knowing the prime factorization of each number. However, if we are trying to find the GCD of very large numbers, then finding the prime factors can get quite cumbersome. We once again turn to Euclid for a solution. Around 300 B.C., Euclid developed an algorithm that calculates the GCD of two numbers. The algorithm constructs a sequence of calculations by repeatedly dividing one number by another and noting the remainders. The last nonzero remainder is the GCD. This algorithm is referred to as **Euclid's Algorithm**, or the *Euclidian Algorithm*.

Steps for Using Euclid's Algorithm

1. Divide the larger of the given numbers by the smaller (the divisor) and note the remainder.
2. Divide the original divisor (the smaller of the given numbers) by the remainder found in Step 1.
3. Continue dividing the previous divisor by the previous remainder until the remainder is 0.
4. The last nonzero remainder is the GCD of the given numbers.

Example 6: Finding the GCD Using Euclid's Algorithm

Use Euclid's Algorithm to find the greatest common divisor of 88 and 300.

Solution

Begin with the two numbers 88 and 300. Divide the larger number by the smaller number and note the remainder.

$$\begin{array}{r} 3\text{R}36 \\ 88\overline{)\,300} \end{array}$$

Our next division will use the remainder that we just found, 36, and the original divisor, 88. Again, divide the larger by the smaller and note the remainder. Continue until we get a remainder of zero.

$$36\overline{)\,88}\quad 2R16$$

$$16\overline{)\,36}\quad 2R4$$

$$4\overline{)\,16}\quad 4R0$$

Since 4 is the last nonzero remainder, it is the GCD of 88 and 300.

Example 7: Relatively Prime Numbers Using Euclid's Algorithm

Use Euclid's Algorithm to show that 88 and 17 are relatively prime.

Solution

Divide the larger number by the smaller number and note the remainder.

$$17\overline{)\,88}\quad 5R3$$

Our next division will use the remainder we just found and the divisor. Repeat this process until the remainder is zero.

$$3\overline{)\,17}\quad 5R2$$

$$2\overline{)\,3}\quad 1R1$$

$$1\overline{)\,2}\quad 2R0$$

The last nonzero remainder is 1, so the GCD of 88 and 17 is 1, which means that these numbers are relatively prime.

As we will see in the last section on public-key encryption, prime numbers are very useful things. Euclid's prime factorization theorem also ensures that we will never run out of them—there are infinitely many prime numbers. We'll finish this section with a generalization of Euclid's proof of this.

Euclid's Proof of Infinitely Many Primes (A Summary)

One of the consequences of Euclid's Algorithm is that there are infinitely many prime numbers. Let's think about why this is true. Suppose that $p_1, p_2, p_3, \ldots, p_k$ is a complete, finite list of all of the prime numbers that exist. Take a look at the number n, the number that you get by multiplying all of those primes together and adding 1.

$$n = p_1 \cdot p_2 \cdot p_3 \cdots p_k + 1$$

As we have seen in the previous equation, if n is composite, then it must have a prime divisor. Since $p_1, p_2, p_3, \ldots, p_k$ are all of the primes that exist, the divisor must be one of the numbers in the list. But, by our definition of n, there is a remainder of 1 when n is divided by any of the primes on the list. (Check this for yourself.) So the divisor cannot be on our finite list of primes. Therefore, n must be an even bigger prime that we left off the list. Because we can continue to play this game, there must be infinitely many primes.

Skill Check Answers

1. It's even, so it's composite.

2.

14.1   Exercises

Solve each problem.

1. Using the sieve of Eratosthenes, write out the prime numbers up to 202.

2. True or false: When using a sieve to find prime numbers, if you begin with the largest number and work backwards by crossing out divisors, you establish the same set of prime numbers as starting with the smallest number and working up.

Determine whether each of the numbers is prime or composite.

3. 245

4. 939

5. 149

6. 4372

7. 68,045,800

8. 113

Answer each question thoughtfully.

9. Which possible divisors must be checked to see if 283 is prime?

10. Which possible divisors must be checked to see if 1291 is prime?

11. Suppose you are choosing the size of an elementary school class. It must be between 15 and 20 students. Knowing that teachers regularly break the class into smaller groups for projects and activities, answer the following questions.

    a. List the advantages and disadvantages for each of the possible class sizes between 15 and 20.

    b. Which size would you say is best?

Determine the prime factorization of each number.

12. 16

13. 240

14. 162

15. 630

Use a factor tree to determine the greatest common divisor of each pair of numbers.

16. 35 and 14

17. 28 and 42

18. 90 and 225

19. 350 and 217

Use Euclid's Algorithm to find the greatest common divisor of each pair of numbers.

20. 357 and 217

21. 350 and 140

22. 1235 and 5687

23. 1556 and 236

Use Euclid's Algorithm to determine whether each pair of numbers is relatively prime.

24. 67 and 17

25. 1231 and 5673

26. 351 and 141

Solve each problem.

27. There are 16 cellists, 32 violinists, 24 flautists, and 16 violists at a summer classical music camp.

   a. What is the largest number of identical groups that can be made from all of the musicians?

   b. How many violinists will be in each group?

   c. How many campers will be left without a group?

28. Berrylin was asked to make flower arrangements for the tables at a sports banquet. She was given 36 tulips, 27 daisies, and 18 carnations.

   a. What is the largest number of identical arrangements that Berrylin can make from all of the flowers?

   b. How many of each flower will be put into each arrangement?

29. Phillip's Bakery gives away any remaining items at the end of the day to local hunger organizations. Today they had 39 loaves of bread and 86 cookies left. Bill is allowed to take some of the food home himself because he volunteers to box the goods each day. If he plans to take 3 loaves of bread and 10 cookies home, how many boxes will Bill fill given that each box must contain the same number of loaves of bread and cookies?

30. The library received 117 used books for their summer used book drive this year. There were 45 hardbacks and 72 paperbacks. The library delivers the books in bundles to local shelters.

   a. How many bundles can the library make if each bundle must have the same number of hardback and paperback books?

   b. How many paperback books will be in each bundle?

31. Two girl scout troops are planting gardens side-by-side on donated land. Based on the number of girls, Troop A is planning a garden that will cover 180 square feet, and Troop B's garden will be 204 square feet. The gardens must have a fence completely surrounding each one. The fencing is only sold in whole foot sections. In order to save money, the troops decide to maximize the shared amount of fence between the gardens.

   a. What is the greatest amount of fence the troops can share?

   b. What will the dimensions of each garden be if the shared fencing is maximized?

14.2   Modular Arithmetic

In Section 14.1, we looked at the divisibility property of numbers. We introduced clever ways to test for divisibility by using the numbers 2 through 10. We also focused on numbers having precisely two divisors, that is, prime numbers.

Suppose we continue the idea of divisibility by creating a fictitious classification of numbers; let's call it "3visible." In our new classification, a number will be 3visible if it can be divided evenly by 3.

Here's a beginning list of 3visible numbers: 3, 6, 9, 12, 15, 18, . . . (notice they are the multiples of 3).

What if a number is not 3visible? If a number is not 3visible, then it must have a remainder when divided by 3. The number can be either 1 greater than 3visible or 2 greater than 3visible. But, if it's 3 greater than 3visible, then it's actually 3visible. Therefore, the remainder must either be a 1 or 2. Table 1 can help visualize this by listing numbers and their 3visible properties, starting with the number 3.

Table 1		
3visible Properties		
Number	**3visible**	**Remainder When Divided by 3**
3	Yes	0
4	No	1
5	No	2
6	Yes	0
7	No	1
8	No	2

Let's consider the arithmetic properties of 3visible. If we add together two numbers that are 3visible, then the answer is also 3visible. If you add together two numbers that are 1 unit greater than 3visible numbers, then the answer is 2 greater than 3visible. For instance, add together the numbers 4 and 7, which both have a remainder of 1 when divided by 3.

$$4 + 7 = 11$$
$$11 \div 3 = 3 \text{ remainder } 2$$

What do you think happens if you add a number that is 1 greater than 3visible to one that is 2 greater than 3visible? Try the numbers 4 and 5.

$$4 + 5 = 9$$
$$9 \div 3 = 3 \text{ remainder } 0$$

Therefore, the sum of 4 and 5 is 3visible.

In fact, performing 3visible arithmetic is equivalent to adding the remainders together, as long as you remember to go back to 0 every time you get to 3. The classification we created works for 3visible, but what if we wanted to do the same for the number 4, or "4visible"? It would get a bit awkward to have an entire vocabulary of classifications that only we could understand. Thankfully, a system is already in place for this type of classification. It's called **modular arithmetic**.

Here's how it works. The operation modulo (or "mod" for short) is simply the remainder when dividing. For example, 4 mod 3 means "the remainder when 4 is divided by 3". Of course, the answer is 1 in this case, since we already know that when you divide 4 by 3, there is a remainder of 1. We use the following notation in modular arithmetic to denote the procedure.

$$4 \bmod 3 = 1 \quad \text{or} \quad 4 \equiv 1 \ (\bmod 3)$$

Notice that rather than using the equal sign = when working with modular arithmetic, we use the congruency sign ≡ when the modulus is indicated on the right side of the equation. This sign is read "is congruent to."

For instance, $5 \equiv 2 \,(\mathrm{mod}\ 3)$ is read "5 is congruent to 2 mod 3," which means 5 divided by 3 has a remainder of 2.

Modular Arithmetic

In **modular arithmetic**, a number n is congruent to the remainder r when it is divided by a fixed number m. We write $n \equiv r \,(\mathrm{mod}\ m)$. Note that m is referred to as the **modulus**.

Now that we have a notation to check out modular arithmetic, we can understand 3visible more mathematically. Instead of 3visible we can now refer to this classification as modulo 3 (or mod 3 for short). Likewise, 4visible is simply mod 4. Table 2 is a modification of Table 1 with an added column for the congruence of modulo 3.

	Table 2		
	3visible Properties		
Number	**3visible**	**Remainder When Divided by 3**	**Congruence**
3	Yes	0	$3 \equiv 0\ (\mathrm{mod}\ 3)$
4	No	1	$4 \equiv 1\ (\mathrm{mod}\ 3)$
5	No	2	$5 \equiv 2\ (\mathrm{mod}\ 3)$
6	Yes	0	$6 \equiv 0\ (\mathrm{mod}\ 3)$
7	No	1	$7 \equiv 1\ (\mathrm{mod}\ 3)$
8	No	2	$8 \equiv 2\ (\mathrm{mod}\ 3)$

Notice that in modulo 3, all numbers are congruent to either 0, 1, or 2. This cycling pattern is based on the possible remainders when divided by 3. In general, for arithmetic modulo m, the cycle will include the numbers 0, 1, 2, . . . , $m - 1$. For example, the cycle for modulo 6 will consist of the numbers 0, 1, 2, 3, 4, and 5.

Example 1: Completing the Modular Congruence Table

Complete the following table for modulo 5.

	Table 3	
	Modulo 5	
Number	**Remainder When Divided by 5**	**Congruence**
0	0	$0 \equiv 0\ (\mathrm{mod}\ 5)$
1	1	
2	2	
3	3	
4	4	
5	0	
6		
7		
8		
9		

Solution

The cycle of remainders for modulo 5, will consist of 0, 1, 2, 3, and 4. Because the cycle returns to 0 after reaching 4, we can complete the second column with the cycle. The last column of Table 4 is the notation that represents these remainders.

Table 4 Modulo 5		
Number	Remainder When Divided by 5	Congruence
0	0	$0 \equiv 0 \ (\text{mod } 5)$
1	1	$1 \equiv 1 \ (\text{mod } 5)$
2	2	$2 \equiv 2 \ (\text{mod } 5)$
3	3	$3 \equiv 3 \ (\text{mod } 5)$
4	4	$4 \equiv 4 \ (\text{mod } 5)$
5	0	$5 \equiv 0 \ (\text{mod } 5)$
6	1	$6 \equiv 1 \ (\text{mod } 5)$
7	2	$7 \equiv 2 \ (\text{mod } 5)$
8	3	$8 \equiv 3 \ (\text{mod } 5)$
9	4	$9 \equiv 4 \ (\text{mod } 5)$

It is not necessary to make a modular congruency table for all possible modular congruences of a particular number. Often we are asked to find the modular congruence of a single number. Remembering that we only need division, the process is simple.

Example 2: Finding Modular Congruence

Evaluate the following.

a. 15 mod 6

b. 72 mod 13

Solution

a. To find 15 mod 6, we need to divide 15 by 6 and find the remainder.

$$15 \div 6 = 2 \text{ remainder } 3$$
$$\text{So, } 15 \equiv 3 \ (\text{mod } 6).$$

b. To find 72 mod 13, we need to divide 72 by 13 and find the remainder.

$$72 \div 13 = 5 \text{ remainder } 7$$
$$\text{So, } 72 \equiv 7 \ (\text{mod } 13).$$

Most of us use modular congruence every day when we refer to the time of day. The military uses the 24-hour clock, but civilians normally use a 12-hour clock. At 1400 hours military time, we say it's 2:00 p.m. That's because $14 \equiv 2 \ (\text{mod } 12)$. Similarly, 2300 hours is 11:00 p.m. because $23 \equiv 11 \ (\text{mod } 12)$.

Modulo is more than a notation for writing congruent numbers. It extends standard arithmetic to modular arithmetic. It is possible to add, subtract, and multiply with modular congruence. Once again, you are familiar with adding and subtracting in modular arithmetic because of clock time. For instance, 2 hours after 11:00 p.m. is 1:00 a.m., not 13:00 p.m. The following example illustrates addition, subtraction, and multiplication in modular arithmetic.

Example 3: Modular Operations

Evaluate each of the following.

a. $(12 + 7) \bmod 5$

b. $(21 - 18) \bmod 6$

c. $(35 \cdot 22) \bmod 10$

Solution

a. In order to calculate the sum of $(12 + 7)(\bmod 5)$, we can either calculate $(19)(\bmod 5)$ or we can calculate the individual pieces first and then add them together.

$$(12 + 7) \bmod 5 = (12 \bmod 5) + (7 \bmod 5)$$
$$\equiv 2 + 2 \,(\bmod 5)$$
$$\equiv 4 \,(\bmod 5)$$

Since the numbers in this example are small, there is little difference between calculating $(12 \bmod 5) + (7 \bmod 5)$ and 19 mod 5, although both return the same result. However, as we will see later, this way of breaking things down is useful with larger numbers.

b. We can perform subtraction in the same manner as addition by calculating the individual pieces and then subtracting.

$$(21 - 18) \bmod 6 = (21 \bmod 6) - (18 \bmod 6)$$
$$\equiv 3 - 0 \,(\bmod 6)$$
$$\equiv 3 \,(\bmod 6)$$

c. Multiplication is consistent with both addition and subtraction, in that you can find the modular congruences and then multiply.

$$(35 \cdot 22) \bmod 10 = (35 \bmod 10) \cdot (22 \bmod 10)$$
$$\equiv 5 \cdot 2 \,(\bmod 10)$$
$$\equiv 10 \,(\bmod 10)$$
$$\equiv 0 \,(\bmod 10)$$

☞ Helpful Hint

After a congruence symbol, writing "2 + 2 (mod 5)" is the same as writing "(2 + 2) (mod 5)" since the modulus is applied to the entire expression. If written before a congruence symbol, parentheses are needed around the expression.

When performing any of these three operations with a particular modulus, we can either find the individual modulo first, and then carry out the operation, or we can carry out the calculation on the numbers and then find the modular congruence. Both methods provide the same answer. Note, however, that these operations must use the same modulus all the way through the calculation.

Example 4: Modular Addition

Calculate each of the following and compare the answers.

a. $(10 \bmod 4) + (17 \bmod 4)$

b. $(10 + 17) \bmod 4$

Solution

a. In order to find $(10 \bmod 4) + (17 \bmod 4)$, we need to find each modulo first as we did in Example 3.

$$(10 \bmod 4) + (17 \bmod 4) \equiv 2 + 1 \,(\bmod 4)$$
$$\equiv 3\,(\bmod 4)$$

b. For $(10 + 17) \bmod 4$, we can either add the numbers in the brackets together first and then apply the modulus or we can find each individual modulus and then add them together as we did in part **a.** We will add the numbers first since they are small.

$$(10 + 17) \bmod 4 = 27\,(\bmod 4)$$
$$\equiv 3\,(\bmod 4)$$

Comparing answers in parts **a.** and **b.**, we can see that they provide equivalent solutions using modular arithmetic.

Skill Check #1

Show that $(16 \bmod 2) \cdot (3 \bmod 2)$ is the same as $(16 \cdot 3) \bmod 2$.

Modular Arithmetic and Bar Codes

Finding modular arithmetic in action is as easy as picking up a book and finding its bar code. The bar code on a book contains a string of numbers called the International Standard Book Number (ISBN) that is unique to each book title and conveys such information as the publisher and country of origin. If the book was assigned an ISBN before 2007, the ISBN is 10 digits long. After 2007, the ISBN changed to 13 digits. Either way, the numbers in an ISBN are not assigned at random; modular arithmetic plays an integral part in the construction of either the 10-digit or 13-digit ISBN.

Figure 1: ISBN of a Textbook

☞ **Helpful Hint**

The top number on this bar code is the 10-digit ISBN for the book. The bottom number is the 13-digit ISBN, which is the new standard as of 2007.

Let's look at how the 10-digit ISBN is constructed. All 10-digit ISBNs are constructed from the numbers 0 through 9 as well as the letter X. The X stands for the number 10. The first 9 digits are numbers that represent the language group, publisher, and title. The 10th digit is assigned such that when added to certain multiples of the other 9 digits, the sum is 0 modulo 11. This 10th digit is referred to as the **check-sum digit**. Figure 2 shows the breakdown for a 10-digit ISBN.

In order to determine the check-sum digit for a particular 10-digit ISBN, a process of multiplication, addition, and modular arithmetic takes place. The following outlines how the process works once the initial 9 digits are assigned.

14

Process for Finding the Check-Sum Digit for a 10-digit ISBN

1.   Multiply the 1st digit by 10.

2.   Multiply the 2nd digit by 9.

3.   Multiply the 3rd digit by 8.

4.   Multiply the 4th digit by 7.

5.   Multiply the 5th digit by 6.

6.   Multiply the 6th digit by 5.

7.   Multiply the 7th digit by 4.

8.   Multiply the 8th digit by 3.

9.   Multiply the 9th digit by 2.

10. Add the multiples together.

11. The check-sum digit (or 10th digit) is then chosen so that the total of all of these numbers is 0 modulo 11.

Figure 2: ISBN Breakdown

Example 5 validates that a particular ISBN meets the constraints given by the process for finding the check-sum digit.

Example 5: Verifying the Validity of a 10-digit ISBN

Verify the validity of the ISBN shown here by performing the process described for finding the check-sum digit.

Solution

To verify the ISBN, we need to find the various multiples of the first 9 digits in the number, and add them together with the check-sum digit. Remember, the sum should be congruent to 0 mod 11.

First, we'll find the multiples of the first 9 digits.

$$(10 \cdot 0) + (9 \cdot 0) + (8 \cdot 7) + (7 \cdot 3) + (6 \cdot 3) + (5 \cdot 8) + (4 \cdot 1) + (3 \cdot 2) + (2 \cdot 5) = 0 + 0 + 56 + 21 + 18 + 40 + 4 + 6 + 10$$
$$= 155$$

Next, we add 155 and 10, since X represents the number 10.

$$155 + 10 = 165$$

Finally, we need to find $165 \pmod{11}$.

$$165 \equiv 0 \pmod{11}$$

Because the sum is congruent to $0 \pmod{11}$, the ISBN is indeed valid.

Skill Check #2

Verify that the following is a valid 10-digit ISBN.

ISBN 0-306-40615-2

Example 6: Finding a Missing Digit of a 10-digit ISBN

What should the 3rd digit of the ISBN number be for the following barcode? The check-sum number is 5.

ISBN 0-7?56-2153-5

9 780735 621534

Solution

The ISBN we were given is 07?5621535.

We can set up the equation for the multiplication as follows, using x to represent the missing digit.

$$(10 \cdot 0) + (9 \cdot 7) + (8 \cdot x) + (7 \cdot 5) + (6 \cdot 6) + (5 \cdot 2) + (4 \cdot 1) + (3 \cdot 5) + (2 \cdot 3) + (1 \cdot 5) \equiv 0 \,(\text{mod}\,11)$$

Simplifying, we have

$$0 + 63 + 8x + 35 + 36 + 10 + 4 + 15 + 6 + 5 \equiv 0 \,(\text{mod}\,11)$$
$$8x + 174 \equiv 0 \,(\text{mod}\,11).$$

To make things easier on ourselves here, we can go ahead and evaluate 174 modulo 11 and substitute that value into our equation.

$$174 \equiv 9 \,(\text{mod}\,11)$$

So, now we have $8x + 9 \equiv 0 \,(\text{mod}\,11)$.

This tells us that the third digit in the ISBN number x, will need to be multiplied by 8. This number, when added to 9, must be congruent to $0 \,(\text{mod}\,11)$.

Now, we need to use a bit of trial and error to find the missing digit. We know that the digit must be a number between 0 and 10, inclusive. So, we can begin trying numbers, starting with 0. Substituting $x = 0$ in our equation gives the following.

$$8x + 174 = 8(0) + 174$$
$$= 0 + 174$$
$$\equiv 0 \,(\text{mod}\,11) + 9 \,(\text{mod}\,11)$$
$$\equiv 9 \,(\text{mod}\,11)$$

Since this is not congruent to $0 \,(\text{mod}\,11)$, the third digit cannot be a 0.

Next, suppose the third digit was a 1. Let's substitute $x = 1$, in our equation.

$$8x + 174 = 8(1) + 174$$
$$= 8 + 174$$
$$\equiv 8 \ (\text{mod } 11) + 9 \ (\text{mod } 11)$$
$$\equiv 17 \ (\text{mod } 11)$$
$$\equiv 6 \ (\text{mod } 11)$$

Since this is not congruent to $0 \ (\text{mod } 11)$, the third digit cannot be a 1.

Now let $x = 2$.

$$8x + 174 = 8(2) + 174$$
$$= 16 + 174$$
$$\equiv 5 \ (\text{mod } 11) + 9 \ (\text{mod } 11)$$
$$\equiv 14 \ (\text{mod } 11)$$
$$\equiv 3 \ (\text{mod } 11)$$

Again, we do not have a value equivalent to $0 \ (\text{mod } 11)$, so we still have not found our number. Let $x = 3$.

$$8x + 174 = 8(3) + 174$$
$$= 24 + 174$$
$$\equiv 2 \ (\text{mod } 11) + 9 \ (\text{mod } 11)$$
$$\equiv 11 \ (\text{mod } 11)$$
$$\equiv 0 \ (\text{mod } 11)$$

Since 11 modulo 11 is congruent to 0, we have found our missing digit!

The third digit in the ISBN number is **3**.

One might wonder why an elaborate system of numbers, which includes a check-sum number and modular arithmetic, is needed for books. Although it might take us a few minutes to confirm the validity of an ISBN by hand, in our world of computers, this process only takes milliseconds to compute. A computer can quickly identify if a number has been entered incorrectly or even if a particular ISBN is legitimate. We can take one step further and apply this technique to credit card numbers. Since one of the 16 numbers in a credit card number is a check-sum number. You can imagine how computers can immediately identify when you've accidently typed in the wrong number.

Skill Check Answers

1. $(16 \bmod 2) \cdot (3 \bmod 2) \equiv 0 \cdot 1 \ (\text{mod } 2)$
$$\equiv 0 \ (\text{mod } 2)$$
$$(16 \cdot 3) \bmod 2 \equiv 48 \ (\text{mod } 2)$$
$$\equiv 0 \ (\text{mod } 2)$$

2. $(10 \cdot 0) + (9 \cdot 3) + (8 \cdot 0) + (7 \cdot 6) + (6 \cdot 4) + (5 \cdot 0) + (4 \cdot 6) + (3 \cdot 1) + (2 \cdot 5) = 130$
$130 + 2 = 132$
$132 \equiv 0 \ (\text{mod } 11)$

14.2   Exercises

Compute each value.

1. $12 \equiv$ ___ $(\bmod 5)$

2. $13 \equiv$ ___ $(\bmod 4)$

3. $120 \equiv$ ___ $(\bmod 11)$

4. $84 \equiv$ ___ $(\bmod 3)$

5. $5^2 \equiv$ ___ $(\bmod 2)$

6. $4328 \bmod 10$

7. $60{,}002 \bmod 6$

8. $311 \bmod 4$

9. $44 \bmod 12$

10. $113 \bmod 12$

Convert each 24-hour clock time to an equivalent 12-hour clock time.

11.   0800 hours

12.   1300 hours

13.   2100 hours

14.   0000 hours

Compute each value.

15. $(56 + 87) \bmod 13$

16. $\left[12(34 + 6) \right] \bmod 7$

17. $\left[128 - (15 + 8) \cdot 23 \right] \bmod 11$

18. $\left(2^5 - 17 \right) \bmod 3$

Determine whether each statement is true or false.

19. $2^3 \equiv 3 (\bmod 5)$

20. $\sqrt{81} \equiv 3 (\bmod 2)$

21. $(3 \cdot 4) \equiv 0 (\bmod 6)$

22.   Any two even numbers x and y are equivalent to each other mod 2; that is, $x \equiv y (\bmod 2)$.

23.   Any two odd numbers a and b are equivalent to each other mod 2; that is, $a \equiv b (\bmod 2)$.

24. $645{,}234 \equiv 111{,}111{,}111 (\bmod 3)$ (**Hint:** Two numbers are congruent mod 3 if both are divisible by 3. Use Section 14.1 methods to help you decide.)

25. $9{,}436{,}278{,}463{,}920 \equiv 764{,}283{,}237{,}885 (\bmod 5)$

Determine whether each 10-digit ISBN is valid. If it is not, state what the correct check-sum number should be.

26.   0-392-31123-2

27.   1-103-24582-6

28.   0-332-15573-0

29.   1-02-345678-8

30.   0-022-44668-8

14

Determine the missing digit for each 13-digit ISBN.

For a 13-digit ISBN, each of the first 12 digits, from left to right, is alternately multiplied by 1 or 3. These products are summed and the 13th digit is chosen so that the total of all these numbers is 0 modulo 10.

31. *Emma* by Jane Austin: 978048640648?

32. *Harry Potter and the Sorcerer's Stone* by J.K. Rowling: 978059035342?

33. *The Hunger Games* by Suzanne Collins: 97?0439023528

34. *The Girl with the Dragon Tattoo* by Stieg Larsson: 9?80307949486

Decide whether each barcode is valid. If it is not, state what the correct check-sum number should be.

Each of the following bar codes consists of 12 digits. To find the check-sum digit, the first 11 digits are multiplied by 3 or 1, alternatively. These products are summed and the final digit is chosen so that the total of all these numbers is 0 modulo 10.

35.

36.

37.

Calculate the correct first digit for each barcode.

38.

39.

40.

Credit card companies use the Luhn algorithm to help construct secure numbers. Use the following steps to determine whether each credit card number is valid. If it is not, find the correct check-sum number.

1. The check-sum digit is the last digit in the number, whether the card number is 13, 15, or 16 digits long. Working right to left, starting with the digit to the left of the check-sum number, double the value of every other digit.

2. If a number becomes a 2-digit number after doubling, treat each digit as an individual digit. Finally, sum the digits of all doubled numbers as well as the undoubled numbers including the check-sum number.

3. If the total is congruent to 0 (mod 10), then the number is valid according to the Luhn formula; otherwise it is not valid.

41. 3780 2850 1184 225

42. 300 9255 939 6891

43. 389 841 621 516 22

44. 4929 1175 0198 3180

14.3 Fermat's Little Theorem and Prime Testing

Two 17th-century mathematicians, Pierre de Fermat and his friend and confidant Bernard Frénicle de Bessy, often corresponded by letter. In a letter dated October 18, 1640, Fermat tells Frénicle de Bessy about a marvelous discovery in number theory that he has come across. As he often did in his letters, Fermat provided no proof of his discovery. The proof of the comment in the letter took another 100 years to be rediscovered. Thankfully, it was published by yet another great mathematician, Leonhard Euler, and since then has been known as Fermat's Little Theorem. More than 400 years later, we continue to use this theorem to help us establish whether a number is prime. The theorem is stated as follows.

Fermat's Little Theorem

Let p be any prime number and x be any positive integer. Then, $x^p - x \equiv 0 \pmod{p}$.

This theorem is a statement about a calculation in modular arithmetic. But, rather than talking about addition or multiplication as we did in the previous section, this time exponents are involved. Fermat's Little Theorem says that if you multiply a number by itself p times, and then subtract off that same number, the answer is a multiple of p; that is, the answer is congruent to $0 \pmod{p}$.

To try it out, we need a prime number p, let's choose 5, and another positive integer x, such as 2. Now let's carry out the calculation in the theorem with $p = 5$ and $x = 2$. We'll raise 2 to the 5th power and subtract 2, then show that the answer is a multiple of 5.

$$
\begin{aligned}
x^p - x &= 2^5 - 2 \\
&= 2 \cdot 2 \cdot 2 \cdot 2 \cdot 2 - 2 \\
&= 32 - 2 \\
&= 30 \\
&\equiv 0 \pmod{5}
\end{aligned}
$$

Notice that 30 is congruent to 0 modulo 5 because 30 is a multiple of 5.

Lets try another example.

Example 1: Verifying Fermat's Little Theorem

Verify that $x^p - x \equiv 0 \pmod{p}$, when $p = 7$ and $x = 4$.

Solution

$$
4^7 - 4 = 4 \cdot 4 \cdot 4 \cdot 4 \cdot 4 \cdot 4 \cdot 4 - 4
$$

We'll stop at this point and introduce a useful short cut. You might not be able to calculate 4^7 in your head, but the beauty of modular arithmetic is that you don't have to. Remember that we want our equation to work in modulo 7. So changing any number to its equivalent modulo 7 throughout the calculation is helpful in that it makes the numbers we are manipulating smaller. Notice that $4 \cdot 4 = 16 \equiv 2 \pmod{7}$. By replacing each pair of 4s that are multiplied together with a 2, we can make life easier.

$$
\begin{aligned}
4^7 - 4 &= (4 \cdot 4) \cdot (4 \cdot 4) \cdot (4 \cdot 4) \cdot 4 - 4 \\
&\equiv 2 \cdot 2 \cdot 2 \cdot 4 - 4 \pmod{7} \\
&\equiv 8 \cdot 4 - 4 \pmod{7}
\end{aligned}
$$

Finally, $2^3 = 8$, which is congruent to 1 modulo 7.

$$4^7 - 4 \equiv 8 \cdot 4 - 4 \pmod{7}$$
$$\equiv 1 \cdot 4 - 4 \pmod{7}$$
$$\equiv 4 - 4 \pmod{7}$$
$$\equiv 0 \pmod{7}$$

Therefore, we've shown that $4^7 - 4 \equiv 0 \pmod{7}$.

Skill Check #1

Verify that $x^p - x \equiv 0 \pmod{p}$, when $p = 11$ and $x = 10$.

Recall that one method for establishing primes is to test for possible divisors. As we showed in previous examples, that method works very well for small numbers. But how would you like to test the divisors of the number 4,294,967,297 or the number 18,446,744,073,709,551,617? Although we only need to test the prime numbers as large as the square root of each number, for the second number, that's still more than two hundred million prime numbers to try out one-by-one.

Luckily, we can use Fermat's Little Theorem to provide a more useful method for testing large numbers. The theorem states that, for any prime number p, $x^p - x \equiv 0 \pmod{p}$ is true. Recall from our discussion of logic in Chapter 3 that if a statement is true, then its contrapositive is also true. In other words, if a implies b is true, then not b implies not a is true. So, we can use logic to turn the theorem around and tell us something about prime numbers. Fermat's Little Theorem says that if you raise a number to a prime, then the result after subtraction of the same number is equivalent to 0 mod that prime. So, the contrapositive says that if we perform the operation and DON'T get a value equivalent to $0 \pmod{p}$, then p is not a prime number to begin with. The formal theorem is stated as follows.

Contrapositive of Fermat's Little Theorem

Let x and n be positive integers. If $x^n - x \not\equiv 0 \pmod{n}$, then n is **not** a prime number.

Let's stop and emphasize that the contrapositive says that if we do NOT get 0, then the power we used is NOT prime. However, we **must** be careful to not misinterpret the theorem to imply that obtaining 0 means that the power must be a prime number. In other words, if we can find at least one value for x to use in the formula that results in a number other than 0, we've shown that the number n is not prime. So, this contrapositive allows us to test if a number is not prime simply by trying Fermat's Little Theorem with a single number. Let's try this out for some small manageable numbers before we move on to our bigger example of 4,294,967,297. We know that the number 8 is not prime, so we'll start with that.

Example 2: Prime Testing

Use the contrapositive of Fermat's Little Theorem to verify that the number $n = 8$ is not prime by using the number 2 for x.

Solution

Substituting in $n = 8$ and $x = 2$ into the contrapositive, we have the following.

$$x^n - x = 2^8 - 2$$
$$= (2 \cdot 2 \cdot 2) \cdot (2 \cdot 2 \cdot 2) \cdot (2 \cdot 2) - 2$$
$$\equiv 0 \cdot 0 \cdot 4 - 2 \,(\text{mod } 8)$$
$$\equiv -2 \,(\text{mod } 8)$$
$$\equiv (-2 \text{ mod } 8) + (8 \text{ mod } 8)$$
$$\equiv -2 + 8 \,(\text{mod } 8)$$
$$\equiv 6 \,(\text{mod } 8)$$
$$\not\equiv 0 \,(\text{mod } 8)$$

Helpful Hint

To calculate negative equivalences, you can subtract from the modular base. In other words, −2 mod 8 is congruent to (−2 + 8) (mod 8), or 6 (mod 8).

Notice that this is the first time a negative has appeared at the front of an equivalence in our calculations. To simplify, we added 0 to the equation by adding 8 (mod 8). This allowed us to return to a positive equivalence that we are accustomed to.

Since raising 2 to the 8th power and subtracting 2 does not result in a multiple of 8, we have confirmed that 8 is not prime. It's worth noting that although we are told to start with $x = 2$, we could use any number to show that 8 is not a prime number.

To get the most out of this method, we need to learn how to calculate large exponents in modular arithmetic. Table 1 recalls a few rules of exponents.

Table 1	
Rules of Exponents	
Product Rule	$x^a \cdot x^b = x^{a+b}$
Quotient Rule	$\dfrac{x^a}{x^b} = x^{a-b}$
Power Rule	$\left(x^a\right)^b = x^{ab}$

Skill Check #2

Simplify each of the following.

a. $3^3 \cdot 3^4$ **b.** $\left(4^2\right)^5$ **c.** $\dfrac{2^{10}}{2^4}$

To use the contrapositive of Fermat's Little Theorem for a large number n, we will have to raise a number x to that large exponent. Using standard arithmetic, x to the power n will be a very large number to calculate. However, the beauty of using modular arithmetic is that we can always keep numbers relatively small; that is, less than n. The numbers might still be large, but not too large to handle. Let's try testing a slightly bigger number as an example of how we can use modular arithmetic to make calculations easier. We're going to use a squaring technique to break the exponent we want into manageable pieces.

Example 3: Prime Testing Using Modular Arithmetic

Verify that 39 is not prime using the contrapositive of Fermat's Little Theorem and $x = 2$.

Solution

We want to show 39 is not prime using the number 2. So, we have $x = 2$ and $n = 39$. Using $x^n - x \not\equiv 0 \,(\text{mod } n)$, we can see that we have to reduce $2^{39} \text{ mod } 39$. However, since 2^{39} becomes a very large number very quickly, we can slowly build up to 2^{39} piece-by-piece by repeatedly squaring.

14

Notice that $2^{39} = 2^{32} \cdot 2^4 \cdot 2^2 \cdot 2 \cdot 1$. So, if we can find these powers of 2, we can combine them to find 2^{39}. The following steps show the most efficient way to find these powers.

$$2^1 \equiv 2 \ (\text{mod } 39)$$
$$2^2 \equiv 4 \ (\text{mod } 39)$$
$$2^4 = 2^2 \cdot 2^2 \equiv 4 \cdot 4 \equiv 16 \ (\text{mod } 39)$$
$$2^8 = 2^4 \cdot 2^4 \equiv 16 \cdot 16 = 256 \equiv 22 \ (\text{mod } 39) \ \left(\text{verify this yourself}\right)$$

In standard arithmetic, $2^8 = 256$, and we would need to continue on with our calculations of 2^{39} using 256. However, in arithmetic modulo 39, $2^8 \equiv 22 \left(\text{mod } 39\right)$. This smaller number of 22 will be much easier to work with. This pattern of reducing our calculations modulo 39 keeps the numbers manageable. Notice how the pattern continues in the next steps.

$$2^{16} = 2^8 \cdot 2^8 \equiv 22 \cdot 22 = 484 \equiv 16 \ (\text{mod } 39) \ \left(\text{again, verify this yourself}\right)$$
$$2^{32} = 2^{16} \cdot 2^{16} \equiv 16 \cdot 16 = 256 \equiv 22 \ (\text{mod } 39)$$

From these powers of 2, we have all we need to calculate $2^{39} - 2$.

$$2^{39} - 2 = 2^{32} \cdot 2^4 \cdot 2^2 \cdot 2^1 - 2$$
$$\equiv 22 \cdot 16 \cdot 4 \cdot 2 - 2 \ (\text{mod } 39)$$
$$\equiv 2816 - 2 \ (\text{mod } 39)$$
$$\equiv 2814 \ (\text{mod } 39)$$
$$\equiv 6 \ (\text{mod } 39)$$

Verify for yourself that $2814 \equiv 6 \ (\text{mod } 39)$. Once again, since this is not equivalent to $0 \left(\text{mod } 39\right)$, we have established that 39 is not prime. Notice that we did the calculation without ever multiplying two numbers that were bigger than 39.

So far, we have only confirmed that a few relatively small numbers are not prime, which may have been obvious to some. To demonstrate the real power of Fermat's Little of Theorem, we can show that the monstrous number 4,294,967,297 is not prime using the same technique we developed in Example 3. To use Fermat's Little Theorem technique, we need to calculate

$$x^{4,294,967,297} - x \left(\text{mod } 4,294,967,297\right) \text{ for some value of } x.$$

Clearly, there is no way we can just type $x^{4,294,967,297}$ into a calculator no matter what x is, because this would require the calculator to show a number that is more than one billion digits long. (Think for a second what a number this large might look like.) However, we can employ the successive squaring technique we have just developed by using exponent rules once again. This large number happens to be $2^{32} + 1$, so we have the following.

$$x^{4,294,967,297} = x^{\left(2^{32}+1\right)}$$
$$= x^{\left(2^{32}\right)} \cdot x^1$$

We can calculate any number to the power 4,294,967,297 by using the squaring method again to help us break down the calculation of the exponent. Now, we need to choose the number x. There's no reason not to choose the smallest possibility, 2. Therefore, we'll let $x = 2$. Let's begin the process of squaring x.

$$2 \equiv 2 \,(\text{mod } 4{,}294{,}967{,}297)$$

$$2^2 = 2^{(2^1)} = 2 \cdot 2 \equiv 4 \,(\text{mod } 4{,}294{,}967{,}297)$$

$$2^4 = 2^{(2^2)} = 2^2 \cdot 2^2 \equiv 4 \cdot 4 \equiv 16 \,(\text{mod } 4{,}294{,}967{,}297)$$

$$2^8 = 2^{(2^3)} = 2^4 \cdot 2^4 \equiv 16 \cdot 16 \equiv 256 \,(\text{mod } 4{,}294{,}967{,}297)$$

$$2^{16} = 2^{(2^4)} = 2^8 \cdot 2^8 \equiv 256 \cdot 256 \equiv 65{,}536 \,(\text{mod } 4{,}294{,}967{,}297)$$

$$2^{32} = 2^{(2^5)} = 2^{16} \cdot 2^{16} \equiv 65{,}536 \cdot 65{,}536 \equiv 4{,}294{,}967{,}296 \,(\text{mod } 4{,}294{,}967{,}297)$$

$$2^{64} = 2^{(2^6)} = 2^{32} \cdot 2^{32} \equiv (4{,}294{,}967{,}296) \cdot (4{,}294{,}967{,}296)$$
$$= 18{,}446{,}744{,}073{,}709{,}551{,}616 \equiv 1 \,(\text{mod } 4{,}294{,}967{,}297)$$

$$2^{128} = 2^{(2^7)} = 2^{64} \cdot 2^{64} \equiv 1 \cdot 1 \equiv 1 \,(\text{mod } 4{,}294{,}967{,}297)$$

$$2^{256} = 2^{(2^8)} = 2^{128} \cdot 2^{128} \equiv 1 \cdot 1 \equiv 1 \,(\text{mod } 4{,}294{,}967{,}297)$$

$$2^{512} = 2^{(2^9)} = 2^{256} \cdot 2^{256} \equiv 1 \cdot 1 \equiv 1 \,(\text{mod } 4{,}294{,}967{,}297)$$

Now, we can see that the pattern developing means we will get a value equivalent to $1 \,(\text{mod } 4{,}294{,}967{,}297)$ every time we square from now on. We will get $1 \,(\text{mod } 4{,}294{,}967{,}297)$ for 2^{1024}, 2^{2048}, ..., and all the way up to $2^{4{,}294{,}969{,}296}$.

So now we are ready to calculate $2^{4{,}294{,}967{,}297} - 2 \,(\text{mod } 4{,}294{,}967{,}297)$.

$$2^{4{,}294{,}967{,}297} - 2 = 2^{4{,}294{,}967{,}296 + 1} - 2$$
$$= 2^{4{,}294{,}967{,}296} \cdot 2^1 - 2$$
$$\equiv 1 \cdot 2^1 - 2 \,(\text{mod } 4{,}294{,}967{,}297)$$
$$\equiv 2 - 2 \,(\text{mod } 4{,}294{,}967{,}297)$$
$$\equiv 0 \,(\text{mod } 4{,}294{,}967{,}297)$$

Now, let's think carefully about what we just found out. Remember, as we stated before, the theorem says that if we do NOT get $0 \,(\text{mod } n)$, then the number is NOT prime. It does **not** say that getting $0 \,(\text{mod } n)$ means it is prime. So, we have not found out anything helpful in determining if our number is prime *yet*. We can try a different value for x to see if that helps. Let's try $x = 3$ instead. Don't get discouraged here. We know that these calculations are very large, but in bite size steps they are manageable.

$$3 \equiv 3 \,(\text{mod } 4{,}294{,}967{,}297)$$

$$3^2 = 3^{(2^1)} \equiv 3 \cdot 3 \equiv 9 \,(\text{mod } 4{,}294{,}967{,}297)$$

$$3^4 = 3^{(2^2)} \equiv 9 \cdot 9 \equiv 81 \,(\text{mod } 4{,}294{,}967{,}297)$$

$$3^8 = 3^{(2^3)} \equiv 81 \cdot 81 \equiv 6561 \,(\text{mod } 4{,}294{,}967{,}297)$$

$$3^{16} = 3^{(2^4)} \equiv 6561 \cdot 6561 \equiv 43{,}046{,}721 \,(\text{mod } 4{,}294{,}967{,}297)$$

$$3^{32} = 3^{(2^5)} \equiv 43{,}046{,}721 \cdot 43{,}046{,}721 \equiv 3{,}793{,}201{,}458 \,(\text{mod } 4{,}294{,}967{,}297)$$

$$3^{64} = 3^{(2^6)} \equiv 3{,}793{,}201{,}458 \cdot 3{,}793{,}201{,}458 \equiv 1{,}461{,}798{,}105 \,(\text{mod } 4{,}294{,}967{,}297)$$

$$3^{128} = 3^{(2^7)} \equiv 1{,}461{,}798{,}105 \cdot 1{,}461{,}798{,}105 \equiv 852{,}385{,}491 \,(\text{mod } 4{,}294{,}967{,}297)$$

$$3^{256} = 3^{(2^8)} \equiv 852{,}385{,}491 \cdot 852{,}385{,}491 \equiv 547{,}249{,}794 \,(\text{mod } 4{,}294{,}967{,}297)$$

$$3^{512} = 3^{(2^9)} \equiv 547{,}249{,}794 \cdot 547{,}249{,}794 \equiv 1{,}194{,}573{,}931 \,(\text{mod } 4{,}294{,}967{,}297)$$

$$3^{1024} = 3^{(2^{10})} \equiv 1{,}194{,}573{,}931 \cdot 1{,}194{,}573{,}931 \equiv 2{,}171{,}923{,}848 \,(\text{mod } 4{,}294{,}967{,}297)$$

$$\vdots$$

$$3^{2{,}147{,}483{,}648} = 3^{(2^{31})} \equiv 10{,}324{,}303 \,(\text{mod } 4{,}294{,}967{,}297)$$

$$3^{4{,}294{,}967{,}296} = 3^{(2^{32})} \equiv 3{,}029{,}026{,}160 \,(\text{mod } 4{,}294{,}967{,}297)$$

Notice that between 3^{1024} and $3^{2{,}147{,}483{,}648}$ we've skipped over several steps. In fact, we need to continue doubling 21 times between the steps listed. For sake of space, we've taken the liberty to do the work for you and leave you to verify these steps on your own.

So, now to use the theorem, we calculate the following.

$$3^{4,294,967,297} - 3 = 3^{4,294,967,296+1} - 3$$
$$= 3^{4,294,967,296} \cdot 3^1 - 3$$
$$\equiv 3,029,026,160 \cdot 3 - 3 \pmod{4,294,967,297}$$
$$\equiv 9,087,078,480 - 3 \pmod{4,294,967,297}$$
$$\equiv 9,087,078,477 \pmod{4,294,967,297}$$
$$\equiv 497,143,883 \pmod{4,294,967,297}$$

Thankfully, this is not 0. So 4,294,967,297 is indeed not prime, but composite.

This is quite a lot of work to do by hand with a calculator, but certainly far less than the hundreds of millions of divisions we would otherwise have had to check by the divisor method. Even computers can't handle the divisor method for numbers this big, but they can use modular arithmetic to apply Fermat's Little Theorem. The following Tech Training shows one of the websites that calculates these enormous values in no time at all. Now we have a healthy appreciation for what goes on behind the scenes for this computation.

TECH TRAINING

To perform these calculations using Wolfram|Alpha, go to www.wolframalpha.com, and type in the mathematical phrase you would like it to solve for you.

Note that typing in "3^4,294,967,297 − 3 (mod 4,294,967,297)" is not a valid command. In fact, typing in "3^4,294,967,297" returns a number with over 2 billion digits. So, the best we can do is to find $3^{4,294,967,297} \pmod{4,294,967,297}$.

Type "3^4,294,967,297 (mod 4,294,967,297)" into the input bar then click the = button. Wolfram|Alpha will return the following result. [1]

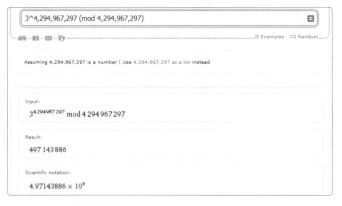

Therefore, we know that $3^{4,294,967,297} \pmod{4,294,967,297}$ is congruent to 497,143,886.

$$3^{4,294,967,297} - 3 = 497,143,886 - 3 \pmod{4,294,967,297}$$
$$= 497,143,883 \pmod{4,294,967,297}$$

---

1 Wolfram Alpha LLC, http://www.wolframalpha.com

Skill Check Answers

1. $10^{11} - 10 = (10 \cdot 10) \cdot (10 \cdot 10) \cdot (10 \cdot 10) \cdot (10 \cdot 10) \cdot (10 \cdot 10) \cdot 10 - 10$

$\equiv 1 \cdot 1 \cdot 1 \cdot 1 \cdot 1 \cdot 10 - 10 \ (\text{mod } 11)$

$\equiv 10 - 10$

$\equiv 0 \ (\text{mod } 11)$

2. a. 3^7;   **b.** 4^{10};   **c.** 2^6

14.3   Exercises

Verify that Fermat's Little Theorem holds true for each prime number using the value of x given.

1.   The prime number 11 and $x = 6$

2. $x = 2$ with the prime 23

3.   The third prime number and $x = 10$

4. $p = 79$ and $x = 3$

Use the contrapositive of Fermat's Little Theorem to show that each given number is composite.

5.   63,571 using $x = 3$

6.   12,731 using $x = 3$

7.   65,476,751 using $x = 2$

Use a resource such as Wolfram|Alpha to confirm that each given number is composite using the contrapositive of Fermat's Little Theorem.

8.   2427

9.   24,271

10.   245,127

11.   10,050,207

14.4   *Fermat's Little Theorem and Public-Key Encryption*

Secret codes, or encryptions, have been around since the beginning of history. One of the earliest and simplest was one that is said to have been used by Julius Caesar to communicate with his military. Now known as a Caesar cipher, or cipher shift, messages were encoded by substituting each letter in the alphabet by another letter some fixed number of letters down the alphabet. For example, in a right shift of 3, the letter A would be replaced by the letter D, and the letter B would be replaced by E, and so on.

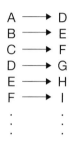

Figure 1: Cipher Shift

Once a message was sent to a general in the army, all the general needed was the size of the shift down the alphabet and he could know Caesar's wishes. Although effective, you can see that if someone knew the size (and direction) of the cipher shift, they could easily decode an intercepted message.

Over time, codes have become more and more advanced and harder to break. One of the most widely used coding techniques today is called public-key encryption. Its existence is crucial in our computer-driven world as the most widely used system for encrypting and protecting Internet transactions. Its beauty lies in the fact that to "unlock" a message, 2 keys are required—one that is secret and another that is public—hence its name.

One of the best explanations for public-key encryption comes from an article in *The Chronicle of Higher Education* by Konrad M. Lawson. He tells a story of two professors, *Professor More* and *Professor Erasmus*, who want to exchange private documents via the public mail system. To do this, both professors buy a padlock. Then they each send the open padlocks (without keys) to the other. When it's time to send a private document, Professor More puts the document in a box and locks it with Professor Erasmus' lock. Upon receiving the package, Professor Erasmus can use his own key, which he kept, to unlock it. If he chooses to send something back, Professor Erasmus uses Professor More's lock to secure the document. Neither professor has risked someone intercepting the package or stealing the key, since the keys never left the possession of the owner. The padlocks are analogous to the "public key" that only allows someone to encode a message, that is, "lock the box." The private keys are the essential bits used to decode the message, or "unlock the box." [1]

The idea for a public-key cipher was first publicly described by three mathematicians/computer scientists, Ron Rivest, Adi Shamir, and Leonard Aldeman in 1977. This cipher, called RSA after the three men, is built into almost all secure computer systems in some way, and luckily for them, they patented their idea (US Patent 4,405,829). Despite their patent and independent work, they were not the first to come up with the idea. Clifford Cocks, an English mathematician discovered this idea in 1973. But because he was working for the UK secret service, GCHQ, at the time of his discovery, his work remained classified until 1998.

The idea for the cipher revolves around a theorem attributed to Leonhard Euler. His theorem is a slight extension of Fermat's Little Theorem from the use of a prime number to that of any number. Remember, Fermat's Little Theorem says that for any prime number p, $x^p - x \equiv 0 \pmod{p}$ for any x. For

---

1 Konrad M. Lawson, *The Chronicle of Higher Education*, http://chronicle.com

Euler's Theorem, though, a number need not be prime; it just needs to be the product of at least two primes. Let's explore this theorem to understand what Euler is saying and then see what connection this has with encryptions.

First, we'll state the theorem formally when n is a product of two primes. Although the formula itself might look a bit intimidating at first, see if you can recognize the similarity in structure to Fermat's Little Theorem.

Euler's Theorem

$$x^{a(p-1)(q-1)+1} - x \equiv 0 \,(\text{mod } n),$$

where p and q are prime numbers, $n = pq$, and x and a are any positive integers.

Did you notice that the basic format of the equation in Euler's Theorem looks similar to the one in Fermat's Little Theorem? Both are of the general form

$$x^{number} - x \equiv 0 \,(\text{mod modulus}).$$

In Fermat's Little Theorem, the exponent and the modulus were the same prime number. In Euler's Theorem, that is not the case. Example 1, uses a set of prime numbers to demonstrate this.

Example 1: Verifying Euler's Theorem

Verify Euler's Theorem with the prime numbers $p = 3$ and $q = 5$. Let $x = 2$ and $a = 1$.

Solution

We are told to use the prime numbers 3 and 5. Therefore, $n = 3 \cdot 5 = 15$. We can substitute these values into the formula as follows.

$$2^{a(p-1)(q-1)+1} - 2 = 2^{1(3-1)(5-1)+1} - 2$$
$$= 2^9 - 2$$
$$= 512 - 2$$
$$= 510$$
$$\equiv 0 \,(\text{mod } 15)$$

Just as with Fermat's Little Theorem in the previous section, there is nothing special about the numbers we chose, other than the fact that 3 and 5 are both prime.

Skill Check #1

Verify Euler's Theorem with the prime numbers $p = 2$ and $q = 3$. Let $x = 2$ and $a = 1$.

The way that we presented both Fermat's Little Theorem and Euler's Theorem illustrates how the two are similar. We've written them side-by-side here so that you can see for yourself.

Table 1	
Fermat's Little Theorem vs. Euler's Theorem	
Fermat's Little Theorem	**Euler's Theorem**
$x^p - x \equiv 0 \,(\text{mod } p)$	$x^{a(p-1)(q-1)+1} - x \equiv 0 \,(\text{mod } n)$
p is prime	p and q are prime, $n = pq$

As you can see, both theorems involve raising x to some power and then subtracting that same x. The results are surprisingly similar in that they are both equivalent to 0 modulo something that is connected to the exponent. Both of these theorems depend on certain numbers being prime. Public-key encryption also relies on prime numbers.

Here's where Euler's Theorem comes into play. To help us better understand public encryption codes, we're going to rearrange Euler's Theorem a bit. Although we are not changing the meaning behind the theorem, changing the way it is written can help us see how to use it in encryption codes. Here's the rearranged version.

> ## Euler's Theorem
>
> $$x^{a(p-1)(q-1)+1} \equiv x \ (\text{mod } n)$$
>
> where p and q are prime numbers, $n = pq$, and x and a are any positive integers.

Think of the new arrangement of the formula in the following way. Suppose you have any number and you raise that number to a special power involving two primes, p and q. What you get is congruent to itself, modulo the product of p and q. In other words, you can manipulate a number exponentially with primes and get right back where you started using modular arithmetic.

Using the formula in this format is beneficial in public encryption. Suppose we take your credit card number and encode it by raising it to a power. Now we have your new encoded credit card number. If I wanted to, I can give your encoded number along with the directions of how I encoded it, to anyone without resulting in a breach in your credit card security. Why? Because I'm the only one who knows the exact two prime numbers that are used for the decoding.

Now, you might think that it would be easy for someone to just "figure out" what the two prime numbers are. That, however, is not so easy if the prime numbers chosen are very, very large—say 200 digits each!

Remember, it is crucial to know the prime numbers p and q in order to decode the number. Usually for these purposes p, q, and n are very large numbers, often hundreds of digits long, and at present, no one knows how to carry out the search for prime numbers in any practical way. A good analogy is to suppose that I were to give you a 200-digit number n and its factorization into two 100-digit primes written down on a piece of paper. Then, you lost the primes by accidentally shredding the piece of paper. It would be quicker for you to search for the shreds of paper in the public dump and try to reconstruct the primes on the paper rather than to try to find the factors by trial and error. It's like finding a needle in a haystack.

Here's how the process works in encryption. The formula requires that we raise a number to a power and then perform a modular reduction to get back to the original number. For encryption purposes, the sender will do part of the raising to a power and the receiver will finish off the raising to a power. Both will perform modular arithmetic using the product of two prime numbers.

Let's use Samantha and Carlos to help us explain. After choosing her prime numbers, p and q, Samantha multiplies them together with another number a in the following manner so that she forms the exponent in the formula from Euler's Theorem.

$$a(p-1)(q-1)+1$$

She then takes this number and factors it into two numbers e and d.

$$a(p-1)(q-1)+1 = ed$$

She tells Carlos the numbers e and n and instructs him to encrypt his secret number by raising it to the power e and finding its congruence modulo n. (Notice that by telling Carlos e and n, where n is the product of p and q, Samantha has not actually given away what p and q are.)

So, Carlos does the following and sends it to Samantha.

$$\left(\text{secret number}\right)^e \left(\text{mod } n\right)$$

Now, when Samantha gets the coded number from Carlos, she raises it to the power d.

$$\left(\text{secret number}^e\right)^d \left(\text{mod } n\right)$$

Now Samantha has: $\left(\text{secret number}\right)^{ed}$. Since Samantha chose e and d based on her prime number calculation, this is the same as $\left(\text{secret number}\right)^{a(p-1)(q-1)+1}$.

Because of Euler's Theorem, she can then perform modular arithmetic with modulus $n = pq$, and have the original secret number Carlos started with.

$$\left(\text{secret number}\right)^{ed} = \left(\text{secret number}\right)^{a(p-1)(q-1)+1}$$
$$\equiv \left(\text{secret number}\right)\left(\text{mod } n\right)$$

With Samantha and Carlos' help, we can see that the public part of the encryption code, called the public key, consists of the numbers e and n. The private part of the process, which is used to decode the number, is d along with p and q.

The following table gives a summary of the variables used in public-key encryption.

Table 2	
Summary of the Variables Used in Public-Key Encryption	
Variable	**Definition**
p and q	prime numbers
a	any positive integer
e and d	factors of $a(p-1)(q-1)+1$
n	the product pq
e and n	the public key
p, q, and d	the private key

The next examples take us through encoding and decoding a message using public-key encryption.

Example 2: Public-Key Encryption

Let $C = 2$ be your simple credit card number, which is the secret number. Encode C using the following public key. Let M be the new encoded credit card number.

Public Key: $n = 115$ and $e = 17$

Solution

To encode C, we have to calculate $C^e \left(\text{mod } n\right)$.

$$C^e \left(\text{mod } n\right) \equiv 2^{17} \left(\text{mod } 115\right)$$
$$\equiv 131,072 \left(\text{mod } 115\right)$$
$$\equiv 87 \left(\text{mod } 115\right)$$

So, our encoded credit card number is now $M = 87$.

828 Chapter 14 Number Theory

Skill Check #2

Using the same public-key information in Example 2, encode the number $C = 3$.

Let's now try to decode the newly transmitted credit card number from Example 2.

Example 3: Decoding Using Private Key Decryption

Decode the number $M = 87$, which is the secret number to the power e, using the private key $d = 57$ and $n = 115$.

Solution

To decode, we calculate the following.

$$M^d \pmod{n} = 87^{57} \pmod{115}$$

As it stands, these numbers are too large for direct computation without an engine such as Wolfram|Alpha, so we will use the squaring method we introduced in the last section. Verify for yourself each step of the process with modular arithmetic.

$$87^1 = 87 \pmod{115}$$
$$87^2 = 87^1 \cdot 87^1 = 87 \cdot 87 = 7569 \equiv 94 \pmod{115}$$
$$87^4 = 87^2 \cdot 87^2 \equiv 94 \cdot 94 = 8836 \equiv 96 \pmod{115}$$
$$87^8 = 87^4 \cdot 87^4 \equiv 96 \cdot 96 = 9216 \equiv 16 \pmod{115}$$
$$87^{16} = 87^8 \cdot 87^8 \equiv 16 \cdot 16 = 256 \equiv 26 \pmod{115}$$
$$87^{32} = 87^{16} \cdot 87^{16} \equiv 26 \cdot 26 = 676 \equiv 101 \pmod{115}$$

Now, we can once again use some rules of exponents to give us 87^{57}. Since we can use a combination of the numbers we have calculated, we have the following.

$$87^{57} = 87^{32} \cdot 87^{16} \cdot 87^8 \cdot 87^1$$
$$\equiv 101 \cdot 26 \cdot 16 \cdot 87 \pmod{115}$$
$$\equiv 3,655,392 \pmod{115}$$
$$\equiv 2 \pmod{115}$$

And, just as we knew, the original credit card number is 2.

Although we have shown this long computation by hand, $87^{57} \pmod{115}$ can be evaluated in one step using WolframAlpha.com.

To finish off the chapter, Example 4 uses Wolfram|Alpha to help with both encoding and decoding of messages using public-key encryption.

TECH Example 4: Encoding and Decoding Using Technology

ECD Technologies uses public-key encryption to receive credit card numbers via the Internet. When an order is placed on its website, the server uses the public key $n = 999,985,999,949$ and $e = 41$ to encode the number. When ECD receives the transmission, it uses the private key $d = 73,169,560,973$ and $n = 999,985,999,949$ to retrieve the original credit card number.

a. Use Wolfram|Alpha to code the credit card number 123 456 7890 to place an order with ECD Technologies online.

b. Use Wolfram|Alpha to decode the transmitted number 257,646,382,798.

Solution

a. In order to encode the credit card number, we need to raise it to the power e modulo n. In other words, we need to calculate $1234567890^{41} \pmod{999,985,999,949}$.

To calculate this value, go to www.wolframalpha.com and type "1234567890^41 $\pmod{999,985,999,949}$" into the input bar. Then, press the = button.

This gives us the following encoded number, as shown in the screenshot: 597,638,266,866. [2]

b. To decode a transmitted number, we need to raise it to the power d modulo n. In other words, we need to calculate $257,646,382,798^{73,169,560,973} \pmod{999,985,999,949}$.

To enter this into Wolfram|Alpha, type "257,646,382,798^73,169,560,973 $\pmod{999,985,999,949}$" into the input bar. Then, click the = button.

This gives us the following decoded number as shown in the screen shot: 9,876,543,210. [3]

Skill Check Answers

1. $2^{1(2-1)(3-1)+1} - 2 = 2^3 - 2$
   $= 8 - 2$
   $= 6$
   $\equiv 0 \pmod{6}$

2. $M = 108$

---

2 Wolfram Alpha LLC, http://www.wolframalpha.com

3 Wolfram Alpha LLC, http://www.wolframalpha.com

14.4   Exercises

Verify Euler's Theorem for each given number.

1. $n = 77, p = 7, q = 11, x = 2$, and $a = 1$

2. $n = 91, p = 13, q = 7, x = 3$, and $a = 2$

3. $n = 155, p = 31, q = 5, x = 4$, and $a = 6$

Encode each number using the given public key.

4.   Encode the number $M = 14$ using the public key $n = 77$ and $e = 13$.

5.   Encode the number $T = 49$ using the public key $n = 77$ and $e = 37$.

6.   Encode the number $CC = 101$ using the public key $n = 119$ and $e = 11$.

7.   Encode the number $K = 55$ using the public key $n = 119$ and $e = 35$.

Decode each number using the private key given.

8.   Decode the number $PR = 15$ using the private key $d = 53$ and $n = 77$.

9.   Decode the number $TD = 38$ using the private key $d = 17$ and $n = 77$.

10.   Decode the number $C = 64$ using the private key $d = 35$ and $n = 221$.

11.   Decode the number $W = 219$ using the private key $d = 151$ and $n = 781$.

Use Wolfram|Alpha to break each code.

12. Suppose $p = 23$ and $q = 71$. Calculate what d must be if $e = 51$ and $a = 5$.

13. Suppose $p = 17$ and $q = 61$. Calculate what d must be if $e = 71$ and $a = 23$.

14. Suppose $n = 187$ and $e = 19$. Calculate what d must be as well as the two primes p and q, if $a = 7$.

15. Suppose $n = 589$ and $e = 23$. Calculate what d must be as well as the two primes p and q, if $a = 2$.

Chapter 14   Summary

Definitions

Prime

A prime number is a positive integer that has precisely two divisors: 1 and itself.

Composite

A composite number is a positive integer that has more than two divisors.

Divisibility Rules

Divisibility Rules	
Divisor	Test
2	The number is even.
3	When the digits of the number are added together, the resulting number is divisible by 3.
4	The last 2 digits of the number form a number divisible by 4.
5	The number ends in a 0 or a 5.
6	The number is divisible by both 2 and 3.
7	Double the last digit, then subtract it from the remaining digits of the number. If the answer is divisible by 7, so is the original number.
8	The last 3 digits of the number form a number divisible by 8.
9	When the digits of the number are added together, the resulting number is divisible by 9.
10	The number ends in a 0.

Fundamental Theorem of Arithmetic

Every positive integer greater than 1 is either a prime number or can be written as a unique product of prime numbers. This unique product of prime numbers is called its prime factorization.

Greatest Common Divisor (GCD)

The largest integer that divides two numbers without a remainder is called the greatest common divisor (GCD). If the GCD = 1 for a pair of numbers, the numbers are said to be relatively prime.

Process

Euclid's Algorithm

1. Divide the larger of the given numbers by the smaller (the divisor) and note the remainder.

2. Divide the original divisor (the smaller of the given numbers) by the remainder found in Step 1.

3. Continue dividing the previous divisor by the previous remainder until the remainder is 0.

4. The last nonzero remainder is the GCD of the given numbers.

Definition

Modular Arithmetic

In modular arithmetic, a number n is congruent to the remainder r when it is divided by a fixed number m. We write $n \equiv r \pmod{m}$. Note that m is referred to as the modulus.

Process

Finding the Check-Sum Digit for a 10-Digit ISBN

1.   Multiply the 1st digit by 10.

2.   Multiply the 2nd digit by 9.

3.   Multiply the 3rd digit by 8.

4.   Multiply the 4th digit by 7.

5.   Multiply the 5th digit by 6.

6.   Multiply the 6th digit by 5.

7.   Multiply the 7th digit by 4.

8.   Multiply the 8th digit by 3.

9.   Multiply the 9th digit by 2.

10. Add the multiples together.

11. The check-sum digit (or 10th digit) is then chosen so that the total of all of these numbers is 0 modulo 11.

Section 14.3   Fermat's Little Theorem and Prime Testing

Definitions

Fermat's Little Theorem

Let p be any prime number and x be any positive integer. Then $x^p - x \equiv 0 \pmod{p}$.

Contrapositive of Fermat's Little Theorem

Let x and n be positive integers. If $x^n - x \not\equiv 0 \pmod{n}$, then n is *not* a prime number.

Section 14.4   Fermat's Little Theorem and Public-Key Encryption

Definition

Euler's Theorem

$x^{a(p-1)(q-1)+1} - x \equiv 0 \pmod{n}$, where p and q are prime numbers, $n = pq$, and x and a are any positive integers.

Euler's Theorem can also be rewritten as $x^{a(p-1)(q-1)+1} \equiv x \pmod{n}$

Variables Used in Public-Key Encryption

Variables Used in Public-Key Encryption	
Variable	**Definition**
p and q	prime numbers
a	any positive integer
e and d	factors of $a(p-1)(q-1)+1$
n	the product pq
e and n	the public key
p, q, and d	the private key

Chapter 14 Exercises

Solve each problem.

1. The student government body consists of 16 females and 12 males. What is the greatest number of subgroups that can be made if each group must have the same number of males and females?

2. As part of a service project, Brooke, Madison, and Collin have collected school supplies (pens, pencils, glue, and notebooks) along with toothbrushes and toothpaste to send to school kids in Haiti. They collected the following.

    60 glue sticks   122 pencils   180 pens   90 notebooks
    70 toothbrushes   40 tubes of toothpaste

    a. If they want to make identical care packages using the supplies they collected, what is the maximum number of care packages they can make?

    b. How many of each item will be in each package?

3. What is the largest number that divides both 691 and 861 leaving a remainder of 11 each for each number.

4. The Food Mission is a local nonprofit organization that teaches high school students how to grow a sustainable garden. They donate all the food grown to local hunger organizations. The land they have must be divided into two pieces so that they can rotate through spring and summer crops. The first garden needs to cover 400 square feet, and the second garden needs to be 340 square feet. In order to stretch donated supplies, the shared boundary between the two gardens needs to be maximized.

    a. What is the maximum boundary the gardens can share?

    b. What will the dimensions of each garden be if the shared boundary is maximized?

5. Find the GCD of the numbers 1785 and 546. Are the numbers relatively prime?

Determine the missing digit for each 10-digit ISBN.

6. 0–337–25?16–9

7. 1–00–?82634–1

8. 0–07–7734?1–1

9. ?–271–53981–1

Decide whether each barcode is valid. If it is not, state what the correct check-sum number should be.

Each of the following bar codes consists of 12 digits. To find the check-sum digit, the first 11 digits are multiplied by 3 and 1, alternatively. These multiples are summed with the last digit so that the result is 0 (mod 10).

10.

1 2 3 4 5 2 3 7 3 9 2 5

11.

3 2 8 9 5 4 3 1 0 0 4 3

12.

2 6 0 0 1 7 0 3 0 1 1 1

Find the missing digit for each credit card number.

13. 4485 4217 6305 8?81

14. 6011 5341 1874 008?

15. ?465 9337 2511 334

Encode each number using the given public key.

16. Encode the number $K = 21$ using the public key $n = 77$ and $e = 13$.

17. Encode the number $T = 28{,}462$ using the public key $n = 5{,}298{,}463$ and $e = 99$.

Decode each number using the private key given.

18. Decode the number 5,031,323 using the private key $d = 105{,}019$ and $n = 5{,}298{,}463$.

19. Decode the number 51,189,234 using the private key $d = 27{,}743$ and $n = 152{,}472{,}479$.

Bibliography

14.3

1. Wolfram Alpha LLC. 2009. Wolfram|Alpha. http://www.wolframalpha.com/input/?i=3%5E4%2C294%2C967%2C297+%28mod+4%2C294%2C967%2C297%29 (accessed July 3, 2014).

14.4

1. Lawson, Konrad M. *The Chronicle of Higher Education*. "Secure Communication with Public-Key Encryption." http://chronicle.com/blogs/profhacker/secure-communication-with-public-key-encryption/30504

2. Wolfram Alpha LLC. 2009. Wolfram|Alpha. http://www.wolframalpha.com/input/?i=1234567890%5E41+%28mod+999%2C985%2C999%2C949%29 (accessed July 3, 2014).

3. Wolfram Alpha LLC. 2009. Wolfram|Alpha. http://www.wolframalpha.com/input/?i=257%2C646%2C382%2C798%5E73%2C169%2C560%2C973+%28mod+999%2C985%2C999%2C949%29 (accessed July 3, 2014).

Appendix A: Critical Values of the Pearson Correlation Coefficient

Appendix A: Table A		
Critical Values of the Pearson Correlation Coefficient		
n	$\alpha = 0.05$	$\alpha = 0.01$
4	0.950	0.990
5	0.878	0.959
6	0.811	0.917
7	0.754	0.875
8	0.707	0.834
9	0.666	0.798
10	0.632	0.765
11	0.602	0.735
12	0.576	0.708
13	0.553	0.684
14	0.532	0.661
15	0.514	0.641
16	0.497	0.623
17	0.482	0.606
18	0.468	0.590
19	0.456	0.575
20	0.444	0.561
21	0.433	0.549
22	0.423	0.537
23	0.413	0.526
24	0.404	0.515
25	0.396	0.505
26	0.388	0.496
27	0.381	0.487
28	0.374	0.479
29	0.367	0.471
30	0.361	0.463
35	0.334	0.430
40	0.312	0.403
45	0.294	0.380
50	0.279	0.361
55	0.266	0.345
60	0.254	0.330
65	0.244	0.317
70	0.235	0.306
75	0.227	0.296
80	0.220	0.286
85	0.213	0.278
90	0.207	0.270
95	0.202	0.268
100	0.197	0.256

Note: r is statistically significant if $|r|$ is greater than or equal to the value given in the table.

Appendix B: Supplemental Calculator Instruction

Throughout this textbook, we have given you instructions on how to use a TI-83/84 Plus graphing calculator and a TI-30XII S/B scientific calculator. You may be using a different brand or model, so the keystrokes for your particular calculator may be different than those listed. If the given instructions do not work for your calculator, you will want to consult your user manual to troubleshoot the issue. A digital form of most user manuals can be found online. Below are websites that contain a list of most Texas Instruments and Casio calculator user manuals.

Texas Instruments: http://education.ti.com/en/us/guidebook/search

Casio: http://world.casio.com/manual/calc/

Specific Instructions for a TI-30 XS/B Multiview Calculator

Finding the Mean, Median, and Standard Deviation of a Data Set

When calculating the mean, median, and standard deviation using a TI-30XS/B Multiview calculator, always remember to clear the data list first using the following commands.

1. Press DATA. You should see three data lists.
2. Press DATA again. You should see an option to clear each list (1:Clear L1, 2:Clear L2, 3:Clear L3).
3. Choose Clear L1.

After clearing the data list, enter the data into the calculator using the following commands.

1. Press DATA to enter the data lists (if you are not already there).
2. Enter your data into L1 by entering each data value and pressing ENTER.

To calculate the mean, median, and standard deviation, use the following commands.

1. Press 2ND DATA.
2. Choose option 1:1-Var Stats and press ENTER.
3. Since the data are in L1, select L1 next to DATA: and leave FRQ set to ONE. Select CALC and press ENTER.

The screen will display a list of values that the calculator computed using the data entered. Use the down arrow key to scroll through the list. The mean of the data will be listed as $\bar{x}$. The sample standard deviation will be listed as Sx and the population standard deviation will be listed as σx. The median will be listed as Med.

Finding the Pearson Correlation Coefficient and Regression Line

When calculating the Pearson correlation coefficient and regression line using a TI-30XS/B Multiview calculator, always remember to clear the data lists first using the following commands.

1. Press DATA. You should see three data lists.
2. Press DATA again. You should see an option to clear each list (1:Clear L1, 2:Clear L2, 3:Clear L3).
3. Choose Clear L1.
4. Repeat Steps 1 and 2, then choose Clear L2.

After clearing the data lists, enter the data into the calculator using the following commands.

1. Press $\boxed{\text{DATA}}$ to enter the data lists (if you are not already there).
2. Enter your data for the variable x into L1 and the data for y into L2 by entering each data value and pressing $\boxed{\text{ENTER}}$.

To calculate the Pearson correlation coefficient and regression line, use the following commands.

1. Press $\boxed{\text{2ND}}$ $\boxed{\text{DATA}}$.
2. Choose option 2:2-Var Stats and press $\boxed{\text{ENTER}}$.
3. Since the data are in L1 and L2, select L1 next to xDATA: and L2 next to yDATA:. Select CALC and press $\boxed{\text{ENTER}}$.

The screen will display a list of values that the calculator computed using the data entered. Use the down arrow key to scroll through the list to the Pearson correlation coefficient, which will be listed as r. The slope of the regression line will be listed as a and the y-intercept of the regression line will be listed as b.

Finding Continuously Compounded Interest

Use the following keystrokes on a TI-30XS/B Multiview calculator for the calculation of continuously compounded interest in Section 9.2 Example 6. Note that you need to use the right arrow key to exit the exponent when you enter the expression.

2500 $\boxed{\times}$ $\boxed{\text{2ND}}$ $\boxed{\text{LN}}$ 0.06 $\boxed{\times}$ 10 $\boxed{\text{ENTER}}$

Finding the Future Value for an Annuity

To perform the calculation from Section 9.3 Example 3 on a TI-30XS/B Multiview calculator, use the following keystrokes. Note that you need to use the right arrow key to exit the exponent when you enter the expression.

50 $\boxed{\times}$ $\boxed{(}$ $\boxed{(}$ 1 $\boxed{+}$ 0.12 $\boxed{\div}$ 12 $\boxed{)}$ $\boxed{\wedge}$ 12 $\boxed{\times}$ 0.5
$\boxed{\blacktriangleright}$ $\boxed{-}$ 1 $\boxed{)}$ $\boxed{\div}$ $\boxed{(}$ 0.12 $\boxed{\div}$ 12 $\boxed{)}$ $\boxed{\text{ENTER}}$

Finding the Future Value for an IRA

To perform the calculation from Section 9.3 Example 4 on a TI-30XS/B Multiview calculator, use the following keystrokes. Note that you need to use the right arrow key to exit the exponent when you enter the expression.

200 $\boxed{\times}$ $\boxed{(}$ $\boxed{(}$ 1 $\boxed{+}$ 0.08 $\boxed{\div}$ 12 $\boxed{)}$ $\boxed{\wedge}$ 12 $\boxed{\times}$ 40
$\boxed{\blacktriangleright}$ $\boxed{-}$ 1 $\boxed{)}$ $\boxed{\div}$ $\boxed{(}$ 0.08 $\boxed{\div}$ 12 $\boxed{)}$ $\boxed{\text{ENTER}}$

Finding Monthly Payments

To perform the calculation from Section 9.3 Example 5 on a TI-30XS/B Multiview calculator, use the following keystrokes. Note that you need to use the right arrow key to exit the exponent when you enter the expression.

3000 $\boxed{\times}$ $\boxed{(}$ 0.0125 $\boxed{\div}$ 12 $\boxed{)}$ $\boxed{\div}$ $\boxed{(}$ $\boxed{(}$ 1 $\boxed{+}$ 0.0125
$\boxed{\div}$ 12 $\boxed{)}$ $\boxed{\wedge}$ 12 $\boxed{\times}$ 3 $\boxed{\blacktriangleright}$ $\boxed{-}$ 1 $\boxed{)}$ $\boxed{\text{ENTER}}$

Finding Maximum Purchase Price

To perform the calculation from Section 9.4 Example 5a. on a TI–30XS/B Multiview calculator, use the following keystrokes. Note that you need to use the right arrow key to exit the exponent when you enter the expression.

805 $\boxed{\times}$ $\boxed{(}$ 1 $\boxed{-}$ $\boxed{(}$ 1 $\boxed{+}$ 0.0337 $\boxed{\div}$ 12 $\boxed{)}$ $\boxed{\wedge}$ $\boxed{(-)}$ 12
$\boxed{\times}$ 15 $\boxed{\blacktriangleright}$ $\boxed{)}$ $\boxed{\div}$ $\boxed{(}$ 0.0337 $\boxed{\div}$ 12 $\boxed{)}$ $\boxed{\text{ENTER}}$

Answer Key

Chapter 1   Critical Thinking and Problem Solving

Section 1.1

1. Answers will vary. There are seven days in a week, so if the first Tuesday of the month was the 2nd (an even day), then seven days later that Tuesday's date would be the 9th (an odd day). Therefore, not every Tuesday in the month is an even day.

3. Answers will vary. $5 - 3 = 2$. The difference between any two odd numbers is always an even number. Therefore, if the difference between two numbers is even, the numbers do not have to both be even.

5. Answers will vary. If $a = 5$, $b = 1$, and $c = 2$ then $a > b$ and $a > c$, but $b < c$.

7. 17, 21, 25; arithmetic; common difference = 4

9. 80, 160, 320; geometric; common ratio = 2

11. 23, 30, 38; neither

13. Geometric; common ratio = 2. (A figure with 24 sides)

15. Deductive reasoning

17. Deductive reasoning

19. Deductive reasoning

21. Inductive reasoning

23. a. 1,333,332   **b.** 1,333,332   **c.** Inductive reasoning

25. a. 31 push-ups   **b.** $2n + 1$ push-ups   **c.** Answers will vary. After day 10, Jessica will be doing more push-ups per day.

Section 1.2

1.

16	2	3	13
5	11	10	8
9	7	6	12
4	14	15	1

3. Three 41¢ stamps and six 8¢ stamps

5. 20

7. 607.5

9. 41 and 22

11. 140 total games

13. 609 students

15. 4.5 ounces

17.
```
    1
  5 6
3 4 2
```

19. Answers will vary. For example:
```
    2            1            2
3 1 4        2 3 4        1 5 4
  5            5            3
```

21. Apples = $0.14, pears = $0.18

23. a. $72 \cdot 1$; $36 \cdot 2$; $24 \cdot 3$; $18 \cdot 4$; $9 \cdot 8$; $8 \cdot 9$; $4 \cdot 18$; $3 \cdot 24$; $2 \cdot 36$; $1 \cdot 72$ **b.** $9 \cdot 8$ or $8 \cdot 9$

25. 62,750

27. a.
```
      1 7 21 35 35 21 7 1
     1 8 28 56 70 56 28 8 1
  1 9 36 84 126 126 84 36 9 1
```

b. The sums of the first six rows, in order, are 1, 2, 4, 8, 16, and 32   **c.** Sum of seventh row = 64; sum of eighth row = 128; sum of ninth row = 256; The sum is equal to two multiplied by the sum of the previous row.

29. $21 + 23 + 25 + 27 + 29 = 125 = 5^3$;
$31 + 33 + 35 + 37 + 39 + 41 = 216 = 6^3$;
$43 + 45 + 47 + 49 + 51 + 53 + 55 = 343 = 7^3$

Section 1.3

1. 820

3. 12,000

5. 100,000

7. Answers will vary. For example, $50 \cdot 30 = 1500$

9. Answers will vary. For example, $5000 \cdot 400 = 2,000,000$

11. Answers will vary. For example, $120/hr \cdot 100$ hrs $+ 2(\$90/hr \cdot 50$ hrs$) + 3(\$120/hr \cdot 100$ hrs$) = \$57,000$

13. a. Approximately $4   **b.** 1968   **c.** Approximate time periods are 1968–1975; 1978–1990; 1991–1996; 1997–2007

15. Answers will vary. Bills come to about $180 + $140 + $180 = $500 per month. So leftover income will be about $1700 - $500 = $1200.

17. Answers will vary.
$2.5 \div 2 + 100/20 + 30/4 = 13.75$ hours

19. Answers will vary. $(\$25,000 - \$7500) \div 4 = \$4375$

21. b. $432

23. a. $2500   **b.** $3600   **c.** $6500   **d.** 2009

Chapter 1 Exercises

1. Answers will vary. Most Super Bowls have been played in January. For example, Super Bowl XXXVII was played on January 26, 2003.

3. Answers will vary. The sum of two odd numbers is always even, so if the sum is odd, then the numbers cannot both be odd. For example, $4 + 5 = 9$, and 4 is even, not odd.

5. 13, 16, 19; arithmetic; common difference = 3

7. 108, 216, 648; neither

9. $\dfrac{1}{81}, \dfrac{1}{243}, \dfrac{1}{729}$; geometric; common difference = $\dfrac{1}{3}$

11. $12{,}345{,}678 \cdot 8 + 8 = 98{,}765{,}432$

13. 11 kittens and 12 parakeets

15. 12

17. $195

19. a. $1 \cdot 120$; $2 \cdot 60$; $3 \cdot 40$; $4 \cdot 30$; $5 \cdot 24$; $6 \cdot 20$; $8 \cdot 15$; $10 \cdot 12$; $12 \cdot 10$; $15 \cdot 8$; $20 \cdot 6$; $24 \cdot 5$; $30 \cdot 4$; $40 \cdot 3$; $60 \cdot 2$; $120 \cdot 1$ **b.** $10 \cdot 12$ or $12 \cdot 10$

21. 4; 7; 10; 13; 49

23. About 8

Chapter 2   Set Theory

Section 2.1

1. False; the set may be an infinite set.

3. True

5. True

7. True

9. True

11. $B = \{$Nebraska, Nevada, New Hampshire, New Jersey, New Mexico, New York, North Dakota, North Carolina$\}$

13. Answers will vary.

15. $F = \{$Monday, Tuesday, Wednesday, Thursday, Friday$\}$

17. Answers will vary.

19. $H = \{x \mid x \in \mathbb{N}, x \le 50\}$

21. $K = \{x \mid x \in U, x \text{ is an athlete}\}$

23. K is the set of US paper currency that is at most $100.

25. N is the set of months beginning with the letter J.

27. $B = \{3, 6, 9, 12, 15\}$

29. $D = \{$Wyoming, Nebraska, Kansas, New Mexico, Oklahoma, Utah, Arizona$\}$

31. $Y = \{$CHN, SWE, ITA, FRA, GER, GBR, RUS, GDR$\}$

33. $G = \{$CHN, ITA, FRA, GER, GBR, URS, USA$\}$

35. No; they have a different cardinal numbers, that is, a different number of elements.

37. Yes; they have the same elements, just in a different order.

39. Yes; P, R, and S are all equivalent. They all contain 5 elements. Q is not equivalent to any of them.

41. No; they have different elements.

43. B is equivalent to C since they each have 5 elements. A and D are not equivalent to the other sets.

45. $|X| = 9$

47. $|Y| = 43$, as of printing

49. $A' = \{$c, d, f, g, h, i, j, l, m, n, o, p, q, r, u, v, w, x, y, z$\}$

51. $A' = \{$c, d, f, g, h, i, j, l, m, n, o, p, q, r, u, v, w, x, y, z, A, B, C, D, E, F, G, H, I, J, K, L, M, N, O, P, Q, R, S, T, U, V, W, X, Y, Z$\}$

53. 4

55. 6

57. 15

59. Answers will vary. Examples may include $U = \{\pi\}$; $U = \mathbb{R}$; $U = \{x \mid x \in \mathbb{R}, x > 0\}$

Section 2.2

1. $A = \{$Greg, Anna, Bethany, Clair, Oscar$\}$; $B = \{$Alex, Georgina, Charles, Olivia$\}$

3. $U = \{$Greg, Anna, Bethany, Clair, Oscar, Alex, Georgina, Charles, Olivia, Richard, Karl, Rhonda, Matthew$\}$

5.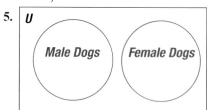

7.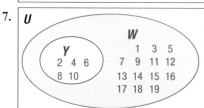

9. Yes

11. Red is not a set, so it cannot be a subset.

13. $\varnothing$, $\{Avatar\}$, $\{Titanic\}$, $\{Marvel's\ The\ Avengers\}$, $\{Avatar, Titanic\}$, $\{Avatar, Marvel's\ The\ Avengers\}$, $\{Titanic, Marvel's\ The\ Avengers\}$

15. True. The empty set has $2^0 - 1 = 0$ proper subsets, which is an even number.

17. Subsets of W: $2^5 = 32$; proper subsets of W: $2^5 - 1 = 31$

19. Proper subsets of Y: $2^{12} = 4096$

21. 7

23. $2^{12} = 4096$ (Some may like a no-topping pizza, also known as bread.) It is possible to qualify for the lifetime offer in less than 12 years if you order one pizza a day.

25.

# of Elements	Proper Subsets	# of Proper Subsets	# of Subsets
0	None	0	1
1	∅	1	2
2	∅, {1}, {2}	3	4
3	∅, {1}, {2}, {3}, {1, 2}, {2, 3}, {1,3}	7	8
4	∅, {1}, {2}, {3}, {4}, {1, 2}, {1, 3}, {1,4}, {2, 3}, {2, 4}, {3, 4}, {1, 2, 3}, {1, 2, 4}, {1, 3, 4}, {2, 3, 4}	15	16
n	All possible subsets except {1, 2, 3, . . . , $n-1$, n}	$2^n - 1$	2^n

Section 2.3

1. $\{1, 2, 3, 4, 5, 6, 7, 8, 10, 12\}$

3. $\varnothing$ or $\{\ \}$

5. $(A \cup B)' = \{9, 11, 13, 14, 15, 16, 17, 18, 19, 20\}$ and $A' = \{9, 10, 11, 12, 13, 14, 15, 16, 17, 18, 19, 20\}$ and $B' = \{1, 3, 5, 7, 9, 11, 13, 14, 15, 16, 17, 18, 19, 20\}$. So $A' \cap B' = \{9, 11, 13, 14, 15, 16, 17, 18, 19, 20\}$.

7. $\{n, u, m, b, e, r, s, l\}$

9. 3

11. $A \cap B = \{r, u, e\}$, $(A \cap B)' = \{a, b, c, d, f, g, h, i, j, k, l, m, n, o, p, q, s, t, v, w, x, y, z\}$, $A' = \{a, c, d, f, g, h, i, j, k, l, o, p, q, t, v, w, x, y, z\}$. $B' = \{a, b, c, d, f, g, h, i, j, k, m, n, o, p, q, s, t, v, w, x, y, z\}$, so $A' \cup B' = \{a, b, c, d, f, g, h, i, j, k, l, m, n, o, p, q, s, t, v, w, x, y, z\}$.

13. $\{C, E\}$

15. $A \cup B = \{I, C, E, U, B\}$, $(A \cup B)' = \{A, D, F, G, H, J, K, L, M, N, O, P, Q, R, S, T, V, W, X, Y, Z\}$, $A' = \{A, B, D, F, G, H, J, K, L, M, N, O, P, Q, R, S, T, U, V, W, X, Y, Z\}$, $B' = \{A, D, F, G, H, I, J, K, L, M, N, O, P, Q, R, S, T, V, W, X, Y, Z\}$, and $A' \cap B' = \{A, D, F, G, H, J, K, L, M, N, O, P, Q, R, S, T, V, W, X, Y, Z\}$

17. $\{A, C, D, F, O, P, R, T, U\}$

19. 4

21. $A \cap B = \{C, O, R, T\}$, $(A \cap B)' = \{A, B, D, E, F, G, H, I, J, K, L, M, N, P, Q, S, U, V, W, X, Y, Z\}$, $A' = \{B, D, E, G, H, I, J, K, L, M, N, P, Q, S, U, V, W, X, Y, Z\}$, $B' = \{A, B, E, F, G, H, I, J, K, L, M, N, Q, S, V, W, X, Y, Z\}$, and $A' \cup B' = \{A, B, D, E, F, G, H, I, J, K, L, M, N, P, Q, S, U, V, W, X, Y, Z\}$

23. $2^{26} = 67,108,864$

25. $\{g, a, t, o, r, b, i, e\}$

27. $M \cup K = \{b, r, i, d, g, e, a, l, s, t\}$, $(M \cup K)' = \{c, f, h, j, k, m, n, o, p, q, u, v, w, x, y, z\}$, $M' = \{a, c, f, h, j, k, l, m, n, o, p, q, s, t, u, v, w, x, y, z\}$, $K' = \{c, d, f, g, h, j, k, m, n, o, p, q, u, v, w, x, y, z\}$, and $M' \cap K' = \{c, f, h, j, k, m, n, o, p, q, u, v, w, x, y, z\}$.

29. 170

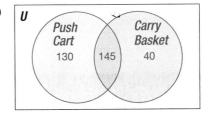

31. Will, David, Kim, Barbara, Alden, Morgan, Ali, Holly, Jessica, Jeff, Kent

33. 11

35. 39

37.

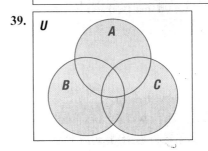

39.

41. $A \cap B'$

Section 2.4

1. $\{1, 2, 3, 4, 5, 6, 7, 8, 9, 10, 11, 12, 13, 15\}$

3. $\{5, 7\}$

5.

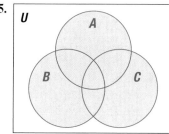

7.

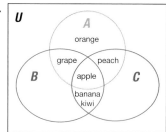

9.

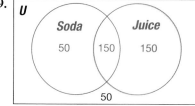

11.

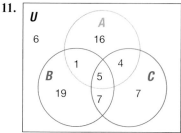

13. 20

15. 3

17. a. 62 **b.** 17 **c.** 5

19. a. 76.6% **b.** 21.7%

Chapter 2 Exercises

1. False; the set containing 3 is not an element of the set.

3. True

5. True

7. False; the cardinal number of the empty set is 0.

9. $A = \{2, 4, 6, 8, 10, 12\}$

11. $C = \{x \mid x \in \mathbb{R}, 100 < x < 1000\}$

13. $|A| = 2, |B| = 3$

15. Proper subsets of $A = \varnothing, \{\text{Felix}\}, \{\text{Amber}\}$

17. No, A has 2 elements and B has 3 elements.

19. Subsets of $G = \varnothing, \{I\}, \{II\}, \{III\}, \{I,II\}, \{I,III\}, \{II, III\}, \{I, II, III\}$, Subsets of $F = \varnothing, \{\text{love}\}, \{\text{joy}\}, \{\text{peace}\}, \{\text{love, joy}\}, \{\text{love, peace}\}, \{\text{joy, peace}\}, \{\text{love, joy, peace}\}$

21. No, they contain different elements.

23. 255

25.

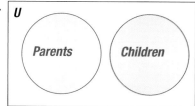

Universal set will vary.

27.

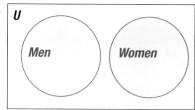

Universal set will vary.

29. $\{1, 3, 5\}$

31. $\{7\}$

33. 3

35. {m, e, d, i, c, a, l, n, k}

37. {n, k}

39. {b, f, h, j, o, p, q, r, s, v, w, x, y, z}

41. {a, b, c, d, e, f, g, h, i, j, k, l, m, n, o, p, q, r, s, t, u, v, w, x, y, z}

43. $|A \cap B| = 2$

45.

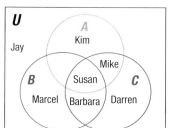

47. a.

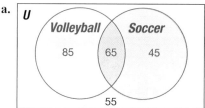

b. 85 **c.** 45 **d.** 55

49. 33%

Chapter 3   Logic

Section 3.1

1. No

3. No

5. No

7. Yes

9. Yes

11. No

13. Driving makes me smile or it is sunny.

15. If it is sunny, then I grill more often than I bake.

17. My video reached 1000 views on YouTube and I have 1000 Facebook friends.

19. I have 1000 Facebook friends or my video reached 1000 views on YouTube.

21. My peddling faster is sufficient for the wheels to turn more slowly.

23. 50% of the class being female is necessary for 50% of the class to be male.

25. Kelsey's website did not have more than 50,000 visits yesterday.

27. I got the job.

29. At least one of the Christmas tree lights is not working.

31. $\sim(m \wedge n)$

33. $a \Rightarrow \sim b$

35. $t \Rightarrow \sim r$

37. $w \Rightarrow z$

39. The moon is full and I've lost my glasses.

41. It is not true that I don't know if it's cloudy or bright outside and I've lost my glasses.

43. If I have not lost my glasses, then I do know if it's cloudy or bright outside.

Section 3.2

1.

Truth Table			
a	b	$\sim b$	$a \wedge \sim b$
T	T	F	F
T	F	T	T
F	T	F	F
F	F	T	F

3.

Truth Table			
c	d	$\sim d$	$c \Rightarrow \sim d$
T	T	F	F
T	F	T	T
F	T	F	T
F	F	T	T

5.

Truth Table				
p	r	q	$p \vee r$	$(p \vee r) \Rightarrow q$
T	T	T	T	T
T	T	F	T	F
T	F	T	T	T
T	F	F	T	F
F	T	T	T	T
F	T	F	T	F
F	F	T	F	T
F	F	F	F	T

7. Tautology

9. Not a tautology

11. p: We will buy a new smart phone.; q: We will buy a new computer.; $p \vee q$

p	q	$p \vee q$
T	T	T
T	F	T
F	T	T
F	F	F

13. r: Ella is on the dance team.; s: Coleman is on the soccer team.; $\sim r \wedge s$

r	s	$\sim r$	$\sim r \wedge s$
T	T	F	F
T	F	F	F
F	T	T	T
F	F	T	F

15. a: Meg passes.; b: Meg will lose her scholarship.; c: Meg will drop out of school.; $\sim a \Rightarrow (b \wedge c)$

a	b	c	$\sim a$	$b \wedge c$	$\sim a \Rightarrow (b \wedge c)$
T	T	T	F	T	T
T	T	F	F	F	T
T	F	T	F	F	T
T	F	F	F	F	T
F	T	T	T	T	T
F	T	F	T	F	F
F	F	T	T	F	F
F	F	F	T	F	F

Section 3.3

1. Converse: If I get popcorn, then I will go to the movies. Inverse: If I do not go to the movies, then I will not get popcorn. Contrapositive: If I do not get popcorn, then I will not go to the movies. Biconditional: I will go to the movies if and only if I get popcorn.

3. Converse: If it is dark, then I got out of my class. Inverse: If I do not get out of my class, then it is not dark. Contrapositive: If it is not dark, then I did not get out of my class. Biconditional: I get out of my class if and only if it is dark outside.

5. $\sim p \Rightarrow \sim q$

7. $\sim q \Rightarrow p$

9. $p \Rightarrow \sim q$ and $\sim(p \wedge q)$ are logically equivalent.

						Truth Table
p	q	$\sim q$	$p \Rightarrow \sim q$	$\sim p \wedge q$	$\sim(p \wedge q)$	$(p \Rightarrow \sim q) \Leftrightarrow \sim(p \wedge q)$
T	T	F	F	T	F	T
T	F	T	T	F	T	T
F	T	F	T	F	T	T
F	F	T	T	F	T	T

11. $\sim p \wedge \sim q$ and $\sim(p \vee q)$ are logically equivalent.

						Truth Table
p	q	$\sim p$	$\sim q$	$\sim p \wedge \sim q$	$p \vee q$	$\sim(p \vee q)$	$(\sim p \wedge \sim q) \Leftrightarrow \sim(p \vee q)$
T	T	F	F	F	T	F	T
T	F	F	T	F	T	F	T
F	T	T	F	F	T	F	T
F	F	T	T	T	F	T	T

13. Not logically equivalent, since **b.** is the converse of **a.**

15. Logically equivalent

p	q	$\sim p$	$\sim q$	$\sim p \vee q$	$p \wedge \sim q$	$\sim(\sim p \vee q)$
T	T	F	F	T	F	F
T	F	F	T	F	T	T
F	T	T	F	T	F	F
F	F	T	T	T	F	F

17. Not logically equivalent, since **b.** is the inverse of **a.**

19.

p	q	$\sim p$	$\sim q$	$p \wedge q$	$\sim(p \wedge q)$	$\sim p \vee \sim q$
T	T	F	F	T	F	F
T	F	F	T	F	T	T
F	T	T	F	F	T	T
F	F	T	T	F	T	T

21. $p \vee \sim q$

23. $\sim(p \vee \sim q)$

25. It is not true that there is not a space available in the 8:00 a.m. Biology class and I am able to make the perfect schedule for next semester.

27. Brooke comes to my room before 2:00 p.m., and I finish my homework before 10:00 p.m.

29. I have a ticket from the Sunday paper, and I do not get a free ice cream at Dairy Dip.

31. $\sim p \wedge \sim a$

33. $(c \wedge d) \wedge \sim b$

Section 3.4

1. Conclusion

3. Premises

5. False dilemma

7. Argumentum ad populum

9. Petitio principii

11. Dicto simpliciter

13. Inductive, invalid

15. Inductive, invalid

17. Deductive, valid

19. Valid

21. Invalid

23. Valid

25. Missing piece: The hand sanitizer was bigger than 3.4 fluid ounces.

27. Missing piece: My guitar is tuned to an open D tuning.

29. Missing piece: Emma is over 16.

31. Missing piece: You did not buy a new car.

33. Premise: We stop burning fossil fuels today. Conclusion: There is enough carbon dioxide in the atmosphere that temperatures will continue to rise for a few hundred years.

35. Premise: Fast food is easily available in grocery shops, gas stations, and dispensers everywhere. Conclusion: Fast food obesity has strikingly increased.

37. Premise: A man is struck down by a heart attack in the street. Conclusion: Americans will care for him whether or not he has insurance.

39. Premises: Penguins are black and white.; Some old TV shows are black and white. Conclusion: Some penguins are old TV shows.

41. Premises: All potatoes have skin; I have skin. Conclusion: I must be a potato.

43. Fallacy: Argumentum ad populum. The spokesman implies that because the candidate is doing the same thing as others, he is doing the right thing. But this may not be the case.

45. Straw man

47. Non sequitur

Chapter 3 Exercises

1. The puppy could keep her eyes open after 10:00.

3. Not all houses have fireplaces (or at least one house does not have a fireplace).

5.

a	b	c	~b	a ∨ ~b	(a ∨ ~b) ⇒ c
T	T	T	F	T	T
T	T	F	F	T	F
T	F	T	T	T	T
T	F	F	T	T	F
F	T	T	F	F	T
F	T	F	F	F	T
F	F	T	T	T	T
F	F	F	T	T	F

7. i: I go to the movies. y: You go to the movies. k: Kathy goes to the movies.

y	k	i	y ∧ k	(y ∧ k) ⇒ i
T	T	T	T	T
T	T	F	T	F
T	F	T	F	T
T	F	F	F	T
F	T	T	F	T
F	T	F	F	T
F	F	T	F	T
F	F	F	F	T

9. Converse: If I enroll in the next course, then my grade in this course will be an A. Inverse: If my grade in this course is not an A, then I cannot enroll in the next course. Contrapositive: If I cannot enroll in the next course, then my grade in this course was not an A. Biconditional: My grade in this course will be an A if and only if I can enroll in the next course.

11.

p	q	p ⇒ q	p ∧ (p ⇒ q)	(p ∧ (p ⇒ q)) ⇒ q
T	T	T	T	T
T	F	F	F	T
F	T	T	F	T
F	F	T	F	T

13. a.

15. Valid argument

17. Valid argument

19. Premise: Criminals support global warming; Conclusion: If you support global warming, you are a criminal

21. False dilemma

Chapter 4 Rates, Ratios, Proportions, and Percentages

Section 4.1

1. $\dfrac{15¢}{1 \text{ min}}$

3. $\dfrac{\$1473.00}{1 \text{ month}}$

5. $\dfrac{65 \text{ hours}}{3 \text{ weeks}}$

7. a.

9. $0.08/mile or 8.04¢/mile

11. 25.17 mpg

13. $0.75/pound or 74.75¢/pound

15. $1.25/eggplant; 0.80 eggplants/$1

17. The 12 batteries for $14.76 is a better buy since each battery costs $1.23 whereas the 3 batteries for $4.80 has a cost per battery of $1.60.

19. a.

21. Sugar = $1\dfrac{1}{2}$ cups, half-and-half = $4\dfrac{1}{2}$ tablespoons

23. $3387

Section 4.2

1. 96

3. a. 10   **b.** 14

5. a. 4.5   **b.** 7.5

7. $\dfrac{1}{2}$ or $1:2$ or 1 to 2

9. $\dfrac{18 \text{ students}}{1.3 \text{ faculty}}$ or 18 students : 1.3 faculty or 18 students to 1.3 faculty

11. 60 locals

13. 1092 uninsured cars

15. 8 ounces

17. 10,400 rodents

19. 12 m

21. $7\dfrac{1}{2}$ cups sifted all-purpose flour, $3\dfrac{3}{4}$ teaspoons baking powder, $1\dfrac{1}{4}$ teaspoons salt, $2\dfrac{1}{2}$ cups white sugar, $2\dfrac{1}{2}$ cups butter, $2\dfrac{1}{2}$ eggs, $7\dfrac{1}{2}$ tablespoons half-and-half, 5 teaspoons vanilla extract

23. 20 wins

25. 9.6 hours per week

27. a. $\dfrac{4.875 \text{ inches}}{870,000 \text{ miles}}$ **b.** 66,923.1 miles

29. 130 points

31. 315 grams

Section 4.3

1. $\dfrac{1}{3}$ are not seniors

3. $\dfrac{4}{5}$ American women have borne a child by the end of their childbearing years

5. 44.44%

7. 66.67%

9. 480

11. a.

13. $\dfrac{51 \text{ female participants}}{49 \text{ male participants}}$

15. one-fifth of 95

17. 170

19. 564 are female

21. 246 nonherbivores

23. 501 students do not live in the county

Section 4.4

1. $3.60

3. $1.43

5. $8.81

7. $109.49

9. $79.52

11. $345.60

13. $8340

15. $95,400

17. $54.36

19. 56; 400% increase

21. 323; 68% decrease

23. 19.22; 31% decrease

25. 20% decrease

27. 37.04 million

29. Company Z has the better absolute change but both companies have the same percent growth.

Chapter 4 Exercises

1. $\dfrac{14}{26} = \dfrac{7}{13}$

3. 360 people

5. a. 40 calories   **b.** 420 calories

7. $11.70

9. $117,600

11. 25%

13. $\dfrac{42 \text{ students}}{3 \text{ advisors}} = \dfrac{14 \text{ students}}{1 \text{ advisors}}$

15. $5.99 for a package of 8, since that is about $0.75 each instead of $1.43 each.

17. Male: 36; female: 45

19. $3\dfrac{3}{4}$ cups water, 5 tablespoons butter, $3\dfrac{3}{4}$ cups milk, 5 cups potato flakes, $1\dfrac{1}{4}$ teaspoons salt

Chapter 5   The Mathematics of Growth

Section 5.1

1. $(-2, -6), (-1, -3), (0, 0), (1, 3), (2, 6)$

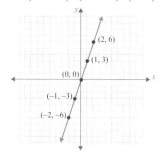

3. $(-2, 7), (-1, 5), (0, 3), (1, 1), (2, -1)$

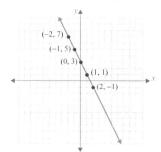

5. $(-2, 2), (-1, -1), (0, -2), (1, -1), (2, 2)$

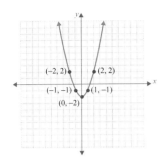

7. $(-2, 9), (-1, 4), (0, 1), (1, 0), (2, 1)$

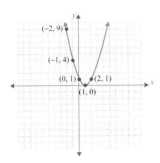

9. $(-2, -1), (-1, -3), (0, -7), (1, -13), (2, -21)$

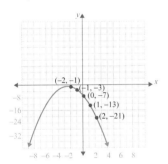

11. $f(x) = x^2 - 3x + 1$

x	$f(x) = x^2 - 3x + 1$
−3	19
−2	11
−1	5
0	1
1	−1
2	−1
3	1

13. $g(x) = (2x - 3)^2$

x	$g(x) = (2x - 3)^2$
−3	81
−2	49
−1	25
0	9
1	1
2	1
3	9

15. $f(x) = 1500 - 150x$ for $x \le 10$; $750

17. $f(x) = 15{,}000 + 575x$ for $x \ge 0$; $18{,}450

19. 4 liters

21. 2 atm

23. 4 meters

Section 5.2

1. $m = 2$; y-intercept: $(0, -7)$

3. $m = -4$; y-intercept: $(0, 0)$

5. $m = -1$; y-intercept: $(0, -7)$

7. $m = -\dfrac{1}{2}$; y-intercept: $\left(0, -\dfrac{5}{4}\right)$

9. $m = 1$; y-intercept: $(0, -4)$

11. $p(h) = 12h$, $h \ge 0$

13. $P(x) = 600x - 450x = 150x$, $x \ge 0$

15. $C(w) = 17.50w + 25$, $w \ge 0$

17. $p(n) = 25n + 1100$, $n \ge 0$; $1187.50

19. $C(m) = 275m + 500$, $m \ge 0$; 40 months

21. $T(x) = 850x + 4750$, $x \ge 0$; $6450

Section 5.3

1.

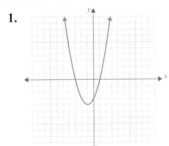

3.

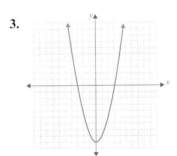

5.

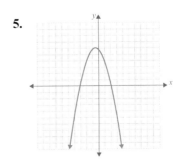

7.

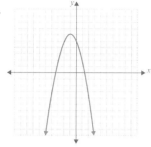

9. 117 feet; 42 feet

11. December of 2094

13. $h(t) = -16t^2 + 1100t + 200$

15. $t = 68.93$ seconds

17. 47.27 feet; 3.44 seconds

19. 8.84 seconds

Section 5.4

1. The growth of an exponential function is proportional to the previous growth while the growth of a linear function is constant.

3. 1,048,576 bacteria

5. $3.61

7. 1211 people

9. 8192 grains

11. 15,728.64 grams

13. Rather get the pennies

Section 5.5

1. $b = 2$

3. $x = 3$

5. $b = 216$

7. $x = 0$

9. $x = 2$

11. 78%

13. 60%

15. About 43.7 months

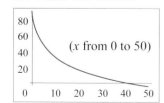

17. pH = 8.89; basic

19. $[H^+] = 10^{-11}$

21. 26 decibels

23. Yes; 127 dB

25. No; 2.99 dB

27. 74%

29. 37 years old

Chapter 5 Exercises

1.

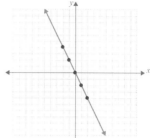

3.

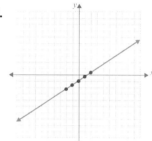

5.

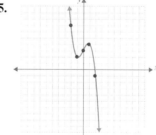

7. Independent variable = years; dependent variable = value of car; $y = -1200x + 19,500$ for $0 \le x \le 16.25$; $13,500

9. Independent variable = number of pizzas delivered; dependent variable = amount of money earned; $y = 0.25x + 31.25$ for $x \ge 0$; $36.50

11. a. 19 miles per gallon   **b.** 30 miles per gallon
c. About 44 miles per gallon

13. $m = \dfrac{3}{4}$; y-intercept $= (0, -3)$

15. $y = 8x - 12,000,000$ for $x \ge 0$

17. $y = 115x + 1500$ for $x \ge 0$; $2075

19. 59°F

21. a. About 24,367 people   **b.** May 2220

23. About 6.85 seconds

25. a. Answers will vary.   **b.** Answers will vary.
c. Graph in part **b.** It will be exponential decay.

27. a. 3,587,707,402   **b.** May 2010   **c.** April 2011
d. Answers will vary depending on the current year.

Chapter 6   Geometry

Section 6.1

1. Right

3. Obtuse

5. Acute

7. $18°$

9. $r = 4$

11. $47°$ and $43°$

13. $84°$ and $96°$

15. $x = 4$

17. $87°$

19. $38°$

21. $121.109°$

23. $41°13'12''$

25. $13°51'36''$

27. 177 miles

29. $AC = 27.7$; $BC = 13.9$

31. 6.23

33. 28.17

35. 25.65

37. 5.50 miles

39. 40.73 miles

41. 0.67 mile

43. 4.92 miles

Section 6.2

1. 320 m

3. 109.2 inches

5. 5

7. $60°, 60°, 120°, 120°$

9. $P = 69.28$ ft; $A = 246$ ft^2

11. $C = 18.84$ in.; $A = 28.26$ in.^{2}

13. $11.42

15. 18 m^2

17. 62.35 ft^2

19. $P = 68.05$ cm; $A = 253.30$ cm^2

21. $600

23. 77.04 cm^2

25. 37.68 in.^{2}

Section 6.3

1. $V = 96$ ft^3; $SA = 136$ ft^2

3. $V = 904.32$ in.^{3}; $SA = 452.16$ in.^{2}

5. $V = 144$ ft^3; $SA = 216$ ft^2

7. $V = 58.88$ in.^{3}

9. $V = 552.64$ m^3

11. $w = 15$ cm; $SA = 6050$ cm^2

13. a. $33,912$ ft^3 **b.** $27,129.6$ bushels   **c.** 169.56 min
(about 2.8 hours)

15. a. $SA = 195,967,400$ mi^2 **b.** $V = 258,023,743,300$ mi^3
c. $SA = 65,322,466.67$ mi^2

17. a. $SA = 65.81$ m^2;   **b.** Cost: $477.13

Chapter 6 Exercises

1. $13°$ and $77°$

3. $78°$ and $102°$

5. $211.103°$

7. $31°20'24''$

9. $14°33'36''$

11. $BC \approx 20.78$; $AC \approx 41.60$

13. 5.70 miles

15. 27.08 mi

17. 525 cm

19. a. $2340°$ **b.** $3600°$ **c.** $5040°$

21. $P = 56$ cm; $A = 84$ cm^2

23. $C = 15.7$ m; $A = 19.63$ m^2

25. $P = 237.25$ cm; $A = 2340.63$ cm^2

27. $P = 43.31$ ft; $A = 102.57$ ft^2

29. 14.13 m^2

31. 328.34 lb

33. 904.3 in.^{3}

35. a. 7.03 in.^{3} **b.** 57.5 in.^{2}

37. About 2 servings

Chapter 7   Probability

Section 7.1

1. $\{$H1, H2, H3, H4, H5, H6, T1, T2, T3, T4, T5, T6$\}$

3. $\{$BC, BG, BY, BR, CB, CG, CY, CR, GB, GC, GY, GR, YB, YC, YG, YR, RB, RC, RG, RY$\}$

5. Empirical

7. Classical

9. a. $\dfrac{8}{64} = 0.125$ **b.** $\dfrac{17}{64} = 0.265625$

c. $\dfrac{49}{64} = 0.765625$

11. $\dfrac{33}{84} \approx 0.392857$

13. $\dfrac{4}{16} = 0.25$

15. a. $\dfrac{2}{6} \approx 0.333333$ **b.** $\dfrac{3}{6} = 0.5$ **c.** $\dfrac{6}{6} = 1$

17. a. $\dfrac{14{,}143.0}{137{,}147.0} \approx 0.103123$

b. $\dfrac{2705 + 5217 + 14{,}341}{137{,}147} \approx 0.162329$

19. $\dfrac{11}{321} \approx 0.034268$

21. $\dfrac{64}{213} \approx 0.300469$

Section 7.2

1. $\{$GGG, GGB, GBG, GBB, BGG, BGB, BBG, BBB$\}$

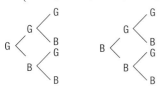

3. $\{$HHH, HHT, HTH, HTT, THH, THT, TTH, TTT$\}$

5. $10 \cdot 10 \cdot 10 = 1000$

7. $26 + 10 = 36$,
so $36 \cdot 36 \cdot 36 \cdot 36 \cdot 36 \cdot 36 = 2{,}176{,}782{,}336$

9. 144 possibilities

11. $3^{11} = 177{,}147$

13. $10^7 = 10{,}000{,}000$

15. $39 \cdot 10 \cdot 8 = 3120$

17. 362,880

19. 1

21. 75,600

23. 5940

25. 56

27. 5

29. 3

31. 1

33. 6

35. 1099

37. Combination; $_{120}C_5 = 190{,}578{,}024$ ways

39. Combination; $_{29}C_3 = 3654$ ways

41. Permutation; $_{18}P_3 = 4896$ ways

43. Permutation; $\dfrac{10!}{3!3!1!2!1!} = 50{,}400$ ways

45. c.

47. Answers will vary; should be
$(\#\text{ of pants}) \cdot (\#\text{ of shirts}) \cdot (\#\text{ of pairs of shoes})$

Section 7.3

1. $\dfrac{5}{5 \cdot 5} = \dfrac{5}{25} = 0.2$

3. a. $_{16}C_3 = 560$ **b.** $\dfrac{1}{560} \approx 0.001786$

5. a. $4! = 24$ **b.** $\dfrac{2 \cdot 3!}{24} = \dfrac{12}{24} = 0.5$

7. a. $_{52}C_5 = 2{,}598{,}960$ **b.** $\dfrac{4}{2{,}598{,}960} \approx 0.000002$

9. a. $10^4 = 10{,}000$ **b.** $_{10}P_4 = 10 \cdot 9 \cdot 8 \cdot 7 = 5040$

c. $\dfrac{1}{4!} = \dfrac{1}{24} \approx 0.041667$

d. $\dfrac{1}{3 \cdot \left(\dfrac{4!}{2!1!1!}\right)} = \dfrac{1}{36} \approx 0.027778$ **e.** Repeat a digit

11. $\dfrac{3}{1000} = 0.003$

13. $\dfrac{1}{26 \cdot 25 \cdot 24} = \dfrac{1}{15600} \approx 0.000064$

15. $\dfrac{_4C_2}{_6C_2} = \dfrac{6}{15} = 0.4$

17. $\{2, 4, 5, 7, 8, 10, 11\}$

19. A card with a face value of ace or 2 through 10 in any suit.

21. $1 - \dfrac{5}{26} \approx 0.807692$

23. $1 - \dfrac{(5,803.7 + 15,004.0)}{137,147.0} \approx 1 - 0.151718 \approx 0.848282$

25. $1 - \dfrac{6}{36} \approx 0.833333$

Section 7.4

1. a. 0.80 **b.** 0.20

3. $\dfrac{8}{11} \approx 0.727273$

5. $\dfrac{12}{46} \approx 0.260870$

7. 0.3

9. $\dfrac{24}{36} = \dfrac{2}{3} \approx 0.666667$

11. $\dfrac{16}{52} = \dfrac{4}{13} \approx 0.307692$

13. $\dfrac{9}{30} = 0.3$

15. $\dfrac{10}{19} \approx 0.526316$

17. These are not independent events. Because we are not allowed to repeat characters, choosing a character omits it from being chosen the next time.

19. Yes

21. 0.312

23. a. $\dfrac{8}{31} \approx 0.258065$ **b.** $\dfrac{3}{31} \cdot \dfrac{4}{30} \approx 0.012903$

25. $0.49^3 \approx 0.117649$

27. $\dfrac{2}{18} = \dfrac{1}{9} \approx 0.111111$

29. 0.385

31. $\dfrac{16}{34} = \dfrac{8}{17} \approx 0.470588$

33. $\dfrac{{}_{18}C_5}{{}_{20}C_5} = \dfrac{8568}{15,504} \approx 0.552632$

Section 7.5

1. Expected value $= 2.15$

3. Expected value $= 16$

5. 0.65 times per week

7. Expected winnings $= -\$0.03$

9. a. 1 ticket: $-\$8$; 3 tickets: $-\$19$; 5 tickets: $-\$30$
b. 1 ticket

11. 32.25%

13. a. 0 **b.** 0

15. $\dfrac{1}{7}$

17. $\dfrac{2}{5}$

19. $\dfrac{5}{7}$

21. a. 0.822 **b.** $\$80.24$

Chapter 7 Exercises

1. $\{$Bankrupt, Lose a Turn, Free Play, $\$250$, $\$300$, $\$350$, $\$400$, $\$500$, $\$700$, $\$750$, $\$800$, $\$850$, $\$900$, $\$1000\}$

3. a. $\dfrac{1}{24} \approx 0.041667$ **b.** $\dfrac{2}{24} \approx 0.083333$

c. $\dfrac{6}{24} = 0.25$ **d.** 0

5. a.

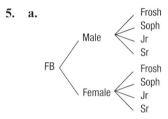
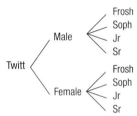
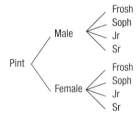

b. $5 \cdot 2 \cdot 5 = 50$

7. a. $\dfrac{1}{100,000} = 0.00001$ **b.** $\dfrac{1}{5!} = \dfrac{1}{120} \approx 0.008333$

c. $\dfrac{1}{4 \cdot {}_5C_4} = \dfrac{1}{20} = 0.05$

9. $\dfrac{6}{11} \approx 0.545455$

11. $\dfrac{50}{100} \cdot \dfrac{25}{100} = \dfrac{1}{8} = 0.125$

13. $\left(\dfrac{50}{100}\right)^2 + \left(\dfrac{10}{100}\right)^2 + \left(\dfrac{15}{100}\right)^2 + \left(\dfrac{25}{100}\right)^2 = \dfrac{69}{200} = 0.345$

15. $\dfrac{10}{100} \cdot \dfrac{9}{99} = \dfrac{1}{110} \approx 0.009091$

17. $\dfrac{5}{36} \approx 0.138889$

19. $\dfrac{18}{36} = 0.5$

21. $\dfrac{22}{52} \approx 0.423077$

23. 0

25. a. $2409.09 **b.** $4000

27. −$48

Chapter 8 Statistics

Section 8.1

1. Census

3. Population; Population parameters

5. Both begin with separating the population into groups, but cluster sampling surveys every member of certain randomly chosen groups, whereas stratified sampling takes a random sample from each group.

7. Population parameter

9. Population parameters based on estimates

11. Population: US real estate for sale; Sample: no information given; Population parameters: 9%, 3.5%, 3.4%, 1%

13. Population: higher education graduates in the United States; Sample: not given; Sample statistics: 4.8% and 6.7%

15. Stratified sampling

17. Convenience sampling

19. Stratified sampling

21. Stratified sample by district or other natural divisions in his area. Possible biases include age, race, and location. Answers will vary.

23. Stratified sampling of four groups: those with a family history of cardiovascular disease, those without cardiovascular disease history, those with family cancer history, and those without. Possible biases to consider are existing risk factors for cardiovascular disease and cancer in the women, as well as their lifestyle—eating habits, exercise habits, whether or not they smoke, if they live in a city or country, if they spend a lot of time indoors or outdoors, etc.—can affect the results. Answers will vary.

25. Answers will vary. No, it is not likely to include all of the American public, since it will not include the section of the public who do not have internet access, and the only people who would respond are people who frequent those news outlet sites. Because some news outlets may have a slight bias one way or the other, each may draw only a certain demographic.

27. Answers will vary. For example, income level, health of residents, happiness of residents, average annual temperature, number of activities in the area.

Section 8.2

1. Histogram

3. Line

5. Frequency distribution

7. a. 6 **b.** 1 kg **c.** 2.00 kg **d.** 4.99 kg
e. 89/194 or 45.9%

9. a.

Class	Freq.	b. Rel. Freq.
0–9	0	0
10–19	6	17.1%
20–29	8	22.9%
30–39	5	14.3%
40–49	4	11.4%
50–59	3	8.6%
60–69	5	14.3%
70–79	4	11.4%

c. 11.4% **d.** 0% **e.** 20–29 calls

11. a. 2 **b.** The western US

13. a. Generally decreasing with brief periods of increase

b. June 2011; 9.2% **c.** February 2013; 7.7%
d. Answers might include making it clearer that these are percentages and that the x-axis is not marked very clearly.

15. a. Approximately 6 hours **b.** Approximately 7 hours
c. There is essentially no difference in the number of hours wives spend sleeping. **d.** A pair of side-by-side bar graphs: one for husbands and one for wives.

Section 8.3

1. Mean: 24; median: 25; mode: 22, 27; range: 17; standard deviation: 4.8; bimodal

3. Mean: 34.3; median: 35; mode: 26; range: 25; standard deviation: 9.0; unimodal

5. Mean: −1.8; median: −3; mode: −7, −3, 3; range: 18; standard deviation: 5.9; multimodal

7. Mean: 147.3 minutes; median: 149 minutes; mode: no mode; range: 31 minutes; standard deviation: 9.8 minutes

9. 86

11. Mode

13. Mode

15. a. 95%   **b.** 16%   **c.** 50%

17. Since the mean is much larger than both the median and the mode, the data is likely skewed to the right.

19. No, it means that all the data points are equal to the mean.

21. Amelia scoring an 82 is just her raw score; it means she got 82% of the questions on the test correct. Amelia scoring in the 82nd percentile tells you that she scored at least as well as 82% of the people who took the test. It tells you more about her relative performance on the test.

23. a. Min: \$100,960; Q_1: \$120,900; Q_2: \$145,850; Q_3: 159,635; Max: \$182,500   **b.** 75%   **c.** \$81,540

25. Min: 310; Q_1: 510; Q_2: 705; Q_3: 1225; Max: 3700;

\$300	\$1550	\$2800	\$4050

27. Answers will vary. Possible examples: large variation: salaries of all employees of a Fortune 500 company; small variation: densities of water from different sources.

29. Yes, the five-number summary can be put back together in numerical order: 9, 11, 13.5, 17.5, 19. So, $Q_1 = 11$.

31. No, percentiles give the relative position in terms of percentages. Without knowing more information about the size of the sample, we cannot compute the size of the sample.

Section 8.4

1. a. $z = 0.55$ **b.** $z = -1.73$ **c.** $z = 0.09$

3. a. $z = -1$ **b.** $z = 0.5$ **c.** $z = -3.17$

5. -0.45

7. 0.41

9. Ava's score of 92

11.

	z	x	μ	σ
a.	1.29	82.1	74.0	6.3
b.	1.05	162.3	153.0	8.9
c.	3.04	49.8	34.5	5.02
d.	−2.73	379	634	93.4

13. 42.47%

15. 99.89%

17. 50.00%

19. 96.93%

21. 0.23%

23. 68.27%

25. 2.14%

27. 49.55%

29. 0.45%

31. 86.41%

33. a. 25.14%   **b.** 15.87%   **c.** 40.82%   **d.** 0.27%

35. 1.29

37. Q_1: −0.67; Q_2: 0; Q_3: 0.67

Section 8.5

1. Positive linear correlation

3. Negative linear correlation

5. Strong negative relationship

7. Weak positive relationship

9. No

11. Yes

13. a. Negative   **b.** $r = -0.538$ **c.** No

15. a. Positive   **b.** $r = 0.903$ **c.** Yes

17. a. 606.04   **b.** 886.14   **c.** 1138.23

19. a. $\hat{y} = 0.344x + 1.134$ **b.** Yes, $r = 0.728$ **c.** 3.886, or about 4 times

21. a. $\hat{y} = 0.429x + 0.4$ **b.** No, $r = 0.407$ **c.** N/A

Chapter 8 Exercises

1. Population: all US college freshmen; Sample: 203,967 freshmen who responded; Sample statistic: 27.6% describe themselves as "liberal;" Sample statistic: 20.7% describe themselves as "conservative;" Sample statistic: 47.4% describe themselves as "middle of the road."

3. Not necessarily; it is convenient, but may not be representative of the campus as a whole.

5. a. Answers will vary. For example, choose one store and survey people as they enter the store; choose one mall entrance and survey people as they enter the mall; set up a booth in the middle of the mall and ask people to participate as they walk by, etc.   **b.** Answers will vary.

7. a. About \$700 billion   **b.** About \$150 billion   **c.** About \$250 billion   **d.** Yes, you can compare total spending and categorical spending easily with one graph.

9. a. Answers will vary. For example, labels on axes, the number of participants, publisher, how data was recorded, when the data was recorded, etc.   **b.** Answers will vary. For example, it is unclear what the percentages represent. It is unclear whether one person was asked about each of the categories and responded with a yes or no, or whether the responses were open-ended and the percentages represent the number of people who included each response.

11. Mean = 1.22; median = 1; mode = 0; range = 6; stdev = 1.66; min = 0; Q_1 = 0; med =1; Q_3 = 2; max = 6

13. a. $z_1 = 6.25$ **b.** $z_2 = -1.5$ **c.** $z_3 = 12.5$

15. Hasef: 0.889; Kimberly: 0.885; Hasef's score was better

17. 38.59%

19. 99.93%

21. 12.10%

23. 99.94%

25. 95.45%

27. 13.59%

29. True; $z > 0$, so it is higher than the mean.

31. a. 40.13%   **b.** 45.03%   **c.** 14.84%

33. Positive linear correlation

35. No linear correlation

37. No relationship

39. Statistically significant

41. a. Positive   **b.** $r = 0.789$ **c.** Statistically significant

43. a. $\hat{y} = 0.638x + 28.902$ **b.** $r = 0.892$; statistically significant   **c.** 74

Chapter 9   Personal Finance

Section 9.1

1. $7200

3. Taxes are $182.24/week; max. car payment is $292.64/month

5. $3890/month

7. $11,700

9. $189.88

11. $117.06

13. $164,285.71

15. $312.50

17. $6125

Section 9.2

1. $340

3. $585

5. $1653.71

7. a. $5484.47   **b.** $1984.47

9. a. $18,325.20   **b.** $12,675.20

11. a. $65,758.59   **b.** $50,758.59

13. a. $387,717.95   **b.** $380,417.95

15. a. $111,624.25   **b.** $111,945.38

17. a. $6000   **b.** $7373.07   **c.** $7358.78

19. a. $1770.36   **b.** $1827.69   **c.** $1831.87

d. $1832.95   **e.** $1833.13

21. a. $3000   **b.** $6000   **c.** $12,000   **d.** $24,000

23. a. $46,970.43   **b.** $16,970.43

25. a. $187,881.72   **b.** $67,881.72

27.

First Bank of Lending Loan APR	
Loan Amount	APY
< $20,000	11.73%
$20,000–$99,000	9.30%
$100,000	5.88%

Section 9.3

1. $11,263.09

3. $34,233.00

5. $782.48

7. $773.91

9. a. $17.91   **b.** $39.39   **c.** $91.19

11. $81,466.12

13. $921.76

15. $978.28

17. a. $257.89   **b.** $46,420.20; I = $28,579.80

19. a. $1,474,258.37   **b.** Blake deposited $232,200 and made $1,242,058.37 in interest.

Section 9.4

1. a. $1240   **b.** $4464

3. 0.9% financing for 48 months will cost $29,492.16 total, and the cash back with 4.75% APR for 48 months will cost $30,752.16 total. The 0.9% financing is the best option.

5. a. $2599.60   **b.** $124,780.80   **c.** $10,780.80

7. a. $125.36   **b.** $3008.64   **c.** $358.64

9. a. $1266.71   **b.** $206,015.60   **c.** $456,015.60
d. $1912.48   **e.** $94,246.40   **f.** $344,246.40

11. The five-year loan with an APR of 7.5% ($400.76 monthly payment, $4045.60 in interest paid)

13. a. $331.15   **b.** $7947.60

15. a. $501.81   **b.** $6021.72

17. a. $168.89   **b.** $6080.04

19. 20.12 months, so 21 months, or about 1.7 years

21. a. $169,024.96   **b.** $247,502.21

Chapter 9 Exercises

1. $5812.50

3. $1916/month

5. $21,155.55

7. $10.09

9. $45.50

11. 9%

13. $1308.31

15. 442%

17. $94,200.47

19. a. 5.12%   **b.** 5.12%   **c.** 5.13%

21. a. $38,481.59   **b.** $8481.59

23. a. $153,926.38   **b.** $33,926.38

25.

Loan Amount	APY
< $20,000	10.11%
$20,000–$99.999	6.13%
$100,000	3.80%

27. a. $160.19   **b.** $385.27   **c.** $968.94

29. a. $330.61   **b.** $71,411.24; $I = $63,588.76

31. $237.08

33. $243,932.24

35. a. $127,628.30   **b.** $78.74   **c.** $45,946,188.00

37. a. $451.57   **b.** $5418.84

39. a. $206.92   **b.** $7449.12

41. a. $343.19   **b.** $510.09   **c.** $11,140.80

Chapter 10   Voting and Apportionment

Section 10.1

1. Plurality

3. Preference ballot

5. Majority rule

7.

	Rankings							
1st	S	N	O	J	O	N	S	S
2nd	O	O	S	S	N	L	L	J
3rd	J	J	J	L	L	S	O	L
4th	L	L	N	N	J	J	N	N
5th	N	S	L	O	S	O	J	O
# Votes	212	133	543	24	179	8	201	11

9.

	Rankings								
1st	M	N	D	S	M	S	M	S	S
2nd	A	D	S	M	A	A	A	D	A
3rd	D	S	M	A	N	M	N	M	M
4th	N	M	A	N	D	N	S	A	D
5th	S	A	N	D	S	D	D	N	N
# Votes	3	7	3	6	5	2	1	1	2

11. a. 720   **b.** 9194   **c.** *Nothing Compares to You* by Sinead O'Connor   **d.** 4598   **e.** No

13. 36

15. Ryan Braun: 388; Matt Kemp: 332; Prince Fielder: 229; Justin Upton: 214; Albert Pujols: 166; Joey Votto: 135; Lance Berkman: 118; Troy Tulowitzki: 69; Roy Halladay: 52; Ryan Howard: 39

17. a. Little   **b.** Little   **c.** Three-way tie between Little, Braugh, and Costa.   **d.** Little; no, same as in part **a.**

19. a. Knoxville   **b.** Knoxville; no, Knoxville had 33 first-place votes, but 50 are needed for a majority. **c.** Knoxville   **d.** Memphis   **e.** Answers will vary. For example, Borda count method, since it represents votes for a bus route, people may still use the route that they ranked second, third, and fourth.

21. Any number greater than 11

23. Any number of votes will satisfy the requirement that the favorite feature among customers is the voice-activated command using the plurality with elimination method.

25. For location to be the top influencer, the number of votes missing must be at least 171.

Section 10.2

1. Answers will vary.

3. a. Susan   **b.** Courtney   **c.** The Condorcet criterion does not apply because no candidate wins all head-to-head comparisons with every other candidate.

5. a. Mechanics   **b.** Heat and Optics   **c.** No, the Borda count method declares Heat and Optics the winner, but Mechanics has a majority of first-place votes, so the majority criterion is not satisfied.

7. a. New Orleans, LA   **b.** New Orleans, LA   **c.** Yes, New Orleans is the preferred destination both before and after Chicago was taken out of the selected options, so the irrelevant alternatives criterion is satisfied.

9. a. H. Kennedy **b.** T. Parchment

Rankings				
1st	H. Kennedy	T. Parchment	H. Kennedy	T. Parchment
2nd	T. Parchment	H. Kennedy	T. Parchment	H. Kennedy
# Votes	8	14	10	11

c. No, Kennedy was the winner before Jones pulled out of the election, but not after, so the Borda count method does not satisfy the alternative criterion here.

11. The majority criterion is satisfied because Johnny Depp wins using both the Borda count method or the majority method.

13. Any x such that x is less than or equal to 11

15. Any x between 42 and 100, inclusive

17. a. Blake **b.** The pairwise comparison method does not produce a winner. Luke, Lauren, and Blake all receive 2 points and therefore there is a tie. **c.** The Condorcet criterion does not apply here since no candidate wins every head-to-head comparison against the other candidates.

Section 10.3

1. $260 million

3. Los Angeles: $412,103,095.59; Merced: $29,893,833.00

5. a. 469.662 **b.** San Francisco: 38.624; Irvine: 59.617

7.

Campus	Electric Vehicles
Berkeley	71
Davis	69
Irvine	59
Los Angeles	79
Merced	6
Riverside	45
San Diego	55
San Francisco	39
Santa Barbara	45
Santa Cruz	32

9. a. SD = 21; SQ: I = 1.667; II = 2.048; III = 2.286 **b.** They should offer two sections of each course with about 21 students per section.

11. They should offer two sections of each course with about 21 students per section.

13. a. Northern: 5; Southern: 6; Eastern: 3; Western: 2 **b.** Northern: 5; Southern: 6; Eastern: 3; Western: 2 **c.** No, the apportionments are the same.

15. $13 + 3 + 3 + 4 + 2 = 25$

17. Campus 1: 20 police officers; campus 2: 18 police officers; campus 3: 2 police officers

19. A: 33 representatives; B: 138 representatives; C: 3 representatives; D: 42 representatives; E: 14 representatives; F: 20 representatives

21. a. History: 18 positions; liberal arts math: 141 positions; English: 38 positions **b.** History: 19 positions; liberal arts math: 140 positions; English: 38 positions **c.** No **d.** History: 19 positions; liberal arts math: 140 positions; English: 38 positions **e.** History: 19 positions; liberal arts math: 139 positions; English: 39 positions **f.** No **g.** History: 19 positions; liberal arts math: 140 positions; English: 38 positions **h.** History: 19 positions; liberal arts math: 139 positions; English: 39 positions **i.** No

Section 10.4

1. Answers will vary. For example, a voter with veto power in a coalition.

3. 120

5. 16

7. $[13: 9, 6, 5, 3, 2]$

9. a. 25 **b.** 2 **c.** No **d.** No

11. No dictator; no dummy player

13. a. 17 **b.** 32

15. 45

17. $\text{BPI}(P_1) = \dfrac{10}{26} \approx 0.385 = 38.5\%$;

$\text{BPI}(P_2) = \text{BPI}(P_3) = \text{BPI}(P_4) = \text{BPI}(P_5) = \dfrac{4}{26} \approx 0.154 = 15.4\%$; no dictators; none have veto power

19. a. When all permanent members and exactly 3 nonpermanent members come before the nonpermanent member. **b.** 0.000259% **c.** 0.00259% **d.** 0.0006475% **e.** 0.0032375%

21. a. $\langle P_2, P_1, P_3, P_4 \rangle$ **b.** $\langle P_1, P_2, P_3, P_4 \rangle$
$\langle P_2, P_1, P_4, P_3 \rangle$ $\langle P_1, P_2, P_4, P_3 \rangle$
$\langle P_3, P_1, P_2, P_4 \rangle$ $\langle P_3, P_4, P_2, P_1 \rangle$
$\langle P_3, P_1, P_4, P_2 \rangle$ $\langle P_4, P_3, P_2, P_1 \rangle$
$\langle P_2, P_3, P_1, P_4 \rangle$ $\langle P_4, P_1, P_2, P_3 \rangle$
$\langle P_3, P_2, P_1, P_4 \rangle$ $\langle P_1, P_4, P_2, P_3 \rangle$
$\langle P_2, P_4, P_1, P_3 \rangle$
$\langle P_4, P_2, P_1, P_3 \rangle$
$\langle P_3, P_4, P_1, P_2 \rangle$
$\langle P_4, P_3, P_1, P_2 \rangle$

c. P_1: 41.67%; P_2: 25%; P_3: 25%; P_4: 8.33%

d. P_1: 41.67%; P_2: 25%; P_3: 25%; P_4: 8.33% **e.** No

23. a. Winning coalitions: $\{S_1,S_2,S_3,S_4,S_5\}$: 500 votes;

$\{S_1,S_2,S_3,S_4\}$: 483 votes;

$\{S_1,S_2,S_3,S_5\}$: 460 votes;

$\{S_1,S_2,S_4,S_5\}$: 380 votes;

$\{S_1,S_3,S_4,S_5\}$: 377 votes;

$\{S_2,S_3,S_4,S_5\}$: 300 votes;

$\{S_1,S_2,S_3\}$: 443 votes;

$\{S_1,S_2,S_4\}$: 363 votes;

$\{S_1,S_2,S_5\}$: 340 votes;

$\{S_1,S_3,S_4\}$: 360 votes;

$\{S_1,S_3,S_5\}$: 337 votes;

$\{S_1,S_4,S_5\}$: 257 votes;

$\{S_2,S_3,S_4\}$: 283 votes;

$\{S_2,S_3,S_5\}$: 260 votes;

$\{S_1,S_2\}$: 323 votes;

$\{S_1,S_3\}$: 320 votes

Losing coalitions: $\{S_2,S_4,S_5\}$: 71 votes;

$\{S_3,S_4,S_5\}$: 74 votes;

$\{S_1,S_4\}$: 11 votes;

$\{S_1,S_5\}$: 34 votes;

$\{S_2,S_3\}$: 8 votes;

$\{S_2,S_4\}$: 88 votes;

$\{S_2,S_5\}$: 111 votes;

$\{S_3,S_4\}$: 91 votes;

$\{S_3,S_5\}$: 114 votes;

$\{S_4,S_5\}$: 194 votes;

$\{S_1\}$: 51 votes;

$\{S_2\}$: 128 votes;

$\{S_3\}$: 131 votes;

$\{S_4\}$: 211 votes;

$\{S_5\}$: 234 votes

b. 7 votes   **c.** 69 votes   **d.** 6; selling 7 votes will cause $\{S_1,S_4,S_5\}$ to lose

Chapter 10 Exercises

1.

					Rankings					
1st	H	P	S	S	K	P	P	J	S	K
2nd	K	J	H	H	J	H	J	P	H	P
3rd	P	K	J	K	P	S	S	H	J	J
4th	J	S	P	P	H	J	H	K	K	S
5th	S	H	K	J	S	K	K	S	P	H
# Votes	5	4	6	3	7	1	4	1	2	2

3. $\dfrac{7(6)}{2} = 21$

5. a. $5! = 120$ **b.** 793   **c.** 388   **d.** Candidate A
e. 397   **f.** No

7. a.

Harris; SGA	20
Icin; Social Work Club	73
Green; ΑΛΠ	86
Lawson; FCA	6
Albert; GSA	132
Roman; ΓΒΘ	89
Switzer; Galois Club	37
Belton; Hispanic Culture Center	10
Finley; History Club	38
Issen; NAEA	33
San Marie; NBS	19
Centus; NTSS	82
Dennis; ΩΨΦ	56
Russan; ΦΑ	130
Hunter; ΦΜΑ	45
Austin; ΣΓΡ	8
Molda; Student Design Group	92
Irene; Voices of Praise	88

b.

1	Albert; GSA	132
2	Russan; ΦΑ	130
3	Molda; Student Design Group	92
4	Roman; ΓΒΘ	89
5	Irene; Voices of Praise	88
6	Green; ΑΛΠ	86
7	Centus; NTSS	82
8	Icin; Social Work Club	73
9	Dennis; ΩΨΦ	56
10	Hunter; ΦΜΑ	45

9. a. Clinton; no   **b.** Clinton; yes
c. No; answers will vary. For example, although it is possible that the popular vote may have changed, the electoral vote would not (since Perot did not receive any electoral votes) and Clinton would still win the election.

11. a. *Django Unchained*   **b.** *Zero Dark Thirty*
c. No; the Condorcet criterion is not satisfied because the plurality method chose *Django Unchained* as the winner, but *Zero Dark Thirty* won the head-to-head comparisons against every other candidate.

13. The irrelevant alternatives criterion would be met using the plurality with elimination method because even if Jacksonville, FL were removed, the winner is still Boston, MA.

15. The Condorcet criterion is satisfied because outdoor living spaces wins using the plurality method and also wins all head-to-head match ups with the other categories.

17. The missing ranking for honesty is 2nd and the missing ranking for sense of humor is 1st.

19.

County	Officers
A	45
B	52
C	55
D	40
E	24
F	34

21. $62,816.18

23. Austin: $4,014,970,172; Dallas: $1,084,773,995

25. a. 183.515 students/vehicle   **b.** San Antonio: 126.409 vehicles; Permian Basin: 14.691 vehicles

27. Yes, the apportionments are different.

Campus	Allocation
Arlington	112
Austin	253
Brownsville	51
Dallas	68
El Paso	89
San Antonio	126
Tyler	26
Permian Basin	15
Pan American	85

29.

Course	Allocation
Phys Sci	2 Sections
Biology	1 Section
Anatomy	2 Sections

31.

Course	Allocation
Comp I	8 Grad Stud
Comp II	5 Grad Stud
Surv. of Lit	2 Grad Stud
Poetry	4 Grad Stud

33. a.

County	Allocation
A	5 Members
B	6 Members
C	7 Members
D	5 Members
E	3 Members
F	4 Members

b.

County	Allocation
A	5 Members
B	5 Members
C	6 Members
D	6 Members
E	4 Members
F	4 Members

c.

County	Allocation
A	5 Members
B	6 Members
C	7 Members
D	5 Members
E	3 Members
F	4 Members

d.

County	Allocation
A	5 Members
B	5 Members
C	6 Members
D	6 Members
E	4 Members
F	4 Members

e. No

35. No

37. No

39. 10

41. a. 25   **b.** 2   **c.** No   **d.** No

e. $\{P_1, P_2, P_3, P_4, P_5, P_6, P_7\}$,
$\{P_1, P_2, P_3, P_4, P_5, P_6\}$,
$\{P_1, P_2, P_3, P_4, P_5, P_7\}$,
$\{P_1, P_2, P_3, P_4, P_6, P_7\}$,
$\{P_1, P_2, P_3, P_5, P_6, P_7\}$,
$\{P_1, P_2, P_3, P_4, P_5\}$,
$\{P_1, P_2, P_3, P_4, P_6\}$

43. 28

45. a. $\langle P_2, P_1, P_3, P_4 \rangle$,   **b.** $\langle P_1, P_2, P_3, P_4 \rangle$,
$\langle P_2, P_1, P_4, P_3 \rangle$, $\langle P_1, P_2, P_4, P_3 \rangle$,
$\langle P_2, P_3, P_1, P_4 \rangle$, $\langle P_1, P_4, P_2, P_3 \rangle$,
$\langle P_2, P_4, P_1, P_3 \rangle$, $\langle P_3, P_4, P_2, P_1 \rangle$,
$\langle P_3, P_1, P_2, P_4 \rangle$, $\langle P_4, P_1, P_2, P_3 \rangle$,
$\langle P_3, P_1, P_4, P_2 \rangle$, $\langle P_4, P_3, P_2, P_1 \rangle$
$\langle P_3, P_2, P_1, P_4 \rangle$,
$\langle P_3, P_4, P_1, P_2 \rangle$,
$\langle P_4, P_2, P_1, P_3 \rangle$,
$\langle P_4, P_3, P_1, P_2 \rangle$

c. $\{P_1, P_2, P_3, P_4\}$,
$\{P_1, P_2, P_3\}$,
$\{P_1, P_2, P_4\}$,
$\{P_1, P_3, P_4\}$,
$\{P_2, P_3, P_4\}, \{P_1, P_2\}$,
$\{P_1, P_3\}$

d. $\text{SSPI}(P_1) = \dfrac{10}{24} = \dfrac{5}{12} \approx 0.417 = 41.7\%;$

  $\text{SSPI}(P_2) = \text{SSPI}(P_3) = \dfrac{6}{24} = \dfrac{1}{4} = 0.25 = 25\%;$

  $\text{SSPI}(P_4) = \dfrac{2}{24} = \dfrac{1}{12} \approx 0.083 = 8.3\%$

e. $\text{BPI}(P_1) = \dfrac{5}{12} \approx 0.417 = 41.7\%;$ **f.** No

  $\text{BPI}(P_2) = \text{BPI}(P_3) = \dfrac{3}{12} = 0.25 = 25\%;$

  $\text{BPI}(P_4) = \dfrac{1}{12} \approx 0.083 = 8.3\%$

Chapter 11   The Arts

Section 11.1

1. 19

3. 29

5. $\dfrac{144}{89} \approx 1.618 \approx \phi$

7. 199, 322; $\dfrac{322}{199} \approx 1.618 \approx \phi$; $\dfrac{F_{11}}{F_{10}} = \dfrac{89}{55} \approx 1.618 \approx \phi$

9. $\dfrac{EH}{EA} = \dfrac{12.94 \text{ cm}}{8 \text{ cm}} \approx 1.618$

11. $\dfrac{EH}{EA} = \dfrac{12.94 \text{ cm}}{8.00 \text{ cm}} \approx 1.618;$

  $\dfrac{EH + EA}{EH} = \dfrac{12.94 \text{ cm} + 8.00 \text{ cm}}{12.94 \text{ cm}} = \dfrac{20.94 \text{ cm}}{12.94 \text{ cm}} \approx 1.618$

13. $\left(\dfrac{1}{\phi}\right)^2 = \left(\dfrac{1}{\dfrac{1+\sqrt{5}}{2}}\right)^2 = \left(\dfrac{2}{1+\sqrt{5}}\right)^2 = \dfrac{4}{1 + 2\sqrt{5} + 5}$

  $= \dfrac{4}{2\sqrt{5}+6} = \dfrac{2}{\sqrt{5}+3} = \dfrac{2}{\sqrt{5}+3} \cdot \dfrac{\sqrt{5}-3}{\sqrt{5}-3}$

  $= \dfrac{2\sqrt{5}-6}{5-9} = \dfrac{2\sqrt{5}-6}{-4} = \dfrac{3-\sqrt{5}}{2}$

  $1 - \dfrac{1}{\phi} = 1 - \dfrac{1}{\dfrac{1+\sqrt{5}}{2}} = 1 - \dfrac{2}{1+\sqrt{5}} = \dfrac{1+\sqrt{5}}{1+\sqrt{5}} - \dfrac{2}{1+\sqrt{5}}$

  $= \dfrac{\sqrt{5}-1}{1+\sqrt{5}} = \dfrac{\sqrt{5}-1}{1+\sqrt{5}} \cdot \dfrac{1-\sqrt{5}}{1-\sqrt{5}} = \dfrac{\sqrt{5}-5+\sqrt{5}-1}{1-5}$

  $= \dfrac{2\sqrt{5}-6}{-4} = \dfrac{3-\sqrt{5}}{2}$

15. 7.4 feet

17. 7.8 miles or 20.4 miles

19. $\dfrac{17.55 \text{ cm}}{10.85 \text{ cm}} \approx 1.618$

21. Answers will vary.

Section 11.2

1. 28, 36, 45

3. Yes, it is the 11th triangular number.

5. $d_n = 3(n-1)$

7.

9.

11. Yes, the four angles of a rhombus must add up to 360 degrees, so they can be used to form a tessellation where all four angles come together at each vertex.

13. Yes, because essentially four squares come together at each vertex.

$$\dfrac{1}{4} + \dfrac{1}{4} + \dfrac{1}{4} + \dfrac{1}{4} = \dfrac{4}{2} - 1$$
$$\dfrac{4}{4} = 2 - 1$$
$$1 = 1$$

15.

17. Answers will vary.

Section 11.3

1. 11 half steps up or 1 half step down

3. E

5. Four octaves higher

7.

Note	Frequency (Hz)
A	110
A♯	116.541
B	123.471
C	130.813
C♯	138.592
D	146.833
D♯	155.564
E	164.814
F	174.614
F♯	184.997
G	195.997
G♯	207.652
A	220

9. a. 1567.980 Hz **b.** 493.884 Hz **c.** 1760 Hz

d. 440 Hz **e.** 2093.002 Hz

11. 155.192 Hz

13. B♭, D, and F

15. a. 391.5 Hz

b.

Note	Frequency (Hz)
C	261 Hz
C♯	276.519 Hz
D	292.962 Hz
D♯	310.382 Hz
E	328.839 Hz
F	348.393 Hz
F♯	369.110 Hz
G	391.057 Hz

c. $\dfrac{3}{2}$

17. C♯ (or D♭)

Chapter 11 Exercises

1. 31

3. 260

5. $F_{11} = 89; \dfrac{1}{F_{11}} = \dfrac{1}{89} \approx 0.011$

7. $\dfrac{1220}{754} \approx 1.618 \approx \phi$

9. $\dfrac{AC}{EA} = \dfrac{8.62}{5.33} \approx 1.617 \approx \phi$

11. $\dfrac{EA}{AC} = \dfrac{5.33 \text{ cm}}{8.62 \text{ cm}} \approx 0.618;$

$\dfrac{AC}{AC + EA} = \dfrac{8.62 \text{ cm}}{8.62 \text{ cm} + 5.33 \text{ cm}} = \dfrac{8.62 \text{ cm}}{13.95 \text{ cm}} \approx 0.618$

13. 9.3 feet

15. 6.1 km or 15.9 km

17. $\dfrac{44.55 \text{ ft}}{27.5 \text{ ft}} = 1.62$

19. No

21. 900

23. 87

25. $t_{14} + t_{15} = 105 + 120$

27. Yes, each angle in a regular hexagon measures 120°, which divides evenly into 360°. So 3 hexagons will meet at each vertex.

29. Yes

31. 696.785 Hz

33. a. 783.990 Hz **b.** 246.942 Hz **c.** 880 Hz

d. 220 Hz **e.** 1046.501 Hz

35. 783 Hz

Chapter 12 Sports

Section 12.1

1. 0.533

3. 0.340

5. 2.182

7. 0.82

9. 0.344

11. 125 singles

13. 0.586

15. Cabrera

17. $\dfrac{159}{195} \approx 0.815$

19. 0.273

21. 0.326

23. 4.512

25. 2.816

27. $\dfrac{57}{251} \approx 0.227$

29. Kershaw

Section 12.2

1. 0.631

3. 0.064

5. 8.375

7. 0.021

9. Brees

11. Roethlisberger

13. 13.741

15. 174.031

17. 8.381

19. 145.345

21. Griffin

23. The Ravens have a scoreability of 17.38 and the Steelers have a bendability of 12.12. Since the scoreability is higher than the bendability, the Ravens are favored to win.

25. 201.313

27. 19.188

29. Kansas City Chiefs

Section 12.3

1. 19.5 points per game

3. 0.200

5. $0.86 = 86\%$

7. 26.0

9. 0.831

11. Bryant

13. 14.833

15. Durant

17. Griffin

19. 0.362

Section 12.4

1. f^6

3. a. 0.408   **b.** 0.334   **c.** 0.303   **d.** 0.203   **e.** 0.368
f. 0.299   **g.** 0.200   **h.** 0.092   **i.** 0.041

5. 0.525 seconds

7. a. 2.756 m/s   **b.** 9.921 km/h   **c.** 9.042 ft/s
d. 6.165 mi/h

9. 2.758 hours

11. An average of at least 6.55 mph

13. a. 40.468%   **b.** 24.178%   **c.** 14.445%   **d.** 36.507%
e. 55.123%   **f.** 44.859%   **g.** 13.031%   **h.** 12.377%
i. 11.755%

Chapter 12 Exercises

1. 0.562

3. 0.334

5. 2.177

7. 0.310

9. 0.336

11. 0.446

13. Posey

15. 1.114

17. 1.004

19. 6.273

21. Lee

23. 250 attempts

25. 0.672

27. 90.5

29. Rodgers

31. 8.539

33. 0.661

35. 0.093

37. Aaron Murray

39. 21.533

41. 1.313

43. 12.119

45. 16.833

47. 0.008

49. f^{12}

51. 0.724 seconds

53. a. 9.009 m/s   **b.** 32.432 km/hr   **c.** 29.557 ft/s
d. 20.153 mi/hr

55. 6.55 mi/h

Chapter 13   Graph Theory

Section 13.1

1. degree

3. edge

5. loop

7. chromatic number

9. cycle

11. Vertices: USA, UK, Algeria, Canada, Mexico; edges: 10, 1, 2, 60, 3

13. Answers will vary. For example,

15. a. v_1, v_2, v_3 **b.** No, it does not start and end at the same vertex. **c.** Answers will vary. The walk must start and end with the same vertex.

17. Answers will vary.

19. The graph on the left is connected, but the graph on the right is not.

21. a. Answers will vary. For example, $a, b, c, b, e, a, e, c,$ d, e, f, d, f, a. **b.** Answers will vary. For example, $u, w, v,$ x, u, y, z, u.

23. a. Answers will vary. For example, c, d, i, j.
b. Answers will vary. For example, d, e, f, a, b, c.
c. Answers will vary. For example, a, f, g, h, i, j, k, l.

25. a. Answers will vary. For example,

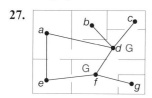

b.

c. This graph doesn't have a Hamilton cycle.

27.

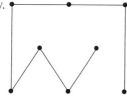

Section 13.2

1. Yes, the graph is a tree.

3. No, the graph has cycles in it and is not connected, and is therefore not a tree.

5. 7

7. 19

9. Answers will vary.

11. Answers will vary.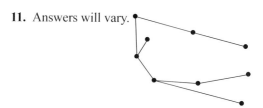

13.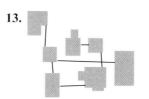

15. a. The traveling salesman route must visit each vertex once. If he visits each destination exactly once, then it's a minimum-weight spanning tree. So the route must have weight at least equal to that of the minimum-weight spanning tree. **b.** Take a minimum-weight spanning tree. Begin at any vertex and walk around the tree crossing each edge twice. This is a route that visits each vertex with weight at most 200. The optimal solution is at most 200.

17. 4

19. 14

21. 8

23. 22 cities

Section 13.3

1. bipartite graph

3. matching

5. Yes;

7. $N(A) = \{x_1, x_2, x_6, y_1, y_2, y_6\}$

9. $N(C) = \{$all vertices in $Q\}$

11. Yes

13. Yes

15. Regular bipartite graph; yes

17. No, the graph is bipartite, but the left-side has more vertices than the right.

19. The graph is regular bipartite, so it has a matching.

21. No, in order for a graph to be bipartite, vertices on one side of the graph can't be joined to one another. Consider the cycle a, b, c, d, e. Label each vertex in the cycle either left-side (L) or right-side (R). Then, we have a (L); b (R); c (L); d (R); e (L). This means that a and e are both on the left, but are connected. Therefore, the graph cannot be bipartite.

23. Answers will vary. For example,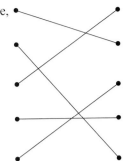

25.

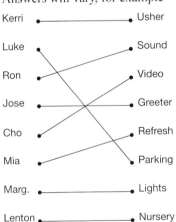

Kerri — Usher
Luke — Sound
Ron — Video
Jose — Greeter
Cho — Refresh
Mia — Parking
Margaret — Lights
Lenton — Nursery

Answers will vary, for example

Kerri ———— Usher

Luke — Sound

Ron — Video

Jose — Greeter

Cho — Refresh

Mia — Parking

Marg. ———— Lights

Lenton ———— Nursery

27. Since there is no restriction on whether vertices lie in the left- or the right-side of a bipartite graph, we can simply switch the sides of the vertices; that is, the right-side becomes the left, and the left-side becomes the right. Now we can apply Hall's Marriage Theorem to see there is a matching.

Section 13.4

1.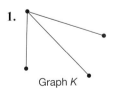

Graph *K*

3. 2

5. 1

7. $v = 5, e = 5, f = 2; 5 + 2 - 5 = 2$ so Euler's formula is satisfied.

9. $v = 15, e = 22, f = 9; 15 + 9 - 22 = 2$ so Euler's formula is satisfied.

11. Yes, the graph is planar.

13. Yes, the graph is planar.

15. If every vertex has degree at least 6 there are at least $\frac{6v}{2} = 3v$ edges. $3v > 3v - 6$, so not planar.

17. $\chi(K_{5,5}) = 2; \chi(K_{k,k}) = 2$

19. The sphere will flatten into the shape of the cube drawn on its surface.

Chapter 13 Exercises

1. No, the graph is not connected.

3. Yes, they are the same graph.

5. Chromatic number = 3; Answers will vary. For example,

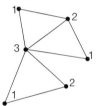

7. No, the graph is not a tree because it contains at least one cycle.

9. 11

11. Answers will vary. For example,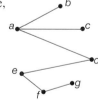

13. a. Yes, it is a spanning tree. A spanning tree will ensure that all cities are connected with optical fiber. **b.** Yes, it is minimum-weight spanning tree.

15. $N(A) = \{\text{Jeff, Isaac, Peg, Sarah}\}$

17. The graph has a matching.

19. Answers will vary. For example,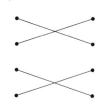

21. 8

23. #21: $f + v - e = 2$; #22: $f + v - e = 2$

$$8 + 9 - 15 \overset{?}{=} 2 \qquad 8 + 12 - 18 \overset{?}{=} 2$$
$$17 - 15 \overset{?}{=} 2 \qquad 20 - 18 \overset{?}{=} 2$$
$$2 = 2 \qquad\qquad 2 = 2$$

25. No, the graph is not planar because it contains K_5 as a minor.

Chapter 14   Number Theory

Section 14.1

1. 2, 3, 5, 7, 11, 13, 17, 19, 23, 29, 31, 37, 41, 43, 47, 53, 59, 61, 67, 71, 73, 79, 83, 89, 97, 101, 103, 107, 109, 113, 127, 131, 137, 139, 149, 151, 157, 163, 167, 173, 179, 181, 191, 193, 197, 199

3. Composite

5. Prime

7. Composite

9. All prime numbers smaller than $\sqrt{283} \approx 16.8226$: 2, 3, 5, 7, 11, and 13

11. a. Answers will vary. Breaking the class into groups of the same size requires the class size to have divisors. 17 and 19 are prime, classes of these sizes could not be broken into even groups. A class size of 15 could only be broken into groups of 3 or 5 students. A class size of 16 could be broken into groups of 2, 4, or 8 students. A class size of 18 would allow groups with 2, 3, 6, or 9 students. A class size of 20 would allow groups of sizes 2, 4, 5, or 10 students.
b. Answers will vary. Consider both the size of the class and the number of group divisions that are possible.

13. $2 \cdot 2 \cdot 2 \cdot 2 \cdot 3 \cdot 5$

15. $2 \cdot 3 \cdot 3 \cdot 5 \cdot 7$

17. 14

19. 7

21. 70

23. 4

25. Yes

27. a. 8 groups   **b.** 4 violinists   **c.** None

29. 4 boxes

31. a. 12 ft   **b.** Troop A: 15 ft × 12 ft; Troop B: 17 ft × 12 ft

Section 14.2

1. 2

3. 10

5. 1

7. $2 \,(\text{mod } 6)$

9. $8 \,(\text{mod } 12)$

11. 8:00 a.m.

13. 9:00 p.m.

15. $0 \,(\text{mod } 13)$

17. $6 \,(\text{mod } 11)$

19. True

21. True

23. True

25. True

27. No, the check-sum digit should be 1.

29. Valid

31. 0

33. 8

35. 5

37. 1

39. 0

41. Yes

43. No, check-sum digit should be 6.

Section 14.3

1. $6^{11} - 6 = (6 \cdot 6) \cdot (6 \cdot 6) \cdot (6 \cdot 6) \cdot (6 \cdot 6) \cdot (6 \cdot 6) \cdot 6 - 6$
$\equiv 3 \cdot 3 \cdot 3 \cdot 3 \cdot 6 - 6 \,(\text{mod } 11)$
$\equiv 6 - 6 \,(\text{mod } 11)$
$\equiv 0 \,(\text{mod } 11)$

3. $10^5 - 10 \equiv 0^5 - 0 \,(\text{mod } 5)$
$\equiv 0 \,(\text{mod } 5)$

5. Answers will vary.

7. Answers will vary.

9. Answers will vary.

11. Answers will vary.

Section 14.4

1. $2^{1(7-1)(11-1)+1} - 2 = 2^{(6)(10)+1} - 2$
   $= 2^{61} - 2$
   $= 2^{16} \cdot 2^{16} \cdot 2^{16} \cdot 2^{10} \cdot 2^3 - 2$
   $\equiv 9 \cdot 9 \cdot 9 \cdot 23 \cdot 8 - 2 \pmod{77}$
   $\equiv 2 - 2 \pmod{77}$
   $\equiv 0 \pmod{77}$

3. $4^{6(31-1)(5-1)+1} - 4 = 4^{6(30)(4)+1} - 4$
   $= 4^{721} - 4$
   $= \left(4^{10}\right)^{72} \cdot 4 - 4$
   $\equiv (1)^{72} \cdot 4 - 4 \pmod{155}$
   $\equiv 4 - 4 \pmod{155}$
   $\equiv 0 \pmod{155}$

5. 14

7. 13

9. 47

11. 120

13. $d = 311$

15. $p = 19, q = 31,$ and $d = 47$

Chapter 14 Exercises

1. 4 groups

3. 170

5. GCD = 21; because the GCD is not 1, the numbers are not relatively prime.

7. 0

9. 2

11. No, the check-sum digit should be 5.

13. 1

15. 3

17. 4,459,580

19. 12,345

Index